New Introduction to Mu
Time Series Analysis

Helmut Lütkepohl

New Introduction to Multiple Time Series Analysis

With 49 Figures
and 36 Tables

Professor Dr. Helmut Lütkepohl
Department of Economics
European University Institute
Villa San Paolo
Via della Piazzola 43
50133 Firenze
Italy
E-mail: helmut.luetkepohl@iue.it

Cataloging-in-Publication Data
Library of Congress Control Number: 2005927322

ISBN 3-540-26239-3 Springer Berlin Heidelberg New York

This work is subject to copyright. All rights are reserved, whether the whole or part of the material is concerned, specifically the rights of translation, reprinting, reuse of illustrations, recitation, broadcasting, reproduction on microfilm or in any other way, and storage in data banks. Duplication of this publication or parts thereof is permitted only under the provisions of the German Copyright Law of September 9, 1965, in its current version, and permission for use must always be obtained from Springer-Verlag. Violations are liable for prosecution under the German Copyright Law.

Springer is a part of Springer Science+Business Media
springeronline.com

© Springer-Verlag Berlin Heidelberg 2005
Printed in Germany

The use of general descriptive names, registered names, trademarks, etc. in this publication does not imply, even in the absence of a specific statement, that such names are exempt from the relevant protective laws and regulations and therefore free for general use.

Cover design: Erich Kirchner
Production: Helmut Petri
Printing: Strauss Offsetdruck

SPIN 11496953   Printed on acid-free paper – 43/3153 – 5 4 3 2 1 0

To Sabine

Preface

When I worked on my *Introduction to Multiple Time Series Analysis* (Lütkepohl (1991)), a suitable textbook for this field was not available. Given the great importance these methods have gained in applied econometric work, it is perhaps not surprising in retrospect that the book was quite successful. Now, almost one and a half decades later the field has undergone substantial development and, therefore, the book does not cover all topics of my own courses on the subject anymore. Therefore, I started to think about a serious revision of the book when I moved to the European University Institute in Florence in 2002. Here in the lovely hills of Toscany I had the time to think about bigger projects again and decided to prepare a substantial revision of my previous book. Because the label *Second Edition* was already used for a previous reprint of the book, I decided to modify the title and thereby hope to signal to potential readers that significant changes have been made relative to my previous multiple time series book.

Although Chapters 1–5 still contain an introduction to the vector autoregressive (VAR) methodology and their structure is largely the same as in Lütkepohl (1991), there have been some adjustments and additions, partly in response to feedback from students and colleagues. In particular, some discussion on multi-step causality and also bootstrap inference for impulse responses has been added. Moreover, the LM test for residual autocorrelation is now presented in addition to the portmanteau test and Chow tests for structural change are discussed on top of the previously considered prediction tests. When I wrote my first book on multiple time series, the cointegration revolution had just started. Hence, only one chapter was devoted to the topic. By now the related models and methods have become far more important for applied econometric work than, for example, vector autoregressive moving average (VARMA) models. Therefore, Part II (Chapters 6–8) is now entirely devoted to VAR models with cointegrated variables. The basic framework in this new part is the vector error correction model (VECM). Chapter 9 is also new. It contains a discussion of structural vector autoregressive and vector error correction models which are by now also standard tools in applied econometric analysis. Chapter 10 on systems of dynamic simultaneous equations maintains much of the contents of the corresponding chapter in Lütkepohl (1991). Some discussion of nonstationary, integrated series has been added, however. Chapters 9 and 10 together constitute Part III. Given that the research activities devoted to VARMA models have been less important than those on cointegration, I have shifted them to Part IV (Chapters 11–15) of the new book. This part also contains a new chapter on cointegrated VARMA models (Chapter 14) and in Chapter 15 on infinite order VAR models, a section on models with cointegrated variables has been added. The last part of the new book contains three chapters on special topics related to multiple time series. One chapter deals with autoregressive conditional heteroskedasticity (Chapter 16) and is new, whereas the other two chapters on periodic models (Chapter 17) and state space models (Chapter 18) are largely taken from Lütkepohl (1991). All chapters have been adjusted to account for the new material and the new structure of the book. In some instances, also the notation has been modified. In Appendix A, some additional matrix results are presented because they are used in the new parts of the text. Also Appendix C has been expanded by sections on unit root asymptotics. These results are important in the more extensive discussion of cointegration. Moreover, the discussion of bootstrap methods in Appendix D has been revised. Generally, I have added many new references and consequently the reference list is now much longer than in the previous version. To keep the length of the book in acceptable bounds, I have also deleted some material from the previous version. For example, stationary reduced rank VAR models are just mentioned as examples of models with nonlinear parameter restrictions and not discussed in detail anymore. Reduced rank models are now more important in the context of cointegration analysis. Also the tables with example time series are not timely anymore and have been eliminated. The example time series are available from my webpage and they can also be downloaded from www.jmulti.de. It is my hope that these revisions make the book more suitable for a modern course on multiple time series analysis.

Although multiple time series analysis is applied in many disciplines, I have prepared the text with economics and business students in mind. The examples and exercises are chosen accordingly. Despite this orientation, I hope that the book will also serve multiple time series courses in other fields. It contains enough material for a one semester course on multiple time series analysis. It may also be combined with univariate times series books or with texts like Fuller (1976) or Hamilton (1994) to form the basis of a one or two semester course on univariate and multivariate time series analysis. Alternatively, it is also possible to select some of the chapters or sections for a special topic of a graduate level econometrics course. For example, Chapters 1–8 could be used for an introduction to stationary and cointegrated VARs. For students already familiar with these topics, Chapter 9 could be a special topic on structural VAR modelling in an advanced econometrics course.

The students using the book must have knowledge of matrix algebra and should also have been introduced to mathematical statistics, for instance, based on textbooks like Mood, Graybill & Boes (1974), Hogg & Craig (1978) or Rohatgi (1976). Moreover, a working knowledge of the Box-Jenkins approach and other univariate time series techniques is an advantage. Although, in principle, it may be possible to use the present text without any prior knowledge of univariate time series analysis if the instructor provides the required motivation, it is clearly an advantage to have some time series background. Also, a previous introduction to econometrics will be helpful. Matrix algebra and an introductory mathematical statistics course plus the multiple regression model are necessary prerequisites.

As the previous book, the present one is meant to be an introductory exposition. Hence, I am not striving for utmost generality. For instance, quite often I use the normality assumption although the considered results hold under more general conditions. The emphasis is on explaining the underlying ideas and not on generality. In Chapters 2–7 a number of results are proven to illustrate some of the techniques that are often used in the multiple time series arena. Most proofs may be skipped without loss of continuity. Therefore the beginning and the end of a proof are usually clearly marked. Many results are summarized in propositions for easy reference.

Exercises are given at the end of each chapter with the exception of Chapter 1. Some of the problems may be too difficult for students without a good formal training, some are just included to avoid details of proofs given in the text. In most chapters empirical exercises are provided in addition to algebraic problems. Solving the empirical problems requires the use of a computer. Matrix oriented software such as GAUSS, MATLAB, or Ox will be most helpful. Most of the empirical exercises can also be done with the easy-to-use software JMulTi (see Lütkepohl & Krätzig (2004)) which is available free of charge at the website www.jmulti.de. The data needed for the exercises are also available at that website, as mentioned earlier.

Many persons have contributed directly or indirectly to this book and I am very grateful to all of them. Many students and colleagues have commented on my earlier book on the topic. Thereby they have helped to improve the presentation and to correct errors. A number of colleagues have commented on parts of the manuscript and have been available for discussions on the topics covered. These comments and discussions have been very helpful for my own understanding of the subject and have resulted in improvements to the manuscript.

Although the persons who have contributed to the project in some way or other are too numerous to be listed here, I wish to express my special gratitude to some of them. Because some parts of the old book are still maintained, it is only fair to mention those who have helped in a special way in the preparation of that book. They include Theo Dykstra who read and commented on a large part of the manuscript during his visit in Kiel in the summer of 1990, Hans-Eggert Reimers who read the entire manuscript, suggested many improvements, and pointed out numerous errors, Wolfgang Schneider who helped with examples and also commented on parts of the manuscript as well as Bernd Theilen who prepared the final versions of most figures, and Knut Haase and Holger Claessen who performed the computations for many of the examples. I deeply appreciate the help of all these collaborators.

Special thanks for comments on parts of the new book go to Pentti Saikkonen for helping with Part II and to Ralf Brüggemann, Helmut Herwartz, and Martin Wagner for reading Chapters 9, 16, and 18, respectively. Christian Kascha prepared some of the new figures and my wife Sabine helped with the preparation of the author index. Of course, I assume full responsibility for any remaining errors, in particular, as I have keyboarded large parts of the manuscript myself. A preliminary LaTeX version of parts of the old book was provided by Springer-Verlag. I thank Martina Bihn for taking charge of the project on the side of Springer-Verlag. Needless to say, I welcome any comments by readers.

Florence and Berlin, *Helmut Lütkepohl*
March 2005

Contents

1 **Introduction** 1
   1.1 Objectives of Analyzing Multiple Time Series 1
   1.2 Some Basics 2
   1.3 Vector Autoregressive Processes 4
   1.4 Outline of the Following Chapters 5

Part I Finite Order Vector Autoregressive Processes

2 **Stable Vector Autoregressive Processes** 13
   2.1 Basic Assumptions and Properties of VAR Processes 13
      2.1.1 Stable VAR(p) Processes 13
      2.1.2 The Moving Average Representation of a VAR Process . 18
      2.1.3 Stationary Processes 24
      2.1.4 Computation of Autocovariances and Autocorrelations
           of Stable VAR Processes 26
   2.2 Forecasting 31
      2.2.1 The Loss Function 32
      2.2.2 Point Forecasts 33
      2.2.3 Interval Forecasts and Forecast Regions 39
   2.3 Structural Analysis with VAR Models 41
      2.3.1 Granger-Causality, Instantaneous Causality, and
           Multi-Step Causality 41
      2.3.2 Impulse Response Analysis 51
      2.3.3 Forecast Error Variance Decomposition 63
      2.3.4 Remarks on the Interpretation of VAR Models 66
   2.4 Exercises 66

3 **Estimation of Vector Autoregressive Processes** 69
   3.1 Introduction 69
   3.2 Multivariate Least Squares Estimation 69

    3.2.1 The Estimator 70
    3.2.2 Asymptotic Properties of the Least Squares Estimator . 73
    3.2.3 An Example 77
    3.2.4 Small Sample Properties of the LS Estimator 80
  3.3 Least Squares Estimation with Mean-Adjusted Data and
      Yule-Walker Estimation 82
    3.3.1 Estimation when the Process Mean Is Known 82
    3.3.2 Estimation of the Process Mean 83
    3.3.3 Estimation with Unknown Process Mean 85
    3.3.4 The Yule-Walker Estimator 85
    3.3.5 An Example 87
  3.4 Maximum Likelihood Estimation 87
    3.4.1 The Likelihood Function 87
    3.4.2 The ML Estimators 89
    3.4.3 Properties of the ML Estimators 90
  3.5 Forecasting with Estimated Processes 94
    3.5.1 General Assumptions and Results 94
    3.5.2 The Approximate MSE Matrix 96
    3.5.3 An Example 98
    3.5.4 A Small Sample Investigation 100
  3.6 Testing for Causality 102
    3.6.1 A Wald Test for Granger-Causality 102
    3.6.2 An Example 103
    3.6.3 Testing for Instantaneous Causality 104
    3.6.4 Testing for Multi-Step Causality 106
  3.7 The Asymptotic Distributions of Impulse Responses and
      Forecast Error Variance Decompositions 109
    3.7.1 The Main Results 109
    3.7.2 Proof of Proposition 3.6 116
    3.7.3 An Example 118
    3.7.4 Investigating the Distributions of the Impulse
          Responses by Simulation Techniques 126
  3.8 Exercises 130
    3.8.1 Algebraic Problems 130
    3.8.2 Numerical Problems 132

4 **VAR Order Selection and Checking the Model Adequacy** .. 135
  4.1 Introduction 135
  4.2 A Sequence of Tests for Determining the VAR Order 136
    4.2.1 The Impact of the Fitted VAR Order on the Forecast
          MSE 136
    4.2.2 The Likelihood Ratio Test Statistic 138
    4.2.3 A Testing Scheme for VAR Order Determination 143
    4.2.4 An Example 145
  4.3 Criteria for VAR Order Selection 146

		4.3.1	Minimizing the Forecast MSE 146

- 4.3.1 Minimizing the Forecast MSE 146
- 4.3.2 Consistent Order Selection 148
- 4.3.3 Comparison of Order Selection Criteria 151
- 4.3.4 Some Small Sample Simulation Results.............. 153
- 4.4 Checking the Whiteness of the Residuals 157
  - 4.4.1 The Asymptotic Distributions of the Autocovariances and Autocorrelations of a White Noise Process 157
  - 4.4.2 The Asymptotic Distributions of the Residual Autocovariances and Autocorrelations of an Estimated VAR Process 161
  - 4.4.3 Portmanteau Tests 169
  - 4.4.4 Lagrange Multiplier Tests 171
- 4.5 Testing for Nonnormality 174
  - 4.5.1 Tests for Nonnormality of a Vector White Noise Process 174
  - 4.5.2 Tests for Nonnormality of a VAR Process 177
- 4.6 Tests for Structural Change.............................. 181
  - 4.6.1 Chow Tests 182
  - 4.6.2 Forecast Tests for Structural Change................ 184
- 4.7 Exercises 189
  - 4.7.1 Algebraic Problems 189
  - 4.7.2 Numerical Problems 191

5 VAR Processes with Parameter Constraints 193
- 5.1 Introduction 193
- 5.2 Linear Constraints...................................... 194
  - 5.2.1 The Model and the Constraints 194
  - 5.2.2 LS, GLS, and EGLS Estimation 195
  - 5.2.3 Maximum Likelihood Estimation 200
  - 5.2.4 Constraints for Individual Equations 201
  - 5.2.5 Restrictions for the White Noise Covariance Matrix.... 202
  - 5.2.6 Forecasting 204
  - 5.2.7 Impulse Response Analysis and Forecast Error Variance Decomposition 205
  - 5.2.8 Specification of Subset VAR Models 206
  - 5.2.9 Model Checking 212
  - 5.2.10 An Example 217
- 5.3 VAR Processes with Nonlinear Parameter Restrictions 221
- 5.4 Bayesian Estimation 222
  - 5.4.1 Basic Terms and Notation 222
  - 5.4.2 Normal Priors for the Parameters of a Gaussian VAR Process 223
  - 5.4.3 The Minnesota or Litterman Prior................... 225
  - 5.4.4 Practical Considerations........................... 227
  - 5.4.5 An Example 227

XIV   Contents

        5.4.6   Classical versus Bayesian Interpretation of $\bar{\alpha}$ in Forecasting and Structural Analysis 228
   5.5   Exercises 230
        5.5.1   Algebraic Exercises 230
        5.5.2   Numerical Problems 231

Part II Cointegrated Processes

6   Vector Error Correction Models 237
   6.1   Integrated Processes 238
   6.2   VAR Processes with Integrated Variables 243
   6.3   Cointegrated Processes, Common Stochastic Trends, and Vector Error Correction Models 244
   6.4   Deterministic Terms in Cointegrated Processes 256
   6.5   Forecasting Integrated and Cointegrated Variables 258
   6.6   Causality Analysis 261
   6.7   Impulse Response Analysis 262
   6.8   Exercises 265

7   Estimation of Vector Error Correction Models 269
   7.1   Estimation of a Simple Special Case VECM 269
   7.2   Estimation of General VECMs 286
        7.2.1   LS Estimation 287
        7.2.2   EGLS Estimation of the Cointegration Parameters 291
        7.2.3   ML Estimation 294
        7.2.4   Including Deterministic Terms 299
        7.2.5   Other Estimation Methods for Cointegrated Systems ... 300
        7.2.6   An Example 302
   7.3   Estimating VECMs with Parameter Restrictions 305
        7.3.1   Linear Restrictions for the Cointegration Matrix 305
        7.3.2   Linear Restrictions for the Short-Run and Loading Parameters 307
        7.3.3   An Example 309
   7.4   Bayesian Estimation of Integrated Systems 309
        7.4.1   The Model Setup 310
        7.4.2   The Minnesota or Litterman Prior 310
        7.4.3   An Example 312
   7.5   Forecasting Estimated Integrated and Cointegrated Systems .. 315
   7.6   Testing for Granger-Causality 316
        7.6.1   The Noncausality Restrictions 316
        7.6.2   Problems Related to Standard Wald Tests 317
        7.6.3   A Wald Test Based on a Lag Augmented VAR 318
        7.6.4   An Example 320
   7.7   Impulse Response Analysis 321

	7.8	Exercises 323
		7.8.1 Algebraic Exercises 323
		7.8.2 Numerical Exercises 324

8 Specification of VECMs 325
- 8.1 Lag Order Selection...................................... 325
- 8.2 Testing for the Rank of Cointegration 327
  - 8.2.1 A VECM without Deterministic Terms 328
  - 8.2.2 A Nonzero Mean Term 330
  - 8.2.3 A Linear Trend 331
  - 8.2.4 A Linear Trend in the Variables and Not in the Cointegration Relations 331
  - 8.2.5 Summary of Results and Other Deterministic Terms ... 332
  - 8.2.6 An Example 335
  - 8.2.7 Prior Adjustment for Deterministic Terms 337
  - 8.2.8 Choice of Deterministic Terms 341
  - 8.2.9 Other Approaches to Testing for the Cointegrating Rank 342
- 8.3 Subset VECMs... 343
- 8.4 Model Diagnostics 345
  - 8.4.1 Checking for Residual Autocorrelation 345
  - 8.4.2 Testing for Nonnormality 348
  - 8.4.3 Tests for Structural Change........................ 348
- 8.5 Exercises 351
  - 8.5.1 Algebraic Exercises 351
  - 8.5.2 Numerical Exercises 352

Part III Structural and Conditional Models

9 Structural VARs and VECMs 357
- 9.1 Structural Vector Autoregressions 358
  - 9.1.1 The A-Model 358
  - 9.1.2 The B-Model 362
  - 9.1.3 The AB-Model 364
  - 9.1.4 Long-Run Restrictions à la Blanchard-Quah 367
- 9.2 Structural Vector Error Correction Models.................. 368
- 9.3 Estimation of Structural Parameters 372
  - 9.3.1 Estimating SVAR Models 372
  - 9.3.2 Estimating Structural VECMs 376
- 9.4 Impulse Response Analysis and Forecast Error Variance Decomposition 377
- 9.5 Further Issues... 383
- 9.6 Exercises 384
  - 9.6.1 Algebraic Problems 384
  - 9.6.2 Numerical Problems 385

10 Systems of Dynamic Simultaneous Equations 387
10.1 Background 387
10.2 Systems with Unmodelled Variables 388
    10.2.1 Types of Variables 388
    10.2.2 Structural Form, Reduced Form, Final Form 390
    10.2.3 Models with Rational Expectations 393
    10.2.4 Cointegrated Variables 394
10.3 Estimation 395
    10.3.1 Stationary Variables 396
    10.3.2 Estimation of Models with $I(1)$ Variables 398
10.4 Remarks on Model Specification and Model Checking 400
10.5 Forecasting 401
    10.5.1 Unconditional and Conditional Forecasts 401
    10.5.2 Forecasting Estimated Dynamic SEMs 405
10.6 Multiplier Analysis 406
10.7 Optimal Control 408
10.8 Concluding Remarks on Dynamic SEMs 411
10.9 Exercises 412

Part IV Infinite Order Vector Autoregressive Processes

11 Vector Autoregressive Moving Average Processes 419
11.1 Introduction 419
11.2 Finite Order Moving Average Processes 420
11.3 VARMA Processes 423
    11.3.1 The Pure MA and Pure VAR Representations of a VARMA Process 423
    11.3.2 A VAR(1) Representation of a VARMA Process 426
11.4 The Autocovariances and Autocorrelations of a VARMA(p,q) Process 429
11.5 Forecasting VARMA Processes 432
11.6 Transforming and Aggregating VARMA Processes 434
    11.6.1 Linear Transformations of VARMA Processes 435
    11.6.2 Aggregation of VARMA Processes 440
11.7 Interpretation of VARMA Models 442
    11.7.1 Granger-Causality 442
    11.7.2 Impulse Response Analysis 444
11.8 Exercises 444

12 Estimation of VARMA Models 447
12.1 The Identification Problem 447
    12.1.1 Nonuniqueness of VARMA Representations 447
    12.1.2 Final Equations Form and Echelon Form 452
    12.1.3 Illustrations 455

12.2 The Gaussian Likelihood Function 459
    12.2.1 The Likelihood Function of an MA(1) Process 459
    12.2.2 The MA(q) Case 461
    12.2.3 The VARMA(1, 1) Case 463
    12.2.4 The General VARMA(p, q) Case..................... 464
12.3 Computation of the ML Estimates........................... 467
    12.3.1 The Normal Equations 468
    12.3.2 Optimization Algorithms 470
    12.3.3 The Information Matrix 473
    12.3.4 Preliminary Estimation 474
    12.3.5 An Illustration 477
12.4 Asymptotic Properties of the ML Estimators................ 479
    12.4.1 Theoretical Results 479
    12.4.2 A Real Data Example................................. 486
12.5 Forecasting Estimated VARMA Processes 487
12.6 Estimated Impulse Responses 490
12.7 Exercises 491

13 Specification and Checking the Adequacy of VARMA Models 493
13.1 Introduction 493
13.2 Specification of the Final Equations Form 494
    13.2.1 A Specification Procedure 494
    13.2.2 An Example 497
13.3 Specification of Echelon Forms 498
    13.3.1 A Procedure for Small Systems 499
    13.3.2 A Full Search Procedure Based on Linear Least
           Squares Computations 501
    13.3.3 Hannan-Kavalieris Procedure 503
    13.3.4 Poskitt's Procedure 505
13.4 Remarks on Other Specification Strategies for VARMA Models 507
13.5 Model Checking 508
    13.5.1 LM Tests.. 508
    13.5.2 Residual Autocorrelations and Portmanteau Tests 510
    13.5.3 Prediction Tests for Structural Change 511
13.6 Critique of VARMA Model Fitting 511
13.7 Exercises 512

14 Cointegrated VARMA Processes 515
14.1 Introduction 515
14.2 The VARMA Framework for $I(1)$ Variables 516
    14.2.1 Levels VARMA Models 516
    14.2.2 The Reverse Echelon Form 518
    14.2.3 The Error Correction Echelon Form 519
14.3 Estimation 521

14.3.1 Estimation of ARMA$_{RE}$ Models.....................521
14.3.2 Estimation of EC-ARMA$_{RE}$ Models 522
14.4 Specification of EC-ARMA$_{RE}$ Models......................523
14.4.1 Specification of Kronecker Indices 523
14.4.2 Specification of the Cointegrating Rank 525
14.5 Forecasting Cointegrated VARMA Processes 526
14.6 An Example 526
14.7 Exercises 528
14.7.1 Algebraic Exercises 528
14.7.2 Numerical Exercises 529

15 Fitting Finite Order VAR Models to Infinite Order Processes 531
15.1 Background..531
15.2 Multivariate Least Squares Estimation 532
15.3 Forecasting 536
15.3.1 Theoretical Results 536
15.3.2 An Example 538
15.4 Impulse Response Analysis and Forecast Error Variance Decompositions 540
15.4.1 Asymptotic Theory 540
15.4.2 An Example 543
15.5 Cointegrated Infinite Order VARs 545
15.5.1 The Model Setup....................................546
15.5.2 Estimation 549
15.5.3 Testing for the Cointegrating Rank 551
15.6 Exercises 552

Part V Time Series Topics

16 Multivariate ARCH and GARCH Models 557
16.1 Background..557
16.2 Univariate GARCH Models 559
16.2.1 Definitions 559
16.2.2 Forecasting 561
16.3 Multivariate GARCH Models 562
16.3.1 Multivariate ARCH..................................563
16.3.2 MGARCH..564
16.3.3 Other Multivariate ARCH and GARCH Models 567
16.4 Estimation 569
16.4.1 Theory..569
16.4.2 An Example 571
16.5 Checking MGARCH Models 576
16.5.1 ARCH-LM and ARCH-Portmanteau Tests............576

		Contents XIX

```
            16.5.2  LM and Portmanteau Tests for Remaining ARCH ..... 577
            16.5.3  Other Diagnostic Tests ............................. 578
            16.5.4  An Example ........................................ 578
     16.6  Interpreting GARCH Models ............................... 579
            16.6.1  Causality in Variance ............................. 579
            16.6.2  Conditional Moment Profiles and Generalized Impulse
                    Responses ......................................... 580
     16.7  Problems and Extensions .................................. 582
     16.8  Exercises ................................................ 584

17  Periodic VAR Processes and Intervention Models .......... 585
     17.1  Introduction ............................................. 585
     17.2  The VAR(p) Model with Time Varying Coefficients .......... 587
            17.2.1  General Properties ................................ 587
            17.2.2  ML Estimation ..................................... 589
     17.3  Periodic Processes ....................................... 591
            17.3.1  A VAR Representation with Time Invariant Coefficients 592
            17.3.2  ML Estimation and Testing for Time Varying
                    Coefficients ...................................... 595
            17.3.3  An Example ........................................ 602
            17.3.4  Bibliographical Notes and Extensions .............. 604
     17.4  Intervention Models ...................................... 604
            17.4.1  Interventions in the Intercept Model .............. 605
            17.4.2  A Discrete Change in the Mean ..................... 606
            17.4.3  An Illustrative Example ........................... 608
            17.4.4  Extensions and References ......................... 609
     17.5  Exercises ................................................ 609

18  State Space Models ........................................ 611
     18.1  Background ............................................... 611
     18.2  State Space Models ....................................... 613
            18.2.1  The Model Setup ................................... 613
            18.2.2  More General State Space Models ................... 624
     18.3  The Kalman Filter ........................................ 625
            18.3.1  The Kalman Filter Recursions ...................... 626
            18.3.2  Proof of the Kalman Filter Recursions ............. 630
     18.4  Maximum Likelihood Estimation of State Space Models ...... 631
            18.4.1  The Log-Likelihood Function ....................... 632
            18.4.2  The Identification Problem ........................ 633
            18.4.3  Maximization of the Log-Likelihood Function ....... 634
            18.4.4  Asymptotic Properties of the ML Estimator ......... 636
     18.5  A Real Data Example ...................................... 637
     18.6  Exercises ................................................ 641
```

Appendix

A Vectors and Matrices 645
  A.1 Basic Definitions 645
  A.2 Basic Matrix Operations 646
  A.3 The Determinant 647
  A.4 The Inverse, the Adjoint, and Generalized Inverses 649
    A.4.1 Inverse and Adjoint of a Square Matrix 649
    A.4.2 Generalized Inverses 650
  A.5 The Rank 651
  A.6 Eigenvalues and -vectors – Characteristic Values and Vectors .. 652
  A.7 The Trace 653
  A.8 Some Special Matrices and Vectors 653
    A.8.1 Idempotent and Nilpotent Matrices 653
    A.8.2 Orthogonal Matrices and Vectors and Orthogonal Complements 654
    A.8.3 Definite Matrices and Quadratic Forms 655
  A.9 Decomposition and Diagonalization of Matrices 656
    A.9.1 The Jordan Canonical Form 656
    A.9.2 Decomposition of Symmetric Matrices 658
    A.9.3 The Choleski Decomposition of a Positive Definite Matrix 658
  A.10 Partitioned Matrices 659
  A.11 The Kronecker Product 660
  A.12 The vec and vech Operators and Related Matrices 661
    A.12.1 The Operators 661
    A.12.2 Elimination, Duplication, and Commutation Matrices .. 662
  A.13 Vector and Matrix Differentiation 664
  A.14 Optimization of Vector Functions 671
  A.15 Problems 675

B Multivariate Normal and Related Distributions 677
  B.1 Multivariate Normal Distributions 677
  B.2 Related Distributions 678

C Stochastic Convergence and Asymptotic Distributions 681
  C.1 Concepts of Stochastic Convergence 681
  C.2 Order in Probability 684
  C.3 Infinite Sums of Random Variables 685
  C.4 Laws of Large Numbers and Central Limit Theorems 689
  C.5 Standard Asymptotic Properties of Estimators and Test Statistics 692
  C.6 Maximum Likelihood Estimation 693
  C.7 Likelihood Ratio, Lagrange Multiplier, and Wald Tests 694

    C.8 Unit Root Asymptotics................................... 698
        C.8.1 Univariate Processes 698
        C.8.2 Multivariate Processes 703

D Evaluating Properties of Estimators and Test Statistics by Simulation and Resampling Techniques 707
    D.1 Simulating a Multiple Time Series with VAR Generation Process 707
    D.2 Evaluating Distributions of Functions of Multiple Time Series by Simulation 708
    D.3 Resampling Methods..................................... 709

References 713

Index of Notation 733

Author Index 741

Subject Index 747

1
Introduction

1.1 Objectives of Analyzing Multiple Time Series

In making choices between alternative courses of action, decision makers at all structural levels often need predictions of economic variables. If time series observations are available for a variable of interest and the data from the past contain information about the future development of a variable, it is plausible to use as forecast some function of the data collected in the past. For instance, in forecasting the monthly unemployment rate, from past experience a forecaster may know that in some country or region a high unemployment rate in one month tends to be followed by a high rate in the next month. In other words, the rate changes only gradually. Assuming that the tendency prevails in future periods, forecasts can be based on current and past data.

Formally, this approach to forecasting may be expressed as follows. Let y_t denote the value of the variable of interest in period t. Then a forecast for period $T + h$, made at the end of period T, may have the form

$$\widehat{y}_{T+h} = f(y_T, y_{T-1}, \ldots), \tag{1.1.1}$$

where $f(\cdot)$ denotes some suitable function of the past observations y_T, y_{T-1}, ... For the moment it is left open how many past observations enter into the forecast. One major goal of univariate time series analysis is to specify sensible forms of functions $f(\cdot)$. In many applications, linear functions have been used so that, for example,

$$\widehat{y}_{T+h} = \nu + \alpha_1 y_T + \alpha_2 y_{T-1} + \cdots.$$

In dealing with economic variables, often the value of one variable is not only related to its predecessors in time but, in addition, it depends on past values of other variables. For instance, household consumption expenditures may depend on variables such as income, interest rates, and investment expenditures. If all these variables are related to the consumption expenditures it makes sense to use their possible additional information content in forecasting consumption expenditures. In other words, denoting the related variables by $y_{1t}, y_{2t}, \ldots, y_{Kt}$, the forecast of $y_{1,T+h}$ at the end of period T may be of the form

$$\widehat{y}_{1,T+h} = f_1(y_{1,T}, y_{2,T}, \ldots, y_{K,T}, y_{1,T-1}, y_{2,T-1}, \ldots, y_{K,T-1}, y_{1,T-2}, \ldots).$$

Similarly, a forecast for the second variable may be based on past values of all variables in the system. More generally, a forecast of the k-th variable may be expressed as

$$\widehat{y}_{k,T+h} = f_k(y_{1,T}, \ldots, y_{K,T}, y_{1,T-1}, \ldots, y_{K,T-1}, \ldots). \tag{1.1.2}$$

A set of time series y_{kt}, $k = 1, \ldots, K$, $t = 1, \ldots, T$, is called a *multiple time series* and the previous formula expresses the forecast $\widehat{y}_{k,T+h}$ as a function of a multiple time series. In analogy with the univariate case, it is one major objective of multiple time series analysis to determine suitable functions $f_1, \ldots, f_K$ that may be used to obtain forecasts with good properties for the variables of the system.

It is also often of interest to learn about the dynamic interrelationships between a number of variables. For instance, in a system consisting of investment, income, and consumption one may want to know about the likely impact of a change in income. What will be the present and future implications of such an event for consumption and for investment? Under what conditions can the effect of an increase in income be isolated and traced through the system? Alternatively, given a particular subject matter theory, is it consistent with the relations implied by a multiple time series model which is developed with the help of statistical tools? These and other questions regarding the structure of the relationships between the variables involved are occasionally investigated in the context of multiple time series analysis. Thus, obtaining insight into the dynamic structure of a system is a further objective of multiple time series analysis.

1.2 Some Basics

In the following chapters, we will regard the values that a particular economic variable has assumed in a specific period as realizations of random variables. A time series will be assumed to be generated by a stochastic process. Although the reader is assumed to be familiar with these terms, it may be useful to briefly review some of the basic definitions and expressions at this point, in order to make the underlying concepts precise.

Let $(\Omega, \mathcal{F}, \Pr)$ be a *probability space*, where Ω is the set of all elementary events (sample space), $\mathcal{F}$ is a sigma-algebra of events or subsets of Ω and $\Pr$ is a probability measure defined on $\mathcal{F}$. A *random variable* y is a real valued function defined on Ω such that for each real number c, $A_c = \{\omega \in \Omega | y(\omega) \leq$

$c\} \in \mathcal{F}$. In other words, A_c is an event for which the probability is defined in terms of Pr. The function $F : \mathbb{R} \to [0,1]$, defined by $F(c) = \Pr(A_c)$, is the distribution function of y.

A K-dimensional *random vector* or a K-dimensional *vector of random variables* is a function y from Ω into the K-dimensional Euclidean space $\mathbb{R}^K$, that is, y maps $\omega \in \Omega$ on $y(\omega) = (y_1(\omega), \ldots, y_K(\omega))'$ such that for each $c = (c_1, \ldots, c_K)' \in \mathbb{R}^K$,

$$A_c = \{\omega | y_1(\omega) \leq c_1, \ldots, y_K(\omega) \leq c_K\} \in \mathcal{F}.$$

The function $F : \mathbb{R}^K \to [0,1]$ defined by $F(c) = \Pr(A_c)$ is the joint *distribution function* of y.

Suppose Z is some index set with at most countably many elements like, for instance, the set of all integers or all positive integers. A (discrete) *stochastic process* is a real valued function

$$y : Z \times \Omega \to \mathbb{R}$$

such that for each fixed $t \in Z$, $y(t, \omega)$ is a random variable. The random variable corresponding to a fixed t is usually denoted by y_t in the following. The underlying probability space will usually not even be mentioned. In that case, it is understood that all the members y_t of a stochastic process are defined on the same probability space. Usually the stochastic process will also be denoted by y_t if the meaning of the symbol is clear from the context.

A stochastic process may be described by the joint distribution functions of all finite subcollections of y_t's, $t \in S \subset Z$. In practice, the complete system of distributions will often be unknown. Therefore, in the following chapters, we will often be concerned with the first and second moments of the distributions. In other words, we will be concerned with the means $E(y_t) = \mu_t$, the variances $E[(y_t - \mu_t)^2]$ and the covariances $E[(y_t - \mu_t)(y_s - \mu_s)]$.

A K-dimensional *vector stochastic process* or *multivariate stochastic process* is a function

$$y : Z \times \Omega \to \mathbb{R}^K,$$

where, for each fixed $t \in Z$, $y(t, \omega)$ is a K-dimensional random vector. Again we usually use the symbol y_t for the random vector corresponding to a fixed $t \in Z$. For simplicity, we also often denote the complete process by y_t. The particular meaning of the symbol should be clear from the context. With respect to the stochastic characteristics the same applies as for univariate processes. That is, the stochastic characteristics are summarized in the joint distribution functions of all finite subcollections of random vectors y_t. In practice, interest will often focus on the first and second moments of all random variables involved.

A realization of a (vector) stochastic process is a sequence (of vectors) $y_t(\omega)$, $t \in Z$, for a fixed ω. In other words, a realization of a stochastic process is a function $Z \to \mathbb{R}^K$ where $t \to y_t(\omega)$. A (multiple) time series is regarded as such a realization or possibly a finite part of such a realization, that is, it consists, for instance, of values (vectors) $y_1(\omega), \ldots, y_T(\omega)$. The underlying stochastic process is said to have *generated* the (multiple) time series or it is called the *generating* or *generation process* of the time series or the *data generation process* (DGP). A time series $y_1(\omega), \ldots, y_T(\omega)$ will usually be denoted by $y_1, \ldots, y_T$ or simply by y_t just like the underlying stochastic process, if no confusion is possible. The number of observations, T, is called the *sample size* or *time series length*. With this terminology at hand, we may now return to the problem of specifying forecast functions.

1.3 Vector Autoregressive Processes

Because linear functions are relatively easy to deal with, it makes sense to begin with forecasts that are linear functions of past observations. Let us consider a univariate time series y_t and a forecast $h = 1$ period into the future. If $f(\cdot)$ in (1.1.1) is a linear function, we have

$$\widehat{y}_{T+1} = \nu + \alpha_1 y_T + \alpha_2 y_{T-1} + \cdots.$$

Assuming that only a finite number p, say, of past y values are used in the prediction formula, we get

$$\widehat{y}_{T+1} = \nu + \alpha_1 y_T + \alpha_2 y_{T-1} + \cdots + \alpha_p y_{T-p+1}. \tag{1.3.1}$$

Of course, the true value y_{T+1} will usually not be exactly equal to the forecast $\widehat{y}_{T+1}$. Let us denote the forecast error by $u_{T+1} := y_{T+1} - \widehat{y}_{T+1}$ so that

$$y_{T+1} = \widehat{y}_{T+1} + u_{T+1} = \nu + \alpha_1 y_T + \cdots + \alpha_p y_{T-p+1} + u_{T+1}. \tag{1.3.2}$$

Now, assuming that our numbers are realizations of random variables and that the same data generation law prevails in each period T, (1.3.2) has the form of an *autoregressive process*,

$$y_t = \nu + \alpha_1 y_{t-1} + \cdots + \alpha_p y_{t-p} + u_t, \tag{1.3.3}$$

where the quantities $y_t, y_{t-1}, \ldots, y_{t-p}$, and u_t are now random variables. To actually get an autoregressive (AR) process we assume that the forecast errors u_t for different periods are uncorrelated, that is, u_t and u_s are uncorrelated for $s \neq t$. In other words, we assume that all useful information in the past y_t's is used in the forecasts so that there are no systematic forecast errors.

If a multiple time series is considered, an obvious extension of (1.3.1) would be

$$\begin{aligned}\widehat{y}_{k,T+1} &= \nu + \alpha_{k1,1} y_{1,T} + \alpha_{k2,1} y_{2,T} + \cdots + \alpha_{kK,1} y_{K,T} \\ &\quad + \cdots + \alpha_{k1,p} y_{1,T-p+1} + \cdots + \alpha_{kK,p} y_{K,T-p+1}, \\ &\qquad k = 1, \ldots, K.\end{aligned} \tag{1.3.4}$$

To simplify the notation, let $y_t := (y_{1t}, \ldots, y_{Kt})'$, $\widehat{y}_t := (\widehat{y}_{1t}, \ldots, \widehat{y}_{Kt})'$, $\nu := (\nu_1, \ldots, \nu_K)'$ and

$$A_i := \begin{bmatrix} \alpha_{11,i} & \cdots & \alpha_{1K,i} \\ \vdots & \ddots & \vdots \\ \alpha_{K1,i} & \cdots & \alpha_{KK,i} \end{bmatrix}.$$

Then (1.3.4) can be written compactly as

$$\widehat{y}_{T+1} = \nu + A_1 y_T + \cdots + A_p y_{T-p+1}. \tag{1.3.5}$$

If the y_t's are regarded as random vectors, this predictor is just the optimal forecast obtained from a vector autoregressive model of the form

$$y_t = \nu + A_1 y_{t-1} + \cdots + A_p y_{t-p} + u_t, \tag{1.3.6}$$

where the $u_t = (u_{1t}, \ldots, u_{Kt})'$ form a sequence of independently identically distributed random K-vectors with zero mean vector.

Obviously such a model represents a tremendous simplification compared with the general form (1.1.2). Because of its simple structure, it enjoys great popularity in applied work. We will study this particular model in the following chapters in some detail.

1.4 Outline of the Following Chapters

In Part I of the book, consisting of the next four chapters, we will investigate some basic properties of stationary vector autoregressive (VAR) processes such as (1.3.6). Forecasts based on these processes are discussed and it is shown how VAR processes may be used for analyzing the dynamic structure of a system of variables. Throughout Chapter 2, it is assumed that the process under study is completely known including its coefficient matrices. In practice, for a given multiple time series, first a model of the DGP has to be specified and its parameters have to be estimated. Then the adequacy of the model is checked by various statistical tools and then the estimated model can be used for forecasting and dynamic or structural analysis. The main steps of a VAR analysis are presented in Figure 1.1 in a schematic way. Estimation and model specification are discussed in Chapters 3 and 4, respectively. In the former chapter the estimation of the VAR coefficients is considered and the consequences of using estimated rather than known processes for forecasting and economic analysis are explored. In Chapter 4, the specification and model checking stages of an analysis are considered. Criteria for determining the order p of a VAR process are given and possibilities for checking the assumptions underlying a VAR analysis are discussed.

In systems with many variables and/or large VAR order p, the number of coefficients is quite substantial. As a result the estimation precision will

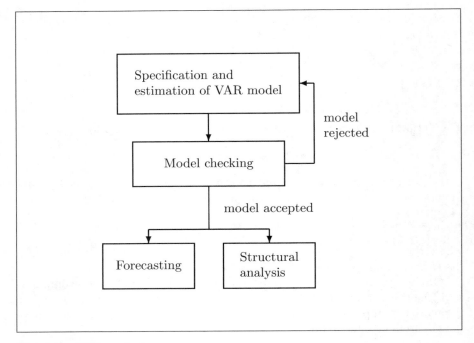

Fig. 1.1. VAR analysis.

be low if estimation is based on time series of the size typically available in economic applications. In order to improve the estimation precision, it is useful to place restrictions from nonsample sources on the parameters and thereby reduce the number of coefficients to be estimated. In Chapter 5, VAR processes with parameter constraints and restricted estimation are discussed. Zero restrictions, nonlinear constraints, and Bayesian estimation are treated.

In Part I, stationary processes are considered which have time invariant expected values, variances, and covariances. In other words, the first and second moments of the random variables do not change over time. In practice many time series have a trending behavior which is not compatible with such an assumption. This fact is recognized in Part II, where VAR processes with stochastic and deterministic trends are considered. Processes with stochastic trends are often called integrated and if two or more variables are driven by the same stochastic trend, they are called cointegrated. Cointegrated VAR processes have quite different properties from stationary ones and this has to be taken into account in the statistical analysis. The specific estimation, specification, and model checking procedures are discussed in Chapters 6–8.

The models discussed in Parts I and II are essentially reduced form models which capture the dynamic properties of the variables and are useful forecasting tools. For structural economic analysis, these models are often insufficient because different economic theories may be compatible with the same statistical reduced form model. In Chapter 9, it is discussed how to integrate structural information in stationary and cointegrated VAR models. In many econometric applications it is assumed that some of the variables are determined outside the system under consideration. In other words, they are exogenous or unmodelled variables. VAR processes with exogenous variables are considered in Chapter 10. In the econometrics literature such systems are often called systems of dynamic simultaneous equations. In the time series literature they are sometimes referred to as multivariate transfer function models. Together Chapters 9 and 10 constitute Part III of this volume.

In Part IV of the book, it is recognized that an upper bound p for the VAR order is often not known with certainty. In such a case, one may not want to impose any upper bound and allow for an infinite VAR order. There are two ways to make the estimation problem for the potentially infinite number of parameters tractable. First, it may be assumed that they depend on a finite set of parameters. This assumption leads to vector autoregressive moving average (VARMA) processes. Some properties of these processes, parameter estimation and model specification are discussed in Chapters 11–13 for the stationary case and in Chapter 14 for cointegrated systems. In the second approach for dealing with infinite order VAR processes, it is assumed that finite order VAR processes are fitted and that the VAR order goes to infinity with the sample size. This approach and its consequences for the estimators, forecasts, and structural analysis are discussed in Chapter 15 for both the stationary and the cointegrated cases.

In Part V, some special models and issues for multiple time series are studied. In Chapter 16, models for conditionally heteroskedastic series are considered and, in particular, multivariate generalized autoregressive conditionally heteroskedastic (MGARCH) processes are presented and analyzed. In Chapter 17, VAR processes with time varying coefficients are considered. The coefficient variability may be due to a one-time intervention from outside the system or it may result from seasonal variation. Finally, in Chapter 18, so-called state space models are introduced. The models represent a very general class which encompasses most of the models previously discussed and includes in addition VAR models with stochastically varying coefficients. A brief review of these and other important models for multiple time series is given. The Kalman filter is presented as an important tool for dealing with state space models.

The reader is assumed to be familiar with vectors and matrices. The rules used in the text are summarized in Appendix A. Some results on the multivariate normal and related distributions are listed in Appendix B and stochastic convergence and some asymptotic distribution theory are reviewed in Appendix C. In Appendix D, a brief outline is given of the use of simulation techniques in evaluating properties of estimators and test statistics. Although it is not necessary for the reader to be familiar with all the particular rules and propositions listed in the appendices, it is implicitly assumed in the following chapters that the reader has knowledge of the basic terms and results.

Part I

Finite Order Vector Autoregressive Processes

In the four chapters of this part, finite order, stationary vector autoregressive (VAR) processes and their uses are discussed. Chapter 2 is dedicated to processes with known coefficients. Some of their basic properties are derived, their use for prediction and analysis purposes is considered. Unconstrained estimation is discussed in Chapter 3, model specification and checking the model adequacy are treated in Chapter 4, and estimation with parameter restrictions is the subject of Chapter 5.

2

Stable Vector Autoregressive Processes

In this chapter, the basic, stationary finite order vector autoregressive (VAR) model will be introduced. Some important properties will be discussed. The main uses of vector autoregressive models are forecasting and structural analysis. These two uses will be considered in Sections 2.2 and 2.3. Throughout this chapter, the model of interest is assumed to be known. Although this assumption is unrealistic in practice, it helps to see the problems related to VAR models without contamination by estimation and specification issues. The latter two aspects of an analysis will be treated in detail in subsequent chapters.

2.1 Basic Assumptions and Properties of VAR Processes

2.1.1 Stable VAR(p) Processes

The object of interest in the following is the VAR(p) model (VAR model of order p),

$$y_t = \nu + A_1 y_{t-1} + \cdots + A_p y_{t-p} + u_t, \quad t = 0, \pm 1, \pm 2, \ldots, \tag{2.1.1}$$

where $y_t = (y_{1t}, \ldots, y_{Kt})'$ is a $(K \times 1)$ random vector, the A_i are fixed $(K \times K)$ coefficient matrices, $\nu = (\nu_1, \ldots, \nu_K)'$ is a fixed $(K \times 1)$ vector of intercept terms allowing for the possibility of a nonzero mean $E(y_t)$. Finally, $u_t = (u_{1t}, \ldots, u_{Kt})'$ is a K-dimensional *white noise* or *innovation process*, that is, $E(u_t) = 0$, $E(u_t u_t') = \Sigma_u$ and $E(u_t u_s') = 0$ for $s \neq t$. The covariance matrix Σ_u is assumed to be nonsingular if not otherwise stated.

At this stage, it may be worth thinking a little more about which process is described by (2.1.1). In order to investigate the implications of the model let us consider the VAR(1) model

$$y_t = \nu + A_1 y_{t-1} + u_t. \tag{2.1.2}$$

If this generation mechanism starts at some time $t = 1$, say, we get

$$\begin{aligned} y_1 &= \nu + A_1 y_0 + u_1, \\ y_2 &= \nu + A_1 y_1 + u_2 = \nu + A_1(\nu + A_1 y_0 + u_1) + u_2 \\ &= (I_K + A_1)\nu + A_1^2 y_0 + A_1 u_1 + u_2, \\ &\vdots \\ y_t &= (I_K + A_1 + \cdots + A_1^{t-1})\nu + A_1^t y_0 + \sum_{i=0}^{t-1} A_1^i u_{t-i} \\ &\vdots \end{aligned} \qquad (2.1.3)$$

Hence, the vectors $y_1, \ldots, y_t$ are uniquely determined by $y_0, u_1, \ldots, u_t$. Also, the joint distribution of $y_1, \ldots, y_t$ is determined by the joint distribution of $y_0, u_1, \ldots, u_t$.

Although we will sometimes assume that a process is started in a specified period, it is often convenient to assume that it has been started in the infinite past. This assumption is in fact made in (2.1.1). What kind of process is consistent with the mechanism (2.1.1) in that case? To investigate this question we consider again the VAR(1) process (2.1.2). From (2.1.3) we have

$$\begin{aligned} y_t &= \nu + A_1 y_{t-1} + u_t \\ &= (I_K + A_1 + \cdots + A_1^j)\nu + A_1^{j+1} y_{t-j-1} + \sum_{i=0}^{j} A_1^i u_{t-i}. \end{aligned}$$

If all eigenvalues of A_1 have modulus less than 1, the sequence A_1^i, $i = 0, 1, \ldots$, is absolutely summable (see Appendix A, Section A.9.1). Hence, the infinite sum

$$\sum_{i=1}^{\infty} A_1^i u_{t-i}$$

exists in mean square (Appendix C, Proposition C.9). Moreover,

$$(I_K + A_1 + \cdots + A_1^j)\nu \xrightarrow[j \to \infty]{} (I_K - A_1)^{-1} \nu$$

(Appendix A, Section A.9.1). Furthermore, A_1^{j+1} converges to zero rapidly as $j \to \infty$ and, thus, we ignore the term $A_1^{j+1} y_{t-j-1}$ in the limit. Hence, if all eigenvalues of A_1 have modulus less than 1, by saying that y_t is the VAR(1) process (2.1.2) we mean that y_t is the well-defined stochastic process

$$y_t = \mu + \sum_{i=0}^{\infty} A_1^i u_{t-i}, \quad t = 0, \pm 1, \pm 2, \ldots, \qquad (2.1.4)$$

where

2.1 Basic Assumptions and Properties of VAR Processes 15

$$\mu := (I_K - A_1)^{-1}\nu.$$

The distributions and joint distributions of the y_t's are uniquely determined by the distributions of the u_t process. From Appendix C.3, Proposition C.10, the first and second moments of the y_t process are seen to be

$$E(y_t) = \mu \quad \text{for all } t \tag{2.1.5}$$

and

$$\begin{aligned}
\Gamma_y(h) &:= E(y_t - \mu)(y_{t-h} - \mu)' \\
&= \lim_{n \to \infty} \sum_{i=0}^{n} \sum_{j=0}^{n} A_1^i E(u_{t-i} u'_{t-h-j})(A_1^j)' \\
&= \lim \sum_{i=0}^{n} A_1^{h+i} \Sigma_u A_1^{i'} = \sum_{i=0}^{\infty} A_1^{h+i} \Sigma_u A_1^{i'},
\end{aligned} \tag{2.1.6}$$

because $E(u_t u'_s) = 0$ for $s \neq t$ and $E(u_t u'_t) = \Sigma_u$ for all t.

Because the condition for the eigenvalues of the matrix A_1 is of importance, we call a VAR(1) process *stable* if all eigenvalues of A_1 have modulus less than 1. By Rule (7) of Appendix A.6, the condition is equivalent to

$$\det(I_K - A_1 z) \neq 0 \quad \text{for } |z| \leq 1. \tag{2.1.7}$$

It is perhaps worth pointing out that the process y_t for $t = 0, \pm 1, \pm 2, \ldots$ may also be defined if the *stability condition* (2.1.7) is not satisfied. We will not do so here because we will always assume stability of processes defined for all $t \in \mathbb{Z}$.

The previous discussion can be extended easily to VAR(p) processes with $p > 1$ because any VAR(p) process can be written in VAR(1) form. More precisely, if y_t is a VAR(p) as in (2.1.1), a corresponding Kp-dimensional VAR(1)

$$Y_t = \boldsymbol{\nu} + \mathbf{A} Y_{t-1} + U_t \tag{2.1.8}$$

can be defined, where

$$Y_t := \underbrace{\begin{bmatrix} y_t \\ y_{t-1} \\ \vdots \\ y_{t-p+1} \end{bmatrix}}_{(Kp \times 1)}, \quad \boldsymbol{\nu} := \underbrace{\begin{bmatrix} \nu \\ 0 \\ \vdots \\ 0 \end{bmatrix}}_{(Kp \times 1)},$$

$$\mathbf{A} := \underbrace{\begin{bmatrix} A_1 & A_2 & \cdots & A_{p-1} & A_p \\ I_K & 0 & \cdots & 0 & 0 \\ 0 & I_K & & 0 & 0 \\ \vdots & & \ddots & \vdots & \vdots \\ 0 & 0 & \cdots & I_K & 0 \end{bmatrix}}_{(Kp \times Kp)}, \quad U_t := \underbrace{\begin{bmatrix} u_t \\ 0 \\ \vdots \\ 0 \end{bmatrix}}_{(Kp \times 1)}.$$

Following the foregoing discussion, Y_t is *stable* if

$$\det(I_{Kp} - \mathbf{A}z) \neq 0 \quad \text{for } |z| \leq 1. \tag{2.1.9}$$

Its mean vector is

$$\boldsymbol{\mu} := E(Y_t) = (I_{Kp} - \mathbf{A})^{-1}\boldsymbol{\nu}$$

and the autocovariances are

$$\Gamma_Y(h) = \sum_{i=0}^{\infty} \mathbf{A}^{h+i} \Sigma_U (\mathbf{A}^i)', \tag{2.1.10}$$

where $\Sigma_U := E(U_t U_t')$. Using the $(K \times Kp)$ matrix

$$J := [I_K : 0 : \cdots : 0], \tag{2.1.11}$$

the process y_t is obtained as $y_t = JY_t$. Because Y_t is a well-defined stochastic process, the same is true for y_t. Its mean is $E(y_t) = J\boldsymbol{\mu}$ which is constant for all t and the autocovariances $\Gamma_y(h) = J\Gamma_Y(h)J'$ are also time invariant.

It is easy to see that

$$\det(I_{Kp} - \mathbf{A}z) = \det(I_K - A_1 z - \cdots - A_p z^p)$$

(see Problem 2.1). Given the definition of the characteristic polynomial of a matrix, we call this polynomial the *reverse characteristic polynomial* of the VAR(p) process. Hence, the process (2.1.1) is *stable* if its reverse characteristic polynomial has no roots in and on the complex unit circle. Formally y_t is stable if

$$\det(I_K - A_1 z - \cdots - A_p z^p) \neq 0 \quad \text{for } |z| \leq 1. \tag{2.1.12}$$

This condition is called the *stability condition*.

In summary, we say that y_t is a stable VAR(p) process if (2.1.12) holds and

$$y_t = JY_t = J\boldsymbol{\mu} + J \sum_{i=0}^{\infty} \mathbf{A}^i U_{t-i}. \tag{2.1.13}$$

Because the $U_t := (u_t', 0, \ldots, 0)'$ involve the white noise process u_t, the process y_t is seen to be determined by its white noise or innovation process. Often specific assumptions regarding u_t are made which determine the process y_t by the foregoing convention. An important example is the assumption that u_t is *Gaussian white noise*, that is, $u_t \sim \mathcal{N}(0, \Sigma_u)$ for all t and u_t and u_s are independent for $s \neq t$. In that case, it can be shown that y_t is a *Gaussian process*, that is, subcollections $y_t, \ldots, y_{t+h}$ have multivariate normal distributions for all t and h.

2.1 Basic Assumptions and Properties of VAR Processes

The condition (2.1.12) provides an easy tool for checking the stability of a VAR process. Consider, for instance, the three-dimensional VAR(1) process

$$y_t = \nu + \begin{bmatrix} .5 & 0 & 0 \\ .1 & .1 & .3 \\ 0 & .2 & .3 \end{bmatrix} y_{t-1} + u_t. \tag{2.1.14}$$

For this process the reverse characteristic polynomial is

$$\det\left(\begin{bmatrix} 1 & 0 & 0 \\ 0 & 1 & 0 \\ 0 & 0 & 1 \end{bmatrix} - \begin{bmatrix} .5 & 0 & 0 \\ .1 & .1 & .3 \\ 0 & .2 & .3 \end{bmatrix} z\right)$$

$$= \det \begin{bmatrix} 1 - .5z & 0 & 0 \\ -.1z & 1 - .1z & -.3z \\ 0 & -.2z & 1 - .3z \end{bmatrix}$$

$$= (1 - .5z)(1 - .4z - .03z^2).$$

The roots of this polynomial are easily seen to be

$$z_1 = 2, \qquad z_2 = 2.1525, \qquad z_3 = -15.4858.$$

They are obviously all greater than 1 in absolute value. Therefore the process (2.1.14) is stable.

As another example consider the bivariate (two-dimensional) VAR(2) process

$$y_t = \nu + \begin{bmatrix} .5 & .1 \\ .4 & .5 \end{bmatrix} y_{t-1} + \begin{bmatrix} 0 & 0 \\ .25 & 0 \end{bmatrix} y_{t-2} + u_t. \tag{2.1.15}$$

Its reverse characteristic polynomial is

$$\det\left(\begin{bmatrix} 1 & 0 \\ 0 & 1 \end{bmatrix} - \begin{bmatrix} .5 & .1 \\ .4 & .5 \end{bmatrix} z - \begin{bmatrix} 0 & 0 \\ .25 & 0 \end{bmatrix} z^2\right) = 1 - z + .21z^2 - .025z^3.$$

The roots of this polynomial are

$$z_1 = 1.3, \qquad z_2 = 3.55 + 4.26i, \qquad \text{and} \qquad z_3 = 3.55 - 4.26i.$$

Here $i := \sqrt{-1}$ denotes the imaginary unit. Note that the modulus of z_2 and z_3 is $|z_2| = |z_3| = \sqrt{3.55^2 + 4.26^2} = 5.545$. Thus, the process (2.1.15) satisfies the stability condition (2.1.12) because all roots are outside the unit circle. Although the roots for higher dimensional and higher order processes are often difficult to compute by hand, efficient computer programs exist that do the job.

To understand the implications of the stability assumption, it may be helpful to visualize time series generated by stable processes and contrast them with realizations from unstable VAR processes. In Figure 2.1 three pairs of time series generated by three different stable bivariate (two-dimensional) VAR processes are depicted. Although they differ considerably, a common feature is that they fluctuate around constant means and their variability (variance) does not change as they wander along. In contrast, the pairs of series plotted in Figures 2.2 and 2.3 are generated by unstable, bivariate VAR processes. The time series in Figure 2.2 have a trend and those in Figure 2.3 exhibit quite pronounced seasonal fluctuations. Both shapes are typical of certain instabilities although they are quite common in practice. Hence, the stability assumption excludes many series of practical interest. We shall therefore discuss unstable processes in more detail in Part II. For that analysis understanding the stable case first is helpful.

2.1.2 The Moving Average Representation of a VAR Process

In the previous subsection we have considered the VAR(1) representation

$$Y_t = \boldsymbol{\nu} + \mathbf{A} Y_{t-1} + U_t$$

of the VAR(p) process (2.1.1). Under the stability assumption, the process Y_t has a representation

$$Y_t = \boldsymbol{\mu} + \sum_{i=0}^{\infty} \mathbf{A}^i U_{t-i}. \tag{2.1.16}$$

This form of the process is called the *moving average* (MA) *representation*, where Y_t is expressed in terms of past and present error or innovation vectors U_t and the mean term $\boldsymbol{\mu}$. This representation can be used to determine the autocovariances of Y_t and the mean and autocovariances of y_t can be obtained as outlined in Section 2.1.1. Moreover, an MA representation of y_t can be found by premultiplying (2.1.16) by the ($K \times Kp$) matrix $J := [I_K : 0 : \cdots : 0]$ (defined in (2.1.11)),

$$\begin{aligned} y_t &= J Y_t = J\boldsymbol{\mu} + \sum_{i=0}^{\infty} J \mathbf{A}^i J' J U_{t-i} \\ &= \mu + \sum_{i=0}^{\infty} \Phi_i u_{t-i}. \end{aligned} \tag{2.1.17}$$

Here $\mu := J\boldsymbol{\mu}$, $\Phi_i := J \mathbf{A}^i J'$ and, due to the special structure of the white noise process U_t, we have $U_t = J' J U_t$ and $J U_t = u_t$. Because the $\mathbf{A}^i$ are absolutely summable, the same is true for the Φ_i.

Later we will also consider other MA representations of a stable VAR(p) process. The unique feature of the present representation is that the zero order coefficient matrix $\Phi_0 = I_K$ and the white noise process involved consists of the error terms u_t of the VAR representation (2.1.1). In Section 2.2.2, the u_t will be seen to be the errors of optimal forecasts made in period $t - 1$. Therefore,

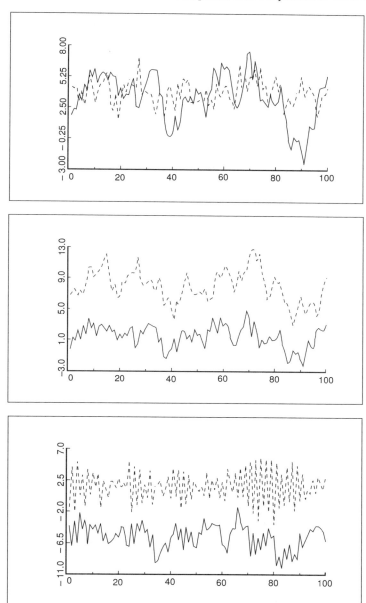

Fig. 2.1. Bivariate time series generated by stable processes.

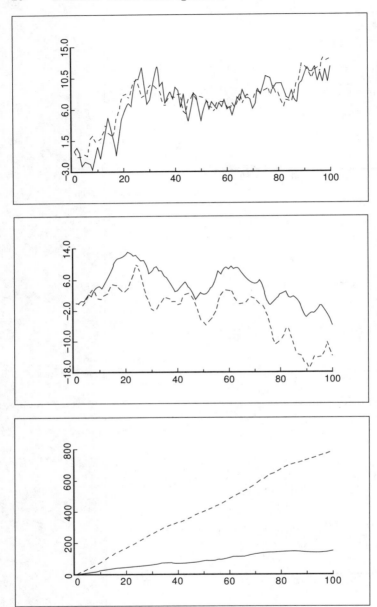

Fig. 2.2. Bivariate time series generated by unstable VAR processes.

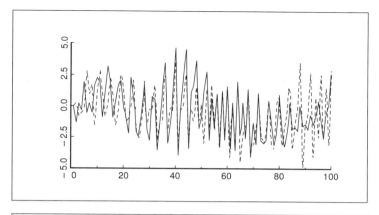

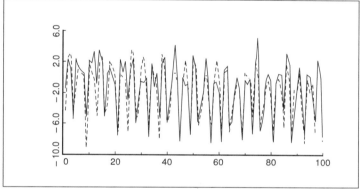

Fig. 2.3. Unstable seasonal time series.

to distinguish the present representation from other MA representations, we will sometimes refer to it as the *canonical* or *fundamental* or *prediction error representation*.

Using Proposition C.10 of Appendix C.3, the representation (2.1.17) provides a possibility for determining the mean and autocovariances of y_t:

$$E(y_t) = \mu$$

and

$$\begin{aligned}
\Gamma_y(h) &= E[(y_t - \mu)(y_{t-h} - \mu)'] \\
&= E\left[\left(\sum_{i=0}^{h-1} \Phi_i u_{t-i} + \sum_{i=0}^{\infty} \Phi_{h+i} u_{t-h-i}\right)\left(\sum_{i=0}^{\infty} \Phi_i u_{t-h-i}\right)'\right] \\
&= \sum_{i=0}^{\infty} \Phi_{h+i} \Sigma_u \Phi_i'.
\end{aligned} \quad (2.1.18)$$

There is no need to compute the MA coefficient matrices Φ_i via the VAR(1) representation corresponding to y_t as in the foregoing derivation. A more direct way for determining these matrices results from writing the VAR(p) process in *lag operator* notation. The lag operator L is defined such that $Ly_t = y_{t-1}$, that is, it lags (shifts back) the index by one period. Because of this property it is sometimes called *backshift operator*. Using this operator, (2.1.1) can be written as

$$y_t = \nu + (A_1 L + \cdots + A_p L^p) y_t + u_t$$

or

$$A(L) y_t = \nu + u_t, \tag{2.1.19}$$

where

$$A(L) := I_K - A_1 L - \cdots - A_p L^p.$$

Let

$$\Phi(L) := \sum_{i=0}^{\infty} \Phi_i L^i$$

be an operator such that

$$\Phi(L) A(L) = I_K. \tag{2.1.20}$$

Premultiplying (2.1.19) by $\Phi(L)$ gives

$$\begin{aligned} y_t &= \Phi(L)\nu + \Phi(L) u_t \\ &= \left(\sum_{i=0}^{\infty} \Phi_i \right) \nu + \sum_{i=0}^{\infty} \Phi_i u_{t-i}. \end{aligned} \tag{2.1.21}$$

The operator $\Phi(L)$ is the *inverse* of $A(L)$ and it is therefore sometimes denoted by $A(L)^{-1}$. Generally, we call the operator $A(L)$ *invertible* if $|A(z)| \neq 0$ for $|z| \leq 1$. If this condition is satisfied, the coefficient matrices of $\Phi(L) = A(L)^{-1}$ are absolutely summable and, hence, the process $\Phi(L) u_t = A(L)^{-1} u_t$ is well-defined (see Appendix C.3). The coefficient matrices Φ_i can be obtained from (2.1.20) using the relations

$$\begin{aligned} I_K &= (\Phi_0 + \Phi_1 L + \Phi_2 L^2 + \cdots)(I_K - A_1 L - \cdots - A_p L^p) \\ &= \Phi_0 + (\Phi_1 - \Phi_0 A_1) L + (\Phi_2 - \Phi_1 A_1 - \Phi_0 A_2) L^2 + \cdots \\ &\quad + \left(\Phi_i - \sum_{j=1}^{i} \Phi_{i-j} A_j \right) L^i + \cdots \end{aligned}$$

or

$$
\begin{aligned}
I_K &= \Phi_0 \\
0 &= \Phi_1 - \Phi_0 A_1 \\
0 &= \Phi_2 - \Phi_1 A_1 - \Phi_0 A_2 \\
&\vdots \\
0 &= \Phi_i - \sum_{j=1}^{i} \Phi_{i-j} A_j \\
&\vdots
\end{aligned}
$$

where $A_j = 0$ for $j > p$. Hence, the Φ_i can be computed recursively using

$$
\Phi_0 = I_K,
$$
$$
\Phi_i = \sum_{j=1}^{i} \Phi_{i-j} A_j, \quad i = 1, 2, \ldots. \tag{2.1.22}
$$

The mean μ of y_t can be obtained as follows:

$$
\mu = \Phi(1)\nu = A(1)^{-1}\nu = (I_K - A_1 - \cdots - A_p)^{-1}\nu. \tag{2.1.23}
$$

For a VAR(1) process, the recursions (2.1.22) imply that $\Phi_0 = I_K$, $\Phi_1 = A_1, \ldots, \Phi_i = A_1^i, \ldots$. This result is in line with (2.1.4). For the example VAR(1) process (2.1.14), we get $\Phi_0 = I_3$,

$$
\Phi_1 = \begin{bmatrix} .5 & 0 & 0 \\ .1 & .1 & .3 \\ 0 & .2 & .3 \end{bmatrix}, \quad \Phi_2 = \begin{bmatrix} .25 & 0 & 0 \\ .06 & .07 & .12 \\ .02 & .08 & .15 \end{bmatrix},
$$
$$
\Phi_3 = \begin{bmatrix} .125 & 0 & 0 \\ .037 & .031 & .057 \\ .018 & .038 & .069 \end{bmatrix}, \tag{2.1.24}
$$

etc. For a VAR(2), the recursions (2.1.22) result in

$$
\begin{aligned}
\Phi_1 &= A_1 \\
\Phi_2 &= \Phi_1 A_1 + A_2 = A_1^2 + A_2 \\
\Phi_3 &= \Phi_2 A_1 + \Phi_1 A_2 = A_1^3 + A_2 A_1 + A_1 A_2 \\
&\vdots \\
\Phi_i &= \Phi_{i-1} A_1 + \Phi_{i-2} A_2 \\
&\vdots
\end{aligned}
$$

Thus, for the example VAR(2) process (2.1.15), we get the MA coefficient matrices $\Phi_0 = I_2$,

$$
\Phi_1 = \begin{bmatrix} .5 & .1 \\ .4 & .5 \end{bmatrix}, \quad \Phi_2 = \begin{bmatrix} .29 & .1 \\ .65 & .29 \end{bmatrix}, \quad \Phi_3 = \begin{bmatrix} .21 & .079 \\ .566 & .21 \end{bmatrix}, \tag{2.1.25}
$$

etc. For both example processes, the Φ_i matrices approach zero as $i \to \infty$. This property is a consequence of the stability of the two processes.

It may be worth noting that the MA representation of a stable VAR(p) process is not necessarily of infinite order. That is, the Φ_i may all be zero for i greater than some finite integer q. For instance, for the bivariate VAR(1)

$$y_t = \nu + \begin{bmatrix} 0 & \alpha \\ 0 & 0 \end{bmatrix} y_{t-1} + u_t,$$

the MA representation is easily seen to be

$$y_t = \mu + u_t + \begin{bmatrix} 0 & \alpha \\ 0 & 0 \end{bmatrix} u_{t-1},$$

because

$$\begin{bmatrix} 0 & \alpha \\ 0 & 0 \end{bmatrix}^i = 0$$

for $i > 1$.

2.1.3 Stationary Processes

A stochastic process is *stationary* if its first and second moments are time invariant. In other words, a stochastic process y_t is stationary if

$$E(y_t) = \mu \qquad \text{for all } t \tag{2.1.26a}$$

and

$$E[(y_t - \mu)(y_{t-h} - \mu)'] = \Gamma_y(h) = \Gamma_y(-h)' \qquad \text{for all } t \text{ and } h = 0, 1, 2, \ldots. \tag{2.1.26b}$$

Condition (2.1.26a) means that all y_t have the same finite mean vector μ and (2.1.26b) requires that the autocovariances of the process do not depend on t but just on the time period h the two vectors y_t and y_{t-h} are apart. Note that, if not otherwise stated, all quantities are assumed to be finite. For instance, μ is a vector of finite mean terms and $\Gamma_y(h)$ is a matrix of finite covariances. Other definitions of stationarity are often used in the literature. For example, the joint distribution of n consecutive vectors may be assumed to be time invariant for all n. We shall, however, use the foregoing definition in the following. We call a process *strictly stationary* if the joint distributions of n consecutive variables are time invariant and there is a reason to distinguish between our notion of stationarity and the stricter form. By our definition, the white noise process u_t used in (2.1.1) is an obvious example of a stationary process. Also, from (2.1.18) we know that a stable VAR(p) process is stationary. We state this fact as a proposition.

2.1 Basic Assumptions and Properties of VAR Processes

Proposition 2.1 (*Stationarity Condition*)
A stable VAR(p) process y_t, $t = 0, \pm 1, \pm 2, \ldots$, is stationary. ∎

Because stability implies stationarity, the stability condition (2.1.12) is often referred to as *stationarity condition* in the time series literature. The converse of Proposition 2.1 is not true. In other words, an unstable process is not necessarily nonstationary. Because unstable stationary processes are not of interest in the following, we will not discuss this possibility here.

At this stage, it may be worth thinking about the generality of the VAR(p) processes considered in this and many other chapters. In this context, an important result due to Wold (1938) is of interest. He has shown that every stationary process x_t can be written as the sum of two uncorrelated processes z_t and y_t,

$$x_t = z_t + y_t,$$

where z_t is a deterministic process that can be forecast perfectly from its own past and y_t is a process with MA representation

$$y_t = \sum_{i=0}^{\infty} \Phi_i u_{t-i}, \tag{2.1.27}$$

where $\Phi_0 = I_K$, the u_t constitute a white noise process and the infinite sum is defined as a limit in mean square although the Φ_i are not necessarily absolutely summable (Hannan (1970, Chapter III)). The term "deterministic" will be explained more formally in Section 2.2. This result is often called *Wold's Decomposition Theorem*. If we assume that in the system of interest the only deterministic component is the mean term, the theorem states that the system has an MA representation. Suppose the Φ_i are absolutely summable and there exists an operator $A(L)$ with absolutely summable coefficient matrices satisfying $A(L)\Phi(L) = I_K$. Then $\Phi(L)$ is invertible ($A(L) = \Phi(L)^{-1}$) and y_t has a VAR representation of possibly infinite order,

$$y_t = \sum_{i=1}^{\infty} A_i y_{t-i} + u_t, \tag{2.1.28}$$

where

$$A(z) := I_K - \sum_{i=1}^{\infty} A_i z^i = \left(\sum_{i=0}^{\infty} \Phi_i z^i \right)^{-1} \quad \text{for } |z| \leq 1.$$

The A_i can be obtained from the Φ_i by recursions similar to (2.1.22).

The absolute summability of the A_i implies that the VAR coefficient matrices converge to zero rapidly. In other words, under quite general conditions, every stationary, purely nondeterministic process (a process without a deterministic component) can be approximated well by a finite order VAR process.

This is a very powerful result which demonstrates the generality of the processes under study. Note that economic variables can rarely be predicted without error. Thus the assumption of having a nondeterministic system except perhaps for a mean term is not a very restrictive one. The crucial and restrictive condition is the stationarity of the system, however. We will consider nonstationary processes later. For that discussion it is useful to understand the stationary case first.

An important implication of Wold's Decomposition Theorem is worth noting at this point. The theorem implies that any *subprocess* of a purely nondeterministic, stationary process y_t consisting of any subset of the components of y_t also has an MA representation. Suppose, for instance, that interest centers on the first M components of the K-dimensional process y_t, that is, we are interested in $x_t = F y_t$, where $F = [I_M : 0]$ is an $(M \times K)$ matrix. Then $E(x_t) = F E(y_t) = F \mu$ and $\Gamma_x(h) = F \Gamma_y(h) F'$ and, thus, x_t is stationary. Application of Wold's theorem then implies that x_t has an MA representation.

2.1.4 Computation of Autocovariances and Autocorrelations of Stable VAR Processes

Although the autocovariances of a stationary, stable VAR(p) process can be given in terms of its MA coefficient matrices as in (2.1.18), that formula is unattractive in practice, because it involves an infinite sum. For practical purposes it is easier to compute the autocovariances directly from the VAR coefficient matrices. In this section, we will develop the relevant formulas.

Autocovariances of a VAR(1) Process

In order to illustrate the computation of the autocovariances when the process coefficients are given, suppose that y_t is a stationary, stable VAR(1) process

$$y_t = \nu + A_1 y_{t-1} + u_t$$

with white noise covariance matrix $E(u_t u_t') = \Sigma_u$. Alternatively, the process may be written in mean-adjusted form as

$$y_t - \mu = A_1(y_{t-1} - \mu) + u_t, \qquad (2.1.29)$$

where $\mu = E(y_t)$, as before. Postmultiplying by $(y_{t-h} - \mu)'$ and taking expectations gives

$$E[(y_t - \mu)(y_{t-h} - \mu)'] = A_1 E[(y_{t-1} - \mu)(y_{t-h} - \mu)'] + E[u_t(y_{t-h} - \mu)'].$$

Thus, for $h = 0$,

$$\Gamma_y(0) = A_1 \Gamma_y(-1) + \Sigma_u = A_1 \Gamma_y(1)' + \Sigma_u \qquad (2.1.30)$$

and for $h > 0$,

2.1 Basic Assumptions and Properties of VAR Processes   27

$$\Gamma_y(h) = A_1 \Gamma_y(h-1). \tag{2.1.31}$$

These equations are usually referred to as *Yule-Walker equations*. If A_1 and the covariance matrix $\Gamma_y(0) = \Sigma_y$ of y_t are known, the $\Gamma_y(h)$ can be computed recursively using (2.1.31).

If A_1 and Σ_u are given, $\Gamma_y(0)$ can be determined as follows. For $h = 1$, we get from (2.1.31), $\Gamma_y(1) = A_1 \Gamma_y(0)$. Substituting $A_1 \Gamma_y(0)$ for $\Gamma_y(1)$ in (2.1.30) gives

$$\Gamma_y(0) = A_1 \Gamma_y(0) A_1' + \Sigma_u$$

or

$$\begin{aligned} \operatorname{vec} \Gamma_y(0) &= \operatorname{vec}(A_1 \Gamma_y(0) A_1') + \operatorname{vec} \Sigma_u \\ &= (A_1 \otimes A_1) \operatorname{vec} \Gamma_y(0) + \operatorname{vec} \Sigma_u. \end{aligned}$$

(For the definition of the Kronecker product $\otimes$, the vec operator and the rules used here, see Appendix A). Hence,

$$\operatorname{vec} \Gamma_y(0) = (I_{K^2} - A_1 \otimes A_1)^{-1} \operatorname{vec} \Sigma_u. \tag{2.1.32}$$

Note that the invertibility of $I_{K^2} - A_1 \otimes A_1$ follows from the stability of y_t because the eigenvalues of $A_1 \otimes A_1$ are the products of the eigenvalues of A_1 (see Appendix A). Hence, the eigenvalues of $A_1 \otimes A_1$ have modulus less than 1. Consequently, $\det(I_{K^2} - A_1 \otimes A_1) \neq 0$ (see Appendix A.9.1).

Using, for instance,

$$\Sigma_u = \begin{bmatrix} 2.25 & 0 & 0 \\ 0 & 1.0 & .5 \\ 0 & .5 & .74 \end{bmatrix}, \tag{2.1.33}$$

we get for the example process (2.1.14),

$$\operatorname{vec} \Gamma_y(0) = (I_9 - A_1 \otimes A_1)^{-1} \operatorname{vec} \Sigma_u$$

$$= \begin{bmatrix} .75 & 0 & 0 & 0 & 0 & 0 & 0 & 0 & 0 \\ -.05 & .95 & -.15 & 0 & 0 & 0 & 0 & 0 & 0 \\ 0 & -.10 & .85 & 0 & 0 & 0 & 0 & 0 & 0 \\ -.05 & 0 & 0 & .95 & 0 & 0 & -.15 & 0 & 0 \\ -.01 & -.01 & -.03 & -.01 & .99 & -.03 & -.03 & -.03 & -.09 \\ 0 & -.02 & -.03 & 0 & -.02 & .97 & 0 & -.06 & -.09 \\ 0 & 0 & 0 & 0 & -.01 & 0 & 0 & .85 & 0 & 0 \\ 0 & 0 & 0 & -.02 & -.02 & -.06 & -.03 & .97 & -.09 \\ 0 & 0 & 0 & 0 & -.04 & -.06 & 0 & -.06 & .91 \end{bmatrix}^{-1} \begin{bmatrix} 2.25 \\ 0 \\ 0 \\ 0 \\ 1.0 \\ .5 \\ 0 \\ .5 \\ .74 \end{bmatrix}$$

$$= \begin{bmatrix} 3.000 \\ .161 \\ .019 \\ .161 \\ 1.172 \\ .674 \\ .019 \\ .674 \\ .954 \end{bmatrix}.$$

It follows that

$$\Gamma_y(0) = \begin{bmatrix} 3.000 & .161 & .019 \\ .161 & 1.172 & .674 \\ .019 & .674 & .954 \end{bmatrix},$$

$$\Gamma_y(1) = A_1 \Gamma_y(0) = \begin{bmatrix} 1.500 & .080 & .009 \\ .322 & .335 & .355 \\ .038 & .437 & .421 \end{bmatrix}, \tag{2.1.34}$$

$$\Gamma_y(2) = A_1 \Gamma_y(1) = \begin{bmatrix} .750 & .040 & .005 \\ .194 & .173 & .163 \\ .076 & .198 & .197 \end{bmatrix}.$$

Note that the results are rounded after the computation. A higher precision has been used in intermediate steps.

Autocovariances of a Stable VAR(p) Process

For a higher order VAR(p) process,

$$y_t - \mu = A_1(y_{t-1} - \mu) + \cdots + A_p(y_{t-p} - \mu) + u_t, \tag{2.1.35}$$

the *Yule-Walker equations* are also obtained by postmultiplying with $(y_{t-h} - \mu)'$ and taking expectations. For $h = 0$, using $\Gamma_y(i) = \Gamma_y(-i)'$,

$$\begin{aligned} \Gamma_y(0) &= A_1 \Gamma_y(-1) + \cdots + A_p \Gamma_y(-p) + \Sigma_u \\ &= A_1 \Gamma_y(1)' + \cdots + A_p \Gamma_y(p)' + \Sigma_u, \end{aligned} \tag{2.1.36}$$

and for $h > 0$,

$$\Gamma_y(h) = A_1 \Gamma_y(h-1) + \cdots + A_p \Gamma_y(h-p). \tag{2.1.37}$$

These equations may be used to compute the $\Gamma_y(h)$ recursively for $h \geq p$, if $A_1, \ldots, A_p$ and $\Gamma_y(p-1), \ldots, \Gamma_y(0)$ are known.

The initial autocovariance matrices for $|h| < p$ can be determined using the VAR(1) process that corresponds to (2.1.35),

$$Y_t - \boldsymbol{\mu} = \mathbf{A}(Y_{t-1} - \boldsymbol{\mu}) + U_t, \tag{2.1.38}$$

2.1 Basic Assumptions and Properties of VAR Processes

where Y_t, $\mathbf{A}$, and U_t are as in (2.1.8) and $\boldsymbol{\mu} := (\mu', \ldots, \mu')' = E(Y_t)$. Proceeding as in the VAR(1) case gives

$$\Gamma_Y(0) = \mathbf{A}\Gamma_Y(0)\mathbf{A}' + \Sigma_U,$$

where $\Sigma_U = E(U_t U_t')$ and

$$\Gamma_Y(0) = E\left(\begin{bmatrix} y_t - \mu \\ \vdots \\ y_{t-p+1} - \mu \end{bmatrix} [(y_t - \mu)', \ldots, (y_{t-p+1} - \mu)']\right)$$

$$= \begin{bmatrix} \Gamma_y(0) & \Gamma_y(1) & \cdots & \Gamma_y(p-1) \\ \Gamma_y(-1) & \Gamma_y(0) & \cdots & \Gamma_y(p-2) \\ \vdots & \vdots & \ddots & \vdots \\ \Gamma_y(-p+1) & \Gamma_y(-p+2) & \cdots & \Gamma_y(0) \end{bmatrix}.$$

Thus, the $\Gamma_y(h)$, $h = -p+1, \ldots, p-1$, are obtained from

$$\text{vec}\,\Gamma_Y(0) = (I_{(Kp)^2} - \mathbf{A} \otimes \mathbf{A})^{-1} \text{vec}\,\Sigma_U. \tag{2.1.39}$$

For instance, for the example VAR(2) process (2.1.15) we get

$$\mathbf{A} = \begin{bmatrix} .5 & .1 & 0 & 0 \\ .4 & .5 & .25 & 0 \\ 1 & 0 & 0 & 0 \\ 0 & 1 & 0 & 0 \end{bmatrix} \tag{2.1.40}$$

and, assuming

$$\Sigma_u = \begin{bmatrix} .09 & 0 \\ 0 & .04 \end{bmatrix}, \tag{2.1.41}$$

we have

$$\Sigma_U = \begin{bmatrix} \Sigma_u & 0 \\ 0 & 0 \end{bmatrix} = \begin{bmatrix} .09 & 0 & 0 & 0 \\ 0 & .04 & 0 & 0 \\ 0 & 0 & 0 & 0 \\ 0 & 0 & 0 & 0 \end{bmatrix}.$$

Hence, using (2.1.39) and

$$\Gamma_Y(0) = \begin{bmatrix} \Gamma_y(0) & \Gamma_y(1) \\ \Gamma_y(1)' & \Gamma_y(0) \end{bmatrix}$$

gives

$$\Gamma_y(0) = \begin{bmatrix} .131 & .066 \\ .066 & .181 \end{bmatrix}, \quad \Gamma_y(1) = \begin{bmatrix} .072 & .051 \\ .104 & .143 \end{bmatrix},$$

$$\Gamma_y(2) = A_1\Gamma_y(1) + A_2\Gamma_y(0) = \begin{bmatrix} .046 & .040 \\ .113 & .108 \end{bmatrix}, \tag{2.1.42}$$

$$\Gamma_y(3) = A_1\Gamma_y(2) + A_2\Gamma_y(1) = \begin{bmatrix} .035 & .031 \\ .093 & .083 \end{bmatrix},$$

and so on. A method for computing $\Gamma_y(0)$ without explicitly inverting $(I - \mathbf{A} \otimes \mathbf{A})$ is given by Barone (1987).

The *autocovariance function* of a stationary VAR(p) process is *positive semidefinite*, that is,

$$\sum_{j=0}^{n}\sum_{i=0}^{n} a_j' \Gamma_y(i-j) a_i$$

$$= (a_0', \ldots, a_n') \begin{bmatrix} \Gamma_y(0) & \Gamma_y(1) & \cdots & \Gamma_y(n) \\ \Gamma_y(-1) & \Gamma_y(0) & \cdots & \Gamma_y(n-1) \\ \vdots & \vdots & \ddots & \vdots \\ \Gamma_y(-n) & \Gamma_y(-n+1) & \cdots & \Gamma_y(0) \end{bmatrix} \begin{bmatrix} a_0 \\ a_1 \\ \vdots \\ a_n \end{bmatrix} \geq 0 \tag{2.1.43}$$

for any $n \geq 0$. Here the a_i are arbitrary $(K \times 1)$ vectors. This result follows because (2.1.43) is just the variance of

$$(a_0', \ldots, a_n') \begin{bmatrix} y_t \\ y_{t-1} \\ \vdots \\ y_{t-n} \end{bmatrix}$$

which is always nonnegative.

Autocorrelations of a Stable VAR(p) Process

Because the autocovariances depend on the unit of measurement used for the variables of the system, they are sometimes difficult to interpret. Therefore, the autocorrelations

$$R_y(h) = D^{-1}\Gamma_y(h)D^{-1} \tag{2.1.44}$$

are usually more convenient to work with as they are scale invariant measures of the linear dependencies among the variables of the system. Here D is a diagonal matrix with the standard deviations of the components of y_t on the main diagonal. That is, the diagonal elements of D are the square roots of the diagonal elements of $\Gamma_y(0)$. Denoting the covariance between $y_{i,t}$ and $y_{j,t-h}$ by $\gamma_{ij}(h)$ (i.e., $\gamma_{ij}(h)$ is the ij-th element of $\Gamma_y(h)$) the diagonal elements $\gamma_{11}(0), \ldots, \gamma_{KK}(0)$ of $\Gamma_y(0)$ are the variances of $y_{1t}, \ldots, y_{Kt}$. Thus,

$$D^{-1} = \begin{bmatrix} 1/\sqrt{\gamma_{11}(0)} & & 0 \\ & \ddots & \\ 0 & & 1/\sqrt{\gamma_{KK}(0)} \end{bmatrix}$$

and the correlation between $y_{i,t}$ and $y_{j,t-h}$ is

$$\rho_{ij}(h) = \frac{\gamma_{ij}(h)}{\sqrt{\gamma_{ii}(0)}\sqrt{\gamma_{jj}(0)}} \qquad (2.1.45)$$

which is just the ij-th element of $R_y(h)$.

For the VAR(1) example process (2.1.14) we get from (2.1.34),

$$D = \begin{bmatrix} \sqrt{3.000} & 0 & 0 \\ 0 & \sqrt{1.172} & 0 \\ 0 & 0 & \sqrt{.954} \end{bmatrix} = \begin{bmatrix} 1.732 & 0 & 0 \\ 0 & 1.083 & 0 \\ 0 & 0 & .977 \end{bmatrix}$$

and

$$R_y(0) = D^{-1}\Gamma_y(0)D^{-1} = \begin{bmatrix} 1 & .086 & .011 \\ .086 & 1 & .637 \\ .011 & .637 & 1 \end{bmatrix},$$

$$R_y(1) = D^{-1}\Gamma_y(1)D^{-1} = \begin{bmatrix} .500 & .043 & .005 \\ .172 & .286 & .336 \\ .022 & .413 & .441 \end{bmatrix}, \qquad (2.1.46)$$

$$R_y(2) = D^{-1}\Gamma_y(2)D^{-1} = \begin{bmatrix} .250 & .021 & .003 \\ .103 & .148 & .154 \\ .045 & .187 & .206 \end{bmatrix}.$$

A plot of some autocorrelations is shown in Figure 2.4. Assuming that the three variables of the system represent rates of change of investment, income, and consumption, respectively, it can, for instance, be seen that the contemporaneous and intertemporal correlations between consumption and investment are quite small, while the patterns of the autocorrelations of the individual series are similar.

2.2 Forecasting

We have argued in the introduction that forecasting is one of the main objectives of multiple time series analysis. Therefore, we will now discuss predictors based on VAR processes. Point forecasts and interval forecasts will be considered in turn. Before discussing particular predictors or forecasts (the two terms will be used interchangeably) we comment on the prediction problem in general.

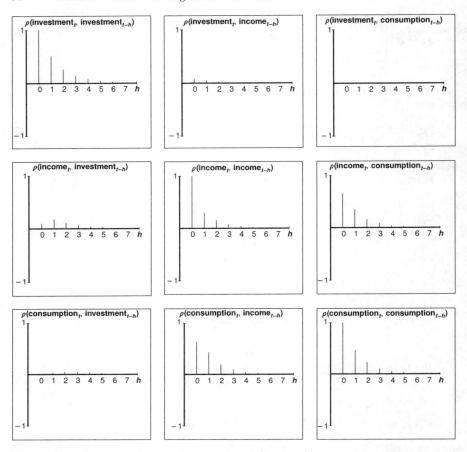

Fig. 2.4. Autocorrelations of the investment/income/consumption system.

2.2.1 The Loss Function

The forecaster usually finds himself in a situation where in a particular period t he has to make statements about the future values of variables $y_1, \ldots, y_K$. For this purpose he has available a model for the data generation process and an information set, say Ω_t, containing the available information in period t. The data generation process may, for instance, be a VAR(p) process and Ω_t may contain the past and present variables of the system under consideration, that is, $\Omega_t = \{y_s | s \leq t\}$, where $y_s = (y_{1s}, \ldots, y_{Ks})'$. The period t, where the forecast is made, is the *forecast origin* and the number of periods into the future for which a forecast is desired is the *forecast horizon*. A predictor, h periods ahead, is an *h-step predictor*.

If forecasts are desired for a particular purpose, a specific cost function may be associated with the forecast errors. A forecast will be optimal if it minimizes the cost. To find a forecast that is optimal in this sense is usually too ambitious a goal to be attainable in practice. Therefore, minimizing the *expected* cost or loss is often used as an objective. In general, it will depend on the particular loss function which forecast is optimal. On the other hand, forecasts of economic variables are often published for general use. In that case, the specific cost or loss function of all potential users cannot be taken into account in computing a forecast. In this situation, the statistical properties of the forecasts and perhaps interval forecasts are of interest to enable the user to draw proper conclusions for his or her particular needs. It may also be desirable to choose the forecast such that it minimizes a wide range of plausible loss functions.

In the context of VAR models, predictors that minimize the forecast mean squared errors (MSEs) are the most widely used ones. Arguments in favor of using the MSE as loss function are given by Granger (1969b) and Granger & Newbold (1986). They show that minimum MSE forecasts also minimize a range of loss functions other than the MSE. Moreover, for many loss functions the optimal predictors are simple functions of minimum MSE predictors. Furthermore, for an unbiased predictor, the MSE is the forecast error variance which is useful in setting up interval forecasts. Therefore, minimum MSE predictors will be of major interest in the following. If not otherwise stated, the information set Ω_t is assumed to contain the variables of the system under consideration up to and including period t.

2.2.2 Point Forecasts

Conditional Expectation

Suppose $y_t = (y_{1t}, \ldots, y_{Kt})'$ is a K-dimensional stable VAR(p) process as in (2.1.1). Then, the minimum MSE predictor for forecast horizon h at forecast origin t is the conditional expected value

$$E_t(y_{t+h}) := E(y_{t+h}|\Omega_t) = E(y_{t+h}|\{y_s|s \leq t\}). \qquad (2.2.1)$$

This predictor minimizes the MSE of each component of y_t. In other words, if $\bar{y}_t(h)$ is any h-step predictor at origin t,

$$\begin{aligned}\text{MSE}[\bar{y}_t(h)] &= E[(y_{t+h} - \bar{y}_t(h))(y_{t+h} - \bar{y}_t(h))'] \\ &\geq \text{MSE}[E_t(y_{t+h})] = E[(y_{t+h} - E_t(y_{t+h}))(y_{t+h} - E_t(y_{t+h}))'], \end{aligned} \qquad (2.2.2)$$

where the inequality sign $\geq$ between two matrices means that the difference between the left-hand and the right-hand matrix is positive semidefinite. Equivalently, for any $(K \times 1)$ vector c,

$$\text{MSE}[c'\bar{y}_t(h)] \geq \text{MSE}[c'E_t(y_{t+h})].$$

The optimality of the conditional expectation can be seen by noting that

$$\text{MSE}[\bar{y}_t(h)] = E\{[y_{t+h} - E_t(y_{t+h}) + E_t(y_{t+h}) - \bar{y}_t(h)]$$
$$\times [y_{t+h} - E_t(y_{t+h}) + E_t(y_{t+h}) - \bar{y}_t(h)]'\}$$
$$= \text{MSE}[E_t(y_{t+h})]$$
$$+ E\{[E_t(y_{t+h}) - \bar{y}_t(h)][E_t(y_{t+h}) - \bar{y}_t(h)]'\},$$

where $E\{[y_{t+h} - E_t(y_{t+h})][E_t(y_{t+h}) - \bar{y}_t(h)]'\} = 0$ has been used. The latter result holds because $[y_{t+h} - E_t(y_{t+h})]$ is a function of innovations after period t which are uncorrelated with the terms contained in $[E_t(y_{t+h}) - \bar{y}_t(h)]$ which are functions of y_s, $s \leq t$.

The optimality of the conditional expectation implies that

$$E_t(y_{t+h}) = \nu + A_1 E_t(y_{t+h-1}) + \cdots + A_p E_t(y_{t+h-p}) \qquad (2.2.3)$$

is the optimal h-step predictor of a VAR(p) process y_t, provided u_t is *independent white noise* so that u_t and u_s are independent for $s \neq t$ and, hence, $E_t(u_{t+h}) = 0$ for $h > 0$.

The formula (2.2.3) can be used for recursively computing the h-step predictors starting with $h = 1$:

$$E_t(y_{t+1}) = \nu + A_1 y_t + \cdots + A_p y_{t-p+1},$$
$$E_t(y_{t+2}) = \nu + A_1 E_t(y_{t+1}) + A_2 y_t + \cdots + A_p y_{t-p+2},$$
$$\vdots$$

By these recursions we get for a VAR(1) process

$$E_t(y_{t+h}) = (I_K + A_1 + \cdots + A_1^{h-1})\nu + A_1^h y_t.$$

Assuming $y_t = (-6, 3, 5)'$ and $\nu = (0, 2, 1)'$, the following forecasts are obtained for the VAR(1) example process (2.1.14):

$$E_t(y_{t+1}) = \begin{bmatrix} 0 \\ 2 \\ 1 \end{bmatrix} + \begin{bmatrix} .5 & 0 & 0 \\ .1 & .1 & .3 \\ 0 & .2 & .3 \end{bmatrix} \begin{bmatrix} -6 \\ 3 \\ 5 \end{bmatrix} = \begin{bmatrix} -3.0 \\ 3.2 \\ 3.1 \end{bmatrix}, \qquad (2.2.4a)$$

$$E_t(y_{t+2}) = (I_3 + A_1)\nu + A_1^2 y_t = \begin{bmatrix} -1.50 \\ 2.95 \\ 2.57 \end{bmatrix}, \qquad (2.2.4b)$$

etc. Similarly, we get for the VAR(2) process (2.1.15) with $\nu = (.02, .03)'$, $y_t = (.06, .03)'$ and $y_{t-1} = (.055, .03)'$,

$$E_t(y_{t+1}) = \begin{bmatrix} .02 \\ .03 \end{bmatrix} + \begin{bmatrix} .5 & .1 \\ .4 & .5 \end{bmatrix} \begin{bmatrix} .06 \\ .03 \end{bmatrix} + \begin{bmatrix} 0 & 0 \\ .25 & 0 \end{bmatrix} \begin{bmatrix} .055 \\ .03 \end{bmatrix}$$
$$= \begin{bmatrix} .053 \\ .08275 \end{bmatrix},$$

$$E_t(y_{t+2}) = \begin{bmatrix} .02 \\ .03 \end{bmatrix} + \begin{bmatrix} .5 & .1 \\ .4 & .5 \end{bmatrix} \begin{bmatrix} .053 \\ .08275 \end{bmatrix} + \begin{bmatrix} 0 & 0 \\ .25 & 0 \end{bmatrix} \begin{bmatrix} .06 \\ .03 \end{bmatrix}$$
$$= \begin{bmatrix} .0548 \\ .1076 \end{bmatrix}. \qquad (2.2.5)$$

The conditional expectation has the following properties:

(1) It is an unbiased predictor, that is, $E[y_{t+h} - E_t(y_{t+h})] = 0$.
(2) If u_t is independent white noise, $\text{MSE}[E_t(y_{t+h})] = \text{MSE}[E_t(y_{t+h})|y_t, y_{t-1}, \ldots]$, that is, the MSE of the predictor equals the conditional MSE given $y_t, y_{t-1}, \ldots$.

The latter property follows by similar arguments as the optimality of the predictor $E_t(y_{t+h})$.

It must be emphasized that the prediction formula (2.2.3) relies on u_t being independent white noise. If u_t and u_s are not independent but just uncorrelated, $E_t(u_{t+h})$ will be nonzero in general. As an example consider the univariate AR(1) process $y_t = \nu + \alpha y_{t-1} + u_t$ with

$$u_t = \begin{cases} e_t & \text{for } t = 0, \pm 2, \pm 4, \ldots, \\ (e_{t-1}^2 - 1)/\sqrt{2} & \text{for } t = \pm 1, \pm 3, \ldots, \end{cases}$$

where the e_t are independent standard normal ($\mathcal{N}(0,1)$) random variables (see also Fuller (1976, Chapter 2, Exercise 16)). The process u_t is easily seen to be uncorrelated but not independent white noise. For even t,

$$E_t(u_{t+1}) = E[(e_t^2 - 1)/\sqrt{2}|y_t, y_{t-1}, \ldots]$$
$$= (e_t^2 - 1)/\sqrt{2},$$

because $e_t = y_t - \nu - \alpha y_{t-1}$.

Linear Minimum MSE Predictor

If u_t is not independent white noise, additional assumptions are usually required to find the optimal predictor (conditional expectation) of a VAR(p) process. Without such assumptions we can achieve the less ambitious goal of finding the minimum MSE predictors among those that are *linear* functions of $y_t, y_{t-1}, \ldots$. Let us consider a zero mean VAR(1) process

$$y_t = A_1 y_{t-1} + u_t \qquad (2.2.6)$$

first. As in (2.1.3), it follows that

$$y_{t+h} = A_1^h y_t + \sum_{i=0}^{h-1} A_1^i u_{t+h-i}.$$

Thus, for a predictor

$$y_t(h) = B_0 y_t + B_1 y_{t-1} + \cdots,$$

where the B_i's are $(K \times K)$ coefficient matrices, we get a forecast error

$$y_{t+h} - y_t(h) = \sum_{i=0}^{h-1} A_1^i u_{t+h-i} + (A_1^h - B_0) y_t - \sum_{i=1}^{\infty} B_i y_{t-i}.$$

Using that u_{t+j}, for $j > 0$, is uncorrelated with y_{t-i}, for $i \geq 0$, we get

$$\text{MSE}[y_t(h)]$$
$$= E\left[\left(\sum_{i=0}^{h-1} A_1^i u_{t+h-i}\right)\left(\sum_{i=0}^{h-1} A_1^i u_{t+h-i}\right)'\right]$$
$$+ E\left\{\left[(A_1^h - B_0)y_t - \sum_{i=1}^{\infty} B_i y_{t-i}\right]\left[(A_1^h - B_0)y_t - \sum_{i=1}^{\infty} B_i y_{t-i}\right]'\right\}.$$

Obviously, this MSE matrix is minimal for $B_0 = A_1^h$ and $B_i = 0$. Thus, the optimal (*linear* minimum MSE) predictor for this special case is

$$y_t(h) = A_1^h y_t = A_1 y_t(h-1).$$

The forecast error is

$$\sum_{i=0}^{h-1} A_1^i u_{t+h-i}$$

and the MSE or forecast error covariance matrix is

$$\Sigma_y(h) := \text{MSE}[y_t(h)] = E\left[\left(\sum_{i=0}^{h-1} A_1^i u_{t+h-i}\right)\left(\sum_{i=0}^{h-1} A_1^i u_{t+h-i}\right)'\right]$$
$$= \sum_{i=0}^{h-1} A_1^i \Sigma_u (A_1^i)' = \text{MSE}[y_t(h-1)] + A_1^{h-1} \Sigma_u (A_1^{h-1})'.$$

A general VAR(p) process with zero mean,

$$y_t = A_1 y_{t-1} + \cdots + A_p y_{t-p} + u_t,$$

has a VAR(1) counterpart,

$$Y_t = \mathbf{A} Y_{t-1} + U_t,$$

where Y_t, $\mathbf{A}$, and U_t are as defined in (2.1.8). Using the same arguments as above, the optimal predictor of Y_{t+h} is seen to be

$$Y_t(h) = \mathbf{A}^h Y_t = \mathbf{A} Y_t(h-1).$$

It is easily seen by induction with respect to h that

$$Y_t(h) = \begin{bmatrix} y_t(h) \\ y_t(h-1) \\ \vdots \\ y_t(h-p+1) \end{bmatrix},$$

where $y_t(j) := y_{t+j}$ for $j \leq 0$. Defining the $(K \times Kp)$ matrix $J := [I_K : 0 : \cdots : 0]$ as in (2.1.11), we get the optimal h-step predictor of the process y_t at origin t as

$$\begin{aligned} y_t(h) &= J\mathbf{A} Y_t(h-1) = [A_1, \ldots, A_p] Y_t(h-1) \\ &= A_1 y_t(h-1) + \cdots + A_p y_t(h-p). \end{aligned} \quad (2.2.7)$$

This formula may be used for recursively computing the forecasts. Obviously, $y_t(h)$ is the conditional expectation $E_t(y_{t+h})$ if u_t is independent white noise because the recursion in (2.2.3) is the same as the one obtained here for a zero mean process with $\nu = 0$.

If the process y_t has nonzero mean, that is,

$$y_t = \nu + A_1 y_{t-1} + \cdots + A_p y_{t-p} + u_t,$$

we define $x_t := y_t - \mu$, where $\mu := E(y_t) = (I - A_1 - \cdots - A_p)^{-1} \nu$. The process x_t has zero mean and the optimal h-step predictor is

$$x_t(h) = A_1 x_t(h-1) + \cdots + A_p x_t(h-p).$$

Adding μ to both sides of this equation gives the optimal linear predictor of y_t,

$$\begin{aligned} y_t(h) &= x_t(h) + \mu = \mu + A_1(y_t(h-1) - \mu) + \cdots + A_p(y_t(h-p) - \mu) \\ &= \nu + A_1 y_t(h-1) + \cdots + A_p y_t(h-p). \end{aligned} \quad (2.2.8)$$

Henceforth, we will refer to $y_t(h)$ as the *optimal predictor* irrespective of the properties of the white noise process u_t, that is, even if u_t is not independent but just uncorrelated white noise.

Using

$$Y_{t+h} = \mathbf{A}^h Y_t + \sum_{i=0}^{h-1} \mathbf{A}^i U_{t+h-i}$$

for a zero mean process, we get the forecast error

$$\begin{aligned} y_{t+h} - y_t(h) &= J[Y_{t+h} - Y_t(h)] = J \left[\sum_{i=0}^{h-1} \mathbf{A}^i U_{t+h-i} \right] \\ &= \sum_{i=0}^{h-1} J\mathbf{A}^i J' J U_{t+h-i} = \sum_{i=0}^{h-1} \Phi_i u_{t+h-i}, \end{aligned} \quad (2.2.9)$$

where the Φ_i are the MA coefficient matrices from (2.1.17). The forecast error is unchanged if y_t has nonzero mean because the mean term cancels. The forecast error representation (2.2.9) shows that the predictor $y_t(h)$ can also be expressed in terms of the MA representation (2.1.17),

$$y_t(h) = \mu + \sum_{i=h}^{\infty} \Phi_i u_{t+h-i} = \mu + \sum_{i=0}^{\infty} \Phi_{h+i} u_{t-i}. \tag{2.2.10}$$

From (2.2.9) the forecast error covariance or MSE matrix is easy to obtain,

$$\Sigma_y(h) := \text{MSE}[y_t(h)] = \sum_{i=0}^{h-1} \Phi_i \Sigma_u \Phi_i' = \Sigma_y(h-1) + \Phi_{h-1} \Sigma_u \Phi_{h-1}'. \tag{2.2.11}$$

Hence, the MSEs are monotonically nondecreasing and, for $h \to \infty$, the MSE matrices approach the covariance matrix of y_t,

$$\Gamma_y(0) = \Sigma_y = \sum_{i=0}^{\infty} \Phi_i \Sigma_u \Phi_i'$$

(see (2.1.18)). That is,

$$\Sigma_y(h) \xrightarrow[h \to \infty]{} \Sigma_y. \tag{2.2.12}$$

If the process mean μ is used as a forecast, the MSE matrix of that predictor is just the covariance matrix Σ_y of y_t. Hence, the optimal long range forecast ($h \to \infty$) is the process mean. In other words, the past of the process contains no information on the development of the process in the distant future. Zero mean processes with this property are *purely nondeterministic*, that is, $y_t - \mu$ is purely nondeterministic if the forecast MSEs satisfy (2.2.12).

For the example VAR(1) process (2.1.14) with Σ_u as in (2.1.33), using the MA coefficient matrices from (2.1.24), the forecast MSE matrices

$$\Sigma_y(1) = \Sigma_u = \begin{bmatrix} 2.25 & 0 & 0 \\ 0 & 1.0 & .5 \\ 0 & .5 & .74 \end{bmatrix},$$

$$\Sigma_y(2) = \Sigma_u + \Phi_1 \Sigma_u \Phi_1' = \begin{bmatrix} 2.813 & .113 & 0 \\ .113 & 1.129 & .632 \\ 0 & .632 & .907 \end{bmatrix}, \tag{2.2.13}$$

$$\Sigma_y(3) = \Sigma_y(2) + \Phi_2 \Sigma_u \Phi_2' = \begin{bmatrix} 2.953 & .146 & .011 \\ .146 & 1.161 & .663 \\ .011 & .663 & .943 \end{bmatrix}$$

are obtained. Similarly, for the VAR(2) example process (2.1.15) with white noise covariance matrix (2.1.41), we get with Φ_1 from (2.1.25),

$$\Sigma_y(1) = \Sigma_u = \begin{bmatrix} .09 & 0 \\ 0 & .04 \end{bmatrix},$$

$$\Sigma_y(2) = \Sigma_u + \Phi_1 \Sigma_u \Phi_1' = \begin{bmatrix} .1129 & .02 \\ .02 & .0644 \end{bmatrix}. \qquad (2.2.14)$$

2.2.3 Interval Forecasts and Forecast Regions

In order to set up interval forecasts or forecast intervals, we make an assumption about the distributions of the y_t or the u_t. It is most common to consider *Gaussian processes* where $y_t, y_{t+1}, \ldots, y_{t+h}$ have a multivariate normal distribution for any t and h. Equivalently, it may be assumed that u_t is Gaussian, that is, the u_t are multivariate normal, $u_t \sim \mathcal{N}(0, \Sigma_u)$, and u_t and u_s are independent for $s \neq t$.

Under these conditions the forecast errors are also normally distributed as linear transformations of normal vectors,

$$y_{t+h} - y_t(h) = \sum_{i=0}^{h-1} \Phi_i u_{t+h-i} \sim \mathcal{N}(0, \Sigma_y(h)). \qquad (2.2.15)$$

This result implies that the forecast errors of the individual components are normal so that

$$\frac{y_{k,t+h} - y_{k,t}(h)}{\sigma_k(h)} \sim \mathcal{N}(0, 1), \qquad (2.2.16)$$

where $y_{k,t}(h)$ is the k-th component of $y_t(h)$ and $\sigma_k(h)$ is the square root of the k-th diagonal element of $\Sigma_y(h)$. Denoting by $z_{(\alpha)}$ the upper $\alpha 100$ percentage point of the normal distribution, we get

$$\begin{aligned} 1 - \alpha &= \Pr\left\{-z_{(\alpha/2)} \leq \frac{y_{k,t+h} - y_{k,t}(h)}{\sigma_k(h)} \leq z_{(\alpha/2)}\right\} \\ &= \Pr\left\{y_{k,t}(h) - z_{(\alpha/2)}\sigma_k(h) \leq y_{k,t+h} \leq y_{k,t}(h) + z_{(\alpha/2)}\sigma_k(h)\right\}. \end{aligned}$$

Hence, a $(1-\alpha)100\%$ interval forecast, h periods ahead, for the k-th component of y_t is

$$y_{k,t}(h) \pm z_{(\alpha/2)}\sigma_k(h) \qquad (2.2.17a)$$

or

$$[y_{k,t}(h) - z_{(\alpha/2)}\sigma_k(h), y_{k,t}(h) + z_{(\alpha/2)}\sigma_k(h)]. \qquad (2.2.17b)$$

If forecast intervals of this type are computed repeatedly from a large number of time series (realizations of the considered process), then about $(1-\alpha)100\%$ of the intervals will contain the actual value of the random variable $y_{k,t+h}$.

Using (2.2.4a), (2.2.4b) and (2.2.13), 95% forecast intervals for the components of the example VAR(1) process (2.1.14) are

$$y_{1,t}(1) \pm 1.96\sqrt{2.25} \quad \text{or} \quad -3.0 \pm 2.94,$$
$$y_{2,t}(1) \pm 1.96\sqrt{1.0} \quad \text{or} \quad 3.2 \pm 1.96,$$
$$y_{3,t}(1) \pm 1.96\sqrt{.74} \quad \text{or} \quad 3.1 \pm 1.69, \qquad (2.2.18)$$
$$y_{1,t}(2) \pm 1.96\sqrt{2.813} \quad \text{or} \quad -1.50 \pm 3.29,$$
$$y_{2,t}(2) \pm 1.96\sqrt{1.129} \quad \text{or} \quad 2.95 \pm 2.08,$$
$$y_{3,t}(2) \pm 1.96\sqrt{.907} \quad \text{or} \quad 2.57 \pm 1.87.$$

The result in (2.2.15) can also be used to establish joint forecast regions for two or more variables. For instance, if a joint forecast region for the first N components is desired, we define the $(N \times K)$ matrix $F := [I_N : 0]$ and note that

$$[y_{t+h} - y_t(h)]'F'[F\Sigma_y(h)F']^{-1}F[y_{t+h} - y_t(h)] \sim \chi^2(N) \qquad (2.2.19)$$

by a well-known result for multivariate normal vectors (see Appendix B). Hence, the $\chi^2(N)$-distribution can be used to determine a $(1-\alpha)100\%$ forecast ellipsoid for the first N components of the process.

In practice, the construction of the ellipsoid is quite demanding if N is greater than two or three. Therefore, a more practical approach is to use *Bonferroni's method* for constructing joint confidence regions. It is based on the fact that for events $E_1, \ldots, E_N$ the following probability inequality holds:

$$\Pr(E_1 \cup \cdots \cup E_N) \leq \Pr(E_1) + \cdots + \Pr(E_N).$$

Hence,

$$\Pr\left(\bigcap_{i=1}^{N} E_i\right) \geq 1 - \sum_{i=1}^{N} \Pr(\bar{E}_i),$$

where $\bar{E}_i$ denotes the complement of E_i. Consequently, if E_i is the event that $y_{i,t+h}$ falls within an interval H_i,

$$\Pr(Fy_{t+h} \in H_1 \times \cdots \times H_N) \geq 1 - \sum_{i=1}^{N} \Pr(\bar{E}_i). \qquad (2.2.20)$$

In other words, if we choose a $\left(1 - \frac{\alpha}{N}\right)100\%$ forecast interval for each of the N components, the resulting joint forecast region has probability at least $(1-\alpha)100\%$ of containing all N variables jointly. For instance, for the VAR(1) example process considered previously,

$$\{(y_1, y_2) | -3.0 - 2.94 \leq y_1 \leq -3.0 + 2.94,\ 3.2 - 1.96 \leq y_2 \leq 3.2 + 1.96\}$$

is a joint forecast region of $(y_{1,t+1}, y_{2,t+1})$ with probability content at least 90%.

By the same method joint forecast regions for different horizons h can be obtained. For instance, a joint forecast region with probability content of at least $(1-\alpha)100\%$ for $y_{k,t+1}, \ldots, y_{k,t+h}$ is

$$\{(y_{k,1}, \ldots, y_{k,h}) | y_{k,t}(i) - z_{(\alpha/2h)}\sigma_k(i) \leq y_{k,i} \leq y_{k,t}(i) + z_{(\alpha/2h)}\sigma_k(i),$$
$$i = 1, \ldots, h\}.$$
(2.2.21)

Thus, for the example, a joint forecast region for $y_{2,t+1}, y_{2,t+2}$ with probability content of at least 90% is given by

$$\{(y_{2,1}, y_{2,2}) | 1.24 \leq y_{2,1} \leq 5.16, \ .87 \leq y_{2,2} \leq 5.03\}.$$

Under our assumption of a Gaussian process, the distribution of the forecasts and forecast errors is known and, consequently, forecast intervals are easy to set up. If the underlying process has a different and potentially unknown distribution, considering the forecast distribution becomes more difficult. Even then methods are available to determine more than just point forecasts. A survey of density forecasting is given by Tay & Wallis (2002).

2.3 Structural Analysis with VAR Models

Because VAR models represent the correlations among a set of variables, they are often used to analyze certain aspects of the relationships between the variables of interest. In the following, three ways to interpret a VAR model will be discussed. They are all closely related and they are all beset with problems that will be pointed out subsequently.

2.3.1 Granger-Causality, Instantaneous Causality, and Multi-Step Causality

Definitions of Causality

Granger (1969a) has defined a concept of causality which, under suitable conditions, is fairly easy to deal with in the context of VAR models. Therefore it has become quite popular in recent years. The idea is that a cause cannot come after the effect. Thus, if a variable x affects a variable z, the former should help improving the predictions of the latter variable.

To formalize this idea, suppose that Ω_t is the information set containing all the relevant information in the universe available up to and including period t. Let $z_t(h|\Omega_t)$ be the optimal (minimum MSE) h-step predictor of the process z_t at origin t, based on the information in Ω_t. The corresponding forecast MSE will be denoted by $\Sigma_z(h|\Omega_t)$. The process x_t is said to *cause z_t in Granger's sense* if

$$\Sigma_z(h|\Omega_t) < \Sigma_z(h|\Omega_t \setminus \{x_s|s \leq t\}) \quad \text{for at least one } h = 1, 2, \ldots. \quad (2.3.1)$$

Alternatively, we will say that x_t *Granger-causes* (or briefly causes) z_t or x_t is *Granger-causal* for z_t if (2.3.1) holds. In (2.3.1) $\Omega_t \setminus \{x_s|s \leq t\}$ is the set containing all the relevant information in the universe except for the information in the past and present of the x_t process. In other words, if z_t can be predicted more efficiently if the information in the x_t process is taken into account in addition to all other information in the universe, then x_t is *Granger-causal* for z_t.

The definition extends immediately to the case where z_t and x_t are M- and N-dimensional processes, respectively. In that case, x_t is said to Granger-cause z_t if

$$\Sigma_z(h|\Omega_t) \neq \Sigma_z(h|\Omega_t \setminus \{x_s|s \leq t\}) \quad (2.3.2)$$

for some t and h. Alternatively, this could be expressed by requiring the two MSEs to be different and

$$\Sigma_z(h|\Omega_t) \leq \Sigma_z(h|\Omega_t \setminus \{x_s|s \leq t\})$$

(i.e., the difference between the right-hand and the left-hand matrix is positive semidefinite). Because the null matrix is also positive semidefinite, it is necessary to require in addition that the two matrices are not identical. If x_t causes z_t and z_t also causes x_t the process $(z_t', x_t')'$ is called a *feedback system*.

Sometimes the term "instantaneous causality" is used in economic analyses. We say that there is *instantaneous causality* between z_t and x_t if

$$\Sigma_z(1|\Omega_t \cup \{x_{t+1}\}) \neq \Sigma_z(1|\Omega_t). \quad (2.3.3)$$

In other words, in period t, adding x_{t+1} to the information set helps to improve the forecast of z_{t+1}. We will see shortly that this concept of causality is really symmetric, that is, if there is instantaneous causality between z_t and x_t, then there is also instantaneous causality between x_t and z_t (see Proposition 2.3). Therefore we do not use the notion "instantaneous causality *from x_t to z_t*" in the foregoing definition.

A possible criticism of the foregoing definitions could relate to the choice of the MSE as a measure of the forecast precision. Of course, the choice of another measure could lead to a different definition of causality. However, in the situations of interest in the following, equality of the MSEs will imply equality of the corresponding predictors. In that case a process z_t is not Granger-caused by x_t if the optimal predictor of z_t does not use information from the x_t process. This result is intuitively appealing.

A more serious practical problem is the choice of the information set Ω_t. Usually all the relevant information in the universe is not available to a forecaster and, thus, the optimal predictor given Ω_t cannot be determined. Therefore a less demanding definition of causality is often used in practice. Instead

2.3 Structural Analysis with VAR Models   43

of all the information in the universe, only the information in the past and present of the process under study is considered relevant and Ω_t is replaced by $\{z_s, x_s | s \leq t\}$. Furthermore, instead of *optimal* predictors, *optimal linear* predictors are compared. In other words, $z_t(h|\Omega_t)$ is replaced by the linear minimum MSE h-step predictor based on the information in $\{z_s, x_s | s \leq t\}$ and $z_t(h|\Omega_t \setminus \{x_s | s \leq t\})$ is replaced by the linear minimum MSE h-step predictor based on $\{z_s | s \leq t\}$. In the following, when the terms "Granger-causality" and "instantaneous causality" are used, these restrictive assumptions are implicitly used if not otherwise noted.

Characterization of Granger-Causality

In order to determine the Granger-causal relationships between the variables of the K-dimensional VAR process y_t, suppose it has the canonical MA representation

$$y_t = \mu + \sum_{i=0}^{\infty} \Phi_i u_{t-i} = \mu + \Phi(L)u_t, \quad \Phi_0 = I_K, \tag{2.3.4}$$

where u_t is a white noise process with nonsingular covariance matrix Σ_u. Suppose that y_t consists of the M-dimensional process z_t and the $(K-M)$-dimensional process x_t and the MA representation is partitioned accordingly,

$$y_t = \begin{bmatrix} z_t \\ x_t \end{bmatrix} = \begin{bmatrix} \mu_1 \\ \mu_2 \end{bmatrix} + \begin{bmatrix} \Phi_{11}(L) & \Phi_{12}(L) \\ \Phi_{21}(L) & \Phi_{22}(L) \end{bmatrix} \begin{bmatrix} u_{1t} \\ u_{2t} \end{bmatrix}. \tag{2.3.5}$$

Using the prediction formula (2.2.10), the optimal 1-step forecast of z_t based on y_t is

$$z_t(1|\{y_s | s \leq t\}) = [I_M : 0]y_t(1) \tag{2.3.6}$$

$$= \mu_1 + \sum_{i=1}^{\infty} \Phi_{11,i} u_{1,t+1-i} + \sum_{i=1}^{\infty} \Phi_{12,i} u_{2,t+1-i}.$$

Hence the forecast error is

$$z_{t+1} - z_t(1|\{y_s | s \leq t\}) = u_{1,t+1}. \tag{2.3.7}$$

As mentioned in Section 2.1.3, a subprocess of a stationary process also has a prediction error MA representation. Thus,

$$z_t = \mu_1 + \sum_{i=0}^{\infty} \Phi_{11,i} u_{1,t-i} + \sum_{i=1}^{\infty} \Phi_{12,i} u_{2,t-i}$$

$$= \mu_1 + \sum_{i=0}^{\infty} F_i v_{t-i}, \tag{2.3.8}$$

where $F_0 = I_M$ and the last expression is a prediction error MA representation. Thus, the optimal 1-step predictor based on z_t only is

44   2 Stable Vector Autoregressive Processes

$$z_t(1|\{z_s|s \leq t\}) = \mu_1 + \sum_{i=1}^{\infty} F_i v_{t+1-i} \qquad (2.3.9)$$

and the corresponding forecast error is

$$z_{t+1} - z_t(1|\{z_s|s \leq t\}) = v_{t+1}. \qquad (2.3.10)$$

Consequently, the predictors (2.3.6) and (2.3.9) are identical if and only if $v_t = u_{1,t}$ for all t. In other words, equality of the predictors is equivalent to z_t having the MA representation

$$\begin{aligned}
z_t &= \mu_1 + \sum_{i=0}^{\infty} F_i u_{1,t-i} = \mu_1 + \sum_{i=0}^{\infty} [F_i : 0] u_{t-i} \\
&= \mu_1 + \sum_{i=0}^{\infty} [\Phi_{11,i} : \Phi_{12,i}] u_{t-i} \\
&= \mu_1 + \sum_{i=0}^{\infty} \Phi_{11,i} u_{1,t-i} + \sum_{i=1}^{\infty} \Phi_{12,i} u_{2,t-i}.
\end{aligned}$$

Uniqueness of the canonical MA representation implies that $F_i = \Phi_{11,i}$ and $\Phi_{12,i} = 0$ for $i = 1, 2, \ldots$. Hence, we get the following proposition.

Proposition 2.2 (*Characterization of Granger-Noncausality*)
Let y_t be a VAR process as in (2.3.4)/(2.3.5) with canonical MA operator $\Phi(z)$. Then

$$z_t(1|\{y_s|s \leq t\}) = z_t(1|\{z_s|s \leq t\}) \quad \Leftrightarrow \quad \Phi_{12,i} = 0 \quad \text{for } i = 1, 2, \ldots.$$
(2.3.11)

■

Because we have just used the MA representation (2.3.4) and not its finite order VAR form, the proposition is not only valid for VAR processes but more generally for processes having a canonical MA representation such as (2.3.4). From (2.2.10) it is obvious that equality of the 1-step predictors implies equality of the h-step predictors for $h = 2, 3, \ldots$. Hence, the proposition provides a necessary and sufficient condition for x_t being not Granger-causal for z_t, that is, z_t is not Granger-caused by x_t if and only if $\Phi_{12,i} = 0$ for $i = 1, 2, \ldots$. Thus, Granger-noncausality can be checked easily by looking at the MA representation of y_t. Because we are mostly concerned with VAR processes, it is worth noting that for a stationary, stable VAR(p) process

$$y_t = \begin{bmatrix} z_t \\ x_t \end{bmatrix} = \begin{bmatrix} \nu_1 \\ \nu_2 \end{bmatrix} + \begin{bmatrix} A_{11,1} & A_{12,1} \\ A_{21,1} & A_{22,1} \end{bmatrix} \begin{bmatrix} z_{t-1} \\ x_{t-1} \end{bmatrix} + \cdots$$

$$+ \begin{bmatrix} A_{11,p} & A_{12,p} \\ A_{21,p} & A_{22,p} \end{bmatrix} \begin{bmatrix} z_{t-p} \\ x_{t-p} \end{bmatrix} + \begin{bmatrix} u_{1t} \\ u_{2t} \end{bmatrix}, \qquad (2.3.12)$$

the condition (2.3.11) is satisfied if and only if

$$A_{12,i} = 0 \text{ for } i = 1, \ldots, p.$$

This result follows from the recursions in (2.1.22) or, alternatively, because the inverse of

$$\begin{bmatrix} \Phi_{11}(L) & 0 \\ \Phi_{21}(L) & \Phi_{22}(L) \end{bmatrix}$$

is

$$\begin{bmatrix} \Phi_{11}(L)^{-1} & 0 \\ -\Phi_{22}(L)^{-1}\Phi_{21}(L)\Phi_{11}(L)^{-1} & \Phi_{22}(L)^{-1} \end{bmatrix}$$

Thus, we have the following result.

Corollary 2.2.1
If y_t is a stable VAR(p) process as in (2.3.12) with nonsingular white noise covariance matrix Σ_u, then

$$z_t(h|\{y_s|s \leq t\}) = z_t(h|\{z_s|s \leq t\}), \quad h = 1, 2, \ldots$$
$$\Leftrightarrow A_{12,i} = 0 \quad \text{for } i = 1, \ldots, p. \tag{2.3.13}$$

Alternatively,

$$x_t(h|\{y_s|s \leq t\}) = x_t(h|\{x_s|s \leq t\}), \quad h = 1, 2, \ldots$$
$$\Leftrightarrow A_{21,i} = 0 \quad \text{for } i = 1, \ldots, p. \tag{2.3.14}$$

■

This corollary implies that noncausalities can be determined by just looking at the VAR representation of the system. For instance, for the example process (2.1.14),

$$\begin{bmatrix} y_{1,t} \\ y_{2,t} \\ y_{3,t} \end{bmatrix} = \nu + \begin{bmatrix} .5 & 0 & 0 \\ .1 & .1 & .3 \\ 0 & .2 & .3 \end{bmatrix} \begin{bmatrix} y_{1,t-1} \\ y_{2,t-1} \\ y_{3,t-1} \end{bmatrix} + u_t,$$

$x_t := (y_{2t}, y_{3t})'$ does not Granger-cause $z_t := y_{1t}$ because $A_{12,1} = 0$ if the coefficient matrix is partitioned according to (2.3.12). On the other hand, z_t Granger-causes x_t. To give this discussion economic content let us assume that the variables in the system are rates of change of investment (y_1), income (y_2), and consumption (y_3). With these specifications, the previous discussion shows that investment Granger-causes the consumption/income system whereas the converse is not true. It is also easy to check that consumption causes the income/investment system and vice versa. Note that so far we have defined Granger-causality only in terms of two groups of variables. Therefore, at this stage, we cannot talk about the Granger-causal relationship between consumption and income in the three-dimensional investment/income/consumption system.

Let us assume that the variables in the example VAR(2) process (2.1.15),

$$\begin{bmatrix} y_{1,t} \\ y_{2,t} \end{bmatrix} = \nu + \begin{bmatrix} .5 & .1 \\ .4 & .5 \end{bmatrix} \begin{bmatrix} y_{1,t-1} \\ y_{2,t-1} \end{bmatrix} + \begin{bmatrix} 0 & 0 \\ .25 & 0 \end{bmatrix} \begin{bmatrix} y_{1,t-2} \\ y_{2,t-2} \end{bmatrix} + u_t,$$

represent the inflation rate (y_1), and some interest rate (y_2). Using Corollary 2.2.1, it is immediately obvious that inflation causes the interest rate and vice versa. Hence the system is a feedback system. In the following we will refer to (2.1.15) as the inflation/interest rate system.

Characterization of Instantaneous Causality

In order to study the concept of instantaneous causality in the framework of the MA process (2.3.5), it is useful to rewrite that representation. Note that the positive definite symmetric matrix Σ_u can be written as the product $\Sigma_u = PP'$, where P is a lower triangular nonsingular matrix with positive diagonal elements (see Appendix A.9.3). Thus, (2.3.5) can be represented as

$$y_t = \mu + \sum_{i=0}^{\infty} \Phi_i P P^{-1} u_{t-i} = \mu + \sum_{i=0}^{\infty} \Theta_i w_{t-i}, \tag{2.3.15}$$

where $\Theta_i := \Phi_i P$ and $w_t := P^{-1} u_t$ is white noise with covariance matrix

$$\Sigma_w = P^{-1} \Sigma_u (P^{-1})' = I_K. \tag{2.3.16}$$

Because the white noise errors w_t have uncorrelated components, they are often called *orthogonal* residuals or innovations.

Partitioning the representation (2.3.15) according to the partitioning of $y_t = (z_t', x_t')'$ gives

$$\begin{bmatrix} z_t \\ x_t \end{bmatrix} = \begin{bmatrix} \mu_1 \\ \mu_2 \end{bmatrix} + \begin{bmatrix} \Theta_{11,0} & 0 \\ \Theta_{21,0} & \Theta_{22,0} \end{bmatrix} \begin{bmatrix} w_{1,t} \\ w_{2,t} \end{bmatrix}$$

$$+ \begin{bmatrix} \Theta_{11,1} & \Theta_{12,1} \\ \Theta_{21,1} & \Theta_{22,1} \end{bmatrix} \begin{bmatrix} w_{1,t-1} \\ w_{2,t-1} \end{bmatrix} + \cdots.$$

Hence,

$$z_{t+1} = \mu_1 + \Theta_{11,0} w_{1,t+1} + \Theta_{11,1} w_{1,t} + \Theta_{12,1} w_{2,t} + \cdots$$

and

$$x_{t+1} = \mu_2 + \Theta_{21,0} w_{1,t+1} + \Theta_{22,0} w_{2,t+1} + \Theta_{21,1} w_{1,t} + \Theta_{22,1} w_{2,t} + \cdots.$$

2.3 Structural Analysis with VAR Models

The optimal 1-step predictor of x_t based on $\{y_s|s \leq t\}$ and, in addition, on z_{t+1}, is equal to the 1-step predictor of x_t based on $\{w_s|s \leq t\} \cup \{w_{1,t+1}\}$, that is,

$$\begin{aligned}
x_t(1|\{y_s|s \leq t\} \cup \{z_{t+1}\}) &= x_t(1|\{w_s = (w'_{1,s}, w'_{2,s})'|s \leq t\} \cup \{w_{1,t+1}\}) \\
&= \Theta_{21,0} w_{1,t+1} + x_t(1|\{y_s|s \leq t\}). \quad (2.3.17)
\end{aligned}$$

Consequently,

$$x_t(1|\{y_s|s \leq t\} \cup \{z_{t+1}\}) = x_t(1|\{y_s|s \leq t\})$$

if and only if $\Theta_{21,0} = 0$. This condition, in turn, is easily seen to hold if and only if the covariance matrix Σ_u is block diagonal with a $((K-M) \times M)$ block of zeros in the lower left-hand corner and an $(M \times (K-M))$ block of zeros in the upper right-hand corner. Of course, this means that u_{1t} and u_{2t} in (2.3.5) have to be uncorrelated, i.e., $E(u_{1t} u'_{2t}) = 0$. Thereby the following proposition is proven.

Proposition 2.3 (*Characterization of Instantaneous Causality*)
Let y_t be as in (2.3.5)/(2.3.15) with nonsingular innovation covariance matrix Σ_u. Then there is no instantaneous causality between z_t and x_t if and only if

$$E(u_{1t} u'_{2t}) = 0. \quad (2.3.18)$$

∎

This proposition provides a condition for instantaneous causality which is easy to check if the process is given in MA or VAR form. For instance, for the investment/income/consumption system with white noise covariance matrix (2.1.33),

$$\Sigma_u = \begin{bmatrix} 2.25 & 0 & 0 \\ 0 & 1.0 & .5 \\ 0 & .5 & .74 \end{bmatrix},$$

there is no instantaneous causality between (income, consumption) and investment.

From Propositions 2.2 and 2.3 it follows that $y_t = (z'_t, x'_t)'$ has a representation with orthogonal innovations as in (2.3.15) of the form

$$\begin{aligned}
\begin{bmatrix} z_t \\ x_t \end{bmatrix} &= \begin{bmatrix} \mu_1 \\ \mu_2 \end{bmatrix} + \begin{bmatrix} \Theta_{11,0} & 0 \\ 0 & \Theta_{22,0} \end{bmatrix} \begin{bmatrix} w_{1,t} \\ w_{2,t} \end{bmatrix} \\
&\quad + \begin{bmatrix} \Theta_{11,1} & 0 \\ \Theta_{21,1} & \Theta_{22,1} \end{bmatrix} \begin{bmatrix} w_{1,t-1} \\ w_{2,t-1} \end{bmatrix} + \cdots \\
&= \begin{bmatrix} \mu_1 \\ \mu_2 \end{bmatrix} + \begin{bmatrix} \Theta_{11}(L) & 0 \\ \Theta_{21}(L) & \Theta_{22}(L) \end{bmatrix} \begin{bmatrix} w_{1,t} \\ w_{2,t} \end{bmatrix}, \quad (2.3.19)
\end{aligned}$$

if x_t does not Granger-cause z_t and, furthermore, there is no instantaneous causation between x_t and z_t. In the absence of instantaneous causality, a similar representation with $\Theta_{21}(L) \equiv 0$ is obtained if z_t is not Granger-causal for x_t.

Discussion of Instantaneous and Granger-Causality

At this point, some words of caution seem appropriate. The term "causality" suggests a cause and effect relationship between two sets of variables. Proposition 2.3 shows that such an interpretation is problematic with respect to instantaneous causality because this term only describes a nonzero correlation between two sets of variables. It does not say anything about the cause and effect relation. The direction of instantaneous causation cannot be derived from the MA or VAR representation of the process but must be obtained from further knowledge on the relationship between the variables. Such knowledge may exist in the form of an economic theory.

Although a direction of causation has been defined in relation with Granger-causality it is problematic to interpret the absence of causality from x_t to z_t in the sense that variations in x_t will have no effect on z_t. To see this consider, for instance, the stable bivariate VAR(1) system

$$\begin{bmatrix} z_t \\ x_t \end{bmatrix} = \begin{bmatrix} \alpha_{11} & 0 \\ \alpha_{21} & \alpha_{22} \end{bmatrix} \begin{bmatrix} z_{t-1} \\ x_{t-1} \end{bmatrix} + \begin{bmatrix} u_{1t} \\ u_{2t} \end{bmatrix}. \qquad (2.3.20)$$

In this system, x_t does not Granger-cause z_t by Corollary 2.2.1. However, the system may be multiplied by some nonsingular matrix

$$B = \begin{bmatrix} 1 & \beta \\ 0 & 1 \end{bmatrix}$$

so that

$$\begin{bmatrix} z_t \\ x_t \end{bmatrix} = \begin{bmatrix} 0 & -\beta \\ 0 & 0 \end{bmatrix} \begin{bmatrix} z_t \\ x_t \end{bmatrix} + \begin{bmatrix} \gamma_{11} & \gamma_{12} \\ \gamma_{21} & \gamma_{22} \end{bmatrix} \begin{bmatrix} z_{t-1} \\ x_{t-1} \end{bmatrix} + \begin{bmatrix} v_{1t} \\ v_{2t} \end{bmatrix}, \qquad (2.3.21)$$

where $\gamma_{11} := \alpha_{11} + \alpha_{21}\beta$, $\gamma_{12} := \alpha_{22}\beta$, $\gamma_{21} := \alpha_{21}$, $\gamma_{22} := \alpha_{22}$ and $(v_{1t}, v_{2t})' := B(u_{1t}, u_{2t})'$. Note that this is just another representation of the process $(z_t, x_t)'$ and not another process. (The reader may check that the process (2.3.21) has the same means and autocovariances as the one in (2.3.20).)

In other words, the stochastic interrelationships between the random variables of the system can either be characterized by (2.3.20) or by (2.3.21) although the two representations have quite different physical interpretations. If (2.3.21) happens to represent the actual ongoings in the system, changes in x_t may affect z_t through the term with the coefficient $-\beta$ in the first equation. Thus, the lack of a Granger-causal relationship from one group of variables to the remaining variables cannot necessarily be interpreted as lack of a cause and effect relationship. It must be remembered that a VAR or MA representation characterizes the joint distribution of sets of random variables. In order to derive cause and effect relationships from it, usually requires further assumptions regarding the relationship between the variables involved. We will return to this problem in the following subsections.

Further problems related to the interpretation of Granger-causality result from restricting the information set to contain only past and present variables of the system rather than all information in the universe. Only if all other information in the universe is irrelevant for the problem at hand, the reduction of the information set is of no consequence. Some related problems will be discussed in the following.

Changing the Information Set

So far it has been assumed that the information set contains the variables or groups of variables only for which we want to analyze the causal links. Often we are interested in the causal links between two variables in a higher dimensional system. In other words, we are interested in analyzing Granger-causality in a framework where the information set contains more than just the variables of direct interest. In the bivariate framework when the information set is limited to the two variables of interest, it was seen that if the 1-step ahead forecasts of one variable cannot be improved by using the information in the other variable, the same holds for all h-step forecasts, $h = 1, 2, \ldots$. This result does not hold anymore if the information set contains additional variables, as pointed out by Lütkepohl (1993).

To be more explicit, suppose the vector time series z_t, y_t, x_t with dimensions K_z, K_y, K_x, respectively, are jointly generated by a VAR(p) process

$$\begin{bmatrix} z_t \\ y_t \\ x_t \end{bmatrix} = \sum_{i=1}^{p} A_i \begin{bmatrix} z_{t-i} \\ y_{t-i} \\ x_{t-i} \end{bmatrix} + u_t, \qquad (2.3.22)$$

where

$$A_i = \begin{bmatrix} A_{zz,i} & A_{zy,i} & A_{zx,i} \\ A_{yz,i} & A_{yy,i} & A_{yx,i} \\ A_{xz,i} & A_{xy,i} & A_{xx,i} \end{bmatrix}, \quad i = 1, \ldots, p,$$

with $A_{kl,i}$ having dimension $(K_k \times K_l)$ and u_t is zero mean white noise, as usual. In this process, if $A_{zy,i} = 0$, $i = 1, 2, \ldots, p$, it is not difficult to see that the information in y_t cannot be used to improve the 1-step ahead forecasts of z_t but it is still possible that it can be used to improve the h-step forecasts for $h = 2, 3, \ldots$. In other words, if y_t is 1-step noncausal for z_t, it may still be h-step causal for $h > 1$. Consequently, it makes sense to define more refined concepts of causality which refer explicitly to the forecast horizon. For instance, y_t may be called h-step noncausal for z_t ($y_t \not\to_{(h)} z_t$) for $h = 1, 2, \ldots$, if the j-step ahead forecasts of z_t cannot be improved for $j \leq h$ by taking into account the information in past and present y_t. Now the original concept of Granger-causality corresponds to infinite-step causality.

The corresponding restrictions of multi-step causality on the VAR coefficients have been considered by Dufour & Renault (1998). Unlike in the bivariate setting explored earlier, now nonlinear restrictions on the VAR coefficients are obtained which make it more difficult to check for h-step causality if the information set is expanded by additional variables.

To state the restrictions formally, let $\mathbf{A}$ be defined as in the VAR(1) representation (2.1.8), let $J = [I_K : 0 : \cdots : 0]$ be a $(K \times Kp)$ matrix as before and define $A^{(j)} = J\mathbf{A}^j$ and $\boldsymbol{\alpha}^{(j)} = \text{vec}(A^{(j)})$. Dufour & Renault (1998) show that in the process (2.3.22), $y_t \not\to_{(h)} z_t$ if and only if

$$R\boldsymbol{\alpha}^{(j)} = 0 \quad \text{for} \quad j = 1, \ldots, h, \qquad (2.3.23)$$

and $y_t \not\to_{(\infty)} z_t$ if and only if

$$R\boldsymbol{\alpha}^{(j)} = 0 \quad \text{for} \quad j = 1, \ldots, pK_x + 1. \qquad (2.3.24)$$

Here the restriction matrix R is such that $R\text{vec}[A_1, \ldots, A_p] = \text{vec}[A_{zy,1}, \ldots, A_{zy,p}]$, that is, it collects the elements of the second block in the first row of each of the coefficient matrices.

As an example consider again the 3-dimensional VAR(1) process (2.1.14). For infinite-step causality or noncausality from y_{2t} to y_{1t} we need to check the relevant elements of the coefficient matrix and its second power:

$$A_1 = \begin{bmatrix} .5 & 0 & 0 \\ .1 & .1 & .3 \\ 0 & .2 & .3 \end{bmatrix}, \qquad A_1^2 = \begin{bmatrix} .25 & 0 & 0 \\ .06 & .07 & .12 \\ .02 & .08 & .15 \end{bmatrix}.$$

Clearly, $y_{2t} \not\to_{(1)} y_{1t}$ holds because $\alpha_{12,1} = 0$ and also the restrictions for $y_{2t} \not\to_{(\infty)} y_{1t}$ are satisfied in this case because the (1,2)-th element of A_1^2 is also zero. In contrast, $y_{1t} \not\to_{(1)} y_{3t}$ holds, while $y_{1t} \not\to_{(\infty)} y_{3t}$ does not, because the lower left-hand element of A_1^2 is nonzero. Notice that the definition and characterizations of multi-step causality are given for the first two sets of subvectors with the third one containing the extra variables. For applying the definition and results in the present example, the variables may just be rearranged accordingly.

In addition to these extensions related to increasing the information set, there are also other problems which may make it difficult to interpret Granger-causal relations even in a bivariate setting. Let us discuss some of them in terms of an inflation/interest rate system. For example, it may make a difference whether the information set contains monthly, quarterly or annual data. If a quarterly system is considered and no causality is found from the interest rate to inflation it does not follow that a corresponding monthly interest rate has no impact on the monthly inflation rate. In other words, the interest rate may Granger-cause inflation in a monthly system even if it does not in a quarterly system.

Furthermore, putting seasonally adjusted variables in the information set is not the same as using unadjusted variables. Consequently, if Granger-causality is found for the seasonally adjusted variables, it is still possible that in the actual seasonal system the interest rate is not Granger-causal for inflation. Similar comments apply in the presence of measurement errors. Finally, causality analyses are usually based on estimated rather than known systems. Additional problems result in that case. We will return to them in the next chapter.

The previous critical remarks are meant to caution the reader and multiple time series analyst against overinterpreting the evidence from a VAR model. Still, causality analyses are useful tools in practice if these critical points are kept in mind. At the very least, a Granger-causality analysis tells the analyst whether a set of variables contains useful information for improving the predictions of another set of variables. Further discussions of causality issues and many further references may be found in Geweke (1984) and Granger (1982).

2.3.2 Impulse Response Analysis

In the previous subsection, we have seen that Granger-causality may not tell us the complete story about the interactions between the variables of a system. In applied work, it is often of interest to know the *response* of *one* variable to an *impulse* in another variable in a system that involves a number of further variables as well. Thus, one would like to investigate the impulse response relationship between two variables in a higher dimensional system. Of course, if there is a reaction of one variable to an impulse in another variable we may call the latter causal for the former. In this subsection, we will study this type of causality by tracing out the effect of an exogenous shock or innovation in one of the variables on some or all of the other variables. This kind of impulse response analysis is sometimes called *multiplier analysis.* For instance, in a system consisting of an inflation rate and an interest rate, the effect of an increase in the inflation rate may be of interest. In the real world, such an increase may be induced exogenously from outside the system by events like the increase of the oil price in 1973/74 when the OPEC agreed on a joint action to raise prices. Alternatively, an increase or reduction in the interest rate may be administered by the central bank for reasons outside the simple two variable system under study.

Responses to Forecast Errors

Suppose the effect of an innovation in investment in a system containing investment (y_1), income (y_2), and consumption (y_3) is of interest. To isolate such an effect, suppose that all three variables assume their mean value prior to time $t = 0$, $y_t = \mu$ for $t < 0$, and investment increases by one unit in period $t = 0$, that is, $u_{1,0} = 1$. Now we can trace out what happens to the system during periods $t = 1, 2, \ldots$ if no further shocks occur, that is, $u_{2,0} = u_{3,0} = 0$, $u_1 = 0$, $u_2 = 0, \ldots$. Because we are not interested in the mean of the system in such an exercise but just in the variations of the variables around their means, we assume that all three variables have mean zero and set $\nu = 0$ in (2.1.14). Hence, $y_t = A_1 y_{t-1} + u_t$ or, more precisely,

$$\begin{bmatrix} y_{1,t} \\ y_{2,t} \\ y_{3,t} \end{bmatrix} = \begin{bmatrix} .5 & 0 & 0 \\ .1 & .1 & .3 \\ 0 & .2 & .3 \end{bmatrix} \begin{bmatrix} y_{1,t-1} \\ y_{2,t-1} \\ y_{3,t-1} \end{bmatrix} + \begin{bmatrix} u_{1,t} \\ u_{2,t} \\ u_{3,t} \end{bmatrix}. \qquad (2.3.25)$$

Tracing a unit shock in the first variable in period $t = 0$ in this system we get

$$y_0 = \begin{bmatrix} y_{1,0} \\ y_{2,0} \\ y_{3,0} \end{bmatrix} = \begin{bmatrix} u_{1,0} \\ u_{2,0} \\ u_{3,0} \end{bmatrix} = \begin{bmatrix} 1 \\ 0 \\ 0 \end{bmatrix},$$

$$y_1 = \begin{bmatrix} y_{1,1} \\ y_{2,1} \\ y_{3,1} \end{bmatrix} = A_1 y_0 = \begin{bmatrix} .5 \\ .1 \\ 0 \end{bmatrix},$$

$$y_2 = \begin{bmatrix} y_{1,2} \\ y_{2,2} \\ y_{3,2} \end{bmatrix} = A_1 y_1 = A_1^2 y_0 = \begin{bmatrix} .25 \\ .06 \\ .02 \end{bmatrix}.$$

Continuing the procedure, it turns out that $y_i = (y_{1,i}, y_{2,i}, y_{3,i})'$ is just the first column of A_1^i. An analogous line of arguments shows that a unit shock in y_{2t} (y_{3t}) at $t = 0$, after i periods, results in a vector y_i which is just the second (third) column of A_1^i. Thus, the elements of A_1^i represent the effects of unit shocks in the variables of the system after i periods. Therefore they are called impulse responses or dynamic multipliers.

Recall that $A_1^i = \Phi_i$ is just the i-th coefficient matrix of the MA representation of a VAR(1) process. Consequently, the MA coefficient matrices contain the impulse responses of the system. This result holds more generally for higher order VAR(p) processes as well. To see this, suppose that y_t is a stationary VAR(p) process as in (2.1.1) with $\nu = 0$. This process has a corresponding VAR(1) process $Y_t = \mathbf{A} Y_{t-1} + U_t$ as in (2.1.8) with $\nu = 0$. Under the assumptions of the previous example, $y_t = 0$ for $t < 0$, $u_t = 0$ for $t > 0$ and $y_0 = u_0$ is a K-dimensional unit vector e_k, say, with a one as the k-th coordinate and zeros elsewhere. It follows that $Y_0 = (e_k', 0, \ldots, 0)'$ and $Y_i = \mathbf{A}^i Y_0$. Hence, the impulse responses are the elements of the upper left-hand $(K \times K)$ block of $\mathbf{A}^i$. This matrix, however, was shown to be the i-th coefficient matrix Φ_i of the MA representation (2.1.17) of y_t, i.e., $\Phi_i = J \mathbf{A}^i J'$ with $J := [I_K : 0 : \cdots : 0]$ a $(K \times Kp)$ matrix. In other words, $\phi_{jk,i}$, the jk-th element of Φ_i, represents the reaction of the j-th variable of the system to a unit shock in variable k, i periods ago, provided, of course, the effect is not contaminated by other shocks to the system. Because the u_t are just the 1-step ahead forecast errors of the VAR process, the shocks considered here may be regarded as forecast errors and the impulse responses are sometimes referred to as *forecast error impulse responses*.

The response of variable j to a unit shock (forecast error) in variable k is sometimes depicted graphically to get a visual impression of the dynamic interrelationships within the system. Impulse responses of the investment/income/consumption system are plotted in Figure 2.5 and the dynamic

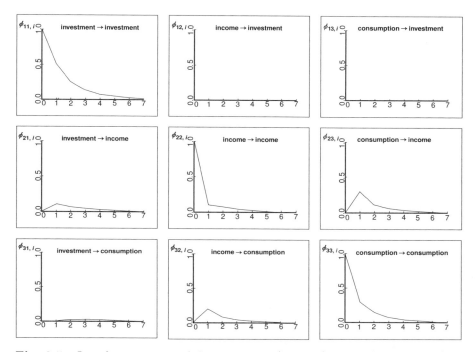

Fig. 2.5. Impulse responses of the investment/income/consumption system (impulse → response).

responses of the inflation/interest rate system are depicted in Figure 2.6. For instance, in the latter figure an inflation innovation is seen to induce the interest rate to increase for two periods and then it tapers off to zero. In both systems the effect of a unit shock in any of the variables dies away quite rapidly due to the stability of the systems.

If the variables have different scales, it is sometimes useful to consider innovations of one standard deviation rather than unit shocks. For instance, instead of tracing an unexpected unit increase in investment in the investment/income/consumption system with white noise covariance matrix (2.1.33), one may follow up on a shock of $\sqrt{2.25} = 1.5$ units because the standard deviation of u_{1t} is 1.5. Of course, this is just a matter of rescaling the impulse responses. In Figures 2.5 and 2.6, it suffices to choose the units at the vertical axes equal to the standard deviations of the residuals corresponding to the variables whose effects are considered. Such a rescaling may sometimes give a better picture of the dynamic relationships because the average size of the innovations occurring in a system depends on their standard deviation.

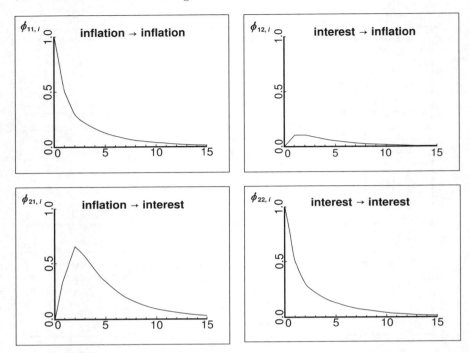

Fig. 2.6. Impulse responses of the inflation/interest rate system (impulse → response).

It follows from Proposition 2.2 that the impulse responses are zero if one of the variables does not Granger-cause the other variables taken as a group. More precisely, an innovation in variable k has no effect on the other variables if the former variable does not Granger-cause the set of the remaining variables. As we have mentioned previously, in applied work it is often of foremost interest whether one variable has an impact on a specific other variable. That is, one would like to know whether, for some $k \neq j$, $\phi_{jk,i} = 0$ for $i = 1, 2, \ldots$. If the $\phi_{jk,i}$ represent the actual reactions of variable j to a unit shock in variable k, we may call the latter *noncausal* for the j-th variable if $\phi_{jk,i} = 0$ for $i = 1, 2, \ldots$. In order to check the latter condition, it is not necessary to compute infinitely many Φ_i matrices. The following proposition shows that it suffices to check the first $p(K-1)$ Φ_i matrices.

Proposition 2.4 (*Zero Impulse Responses*)
If y_t is a K-dimensional stable VAR(p) process, then, for $j \neq k$,

$$\phi_{jk,i} = 0 \qquad \text{for } i = 1, 2, \ldots$$

is equivalent to

$$\phi_{jk,i} = 0 \qquad \text{for } i = 1, \ldots, p(K-1).$$

■

In other words, the proposition asserts that for a K-dimensional, stationary, stable VAR(p), if the first $pK - p$ responses of variable j to an impulse in variable k are zero, all the following responses must also be zero. For instance, in the investment/income/consumption VAR(1) system, because the responses of investment for the next two periods after a consumption impulse are zero, we know that investment will not react at all to such an impulse. Note, that in a VAR(1) system of dimension greater than 2, it does not suffice to check, say, the upper right-hand corner element of the coefficient matrix in order to determine whether the last variable is noncausal for the first variable. Notice that Proposition 2.4 is related to the conditions for multi-step causality in (2.3.23) and (2.3.24). In general, the conditions are not identical, however, because the two concepts differ. Proposition 2.4 will be helpful when testing of impulse response relations is discussed in the next chapter. We will now prove the proposition.

Proof of Proposition 2.4:
Returning to the lag operator notation of Section 2.1.2, we have

$$\Phi(L) = (\phi_{jk}(L))_{j,k} = A(L)^{-1} = A(L)^{adj}/\det(A(L)),$$

where $A(L)^{adj} = (A_{jk}(L))_{j,k}$ is the adjoint of $A(L) = I_K - A_1 L - \cdots - A_p L^p$ (see Appendix A.4.1). Obviously, $\phi_{jk}(L) \equiv 0$ is equivalent to $A_{jk}(L) \equiv 0$. From the definition of a cofactor of a matrix in Appendix A.3, it is easy to see that $A_{jk}(L)$ has degree not greater than $pK - p$. Defining $\gamma(L) = [\det A(L)]^{-1}$, we get for $k \neq j$,

$$\begin{aligned}
\phi_{jk}(L) &= \phi_{jk,1} L + \phi_{jk,2} L^2 + \cdots \\
&= A_{jk}(L)\gamma(L) \\
&= (A_{jk,1} L + \cdots + A_{jk,pK-p} L^{pK-p})(1 + \gamma_1 L + \cdots).
\end{aligned}$$

Hence,

$$\phi_{jk,1} = A_{jk,1} \qquad \text{and} \qquad \phi_{jk,i} = A_{jk,i} + \sum_{n=1}^{i-1} A_{jk,n}\gamma_{i-n} \qquad \text{for } i > 1,$$

with $A_{jk,n} = 0$ for $n > pK - p$. Consequently, $A_{jk,i} = 0$ for $i = 1, \ldots, pK - p$, is equivalent to $\phi_{jk,i} = 0$ for $i = 1, 2, \ldots, pK - p$, which proves the proposition. ■

Sometimes interest centers on the accumulated effect over several or more periods of a shock in one variable. This effect may be determined by summing up the MA coefficient matrices. For instance, the k-th column of $\Psi_n := \sum_{i=0}^{n} \Phi_i$ contains the *accumulated responses* over n periods to a unit shock in the k-th variable of the system. These quantities are sometimes called n-th *interim multipliers*. The total accumulated effects for all future periods are obtained by summing up all the MA coefficient matrices. $\Psi_\infty := \sum_{i=0}^{\infty} \Phi_i$ is sometimes called the matrix of *long-run effects* or *total multipliers*. Because the MA operator $\Phi(z)$ is the inverse of the VAR operator $A(z) = I_K - A_1 z - \cdots - A_p z^p$, the long-run effects are easily obtained as

$$\Psi_\infty = \Phi(1) = (I_K - A_1 - \cdots - A_p)^{-1}. \tag{2.3.26}$$

As an example, accumulated responses for the investment/income/consumption system are depicted in Figure 2.7. Similarly, interim and total multipliers of the inflation/interest rate system are shown in Figure 2.8.

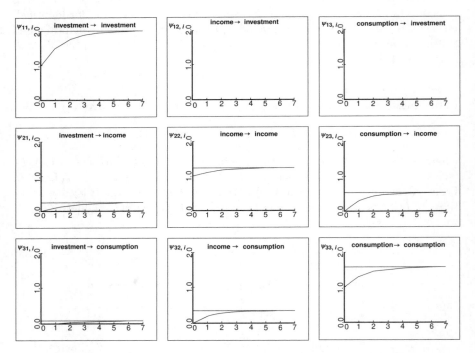

Fig. 2.7. Accumulated and long-run responses of the investment/income/consumption system (impulse $\rightarrow$ response).

Responses to Orthogonal Impulses

A problematic assumption in this type of impulse response analysis is that a shock occurs only in one variable at a time. Such an assumption may be

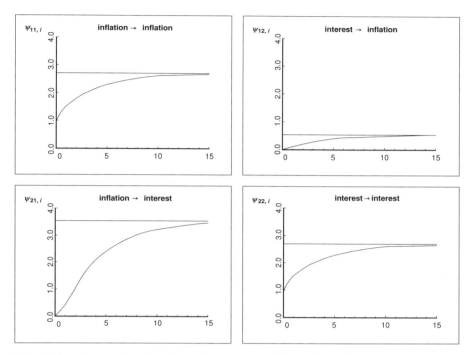

Fig. 2.8. Accumulated and total responses of the inflation/interest rate system (impulse → response).

reasonable if the shocks in different variables are independent. If they are not independent one may argue that the error terms consist of all the influences and variables that are not directly included in the set of y variables. Thus, in addition to forces that affect all the variables, there may be forces that affect variable 1, say, only. If a shock in the first variable is due to such forces it may again be reasonable to interpret the Φ_i coefficients as dynamic responses. On the other hand, correlation of the error terms may indicate that a shock in one variable is likely to be accompanied by a shock in another variable. In that case, setting all other residuals to zero may provide a misleading picture of the actual dynamic relationships between the variables. For example, in the investment/income/consumption system, the white noise or innovation covariance matrix is given in (2.1.33),

$$\Sigma_u = \begin{bmatrix} 2.25 & 0 & 0 \\ 0 & 1.0 & .5 \\ 0 & .5 & .74 \end{bmatrix}.$$

Obviously, there is a quite strong positive correlation between $u_{2,t}$ and $u_{3,t}$, the residuals of the income and consumption equations, respectively. Consequently, a shock in income may be accompanied by a shock in consumption in the same period. Therefore, forcing the consumption innovation to zero when the effect of an income shock is traced, as in the previous analysis, may in fact obscure the actual relation between the variables.

This is the reason why impulse response analysis is often performed in terms of the MA representation (2.3.15),

$$y_t = \sum_{i=0}^{\infty} \Theta_i w_{t-i}, \qquad (2.3.27)$$

where the components of $w_t = (w_{1t}, \ldots, w_{Kt})'$ are uncorrelated and have unit variance, $\Sigma_w = I_K$. The mean term is dropped again because it is of no interest in the present analysis. Recall that the representation (2.3.27) is obtained by decomposing Σ_u as $\Sigma_u = PP'$, where P is a lower triangular matrix, and defining $\Theta_i = \Phi_i P$ and $w_t = P^{-1} u_t$. In (2.3.27) it is reasonable to assume that a change in one component of w_t has no effect on the other components because the components are orthogonal (uncorrelated). Moreover, the variances of the components are one. Thus, a unit innovation is just an innovation of size one standard deviation. The elements of the Θ_i are interpreted as responses of the system to such innovations. More precisely, the jk-th element of Θ_i is assumed to represent the effect on variable j of a unit innovation in the k-th variable that has occurred i periods ago.

To relate these impulse responses to a VAR model, we consider the zero mean VAR(p) process

$$y_t = A_1 y_{t-1} + \cdots + A_p y_{t-p} + u_t. \qquad (2.3.28)$$

This process can be rewritten in such a way that the residuals of different equations are uncorrelated. For this purpose, we choose a decomposition of the white noise covariance matrix $\Sigma_u = W \Sigma_\varepsilon W'$, where Σ_ε is a diagonal matrix with positive diagonal elements and W is a lower triangular matrix with unit diagonal. This decomposition is obtained from the Choleski decomposition $\Sigma_u = PP'$ by defining a diagonal matrix D which has the same main diagonal as P and by specifying $W = PD^{-1}$ and $\Sigma_\varepsilon = DD'$.

Premultiplying (2.3.28) by $\mathsf{A} := W^{-1}$ gives

$$\mathsf{A} y_t = A_1^* y_{t-1} + \cdots + A_p^* y_{t-p} + \varepsilon_t, \qquad (2.3.29)$$

where $A_i^* := \mathsf{A} A_i$, $i = 1, \ldots, p$, and $\varepsilon_t = (\varepsilon_{1t}, \ldots, \varepsilon_{Kt})' := \mathsf{A} u_t$ has diagonal covariance matrix,

$$\Sigma_\varepsilon = E(\varepsilon_t \varepsilon_t') = \mathsf{A} E(u_t u_t') \mathsf{A}' = \mathsf{A} \Sigma_u \mathsf{A}'.$$

Adding $(I_K - \mathsf{A}) y_t$ to both sides of (2.3.29) gives

$$y_t = A_0^* y_t + A_1^* y_{t-1} + \cdots + A_p^* y_{t-p} + \varepsilon_t, \qquad (2.3.30)$$

where $A_0^* := I_K - \mathsf{A}$. Because W is lower triangular with unit diagonal, the same is true for A. Hence,

$$A_0^* = I_K - \mathsf{A} = \begin{bmatrix} 0 & 0 & \cdots & 0 & 0 \\ \beta_{21} & 0 & \cdots & 0 & 0 \\ \vdots & \ddots & \ddots & & \vdots \\ \vdots & & \ddots & \ddots & \vdots \\ \beta_{K1} & \beta_{K2} & \cdots & \beta_{K,K-1} & 0 \end{bmatrix}$$

is a lower triangular matrix with zero diagonal and, thus, in the representation (2.3.30) of our VAR(p) process, the first equation contains no instantaneous y's on the right-hand side. The second equation may contain y_{1t} and otherwise lagged y's on the right-hand side. More generally, the k-th equation may contain $y_{1t}, \ldots, y_{k-1,t}$ and not $y_{kt}, \ldots, y_{Kt}$ on the right-hand side. Thus, if (2.3.30) reflects the actual ongoings in the system, y_{st} cannot have an instantaneous impact on y_{kt} for $k < s$. In the econometrics literature such a system is called a *recursive model* (see Theil (1971, Section 9.6)). Herman Wold has advocated these models where the researcher has to specify the instantaneous "causal" ordering of the variables. This type of causality is therefore sometimes referred to as *Wold-causality*. If we trace ε_{it} innovations of size one standard error through the system (2.3.30), we just get the Θ impulse responses. This can be seen by solving the system (2.3.30) for y_t,

$$y_t = (I_K - A_0^*)^{-1} A_1^* y_{t-1} + \cdots + (I_K - A_0^*)^{-1} A_p^* y_{t-p} + (I_K - A_0^*)^{-1} \varepsilon_t.$$

Noting that $(I_K - A_0^*)^{-1} = W = PD^{-1}$ shows that the instantaneous effects of one-standard deviation shocks (ε_{it}'s of size one standard deviation) to the system are represented by the elements of $WD = P = \Theta_0$ because the diagonal elements of D are just standard deviations of the components of ε_t. The Θ_i may then be obtained by tracing these effects through the system.

The Θ_i's may provide response functions that are quite different from the Φ_i responses. For the example VAR(1) system (2.3.25) with Σ_u as in (2.1.33) we get

$$\Theta_0 = P = \begin{bmatrix} 1.5 & 0 & 0 \\ 0 & 1 & 0 \\ 0 & .5 & .7 \end{bmatrix},$$

$$\Theta_1 = \Phi_1 P = \begin{bmatrix} .75 & 0 & 0 \\ .15 & .25 & .21 \\ 0 & .35 & .21 \end{bmatrix}, \quad (2.3.31)$$

$$\Theta_2 = \Phi_2 P = \begin{bmatrix} .375 & 0 & 0 \\ .090 & .130 & .084 \\ .030 & .055 & .105 \end{bmatrix},$$

and so on. Some more innovation responses are depicted in Figure 2.9. Although they are similar to those given in Figure 2.5, there is an obvious difference in the response of consumption to an income innovation. While consumption responds with a time lag of one period in Figure 2.5, there is an instantaneous effect in Figure 2.9.

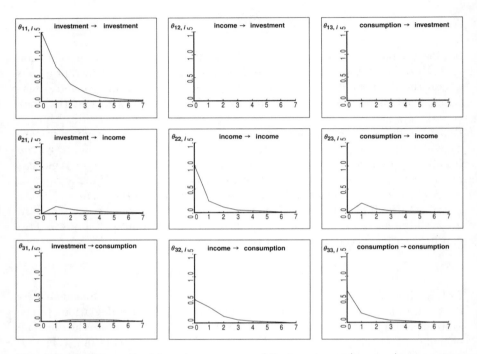

Fig. 2.9. Orthogonalized impulse responses of the investment/income/consumption system (impulse → response).

Note that $\Theta_0 = P$ is lower triangular and some elements below the diagonal will be nonzero if Σ_u has nonzero off-diagonal elements. For instance, for the investment/income/consumption example Θ_0 indicates that an income (y_2) innovation has an immediate impact on consumption (y_3). If the white noise covariance matrix Σ_u contains zeros, some components of $u_t = (u_{1t}, \ldots, u_{Kt})'$ are contemporaneously uncorrelated. Suppose, for instance, that u_{1t} is uncorrelated with u_{it} for $i = 2, \ldots, K$. In this case, $\mathsf{A} = W^{-1}$ and, thus, A_0^* has a block of zeros so that y_1 has no instantaneous effect on y_i, $i = 2, \ldots, K$. In the example, investment has no instantaneous impact on income and consumption because

$$\Sigma_u = \begin{bmatrix} 2.25 & 0 & 0 \\ 0 & 1.0 & .5 \\ 0 & .5 & .74 \end{bmatrix}$$

and, hence, u_{1t} is uncorrelated with u_{2t} and u_{3t}. This, of course, is reflected in the matrix of instantaneous effects Θ_0 given in (2.3.31). Because the elements of $P = \Theta_0$ represent the immediate responses of the system to unit innovations they are sometimes called *impact multipliers*.

In order to determine whether there is no response at all of one variable to an impulse in one of the other variables, it suffices to consider the first $pK - p$ response coefficients and the immediate effect. This result is stated formally in the next proposition where $\theta_{jk,i}$ denotes the jk-th element of Θ_i.

Proposition 2.5 (*Zero Orthogonalized Impulse Responses*)
If y_t is a K-dimensional stable VAR(p) process, then, for $j \neq k$,

$$\theta_{jk,i} = 0 \qquad \text{for } i = 0, 1, 2, \ldots$$

is equivalent to

$$\theta_{jk,i} = 0 \qquad \text{for } i = 0, 1, \ldots, p(K-1).$$

∎

The proof of this result is analogous to that of Proposition 2.4 and is left as an exercise (see Problem 2.2).

The fact that Θ_0 is lower triangular shows that the ordering of the variables is of importance, that is, it is important which of the variables is called y_1 and which one is called y_2 and so on. One problem with this type of impulse response analysis is that the ordering of the variables cannot be determined with statistical methods but has to be specified by the analyst. The ordering has to be such that the first variable is the only one with a potential immediate impact on all other variables. The second variable may have an immediate impact on the last $K - 2$ components of y_t but not on y_{1t} and so on. To establish such an ordering may be a quite difficult exercise in practice. The choice of the ordering, the Wold causal ordering, may, to a large extent, determine the impulse responses and is therefore critical for the interpretation of the system. Currently we are dealing with known systems only. In this situation, assuming that the ordering is known may not be a great restriction. For the investment/income/consumption example it may be reasonable to assume that an increase in income has an immediate effect on consumption while increased consumption stimulates the economy and, hence, income with some time lag.

Our interpretation of orthogonalized impulse responses is based on the representation (2.3.30) and the impulses are viewed as changes in the observed variables. Sometimes it is more plausible to focus on impulses which cannot be associated easily with changes in a specific observed variable within the system. In that case, it may be more logical to base the interpretation on the MA representation (2.3.27) which decomposes the variables in contributions of the w_{kt} innovations. If these innovations can be associated with a specific impulse to the system, the orthogonalized impulse responses reflect the reactions of the variables to such possibly unobserved innovations. In that case, a specific impulse or shock to the system can have an instantaneous impact on several variables while some other impulse may only have an instantaneous effect on one specific variable and may effect the other variables only with some delay. By decomposing $\Sigma_u = PP'$ with some non-triangular P matrix, it is also possible that all shocks have instantaneous effects on all observed variables of the system. In this kind of interpretation, finding the decomposition matrix P and, hence, the innovations w_t which actually can be associated with shocks of interest, is often a difficult part of the analysis. We will provide a more in-depth discussion of the related problems in Chapter 9 which deals with structural VAR models.

Critique of Impulse Response Analysis

Besides specifying the relevant impulses to a system, there are a number of further problems that render the interpretation of impulse responses difficult. We have mentioned some of them in the context of Granger-causality. A major limitation of our systems is their potential incompleteness. Although in real economic systems almost everything depends on everything else, we will usually work with low-dimensional VAR systems. All effects of omitted variables are assumed to be in the innovations. If important variables are omitted from the system, this may lead to major distortions in the impulse responses and makes them worthless for structural interpretations. The system may still be useful for prediction, though.

To see the related problems more clearly, consider a system y_t which is partitioned in vectors z_t and x_t as in (2.3.5). If the z_t variables are considered only and the x_t variables are omitted from the analysis, we get a system

$$\begin{aligned} z_t &= \mu_1 + \sum_{i=0}^{\infty} \Phi_{11,i} u_{1,t-i} + \sum_{i=1}^{\infty} \Phi_{12,i} u_{2,t-i} \\ &= \mu_1 + \sum_{i=0}^{\infty} F_i v_{t-i}, \end{aligned} \qquad (2.3.32)$$

as in (2.3.8). The actual reactions of the z_t components to innovations u_{1t} may be given by the $\Phi_{11,i}$ matrices. On the other hand, the F_i or corresponding orthogonalized "impulse responses" are likely to be interpreted as impulse responses if the analyst does not realize that important variables have been omitted. As we have seen in Section 2.3.1, the F_i will be equal to the $\Phi_{11,i}$ if and only if x_t does not Granger-cause z_t.

Further problems related to the interpretation of the MA coefficients as dynamic multipliers or impulse responses result from measurement errors and the use of seasonally adjusted or temporally and/or contemporaneously aggregated variables. A detailed account of the aggregation problem is given by Lütkepohl (1987). We will discuss these problems in more detail in Chapter 11 in the context of more general models. These problems severely limit the interpretability of the MA coefficients of a VAR system as impulse responses. In the next subsection a further possibility to interpret VAR models will be considered.

2.3.3 Forecast Error Variance Decomposition

If the innovations which actually drive the system can be identified, a further tool for interpreting VAR models is available. Suppose a recursive identification scheme is available so that the MA representation (2.3.15) with orthogonal white noise innovations may be considered. In the context of the representation

$$y_t = \mu + \sum_{i=0}^{\infty} \Theta_i w_{t-i} \qquad (2.3.33)$$

with $\Sigma_w = I_K$, the error of the optimal h-step forecast is

$$\begin{aligned} y_{t+h} - y_t(h) &= \sum_{i=0}^{h-1} \Phi_i u_{t+h-i} = \sum_{i=0}^{h-1} \Phi_i P P^{-1} u_{t+h-i} \\ &= \sum_{i=0}^{h-1} \Theta_i w_{t+h-i}. \end{aligned} \qquad (2.3.34)$$

Denoting the mn-th element of Θ_i by $\theta_{mn,i}$ as before, the h-step forecast error of the j-th component of y_t is

$$\begin{aligned} y_{j,t+h} - y_{j,t}(h) &= \sum_{i=0}^{h-1} (\theta_{j1,i} w_{1,t+h-i} + \cdots + \theta_{jK,i} w_{K,t+h-i}) \\ &= \sum_{k=1}^{K} (\theta_{jk,0} w_{k,t+h} + \cdots + \theta_{jk,h-1} w_{k,t+1}). \end{aligned} \qquad (2.3.35)$$

Thus, the forecast error of the j-th component potentially consists of all the innovations $w_{1t}, \ldots, w_{Kt}$. Of course, some of the $\theta_{mn,i}$ may be zero so that some components may not appear in (2.3.35). Because the $w_{k,t}$'s are uncorrelated and have unit variances, the MSE of $y_{j,t}(h)$ is

$$E(y_{j,t+h} - y_{j,t}(h))^2 = \sum_{k=1}^{K} (\theta_{jk,0}^2 + \cdots + \theta_{jk,h-1}^2).$$

Therefore,

$$\theta_{jk,0}^2 + \theta_{jk,1}^2 + \cdots + \theta_{jk,h-1}^2 = \sum_{i=0}^{h-1}(e_j'\Theta_i e_k)^2 \qquad (2.3.36)$$

is sometimes interpreted as the contribution of innovations in variable k to the forecast error variance or MSE of the h-step forecast of variable j. Here e_k is the k-th column of I_K. Dividing (2.3.36) by

$$\text{MSE}[y_{j,t}(h)] = \sum_{i=0}^{h-1}\sum_{k=1}^{K}\theta_{jk,i}^2$$

gives

$$\omega_{jk,h} = \sum_{i=0}^{h-1}(e_j'\Theta_i e_k)^2 / \text{MSE}[y_{j,t}(h)] \qquad (2.3.37)$$

which is the proportion of the h-step forecast error variance of variable j, accounted for by w_{kt} innovations. If w_{kt} can be associated with variable k, $\omega_{jk,h}$ represents the proportion of the h-step forecast error variance accounted for by innovations in variable k. Thereby, the forecast error variance is decomposed into components accounted for by innovations in the different variables of the system. From (2.3.34), the h-step forecast MSE matrix is seen to be

$$\Sigma_y(h) = \text{MSE}[y_t(h)] = \sum_{i=0}^{h-1}\Theta_i\Theta_i' = \sum_{i=0}^{h-1}\Phi_i\Sigma_u\Phi_i'.$$

The diagonal elements of this matrix are the MSEs of the y_{jt} variables which may be used in (2.3.37).

For the investment/income/consumption example, forecast error variance decompositions of all three variables are given in Table 2.1. For instance, about 66% of the 1-step forecast error variance of consumption is accounted for by own innovations and about 34% is accounted for by income innovations. For long term forecasts, 57.5% and 42.3% of the error variance is accounted for by consumption and income innovations, respectively. For any forecast horizon, investment innovations contribute less than 1% to the forecast error variance of consumption. Moreover, only small fractions (less than 10%) of the forecast error variances of income are accounted for by innovations in the other variables of the system. This kind of analysis is sometimes called *innovation accounting*.

From Proposition 2.5, it is obvious that for a stationary, stable, K-dimensional VAR(p) process y_t all forecast error variance proportions of variable j, accounted for by innovations in variable k, will be zero if $\omega_{jk,h} = 0$ for $h = pK - p + 1$. In this context it is perhaps worth pointing out the relationship between Granger-causality and forecast error variance components. For that purpose we consider a bivariate system $y_t = (z_t, x_t)'$ first. In such a system, if z_t does not Granger-cause x_t, the proportions of forecast error variances of x_t

2.3 Structural Analysis with VAR Models

Table 2.1. Forecast error variance decomposition of the investment/income/consumption system

forecast error in	forecast horizon h	proportions of forecast error variance h periods ahead accounted for by innovations in		
		investment	income	consumption
investment	1	1	0	0
	2	1	0	0
	3	1	0	0
	4	1	0	0
	5	1	0	0
	10	1	0	0
	∞	1	0	0
income	1	0	1	0
	2	.020	.941	.039
	3	.026	.930	.044
	4	.029	.926	.045
	5	.030	.925	.045
	10	.030	.925	.045
	∞	.030	.925	.045
consumption	1	0	.338	.662
	2	0	.411	.589
	3	.001	.421	.578
	4	.002	.423	.576
	5	.002	.423	.575
	10	.002	.423	.575
	∞	.002	.423	.575

accounted for by innovations in z_t may still be nonzero. This property follows directly from the definition of the Θ_i in (2.3.15). Granger-noncausality, by Proposition 2.2, implies zero constraints on the Φ_i which may disappear in the Θ_i if the error covariance matrix Σ_u is not diagonal. On the other hand, if Σ_u is diagonal, so that there is no instantaneous causation between z_t and x_t and if, in addition, z_t is not Granger-causal for x_t the lower left-hand elements of the Θ_i will be zero (see (2.3.19)). Therefore, the proportion of forecast error variance of x_t accounted for by z_t innovations will also be zero.

In a higher dimensional system, suppose a set of variables z_t does not Granger-cause the remaining variables x_t and there is also no instantaneous causality between the two sets of variables. In that case, the forecast MSE proportions of all x_t variables accounted for by z_t variables will be zero.

It is important to understand, however, that Granger-causality and forecast error variance decompositions are quite different concepts because Granger-causality and instantaneous causality are different concepts. While Granger-causality is a uniquely defined property of two subsets of variables of a given process, the forecast error variance decomposition is not unique as it depends on the Θ_i matrices and, thus, on the choice of the transformation matrix

P. Therefore, the interpretation of a forecast error variance decomposition is subject to similar criticisms as the interpretation of impulse responses. In addition, all the critical points raised in the context of Granger-causality apply. That is, the forecast error variance components are conditional on the system under consideration. They may change if the system is expanded by adding further variables or if variables are deleted from the system. Also measurement errors, seasonal adjustment and the use of aggregates may contaminate the forecast error variance decompositions.

2.3.4 Remarks on the Interpretation of VAR Models

Innovation accounting and impulse response analysis in the framework of VAR models have been pioneered by Sims (1980, 1981) and others as an alternative to classical macroeconomic analyses. Sims' main criticism of the latter type of analysis is that macroeconometric models are often not based on sound economic theories or the available theories are not capable of providing a completely specified model. If economic theories are not available to specify the model, statistical tools must be applied. In this approach, a fairly loose model is set up which does not impose rigid a priori restrictions on the data generation process. Statistical tools are then used to determine possible constraints. VAR models represent a class of loose models that may be used in such an approach. Of course, in order to interpret these models, some restrictive assumptions need to be made. In particular, the ordering of the variables may be essential for interpretations of the types discussed in the previous subsections. Sims (1981) suggests to try different orderings and investigate the sensitivity of the corresponding orthogonalized impulse responses and the related conclusions to the ordering of the variables.

So far we have assumed that a VAR model is given to us. Under this assumption we have discussed forecasting and interpretation of the system. In this situation it is of course unnecessary to use statistical tools in order to determine constraints for the system because all constraints are known. In practice, we will virtually never be in such a fortunate situation but we have to determine the model from a given set of time series data. This problem will be treated in subsequent chapters. The purpose of this chapter is to identify some problems that are not related to estimation and model specification but are inherent to the types of models considered.

2.4 Exercises

Problem 2.1
Show that

$$\det(I_{Kp} - \mathbf{A}z) = \det(I_K - A_1 z - \cdots - A_p z^p)$$

where A_i, $i = 1, \ldots, p$, and $\mathbf{A}$ are as in (2.1.1) and (2.1.8), respectively.

Problem 2.2
Prove Proposition 2.5.
(Hint: $\Theta(L) = \Phi(L)P = A(L)^{adj} P/\det A(L)$.)

Problem 2.3
In the United States of Wonderland the growth rates of income (GNP) and the money stock (M2) as well as an interest rate (IR) are related as in the following VAR(2) model:

$$\begin{bmatrix} \text{GNP}_t \\ \text{M2}_t \\ \text{IR}_t \end{bmatrix} = \begin{bmatrix} 2 \\ 1 \\ 0 \end{bmatrix} + \begin{bmatrix} .7 & .1 & 0 \\ 0 & .4 & .1 \\ .9 & 0 & .8 \end{bmatrix} \begin{bmatrix} \text{GNP}_{t-1} \\ \text{M2}_{t-1} \\ \text{IR}_{t-1} \end{bmatrix}$$

$$+ \begin{bmatrix} -.2 & 0 & 0 \\ 0 & .1 & .1 \\ 0 & 0 & 0 \end{bmatrix} \begin{bmatrix} \text{GNP}_{t-2} \\ \text{M2}_{t-2} \\ \text{IR}_{t-2} \end{bmatrix} + \begin{bmatrix} u_{1t} \\ u_{2t} \\ u_{3t} \end{bmatrix},$$

$$\Sigma_u = \begin{bmatrix} .26 & .03 & 0 \\ .03 & .09 & 0 \\ 0 & 0 & .81 \end{bmatrix} = PP', \quad P = \begin{bmatrix} .5 & .1 & 0 \\ 0 & .3 & 0 \\ 0 & 0 & .9 \end{bmatrix}. \quad (2.4.1)$$

(a) Show that the process $y_t = (\text{GNP}_t, \text{M2}_t, \text{IR}_t)'$ is stable.
(b) Determine the mean vector of y_t.
(c) Write the process y_t in VAR(1) form.
(d) Compute the coefficient matrices $\Phi_1, \ldots, \Phi_5$ of the MA representation (2.1.17) of y_t.

Problem 2.4
Determine the autocovariances $\Gamma_y(0), \Gamma_y(1), \Gamma_y(2), \Gamma_y(3)$ of the process defined in (2.4.1). Compute and plot the autocorrelations $R_y(0), R_y(1), R_y(2), R_y(3)$.

Problem 2.5
Consider again the process (2.4.1).

(a) Suppose that

$$y_{2000} = \begin{bmatrix} .7 \\ 1.0 \\ 1.5 \end{bmatrix} \quad \text{and} \quad y_{1999} = \begin{bmatrix} 1.0 \\ 1.5 \\ 3.0 \end{bmatrix}$$

and forecast y_{2001}, y_{2002}, and y_{2003}.
(b) Determine the MSE matrices for forecast horizons $h = 1, 2, 3$.
(c) Assume that y_t is a Gaussian process and construct 90% and 95% forecast intervals for $t = 2001, 2002, 2003$.
(d) Use the Bonferroni method to determine a joint forecast region for $\text{GNP}_{2001}, \text{GNP}_{2002}, \text{GNP}_{2003}$ with probability content at least 97%.

Problem 2.6
Answer the following questions for the process (2.4.1).

(a) Is M2 Granger-causal for (GNP, IR)?
(b) Is IR Granger-causal for (GNP, M2)?
(c) Is there instantaneous causality between M2 and (GNP, IR)?
(d) Is there instantaneous causality between IR and (GNP, M2)?
(e) Is IR 2-step causal for GNP?

Problem 2.7
Plot the effect of a unit innovation in the interest rate (IR) on the three variables of the system (2.4.1) in terms of the MA representation (2.1.17). Consider only 5 periods following the innovation. Plot also the accumulated responses and interpret the plots.

Problem 2.8
For the system (2.4.1), derive the coefficient matrices $\Theta_0, \ldots, \Theta_5$ of the MA representation (2.3.15) using the *upper* triangular P matrix given in (2.4.1). Plot the effects of a unit innovation in IR in terms of that representation. Compare to the plots obtained in Problem 2.7 and interpret. Repeat the analysis with a lower triangular P matrix and comment on the results.

Problem 2.9
Decompose the MSE of the forecast $\text{GNP}_t(5)$ into the proportions accounted for by its own innovations and innovations in M2 and IR.

3
Estimation of Vector Autoregressive Processes

3.1 Introduction

In this chapter, it is assumed that a K-dimensional multiple time series $y_1, \ldots, y_T$ with $y_t = (y_{1t}, \ldots, y_{Kt})'$ is available that is known to be generated by a stationary, stable VAR(p) process

$$y_t = \nu + A_1 y_{t-1} + \cdots + A_p y_{t-p} + u_t. \tag{3.1.1}$$

All symbols have their usual meanings, that is, $\nu = (\nu_1, \ldots, \nu_K)'$ is a ($K \times 1$) vector of intercept terms, the A_i are ($K \times K$) coefficient matrices and u_t is white noise with nonsingular covariance matrix Σ_u. In contrast to the assumptions of the previous chapter, the coefficients $\nu, A_1, \ldots, A_p$, and Σ_u are assumed to be unknown in the following. The time series data will be used to estimate the coefficients. Note that notationwise we do not distinguish between the stochastic process and a time series as a realization of a stochastic process. The particular meaning of a symbol should be obvious from the context.

In the next three sections, different possibilities for estimating a VAR(p) process are discussed. In Section 3.5, the consequences of forecasting with estimated processes will be considered and, in Section 3.6, tests for causality are described. The distribution of impulse responses obtained from estimated processes is considered in Section 3.7.

3.2 Multivariate Least Squares Estimation

In this section, multivariate least squares (LS) estimation is discussed. The estimator obtained for the standard form (3.1.1) of a VAR(p) process is considered in Section 3.2.1. Some properties of the estimator are derived in Sections 3.2.2 and 3.2.4 and an example is given in Section 3.2.3.

3.2.1 The Estimator

It is assumed that a time series $y_1, \ldots, y_T$ of the y variables is available, that is, we have a sample of size T for each of the K variables for the same sample period. In addition, p presample values for each variable, $y_{-p+1}, \ldots, y_0$, are assumed to be available. Partitioning a multiple time series in sample and presample values is convenient in order to simplify the notation. We define

$$
\begin{aligned}
Y &:= (y_1, \ldots, y_T) & (K \times T), \\
B &:= (\nu, A_1, \ldots, A_p) & (K \times (Kp+1)), \\
Z_t &:= \begin{bmatrix} 1 \\ y_t \\ \vdots \\ y_{t-p+1} \end{bmatrix} & ((Kp+1) \times 1), \\
Z &:= (Z_0, \ldots, Z_{T-1}) & ((Kp+1) \times T), \\
U &:= (u_1, \ldots, u_T) & (K \times T), \\
\mathbf{y} &:= \text{vec}(Y) & (KT \times 1), \\
\boldsymbol{\beta} &:= \text{vec}(B) & ((K^2 p + K) \times 1), \\
\mathbf{b} &:= \text{vec}(B') & ((K^2 p + K) \times 1), \\
\mathbf{u} &:= \text{vec}(U) & ((KT \times 1).
\end{aligned}
\tag{3.2.1}
$$

Here vec is the column stacking operator as defined in Appendix A.12.

Using this notation, for $t = 1, \ldots, T$, the VAR(p) model (3.1.1) can be written compactly as

$$Y = BZ + U \tag{3.2.2}$$

or

$$
\begin{aligned}
\text{vec}(Y) &= \text{vec}(BZ) + \text{vec}(U) \\
&= (Z' \otimes I_K) \text{vec}(B) + \text{vec}(U)
\end{aligned}
$$

or

$$\mathbf{y} = (Z' \otimes I_K)\boldsymbol{\beta} + \mathbf{u}. \tag{3.2.3}$$

Note that the covariance matrix of $\mathbf{u}$ is

$$\Sigma_{\mathbf{u}} = I_T \otimes \Sigma_u. \tag{3.2.4}$$

Thus, multivariate LS estimation (or GLS estimation) of $\boldsymbol{\beta}$ means to choose the estimator that minimizes

$$\begin{aligned}
S(\boldsymbol{\beta}) &= \mathbf{u}'(I_T \otimes \Sigma_u)^{-1}\mathbf{u} = \mathbf{u}'(I_T \otimes \Sigma_u^{-1})\mathbf{u} \\
&= [\mathbf{y} - (Z' \otimes I_K)\boldsymbol{\beta}]'(I_T \otimes \Sigma_u^{-1})[\mathbf{y} - (Z' \otimes I_K)\boldsymbol{\beta}] \\
&= \text{vec}(Y - BZ)'(I_T \otimes \Sigma_u^{-1}) \, \text{vec}(Y - BZ) \\
&= \text{tr}\left[(Y - BZ)'\Sigma_u^{-1}(Y - BZ)\right].
\end{aligned} \quad (3.2.5)$$

In order to find the minimum of this function we note that

$$\begin{aligned}
S(\boldsymbol{\beta}) &= \mathbf{y}'(I_T \otimes \Sigma_u^{-1})\mathbf{y} + \boldsymbol{\beta}'(Z \otimes I_K)(I_T \otimes \Sigma_u^{-1})(Z' \otimes I_K)\boldsymbol{\beta} \\
&\quad - 2\boldsymbol{\beta}'(Z \otimes I_K)(I_T \otimes \Sigma_u^{-1})\mathbf{y} \\
&= \mathbf{y}'(I_T \otimes \Sigma_u^{-1})\mathbf{y} + \boldsymbol{\beta}'(ZZ' \otimes \Sigma_u^{-1})\boldsymbol{\beta} - 2\boldsymbol{\beta}'(Z \otimes \Sigma_u^{-1})\mathbf{y}.
\end{aligned}$$

Hence,

$$\frac{\partial S(\boldsymbol{\beta})}{\partial \boldsymbol{\beta}} = 2(ZZ' \otimes \Sigma_u^{-1})\boldsymbol{\beta} - 2(Z \otimes \Sigma_u^{-1})\mathbf{y}.$$

Equating to zero gives the *normal equations*

$$(ZZ' \otimes \Sigma_u^{-1})\widehat{\boldsymbol{\beta}} = (Z \otimes \Sigma_u^{-1})\mathbf{y} \quad (3.2.6)$$

and, consequently, the LS estimator is

$$\begin{aligned}
\widehat{\boldsymbol{\beta}} &= ((ZZ')^{-1} \otimes \Sigma_u)(Z \otimes \Sigma_u^{-1})\mathbf{y} \\
&= ((ZZ')^{-1}Z \otimes I_K)\mathbf{y}.
\end{aligned} \quad (3.2.7)$$

The Hessian of $S(\boldsymbol{\beta})$,

$$\frac{\partial^2 S}{\partial \boldsymbol{\beta} \partial \boldsymbol{\beta}'} = 2(ZZ' \otimes \Sigma_u^{-1}),$$

is positive definite which confirms that $\widehat{\boldsymbol{\beta}}$ is indeed a minimizing vector. Strictly speaking, for these results to hold, it has to be assumed that ZZ' is nonsingular. This result will hold with probability 1 if y_t has a continuous distribution which will always be assumed in the following.

It may be worth noting that the multivariate LS estimator $\widehat{\boldsymbol{\beta}}$ is identical to the ordinary LS (OLS) estimator obtained by minimizing

$$\bar{S}(\boldsymbol{\beta}) = \mathbf{u}'\mathbf{u} = [\mathbf{y} - (Z' \otimes I_K)\boldsymbol{\beta}]'[\mathbf{y} - (Z' \otimes I_K)\boldsymbol{\beta}] \quad (3.2.8)$$

(see Problem 3.1). This result is due to Zellner (1962) who showed that GLS and LS estimation in a multiple equation model are identical if the regressors in all equations are the same.

The LS estimator can be written in different ways that will be useful later on:

$$\begin{aligned}
\widehat{\boldsymbol{\beta}} &= ((ZZ')^{-1}Z \otimes I_K)[(Z' \otimes I_K)\boldsymbol{\beta} + \mathbf{u}] \\
&= \boldsymbol{\beta} + ((ZZ')^{-1}Z \otimes I_K)\mathbf{u}
\end{aligned} \quad (3.2.9)$$

or

$$\text{vec}(\widehat{B}) = \widehat{\boldsymbol{\beta}} = ((ZZ')^{-1}Z \otimes I_K)\,\text{vec}(Y)$$
$$= \text{vec}(YZ'(ZZ')^{-1}).$$

Thus,
$$\widehat{B} = YZ'(ZZ')^{-1}$$
$$= (BZ+U)Z'(ZZ')^{-1}$$
$$= B + UZ'(ZZ')^{-1}. \tag{3.2.10}$$

Another possibility for deriving this estimator results from postmultiplying
$$y_t = BZ_{t-1} + u_t$$
by Z'_{t-1} and taking expectations:
$$E(y_t Z'_{t-1}) = BE(Z_{t-1}Z'_{t-1}). \tag{3.2.11}$$

Estimating $E(y_t Z'_{t-1})$ by
$$\frac{1}{T}\sum_{t=1}^{T} y_t Z'_{t-1} = \frac{1}{T} YZ'$$

and $E(Z_{t-1}Z'_{t-1})$ by
$$\frac{1}{T}\sum_{t=1}^{T} Z_{t-1}Z'_{t-1} = \frac{1}{T} ZZ',$$

we obtain the normal equations
$$\frac{1}{T}YZ' = \widehat{B}\frac{1}{T}ZZ'$$

and, hence, $\widehat{B} = YZ'(ZZ')^{-1}$. Note that (3.2.11) is similar but not identical to the system of Yule-Walker equations in (2.1.37). While central moments about the expectation $\mu = E(y_t)$ are considered in (2.1.37), moments about zero are used in (3.2.11).

Yet another possibility to write the LS estimator is
$$\widehat{\mathbf{b}} = \text{vec}(\widehat{B}') = (I_K \otimes (ZZ')^{-1}Z)\,\text{vec}(Y'). \tag{3.2.12}$$

In this form, it is particularly easy to see that multivariate LS estimation is equivalent to OLS estimation of each of the K equations in (3.1.1) separately. Let b'_k be the k-th row of B, that is, b_k contains all the parameters of the k-th equation. Obviously $\mathbf{b}' = (b'_1, \ldots, b'_K)$. Furthermore, let $y_{(k)} = (y_{k1}, \ldots, y_{kT})'$ be the time series available for the k-th variable, so that

$$\text{vec}(Y') = \begin{bmatrix} y_{(1)} \\ \vdots \\ y_{(K)} \end{bmatrix}.$$

With this notation $\widehat{b}_k = (ZZ')^{-1}Zy_{(k)}$ is the OLS estimator of the model $y_{(k)} = Z'b_k + u_{(k)}$, where $u_{(k)} = (u_{k1}, \ldots, u_{kT})'$ and $\widehat{\mathbf{b}}' = (\widehat{b}'_1, \ldots, \widehat{b}'_K)$.

3.2.2 Asymptotic Properties of the Least Squares Estimator

Because small sample properties of the LS estimator are difficult to derive analytically, we focus on asymptotic properties. Consistency and asymptotic normality of the LS estimator are easily established if the following results hold:

$$\Gamma := \operatorname{plim} ZZ'/T \text{ exists and is nonsingular} \tag{3.2.13}$$

and

$$\frac{1}{\sqrt{T}} \sum_{t=1}^{T} \operatorname{vec}(u_t Z'_{t-1}) = \frac{1}{\sqrt{T}} \operatorname{vec}(UZ') = \frac{1}{\sqrt{T}}(Z \otimes I_K)\mathbf{u} \tag{3.2.14}$$
$$\xrightarrow[T \to \infty]{d} \mathcal{N}(0, \Gamma \otimes \Sigma_u),$$

where, as usual, $\xrightarrow{d}$ denotes convergence in distribution. It follows from a theorem due to Mann & Wald (1943) that these results are true under suitable conditions for u_t, if y_t is a stationary, stable VAR(p). For instance, the conditions stated in the following definition are sufficient.

Definition 3.1 (*Standard White Noise*)
A white noise process $u_t = (u_{1t}, \ldots, u_{Kt})'$ is called *standard white noise* if the u_t are continuous random vectors satisfying $E(u_t) = 0$, $\Sigma_u = E(u_t u'_t)$ is nonsingular, u_t and u_s are independent for $s \neq t$, and, for some finite constant c,

$$E|u_{it} u_{jt} u_{kt} u_{mt}| \leq c \quad \text{for } i, j, k, m = 1, \ldots, K, \text{ and all } t.$$

■

The last condition means that all fourth moments exist and are bounded. Obviously, if the u_t are normally distributed (Gaussian) they satisfy the moment requirements. With this definition it is easy to state conditions for consistency and asymptotic normality of the LS estimator. The following lemma will be essential in proving these large sample results.

Lemma 3.1
If y_t is a stable, K-dimensional VAR(p) process as in (3.1.1) with standard white noise residuals u_t, then (3.2.13) and (3.2.14) hold. ■

Proof: See Theorem 8.2.3 of Fuller (1976, p. 340). ■

The lemma holds also for other definitions of standard white noise. For example, the convergence result in (3.2.14) follows from a central limit theorem for martingale differences or martingale difference arrays (see Proposition C.13) by noting that $w_t = \operatorname{vec}(u_t Z'_{t-1})$ is a martingale difference sequence under quite general conditions. The convergence result in (3.2.13) may then be obtained from a suitable weak law of large numbers (see Proposition C.12). In the next proposition the resulting asymptotic properties of the LS estimator are stated formally.

Proposition 3.1 (*Asymptotic Properties of the LS Estimator*)
Let y_t be a stable, K-dimensional VAR(p) process as in (3.1.1) with standard white noise residuals, $\widehat{B} = YZ'(ZZ')^{-1}$ is the LS estimator of the VAR coefficients B and all symbols are as defined in (3.2.1). Then,

$$\text{plim } \widehat{B} = B$$

and

$$\sqrt{T}(\widehat{\boldsymbol{\beta}} - \boldsymbol{\beta}) = \sqrt{T} \text{ vec}(\widehat{B} - B) \xrightarrow{d} \mathcal{N}(0, \Gamma^{-1} \otimes \Sigma_u) \qquad (3.2.15)$$

or, equivalently,

$$\sqrt{T}(\widehat{\mathbf{b}} - \mathbf{b}) = \sqrt{T} \text{ vec}(\widehat{B}' - B') \xrightarrow{d} \mathcal{N}(0, \Sigma_u \otimes \Gamma^{-1}), \qquad (3.2.16)$$

where $\Gamma = \text{plim } ZZ'/T$. ∎

Proof: Using (3.2.10),

$$\text{plim}(\widehat{B} - B) = \text{plim}\left(\frac{UZ'}{T}\right) \text{plim}\left(\frac{ZZ'}{T}\right)^{-1} = 0$$

by Lemma 3.1, because (3.2.14) implies plim $UZ'/T = 0$. Thus, the consistency of $\widehat{B}$ is established.
Using (3.2.9),

$$\begin{aligned}
\sqrt{T}(\widehat{\boldsymbol{\beta}} - \boldsymbol{\beta}) &= \sqrt{T}((ZZ')^{-1}Z \otimes I_K)\mathbf{u} \\
&= \left(\left(\frac{1}{T}ZZ'\right)^{-1} \otimes I_K\right) \frac{1}{\sqrt{T}}(Z \otimes I_K)\mathbf{u}.
\end{aligned}$$

Thus, by Proposition C.2(4) of Appendix C, $\sqrt{T}(\widehat{\boldsymbol{\beta}} - \boldsymbol{\beta})$ has the same asymptotic distribution as

$$\left[\text{plim}\left(\frac{1}{T}ZZ'\right)^{-1} \otimes I_K\right] \frac{1}{\sqrt{T}}(Z \otimes I_K)\mathbf{u} = (\Gamma^{-1} \otimes I_K)\frac{1}{\sqrt{T}}(Z \otimes I_K)\mathbf{u}.$$

Hence, the asymptotic distribution of $\sqrt{T}(\widehat{\boldsymbol{\beta}} - \boldsymbol{\beta})$ is normal by Lemma 3.1 and the covariance matrix is

$$(\Gamma^{-1} \otimes I_K)(\Gamma \otimes \Sigma_u)(\Gamma^{-1} \otimes I_K) = \Gamma^{-1} \otimes \Sigma_u.$$

The result (3.2.16) can be established with similar arguments (see Problem 3.2). ∎

As mentioned previously, if u_t is *Gaussian* (normally distributed) white noise, it satisfies the conditions of Proposition 3.1 so that consistency and asymptotic normality of the LS estimator are ensured for stable Gaussian (normally distributed) VAR(p) processes y_t. Note that normality of u_t implies normality of the y_t for stable processes.

In order to assess the asymptotic dispersion of the LS estimator, we need to know the matrices Γ and Σ_u. From (3.2.13) an obvious consistent estimator of Γ is

$$\widehat{\Gamma} = ZZ'/T. \tag{3.2.17}$$

Because $\Sigma_u = E(u_t u_t')$, a plausible estimator for this matrix is

$$\begin{aligned}
\widetilde{\Sigma}_u &= \frac{1}{T}\sum_{t=1}^T \widehat{u}_t \widehat{u}_t' = \frac{1}{T}\widehat{U}\widehat{U}' = \frac{1}{T}(Y - \widehat{B}Z)(Y - \widehat{B}Z)' \\
&= \frac{1}{T}[Y - YZ'(ZZ')^{-1}Z][Y - YZ'(ZZ')^{-1}Z]' \\
&= \frac{1}{T}Y[I_T - Z'(ZZ')^{-1}Z][I_T - Z'(ZZ')^{-1}Z]'Y' \\
&= \frac{1}{T}Y(I_T - Z'(ZZ')^{-1}Z)Y'. \tag{3.2.18}
\end{aligned}$$

Often an adjustment for degrees of freedom is desired because in a regression with fixed, nonstochastic regressors this leads to an unbiased estimator of the covariance matrix. Thus, an estimator

$$\widehat{\Sigma}_u = \frac{T}{T - Kp - 1}\widetilde{\Sigma}_u \tag{3.2.19}$$

may be considered. Note that there are $Kp + 1$ parameters in each of the K equations of (3.1.1) and, hence, there are $Kp+1$ parameters in each equation of the system (3.2.2). Of course, $\widehat{\Sigma}_u$ and $\widetilde{\Sigma}_u$ are asymptotically equivalent. They are consistent estimators of Σ_u if the conditions of Proposition 3.1 hold. In fact, a bit more can be shown.

Proposition 3.2 (*Asymptotic Properties of the White Noise Covariance Matrix Estimators*)

Let y_t be a stable, K-dimensional VAR(p) process as in (3.1.1) with standard white noise innovations and let $\bar{B}$ be an estimator of the VAR coefficients B so that $\sqrt{T}\,\text{vec}(\bar{B} - B)$ converges in distribution. Furthermore, using the symbols from (3.2.1), suppose that

$$\bar{\Sigma}_u = (Y - \bar{B}Z)(Y - \bar{B}Z)'/(T - c),$$

where c is a fixed constant. Then

$$\text{plim}\sqrt{T}(\bar{\Sigma}_u - UU'/T) = 0. \tag{3.2.20}$$

■

76   3 Estimation of Vector Autoregressive Processes

Proof:

$$\frac{1}{T}(Y - \bar{B}Z)(Y - \bar{B}Z)' = (B - \bar{B})\left(\frac{ZZ'}{T}\right)(B - \bar{B})' + (B - \bar{B})\frac{ZU'}{T}$$
$$+ \frac{UZ'}{T}(B - \bar{B})' + \frac{UU'}{T}.$$

Under the conditions of the proposition, $\text{plim}(B - \bar{B}) = 0$. Hence, by Lemma 3.1,

$$\text{plim } (B - \bar{B})ZU'/\sqrt{T} = 0$$

and

$$\text{plim } \left[(B - \bar{B})\frac{ZZ'}{T}\sqrt{T}(B - \bar{B})'\right] = 0$$

(see Appendix C.1). Thus,

$$\text{plim } \sqrt{T}\left[(Y - \bar{B}Z)(Y - \bar{B}Z)'/T - UU'/T\right] = 0.$$

Therefore, the proposition follows by noting that $T/(T - c) \to 1$ as $T \to \infty$. ■

The proposition covers both estimators $\widehat{\Sigma}_u$ and $\widetilde{\Sigma}_u$. It implies that the feasible estimators $\widetilde{\Sigma}_u$ and $\widehat{\Sigma}_u$ have the same asymptotic properties as the estimator

$$\frac{UU'}{T} = \frac{1}{T}\sum_{t=1}^{T} u_t u_t'$$

which is based on the unknown true residuals and is therefore not feasible in practice. In particular, if $\sqrt{T} \text{ vec}(UU'/T - \Sigma_u)$ converges in distribution, $\sqrt{T} \text{ vec}(\widehat{\Sigma}_u - \Sigma_u)$ and $\sqrt{T} \text{ vec}(\widetilde{\Sigma}_u - \Sigma_u)$ will have the same limiting distribution (see Proposition C.2 of Appendix C.1). Moreover, it can be shown that the asymptotic distributions are independent of the limiting distribution of the LS estimator $\widehat{B}$. Another immediate implication of Proposition 3.2 is that $\widetilde{\Sigma}_u$ and $\widehat{\Sigma}_u$ are consistent estimators of Σ_u. This result is established next.

Corollary 3.2.1

Under the conditions of Proposition 3.2,

$$\text{plim } \widetilde{\Sigma}_u = \text{plim } \widehat{\Sigma}_u = \text{plim } UU'/T = \Sigma_u.$$

■

Proof: By Proposition 3.2, it suffices to show that $\text{plim } UU'/T = \Sigma_u$ which follows from Proposition C.12(4) because

$$E\left(\frac{1}{T}UU'\right) = \frac{1}{T}\sum_{t=1}^{T} E(u_t u_t') = \Sigma_u$$

and

$$\text{Var}\left(\frac{1}{T}\text{vec}(UU')\right) = \frac{1}{T^2}\sum_{t=1}^{T} \text{Var}[\text{vec}(u_t u_t')] \le \frac{T}{T^2}g \xrightarrow[T\to\infty]{} 0,$$

where g is a constant upper bound for $\text{Var}[\text{vec}(u_t u_t')]$. This bound exists because the fourth moments of u_t are bounded by Definition 3.1. ∎

If y_t is stable with standard white noise, Proposition 3.1 and Corollary 3.2.1 imply that $(\widehat{\beta}_i - \beta_i)/\widehat{s}_i$ has an asymptotic standard normal distribution. Here β_i ($\widehat{\beta}_i$) is the i-th component of β ($\widehat{\beta}$) and $\widehat{s}_i$ is the square root of the i-th diagonal element of

$$(ZZ')^{-1} \otimes \widehat{\Sigma}_u. \tag{3.2.21}$$

This result means that we can use the "t-ratios" provided by common regression programs in setting up confidence intervals and tests for individual coefficients. The critical values and percentiles may be based on the asymptotic standard normal distribution. Because it was found in simulation studies that the small sample distributions of the "t-ratios" have fatter tails than the standard normal distribution, one may want to approximate the small sample distribution by some t-distribution. The question is then what number of degrees of freedom (d.f.) should be used. The overall model (3.2.3) may suggest a choice of d.f. $= KT - K^2p - K$ because in a standard regression model with nonstochastic regressors the d.f. of the "t-ratios" are equal to the sample size minus the number of estimated parameters. In the present case, it seems also reasonable to use d.f. $= T - Kp - 1$ because the multivariate LS estimator is identical to the LS estimator obtained for each of the K equations in (3.2.2) separately. In a separate regression for each individual equation, we would have T observations and $Kp+1$ parameters. If the sample size T is large and, thus, the number of degrees of freedom is large, the corresponding t-distribution will be very close to the standard normal so that the choice between the two becomes irrelevant for large samples. Before we look a little further into the problem of choosing appropriate critical values, let us illustrate the foregoing results by an example.

3.2.3 An Example

As an example, we consider a three-dimensional system consisting of first differences of the logarithms of quarterly, seasonally adjusted West German fixed investment (y_1), disposable income (y_2), and consumption expenditures (y_3) from File E1 of the data sets associated with this book. We use only data from 1960–1978 and reserve the data for 1979–1982 for a subsequent analysis. The original data and first differences of logarithms are plotted in Figures 3.1 and 3.2, respectively. The original data have a trend and are thus considered to be nonstationary. The trend is removed by taking first differences of logarithms. We will discuss this issue in some more detail in Part II. Note that the value for 1960.1 is lost in the differenced series.

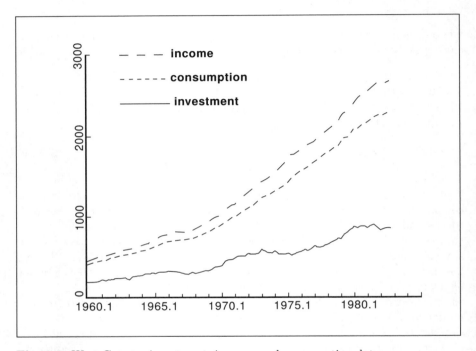

Fig. 3.1. West German investment, income, and consumption data.

Let us assume that the data have been generated by a VAR(2) process. The choice of the VAR order $p = 2$ is arbitrary at this point. In the next chapter, criteria for choosing the VAR order will be considered. Because the VAR order is two, we keep the first two observations of the differenced series as presample values and use a sample size of $T = 73$. Thus, we have a (3×73) matrix Y, $B = (\nu, A_1, A_2)$ is (3×7), Z is (7×73) and β and $\mathbf{b}$ are both (21×1) vectors.

The LS estimates are

$$\widehat{B} = (\widehat{\nu}, \widehat{A}_1, \widehat{A}_2) = YZ'(ZZ')^{-1}$$

$$= \begin{bmatrix} -.017 & -.320 & .146 & .961 & -.161 & .115 & .934 \\ .016 & .044 & -.153 & .289 & .050 & .019 & -.010 \\ .013 & -.002 & .225 & -.264 & .034 & .355 & -.022 \end{bmatrix}. \quad (3.2.22)$$

3.2 Multivariate Least Squares Estimation 79

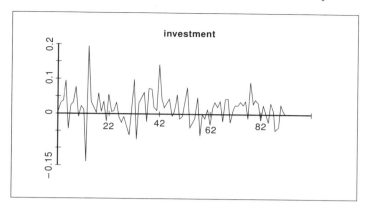

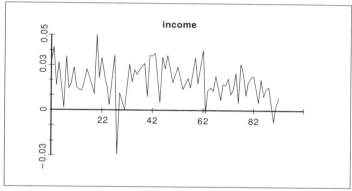

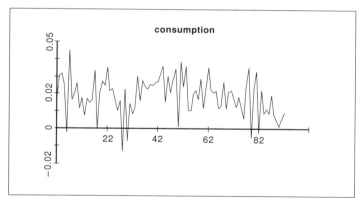

Fig. 3.2. First differences of logarithms of West German investment, income, and consumption.

To check the stability of the estimated process, we determine the roots of the polynomial $\det(I_3 - \widehat{A}_1 z - \widehat{A}_2 z^2)$ which is easily seen to have degree 6. Its roots are

80   3 Estimation of Vector Autoregressive Processes

$$z_1 = 1.753, \ z_2 = -2.694, \ z_{3/4} = -0.320 \pm 2.008i, \ z_{5/6} = -1.285 \pm 1.280i.$$

Note that these roots have been computed using higher precision than the three digits in (3.2.22). They all have modulus greater than 1 and, hence, the stability condition is satisfied.

We get

$$\widehat{\Sigma}_u = \frac{1}{T - Kp - 1}(YY' - YZ'(ZZ')^{-1}ZY')$$

$$= \begin{bmatrix} 21.30 & .72 & 1.23 \\ .72 & 1.37 & .61 \\ 1.23 & .61 & .89 \end{bmatrix} \times 10^{-4} \qquad (3.2.23)$$

as estimate of the residual covariance matrix Σ_u. Furthermore,

$$\widehat{\Gamma}^{-1} = (ZZ'/T)^{-1}$$

$$= T \begin{bmatrix} .14 & .17 & -.69 & -2.51 & .10 & -.67 & -2.57 \\ \bullet & 7.39 & 1.24 & -10.56 & 1.80 & 1.08 & -8.70 \\ \bullet & \bullet & 139.81 & -87.40 & -4.58 & 30.21 & -50.88 \\ \bullet & \bullet & \bullet & 207.22 & .84 & -55.35 & 73.82 \\ \bullet & \bullet & \bullet & \bullet & 7.33 & -.03 & -9.31 \\ \bullet & \bullet & \bullet & \bullet & \bullet & 134.19 & -82.64 \\ \bullet & \bullet & \bullet & \bullet & \bullet & \bullet & 207.71 \end{bmatrix}.$$

Dividing the elements of $\widehat{B}$ by square roots of the corresponding diagonal elements of $(ZZ')^{-1} \otimes \widehat{\Sigma}_u$ we get the matrix of t-ratios:

$$\begin{bmatrix} -0.97 & -2.55 & 0.27 & 1.45 & -1.29 & 0.21 & 1.41 \\ 3.60 & 1.38 & -1.10 & 1.71 & 1.58 & 0.14 & -0.06 \\ 3.67 & -0.09 & 2.01 & -1.94 & 1.33 & 3.24 & -0.16 \end{bmatrix}. \qquad (3.2.24)$$

We may compare these quantities with critical values from a t-distribution with d.f. $= KT - K^2p - K = 198$ or d.f. $= T - Kp - 1 = 66$. In both cases, we get critical values of approximately ± 2 for a two-tailed test with significance level 5%. Thus, the critical values are approximately the same as those from a standard normal distribution.

Apparently quite a few coefficients are not significant under this criterion. This observation suggests that the model contains unnecessarily many free parameters. In subsequent chapters, we will discuss the problem of choosing the VAR order and possible restrictions for the coefficients. Also, before an estimated model is used for forecasting and analysis purposes, the assumptions underlying the analysis should be checked carefully. Checking the model adequacy will be treated in greater detail in Chapter 4.

3.2.4 Small Sample Properties of the LS Estimator

As mentioned earlier, it is difficult to derive small sample properties of the LS estimator analytically. In such a case it is sometimes helpful to use *Monte*

Carlo methods to get some idea about the small sample properties. In a Monte Carlo analysis, specific processes are used to artificially generate a large number of time series. Then a set of estimates is computed for each multiple time series generated and the properties of the resulting empirical distributions of these estimates are studied (see Appendix D). Such an approach usually permits rather limited conclusions only because the findings may depend on the particular processes used for generating the time series. Nevertheless, such exercises give some insight into the small sample properties of estimators.

In the following, we use the bivariate VAR(2) example process (2.1.15),

$$y_t = \begin{bmatrix} .02 \\ .03 \end{bmatrix} + \begin{bmatrix} .5 & .1 \\ .4 & .5 \end{bmatrix} y_{t-1} + \begin{bmatrix} 0 & 0 \\ .25 & 0 \end{bmatrix} y_{t-2} + u_t \qquad (3.2.25)$$

with error covariance matrix

$$\Sigma_u = \begin{bmatrix} 9 & 0 \\ 0 & 4 \end{bmatrix} \times 10^{-4} \qquad (3.2.26)$$

to investigate the small sample properties of the multivariate LS estimator. With this process we have generated 1000 bivariate time series of length $T = 30$ plus 2 presample values using independent standard normal errors, that is, $u_t \sim \mathcal{N}(0, \Sigma_u)$. Thus the 1000 bivariate time series are generated by a stable Gaussian process so that Propositions 3.1 and 3.2 provide the asymptotic properties of the LS estimators.

In Table 3.1, some empirical results are given. In particular, the empirical mean, variance, and mean squared error (MSE) of each parameter estimator are given. Obviously, the empirical means differ from the actual values of the coefficients. However, measuring the estimation precision by the empirical variance (average squared deviation from the mean in 1000 samples) or MSE (average squared deviation from the true parameter value), the coefficients are seen to be estimated quite precisely even with a sample size as small as $T = 30$. This is partly a consequence of the special properties of the process.

In Table 3.1, empirical percentiles of the t-ratios are also given together with the corresponding percentiles from the t- and standard normal distributions (d.f. $= \infty$). Even with the presently considered relatively small sample size the percentiles of the three distributions that might be used for inference do not differ much. Consequently, it does not matter much which of the theoretical percentiles are used, in particular, because the empirical percentiles, in many cases, differ quite a bit from the corresponding theoretical quantities. This example shows that the asymptotic results have to be used cautiously in setting up small sample tests and confidence intervals. On the other hand, this example also demonstrates that the asymptotic theory does provide some guidance for inference. For example, the empirical 95th percentiles of all coefficients lie between the 90th and the 99th percentile of the standard normal distribution given in the last row of the table. Of course, this is just one example and not a general finding.

Table 3.1. Empirical percentiles of t-ratios of parameter estimates for the example process and actual percentiles of t-distributions for sample size $T = 30$

parameter	empirical			empirical percentiles of t-ratios						
	mean	variance	MSE	1.	5.	10.	50.	90.	95.	99.
$\nu_1 = .02$	.041	.0011	.0015	−1.91	−1.04	−0.64	0.62	1.92	2.29	3.12
$\nu_2 = .03$	.038	.0005	.0006	−2.30	−1.40	−1.02	0.25	1.65	2.11	2.83
$\alpha_{11,1} = .5$	.41	.041	.049	−2.78	−2.18	−1.74	−0.43	0.92	1.28	2.01
$\alpha_{21,1} = .4$	.40	.018	.018	−2.61	−1.74	−1.28	0.04	1.28	1.71	2.65
$\alpha_{12,1} = .1$	.10	.078	.078	−2.27	−1.67	−1.35	−0.03	1.29	1.67	2.38
$\alpha_{22,1} = .5$	.44	.030	.034	−2.69	−1.97	−1.59	−0.35	0.89	1.30	2.06
$\alpha_{11,2} = 0$	−.05	.056	.058	−2.75	−1.93	−1.50	−0.24	1.02	1.38	2.09
$\alpha_{21,2} = .25$	.29	.023	.024	−1.99	−1.32	−0.99	0.20	1.45	1.81	2.48
$\alpha_{12,2} = 0$	−.07	.053	.058	−2.48	−1.91	−1.61	−0.28	0.97	1.39	2.03
$\alpha_{22,2} = 0$	−.01	.023	.024	−2.71	−1.72	−1.36	−0.03	1.18	1.53	2.18
degrees of freedom (d.f.)				percentiles of t-distributions						
				1.	5.	10.	50.	90.	95.	99.
$T − Kp − 1 = 25$				−2.49	−1.71	−1.32	0	1.32	1.71	2.49
$K(T − Kp − 1) = 50$				−2.41	−1.68	−1.30	0	1.30	1.68	2.41
∞ (normal distribution)				−2.33	−1.65	−1.28	0	1.28	1.65	2.33

In an extensive study, Nankervis & Savin (1988) investigated the small sample distribution of the "t-statistic" for the parameter of a *univariate* AR(1) process. They found that it differs quite substantially from the corresponding t-distribution, especially if the sample size is small ($T < 100$) and the parameter lies close to the instability region. Analytical results on the bias in estimating VAR models were derived by Nicholls & Pope (1988) and Tjøstheim & Paulsen (1983). What should be learned from our Monte Carlo investigation and these remarks is that asymptotic distributions in the present context can only be used as rough guidelines for small sample inference. That, however, is much better than having no guidance at all.

3.3 Least Squares Estimation with Mean-Adjusted Data and Yule-Walker Estimation

3.3.1 Estimation when the Process Mean Is Known

Occasionally a VAR(p) model is given in *mean-adjusted form*,

$$(y_t - \mu) = A_1(y_{t-1} - \mu) + \cdots + A_p(y_{t-p} - \mu) + u_t. \tag{3.3.1}$$

Multivariate LS estimation of this model form is straightforward if the mean vector μ is known. Defining

$$Y^0 := (y_1 - \mu, \ldots, y_T - \mu) \quad (K \times T),$$
$$A := (A_1, \ldots, A_p) \quad (K \times Kp),$$
$$Y_t^0 := \begin{bmatrix} y_t - \mu \\ \vdots \\ y_{t-p+1} - \mu \end{bmatrix} \quad (Kp \times 1), \tag{3.3.2}$$
$$X := (Y_0^0, \ldots, Y_{T-1}^0) \quad (Kp \times T),$$
$$\mathbf{y}^0 := \text{vec}(Y^0) \quad (KT \times 1),$$
$$\boldsymbol{\alpha} := \text{vec}(A) \quad (K^2 p \times 1),$$

we can write (3.3.1), for $t = 1, \ldots, T$, compactly as

$$Y^0 = AX + U \tag{3.3.3}$$

or

$$\mathbf{y}^0 = (X' \otimes I_K)\boldsymbol{\alpha} + \mathbf{u}, \tag{3.3.4}$$

where U and $\mathbf{u}$ are defined as in (3.2.1). The LS estimator is easily seen to be

$$\widehat{\boldsymbol{\alpha}} = ((XX')^{-1}X \otimes I_K)\mathbf{y}^0 \tag{3.3.5}$$

or

$$\widehat{A} = Y^0 X'(XX')^{-1}. \tag{3.3.6}$$

If y_t is stable and u_t is standard white noise, it can be shown that

$$\sqrt{T}(\widehat{\boldsymbol{\alpha}} - \boldsymbol{\alpha}) \xrightarrow{d} \mathcal{N}(0, \Sigma_{\widehat{\boldsymbol{\alpha}}}), \tag{3.3.7}$$

where

$$\Sigma_{\widehat{\boldsymbol{\alpha}}} = \Gamma_Y(0)^{-1} \otimes \Sigma_u \tag{3.3.8}$$

and $\Gamma_Y(0) := E(Y_t^0 Y_t^{0\prime})$.

3.3.2 Estimation of the Process Mean

Usually μ will not be known in advance. In that case, it may be estimated by the vector of sample means,

$$\bar{y} = \frac{1}{T} \sum_{t=1}^{T} y_t. \tag{3.3.9}$$

Using (3.3.1), $\bar{y}$ can be written as

$$\bar{y} = \mu + A_1 \left[\bar{y} + \frac{1}{T}(y_0 - y_T) - \mu \right] + \cdots$$

$$+ A_p \left[\bar{y} + \frac{1}{T}(y_{-p+1} + \cdots + y_0 - y_{T-p+1} - \cdots - y_T) - \mu \right]$$

$$+ \frac{1}{T} \sum_{t=1}^{T} u_t.$$

Hence,

$$(I_K - A_1 - \cdots - A_p)(\bar{y} - \mu) = \frac{1}{T} z_T + \frac{1}{T} \sum_t u_t, \qquad (3.3.10)$$

where

$$z_T = \sum_{i=1}^{p} A_i \left[\sum_{j=0}^{i-1} (y_{0-j} - y_{T-j}) \right].$$

Evidently,

$$E(z_T/\sqrt{T}) = \frac{1}{\sqrt{T}} E(z_T) = 0$$

and

$$\mathrm{Var}(z_T/\sqrt{T}) = \frac{1}{T} \mathrm{Var}(z_T) \xrightarrow[T \to \infty]{} 0$$

because y_t is stable. In other words, $z_T/\sqrt{T}$ converges to zero in mean square. It follows that $\sqrt{T}(I_K - A_1 - \cdots - A_p)(\bar{y} - \mu)$ has the same asymptotic distribution as $\sum u_t/\sqrt{T}$ (see Appendix C, Proposition C.2). Hence, noting that, by a central limit theorem (e.g., Fuller (1976) or Proposition C.13),

$$\frac{1}{\sqrt{T}} \sum_{t=1}^{T} u_t \xrightarrow{d} \mathcal{N}(0, \Sigma_u), \qquad (3.3.11)$$

if u_t is standard white noise, we get the following result:

Proposition 3.3 (*Asymptotic Properties of the Sample Mean*)
If the VAR(p) process y_t given in (3.3.1) is stable and u_t is standard white noise, then

$$\sqrt{T}(\bar{y} - \mu) \xrightarrow{d} \mathcal{N}(0, \Sigma_{\bar{y}}), \qquad (3.3.12)$$

where

$$\Sigma_{\bar{y}} = (I_K - A_1 - \cdots - A_p)^{-1} \Sigma_u (I_K - A_1 - \cdots - A_p)'^{-1}.$$

In particular, plim $\bar{y} = \mu$. ∎

The proposition follows from (3.3.10), (3.3.11), and Proposition C.15 of Appendix C. The limiting distribution in (3.3.11) holds even in small samples for Gaussian white noise u_t.

Because $\mu = (I_K - A_1 - \cdots - A_p)^{-1}\nu$ (see Chapter 2, Section 2.1), an alternative estimator for the process mean is obtained from the LS estimator of the previous section:

$$\widehat{\mu} = (I_k - \widehat{A}_1 - \cdots - \widehat{A}_p)^{-1}\widehat{\nu}. \tag{3.3.13}$$

Using again Proposition C.15 of Appendix C, this estimator is also consistent and has an asymptotic normal distribution,

$$\sqrt{T}(\widehat{\mu} - \mu) \xrightarrow{d} \mathcal{N}\left(0, \frac{\partial \mu}{\partial \boldsymbol{\beta}'}(\Gamma^{-1} \otimes \Sigma_u)\frac{\partial \mu'}{\partial \boldsymbol{\beta}}\right), \tag{3.3.14}$$

provided the conditions of Proposition 3.1 are satisfied. It can be shown that

$$\frac{\partial \mu}{\partial \boldsymbol{\beta}'}(\Gamma^{-1} \otimes \Sigma_u)\frac{\partial \mu'}{\partial \boldsymbol{\beta}} = \Sigma_{\bar{y}} \tag{3.3.15}$$

and, hence, the estimators $\widehat{\mu}$ and $\bar{y}$ for μ are asymptotically equivalent (see Section 3.4). This result suggests that it does not matter asymptotically whether the mean is estimated separately or jointly with the other VAR coefficients. While this holds asymptotically, it will usually matter in small samples which estimator is used. An example will be given shortly.

3.3.3 Estimation with Unknown Process Mean

If the mean vector μ is unknown, it may be replaced by $\bar{y}$ in the vectors and matrices in (3.3.2) giving $\widehat{X}, \widehat{Y}^0$ and so on. The resulting LS estimator,

$$\widehat{\widehat{\alpha}} = ((\widehat{X}\widehat{X}')^{-1}\widehat{X} \otimes I_K)\widehat{\mathbf{y}}^0,$$

is asymptotically equivalent to $\widehat{\alpha}$. More precisely, it can be shown that, under the conditions of Proposition 3.3,

$$\sqrt{T}(\widehat{\widehat{\alpha}} - \boldsymbol{\alpha}) \xrightarrow{d} \mathcal{N}(0, \Gamma_Y(0)^{-1} \otimes \Sigma_u), \tag{3.3.16}$$

where $\Gamma_Y(0) := E(Y_t^0 Y_t^{0\prime})$. This result will be discussed further in the next section on maximum likelihood estimation for Gaussian processes.

3.3.4 The Yule-Walker Estimator

The LS estimator can also be derived from the Yule-Walker equations given in Chapter 2, (2.1.37). They imply

$$\Gamma_y(h) = [A_1, \ldots, A_p] \begin{bmatrix} \Gamma_y(h-1) \\ \vdots \\ \Gamma_y(h-p) \end{bmatrix}, \quad h > 0,$$

or

$$\begin{aligned}[][\Gamma_y(1), \ldots, \Gamma_y(p)] &= [A_1, \ldots, A_p] \begin{bmatrix} \Gamma_y(0) & \cdots & \Gamma_y(p-1) \\ \vdots & \ddots & \vdots \\ \Gamma_y(-p+1) & \cdots & \Gamma_y(0) \end{bmatrix} \\ &= A\Gamma_Y(0) \end{aligned} \quad (3.3.17)$$

and, hence,

$$A = [\Gamma_y(1), \ldots, \Gamma_y(p)] \Gamma_Y(0)^{-1}.$$

Estimating $\Gamma_Y(0)$ by $\widehat{X}\widehat{X}'/T$ and $[\Gamma_y(1), \ldots, \Gamma_y(p)]$ by $\widehat{Y}^0\widehat{X}'/T$, the resulting estimator is just the LS estimator,

$$\widehat{\widehat{A}} = \widehat{Y}^0 \widehat{X}'(\widehat{X}\widehat{X}')^{-1}. \quad (3.3.18)$$

Alternatively, the moment matrices $\Gamma_y(h)$ may be estimated using as many data as are available, including the presample values. Thus, if a sample $y_1, \ldots, y_T$ and p presample observations $y_{-p+1}, \ldots, y_0$ are available, μ may be estimated as

$$\overline{y}^* = \frac{1}{T+p} \sum_{t=-p+1}^{T} y_t$$

and $\Gamma_y(h)$ may be estimated as

$$\widehat{\Gamma}_y(h) = \frac{1}{T+p-h} \sum_{t=-p+h+1}^{T} (y_t - \overline{y}^*)(y_{t-h} - \overline{y}^*)'. \quad (3.3.19)$$

Using these estimators in (3.3.17), the so-called *Yule-Walker estimator* for A is obtained. For stable processes, this estimator has the same asymptotic properties as the LS estimator. However, it may have less attractive small sample properties (e.g., Tjøstheim & Paulsen (1983)).

The Yule-Walker estimator always produces estimates in the stability region (see Brockwell & Davis (1987, §8.1) for a discussion of the univariate case). In other words, the estimated process is always stable. This property is sometimes regarded as an advantage of the Yule-Walker estimator. It is responsible for possibly considerable bias of the estimator, however. Also, in practice, it may not be known a priori whether the data generation process of a given multiple time series is stable. In the unstable case, LS and Yule-Walker estimation are not asymptotically equivalent anymore (see also the discussion in Reinsel (1993, Section 4.4)). Therefore, enforcing stability may not be a good strategy in practice. The LS estimator is usually used in the following.

3.3.5 An Example

To illustrate the results of this section, we use again the West German investment, income, and consumption data. The variables y_1, y_2, and y_3 are defined as in Section 3.2.3, the sample period ranges from 1960.4 to 1978.4, that is, $T = 73$ and the data for 1960.2 and 1960.3 are used as presample values. Using only the sample values we get

$$\bar{y} = \begin{bmatrix} .018 \\ .020 \\ .020 \end{bmatrix} \qquad (3.3.20)$$

which is different, though not substantially so, from

$$\widehat{\mu} = (I_3 - \widehat{A}_1 - \widehat{A}_2)^{-1}\widehat{\nu} = \begin{bmatrix} .017 \\ .020 \\ .020 \end{bmatrix} \qquad (3.3.21)$$

as obtained from the LS estimates in (3.2.22).

Subtracting the sample means from the data we get, based on (3.3.18),

$$\widehat{A} = (\widehat{A}_1, \widehat{A}_2) = \begin{bmatrix} -.319 & .143 & .960 & -.160 & .112 & .933 \\ .044 & -.153 & .288 & .050 & .019 & -.010 \\ -.002 & .224 & -.264 & .034 & .354 & -.023 \end{bmatrix}. \qquad (3.3.22)$$

This estimate is clearly distinct from the corresponding part of (3.2.22), although the two estimates do not differ dramatically.

If the two presample values are used in estimating the process means and moment matrices we get

$$\widehat{A}_{YW} = \begin{bmatrix} -.319 & .147 & .959 & -.160 & .115 & .932 \\ .044 & -.152 & .286 & .050 & .020 & -.012 \\ -.002 & .225 & -.264 & .034 & .355 & -.022 \end{bmatrix} \qquad (3.3.23)$$

which is the Yule-Walker estimate. Although the sample size is moderate, there is a slight difference between the estimates in (3.3.22) and (3.3.23).

3.4 Maximum Likelihood Estimation

3.4.1 The Likelihood Function

Assuming that the distribution of the process is known, maximum likelihood (ML) estimation is an alternative to LS estimation. We will consider ML estimation under the assumption that the VAR(p) process y_t is Gaussian. More precisely,

3 Estimation of Vector Autoregressive Processes

$$\mathbf{u} = \text{vec}(U) = \begin{bmatrix} u_1 \\ \vdots \\ u_T \end{bmatrix} \sim \mathcal{N}(0, I_T \otimes \Sigma_u). \tag{3.4.1}$$

In other words, the probability density of $\mathbf{u}$ is

$$f_{\mathbf{u}}(\mathbf{u}) = \frac{1}{(2\pi)^{KT/2}} |I_T \otimes \Sigma_u|^{-1/2} \exp\left[-\frac{1}{2}\mathbf{u}'(I_T \otimes \Sigma_u^{-1})\mathbf{u}\right]. \tag{3.4.2}$$

Moreover,

$$\mathbf{u} = \begin{bmatrix} I_K & 0 & \cdots & 0 & \cdots & \cdots & 0 \\ -A_1 & I_K & & 0 & \cdots & \cdots & 0 \\ \vdots & \vdots & \ddots & \vdots & & & \vdots \\ -A_p & -A_{p-1} & \cdots & I_K & & & 0 \\ 0 & -A_p & & & \ddots & & \vdots \\ \vdots & & \ddots & & & \ddots & \vdots \\ 0 & 0 & \cdots & -A_p & \cdots & \cdots & I_K \end{bmatrix} (\mathbf{y} - \boldsymbol{\mu}^*)$$

$$+ \begin{bmatrix} -A_1 & -A_2 & \cdots & -A_p \\ -A_2 & -A_3 & \cdots & 0 \\ \vdots & & & \vdots \\ -A_p & 0 & \cdots & 0 \\ \vdots & & & \vdots \\ 0 & 0 & \cdots & 0 \end{bmatrix} (Y_0 - \boldsymbol{\mu}), \tag{3.4.3}$$

where $\mathbf{y} := \text{vec}(Y)$ and $\boldsymbol{\mu}^* := (\mu', \ldots, \mu')'$ are $(TK \times 1)$ vectors and $Y_0 := (y_0', \ldots, y_{-p+1}')'$ and $\boldsymbol{\mu} := (\mu', \ldots, \mu')'$ are $(Kp \times 1)$. Consequently, $\partial \mathbf{u}/\partial \mathbf{y}'$ is a lower triangular matrix with unit diagonal which has unit determinant. Hence, using that $\mathbf{u} = \mathbf{y} - \boldsymbol{\mu}^* - (X' \otimes I_K)\boldsymbol{\alpha}$,

$$\begin{aligned} f_{\mathbf{y}}(\mathbf{y}) &= \left|\frac{\partial \mathbf{u}}{\partial \mathbf{y}'}\right| f_{\mathbf{u}}(\mathbf{u}) \\ &= \frac{1}{(2\pi)^{KT/2}} |I_T \otimes \Sigma_u|^{-1/2} \\ &\quad \times \exp\left[-\frac{1}{2}(\mathbf{y} - \boldsymbol{\mu}^* - (X' \otimes I_K)\boldsymbol{\alpha})'(I_T \otimes \Sigma_u^{-1}) \right. \\ &\quad \left. \times (\mathbf{y} - \boldsymbol{\mu}^* - (X' \otimes I_K)\boldsymbol{\alpha})\right], \end{aligned} \tag{3.4.4}$$

where X and $\boldsymbol{\alpha}$ are as defined in (3.3.2). For simplicity, the initial values Y_0 are assumed to be given fixed numbers. Hence, we get a log-likelihood function

$$\ln l(\boldsymbol{\mu}, \boldsymbol{\alpha}, \Sigma_u)$$

$$= -\frac{KT}{2}\ln 2\pi - \frac{T}{2}\ln|\Sigma_u|$$
$$-\frac{1}{2}[\mathbf{y} - \boldsymbol{\mu}^* - (X' \otimes I_K)\boldsymbol{\alpha}]'(I_T \otimes \Sigma_u^{-1})[\mathbf{y} - \boldsymbol{\mu}^* - (X' \otimes I_K)\boldsymbol{\alpha}]$$

$$= -\frac{KT}{2}\ln 2\pi - \frac{T}{2}\ln|\Sigma_u| - \frac{1}{2}\sum_{t=1}^{T}\left[(y_t - \mu) - \sum_{i=1}^{p}A_i(y_{t-i} - \mu)\right]'$$
$$\times \Sigma_u^{-1}\left[(y_t - \mu) - \sum_{i=1}^{p}A_i(y_{t-i} - \mu)\right]$$

$$= -\frac{KT}{2}\ln 2\pi - \frac{T}{2}\ln|\Sigma_u|$$
$$-\frac{1}{2}\sum_{t}\left(y_t - \sum_{i}A_i y_{t-i}\right)'\Sigma_u^{-1}\left(y_t - \sum_{i}A_i y_{t-i}\right)$$
$$+ \mu'\left(I_K - \sum_{i}A_i\right)'\Sigma_u^{-1}\sum_{t}\left(y_t - \sum_{i}A_i y_{t-i}\right)$$
$$-\frac{T}{2}\mu'\left(I_K - \sum_{i}A_i\right)'\Sigma_u^{-1}\left(I_K - \sum_{i}A_i\right)\mu$$

$$= -\frac{KT}{2}\ln 2\pi - \frac{T}{2}\ln|\Sigma_u| - \frac{1}{2}\text{tr}[(Y^0 - AX)'\Sigma_u^{-1}(Y^0 - AX)], \quad (3.4.5)$$

where $Y^0 := (y_1 - \mu, \ldots, y_T - \mu)$ and $A := (A_1, \ldots, A_p)$ are as defined in (3.3.2). These different expressions of the log-likelihood function will be useful in the following.

3.4.2 The ML Estimators

In order to determine the ML estimators of $\mu, \boldsymbol{\alpha}$, and Σ_u, the system of first order partial derivatives is needed:

$$\frac{\partial \ln l}{\partial \mu} = \left(I_K - \sum_{i}A_i\right)'\Sigma_u^{-1}\sum_{t}\left(y_t - \sum_{i}A_i y_{t-i}\right)$$
$$-T\left(I_K - \sum_{i}A_i\right)'\Sigma_u^{-1}\left(I_K - \sum_{i}A_i\right)\mu$$
$$= [I_K - A(\mathbf{j} \otimes I_K)]'\Sigma_u^{-1}\left[\sum_{t}(y_t - \mu - AY_{t-1}^0)\right], \quad (3.4.6)$$

where Y_t^0 is as defined in (3.3.2) and $\mathbf{j} := (1, \ldots, 1)'$ is a $(p \times 1)$ vector of ones,

$$\frac{\partial \ln l}{\partial \boldsymbol{\alpha}} = (X \otimes I_K)(I_T \otimes \Sigma_u^{-1})[\mathbf{y} - \boldsymbol{\mu}^* - (X' \otimes I_K)\boldsymbol{\alpha}]$$
$$= (X \otimes \Sigma_u^{-1})(\mathbf{y} - \boldsymbol{\mu}^*) - (XX' \otimes \Sigma_u^{-1})\boldsymbol{\alpha}, \qquad (3.4.7)$$

$$\frac{\partial \ln l}{\partial \Sigma_u} = -\frac{T}{2}\Sigma_u^{-1} + \frac{1}{2}\Sigma_u^{-1}(Y^0 - AX)(Y^0 - AX)'\Sigma_u^{-1}. \qquad (3.4.8)$$

Equating to zero gives the system of normal equations which can be solved for the estimators:

$$\widetilde{\boldsymbol{\mu}} = \frac{1}{T}\left(I_K - \sum_i \widetilde{A}_i\right)^{-1} \sum_t \left(y_t - \sum_i \widetilde{A}_i y_{t-i}\right), \qquad (3.4.9)$$

$$\widetilde{\boldsymbol{\alpha}} = ((\widetilde{X}\widetilde{X}')^{-1}\widetilde{X} \otimes I_K)(\mathbf{y} - \widetilde{\boldsymbol{\mu}}^*), \qquad (3.4.10)$$

$$\widetilde{\Sigma}_u = \frac{1}{T}(\widetilde{Y}^0 - \widetilde{A}\widetilde{X})(\widetilde{Y}^0 - \widetilde{A}\widetilde{X})', \qquad (3.4.11)$$

where $\widetilde{X}$ and $\widetilde{Y}^0$ are obtained from X and Y^0, respectively, by replacing μ with $\widetilde{\mu}$.

3.4.3 Properties of the ML Estimators

Comparing these results with the LS estimators obtained in Section 3.3, it turns out that the ML estimators of μ and $\boldsymbol{\alpha}$ are identical to the LS estimators. Thus, $\widetilde{\mu}$ and $\widetilde{\boldsymbol{\alpha}}$ are consistent estimators if y_t is a stationary, stable Gaussian VAR(p) process and $\sqrt{T}(\widetilde{\mu} - \mu)$ and $\sqrt{T}(\widetilde{\boldsymbol{\alpha}} - \boldsymbol{\alpha})$ are asymptotically normally distributed. This result also follows from a more general maximum likelihood theory (see Appendix C.6). In fact, that theory implies that the covariance matrix of the asymptotic distribution of the ML estimators is the limit of T times the inverse information matrix. The information matrix is

$$\mathcal{I}(\boldsymbol{\delta}) = -E\left[\frac{\partial^2 \ln l}{\partial \boldsymbol{\delta} \partial \boldsymbol{\delta}'}\right] \qquad (3.4.12)$$

where $\boldsymbol{\delta}' := (\mu', \boldsymbol{\alpha}', \sigma')$ with $\sigma := \text{vech}(\Sigma_u)$. Note that vech is a column stacking operator that stacks only the elements on and below the main diagonal of Σ_u. It is related to the vec operator by the ($\frac{1}{2}K(K+1) \times K^2$) elimination matrix $\mathbf{L}_K$, that is, $\text{vech}(\Sigma_u) = \mathbf{L}_K \text{vec}(\Sigma_u)$ or, defining $\boldsymbol{\omega} := \text{vec}(\Sigma_u)$, $\sigma = \mathbf{L}_K \boldsymbol{\omega}$ (see Appendix A.12). For instance, for $K = 3$,

$$\boldsymbol{\omega} = \text{vec}(\Sigma_u) = \text{vec}\begin{bmatrix} \sigma_{11} & \sigma_{12} & \sigma_{13} \\ \sigma_{12} & \sigma_{22} & \sigma_{23} \\ \sigma_{13} & \sigma_{23} & \sigma_{33} \end{bmatrix}$$
$$= (\sigma_{11}, \sigma_{12}, \sigma_{13}, \sigma_{12}, \sigma_{22}, \sigma_{23}, \sigma_{13}, \sigma_{23}, \sigma_{33})'$$

and

$$\boldsymbol{\sigma} = \text{vech}(\Sigma_u) = \mathbf{L}_3\, \boldsymbol{\omega} = \begin{bmatrix} \sigma_{11} \\ \sigma_{12} \\ \sigma_{13} \\ \sigma_{22} \\ \sigma_{23} \\ \sigma_{33} \end{bmatrix}. \tag{3.4.13}$$

Note that in $\boldsymbol{\delta}$ we collect only the potentially different elements of Σ_u.

The asymptotic covariance matrix of the ML estimator $\widetilde{\boldsymbol{\delta}}$ is known to be

$$\lim_{T\to\infty} [\mathcal{I}(\boldsymbol{\delta})/T]^{-1}. \tag{3.4.14}$$

In order to determine this matrix, we need the second order partial derivatives of the log-likelihood. From (3.4.6) to (3.4.8) we get

$$\frac{\partial^2 \ln l}{\partial \mu \, \partial \mu'} = -T \left(I_K - \sum_i A_i\right)' \Sigma_u^{-1} \left(I_K - \sum_i A_i\right), \tag{3.4.15}$$

$$\frac{\partial^2 \ln l}{\partial \alpha \, \partial \alpha'} = -(XX' \otimes \Sigma_u^{-1}), \tag{3.4.16}$$

$$\frac{\partial^2 \ln l}{\partial \boldsymbol{\omega} \, \partial \boldsymbol{\omega}'} = \frac{T}{2}(\Sigma_u^{-1} \otimes \Sigma_u^{-1}) - \frac{1}{2}(\Sigma_u^{-1} \otimes \Sigma_u^{-1} UU' \Sigma_u^{-1})$$
$$- \frac{1}{2}(\Sigma_u^{-1} UU' \Sigma_u^{-1} \otimes \Sigma_u^{-1}), \tag{3.4.17}$$

where $\boldsymbol{\omega} = \text{vec}(\Sigma_u)$ (see Problem 3.3),

$$\frac{\partial^2 \ln l}{\partial \mu \, \partial \alpha'} = -[I_K - (\mathbf{j}' \otimes I_K) A'] \Sigma_u^{-1} \sum_t Y_{t-1}^{0\prime} \otimes I_K$$
$$- \left(\sum_t u_t' \Sigma_u^{-1} \otimes I_K\right) (I_K \otimes \mathbf{j}' \otimes I_K) \frac{\partial \text{vec}(A')}{\partial \alpha'} \tag{3.4.18}$$

(see Problem 3.4),

$$\frac{\partial^2 \ln l}{\partial \boldsymbol{\omega} \, \partial \mu'} = \frac{1}{2}(\Sigma_u^{-1} \otimes \Sigma_u^{-1}) \left[(I_K \otimes U) \frac{\partial \text{vec}(U')}{\partial \mu'} + (U \otimes I_K) \frac{\partial \text{vec}(U)}{\partial \mu'}\right] \tag{3.4.19}$$

(see Problem 3.5), and

$$\frac{\partial^2 \ln l}{\partial \boldsymbol{\omega} \, \partial \alpha'} = -\frac{1}{2}(\Sigma_u^{-1} \otimes \Sigma_u^{-1}) \left[(I_K \otimes UX') \frac{\partial \text{vec}(A')}{\partial \alpha'} + (UX' \otimes I_K)\right] \tag{3.4.20}$$

(see Problem 3.6).

It is obvious from (3.4.18) that

$$\lim \, T^{-1} E \left(\frac{\partial^2 \ln l}{\partial \mu \, \partial \alpha'} \right) = 0 \qquad (3.4.21)$$

because $E(\sum_t Y_{t-1}^0 / T) \to 0$. Furthermore, from (3.4.19), it follows that

$$E \left(\frac{\partial^2 \ln l}{\partial \omega \, \partial \mu'} \right) = 0 \qquad (3.4.22)$$

because $E(U) = 0$ and $\partial \operatorname{vec}(U') / \partial \mu'$ is a matrix of constants. Moreover, from (3.4.20), we have

$$\lim \, T^{-1} E \left(\frac{\partial^2 \ln l}{\partial \omega \, \partial \alpha'} \right) = 0 \qquad (3.4.23)$$

because $E(UX'/T) \to 0$. Thus, $\lim \mathcal{I}(\boldsymbol{\delta})/T$ is block diagonal and we get the asymptotic distributions of $\boldsymbol{\mu}, \boldsymbol{\alpha}$, and $\boldsymbol{\sigma}$ as follows.

Multiplying minus the inverse of (3.4.15) by T gives the asymptotic covariance matrix of the ML estimator for the mean vector $\boldsymbol{\mu}$, that is,

$$\sqrt{T}(\widetilde{\boldsymbol{\mu}} - \boldsymbol{\mu}) \overset{d}{\to} \mathcal{N} \left(0, \left(I_K - \sum_{i=1}^p A_i \right)^{-1} \Sigma_u \left(I_K - \sum_{i=1}^p A_i' \right)^{-1} \right). \qquad (3.4.24)$$

Hence, $\widetilde{\boldsymbol{\mu}}$ has the same asymptotic distribution as $\overline{y}$ (see Proposition 3.3). In other words, the two estimators for μ are asymptotically equivalent and, under the present conditions, this fact implies that $\overline{y}$ is asymptotically efficient because the ML estimator is asymptotically efficient. The asymptotic equivalence of $\widetilde{\boldsymbol{\mu}}$ and $\overline{y}$ can also be seen from (3.4.9) (see the argument prior to Proposition 3.3 and Problem 3.7).

Taking the limit of T^{-1} times the expectation of minus (3.4.16) gives $\Gamma_Y(0) \otimes \Sigma_u^{-1}$. Note that $E(XX'/T)$ is not strictly equal to $\Gamma_Y(0)$ because we have assumed fixed initial values $y_{-p+1}, \ldots, y_0$. However, asymptotically, as T goes to infinity, the impact of the initial values vanishes. Thus, we get

$$\sqrt{T}(\widetilde{\boldsymbol{\alpha}} - \boldsymbol{\alpha}) \overset{d}{\to} \mathcal{N}(0, \Gamma_Y(0)^{-1} \otimes \Sigma_u). \qquad (3.4.25)$$

Of course, this result also follows from the equivalence of the ML and LS estimators.

Noting that $E(UU') = T\Sigma_u$, it follows from (3.4.17) that

$$E \left(\frac{\partial^2 \ln l}{\partial \omega \, \partial \omega'} \right) = -\frac{T}{2} (\Sigma_u^{-1} \otimes \Sigma_u^{-1}). \qquad (3.4.26)$$

Denoting by $\mathbf{D}_K$ the $(K^2 \times \frac{1}{2} K(K+1))$ duplication matrix (see Appendix A.12) so that $\boldsymbol{\omega} = \mathbf{D}_K \boldsymbol{\sigma}$, we get

$$\frac{\partial^2 \ln l}{\partial \boldsymbol{\sigma} \, \partial \boldsymbol{\sigma}'} = \frac{\partial \boldsymbol{\omega}'}{\partial \boldsymbol{\sigma}} \frac{\partial^2 \ln l}{\partial \boldsymbol{\omega} \, \partial \boldsymbol{\omega}'} \frac{\partial \boldsymbol{\omega}}{\partial \boldsymbol{\sigma}'} = \mathbf{D}'_K \frac{\partial^2 \ln l}{\partial \boldsymbol{\omega} \, \partial \boldsymbol{\omega}'} \mathbf{D}_K$$

and, hence,

$$\sqrt{T}(\tilde{\boldsymbol{\sigma}} - \boldsymbol{\sigma}) \xrightarrow{d} \mathcal{N}(0, \Sigma_{\tilde{\boldsymbol{\sigma}}}) \qquad (3.4.27)$$

with

$$\begin{aligned}
\Sigma_{\tilde{\boldsymbol{\sigma}}} &= -TE\left(\frac{\partial^2 \ln l}{\partial \boldsymbol{\sigma} \, \partial \boldsymbol{\sigma}'}\right)^{-1} = 2\left[\mathbf{D}'_K(\Sigma_u^{-1} \otimes \Sigma_u^{-1})\mathbf{D}_K\right]^{-1} \\
&= 2\mathbf{D}_K^+(\Sigma_u \otimes \Sigma_u)\mathbf{D}_K^{+\prime}, \qquad (3.4.28)
\end{aligned}$$

where $\mathbf{D}_K^+ = (\mathbf{D}'_K \mathbf{D}_K)^{-1} \mathbf{D}'_K$ is the Moore-Penrose generalized inverse of the duplication matrix $\mathbf{D}_K$ and Rule (17) from Appendix A.12 has been used. In summary, we get the following proposition.

Proposition 3.4 (*Asymptotic Properties of ML Estimators*)
Let y_t be a stationary, stable Gaussian VAR(p) process as in (3.3.1). Then the ML estimators $\tilde{\mu}, \tilde{\boldsymbol{\alpha}},$ and $\tilde{\boldsymbol{\sigma}} = \text{vech}(\tilde{\Sigma}_u)$ given in (3.4.9)–(3.4.11) are consistent and

$$\sqrt{T}\begin{bmatrix} \tilde{\mu} - \mu \\ \tilde{\boldsymbol{\alpha}} - \boldsymbol{\alpha} \\ \tilde{\boldsymbol{\sigma}} - \boldsymbol{\sigma} \end{bmatrix} \xrightarrow{d} \mathcal{N}\left(0, \begin{bmatrix} \Sigma_{\tilde{\mu}} & 0 & 0 \\ 0 & \Sigma_{\tilde{\boldsymbol{\alpha}}} & 0 \\ 0 & 0 & \Sigma_{\tilde{\boldsymbol{\sigma}}} \end{bmatrix}\right), \qquad (3.4.29)$$

so that $\tilde{\mu}$ is asymptotically independent of $\tilde{\boldsymbol{\alpha}}$ and $\tilde{\Sigma}_u$ and $\tilde{\boldsymbol{\alpha}}$ is asymptotically independent of $\tilde{\mu}$ and $\tilde{\Sigma}_u$. The covariance matrices are

$$\Sigma_{\tilde{\mu}} = \left(I_K - \sum_i A_i\right)^{-1} \Sigma_u \left(I_K - \sum_i A'_i\right)^{-1},$$

$$\Sigma_{\tilde{\boldsymbol{\alpha}}} = \Gamma_Y(0)^{-1} \otimes \Sigma_u,$$

$$\Sigma_{\tilde{\boldsymbol{\sigma}}} = 2\mathbf{D}_K^+(\Sigma_u \otimes \Sigma_u)\mathbf{D}_K^{+\prime}.$$

They may be estimated consistently by replacing the unknown quantities by their ML estimators and estimating $\Gamma_Y(0)$ by $\tilde{X}\tilde{X}'/T$. ∎

In this section, we have chosen to consider the mean-adjusted form of a VAR(p) process. Of course, it is possible to perform a similar derivation for the standard form given in (3.1.1). In that case the ML estimators of ν and $\boldsymbol{\alpha}$ are not asymptotically independent though. Their joint asymptotic distribution is identical to that of $\widehat{\boldsymbol{\beta}}$ given in Proposition 3.1. From Proposition 3.2 we know that the asymptotic distribution of $\tilde{\boldsymbol{\sigma}}$ remains unaltered. In the next section, we will investigate the consequences of forecasting with estimated rather than known processes.

94   3 Estimation of Vector Autoregressive Processes

3.5 Forecasting with Estimated Processes

3.5.1 General Assumptions and Results

In Chapter 2, Section 2.2, we have seen that the optimal h-step forecast of the process (3.1.1) is

$$y_t(h) = \nu + A_1 y_t(h-1) + \cdots + A_p y_t(h-p), \tag{3.5.1}$$

where $y_t(j) = y_{t+j}$ for $j \leq 0$. If the true coefficients $B = (\nu, A_1, \ldots, A_p)$ are replaced by estimators $\widehat{B} = (\widehat{\nu}, \widehat{A}_1, \ldots, \widehat{A}_p)$, we get a forecast

$$\widehat{y}_t(h) = \widehat{\nu} + \widehat{A}_1 \widehat{y}_t(h-1) + \cdots + \widehat{A}_p \widehat{y}_t(h-p), \tag{3.5.2}$$

where $\widehat{y}_t(j) = y_{t+j}$ for $j \leq 0$. Thus, the forecast error is

$$\begin{aligned} y_{t+h} - \widehat{y}_t(h) &= [y_{t+h} - y_t(h)] + [y_t(h) - \widehat{y}_t(h)] \\ &= \sum_{i=0}^{h-1} \Phi_i u_{t+h-i} + [y_t(h) - \widehat{y}_t(h)], \end{aligned} \tag{3.5.3}$$

where the Φ_i are the coefficient matrices of the canonical MA representation of y_t (see (2.2.9)). Under quite general conditions for the process y_t, the forecast errors can be shown to have zero mean, $E[y_{t+h} - \widehat{y}_t(h)] = 0$, so that the forecasts are unbiased even if the coefficients are estimated. Because we do not need this result in the following, we refer to Dufour (1985) for the details and a proof. All the u_s in the first term on the right-hand side of the last equality sign in (3.5.3) are attached to periods $s > t$, whereas all the y_s in the second term correspond to periods $s \leq t$, if estimation is done with observations from periods up to time t only. Therefore, the two terms are uncorrelated. Hence, the MSE matrix of the forecast $\widehat{y}_t(h)$ is of the form

$$\begin{aligned} \Sigma_{\widehat{y}}(h) &:= \text{MSE}[\widehat{y}_t(h)] = E\{[y_{t+h} - \widehat{y}_t(h)][y_{t+h} - \widehat{y}_t(h)]'\} \\ &= \Sigma_y(h) + \text{MSE}[y_t(h) - \widehat{y}_t(h)], \end{aligned} \tag{3.5.4}$$

where

$$\Sigma_y(h) = \sum_{i=0}^{h-1} \Phi_i \Sigma_u \Phi_i'$$

(see (2.2.11)). In order to evaluate the last term in (3.5.4), the distribution of the estimator $\widehat{B}$ is needed. Because we have not been able to derive the small sample distributions of the estimators considered in the previous sections but we have derived the asymptotic distributions instead, we cannot hope for more than an asymptotic approximation to the MSE of $y_t(h) - \widehat{y}_t(h)$. Such an approximation will be derived in the following.

There are two alternative assumptions that can be made in order to facilitate the derivation of the desired result:

(1) Only data up to the forecast origin are used for estimation.
(2) Estimation is done using a realization (time series) of a process that is independent of the process used for prediction and has the same stochastic structure (for instance, it is Gaussian and has the same first and second moments as the process used for prediction).

The first assumption is the more realistic one from a practical point of view because estimation and forecasting are usually based on the same data set. In that case, because the sample size is assumed to go to infinity in deriving asymptotic results, either the forecast origin has to go to infinity too or it has to be assumed that more and more data at the beginning of the sample become available. Because the forecast uses only p vectors y_s prior to the forecast period, these variables will be asymptotically independent of the estimator $\widehat{B}$ (they are asymptotically negligible in comparison with all the other observations going into the estimate). Thus, asymptotically the first assumption implies the same results as the second one. In the following, for simplicity, the second assumption will therefore be used. Furthermore, it will be assumed that for $\beta = \text{vec}(B)$ and $\widehat{\beta} = \text{vec}(\widehat{B})$ we have

$$\sqrt{T}(\widehat{\beta} - \beta) \xrightarrow{d} \mathcal{N}(0, \Sigma_{\widehat{\beta}}). \tag{3.5.5}$$

Samaranayake & Hasza (1988) and Basu & Sen Roy (1986) give a formal proof of the result that the MSE approximation obtained in the following remains valid under assumption (1) above.

With the foregoing assumptions it follows that, conditional on a particular realization $Y_t = (y'_t, \ldots, y'_{t-p+1})'$ of the process used for prediction,

$$\sqrt{T}\left[\widehat{y}_t(h) - y_t(h)|Y_t\right] \xrightarrow{d} \mathcal{N}\left(0, \frac{\partial y_t(h)}{\partial \beta'} \Sigma_{\widehat{\beta}} \frac{\partial y_t(h)'}{\partial \beta}\right) \tag{3.5.6}$$

because $y_t(h)$ is a differentiable function of β (see Appendix C, Proposition C.15(3)). Here T is the sample size (time series length) used for estimation. This result suggests the approximation of $\text{MSE}[\widehat{y}_t(h) - y_t(h)]$ by $\Omega(h)/T$, where

$$\Omega(h) = E\left[\frac{\partial y_t(h)}{\partial \beta'} \Sigma_{\widehat{\beta}} \frac{\partial y_t(h)'}{\partial \beta}\right]. \tag{3.5.7}$$

In fact, for a Gaussian process y_t,

$$\sqrt{T}\left[\widehat{y}_t(h) - y_t(h)\right] \xrightarrow{d} \mathcal{N}(0, \Omega(h)). \tag{3.5.8}$$

Hence, we get an approximation

$$\Sigma_{\widehat{y}}(h) = \Sigma_y(h) + \frac{1}{T}\Omega(h) \tag{3.5.9}$$

for the MSE matrix of $\widehat{y}_t(h)$.

From (3.5.7) it is obvious that $\Omega(h)$ and, thus, the approximate MSE $\Sigma_{\widehat{y}}(h)$ can be reduced by using an estimator that is asymptotically more efficient than $\widehat{\boldsymbol{\beta}}$, if such an estimator exists. In other words, efficient estimation is of importance in order to reduce the forecast uncertainty.

3.5.2 The Approximate MSE Matrix

To derive an explicit expression for $\Omega(h)$, the derivatives $\partial y_t(h)/\partial \boldsymbol{\beta}'$ are needed. They can be obtained easily by noting that

$$y_t(h) = J_1 \mathbf{B}^h Z_t, \tag{3.5.10}$$

where $Z_t := (1, y_t', \ldots, y_{t-p+1}')'$,

$$\mathbf{B} := \begin{bmatrix} 1 & 0 & 0 & \cdots & 0 & 0 \\ \nu & A_1 & A_2 & \cdots & A_{p-1} & A_p \\ 0 & I_K & 0 & \cdots & 0 & 0 \\ 0 & 0 & I_K & & 0 & 0 \\ \vdots & \vdots & & \ddots & & \vdots \\ 0 & 0 & 0 & \cdots & I_K & 0 \end{bmatrix}_{[(Kp+1)\times(Kp+1)]} = \begin{bmatrix} 1 & 0 & \cdots & 0 \\ & B & & \\ 0 & I_{K(p-1)} & & 0 \end{bmatrix}$$

and

$$J_1 := [\underbrace{0}_{(K\times 1)} : I_K : \underbrace{0 : \cdots : 0}_{(K\times K(p-1))}] \quad [K \times (Kp+1)].$$

The relation (3.5.10) follows by induction (see Problem 3.8). Using (3.5.10), we get

$$\begin{aligned}
\frac{\partial y_t(h)}{\partial \boldsymbol{\beta}'} &= \frac{\partial \operatorname{vec}(J_1 \mathbf{B}^h Z_t)}{\partial \boldsymbol{\beta}'} = (Z_t' \otimes J_1) \frac{\partial \operatorname{vec}(\mathbf{B}^h)}{\partial \boldsymbol{\beta}'} \\
&= (Z_t' \otimes J_1) \left[\sum_{i=0}^{h-1} (\mathbf{B}')^{h-1-i} \otimes \mathbf{B}^i \right] \frac{\partial \operatorname{vec}(\mathbf{B})}{\partial \boldsymbol{\beta}'} \\
&\qquad \text{(Appendix A.13, Rule (8))} \\
&= (Z_t' \otimes J_1) \left[\sum_{i=0}^{h-1} (\mathbf{B}')^{h-1-i} \otimes \mathbf{B}^i \right] (I_{Kp+1} \otimes J_1') \\
&\qquad \text{(see the definition of } \mathbf{B}\text{)} \\
&= \sum_{i=0}^{h-1} Z_t'(\mathbf{B}')^{h-1-i} \otimes J_1 \mathbf{B}^i J_1' \\
&= \sum_{i=0}^{h-1} Z_t'(\mathbf{B}')^{h-1-i} \otimes \Phi_i, \tag{3.5.11}
\end{aligned}$$

where $\Phi_i = J_1 \mathbf{B}^i J_1'$ follows as in (2.1.17). Using the LS estimator $\hat{\beta}$ with asymptotic covariance matrix $\Sigma_{\hat{\beta}} = \Gamma^{-1} \otimes \Sigma_u$ (see Proposition 3.1), the matrix $\Omega(h)$ is seen to be

$$\begin{aligned}
\Omega(h) &= E\left[\frac{\partial y_t(h)}{\partial \beta'}(\Gamma^{-1} \otimes \Sigma_u)\frac{\partial y_t(h)'}{\partial \beta}\right] \\
&= \sum_{i=0}^{h-1}\sum_{j=0}^{h-1} E(Z_t'(\mathbf{B}')^{h-1-i}\Gamma^{-1}\mathbf{B}^{h-1-j}Z_t) \otimes \Phi_i \Sigma_u \Phi_j' \\
&= \sum_i \sum_j E[\text{tr}(Z_t'(\mathbf{B}')^{h-1-i}\Gamma^{-1}\mathbf{B}^{h-1-j}Z_t)]\Phi_i \Sigma_u \Phi_j' \\
&= \sum_i \sum_j \text{tr}[(\mathbf{B}')^{h-1-i}\Gamma^{-1}\mathbf{B}^{h-1-j}E(Z_t Z_t')]\Phi_i \Sigma_u \Phi_j' \\
&= \sum_{i=0}^{h-1}\sum_{j=0}^{h-1} \text{tr}[(\mathbf{B}')^{h-1-i}\Gamma^{-1}\mathbf{B}^{h-1-j}\Gamma]\Phi_i \Sigma_u \Phi_j', \quad (3.5.12)
\end{aligned}$$

provided y_t is stable so that

$$\Gamma := \text{plim}(ZZ'/T) = E(Z_t Z_t').$$

Here $Z := (Z_0, \ldots, Z_{T-1})$ is the $((Kp+1) \times T)$ matrix defined in (3.2.1). For example, for $h = 1$,

$$\Omega(1) = (Kp+1)\Sigma_u.$$

Hence, the approximation

$$\Sigma_{\hat{y}}(1) = \Sigma_u + \frac{Kp+1}{T}\Sigma_u = \frac{T + Kp + 1}{T}\Sigma_u \quad (3.5.13)$$

of the MSE matrix of the 1-step forecast with estimated coefficients is obtained. This expression shows that the contribution of the estimation variability to the forecast MSE matrix $\Sigma_{\hat{y}}(1)$ depends on the dimension K of the process, the VAR order p, and the sample size T used for estimation. It can be quite substantial if the sample size is small or moderate. For instance, considering a three-dimensional process of order 8 which is estimated from 15 years of quarterly data (i.e., $T = 52$ plus 8 presample values needed for LS estimation), the 1-step forecast MSE matrix Σ_u for known processes is inflated by a factor $(T + Kp + 1)/T = 1.48$. Of course, this approximation is derived from asymptotic theory so that its small sample validity is not guaranteed. We will take a closer look at this problem shortly. Obviously, the inflation factor $(T + Kp + 1)/T \to 1$ for $T \to \infty$. Thus the MSE contribution due to sampling variability vanishes if the sample size gets large. This result is a consequence of estimating the VAR coefficients consistently. An expression for $\Omega(h)$ can also be derived on the basis of the mean-adjusted form of the VAR process (see Problem 3.9).

In practice, for $h > 1$, it will not be possible to evaluate $\Omega(h)$ without knowing the AR coefficients summarized in the matrix B. A consistent estimator $\widehat{\Omega}(h)$ may be obtained by replacing all unknown parameters by their LS estimators, that is, $\mathbf{B}$ is replaced by $\widehat{\mathbf{B}}$ which is obtained by using $\widehat{B}$ for B, Σ_u is replaced by $\widehat{\Sigma}_u$, Φ_i is estimated by $\widehat{\Phi}_i = J_1 \widehat{\mathbf{B}}^i J_1'$, and Γ is estimated by $\widehat{\Gamma} = ZZ'/T$. The resulting estimator of $\Sigma_{\widehat{y}}(h)$ will be denoted by $\widehat{\Sigma}_{\widehat{y}}(h)$ in the following.

The foregoing discussion is of importance in setting up interval forecasts. Assuming that y_t is Gaussian, an approximate $(1 - \alpha)100\%$ interval forecast, h periods ahead, for the k-th component $y_{k,t}$ of y_t is

$$\widehat{y}_{k,t}(h) \pm z_{(\alpha/2)} \widehat{\widehat{\sigma}}_k(h) \qquad (3.5.14)$$

or

$$\left[\widehat{y}_{k,t}(h) - z_{(\alpha/2)} \widehat{\widehat{\sigma}}_k(h), \; \widehat{y}_{k,t}(h) + z_{(\alpha/2)} \widehat{\widehat{\sigma}}_k(h) \right], \qquad (3.5.15)$$

where $z_{(\alpha)}$ is the upper $\alpha 100$-th percentile of the standard normal distribution and $\widehat{\widehat{\sigma}}_k(h)$ is the square root of the k-th diagonal element of $\widehat{\Sigma}_{\widehat{y}}(h)$. Using Bonferroni's inequality, approximate joint confidence regions for a set of forecasts can be obtained just as described in Section 2.2.3 of Chapter 2.

3.5.3 An Example

To illustrate the previous results, we consider again the investment/income/consumption example of Section 3.2.3. Using the VAR(2) model with the coefficient estimates given in (3.2.22) and

$$y_{T-1} = y_{72} = \begin{bmatrix} .02551 \\ .02434 \\ .01319 \end{bmatrix} \quad \text{and} \quad y_T = y_{73} = \begin{bmatrix} .03637 \\ .00517 \\ .00599 \end{bmatrix}$$

results in forecasts

$$\widehat{y}_T(1) = \widehat{\nu} + \widehat{A}_1 y_T + \widehat{A}_2 y_{T-1} = \begin{bmatrix} -.011 \\ .020 \\ .022 \end{bmatrix},$$

$$\widehat{y}_T(2) = \widehat{\nu} + \widehat{A}_1 \widehat{y}_T(1) + \widehat{A}_2 y_T = \begin{bmatrix} .011 \\ .020 \\ .015 \end{bmatrix},$$

$$(3.5.16)$$

and so on.

The estimated forecast MSE matrix for $h = 1$ is

$$\widehat{\Sigma}_{\widehat{y}}(1) = \frac{T + Kp + 1}{T} \widehat{\Sigma}_u = \frac{73 + 6 + 1}{73} \widehat{\Sigma}_u$$

$$= \begin{bmatrix} 23.34 & .785 & 1.351 \\ .785 & 1.505 & .674 \\ 1.351 & .674 & .978 \end{bmatrix} \times 10^{-4}, \qquad (3.5.17)$$

3.5 Forecasting with Estimated Processes   99

where $\widehat{\Sigma}_u$ from (3.2.23) has been used. We need $\widehat{\Phi}_1$ for evaluating

$$\widehat{\Sigma}_{\widehat{y}}(2) = \widehat{\Sigma}_y(2) + \frac{1}{T}\widehat{\Omega}(2),$$

where

$$\widehat{\Sigma}_y(2) = \widehat{\Sigma}_u + \widehat{\Phi}_1\widehat{\Sigma}_u\widehat{\Phi}'_1$$

and

$$\begin{aligned}\widehat{\Omega}(2) &= \sum_{i=0}^{1}\sum_{j=0}^{1}\text{tr}\left[(\widehat{\mathbf{B}}')^{1-i}(ZZ'/T)^{-1}\widehat{\mathbf{B}}^{1-j}(ZZ'/T)\right]\widehat{\Phi}_i\widehat{\Sigma}_u\widehat{\Phi}'_j \\ &= \text{tr}[\widehat{\mathbf{B}}'(ZZ')^{-1}\widehat{\mathbf{B}}ZZ']\widehat{\Sigma}_u + \text{tr}(\widehat{\mathbf{B}}')\widehat{\Sigma}_u\widehat{\Phi}'_1 \\ &\quad + \text{tr}(\widehat{\mathbf{B}})\widehat{\Phi}_1\widehat{\Sigma}_u + \text{tr}(I_{Kp+1})\widehat{\Phi}_1\widehat{\Sigma}_u\widehat{\Phi}'_1.\end{aligned}$$

From (2.1.22) we know that $\Phi_1 = A_1$. Hence, we use $\widehat{\Phi}_1 = \widehat{A}_1$ from (3.2.22). Thus, we get

$$\widehat{\Sigma}_y(2) = \begin{bmatrix} 23.67 & .547 & 1.226 \\ .547 & 1.488 & .554 \\ 1.226 & .554 & .952 \end{bmatrix} \times 10^{-4}$$

and

$$\widehat{\Omega}(2) = \begin{bmatrix} 10.59 & .238 & .538 \\ .238 & .675 & .233 \\ .538 & .233 & .422 \end{bmatrix} \times 10^{-3}.$$

Consequently,

$$\widehat{\Sigma}_{\widehat{y}}(2) = \begin{bmatrix} 25.12 & .580 & 1.300 \\ .580 & 1.581 & .586 \\ 1.300 & .586 & 1.009 \end{bmatrix} \times 10^{-4}. \tag{3.5.18}$$

Assuming that the data are generated by a Gaussian process, we get the following approximate 95% interval forecasts:

$$\begin{aligned}&\widehat{y}_{1,T}(1) \pm 1.96\widehat{\widehat{\sigma}}_1(1) \text{ or } -.011 \pm .095, \\ &\widehat{y}_{2,T}(1) \pm 1.96\widehat{\widehat{\sigma}}_2(1) \text{ or } .020 \pm .024, \\ &\widehat{y}_{3,T}(1) \pm 1.96\widehat{\widehat{\sigma}}_3(1) \text{ or } .022 \pm .019, \\ &\widehat{y}_{1,T}(2) \pm 1.96\widehat{\widehat{\sigma}}_1(2) \text{ or } .011 \pm .098, \\ &\widehat{y}_{2,T}(2) \pm 1.96\widehat{\widehat{\sigma}}_2(2) \text{ or } .020 \pm .025, \\ &\widehat{y}_{3,T}(2) \pm 1.96\widehat{\widehat{\sigma}}_3(2) \text{ or } .015 \pm .020.\end{aligned} \tag{3.5.19}$$

100   3 Estimation of Vector Autoregressive Processes

In Figure 3.3, some more forecasts of the three variables with two-standard error bounds to each side are depicted. The intervals indicated by the dashed bounds may be interpreted as approximate 95% forecast intervals for the individual forecasts. If the region enclosed by the dashed lines is viewed as a joint confidence region for all 4 forecasts, a lower bound for the (approximate) probability content is $(100-4\times5)\% = 80\%$. In the figure it can be seen that for investment and income the actually observed values for 1979 ($t = 77,\ldots,80$) are well inside the forecast regions, whereas two of the four consumption values are outside that region.

3.5.4 A Small Sample Investigation

It is not obvious that the MSE and interval forecast approximations derived in the foregoing are reasonable in small samples because the MSE modification has been based on asymptotic theory. To investigate the small sample behavior of the predictor with estimated coefficients, we have used again 1000 realizations of the bivariate VAR(2) process (3.2.25)/(3.2.26) of Section 3.2.4 and we have computed forecast intervals for the period following the last sample period. In Table 3.2, the proportions of actual values falling in these intervals are reported for sample sizes of $T = 30$ and 100.

Table 3.2. Accuracy of forecast intervals in small samples based on 1000 bivariate time series

MSE used in interval construction	% forecast interval	percent of actual values falling in the forecast interval			
		$T = 30$		$T = 100$	
		y_1	y_2	y_1	y_2
	90	86.5	85.7	89.7	89.4
$\Sigma_y(1)$	95	92.6	91.8	94.5	94.0
	99	98.1	98.0	99.0	98.5
	90	89.3	88.2	90.4	90.0
$\Sigma_{\hat{y}}(1)$	95	94.4	94.1	95.3	94.6
	99	99.0	98.4	99.3	98.8
	90	85.2	84.2	89.6	88.5
$\widehat{\Sigma}_y(1)$	95	90.5	90.4	94.7	93.9
	99	98.4	96.5	98.9	98.3
	90	88.1	86.9	90.3	89.1
$\widehat{\Sigma}_{\hat{y}}(1)$	95	93.4	92.7	95.2	94.0
	99	99.4	97.8	99.1	98.5

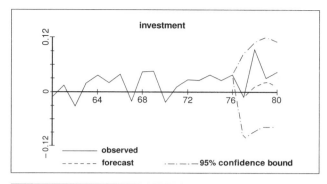
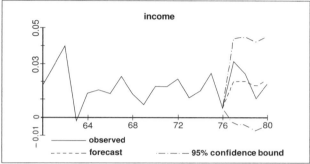
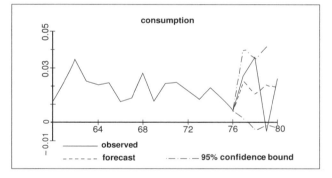

Fig. 3.3. Forecasts of the investment/income/consumption system.

Obviously, for $T = 30$, the theoretical and actual percentages are in best agreement if the approximate MSEs $\Sigma_{\widehat{y}}(h)$ are used in setting up the forecast intervals. On the other hand, only forecast intervals based on $\widehat{\Sigma}_y(h) = \sum_{i=0}^{h-1} \widehat{\Phi}_i \widehat{\Sigma}_u \widehat{\Phi}'_i$ and $\widehat{\Sigma}_{\widehat{y}}(h)$ are feasible in practice when the actual process coefficients are unknown and have to be estimated. Comparing only the results based on these two MSE matrices shows that it pays to use the asymptotic approximation $\widehat{\Sigma}_{\widehat{y}}(h)$.

In Table 3.2, we also give the corresponding results for $T = 100$. Because the estimation uncertainty decreases with increasing sample size, one would expect that now the theoretical and actual percentages are in good agreement for all MSEs. This is precisely what can be observed in the table. Nevertheless, even now the use of the MSE adjustment in $\widehat{\Sigma}_{\hat{y}}(1)$ gives slightly more accurate interval forecasts.

3.6 Testing for Causality

3.6.1 A Wald Test for Granger-Causality

In Chapter 2, Section 2.3.1, we have partitioned the VAR(p) process y_t in subprocesses z_t and x_t, that is, $y_t' = (z_t', x_t')$ and we have defined Granger-causality from x_t to z_t and vice versa. We have seen that this type of causality can be characterized by specific zero constraints on the VAR coefficients (see Corollary 2.2.1). Thus, in an estimated VAR(p) system, if we want to test for Granger-causality, we need to test zero constraints for the coefficients. Given the results of Sections 3.2, 3.3, and 3.4 it is straightforward to derive *asymptotic* tests of such constraints.

More generally we consider testing

$$H_0 : C\boldsymbol{\beta} = c \quad \text{against} \quad H_1 : C\boldsymbol{\beta} \neq c, \qquad (3.6.1)$$

where C is an $(N \times (K^2p + K))$ matrix of rank N and c is an $(N \times 1)$ vector. Assuming that

$$\sqrt{T}(\widehat{\boldsymbol{\beta}} - \boldsymbol{\beta}) \xrightarrow{d} \mathcal{N}(0, \Gamma^{-1} \otimes \Sigma_u) \qquad (3.6.2)$$

as in LS/ML estimation, we get

$$\sqrt{T}(C\widehat{\boldsymbol{\beta}} - C\boldsymbol{\beta}) \xrightarrow{d} \mathcal{N}\left[0, C(\Gamma^{-1} \otimes \Sigma_u)C'\right] \qquad (3.6.3)$$

(see Appendix C, Proposition C.15) and, hence,

$$T(C\widehat{\boldsymbol{\beta}} - c)' \left[C(\Gamma^{-1} \otimes \Sigma_u)C'\right]^{-1} (C\widehat{\boldsymbol{\beta}} - c) \xrightarrow{d} \chi^2(N). \qquad (3.6.4)$$

This statistic is the *Wald statistic* (see Appendix C.7).

Replacing Γ and Σ_u by their usual estimators $\widehat{\Gamma} = ZZ'/T$ and $\widehat{\Sigma}_u$ as given in (3.2.19), the resulting statistic

$$\lambda_W = (C\widehat{\boldsymbol{\beta}} - c)' \left[C((ZZ')^{-1} \otimes \widehat{\Sigma}_u)C'\right]^{-1} (C\widehat{\boldsymbol{\beta}} - c) \qquad (3.6.5)$$

still has an asymptotic χ^2-distribution with N degrees of freedom, provided y_t satisfies the conditions of Proposition 3.2, because under these conditions $[C((ZZ')^{-1} \otimes \widehat{\Sigma}_u)C']^{-1}/T$ is a consistent estimator of $[C(\Gamma^{-1} \otimes \Sigma_u)C']^{-1}$. Hence, we have the following result.

Proposition 3.5 (*Asymptotic Distribution of the Wald Statistic*)
Suppose (3.6.2) holds. Furthermore, $\operatorname{plim}(ZZ'/T) = \Gamma$, $\operatorname{plim} \widehat{\Sigma}_u = \Sigma_u$ are both nonsingular and $H_0 : C\beta = c$ is true, with C being an $(N \times (K^2p + K))$ matrix of rank N. Then

$$\lambda_W = (C\widehat{\beta} - c)'[C((ZZ')^{-1} \otimes \widehat{\Sigma}_u)C']^{-1}(C\widehat{\beta} - c) \xrightarrow{d} \chi^2(N).$$

■

In practice, it may be useful to make adjustments to the statistic or the critical values of the test to compensate for the fact that the matrix $\Gamma^{-1} \otimes \Sigma_u$ is unknown and has been replaced by an estimator. Working in that direction, we note that

$$NF(N,T) \xrightarrow[T \to \infty]{d} \chi^2(N), \tag{3.6.6}$$

where $F(N,T)$ denotes an F random variable with N and T degrees of freedom (d.f.) (Appendix C, Proposition C.3). Because an $F(N,T)$-distribution has a fatter tail than the $\chi^2(N)$-distribution divided by N, it seems reasonable to consider the test statistic

$$\lambda_F = \lambda_W / N \tag{3.6.7}$$

in conjunction with critical values from some F-distribution. The question is then what numbers of degrees of freedom should be used? From the foregoing discussion it is plausible to use N as the numerator degrees of freedom. On the other hand, any sequence that goes to infinity with the sample size qualifies as a candidate for the denominator d.f. The usual F-statistic for a regression model with nonstochastic regressors has denominator d.f. equal to the sample size minus the number of estimated parameters. Therefore we may use this number here too. Note that, in the model (3.2.3), we have a vector **y** with KT observations and β contains $K(Kp+1)$ parameters. Alternatively, we will argue shortly that $T - Kp - 1$ is also a reasonable number for the denominator d.f. Hence, we have the approximate distributions

$$\lambda_F \approx F(N, KT - K^2p - K) \approx F(N, T - Kp - 1). \tag{3.6.8}$$

3.6.2 An Example

To see how this result can be used in a test for Granger-causality, let us consider again our example system from Section 3.2.3. The null hypothesis of no Granger-causality from income/consumption (y_2, y_3) to investment (y_1) may be expressed in terms of the coefficients of the VAR(2) process as

$$H_0 : \alpha_{12,1} = \alpha_{13,1} = \alpha_{12,2} = \alpha_{13,2} = 0. \tag{3.6.9}$$

This null hypothesis may be written as in (3.6.1) by defining the (4×1) vector $c = 0$ and the (4×21) matrix

$$C = \begin{bmatrix} 0\,0\,0 & 0\,0\,0 & 1\,0\,0 & 0\,0\,0 & 0\,0\,0 & 0\,0\,0 & 0\,0\,0 \\ 0\,0\,0 & 0\,0\,0 & 0\,0\,0 & 1\,0\,0 & 0\,0\,0 & 0\,0\,0 & 0\,0\,0 \\ 0\,0\,0 & 0\,0\,0 & 0\,0\,0 & 0\,0\,0 & 0\,0\,0 & 1\,0\,0 & 0\,0\,0 \\ 0\,0\,0 & 0\,0\,0 & 0\,0\,0 & 0\,0\,0 & 0\,0\,0 & 0\,0\,0 & 1\,0\,0 \end{bmatrix}.$$

With this notation, using the estimation results from Section 3.2.3,

$$\lambda_F = \widehat{\boldsymbol{\beta}}'C'\left[C((ZZ')^{-1} \otimes \widehat{\Sigma}_u)C'\right]^{-1}C\widehat{\boldsymbol{\beta}}/4 = 1.59. \qquad (3.6.10)$$

In contrast, the 95th percentile of the $F(4, 3 \cdot 73 - 9 \cdot 2 - 3) = F(4, 198) \approx F(4, 73 - 3 \cdot 2 - 1) = F(4, 66)$-distribution is about 2.5. Thus, in a 5% level test, we cannot reject Granger-noncausality from income/consumption to investment.

In this example, the denominator d.f. are so large (namely 198 or 66) that we could just as well use λ_W in conjunction with a critical value from a $\chi^2(4)$-distribution. The 95th percentile of that distribution is 9.49 and, thus, it is about four times that of the F-test while $\lambda_W = 4\lambda_F$.

In an example of this type it is quite reasonable to use $T - Kp - 1$ denominator d.f. for the F-test because all the restrictions are imposed on coefficients from one equation. Therefore λ_F actually reduces to an F-statistic related to one equation with $Kp + 1$ parameters which are estimated from T observations. The use of $T - Kp - 1$ d.f. may also be justified by arguments that do not rely on the restrictions being imposed on the parameters of one equation only, namely by appealing to the similarity between the λ_F statistic and Hotelling's T^2 (e.g., Anderson (1984)).

Many other tests for Granger-causality have been proposed and investigated (see, e.g., Geweke, Meese & Dent (1983)). In the next chapter, we will return to the testing of hypotheses and then an alternative test will be considered.

3.6.3 Testing for Instantaneous Causality

Tests for instantaneous causality can be developed in the same way as tests for Granger-causality because instantaneous causality can be expressed in terms of zero restrictions for $\boldsymbol{\sigma} = \text{vech}(\Sigma_u)$ (see Proposition 2.3). If y_t is a stable Gaussian VAR(p) process and we wish to test

$$H_0 : C\boldsymbol{\sigma} = 0 \quad \text{against} \quad H_1 : C\boldsymbol{\sigma} \neq 0, \qquad (3.6.11)$$

we may use the asymptotic distribution of the ML estimator given in Proposition 3.4 to set up the Wald statistic

$$\lambda_W = T\widetilde{\boldsymbol{\sigma}}'C'[2C\mathbf{D}_K^+(\widetilde{\Sigma}_u \otimes \widetilde{\Sigma}_u)\mathbf{D}_K^{+\prime}C']^{-1}C\widetilde{\boldsymbol{\sigma}}, \qquad (3.6.12)$$

where $\mathbf{D}_K^+$ is the Moore-Penrose inverse of the duplication matrix $\mathbf{D}_K$ and C is an $(N \times K(K+1)/2)$ matrix of rank N. Under H_0, λ_W has an asymptotic χ^2-distribution with N degrees of freedom.

Alternatively, a Wald test of (3.6.11) could be based on the lower triangular matrix P which is obtained from a Choleski decomposition of Σ_u. Noting that instantaneous noncausality implies zero elements of Σ_u that correspond to zero elements of P, we can write H_0 from (3.6.11) equivalently as

$$H_0: C\text{vech}(P) = 0. \tag{3.6.13}$$

Because $\text{vech}(P)$ is a continuously differentiable function of $\boldsymbol{\sigma}$, the asymptotic distribution of the estimator P obtained from decomposing $\widetilde{\Sigma}_u$ follows from Proposition C.15(3) of Appendix C:

$$\sqrt{T}\,\text{vech}(\widetilde{P} - P) \xrightarrow{d} \mathcal{N}(0, \bar{H}\Sigma_{\widetilde{\sigma}}\bar{H}'), \tag{3.6.14}$$

where

$$\bar{H} = \frac{\partial \text{vech}(P)}{\partial \boldsymbol{\sigma}'} = [\mathbf{L}_K(I_{K^2} + \mathbf{K}_{KK})(P \otimes I_K)\mathbf{L}_K']^{-1}$$

(see Appendix A.13, Rule (10)). Here $\mathbf{K}_{mn}$ is the commutation matrix defined such that $\text{vec}(G) = \mathbf{K}_{mn}\text{vec}(G')$ for any $(m \times n)$ matrix G and $\mathbf{L}_K$ is the $(\frac{1}{2}K(K+1) \times K^2)$ elimination matrix defined such that $\text{vech}(F) = \mathbf{L}_K\,\text{vec}(F)$ for any $(K \times K)$ matrix F (see Appendix A.12.2). A Wald test of (3.6.13) may therefore be based on

$$\lambda_W = T\text{vech}(\widetilde{P})'C'[C\widehat{\bar{H}}\widehat{\Sigma}_{\widetilde{\sigma}}\widehat{\bar{H}}'C']^{-1}C\,\text{vech}(\widetilde{P}) \xrightarrow{d} \chi^2(N), \tag{3.6.15}$$

where hats denote the usual estimators. Although the two tests based on $\widetilde{\boldsymbol{\sigma}}$ and $\widetilde{P}$ are derived from the same asymptotic distribution, they may differ in small samples. Of course, in the previous discussion we may replace $\widetilde{\Sigma}_u$ by the asymptotically equivalent estimator $\widehat{\Sigma}_u$.

In our investment/income/consumption example, suppose we wish to test for instantaneous causality between (income, consumption) and investment. Following Proposition 2.3, the null hypothesis of no causality is

$$H_0: \sigma_{21} = \sigma_{31} = 0 \quad \text{or} \quad C\boldsymbol{\sigma} = 0,$$

where σ_{ij} is a typical element of Σ_u and

$$C = \begin{bmatrix} 0 & 1 & 0 & 0 & 0 & 0 \\ 0 & 0 & 1 & 0 & 0 & 0 \end{bmatrix}.$$

For this hypothesis, the test statistic in (3.6.12) assumes the value $\lambda_W = 5.46$. Alternatively, we may test

$$H_0: p_{21} = p_{31} = 0 \quad \text{or} \quad C\,\text{vech}(P) = 0,$$

where p_{ij} is a typical element of P. The corresponding value of the test statistic from (3.6.15) is $\lambda_W = 5.70$. Both tests are based on asymptotic $\chi^2(2)$-distributions and therefore do not reject the null hypothesis of no instantaneous causality at a 5% level. Note that the critical value for a 5% level test is 5.99.

3.6.4 Testing for Multi-Step Causality

In Section 2.3.1, we have also discussed the possibility of extending the information set and considering causality between two variables in a system that includes further variables. Using the same ideas as in the definition of Granger-causality resulted in the definition of h-step causality. This concept implies nonlinear restrictions for the VAR coefficients for which the usual application of the Wald principle does not result in a valid test. The following example from Lütkepohl & Burda (1997) illustrates the problem.

Consider a three-dimensional VAR(1) process:

$$\begin{bmatrix} z_t \\ y_t \\ x_t \end{bmatrix} = \begin{bmatrix} \alpha_{zz} & \alpha_{zy} & \alpha_{zx} \\ \alpha_{yz} & \alpha_{yy} & \alpha_{yx} \\ \alpha_{xz} & \alpha_{xy} & \alpha_{xx} \end{bmatrix} \begin{bmatrix} z_{t-1} \\ y_{t-1} \\ x_{t-1} \end{bmatrix} + \begin{bmatrix} u_{z,t} \\ u_{y,t} \\ u_{x,t} \end{bmatrix}. \qquad (3.6.16)$$

From (2.3.24) we know that a test of ∞-step noncausality from y_t to z_t ($y_t \not\to_{(\infty)} z_t$) needs to check $h = 2$ restrictions on the VAR coefficient vector. They are of the following nonlinear form:

$$r(\boldsymbol{\alpha}) = \begin{bmatrix} R\boldsymbol{\alpha} \\ R\boldsymbol{\alpha}^{(2)} \end{bmatrix} = (I_2 \otimes R) \begin{bmatrix} \boldsymbol{\alpha} \\ \boldsymbol{\alpha}^{(2)} \end{bmatrix},$$

where

$$R = [0\ 0\ 0\ 1\ 0\ 0\ 0\ 0\ 0],$$

$\boldsymbol{\alpha} = \text{vec}(A_1)$ and $\boldsymbol{\alpha}^{(2)} = \text{vec}(A_1^2)$, with A_1 being the coefficient matrix of the process in (3.6.16). Hence,

$$r(\boldsymbol{\alpha}) = \begin{bmatrix} \alpha_{zy} \\ \alpha_{zz}\alpha_{zy} + \alpha_{zy}\alpha_{yy} + \alpha_{zx}\alpha_{xy} \end{bmatrix} = \begin{bmatrix} 0 \\ 0 \end{bmatrix}. \qquad (3.6.17)$$

Denoting the covariance matrix of the asymptotic distribution of $\sqrt{T}(\widehat{\boldsymbol{\alpha}} - \boldsymbol{\alpha})$ as usual by $\Sigma_{\widehat{\boldsymbol{\alpha}}}$ and a consistent estimator by $\widehat{\Sigma}_{\widehat{\boldsymbol{\alpha}}}$, the Wald statistic for testing these restrictions has the form

$$\lambda_W = Tr(\widehat{\boldsymbol{\alpha}})' \left(\widehat{\frac{\partial r}{\partial \boldsymbol{\alpha}'}} \widehat{\Sigma}_{\widehat{\boldsymbol{\alpha}}} \widehat{\frac{\partial r'}{\partial \boldsymbol{\alpha}}} \right)^{-1} r(\widehat{\boldsymbol{\alpha}}),$$

where $\widehat{\partial r/\partial \boldsymbol{\alpha}'}$ is an estimator of $\partial r/\partial \boldsymbol{\alpha}'$ (see Appendix C.7). The statistic has an asymptotic $\chi^2(2)$-distribution under the null hypothesis, provided the matrix

$$\frac{\partial r}{\partial \boldsymbol{\alpha}'} \Sigma_{\widehat{\boldsymbol{\alpha}}} \frac{\partial r'}{\partial \boldsymbol{\alpha}}$$

is nonsingular. In the present case, the latter condition is unfortunately not satisfied for all relevant parameter values.

To see this, note that the matrix of first order partial derivatives of the function $r(\alpha)$ is

$$\frac{\partial r}{\partial \alpha'} = \begin{bmatrix} 0 & 0 & 0 & 1 & 0 & 0 & 0 & 0 & 0 \\ \alpha_{zy} & 0 & 0 & \alpha_{zz} + \alpha_{yy} & \alpha_{zy} & \alpha_{zx} & \alpha_{xy} & 0 & 0 \end{bmatrix}.$$

The restrictions (3.6.17) are satisfied if

$$\alpha_{zy} = \alpha_{zx} = 0, \alpha_{xy} \neq 0, \tag{3.6.18}$$

or

$$\alpha_{zy} = \alpha_{xy} = 0, \alpha_{zx} \neq 0, \tag{3.6.19}$$

or

$$\alpha_{zy} = \alpha_{zx} = \alpha_{xy} = 0. \tag{3.6.20}$$

Clearly, $\partial r / \partial \alpha'$ has rank 1 only and, thus,

$$\operatorname{rk}\left(\frac{\partial r}{\partial \alpha'} \Sigma_{\widehat{\alpha}} \frac{\partial r'}{\partial \alpha}\right) = 1,$$

if (3.6.20) holds. Hence, the standard Wald statistic will not have its asymptotic $\chi^2(2)$-distribution under the null hypothesis $r(\alpha) = 0$ if (3.6.20) holds.

Lütkepohl & Burda (1997) discussed a possibility to circumvent the problem by simply drawing a random variable from a normal distribution and adding it to the second restriction. Thereby a nonsingular distribution of the modified restriction vector is obtained and a Wald type statistic can be constructed for this vector.

More generally, Lütkepohl & Burda (1997) proposed the following approach for testing the null hypothesis that the K_y-dimensional vector y_t is not h-step causal for the K_z-dimensional vector z_t $(y_t \not\to_{(h)} z_t)$ if additional K_x variables x_t are present in the system of interest. Using the notation from Section 2.3.1, that is, $\mathbf{A}$ is defined as in the VAR(1) representation (2.1.8), $J := [I_K : 0 : \cdots : 0]$ is a $(K \times Kp)$ matrix, $A^{(j)} := J\mathbf{A}^j$, and $\alpha^{(j)} := \operatorname{vec}(A^{(j)})$, the hypotheses of interest can be stated as

$$H_0 : (I_h \otimes R)\, \mathsf{a}^{(h)} = 0 \quad \text{against} \quad H_1 : (I_h \otimes R)\, \mathsf{a}^{(h)} \neq 0, \tag{3.6.21}$$

where R is a $(pK_z K_y \times pK^2)$ matrix, as defined in (2.3.23), and

$$\mathsf{a}^{(h)} = \begin{bmatrix} \alpha \\ \alpha^{(2)} \\ \vdots \\ \alpha^{(h)} \end{bmatrix}.$$

Let $\widehat{\mathsf{a}}^{(h)}$ be the estimator corresponding to $\mathsf{a}^{(h)}$ based on the multivariate LS estimator $\widehat{\boldsymbol{\alpha}}$ of $\boldsymbol{\alpha}$. Furthermore, we denote by $\mathrm{diag}(D)$ a diagonal matrix which has the diagonal elements of the square matrix D on its main diagonal and define the $(hpK_zK_y \times hpK_zK_y)$ matrix

$$\widehat{\Sigma}_w(h) = \begin{bmatrix} 0 & 0 \\ 0 & I_{h-1} \otimes \mathrm{diag}(R\widehat{\Sigma}_{\widehat{\alpha}}R') \end{bmatrix}.$$

Moreover, we define a random vector $w_\lambda^{(h)} \sim \mathcal{N}(0, \lambda \widehat{\Sigma}_w(h))$ which is drawn independently of $\widehat{\boldsymbol{\alpha}}$. Here $\lambda > 0$ is some fixed real number. Lütkepohl & Burda (1997) defined the following modified Wald statistic for testing the pair of hypotheses in (3.6.21):

$$\begin{aligned}
\lambda_W^{mod} &= T \left((I_h \otimes R)\widehat{\mathsf{a}}^{(h)} + \frac{w_\lambda^{(h)}}{\sqrt{T}} \right)' \\
&\quad \times \left[(I_h \otimes R)\widehat{\Sigma}_{\widehat{\mathsf{a}}}(h)(I_h \otimes R') + \lambda \widehat{\Sigma}_w(h) \right]^{-1} \\
&\quad \times \left((I_h \otimes R)\widehat{\mathsf{a}}^{(h)} + \frac{w_\lambda^{(h)}}{\sqrt{T}} \right).
\end{aligned}$$

Here $\widehat{\Sigma}_{\widehat{\mathsf{a}}}(h)$ is a consistent estimator of the asymptotic covariance matrix of $\sqrt{T}(\widehat{\mathsf{a}}^{(h)} - \mathsf{a}^{(h)})$. It can be shown that

$$\lambda_W^{mod} \xrightarrow{d} \chi^2(hpK_zK_y)$$

under H_0. Notice that there is no need to add anything to the first pK_zK_y components of $(I_h \otimes R)\widehat{\mathsf{a}}^{(h)}$ because they are equal to $R\widehat{\boldsymbol{\alpha}}$ which has a nonsingular asymptotic distribution.

Clearly, adding some random term to $\widehat{\mathsf{a}}^{(h)}$ reduces the efficiency of the procedure and is likely to result in a loss in power of the test relative to a procedure which does not use this device. In particular, if the noise term is substantial in relation to the estimated variance, there may be some loss in power. Therefore, the amount of noise (the variance of the noise) is linked to the variance of the estimator through $\Sigma_w(h)$. Moreover, the quantity λ may be chosen close to zero. Thereby the loss in efficiency can be made arbitrarily small.

There are in fact also other possibilities to avoid the problems related to the Wald test. One way to get around it is to impose zero restrictions directly on the VAR coefficients prior to analyzing multi-step causality. The relevant subset models will be discussed in Chapter 5.

3.7 The Asymptotic Distributions of Impulse Responses and Forecast Error Variance Decompositions

3.7.1 The Main Results

In Chapter 2, Section 2.3.2, we have seen that the coefficients of the MA representations

$$y_t = \mu + \sum_{i=0}^{\infty} \Phi_i u_{t-i}, \quad \Phi_0 = I_K, \tag{3.7.1}$$

and

$$y_t = \mu + \sum_{i=0}^{\infty} \Theta_i w_{t-i} \tag{3.7.2}$$

are sometimes interpreted as impulse responses or dynamic multipliers of the system of variables y_t. Here $\mu = E(y_t)$, the $\Theta_i = \Phi_i P$, $w_t = P^{-1} u_t$, and P is the lower triangular Choleski decomposition of Σ_u such that $\Sigma_u = PP'$. Hence, $\Sigma_w = E(w_t w_t') = I_K$. In this section, we will assume that the Φ_i's and Θ_i's are unknown and they are computed from the estimated VAR coefficients and error covariance matrix. We will derive the asymptotic distributions of the resulting estimated Φ_i's and Θ_i's. In these derivations, we will not need the existence of MA representations (3.7.1) and (3.7.2). We will just assume that the Φ_i's are obtained from given coefficient matrices $A_1, \ldots, A_p$ by recursions

$$\Phi_i = \sum_{j=1}^{i} \Phi_{i-j} A_j, \quad i = 1, 2, \ldots,$$

starting with $\Phi_0 = I_K$ and setting $A_j = 0$ for $j > p$. Furthermore, the Θ_i's are obtained from $A_1, \ldots, A_p$, and Σ_u as $\Theta_i = \Phi_i P$, where P is as specified in the foregoing. In addition, the asymptotic distributions of the corresponding accumulated responses

$$\Psi_n = \sum_{i=0}^{n} \Phi_i, \quad \Psi_\infty = \sum_{i=0}^{\infty} \Phi_i = (I_K - A_1 - \cdots - A_p)^{-1} \quad \text{(if it exists)},$$

$$\Xi_n = \sum_{i=0}^{n} \Theta_i, \quad \Xi_\infty = \sum_{i=0}^{\infty} \Theta_i = (I_K - A_1 - \cdots - A_p)^{-1} P \quad \text{(if it exists)},$$

and the forecast error variance components,

$$\omega_{jk,h} = \sum_{i=0}^{h-1} (e_j' \Theta_i e_k)^2 / \text{MSE}_j(h), \tag{3.7.3}$$

will be given. Here e_k is the k-th column of I_K and

$$\mathrm{MSE}_j(h) = \sum_{i=0}^{h-1} e'_j \Phi_i \Sigma_u \Phi'_i e_j$$

is the j-th diagonal element of the MSE matrix $\Sigma_y(h)$ of an h-step forecast (see Chapter 2, Section 2.2.2).

The derivation of the asymptotic distributions is based on the following result from Appendix C, Proposition C.15(3). Suppose $\boldsymbol{\beta}$ is an $(n \times 1)$ vector of parameters and $\widehat{\boldsymbol{\beta}}$ is an estimator such that

$$\sqrt{T}(\widehat{\boldsymbol{\beta}} - \boldsymbol{\beta}) \xrightarrow{d} \mathcal{N}(0, \Sigma_{\widehat{\boldsymbol{\beta}}}),$$

where T, as usual, denotes the sample size (time series length) used for estimation. Let $g(\boldsymbol{\beta})$ be a continuously differentiable function with values in the m-dimensional Euclidean space and suppose that $\partial g_i/\partial \boldsymbol{\beta}' = (\partial g_i/\partial \beta_j)$ is nonzero at the true vector $\boldsymbol{\beta}$, for $i = 1, \ldots, m$. Then,

$$\sqrt{T}\left[g(\widehat{\boldsymbol{\beta}}) - g(\boldsymbol{\beta})\right] \xrightarrow{d} \mathcal{N}\left(0, \frac{\partial g}{\partial \boldsymbol{\beta}'} \Sigma_{\widehat{\boldsymbol{\beta}}} \frac{\partial g'}{\partial \boldsymbol{\beta}}\right).$$

In writing down the asymptotic distributions formally, we use the notation

$$\boldsymbol{\alpha} := \mathrm{vec}(A_1, \ldots, A_p) \qquad (K^2 p \times 1),$$

$$\mathbf{A} := \begin{bmatrix} A_1 & A_2 & \cdots & A_{p-1} & A_p \\ I_K & 0 & \cdots & 0 & 0 \\ 0 & I_K & & 0 & 0 \\ \vdots & & \ddots & \vdots & \vdots \\ 0 & 0 & \cdots & I_K & 0 \end{bmatrix} \qquad (Kp \times Kp),$$

$$\boldsymbol{\sigma} := \mathrm{vech}(\Sigma_u) \qquad (\tfrac{1}{2}K(K+1) \times 1)$$

and the corresponding estimators are furnished with a hat. As before, vec denotes the column stacking operator and vech is the corresponding operator that stacks the elements on and below the main diagonal only. We also use the commutation matrix $\mathbf{K}_{mn}$, defined such that, for any $(m \times n)$ matrix G, $\mathbf{K}_{mn}\mathrm{vec}(G) = \mathrm{vec}(G')$, the $(m^2 \times \tfrac{1}{2}m(m+1))$ duplication matrix $\mathbf{D}_m$, defined such that $\mathbf{D}_m \mathrm{vech}(F) = \mathrm{vec}(F)$, for any symmetric $(m \times m)$ matrix F, and the $(\tfrac{1}{2}m(m+1) \times m^2)$ elimination matrix $\mathbf{L}_m$, defined such that, for any $(m \times m)$ matrix F, $\mathrm{vech}(F) = \mathbf{L}_m \mathrm{vec}(F)$ (see Appendix A.12.2). Furthermore, $J := [I_K : 0 : \cdots : 0]$ is a $(K \times Kp)$ matrix. With this notation, the following proposition from Lütkepohl (1990) can be stated.

Proposition 3.6 (*Asymptotic Distributions of Impulse Responses*)
Suppose

$$\sqrt{T}\begin{bmatrix} \widehat{\boldsymbol{\alpha}} - \boldsymbol{\alpha} \\ \widehat{\boldsymbol{\sigma}} - \boldsymbol{\sigma} \end{bmatrix} \xrightarrow{d} \mathcal{N}\left(0, \begin{bmatrix} \Sigma_{\widehat{\boldsymbol{\alpha}}} & 0 \\ 0 & \Sigma_{\widehat{\boldsymbol{\sigma}}} \end{bmatrix}\right). \tag{3.7.4}$$

Then

3.7 Impulse Responses

$$\sqrt{T} \operatorname{vec}(\widehat{\Phi}_i - \Phi_i) \xrightarrow{d} \mathcal{N}(0, G_i \Sigma_{\widehat{\alpha}} G_i'), \quad i = 1, 2, \ldots, \tag{3.7.5}$$

where

$$G_i := \frac{\partial \operatorname{vec}(\Phi_i)}{\partial \boldsymbol{\alpha}'} = \sum_{m=0}^{i-1} J(\mathbf{A}')^{i-1-m} \otimes \Phi_m.$$

$$\sqrt{T} \operatorname{vec}(\widehat{\Psi}_n - \Psi_n) \xrightarrow{d} \mathcal{N}(0, F_n \Sigma_{\widehat{\alpha}} F_n'), \quad n = 1, 2, \ldots, \tag{3.7.6}$$

where $F_n := G_1 + \cdots + G_n$.

If $(I_K - A_1 - \cdots - A_p)$ is nonsingular,

$$\sqrt{T} \operatorname{vec}(\widehat{\Psi}_\infty - \Psi_\infty) \xrightarrow{d} \mathcal{N}(0, F_\infty \Sigma_{\widehat{\alpha}} F_\infty'), \tag{3.7.7}$$

where $F_\infty := \underbrace{(\Psi_\infty', \ldots, \Psi_\infty')}_{p \text{ times}} \otimes \Psi_\infty$.

$$\sqrt{T} \operatorname{vec}(\widehat{\Theta}_i - \Theta_i) \xrightarrow{d} \mathcal{N}(0, C_i \Sigma_{\widehat{\alpha}} C_i' + \bar{C}_i \Sigma_{\widehat{\sigma}} \bar{C}_i'), \quad i = 0, 1, 2, \ldots, \tag{3.7.8}$$

where

$$C_0 := 0, \ C_i := (P' \otimes I_K) G_i, i = 1, 2, \ldots, \ \bar{C}_i := (I_K \otimes \Phi_i) H, \ i = 0, 1, \ldots,$$

and

$$\begin{aligned} H &:= \frac{\partial \operatorname{vec}(P)}{\partial \boldsymbol{\sigma}'} = \mathbf{L}_K' \{ \mathbf{L}_K [(I_K \otimes P) \mathbf{K}_{KK} + (P \otimes I_K)] \mathbf{L}_K' \}^{-1} \\ &= \mathbf{L}_K' \{ \mathbf{L}_K (I_{K^2} + \mathbf{K}_{KK}) (P \otimes I_K) \mathbf{L}_K' \}^{-1}. \end{aligned}$$

$$\sqrt{T} \operatorname{vec}(\widehat{\Xi}_n - \Xi_n) \xrightarrow{d} \mathcal{N}(0, B_n \Sigma_{\widehat{\alpha}} B_n' + \bar{B}_n \Sigma_{\widehat{\sigma}} \bar{B}_n'), \tag{3.7.9}$$

where $B_n := (P' \otimes I_K) F_n$ and $\bar{B}_n := (I_K \otimes \Psi_n) H$.

If $(I_K - A_1 - \cdots - A_p)$ is nonsingular,

$$\sqrt{T} \operatorname{vec}(\widehat{\Xi}_\infty - \Xi_\infty) \xrightarrow{d} \mathcal{N}(0, B_\infty \Sigma_{\widehat{\alpha}} B_\infty' + \bar{B}_\infty \Sigma_{\widehat{\sigma}} \bar{B}_\infty'), \tag{3.7.10}$$

where $B_\infty := (P' \otimes I_K) F_\infty$ and $\bar{B}_\infty := (I_K \otimes \Psi_\infty) H$.

Finally,

$$\sqrt{T}(\widehat{\omega}_{jk,h} - \omega_{jk,h}) \xrightarrow{d} \mathcal{N}(0, d_{jk,h} \Sigma_{\widehat{\alpha}} d_{jk,h}' + \bar{d}_{jk,h} \Sigma_{\widehat{\sigma}} \bar{d}_{jk,h}')$$
$$j, k = 1, \ldots, K, \quad h = 1, 2, \ldots, \tag{3.7.11}$$

where

$$\begin{aligned} d_{jk,h} &:= \frac{2}{\operatorname{MSE}_j(h)^2} \sum_{i=0}^{h-1} \Big[\operatorname{MSE}_j(h) (e_j' \Phi_i P e_k) (e_k' P' \otimes e_j') G_i \\ &\quad - (e_j' \Phi_i P e_k)^2 \sum_{m=0}^{h-1} (e_j' \Phi_m \Sigma_u \otimes e_j') G_m \Big] \end{aligned}$$

with $G_0 := 0$ and

$$\overline{d}_{jk,h} := \sum_{i=0}^{h-1} \Big[2\,\text{MSE}_j(h)(e'_j \Phi_i P e_k)(e'_k \otimes e'_j \Phi_i) H$$
$$- (e'_j \Phi_i P e_k)^2 \sum_{m=0}^{h-1}(e'_j \Phi_m \otimes e'_j \Phi_m)\mathbf{D}_K \Big] \Big/ \text{MSE}_j(h)^2.$$

∎

In the next subsection, the proof of the proposition is indicated. Some remarks are worthwhile now.

Remark 1 In the proposition, some matrices of partial derivatives may be zero. For instance, if a VAR(1) model is fitted although the true order is zero, that is, y_t is white noise, then

$$G_2 = J\mathbf{A}' \otimes I_K + JI_K \otimes \Phi_1 = 0$$

because $\mathbf{A} = A_1 = 0$ and $\Phi_1 = A_1 = 0$. Hence, a degenerate asymptotic distribution with zero covariance matrix is obtained for $\sqrt{T}\,\text{vec}(\widehat{\Phi}_2 - \Phi_2)$. As explained in Appendix B, we call such a distribution also multivariate normal. Otherwise it would be necessary to distinguish between cases with zero and nonzero partial derivatives or we have to assume that all partial derivatives are such that the covariance matrices have no zeros on the diagonal. Note that estimators of the covariance matrices obtained by replacing unknown quantities by their usual estimators may be problematic when the asymptotic distribution is degenerate. In that case, the usual t-ratios and confidence intervals may not be appropriate.

To illustrate the potential problems resulting from a degenerate asymptotic distribution, we follow Benkwitz, Lütkepohl & Neumann (2000) and consider a univariate AR(1) process $y_t = \alpha y_{t-1} + u_t$. In this case, $\Phi_i = \alpha^i$. Suppose that $\widehat{\alpha}$ is an estimator of α satisfying $\sqrt{T}(\widehat{\alpha} - \alpha) \xrightarrow{d} \mathcal{N}(0, \sigma_{\widehat{\alpha}}^2)$ with $\sigma_{\widehat{\alpha}}^2 \neq 0$. For instance, $\widehat{\alpha}$ may be the LS estimator of α. Then

$$\sqrt{T}(\widehat{\alpha}^2 - \alpha^2) \xrightarrow{d} \mathcal{N}(0, \sigma_{\widehat{\alpha}^2}^2)$$

with $\sigma_{\widehat{\alpha}^2}^2 = 4\alpha^2 \sigma_{\widehat{\alpha}}^2$. This quantity is, of course, zero if $\alpha = 0$. In the latter case, $\sqrt{T}\widehat{\alpha}/\sigma_{\widehat{\alpha}}$ has an asymptotic standard normal distribution and, hence, $T\widehat{\alpha}^2/\sigma_{\widehat{\alpha}}^2$ has an asymptotic $\chi^2(1)$-distribution. Thus, it is clear that in this case $\sqrt{T}\widehat{\alpha}^2$ is asymptotically degenerate.

Because the estimated $\sigma_{\widehat{\alpha}^2}^2$ obtained by replacing α and $\sigma_{\widehat{\alpha}}^2$ by their usual LS estimators is nonzero almost surely, it is tempting to use the quantity $\sqrt{T}(\widehat{\alpha}^2 - \alpha^2)/2\widehat{\alpha}\widehat{\sigma}_{\widehat{\alpha}}$ for constructing a confidence interval, say, for Φ_2. However, for $\alpha = 0$, the t-ratio becomes $\sqrt{T}\widehat{\alpha}/2\widehat{\sigma}_{\widehat{\alpha}}$ which converges to $\mathcal{N}(0, 1/4)$ asymptotically, because $\sqrt{T}\widehat{\alpha}/\widehat{\sigma}_{\widehat{\alpha}} \xrightarrow{d} \mathcal{N}(0,1)$. A confidence interval constructed on the basis of the asymptotic standard normal distribution would therefore be a conservative one. In other words, asymptotic inference which ignores the possible singularity in the asymptotic distribution of the impulse responses may be misleading (see Benkwitz et al. (2000) for further discussion). ∎

Remark 2 In the proposition, it is not explicitly assumed that y_t is stable. While the stability condition is partly introduced in (3.7.7) and (3.7.10) by requiring that $(I_K - A_1 - \cdots - A_p)$ be nonsingular so that

$$\det(I_K - A_1 z - \cdots - A_p z^p) \neq 0 \quad \text{for } z = 1,$$

it is not needed for the other results to hold. The crucial condition is the asymptotic distribution of the process parameters in (3.7.4). Although we have used the stationarity and stability assumptions in Sections 3.2–3.4 in order to derive the asymptotic distribution of the process parameters, we will see in later chapters that asymptotic normality is also obtained for certain nonstationary, unstable processes. Therefore, at least parts of Proposition 3.6 will be useful in a nonstationary environment. ∎

Remark 3 The block-diagonal structure of the covariance matrix of the asymptotic distribution in (3.7.4) is in no way essential for the asymptotic normality of the impulse responses. In fact, the asymptotic distributions in (3.7.5)–(3.7.7) remain unchanged if the asymptotic covariance matrix of the parameter estimators is not block-diagonal. On the other hand, without the block-diagonal structure, the simple additive structure of the asymptotic covariance matrices in (3.7.8)–(3.7.11) is lost. Although these asymptotic distributions are easily generalizable to the case of a general asymptotic covariance matrix of the VAR coefficients in (3.7.4), we have not stated the more general result here because it is not needed in subsequent chapters of this text. ∎

Remark 4 Under the conditions of Proposition 3.4, the covariance matrix of the asymptotic distribution of the parameters has precisely the block-diagonal structure assumed in (3.7.4) with

$$\Sigma_{\widehat{\alpha}} = \Gamma_Y(0)^{-1} \otimes \Sigma_u$$

and

$$\Sigma_{\widehat{\sigma}} = 2\mathbf{D}_K^+ (\Sigma_u \otimes \Sigma_u) \mathbf{D}_K^{+\prime},$$

where $\mathbf{D}_K^+ = (\mathbf{D}_K' \mathbf{D}_K)^{-1} \mathbf{D}_K'$ is the Moore-Penrose inverse of the duplication matrix $\mathbf{D}_K$. Using these expressions in the proposition, some simplifications of the covariance matrices can be obtained. For instance, the covariance matrix in (3.7.5) becomes

$$G_i \Sigma_{\widehat{\alpha}} G_i'$$

$$= \left[\sum_{m=0}^{i-1} J(\mathbf{A}')^{i-1-m} \otimes \Phi_m \right] (\Gamma_Y(0)^{-1} \otimes \Sigma_u) \left[\sum_{n=0}^{i-1} J(\mathbf{A}')^{i-1-n} \otimes \Phi_n \right]'$$

$$= \sum_{m=0}^{i-1} \sum_{n=0}^{i-1} \left[J(\mathbf{A}')^{i-1-m} \Gamma_Y(0)^{-1} \mathbf{A}^{i-1-n} J' \right] \otimes (\Phi_m \Sigma_u \Phi_n')$$

which is computationally convenient because all matrices involved are of a relatively small size. The advantage of the general formulation is that it can be used with other $\Sigma_{\widetilde{\alpha}}$ matrices as well. We will see examples in subsequent chapters. ∎

Remark 5 In practice, to use the asymptotic distributions for inference, the unknown quantities in the covariance matrices in Proposition 3.6 may be replaced by their usual estimators given in Sections 3.2–3.4 for the case of a stationary, stable process y_t (see, however, Remark 1). ∎

Remark 6 Summing the forecast error variance components over k,

$$\sum_{k=1}^{K} \omega_{jk,h} = \sum_{k=1}^{K} \widehat{\omega}_{jk,h} = 1$$

for each j and h. These restrictions are not taken into account in the derivation of the asymptotic distributions in (3.7.11). It is easily checked, however, that for dimension $K = 1$ the standard errors obtained from Proposition 3.6 are zero as they should be, because all forecast error variance components are 1 in that case. A problem in this context is that the asymptotic distribution of $\widehat{\omega}_{jk,h}$ cannot be used in the usual way for tests of significance and setting up confidence intervals if $\omega_{jk,h} = 0$. In that case, from the definitions of $d_{jk,h}$ and $\overline{d}_{jk,h}$, the variance of the asymptotic distribution is easily seen to be zero and, hence, estimating this quantity by replacing unknown parameters by their usual estimators may lead to t-ratios that are not standard normal asymptotically and, hence, cannot be used in the usual way for inference (see Remark 1). This state of affairs is unfortunate from a practical point of view because testing the significance of forecast error variance components is of particular interest in practice. Note, however, that

$$\omega_{jk,h} = 0 \iff \theta_{jk,i} = 0 \quad \text{for} \quad i = 0, \ldots, h.$$

A test of the latter hypothesis may be possible. ∎

Remark 7 Joint confidence regions and test statistics for testing hypotheses that involve several of the response coefficients can be obtained from Proposition 3.6 in the usual way. However, it has to be taken into account that, for instance, the elements of $\widehat{\Phi}_i$ and $\widehat{\Phi}_j$ will not be independent asymptotically. If elements from two or more MA matrices are involved the joint distribution of all the matrices must be determined. This distribution can be derived easily from the results given in the proposition. For instance, the covariance matrix of the joint asymptotic distribution of $\text{vec}(\widehat{\Phi}_i, \widehat{\Phi}_j)$ is

$$\frac{\partial \text{vec}(\Phi_i, \Phi_j)}{\partial \alpha'} \Sigma_{\widehat{\alpha}} \frac{\partial \text{vec}(\Phi_i, \Phi_j)'}{\partial \alpha},$$

where

$$\frac{\partial \text{vec}(\Phi_i, \Phi_j)}{\partial \alpha'} = \begin{bmatrix} \dfrac{\partial \text{vec}(\Phi_i)}{\partial \alpha'} \\ \dfrac{\partial \text{vec}(\Phi_j)}{\partial \alpha'} \end{bmatrix}$$

etc. We have chosen to state the proposition for individual MA coefficient matrices because thereby all required matrices have relatively small dimensions and, hence, are easy to compute. ∎

Remark 8 Denoting the jk-th elements of Φ_i and Θ_i by $\phi_{jk,i}$ and $\theta_{jk,i}$, respectively, hypotheses of obvious interest, for $j \neq k$, are

$$H_0 : \phi_{jk,i} = 0 \quad \text{for} \quad i = 1, 2, \ldots \tag{3.7.12}$$

and

$$H_0 : \theta_{jk,i} = 0 \quad \text{for} \quad i = 0, 1, 2, \ldots \tag{3.7.13}$$

because they can be interpreted as hypotheses on noncausality from variable k to variable j, that is, an impulse in variable k does not induce any response of variable j. From Chapter 2, Propositions 2.4 and 2.5, we know that (3.7.12) is equivalent to

$$H_0 : \phi_{jk,i} = 0 \quad \text{for} \quad i = 1, 2, \ldots, p(K-1) \tag{3.7.14}$$

and (3.7.13) is equivalent to

$$H_0 : \theta_{jk,i} = 0 \quad \text{for} \quad i = 0, 1, \ldots, p(K-1). \tag{3.7.15}$$

Using Bonferroni's inequality (see Chapter 2, Section 2.2.3), a test of (3.7.14) with significance level at most $100\gamma\%$ is obtained by rejecting H_0 if

$$|\sqrt{T}\widehat{\phi}_{jk,i}/\widehat{\sigma}_{\phi_{jk}}(i)| > z_{(\gamma/2p(K-1))} \tag{3.7.16}$$

for at least one $i \in \{1, 2, \ldots, p(K-1)\}$. Here $z_{(\gamma)}$ is the upper 100γ percentage point of the standard normal distribution and $\widehat{\sigma}_{\phi_{jk}}(i)$ is an estimate of the asymptotic standard deviation $\sigma_{\phi_{jk}}(i)$ of $\sqrt{T}\widehat{\phi}_{jk,i}$ obtained via Proposition 3.6. In order to obtain an asymptotic standard normal distribution of the t-ratio $\sqrt{T}\widehat{\phi}_{jk,i}/\widehat{\sigma}_{\phi_{jk}}(i)$, the variance $\sigma^2_{\phi_{jk}}(i)$ must be nonzero, however.

A test of (3.7.15) with significance level at most γ is obtained by rejecting H_0 if

$$|\sqrt{T}\widehat{\theta}_{jk,i}/\widehat{\sigma}_{\theta_{jk}}(i)| \begin{cases} > z_{(\gamma/2(pK-p+1))} & \text{for at least one} \\ & i \in \{0,1,2,\ldots,p(K-1)\} \text{ if } j > k \\ > z_{(\gamma/2(pK-p))} & \text{for at least one} \\ & i \in \{1,2,\ldots,p(K-1)\} \text{ if } j < k. \end{cases} \quad (3.7.17)$$

Here $\widehat{\sigma}_{\theta_{jk}}(i)$ is a consistent estimator of the standard deviation of the asymptotic distribution of $\sqrt{T}\widehat{\theta}_{jk,i}$ obtained from Proposition 3.6 and that standard deviation is assumed to be nonzero.

A test based on Bonferroni's principle may have quite low power because the actual significance level may be much smaller than the given upper bound. Therefore a test based on some χ^2- or F-statistic would be preferable. Unfortunately, such tests are not easily available for the present situation. The problem is similar to the one discussed in Section 3.6.4 in the context of testing for multi-step causality. For more discussion of this point see also Lütkepohl (1990) and for a different approach of representing the uncertainty in estimated impulse responses see Sims & Zha (1999). ∎

3.7.2 Proof of Proposition 3.6

The proof of Proposition 3.6 is a straightforward application of the matrix differentiation rules given in Appendix A.13. It is sketched here for completeness and because it is spread out over a number of publications in the literature. Readers mainly interested in applying the proposition may skip this section without loss of continuity.

To prove (3.7.5), note that $\Phi_i = J\mathbf{A}^i J'$ (see Chapter 2, Section 2.1.2) and apply Rule (8) of Appendix A.13. The expression for F_n in (3.7.6) follows because

$$\frac{\partial \text{vec}(\Psi_n)}{\partial \alpha'} = \sum_{i=1}^{n} \frac{\partial \text{vec}(\Phi_i)}{\partial \alpha'}$$

and

$$\begin{aligned} F_\infty &= \frac{\partial \text{vec}(\Psi_\infty)}{\partial \alpha'} = \frac{\partial \text{vec}(\Psi_\infty)}{\partial \text{vec}(\Psi_\infty^{-1})'} \frac{\partial \text{vec}(\Psi_\infty^{-1})}{\partial \alpha'} \\ &= -(\Psi_\infty' \otimes \Psi_\infty) \frac{\partial \text{vec}(I_K - A_1 - \cdots - A_p)}{\partial \alpha'}. \end{aligned}$$

Furthermore,

$$C_i = \frac{\partial \text{vec}(\Theta_i)}{\partial \alpha'} = \frac{\partial \text{vec}(\Phi_i P)}{\partial \alpha'} = (P' \otimes I_K) \frac{\partial \text{vec}(\Phi_i)}{\partial \alpha'}$$

and
$$\bar{C}_i = \frac{\partial \operatorname{vec}(\Theta_i)}{\partial \boldsymbol{\sigma}'} = (I_K \otimes \Phi_i)\frac{\partial \operatorname{vec}(P)}{\partial \boldsymbol{\sigma}'},$$
where
$$\frac{\partial \operatorname{vec}(P)}{\partial \boldsymbol{\sigma}'} = \mathbf{L}'_K \frac{\partial \operatorname{vech}(P)}{\partial \boldsymbol{\sigma}'} = H,$$

follows from Appendix A.13, Rule (10). The matrices B_n, $\bar{B}_n$, B_∞, and $\bar{B}_\infty$ are obtained in a similar manner, using the relations $\Xi_n = \Psi_n P$ and $\Xi_\infty = \Psi_\infty P$.

Finally, in (3.7.11),

$$\begin{aligned}
d_{jk,h} &= \frac{\partial \omega_{jk,h}}{\partial \boldsymbol{\alpha}'} \\
&= \left[2\sum_{i=0}^{h-1}(e'_j\Phi_i Pe_k)(e'_k P' \otimes e'_j)\frac{\partial \operatorname{vec}(\Phi_i)}{\partial \boldsymbol{\alpha}'}\operatorname{MSE}_j(h) \right. \\
&\quad \left. - \sum_{i=0}^{h-1}(e'_j\Phi_i Pe_k)^2 \frac{\partial \operatorname{MSE}_j(h)}{\partial \boldsymbol{\alpha}'} \right] \Big/ \operatorname{MSE}_j(h)^2,
\end{aligned}$$

$$\begin{aligned}
\frac{\partial \operatorname{MSE}_j(h)}{\partial \boldsymbol{\alpha}'} &= \sum_{m=0}^{h-1} \left[(e'_j\Phi_m \Sigma_u \otimes e'_j)\frac{\partial \operatorname{vec}(\Phi_m)}{\partial \boldsymbol{\alpha}'} \right. \\
&\quad \left. + (e'_j \otimes e'_j\Phi_m\Sigma_u)\frac{\partial \operatorname{vec}(\Phi'_m)}{\partial \boldsymbol{\alpha}'} \right] \\
&= \sum_{m=0}^{h-1} \left[(e'_j\Phi_m\Sigma_u \otimes e'_j) + (e'_j \otimes e'_j\Phi_m\Sigma_u)\mathbf{K}_{KK} \right] \frac{\partial \operatorname{vec}(\Phi_m)}{\partial \boldsymbol{\alpha}'} \\
&= \sum_{m=0}^{h-1} \left[(e'_j\Phi_m\Sigma_u \otimes e'_j) + \mathbf{K}_{11}(e'_j\Phi_m\Sigma_u \otimes e'_j) \right] G_m \\
&= 2\sum_{m=0}^{h-1}(e'_j\Phi_m\Sigma_u \otimes e'_j)G_m,
\end{aligned}$$

(see Appendix A.12.2, Rule (23))

$$\begin{aligned}
\bar{d}_{jk,h} &= \frac{\partial \omega_{jk,h}}{\partial \boldsymbol{\sigma}'} \\
&= \sum_{i=0}^{h-1}\left[2(e'_j\Phi_i Pe_k)(e'_k \otimes e'_j\Phi_i)\frac{\partial \operatorname{vec}(P)}{\partial \boldsymbol{\sigma}'}\operatorname{MSE}_j(h) \right. \\
&\quad \left. - (e'_j\Phi_i Pe_k)^2\frac{\partial \operatorname{MSE}_j(h)}{\partial \boldsymbol{\sigma}'} \right] \Big/ \operatorname{MSE}_j(h)^2,
\end{aligned}$$

and

$$\frac{\partial \mathrm{MSE}_j(h)}{\partial \boldsymbol{\sigma}'} = \sum_{m=0}^{h-1} (e'_j \Phi_m \otimes e'_j \Phi_m) \frac{\partial \operatorname{vec}(\Sigma_u)}{\partial \boldsymbol{\sigma}'}$$

$$= \sum_{m=0}^{h-1} (e'_j \Phi_m \otimes e'_j \Phi_m) \mathbf{D}_K \frac{\partial \operatorname{vech}(\Sigma_u)}{\partial \boldsymbol{\sigma}'}.$$

Thereby Proposition 3.6 is proven. In the next section an example is discussed.

3.7.3 An Example

To illustrate the results of Section 3.7.1, we use again the investment/income/consumption example from Section 3.2.3. Because

$$\widehat{\Phi}_1 = \widehat{A}_1 = \begin{bmatrix} -.320 & .146 & .961 \\ .044 & -.153 & .289 \\ -.002 & .225 & -.264 \end{bmatrix},$$

the elements of $\widehat{\Phi}_1$ must have the same standard errors as the elements of $\widehat{A}_1$. Checking the covariance matrix in (3.7.5), it is seen that the asymptotic covariance matrix of $\widehat{\Phi}_1$ is indeed the upper left-hand ($K^2 \times K^2$) block of $\Sigma_{\widehat{\alpha}}$ because

$$G_1 = J \otimes I_K = [I_{K^2} : 0 : \cdots : 0].$$

Thus, the square roots of the diagonal elements of

$$G_1 \widehat{\Sigma}_{\widehat{\alpha}} G'_1 / T = \frac{1}{T} [I_9 : 0 : \cdots : 0] (\widehat{\Gamma}_Y(0)^{-1} \otimes \widehat{\Sigma}_u) \begin{bmatrix} I_9 \\ 0 \\ \vdots \\ 0 \end{bmatrix}$$

are estimates of the asymptotic standard errors of $\widehat{\Phi}_1$. Note that here and in the following we use the LS estimators from the standard form of the VAR process (see Section 3.2) and not the mean-adjusted form. Accordingly, the estimate $\widehat{\Gamma}_Y(0)^{-1}$ is obtained from $(ZZ'/T)^{-1}$ by deleting the first row and column.

From (2.1.22) we get

$$\widehat{\Phi}_2 = \widehat{\Phi}_1 \widehat{A}_1 + \widehat{A}_2 = \begin{bmatrix} -.054 & .262 & .416 \\ .029 & .114 & -.088 \\ .045 & .261 & .110 \end{bmatrix}.$$

To estimate the corresponding standard errors, we note that

$$G_2 = J\mathbf{A}' \otimes I_K + J \otimes \Phi_1.$$

Replacing the unknown quantities by the usual estimates gives

$$\frac{1}{T}\widehat{G}_2\widehat{\Sigma}_{\widehat{\alpha}}\widehat{G}'_2 = \frac{1}{T}[J\widehat{\mathbf{A}}'\widehat{\Gamma}_Y(0)^{-1}\widehat{\mathbf{A}}J' \otimes \widehat{\Sigma}_u + J\widehat{\mathbf{A}}'\widehat{\Gamma}_Y(0)^{-1}J' \otimes \widehat{\Sigma}_u\widehat{\Phi}'_1$$
$$+J\widehat{\Gamma}_Y(0)^{-1}\widehat{\mathbf{A}}J' \otimes \widehat{\Phi}_1\widehat{\Sigma}_u + J\widehat{\Gamma}_Y(0)^{-1}J' \otimes \widehat{\Phi}_1\widehat{\Sigma}_u\widehat{\Phi}'_1].$$

The square roots of the diagonal elements of this matrix are estimates of the standard deviations of the elements of $\widehat{\Phi}_2$ and so on. Some $\widehat{\Phi}_i$ matrices together with estimated standard errors are given in Table 3.3. In Figures 3.4 and 3.5, some impulse responses are depicted graphically along with two-standard error bounds.

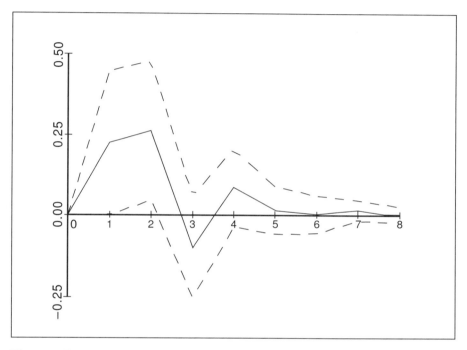

Fig. 3.4. Estimated responses of consumption to a forecast error impulse in income with estimated asymptotic two-standard error bounds.

In Figure 3.4, consumption is seen to increase in response to a unit shock in income. However, under a two-standard error criterion (approximate 95% confidence bounds) only the second response coefficient is significantly different from zero. Of course, the large standard errors of the impulse response coefficients reflect the substantial estimation uncertainty in the VAR coefficient matrices A_1 and A_2.

To check the overall significance of the response coefficients of consumption to an income impulse, we may use the procedure described in Remark 8 of

Table 3.3. Estimates of impulse responses for the investment/income/consumption system with estimated asymptotic standard errors in parentheses

i	$\widehat{\Phi}_i$	$\widehat{\Psi}_i$
1	$\begin{bmatrix} -0.320 & 0.146 & 0.961 \\ (0.125) & (0.562) & (0.657) \\ 0.044 & -0.153 & 0.289 \\ (0.032) & (0.143) & (0.167) \\ -0.002 & 0.225 & -0.264 \\ (0.025) & (0.115) & (0.134) \end{bmatrix}$	$\begin{bmatrix} 0.680 & 0.146 & 0.961 \\ (0.125) & (0.562) & (0.657) \\ 0.044 & 0.847 & 0.289 \\ (0.032) & (0.143) & (0.167) \\ -0.002 & 0.225 & 0.736 \\ (0.025) & (0.115) & (0.134) \end{bmatrix}$
2	$\begin{bmatrix} -0.054 & 0.262 & 0.416 \\ (0.129) & (0.546) & (0.663) \\ 0.029 & 0.114 & -0.088 \\ (0.032) & (0.135) & (0.162) \\ 0.045 & 0.261 & 0.110 \\ (0.026) & (0.108) & (0.131) \end{bmatrix}$	$\begin{bmatrix} 0.626 & 0.408 & 1.377 \\ (0.148) & (0.651) & (0.755) \\ 0.073 & 0.961 & 0.200 \\ (0.043) & (0.192) & (0.222) \\ 0.043 & 0.486 & 0.846 \\ (0.033) & (0.144) & (0.167) \end{bmatrix}$
3	$\begin{bmatrix} 0.119 & 0.353 & -0.408 \\ (0.084) & (0.384) & (0.476) \\ -0.009 & 0.071 & 0.120 \\ (0.016) & (0.078) & (0.094) \\ -0.001 & -0.098 & 0.091 \\ (0.017) & (0.078) & (0.102) \end{bmatrix}$	$\begin{bmatrix} 0.745 & 0.761 & 0.969 \\ (0.099) & (0.483) & (0.550) \\ 0.064 & 1.033 & 0.320 \\ (0.037) & (0.176) & (0.203) \\ 0.042 & 0.388 & 0.937 \\ (0.033) & (0.156) & (0.183) \end{bmatrix}$
∞	0	$\begin{bmatrix} 0.756 & 0.836 & 1.295 \\ (0.133) & (0.661) & (0.798) \\ 0.076 & 1.076 & 0.344 \\ (0.048) & (0.236) & (0.285) \\ 0.053 & 0.505 & 0.964 \\ (0.043) & (0.213) & (0.257) \end{bmatrix}$

Section 3.7.1. That is, we have to check the significance of the first $p(K-1) = 4$ response coefficients. Because one of them is individually significant at an asymptotic 5% level we may reject the null hypothesis of no response of consumption to income impulses at a significance level not greater than $4 \times 5\% = 20\%$. Of course, this is not a significance level we are used to in applied work. However, it becomes clear from Table 3.3 that the second response coefficient $\widehat{\phi}_{32,2}$ is still significant if the individual significance levels are reduced to 2.5%. Note that the upper 1.25 percentage point of the standard normal distribution is $c_{0.0125} = 2.24$. Thus, we may reject the no-response

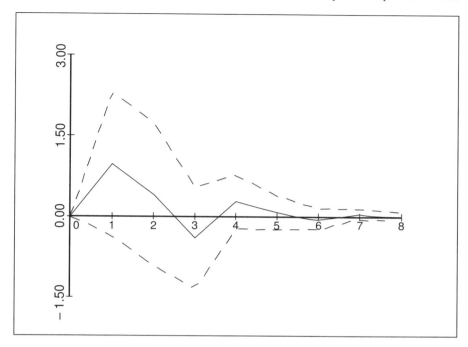

Fig. 3.5. Estimated responses of investment to a forecast error impulse in consumption with estimated asymptotic two-standard error bounds.

hypothesis at an overall $4 \times 2.5\% = 10\%$ level which is clearly a more common size for a test in applied work. Still, in this exercise, the data do not reveal strong evidence for the intuitively appealing hypothesis that consumption responds to income impulses. In later chapters, we will see how the coefficients can potentially be estimated with more precision.

In Figure 3.5, the responses of investment to consumption impulses are depicted. None of them is significant under a two-standard error criterion. This result is in line with the Granger-causality analysis in Section 3.6. In that section, we did not find evidence for Granger-causality from income/consumption to investment. Assuming that the test result describes the actual situation, the $\phi_{13,i}$ must be zero for $i = 1, 2, \ldots$ (see also Chapter 2, Section 2.3.1).

The covariance matrix of

$$\widehat{\Psi}_1 = I_3 + \widehat{\Phi}_1 = \begin{bmatrix} .680 & .146 & .961 \\ .044 & .847 & .289 \\ -.002 & .225 & .736 \end{bmatrix}$$

is, of course, the same as that of $\widehat{\Phi}_1$ and an estimate of the covariance matrix of the elements of

$$\widehat{\Psi}_2 = I_3 + \widehat{\Phi}_1 + \widehat{\Phi}_2 = \begin{bmatrix} .626 & .408 & 1.377 \\ .073 & .961 & .200 \\ .043 & .486 & .846 \end{bmatrix}$$

is obtained as $(G_1 + \widehat{G}_2)\widehat{\Sigma}_{\widehat{\alpha}}(G_1 + \widehat{G}_2)'/T$. Some accumulated impulse responses together with estimated standard errors are also given in Table 3.3 and accumulated responses of consumption to income impulses and of investment to consumption impulses are shown in Figures 3.6 and 3.7, respectively. They reinforce the findings for the individual impulse responses in Figures 3.4 and 3.5.

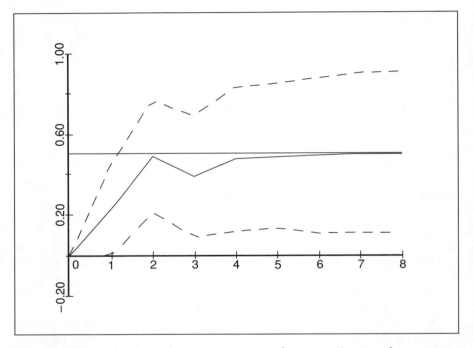

Fig. 3.6. Accumulated and long-run responses of consumption to a forecast error impulse in income with estimated asymptotic two-standard error bounds.

An estimate of the asymptotic covariance matrix of the estimated long-run responses $\widehat{\Psi}_\infty = (I_3 - \widehat{A}_1 - \widehat{A}_2)^{-1}$ is

$$\frac{1}{T}([\widehat{\Psi}'_\infty : \widehat{\Psi}'_\infty] \otimes \widehat{\Psi}_\infty)\widehat{\Sigma}_{\widehat{\alpha}} \left(\begin{bmatrix} \widehat{\Psi}_\infty \\ \widehat{\Psi}_\infty \end{bmatrix} \otimes \widehat{\Psi}'_\infty \right).$$

The matrix $\widehat{\Psi}_\infty$ together with the resulting standard errors is also given in Table 3.3. For instance, the total long-run effect $\widehat{\psi}_{13,\infty}$ of a consumption impulse

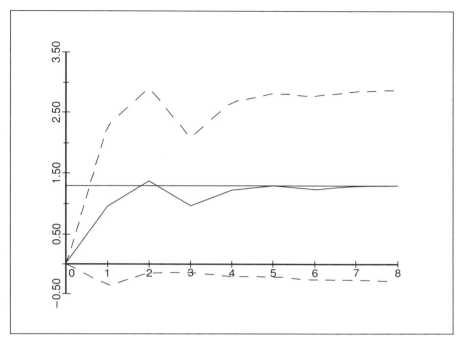

Fig. 3.7. Accumulated and long-run responses of investment to a forecast error impulse in consumption with estimated asymptotic two-standard error bounds.

on investment is 1.295 and its estimated asymptotic standard error is .798. Not surprisingly, $\widehat{\psi}_{13,\infty}$ is not significantly different from zero for any common level of significance (e.g., 10%). On the other hand, $\widehat{\psi}_{32,\infty}$, the long-run effect on consumption due to an impulse in income, is significant at an asymptotic 5% level.

For the interpretation of the $\widehat{\Phi}_i$'s, the critical remarks at the end of Chapter 2 must be kept in mind. As explained there, the $\widehat{\Phi}_i$ and $\widehat{\Psi}_n$ coefficients may not reflect the actual responses of the variables in the system. As an alternative, one may want to determine the responses to orthogonal residuals. In order to obtain the asymptotic covariance matrices of the $\widehat{\Theta}_i$ and $\widehat{\Xi}_i$, a decomposition of $\widehat{\Sigma}_u$ is needed. For our example,

$$\widehat{P} = \begin{bmatrix} 4.61 & 0 & 0 \\ .16 & 1.16 & 0 \\ .27 & .49 & .76 \end{bmatrix} \times 10^{-2}$$

is the lower triangular matrix with positive diagonal elements satisfying $\widehat{P}\widehat{P}' = \widehat{\Sigma}_u$ (Choleski decomposition). The asymptotic covariance matrix of $\text{vec}(\widehat{P}) = \text{vec}(\widehat{\Theta}_0)$ is a (9×9) matrix which is estimated as

$$\frac{1}{T}\widehat{C}_0\widehat{\Sigma}_{\widehat{\sigma}}\widehat{C}'_0 = \frac{2}{T}\widehat{H}\mathbf{D}^+_K(\widehat{\Sigma}_u \otimes \widehat{\Sigma}_u)\mathbf{D}^{+\prime}_K\widehat{H}',$$

where, as usual, $\mathbf{D}^+_K = (\mathbf{D}'_K\mathbf{D}_K)^{-1}\mathbf{D}'_K$ and

$$\widehat{H} = \mathbf{L}'_3\left\{\mathbf{L}_3\left[(I_3 \otimes \widehat{P})\mathbf{K}_{33} + (\widehat{P} \otimes I_3)\right]\mathbf{L}'_3\right\}^{-1}.$$

The resulting estimated asymptotic standard errors of the elements of $\widehat{P}$ are given in Table 3.4. Note that the variances corresponding to elements above the main diagonal of $\widehat{P}$ are all zero because these elements are zero by definition and are not estimated.

The asymptotic covariance matrix of the elements of

$$\widehat{\Theta}_1 = \begin{bmatrix} -1.196 & .644 & .730 \\ .256 & -.035 & .219 \\ -.047 & .131 & -.201 \end{bmatrix} \times 10^{-2}$$

is obtained as the sum of the two matrices

$$\widehat{C}_1\widehat{\Sigma}_{\widehat{\alpha}}\widehat{C}'_1/T = \left[(\widehat{P}' \otimes I_3)G_1\widehat{\Sigma}_{\widehat{\alpha}}G'_1(\widehat{P} \otimes I_3)\right]/T$$

and

$$\widehat{C}_1\widehat{\Sigma}_{\widehat{\sigma}}\widehat{C}'_1/T = (I_3 \otimes \widehat{\Phi}_1)\widehat{H}\widehat{\Sigma}_{\widehat{\sigma}}\widehat{H}'(I_3 \otimes \widehat{\Phi}'_1)/T.$$

The resulting standard errors for the elements of $\widehat{\Theta}_1$ are given in Table 3.4 along with some more $\widehat{\Theta}_i$ and $\widehat{\Xi}_n$ matrices.

Some responses and accumulated responses of consumption to income innovations with two-standard error bounds are depicted in Figures 3.8 and 3.9. The responses in Figures 3.4 and 3.8 are obviously a bit different. Note the (significant) immediate reaction of consumption in Figure 3.8. However, from period 1 onwards the response of consumption in both figures is qualitatively similar. The difference of scales is due to the different sizes of the shocks traced through the system. For instance, Figure 3.4 is based on a unit shock in income while Figure 3.8 is based on an innovation of size one standard deviation due to the transformation of the white noise residuals.

Again, a test of overall significance of the impulse responses in Figure 3.8 could be performed using Bonferroni's principle. Now we have to check the significance of the $\widehat{\theta}_{32,i}$'s for $i = 0, 1, \ldots, 4 = p(K - 1)$. We reject the null hypothesis of no response if at least one of the coefficients is significantly different from zero. In this case, we can reject at an asymptotic 5% level of significance because $\widehat{\theta}_{32,0}$ is significant at the 1% level (see Table 3.4). Thus, we may choose individual significance levels of 1% for each of the 5 coefficients and obtain 5% as an upper bound for the overall level. Of course, all these interpretations are based on the assumption that the actual asymptotic standard errors of the impulse responses are nonzero (see Section 3.7.1, Remark 1).

Table 3.4. Estimates of responses to orthogonal innovations for the investment/income/consumption system with estimated asymptotic standard errors in parentheses

i	$\widehat{\Theta}_i$	$\widehat{\Xi}_i$
0	$\begin{bmatrix} 4.61 & 0 & 0 \\ (.38) & & \\ .16 & 1.16 & 0 \\ (.14) & (.10) & \\ .27 & .49 & .76 \\ (.11) & (.10) & (.06) \end{bmatrix} \times 10^{-2}$	$\begin{bmatrix} 4.61 & 0 & 0 \\ (.38) & & \\ .16 & 1.16 & 0 \\ (.14) & (.10) & \\ .27 & .49 & .76 \\ (.11) & (.10) & (.06) \end{bmatrix} \times 10^{-2}$
1	$\begin{bmatrix} -1.20 & .64 & .73 \\ (.57) & (.56) & (.50) \\ .26 & -.04 & .22 \\ (.14) & (.14) & (.13) \\ -.05 & .13 & -.20 \\ (.12) & (.12) & (.10) \end{bmatrix} \times 10^{-2}$	$\begin{bmatrix} 3.46 & .64 & .73 \\ (.63) & (.56) & (.50) \\ .41 & 1.13 & .22 \\ (.20) & (.17) & (.13) \\ .22 & .62 & .56 \\ (.15) & (.14) & (.11) \end{bmatrix} \times 10^{-2}$
2	$\begin{bmatrix} -.10 & .51 & .32 \\ (.58) & (.57) & (.50) \\ .13 & .09 & -.07 \\ (.14) & (.14) & (.12) \\ .28 & .36 & .08 \\ (.12) & (.12) & (.10) \end{bmatrix} \times 10^{-2}$	$\begin{bmatrix} 3.32 & 1.15 & 1.05 \\ (.74) & (.69) & (.58) \\ .54 & 1.22 & .15 \\ (.24) & (.22) & (.17) \\ .50 & .98 & .64 \\ (.20) & (.18) & (.14) \end{bmatrix} \times 10^{-2}$
∞	0	$\begin{bmatrix} 3.97 & 1.61 & .98 \\ (.82) & (.92) & (.61) \\ .61 & 1.42 & .26 \\ (.31) & (.34) & (.22) \\ .58 & 1.06 & .73 \\ (.28) & (.32) & (.20) \end{bmatrix} \times 10^{-2}$

We have also performed forecast error variance decompositions and we have computed the standard errors on the basis of the results given in Proposition 3.6. For some forecast horizons the decompositions are given in Table 3.5. The standard errors may be regarded as rough indications of the sampling uncertainty. It must be kept in mind, however, that they may be quite misleading if the true forecast error variance components are zero, as explained in Remark 6 of Section 3.7.1. Obviously, this qualification limits their value in

126   3 Estimation of Vector Autoregressive Processes

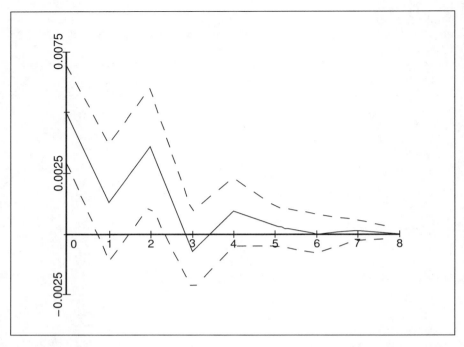

Fig. 3.8. Estimated responses of consumption to an orthogonalized impulse in income with estimated asymptotic two-standard error bounds.

the present example. Students are invited to reproduce the numbers in Table 3.5 and the previous tables of this section.

3.7.4 Investigating the Distributions of the Impulse Responses by Simulation Techniques

In the previous subsections, it was indicated repeatedly that in some cases the small sample validity of the asymptotic results is problematic. In that situation, one possibility is to use Monte Carlo or bootstrapping methods for investigating the sampling properties of the quantities of interest. Although these methods are quite expensive in terms of computer time, they were used in the past for evaluating the properties of impulse response functions (see, e.g., Runkle (1987) and Kilian (1998, 1999)). The general methodology is described in Appendix D.

In the present situation, there are different approaches to simulation. One possibility is to assume a specific distribution of the white noise process, e.g., $u_t \sim \mathcal{N}(0, \widehat{\Sigma}_u)$, and generate a large number of time series realizations based on the estimated VAR coefficients. From these time series, new sets of coefficients are then estimated and the corresponding impulse responses and/or

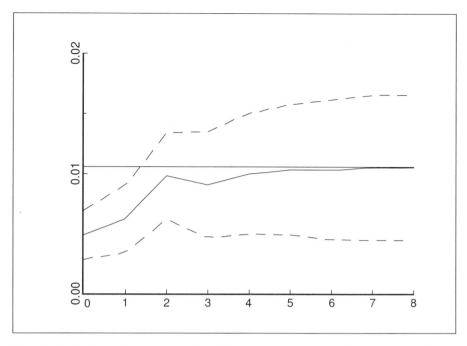

Fig. 3.9. Estimated accumulated and long-run responses of consumption to an orthogonalized impulse in income with estimated asymptotic two-standard error bounds.

forecast error variance components are computed. The empirical distributions obtained in this way may be used to investigate the actual distributions of the quantities of interest.

Alternatively, if an assumption regarding the white noise distribution cannot be made, bootstrap methods may be used and new sets of residuals may be drawn from the estimation residuals. A large number of y_t time series is generated on the basis of these sets of disturbances. The bootstrap multiple time series obtained in this way are then used to compute estimates of the quantities of interest and study their properties. Three different methods for computing bootstrap confidence intervals in the present context are described in Appendix D.3. We have used the standard and the Hall percentile methods to compute confidence intervals for the response of consumption to a forecast error impulse and an orthogonalized impulse in income for our example system. The results are shown in Figures 3.10 and 3.11, respectively.

Some interesting observations can be made. First, for the forecast error impulse responses, the two different methods for establishing confidence intervals produce quite similar results which are also at least qualitatively similar to the asymptotic confidence intervals in Figure 3.4. Second, the situation is a

Table 3.5. Forecast error variance decomposition of the investment/income/consumption system with estimated asymptotic standard errors in parentheses

forecast error in	forecast horizon h	proportions of forecast error variance, h periods ahead, accounted for by innovations in		
		investment $\widehat{\omega}_{j1,h}$	income $\widehat{\omega}_{j2,h}$	consumption $\widehat{\omega}_{j3,h}$
investment	1	1.00(.00)	.00(.00)	.00(.00)
($j=1$)	2	.96(.04)	.02(.03)	.02(.03)
	3	.95(.04)	.03(.03)	.03(.03)
	4	.94(.05)	.03(.03)	.03(.03)
	8	.94(.05)	.03(.03)	.03(.04)
income	1	.02(.03)	.98(.03)	.00(.00)
($j=2$)	2	.06(.05)	.91(.06)	.03(.04)
	3	.07(.06)	.90(.07)	.03(.04)
	4	.07(.06)	.89(.07)	.04(.04)
	8	.07(.06)	.89(.07)	.04(.04)
consumption	1	.08(.06)	.27(.09)	.65(.09)
($j=3$)	2	.08(.06)	.27(.08)	.65(.09)
	3	.13(.08)	.33(.09)	.54(.09)
	4	.13(.08)	.34(.09)	.54(.09)
	8	.13(.08)	.34(.09)	.53(.09)

bit different for the orthogonalized impulse responses in Figure 3.11. Here the two different bootstrap methods produce rather different confidence intervals. These intervals are quite asymmetric in the sense that the estimated impulse responses are not in the middle between the lower and upper bound of the intervals. Thereby they also look quite differently from the asymptotic intervals shown in Figure 3.8. The latter intervals are symmetric around the estimated impulse response coefficients by construction. Again, the qualitative interpretation does not change, however. In other words, the instantaneous and the second coefficient are significantly different from zero, as before. Moreover, the confidence intervals in Figure 3.11 are consistent with a rapidly declining effect of an impulse in income.

It must be emphasized, however, that the bootstrap generally does not solve the problem of a singular asymptotic distribution of the impulse responses and the resulting potentially invalid inference. If the asymptotic distribution is singular, the bootstrap may fail to produce meaningful confidence intervals, for example. Again it may be worth considering a univariate AR(1) process $y_t = \alpha y_{t-1} + u_t$ for illustrative purposes. The second forecast error impulse response coefficient is $\Phi_2 = \alpha^2$. The corresponding estimator $\widehat{\Phi}_2 = \widehat{\alpha}^2$ was found to have a singular asymptotic distribution if $\alpha = 0$ (see Remark 1 in Section 3.7.1). Suppose a bootstrap is used to produce N bootstrap estimates of α, $\widehat{\alpha}^*_{(n)}$, $n = 1, \ldots, N$. Clearly, the corresponding bootstrap estimates

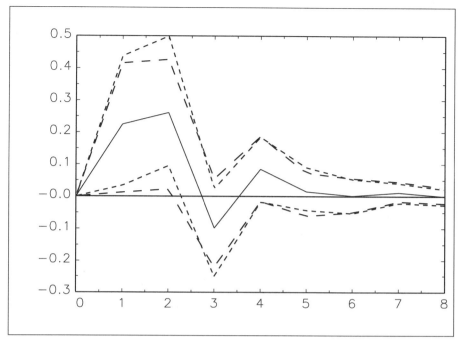

Fig. 3.10. Estimated responses (——) of consumption to a forecast error impulse in income with 95% bootstrap confidence bounds based on 2000 bootstrap replications (— — standard intervals, - - - Hall's percentile intervals).

$\widehat{\Phi}^*_{2(n)} = \widehat{\alpha}^{*2}_{(n)}$ will all be positive with probability one because they are squares. Thus, if the standard $(1-\gamma)100\%$ bootstrap confidence interval is constructed in the usual way by choosing $\widehat{\Phi}^*_{2(N\gamma/2)}$ and $\widehat{\Phi}^*_{2(N(1-\gamma)/2)}$ as lower and upper bound, respectively, the true value of zero will never be within the confidence interval. Hence, in this case the actual confidence level will be zero. Although the Hall confidence intervals may be a bit better in this case, they will also not provide the desired coverage level even in large samples. A more detailed discussion of this problem is given by Benkwitz et al. (2000), where also methods for correct asymptotic inference are considered. One possible solution is to eliminate all points where nonsingularities of the asymptotic distribution may occur by fitting subset models (see Chapter 5). Another possibility to circumvent the problem is to allow the VAR process to be of infinite order and increase the order with growing sample size. This possibility will be discussed in detail in Chapter 15.

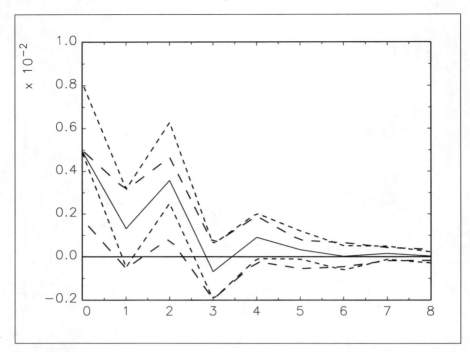

Fig. 3.11. Estimated responses (———) of consumption to an orthogonalized impulse in income with 95% bootstrap confidence bounds based on 2000 bootstrap replications (— — standard intervals, - - - Hall's percentile intervals).

3.8 Exercises

3.8.1 Algebraic Problems

The notation of Sections 3.2–3.5 is used in the following problems.

Problem 3.1
Show that $\widehat{\boldsymbol{\beta}} = ((ZZ')^{-1}Z \otimes I_K)\mathbf{y}$ minimizes

$$\bar{S}(\boldsymbol{\beta}) = \mathbf{u}'\mathbf{u} = [\mathbf{y} - (Z' \otimes I_K)\boldsymbol{\beta}]'[\mathbf{y} - (Z' \otimes I_K)\boldsymbol{\beta}].$$

Problem 3.2
Prove that

$$\sqrt{T}(\widehat{\mathbf{b}} - \mathbf{b}) \xrightarrow{d} \mathcal{N}(0, \Sigma_u \otimes \Gamma^{-1}),$$

if y_t is stable and

$$\frac{1}{\sqrt{T}} \text{vec}(ZU') = \frac{1}{\sqrt{T}}(I_K \otimes Z)\text{vec}(U') \xrightarrow{d} \mathcal{N}(0, \Sigma_u \otimes \Gamma).$$

Problem 3.3
Show (3.4.17). (Hint: Use the product rule for matrix differentiation and $\partial \operatorname{vec}(\Sigma_u^{-1})/\partial \operatorname{vec}(\Sigma_u)' = -\Sigma_u^{-1} \otimes \Sigma_u^{-1}$.)

Problem 3.4
Derive (3.4.18). (Hint: Use the last expression given in (3.4.6).)

Problem 3.5
Show (3.4.19).

Problem 3.6
Derive (3.4.20).

Problem 3.7
Prove that plim $\tilde{z}_T/\sqrt{T} = 0$, where

$$\tilde{z}_T = \sum_{i=1}^{p} \tilde{A}_i \sum_{j=0}^{i-1} (y_{-j} - y_{T-j}).$$

(Hint: Show that $E(\tilde{z}_T/\sqrt{T}) \to 0$ and $\operatorname{Var}(\tilde{z}_T/\sqrt{T}) \to 0$.)

Problem 3.8
Show that Equation (3.5.10) holds.
(Hint: Define

$$Z_t(h) := \begin{bmatrix} 1 \\ y_t(h) \\ \vdots \\ y_t(h-p+1) \end{bmatrix}$$

and show $Z_t(h) = \mathbf{B} Z_t(h-1)$ by induction.)

Problem 3.9
In the context of Section 3.5, suppose that y_t is a stable Gaussian VAR(p) process which is estimated by ML in mean-adjusted form. Show that the forecast MSE correction term has the form

$$\Omega(h) = E\left(\frac{\partial y_t(h)}{\partial \mu'} \Sigma_{\tilde{\mu}} \frac{\partial y_t(h)'}{\partial \mu}\right) + E\left(\frac{\partial y_t(h)}{\partial \alpha'} \Sigma_{\tilde{\alpha}} \frac{\partial y_t(h)'}{\partial \alpha}\right),$$

with

$$\frac{\partial y_t(h)}{\partial \mu'} = I_K - J\mathbf{A}^h \begin{bmatrix} I_K \\ \vdots \\ I_K \end{bmatrix}_{(Kp \times K)}$$

and

$$\frac{\partial y_t(h)}{\partial \boldsymbol{\alpha}'} = \sum_{i=0}^{h-1}(Y_t - \boldsymbol{\mu})'(\mathbf{A}')^{h-1-i} \otimes \Phi_i.$$

Here $\boldsymbol{\mu} := (\mu', \ldots, \mu')'$ is a $(Kp \times 1)$ vector, Y_t and $\mathbf{A}$ are as defined in (2.1.8), $J := [I_K : 0 : \cdots : 0]$ is a $(K \times Kp)$ matrix, and Φ_i is the i-th coefficient matrix of the prediction error MA representation (2.1.17).

Problem 3.10
Derive the ML estimator and its asymptotic distribution for the parameter of a stable AR(1) process, $y_t = \alpha y_{t-1} + u_t$, $u_t \sim$ i.i.d. $\mathcal{N}(0, \sigma_u^2)$.

3.8.2 Numerical Problems

The following problems require the use of a computer. They are based on the two quarterly, seasonally adjusted U.S. investment series given in File E2. Consider the variables

y_1 – first differences of fixed investment,
y_2 – first differences of change in business inventories,

in the following problems. Use the data from 1947 to 1968 only.

Problem 3.11
Plot the two time series y_{1t} and y_{2t} and comment on the stationarity and stability of the series.

Problem 3.12
Estimate the parameters of a VAR(1) model for $(y_{1t}, y_{2t})'$ using multivariate LS, that is, compute $\widehat{B}$ and $\widehat{\Sigma}_u$. Comment on the stability of the estimated process.

Problem 3.13
Use the mean-adjusted form of a VAR(1) model and estimate the coefficients. Assume that the data generation process is Gaussian and estimate the covariance matrix of the asymptotic distribution of the ML estimators.

Problem 3.14
Determine the Yule-Walker estimate of the VAR(1) coefficient matrix and compare it to the LS estimate.

Problem 3.15
Use the LS estimate and compute point forecasts $\widehat{y}_{86}(1), \widehat{y}_{86}(2)$ (that is, the forecast origin is the last quarter of 1968) and the corresponding MSE matrices $\widehat{\Sigma}_y(1), \widehat{\Sigma}_y(2), \widehat{\Sigma}_{\widehat{y}}(1)$, and $\widehat{\Sigma}_{\widehat{y}}(2)$. Use these estimates to set up approximate 95% interval forecasts assuming that the process y_t is Gaussian.

Problem 3.16
Test the hypothesis that y_2 does not Granger-cause y_1.

Problem 3.17
Estimate the coefficient matrices Φ_1 and Φ_2 from the LS estimates of the VAR(1) model for y_t and determine approximate standard errors of the estimates.

Problem 3.18
Determine the *upper* triangular matrix $\widehat{P}$ with positive diagonal for which $\widehat{P}\widehat{P}' = \widehat{\Sigma}_u$. Estimate the covariance matrix of the asymptotic distribution of $\widehat{P}$ under the assumption that y_t is Gaussian. Test the hypothesis that the upper right-hand corner element of the underlying matrix P is zero.

Problem 3.19
Use the results of the previous problems to compute $\widehat{\Theta}_0$, $\widehat{\Theta}_1$, and $\widehat{\Theta}_2$. Determine also estimates of the asymptotic standard errors of the elements of these three matrices.

4

VAR Order Selection and Checking the Model Adequacy

4.1 Introduction

In the previous chapter, we have assumed that we have given a K-dimensional multiple time series $y_1, \ldots, y_T$, with $y_t = (y_{1t}, \ldots, y_{Kt})'$, which is known to be generated by a VAR(p) process,

$$y_t = \nu + A_1 y_{t-1} + \cdots + A_p y_{t-p} + u_t, \qquad (4.1.1)$$

and we have discussed estimation of the parameters $\nu, A_1, \ldots, A_p$, and $\Sigma_u = E(u_t u_t')$. In deriving the properties of the estimators, a number of assumptions were made. In practice, it will rarely be known with certainty whether the conditions hold that are required to derive the consistency and asymptotic normality of the estimators. Therefore statistical tools should be used in order to check the validity of the assumptions made. In this chapter, some such tools will be discussed.

In the next two sections, it will be discussed what to do if the VAR order p is unknown. In practice, the order will usually be unknown. In Chapter 3, we have assumed that a VAR(p) process such as (4.1.1) represents the data generation process. We have not assumed that all the A_i are nonzero. In particular A_p may be zero. In other words, p is just assumed to be an upper bound for the VAR order. On the other hand, from (3.5.13) we know that the approximate MSE matrix of the 1-step ahead predictor will increase with the order p. Thus, choosing p unnecessarily large will reduce the forecast precision of the corresponding estimated VAR(p) model. Also, the estimation precision of the impulse responses depends on the precision of the parameter estimates. Therefore it is useful to have procedures or criteria for choosing an adequate VAR order.

In Sections 4.4–4.6, possibilities are discussed for checking some of the assumptions of the previous chapters. The asymptotic distribution of the residual autocorrelations and so-called portmanteau tests are considered in Section 4.4. The latter tests are popular tools for checking the whiteness of the residuals. More precisely, they are used to test for nonzero residual autocorrelation.

In Section 4.5, tests for nonnormality are considered. The normality assumption was used in Chapter 3 in setting up forecast intervals.

One assumption underlying much of the previous analysis is the stationarity of the systems considered. Nonstationarities may have various forms. Not only trends indicate deviations from stationarity but also changes in the variability or variance of the system. Moreover, exogenous shocks may affect various characteristics of the system. Tests for structural change are presented in Section 4.6.

4.2 A Sequence of Tests for Determining the VAR Order

Obviously, there is not just one correct VAR order for the process (4.1.1). In fact, if (4.1.1) is a correct summary of the characteristics of the process y_t, then the same is true for

$$y_t = \nu + A_1 y_{t-1} + \cdots + A_p y_{t-p} + A_{p+1} y_{t-p-1} + u_t$$

with $A_{p+1} = 0$. In other words, if y_t is a VAR(p) process, in this sense it is also a VAR($p+1$) process. In the assumptions of the previous chapter, the possibility of zero coefficient matrices is not excluded. In this chapter, it is practical to have a unique number that is called the order of the process. Therefore, in the following we will call y_t a VAR(p) process if $A_p \neq 0$ and $A_i = 0$ for $i > p$ so that p is the smallest possible order. This unique number will be called the VAR order.

4.2.1 The Impact of the Fitted VAR Order on the Forecast MSE

If y_t is a VAR(p) process, it is useful to fit a VAR(p) model to the available multiple time series and not, for instance, a VAR($p+i$) because, under a mean square error measure, forecasts from the latter process will be inferior to those based on an estimated VAR(p) model. This result follows from the approximate forecast MSE matrix $\Sigma_{\hat{y}}(h)$ derived in Section 3.5.2 of Chapter 3. For instance, for $h = 1$,

$$\Sigma_{\hat{y}}(1) = \frac{T + Kp + 1}{T} \Sigma_u,$$

if a VAR(p) model is fitted to data generated by a K-dimensional VAR process with order not greater than p. Obviously, $\Sigma_{\hat{y}}(1)$ is an increasing function of the order of the model fitted to the data.

Because the approximate MSE matrix is derived from asymptotic theory, it is of interest to know whether the result remains true in small samples. To get some feeling for the answer to this question, we have generated 1000 Gaussian bivariate time series with a process similar to (3.2.25),

$$y_t = \begin{bmatrix} .02 \\ .03 \end{bmatrix} + \begin{bmatrix} .5 & .1 \\ .4 & .5 \end{bmatrix} y_{t-1} + \begin{bmatrix} 0 & 0 \\ .25 & 0 \end{bmatrix} y_{t-2} + u_t,$$

$$\Sigma_u = \begin{bmatrix} .09 & 0 \\ 0 & .04 \end{bmatrix}. \tag{4.2.1}$$

We have fitted VAR(2), VAR(4), and VAR(6) models to the generated series and we have computed forecasts with the estimated models. Then we have compared these forecasts to generated post-sample values. The resulting average squared forecasting errors for different forecast horizons h and sample sizes T are shown in Table 4.1. Obviously, the forecasts based on estimated VAR(2) models are clearly superior to the VAR(4) and VAR(6) forecasts for sample sizes $T = 30, 50,$ and 100. While the comparative advantage of the VAR(2) models is quite dramatic for $T = 30$, it diminishes with increasing sample size. This, of course, was to be expected given that the approximate forecast MSE matrix of an estimated process approaches that of the known process as the sample size increases (see Section 3.5).

Table 4.1. Average squared forecast errors for the estimated bivariate VAR(2) process (4.2.1) based on 1000 realizations

sample size T	forecast horizon h	VAR(2)		VAR(4)		VAR(6)	
		y_1	y_2	y_1	y_2	y_1	y_2
	1	.111	.052	.132	.062	.165	.075
30	2	.155	.084	.182	.098	.223	.119
	3	.146	.141	.183	.166	.225	.202
	1	.108	.043	.119	.048	.129	.054
50	2	.132	.075	.144	.083	.161	.093
	3	.142	.120	.150	.130	.168	.145
	1	.091	.044	.095	.046	.098	.049
100	2	.120	.064	.125	.067	.130	.069
	3	.130	.108	.135	.113	.140	.113

Of course, the process considered in this example is a very special one. To see whether a similar result is obtained for other processes as well, we have also generated 1000 three-dimensional time series with the VAR(1) process (2.1.14),

$$y_t = \begin{bmatrix} .01 \\ .02 \\ 0 \end{bmatrix} + \begin{bmatrix} .5 & 0 & 0 \\ .1 & .1 & .3 \\ 0 & .2 & .3 \end{bmatrix} y_{t-1} + u_t \quad \text{with} \quad \Sigma_u = \begin{bmatrix} 2.25 & 0 & 0 \\ 0 & 1.0 & .5 \\ 0 & .5 & .74 \end{bmatrix}.$$
$$\tag{4.2.2}$$

138   4 VAR Order Selection and Checking the Model Adequacy

We have fitted VAR(1), VAR(3), and VAR(6) models to these data and we have computed forecasts and forecast errors. Some average squared forecast errors are presented in Table 4.2. Again forecasts from lower order models are clearly superior to higher order models. In fact, in a large scale simulation study involving many more processes, similar results were found (see Lütkepohl (1985)). Thus, it is useful to avoid fitting VAR models with unnecessarily large orders.

Table 4.2. Average squared forecast errors for the estimated three-dimensional VAR(1) process (4.2.2) based on 1000 realizations

sample size T	forecast horizon h	VAR(1)			VAR(3)			VAR(6)		
		y_1	y_2	y_3	y_1	y_2	y_3	y_1	y_2	y_3
30	1	.87	1.14	2.68	1.14	1.52	3.62	2.25	2.78	6.82
	2	1.09	1.21	3.21	1.44	1.67	4.12	2.54	2.98	7.85
	3	1.06	1.31	3.32	1.35	1.58	4.23	2.59	2.79	8.63
50	1	.81	1.03	2.68	.96	1.22	2.97	1.18	1.53	3.88
	2	1.01	1.23	2.92	1.20	1.40	3.47	1.48	1.68	4.38
	3	1.01	1.29	3.11	1.12	1.44	3.48	1.42	1.77	4.66
100	1	.73	.93	2.35	.77	1.00	2.62	.86	1.12	2.91
	2	.94	1.15	2.86	1.00	1.24	3.12	1.12	1.38	3.53
	3	.90	1.15	3.02	.93	1.20	3.23	1.03	1.35	3.51

The question is then what to do if the true order is unknown and an upper bound, say M, for the order is known only. One possibility to check whether certain coefficient matrices may be zero is to set up a significance test. For our particular problem of determining the correct VAR order, we may set up a sequence of tests. First $H_0 : A_M = 0$ is tested. If this null hypothesis cannot be rejected, we test $H_0 : A_{M-1} = 0$ and so on until we can reject a null hypothesis. Before we discuss this procedure in more detail, we will now introduce a possible test statistic.

4.2.2 The Likelihood Ratio Test Statistic

Because we just need to test zero restrictions on the coefficients of a VAR model, we may use the Wald statistic discussed in Section 3.6 in the context of causality tests. To shed some more light on this type of statistic, it may be instructive to consider the likelihood ratio testing principle. It is based on comparing the maxima of the log-likelihood function over the unrestricted and restricted parameter space. Specifically, the likelihood ratio statistic is

$$\lambda_{LR} = 2[\ln l(\widetilde{\boldsymbol{\delta}}) - \ln l(\widetilde{\boldsymbol{\delta}}_r)], \qquad (4.2.3)$$

4.2 A Sequence of Tests for Determining the VAR Order

where $\tilde{\boldsymbol{\delta}}$ is the unrestricted ML estimator for a parameter vector $\boldsymbol{\delta}$ obtained by maximizing the likelihood function over the full feasible parameter space and $\tilde{\boldsymbol{\delta}}_r$ is the restricted ML estimator which is obtained by maximizing the likelihood function over that part of the parameter space where the restrictions of interest are satisfied (see Appendix C.7). For the case of interest here, where we have linear constraints for the coefficients of a VAR process, λ_{LR} can be shown to have an asymptotic χ^2-distribution with as many degrees of freedom as there are distinct linear restrictions.

To obtain this result, let us assume for the moment that y_t is a stable Gaussian (normally distributed) VAR(p) process as in (4.1.1). Using the notation of Section 3.2.1 (as opposed to the mean-adjusted form considered in Section 3.4), the log-likelihood function is

$$\ln l(\boldsymbol{\beta}, \Sigma_u) = -\frac{KT}{2}\ln 2\pi - \frac{T}{2}\ln|\Sigma_u|$$
$$-\frac{1}{2}[\mathbf{y} - (Z' \otimes I_K)\boldsymbol{\beta}]'(I_T \otimes \Sigma_u^{-1})[\mathbf{y} - (Z' \otimes I_K)\boldsymbol{\beta}] \quad (4.2.4)$$

(see (3.4.5)). The first order partial derivatives with respect to $\boldsymbol{\beta}$ are

$$\frac{\partial \ln l}{\partial \boldsymbol{\beta}} = (Z \otimes \Sigma_u^{-1})\mathbf{y} - (ZZ' \otimes \Sigma_u^{-1})\boldsymbol{\beta}. \quad (4.2.5)$$

Equating to zero and solving for $\boldsymbol{\beta}$ gives the unrestricted ML/LS estimator

$$\tilde{\boldsymbol{\beta}} = ((ZZ')^{-1}Z \otimes I_K)\mathbf{y}. \quad (4.2.6)$$

Suppose the restrictions for $\boldsymbol{\beta}$ are given in the form

$$C\boldsymbol{\beta} = c, \quad (4.2.7)$$

where C is a known $(N \times (K^2p+K))$ matrix of rank N and c is a known $(N \times 1)$ vector. Then the restricted ML estimator may be found by a Lagrangian approach (see Appendix A.14). The Lagrange function is

$$\mathcal{L}(\boldsymbol{\beta}, \gamma) = \ln l(\boldsymbol{\beta}) + \gamma'(C\boldsymbol{\beta} - c), \quad (4.2.8)$$

where γ is an $(N \times 1)$ vector of Lagrange multipliers. Of course, $\mathcal{L}$ also depends on Σ_u. Because these parameters are not involved in the restrictions (4.2.7), we have skipped them there. The restricted maximum of the log-likelihood function with respect to $\boldsymbol{\beta}$ is known to be attained at a point where the first order partial derivatives of $\mathcal{L}$ are zero.

$$\frac{\partial \mathcal{L}}{\partial \boldsymbol{\beta}} = (Z \otimes \Sigma_u^{-1})\mathbf{y} - (ZZ' \otimes \Sigma_u^{-1})\boldsymbol{\beta} + C'\gamma, \quad (4.2.9a)$$

$$\frac{\partial \mathcal{L}}{\partial \gamma} = C\boldsymbol{\beta} - c. \quad (4.2.9b)$$

Equating to zero and solving gives

$$\widetilde{\boldsymbol{\beta}}_r = \widetilde{\boldsymbol{\beta}} + \left[(ZZ')^{-1} \otimes \Sigma_u\right] C' \left[C((ZZ')^{-1} \otimes \Sigma_u)C'\right]^{-1} (c - C\widetilde{\boldsymbol{\beta}}) \quad (4.2.10)$$

(see Problem 4.1).

Because for any given coefficient matrix B^0 the maximum of $\ln l$ with respect to Σ_u is obtained for

$$\Sigma_u^0 = \frac{1}{T}(Y - B^0 Z)(Y - B^0 Z)'$$

(see Section 3.4.2, (3.4.8) and (3.4.11)), the maximum for the unrestricted case is attained for

$$\widetilde{\Sigma}_u = \frac{1}{T}(Y - \widetilde{B}Z)(Y - \widetilde{B}Z)' \quad (4.2.11)$$

and for the restricted case we get

$$\widetilde{\Sigma}_u^r = \frac{1}{T}(Y - \widetilde{B}_r Z)(Y - \widetilde{B}_r Z)'. \quad (4.2.12)$$

Here $\widetilde{B}$ and $\widetilde{B}_r$ are the coefficient matrices corresponding to $\widetilde{\boldsymbol{\beta}}$ and $\widetilde{\boldsymbol{\beta}}_r$, respectively, that is, $\widetilde{\boldsymbol{\beta}} = \text{vec}(\widetilde{B})$ and $\widetilde{\boldsymbol{\beta}}_r = \text{vec}(\widetilde{B}_r)$. Thus, for this particular situation, the likelihood ratio statistic becomes

$$\lambda_{LR} = 2[\ln l(\widetilde{\boldsymbol{\beta}}, \widetilde{\Sigma}_u) - \ln l(\widetilde{\boldsymbol{\beta}}_r, \widetilde{\Sigma}_u^r)].$$

This statistic can be shown to have an asymptotic $\chi^2(N)$-distribution. In fact, this result also holds if y_t is not Gaussian, but has a distribution from a larger family. If y_t is not Gaussian, the estimators obtained by maximizing the Gaussian likelihood function in (4.2.4) are called *quasi ML estimators*. We will now state the previous results formally and then present a proof.

Proposition 4.1 (*Asymptotic Distribution of the LR Statistic*)
Let y_t be a stationary, stable VAR(p) process as in (4.1.1) with standard white noise u_t (see Definition 3.1). Suppose the true parameter vector $\boldsymbol{\beta}$ satisfies linear constraints $C\boldsymbol{\beta} = c$, where C is an $(N \times (K^2 p + K))$ matrix of rank N and c is an $(N \times 1)$ vector. Moreover, let $\ln l$ denote the Gaussian log-likelihood function and let $\widetilde{\boldsymbol{\beta}}$ and $\widetilde{\boldsymbol{\beta}}_r$ be the (quasi) ML and restricted (quasi) ML estimators, respectively, with corresponding estimators $\widetilde{\Sigma}_u$ and $\widetilde{\Sigma}_u^r$ of the white noise covariance matrix Σ_u given in (4.2.11) and (4.2.12). Then

$$\begin{aligned}
\lambda_{LR} &= 2\left[\ln l(\widetilde{\boldsymbol{\beta}}, \widetilde{\Sigma}_u) - \ln l(\widetilde{\boldsymbol{\beta}}_r, \widetilde{\Sigma}_u^r)\right] \\
&= T(\ln|\widetilde{\Sigma}_u^r| - \ln|\widetilde{\Sigma}_u|) & (4.2.13a) \\
&= (\widetilde{\boldsymbol{\beta}}_r - \widetilde{\boldsymbol{\beta}})'(ZZ' \otimes \widetilde{\Sigma}_u^{-1})(\widetilde{\boldsymbol{\beta}}_r - \widetilde{\boldsymbol{\beta}}) & (4.2.13b) \\
&= (\widetilde{\boldsymbol{\beta}}_r - \widetilde{\boldsymbol{\beta}})'(ZZ' \otimes (\widetilde{\Sigma}_u^r)^{-1})(\widetilde{\boldsymbol{\beta}}_r - \widetilde{\boldsymbol{\beta}}) + o_p(1) & (4.2.13c) \\
&= (C\widetilde{\boldsymbol{\beta}} - c)' \left[C((ZZ')^{-1} \otimes \widetilde{\Sigma}_u)C'\right]^{-1} (C\widetilde{\boldsymbol{\beta}} - c) + o_p(1) & (4.2.13d) \\
&= (C\widetilde{\boldsymbol{\beta}} - c)' \left[C((ZZ')^{-1} \otimes \widetilde{\Sigma}_u^r)C'\right]^{-1} (C\widetilde{\boldsymbol{\beta}} - c) + o_p(1) & (4.2.13e)
\end{aligned}$$

and

$$\lambda_{LR} \xrightarrow{d} \chi^2(N).$$

Here T is the sample size (time series length) and $Z := (Z_0, \ldots, Z_{T-1})$ with $Z_t' := (1, y_t', \ldots, y_{t-p+1}')$. ∎

In this proposition, the quantity $o_p(1)$ denotes a sequence which converges to zero in probability when the sample size $T \to \infty$ (see Appendix C.2). Note that y_t is not assumed to be Gaussian (normally distributed) in the proposition. It is just assumed that u_t is independent white noise with bounded fourth moments. Thus, $\ln l$ may not really be the log-likelihood function of $\mathbf{y} := \text{vec}(y_1, \ldots, y_T)$. It will only be the actual log-likelihood if $\mathbf{y}$ happens to be multivariate normal. In that case, $\widetilde{\boldsymbol{\beta}}$ and $\widetilde{\boldsymbol{\beta}}_r$ are actual ML and restricted ML estimators. Otherwise they are quasi ML estimators.

The second form of the LR statistic in (4.2.13a) is sometimes convenient for computing the actual test value. It is also useful for comparing the likelihood ratio tests to other procedures for VAR order selection, as we will see in Section 4.3. The expression in (4.2.13b) shows the similarity of the LR statistic to the LM statistic given in (4.2.13c). Using (4.2.5) and

$$\frac{\partial^2 \ln l}{\partial \beta \partial \beta'} = -(ZZ' \otimes \Sigma_u^{-1})$$

gives

$$\lambda_{LM} = \frac{\partial \ln l(\widetilde{\boldsymbol{\beta}}_r)}{\partial \beta'} \frac{\partial^2 \ln l(\widetilde{\boldsymbol{\beta}}_r)}{\partial \beta \partial \beta'} \frac{\partial \ln l(\widetilde{\boldsymbol{\beta}}_r)}{\partial \beta} = (\widetilde{\boldsymbol{\beta}}_r - \widetilde{\boldsymbol{\beta}})'(ZZ' \otimes (\widetilde{\Sigma}_u^r)^{-1})(\widetilde{\boldsymbol{\beta}}_r - \widetilde{\boldsymbol{\beta}})$$

(see Appendix C.7 and Problem 4.5). Notice that in the present case we may ignore the part of the parameter vector which corresponds to Σ_u because its ML estimator is asymptotically independent of the other parameters and the asymptotic covariance matrix is block-diagonal. Therefore, at least asymptotically, the terms related to scores of the covariance parameters vanish from the LM statistic.

Comparing (4.2.13d) to (3.6.5) shows that, for the special case considered here, the LR statistic is also similar to the Wald statistic. In fact, the important difference between the Wald and LR statistics is that the former involves only estimators of the unrestricted model while both unrestricted and restricted estimators enter into λ_{LR} (see also (4.2.13a)). The final form of the statistic given in (4.2.13e) provides another useful expression which is close to both the LR and the LM statistic. It shows that we may use the covariance matrix estimator from the restricted model instead of the unrestricted one.

As in the case of the Wald test, one may consider using the statistic λ_{LR}/N in conjunction with the $F(N, T - Kp - 1)$-distribution in small samples. Another adjustment was suggested by Hannan (1970, p. 341).

4 VAR Order Selection and Checking the Model Adequacy

Proof of Proposition 4.1:
We first show the equivalence of the various forms of the LR statistic given in the proposition. The equality in (4.2.13a) follows by noting that

$$[\mathbf{y} - (Z' \otimes I_K)\boldsymbol{\beta}]' (I_T \otimes \Sigma_u^{-1}) [\mathbf{y} - (Z' \otimes I_K)\boldsymbol{\beta}]$$
$$= \text{tr}[(Y - BZ)'\Sigma_u^{-1}(Y - BZ)]$$
$$= \text{tr}[\Sigma_u^{-1}(Y - BZ)(Y - BZ)'].$$

Replacing the matrices B and Σ_u by $\widetilde{B}$ and $\widetilde{\Sigma}_u$, respectively, gives

$$\ln l(\widetilde{\boldsymbol{\beta}}, \widetilde{\Sigma}_u) = \text{constant} - \frac{T}{2} \ln |\widetilde{\Sigma}_u|.$$

Similarly,

$$\ln l(\widetilde{\boldsymbol{\beta}}_r, \widetilde{\Sigma}_u^r) = \text{constant} - \frac{T}{2} \ln |\widetilde{\Sigma}_u^r|,$$

which gives the desired result.

In order to prove (4.2.13b), we observe that $\ln l$ is a quadratic function in $\boldsymbol{\beta}$. Thus, by Taylor's theorem (Appendix A.13, Proposition A.3), for an arbitrary fixed vector $\boldsymbol{\beta}^0$,

$$\ln l(\boldsymbol{\beta}) = \ln l(\boldsymbol{\beta}^0) + \frac{\partial \ln l(\boldsymbol{\beta}^0)}{\partial \boldsymbol{\beta}'}(\boldsymbol{\beta} - \boldsymbol{\beta}^0)$$
$$+ \frac{1}{2}(\boldsymbol{\beta} - \boldsymbol{\beta}^0)' \frac{\partial^2 \ln l(\boldsymbol{\beta}^0)}{\partial \boldsymbol{\beta} \partial \boldsymbol{\beta}'}(\boldsymbol{\beta} - \boldsymbol{\beta}^0).$$

Choosing $\widetilde{\boldsymbol{\beta}}$ for $\boldsymbol{\beta}^0$ and $\widetilde{\boldsymbol{\beta}}_r$ for $\boldsymbol{\beta}$, $\partial \ln l(\widetilde{\boldsymbol{\beta}})/\partial \boldsymbol{\beta}' = 0$ so that

$$\lambda_{LR} = 2\left[\ln l(\widetilde{\boldsymbol{\beta}}) - \ln l(\widetilde{\boldsymbol{\beta}}_r)\right] = -(\widetilde{\boldsymbol{\beta}}_r - \widetilde{\boldsymbol{\beta}})' \frac{\partial^2 \ln l(\widetilde{\boldsymbol{\beta}})}{\partial \boldsymbol{\beta} \partial \boldsymbol{\beta}'}(\widetilde{\boldsymbol{\beta}}_r - \widetilde{\boldsymbol{\beta}}). \quad (4.2.14)$$

As in Section 3.4, we can derive

$$\frac{\partial^2 \ln l}{\partial \boldsymbol{\beta} \partial \boldsymbol{\beta}'} = -(ZZ' \otimes \Sigma_u^{-1}).$$

Hence, (4.2.13b) follows and (4.2.13c) is an immediate consequence of the fact that the restricted and unrestricted covariance matrix estimators are consistent by Proposition 3.2. Thus, $\text{plim}(\widetilde{\Sigma}_u - \widetilde{\Sigma}_u^r) = 0$ which can be used to show (4.2.13c).

Using (4.2.10) and (4.2.14) gives

$$\lambda_{LR} = (C\widetilde{\boldsymbol{\beta}} - c)' \left[C((ZZ')^{-1} \otimes \Sigma_u)C'\right]^{-1}$$
$$\times C((ZZ')^{-1} \otimes \Sigma_u)(ZZ' \otimes \widetilde{\Sigma}_u^{-1})((ZZ')^{-1} \otimes \Sigma_u)C'$$
$$\times \left[C((ZZ')^{-1} \otimes \Sigma_u)C'\right]^{-1} (C\widetilde{\boldsymbol{\beta}} - c).$$

4.2 A Sequence of Tests for Determining the VAR Order

The result (4.2.13d) follows by replacing Σ_u with $\widetilde{\Sigma}_u$ and noting that this is a consistent estimator of Σ_u. Again by consistency of $\widetilde{\Sigma}_u^r$, using this estimator instead of $\widetilde{\Sigma}_u$ changes the statistic only by a term which vanishes in probability as the sample size increases. Hence, we have (4.2.13e).

The asymptotic $\chi^2(N)$-distribution of λ_{LR} now follows from Proposition C.15(5) of Appendix C because $[C((ZZ'/T)^{-1} \otimes \widetilde{\Sigma}_u^r)C']^{-1}$ is a consistent estimator of $[C(\Gamma^{-1} \otimes \Sigma_u)C']^{-1}$. ∎

In the next subsection a sequential testing scheme based on LR tests is discussed.

4.2.3 A Testing Scheme for VAR Order Determination

Assuming that M is known to be an upper bound for the VAR order, the following sequence of null and alternative hypotheses may be tested using LR tests:

$$
\begin{aligned}
H_0^1 &: A_M = 0 & \text{versus} \quad H_1^1 &: A_M \neq 0 \\
H_0^2 &: A_{M-1} = 0 & \text{versus} \quad H_1^2 &: A_{M-1} \neq 0 \mid A_M = 0 \\
&\vdots \\
H_0^i &: A_{M-i+1} = 0 & \text{versus} \quad H_1^i &: A_{M-i+1} \neq 0 \\
& & & \quad \mid A_M = \cdots = A_{M-i+2} = 0 \\
&\vdots \\
H_0^M &: A_1 = 0 & \text{versus} \quad H_1^M &: A_1 \neq 0 \mid A_M = \cdots = A_2 = 0.
\end{aligned}
\tag{4.2.15}
$$

In this scheme, each null hypothesis is tested conditionally on the previous ones being true. The procedure terminates and the VAR order is chosen accordingly, if one of the null hypotheses is rejected. That is, if H_0^i is rejected, $\widehat{p} = M - i + 1$ will be chosen as estimate of the autoregressive order.

The likelihood ratio statistic for testing the i-th null hypothesis is

$$\lambda_{LR}(i) = T[\ln|\widetilde{\Sigma}_u(M-i)| - \ln|\widetilde{\Sigma}_u(M-i+1)|], \tag{4.2.16}$$

where $\widetilde{\Sigma}_u(m)$ denotes the ML estimator of Σ_u when a VAR(m) model is fitted to a time series of length T. By Proposition 4.1, this statistic has an asymptotic $\chi^2(K^2)$-distribution if H_0^i and all previous null hypotheses are true. Note that K^2 parameters are set to zero in H_0^i. Hence, we have to test K^2 restrictions and we use $\lambda_{LR}(i)$ in conjunction with critical values from a $\chi^2(K^2)$-distribution. Alternatively, one may use $\lambda_{LR}(i)/K^2$ in conjunction with the $F(K^2, T - K(M - i + 1) - 1)$-distribution.

Of course, the order chosen for a particular process will depend on the significance levels used in the tests. In this procedure, it is important to realize that the significance levels of the individual tests must be distinguished from the Type I error of the whole procedure because rejection of H_0^i means that $H_0^{i+1}, \ldots, H_0^M$ are automatically rejected too. Thus, denoting by D_j the event that H_0^j is rejected in the j-th test when it is actually true, the probability of a Type I error for the i-th test in the sequence is

$$\epsilon_i = \Pr(D_1 \cup D_2 \cup \cdots \cup D_i).$$

Because D_j is the event that $\lambda_{LR}(j)$ falls in the rejection region, although H_0^j is true, $\gamma_j = \Pr(D_j)$ is just the significance level of the j-th individual test. It can be shown that for $m \neq j$ and $m, j \leq i$, $\lambda_{LR}(m)$ and $\lambda_{LR}(j)$ are asymptotically independent statistics if $H_0^1, \ldots, H_0^i$ are true (see Paulsen & Tjøstheim (1985, pp. 223–224)). Hence, D_m and D_j are independent events in large samples so that

$$\begin{aligned} \epsilon_i &= \Pr(D_1 \cup \cdots \cup D_{i-1}) + \Pr(D_i) - \Pr\{(D_1 \cup \cdots \cup D_{i-1}) \cap D_i\} \\ &= \epsilon_{i-1} + \gamma_i - \epsilon_{i-1}\gamma_i = \epsilon_{i-1} + \gamma_i(1 - \epsilon_{i-1}), \quad i = 2, 3, \ldots, M. \end{aligned} \quad (4.2.17)$$

Of course, $\epsilon_1 = \gamma_1$. Thus, it is easily seen by induction that

$$\epsilon_i = 1 - (1 - \gamma_1) \cdots (1 - \gamma_i), \quad i = 1, 2, \ldots, M. \tag{4.2.18}$$

If, for example, a 5% significance level is chosen for each individual test ($\gamma_i = .05$), then

$$\epsilon_1 = .05, \quad \epsilon_2 = 1 - .95 \times .95 = .0975, \quad \epsilon_3 = .142625.$$

Hence, the actual rejection probability will become quite substantial if the sequence of null hypotheses to be tested is long.

It is difficult to decide on appropriate significance levels in the testing scheme (4.2.15). Whatever significance levels the researcher decides to use, she or he should keep in mind the distinction between the overall and the individual significance levels. Also, it must be kept in mind that we know the asymptotic distributions of the LR statistics only. Thus, the significance levels chosen will be approximate probabilities of Type I errors only.

Finally, in the literature another testing scheme was also suggested and used. In that scheme the first set of hypotheses ($i = 1$) is as in (4.2.15) and for $i > 1$ the following hypotheses are tested:

$$H_0^i: A_M = \cdots = A_{M-i+1} = 0 \text{ versus } H_1^i: A_M \neq 0 \text{ or} \ldots \text{or } A_{M-i+1} \neq 0.$$

Here H_0^i is not tested conditionally on the previous null hypotheses being true but it is tested against the full VAR(M) model. Unfortunately, the LR statistics to be used in such a sequence will not be independent so that the overall significance level (probability of Type I error) is difficult to determine.

4.2.4 An Example

To illustrate the sequential testing procedure described in the foregoing, we use the investment/income/consumption example from Section 3.2.3. The variables y_1, y_2, and y_3 represent first differences of the logarithms of the investment, income, and consumption data. We assume an upper bound of $M = 4$ for the VAR order and therefore we set aside the first 4 values as presample values. The data up to 1978.4 are used for estimation so that the sample size is $T = 71$ in each test. The estimated error covariance matrices and their determinants are given in Table 4.3. The corresponding χ^2- and F-test values are summarized in Table 4.4. Because the denominator degrees of freedom for the F-statistics are quite large (ranging from 62 to 70), the F-tests are qualitatively similar to the χ^2-tests. Using individual significance levels of .05 in each test, $H_0^3 : A_2 = 0$ is the first null hypothesis that is rejected. Thus, the estimated order from both tests is $\widehat{p} = 2$. This supports the order chosen in the example in Chapter 3. Alternative procedures for choosing VAR orders are considered in the next section.

Table 4.3. ML estimates of the error covariance matrix of the investment/income/consumption system

VAR order m	$\widetilde{\Sigma}_u(m) \times 10^4$	$|\widetilde{\Sigma}_u(m)| \times 10^{11}$
0	$\begin{bmatrix} 21.83 & .410 & 1.228 \\ \cdot & 1.420 & .571 \\ \cdot & \cdot & 1.084 \end{bmatrix}$	2.473
1	$\begin{bmatrix} 20.14 & .493 & 1.173 \\ \cdot & 1.318 & .625 \\ \cdot & \cdot & 1.018 \end{bmatrix}$	1.782
2	$\begin{bmatrix} 19.18 & .617 & 1.126 \\ \cdot & 1.270 & .574 \\ \cdot & \cdot & .821 \end{bmatrix}$	1.255
3	$\begin{bmatrix} 19.08 & .599 & 1.126 \\ \cdot & 1.235 & .543 \\ \cdot & \cdot & .784 \end{bmatrix}$	1.174
4	$\begin{bmatrix} 16.96 & .573 & 1.252 \\ \cdot & 1.234 & .544 \\ \cdot & \cdot & .765 \end{bmatrix}$	.958

Table 4.4. LR statistics for the investment/income/consumption system

i	H_0^i	VAR order under H_0^i	λ_{LR} [a]	$\lambda_{LR}/9$ [b]
1	$A_4 = 0$	3	14.44	1.60
2	$A_3 = 0$	2	4.76	.53
3	$A_2 = 0$	1	24.90	2.77
4	$A_1 = 0$	0	23.25	2.58

[a] Critical value for individual 5% level test: $\chi^2(9)_{.95} = 16.92$.
[b] Critical value for individual 5% level test: $F(9, 71 - 3(5-i) - 1)_{.95} \approx 2$.

4.3 Criteria for VAR Order Selection

Although performing statistical tests is a common strategy for detecting nonzero parameters, the approach described in the previous section is not completely satisfactory if a model is desired for a specific purpose. For instance, a VAR model is often constructed for prediction of the variables involved. In such a case, we are not so much interested in finding the correct order of the underlying data generation process but we are interested in obtaining a good model for prediction. Hence, it seems useful to take the objective of the analysis into account when choosing the VAR order. In the next subsection, we will discuss criteria based on the forecasting objective.

If we really want to know the exact order of the data generation process (e.g., for analysis purposes) it is still questionable whether a testing procedure is the optimal strategy because that strategy has a positive probability of choosing an incorrect order even if the sample size (time series length) is large (see Section 4.3.3). In Section 4.3.2 we will present estimation procedures that choose the correct order with probability 1 at least in large samples.

4.3.1 Minimizing the Forecast MSE

If forecasting is the objective, it makes sense to choose the order such that a measure of forecast precision is minimized. The forecast MSE (mean squared error) is such a measure. Therefore Akaike (1969, 1971) suggested to base the VAR order choice on the approximate 1-step ahead forecast MSE given in Chapter 3, (3.5.13),

$$\Sigma_{\hat{y}}(1) = \frac{T + Km + 1}{T} \Sigma_u,$$

where m denotes the order of the VAR process fitted to the data, T is the sample size, and K is the dimension of the time series. To make this criterion operational, the white noise covariance matrix Σ_u has to be replaced by an estimate. Also, to obtain a unique solution we would like to have a scalar criterion rather than a matrix. Akaike suggested using the LS estimator with degrees of freedom adjustment,

$$\widehat{\Sigma}_u(m) = \frac{T}{T-Km-1}\widetilde{\Sigma}_u(m),$$

for Σ_u and taking the determinant of the resulting expression. Here $\widetilde{\Sigma}_u(m)$ is the ML estimator of Σ_u obtained by fitting a VAR(m) model, as in the previous section. The resulting criterion is called the final prediction error (FPE) criterion, that is,

$$\begin{aligned} \text{FPE}(m) &= \det\left[\frac{T+Km+1}{T}\frac{T}{T-Km-1}\widetilde{\Sigma}_u(m)\right] \\ &= \left[\frac{T+Km+1}{T-Km-1}\right]^K \det\widetilde{\Sigma}_u(m). \end{aligned} \quad (4.3.1)$$

We have written the criterion in terms of the ML estimator of the covariance matrix because in this form the FPE criterion has intuitive appeal. If the order m is increased, $\det\widetilde{\Sigma}_u(m)$ declines while the multiplicative term $(T+Km+1)/(T-Km-1)$ increases. The VAR order estimate is obtained as that value for which the two forces are balanced optimally. Note that the determinant of the LS estimate $\widehat{\Sigma}_u(m)$ may increase with increasing m. On the other hand, it is quite obvious that $|\widetilde{\Sigma}_u(m)|$ cannot become larger when m increases because the maximum of the log-likelihood function is proportional to $-\ln|\widetilde{\Sigma}_u(m)|$ apart from an additive constant and, for $m < n$, a VAR(m) model may be interpreted as a restricted VAR(n) model. Thus, $-\ln|\widetilde{\Sigma}_u(m)| \leq -\ln|\widetilde{\Sigma}_u(n)|$ or $|\widetilde{\Sigma}_u(m)| \geq |\widetilde{\Sigma}_u(n)|$.

Based on the FPE criterion, the estimate $\widehat{p}(\text{FPE})$ of p is chosen such that

$$\text{FPE}[\widehat{p}(\text{FPE})] = \min\{\text{FPE}(m)|m=0,1,\ldots,M\}.$$

That is, VAR models of orders $m = 0, 1, \ldots, M$ are estimated and the corresponding FPE(m) values are computed. The order minimizing the FPE values is then chosen as estimate for p.

Akaike (1973, 1974), based on a quite different reasoning, derived a very similar criterion usually abbreviated by AIC (*Akaike's Information Criterion*). For a VAR(m) process the criterion is defined as

$$\begin{aligned} \text{AIC}(m) &= \ln|\widetilde{\Sigma}_u(m)| + \frac{2}{T}(\text{number of freely estimated parameters}) \\ &= \ln|\widetilde{\Sigma}_u(m)| + \frac{2mK^2}{T}. \end{aligned} \quad (4.3.2)$$

The estimate $\widehat{p}(\text{AIC})$ for p is chosen so that this criterion is minimized. Here the constants in the VAR model may be ignored as freely estimated parameters because counting them would just add a constant to the criterion which does not change the minimizing order.

The similarity of the criteria AIC and FPE can be seen by noting that, for a constant N,

$$\frac{T+N}{T-N} = 1 + \frac{2N}{T} + O(T^{-2}).$$

148   4 VAR Order Selection and Checking the Model Adequacy

The quantity $O(T^{-2})$ denotes a sequence of order T^{-2}, that is, a sequence indexed by T that remains bounded if multiplied by T^2 (see Appendix C.2). Thus, the sequence goes to zero rapidly when $T \to \infty$. Hence,

$$\begin{aligned}\ln \text{FPE}(m) &= \ln|\widetilde{\Sigma}_u(m)| + K \ln\left[(T + Km + 1)/(T - Km - 1)\right] \\ &= \ln|\widetilde{\Sigma}_u(m)| + K \ln\left[1 + 2(Km+1)/T + O(T^{-2})\right] \\ &= \ln|\widetilde{\Sigma}_u(m)| + K\frac{2(Km+1)}{T} + O(T^{-2}) \\ &= \text{AIC}(m) + 2K/T + O(T^{-2}).\end{aligned} \quad (4.3.3)$$

The third equality sign follows from a Taylor series expansion of $\ln(1+x)$ around 1. The term $2K/T$ does not depend on the order m and, hence, AIC(m) + $2K/T$ and AIC(m) assume their minimum for the same value of m. Consequently, AIC and ln FPE differ essentially by a term of order $O(T^{-2})$ and, thus, the two criteria will be about equivalent for moderate and large T.

To illustrate these procedures for VAR order selection, we use again the investment/income/consumption example. The determinants of the residual covariance matrices are given in Table 4.3. Using these determinants, the FPE and AIC values presented in Table 4.5 are obtained. Both criteria reach their minimum for $\widehat{p} = 2$, that is, $\widehat{p}(\text{FPE}) = \widehat{p}(\text{AIC}) = 2$. The other quantities given in the table will be discussed shortly.

Table 4.5. Estimation of the VAR order of the investment/income/consumption system

VAR order m	FPE(m) $\times 10^{11}$	AIC(m)	HQ(m)	SC(m)
0	2.691	−24.42	−24.42*	−24.42*
1	2.500	−24.50	−24.38	−24.21
2	2.272*	−24.59*	−24.37	−24.02
3	2.748	−24.41	−24.07	−23.55
4	2.910	−24.36	−23.90	−23.21

* Minimum.

4.3.2 Consistent Order Selection

If interest centers on the correct VAR order, it makes sense to choose an estimator that has desirable sampling properties. One problem of interest in this context is to determine the statistical properties of order estimators such as $\widehat{p}(\text{FPE})$ and $\widehat{p}(\text{AIC})$. Consistency is a desirable *asymptotic* property of an estimator. As usual, an estimator $\widehat{p}$ of the VAR order p is called *consistent* if

$$\plim_{T \to \infty} \widehat{p} = p \quad \text{or, equivalently,} \quad \lim_{T \to \infty} \Pr\{\widehat{p} = p\} = 1. \quad (4.3.4)$$

The latter definition of the plim may seem to differ slightly from the one given in Appendix C. However, it is easily checked that the two definitions are equivalent for integer valued random variables. Of course, a reasonable estimator for p should be integer valued. The estimator $\hat{p}$ is called *strongly consistent* if

$$\Pr\{\lim \hat{p} = p\} = 1. \tag{4.3.5}$$

Accordingly, a VAR order selection criterion will be called consistent or strongly consistent if the resulting estimator has these properties. The following proposition due to Hannan & Quinn (1979), Quinn (1980), and Paulsen (1984) is useful for investigating the consistency of criteria for order determination.

Proposition 4.2 (*Consistency of VAR Order Estimators*)
Let y_t be a K-dimensional stationary, stable VAR(p) process with standard white noise (that is, u_t is independent white noise with bounded fourth moments). Suppose the maximum order $M \geq p$ and $\hat{p}$ is chosen so as to minimize a criterion

$$\mathrm{Cr}(m) = \ln|\widetilde{\Sigma}_u(m)| + mc_T/T \tag{4.3.6}$$

over $m = 0, 1, \ldots, M$. Here $\widetilde{\Sigma}_u(m)$ denotes the (quasi) ML estimator of Σ_u obtained for a VAR(m) model and c_T is a nondecreasing sequence of real numbers that depends on the sample size T. Then $\hat{p}$ is consistent if and only if

$$c_T \to \infty \quad \text{and} \quad c_T/T \to 0 \quad \text{as} \quad T \to \infty. \tag{4.3.7a}$$

The estimator $\hat{p}$ is a strongly consistent estimator if and only if (4.3.7a) holds and

$$c_T/2\ln\ln T > 1 \tag{4.3.7b}$$

eventually, as $T \to \infty$. ∎

We will not prove this proposition here but refer the reader to Quinn (1980) and Paulsen (1984) for proofs. The basic idea of the proof is to show that, for $p > m$, the quantity $\ln|\widetilde{\Sigma}_u(m)|/\ln|\widetilde{\Sigma}_u(p)|$ will be greater than one in large samples because $\ln|\widetilde{\Sigma}_u(m)|$ is essentially the minimum of minus the Gaussian log-likelihood function for a VAR(m) model. Consequently, because the penalty terms mc_T/T and pc_T/T go to zero as $T \to \infty$, $\mathrm{Cr}(m) > \mathrm{Cr}(p)$ for large T. Thus, the probability of choosing too small an order goes to zero as $T \to \infty$. Similarly, if $m > p$, $\ln|\widetilde{\Sigma}_u(m)|/\ln|\widetilde{\Sigma}_u(p)|$, approaches 1 in probability if $T \to \infty$ and the penalty term of the lower order model is smaller than that of a larger order process. Thus the lower order p will be chosen if the sample size is large. The following corollary is an easy consequence of the proposition.

Corollary 4.2.1

Under the conditions of Proposition 4.2, if $M > p$, $\widehat{p}(\text{FPE})$ and $\widehat{p}(\text{AIC})$ are not consistent. ∎

Proof: Because FPE and AIC are asymptotically equivalent (see (4.3.3)), it suffices to prove the corollary for $\widehat{p}(\text{AIC})$. Equating $\text{AIC}(m)$ and $\text{Cr}(m)$ given in (4.3.6) shows that

$$2mK^2/T = mc_T/T$$

or $c_T = 2K^2$. Obviously, this sequence does not satisfy (4.3.7a). ∎

We will see shortly that the limiting probability for underestimating the VAR order is zero for both $\widehat{p}(\text{AIC})$ and $\widehat{p}(\text{FPE})$ so that asymptotically they overestimate the true order with positive probability. However, Paulsen & Tjøstheim (1985, p. 224) argued that the limiting probability for overestimating the order declines with increasing dimension K and is negligible for $K \geq 5$. In other words, asymptotically AIC and FPE choose the correct order almost with probability one if the underlying multiple time series has large dimension K.

Before we continue the investigation of AIC and FPE, we shall introduce two *consistent* criteria that have been quite popular in recent applied work. The first one is due to Hannan & Quinn (1979) and Quinn (1980). It is often denoted by HQ (*Hannan-Quinn criterion*):

$$\begin{aligned} \text{HQ}(m) &= \ln|\widetilde{\Sigma}_u(m)| + \frac{2\ln\ln T}{T}(\#\text{ freely estimated parameters}) \\ &= \ln|\widetilde{\Sigma}_u(m)| + \frac{2\ln\ln T}{T}mK^2. \end{aligned} \quad (4.3.8)$$

The estimate $\widehat{p}(\text{HQ})$ is the order that minimizes $\text{HQ}(m)$ for $m = 0, 1, \ldots, M$. Comparing this criterion to (4.3.6) shows that $c_T = 2K^2 \ln\ln T$ and, thus, by (4.3.7a), HQ is consistent for univariate processes and by (4.3.7b) it is strongly consistent for $K > 1$, if the conditions of Proposition 4.2 are satisfied for y_t.

Using Bayesian arguments Schwarz (1978) derived the following criterion:

$$\begin{aligned} \text{SC}(m) &= \ln|\widetilde{\Sigma}_u(m)| + \frac{\ln T}{T}(\#\text{ freely estimated parameters}) \\ &= \ln|\widetilde{\Sigma}_u(m)| + \frac{\ln T}{T}mK^2. \end{aligned} \quad (4.3.9)$$

Again the order estimate $\widehat{p}(\text{SC})$ is chosen so as to minimize the value of the criterion. A comparison with (4.3.6) shows that for this criterion $c_T = K^2 \ln T$. Because

$$K^2 \ln T / 2 \ln \ln T$$

approaches infinity for $T \to \infty$, (4.3.7b) is satisfied and SC is seen to be strongly consistent for any dimension K.

Corollary 4.2.2
Under the conditions of Proposition 4.2, SC is strongly consistent and HQ is consistent. If the dimension K of the process is greater than one, both criteria are strongly consistent. ∎

In Table 4.5, the values of HQ and SC for the investment/income/consumption example are given. Both criteria assume the minimum for $m = 0$, that is, $\widehat{p}(\text{HQ}) = \widehat{p}(\text{SC}) = 0$.

4.3.3 Comparison of Order Selection Criteria

It is worth emphasizing that the foregoing results do not necessarily mean that AIC and FPE are inferior to HQ and SC. Only if consistency is the yardstick for evaluating the criteria, the latter two are superior under the conditions of the previous section. So far we have not considered the small sample properties of the estimators. In small samples, AIC and FPE may have better properties (choose the correct order more often) than HQ and SC. Also, the former two criteria are designed for minimizing the forecast error variance. Thus, in small as well as large samples, models based on AIC and FPE may produce superior forecasts although they may not estimate the orders correctly. In fact, Shibata (1980) derived asymptotic optimality properties of AIC and FPE for univariate processes. He showed that, under suitable conditions, they indeed minimize the 1-step ahead forecast MSE asymptotically.

Although it is difficult in general to derive *small sample properties* of the criteria, some such properties can be obtained. The following proposition states small sample relations between the criteria.

Proposition 4.3 (*Small Sample Comparison of AIC, HQ, and SC*)
Let $y_{-M+1}, \ldots, y_0, y_1, \ldots, y_T$ be any K-dimensional multiple time series and suppose that VAR(m) models, $m = 0, 1, \ldots, M$, are fitted to $y_1, \ldots, y_T$. Then the following relations hold:

$$\widehat{p}(\text{SC}) \leq \widehat{p}(\text{AIC}) \quad \text{if } T \geq 8, \tag{4.3.10}$$

$$\widehat{p}(\text{SC}) \leq \widehat{p}(\text{HQ}) \quad \text{for all } T, \tag{4.3.11}$$

$$\widehat{p}(\text{HQ}) \leq \widehat{p}(\text{AIC}) \quad \text{if } T \geq 16. \tag{4.3.12}$$

∎

Note that we do not require stationarity of y_t. In fact, we do not even require that the multiple time series is generated by a VAR process. Moreover, the proposition is valid in small samples and not just asymptotically. The proof is an easy consequence of the following lemma.

Lemma 4.1
Let $a_0, a_1, \ldots, a_M, b_0, b_1, \ldots, b_M$ and $c_0, c_1, \ldots, c_M$ be real numbers. If

$$b_{m+1} - b_m < a_{m+1} - a_m, \quad m = 0, 1, \ldots, M - 1, \tag{4.3.13a}$$

holds and if nonnegative integers n and k are chosen such that

$$c_n + a_n = \min\{c_m + a_m | m = 0, 1, \ldots, M\} \tag{4.3.13b}$$

and

$$c_k + b_k = \min\{c_m + b_m | m = 0, 1, \ldots, M\}, \tag{4.3.13c}$$

then $k \geq n$.[1] ∎

The proof of this lemma is left as an exercise (see Problem 4.2). It is now easy to prove Proposition 4.3.

Proof of Proposition 4.3:
Let $c_m = \ln |\widetilde{\Sigma}_u(m)|$, $b_m = 2mK^2/T$ and $a_m = mK^2 \ln T/T$. Then $\text{AIC}(m) = c_m + b_m$ and $\text{SC}(m) = c_m + a_m$. The sequences a_m, b_m, and c_m satisfy the conditions of the lemma if

$$\begin{aligned} 2K^2/T &= 2(m+1)K^2/T - 2mK^2/T = b_{m+1} - b_m \\ &< a_{m+1} - a_m = (m+1)K^2 \ln T/T - mK^2 \ln T/T = K^2 \ln T/T \end{aligned}$$

or, equivalently, if $\ln T > 2$ or $T > e^2 = 7.39$. Hence, choosing $k = \widehat{p}(\text{AIC})$ and $n = \widehat{p}(\text{SC})$ gives $\widehat{p}(\text{SC}) \leq \widehat{p}(\text{AIC})$ if $T \geq 8$. The relations (4.3.11) and (4.3.12) can be shown analogously. ∎

An immediate consequence of Corollary 4.2.1 and Proposition 4.3 is that AIC and FPE asymptotically overestimate the true order with positive probability and underestimate the true order with probability zero.

Corollary 4.3.1
Under the conditions of Proposition 4.2, if $M > p$,

$$\lim_{T \to \infty} \Pr\{\widehat{p}(\text{AIC}) < p\} = 0 \quad \text{and} \quad \lim \Pr\{\widehat{p}(\text{AIC}) > p\} > 0 \tag{4.3.14}$$

and the same holds for $\widehat{p}(\text{FPE})$. ∎

Proof: By (4.3.10) and Corollary 4.2.2,

$$\Pr\{\widehat{p}(\text{AIC}) < p\} \leq \Pr\{\widehat{p}(\text{SC}) < p\} \to 0.$$

Because AIC is not consistent by Corollary 4.2.1, $\lim \Pr\{\widehat{p}(\text{AIC}) = p\} < 1$. Hence (4.3.14) follows. The same holds for FPE because this criterion is asymptotically equivalent to AIC (see (4.3.3)). ∎

The limitations of the asymptotic theory for the order selection criteria can be seen by considering the criterion obtained by setting c_T equal to $2 \ln \ln T$ in (4.3.6). This results in a criterion

---

[1] I am grateful to Prof. K. Schürger, Universität Bonn, for pointing out the present improvement of the corresponding lemma stated in Lütkepohl (1991).

$$C(m) = \ln|\widetilde{\Sigma}_u(m)| + 2m \ln \ln T / T. \qquad (4.3.15)$$

Under the conditions of Proposition 4.2, it is consistent. Yet, using Lemma 4.1 and the same line of reasoning as in the proof of Proposition 4.3, $\widehat{p}(\text{AIC}) \leq \widehat{p}(C)$ if $2\ln\ln T \leq 2K^2$ or, equivalently, if $T \leq \exp(\exp K^2)$. For instance, for a bivariate process ($K = 2$), $\exp(\exp K^2) \approx 5.14 \times 10^{23}$. Consequently, if $T < 5.14 \times 10^{23}$, the consistent criterion (4.3.15) chooses an order greater than or equal to $\widehat{p}(\text{AIC})$ which in turn has a positive limiting probability for exceeding the true order. This example shows that large sample results sometimes are good approximations only if extreme sample sizes are available. The foregoing result was used by Quinn (1980) as an argument for making c_T a function of the dimension K of the process in the HQ criterion.

It is also of interest to compare the order selection criteria to the sequential testing procedure discussed in the previous section. We have mentioned in Section 4.2 that the order chosen in a sequence of tests will depend on the significance levels used. As a consequence, a testing sequence may give the same order as a selection criterion if the significance levels are chosen accordingly. For instance, AIC chooses an order smaller than the maximum order M if $\text{AIC}(M-1) < \text{AIC}(M)$ or, equivalently, if

$$\lambda_{LR}(1) = T(\ln|\widetilde{\Sigma}_u(M-1)| - \ln|\widetilde{\Sigma}_u(M)|) < 2MK^2 - 2(M-1)K^2 = 2K^2.$$

For $K = 2$, $2K^2 = 8 \approx \chi^2(4)_{.90}$. Thus, for a bivariate process, in order to ensure that AIC chooses an order less than M whenever the LR testing procedure does, we may use approximately a 10% significance level in the first test of the sequence, provided the distribution of $\lambda_{LR}(1)$ is well approximated by a $\chi^2(4)$-distribution.

The sequential testing procedure will not lead to a consistent order estimator if the sequence of individual significance levels is held constant. To see this, note that for $M > p$ and a fixed significance level γ, the null hypothesis $H_0 : A_M = 0$ is rejected with probability γ. In other words, in the testing scheme, M is incorrectly chosen as VAR order with probability γ. Thus, there is a positive probability of choosing too high an order. This problem can be circumvented by letting the significance level go to zero as $T \to \infty$.

4.3.4 Some Small Sample Simulation Results

As mentioned previously, many of the small sample properties of interest in the context of VAR order selection are difficult to derive analytically. Therefore we have performed a small Monte Carlo experiment to get some feeling for the small sample behavior of the estimators. Some results will now be reported.

We have simulated 1000 realizations of the VAR(2) process (4.2.1) and we have recorded the orders chosen by FPE, AIC, HQ, and SC for time series lengths of $T = 30$ and 100 and a maximum VAR order of $M = 6$. In addition, we have determined the order by the sequence of LR tests described in Section

4.2 using a significance level of 5% in each individual test and corresponding critical values from χ^2-distributions. That is, we have used χ^2- rather than F-tests. The frequency distributions obtained with the five different procedures are displayed in Table 4.6. Obviously, for the sample sizes reported, none of the criteria is very successful in estimating the order $p = 2$ correctly. This may be due to the fact that A_2 contains only a single, small nonzero element. The similarity of AIC and FPE derived in (4.3.3) becomes evident for $T = 100$. The orders chosen by the LR testing procedures show that the actual significance levels are quite different from their asymptotic approximations, especially for sample size $T = 30$. If λ_{LR} really had a $\chi^2(4)$-distribution the order $\widehat{p} = M = 6$ should be chosen in about 5% of the cases while in the simulation experiment $\widehat{p} = 6$ is chosen for 25.4% of the realizations. Hence, the $\chi^2(4)$-distribution is hardly a good small sample approximation to the actual distribution of λ_{LR}.

In Table 4.6, we also present the sum of normalized mean squared forecast errors of y_1 and y_2 obtained from post-sample forecasts with the estimated processes. The quantities shown in the table are

$$\frac{1}{N}\sum_{i=1}^{N}(y_{T+h(i)} - \widehat{y}_T(h)_{(i)})'\Sigma_y(h)^{-1}(y_{T+h(i)} - \widehat{y}_T(h)_{(i)}), \quad h = 1,2,3,$$

where N is the number of replications, that is, in this case $N = 1000$, $y_{T+h(i)}$ is the realization in the i-th repetition and $\widehat{y}_T(h)_{(i)}$ is the corresponding forecast. Normalizing with the inverse of the h-step forecast error variance $\Sigma_y(h)$ is useful to standardize the forecast errors in such a way so as to have roughly the same variability and, thus, comparable quantities are averaged. For large sample size T and a large number of replications N, the average normalized squared forecast errors should be roughly equal to the dimension of the process, that is, for the present bivariate process they should be close to 2.

Although in Table 4.6 SC often underestimates the true VAR order $p = 2$, the forecasts obtained with the SC models are generally the best for $T = 30$. The reason is that not restricting the single nonzero coefficient in A_2 to zero does not sufficiently improve the forecasts to offset the additional sampling variability introduced by estimating all four elements of the A_2 coefficient matrix. For $T = 100$, corresponding forecast MSEs obtained with the different criteria and procedures are very similar, although SC chooses the correct order much less often than the other criteria. This result indicates that choosing the correct VAR order and selecting a good forecasting model are objectives that may be reached by different VAR order selection procedures. Specifically, in this example, slight underestimation of the VAR order is not harmful to the forecast precision. In fact, for $T = 30$, the most parsimonious criterion which underestimates the true VAR order in more than 80% of the realizations of our VAR(2) process provides forecasts with the smallest normalized average squared forecast errors. In fact, the LR tests which choose larger orders quite frequently, produce clearly the worst forecasts.

Table 4.6. Simulation results based on 1000 realizations of the bivariate VAR(2) process (4.2.1)

	FPE	AIC	HQ	SC	LR
VAR order	$T=30$				
	frequency distributions of estimated VAR orders in %				
0	0.1	0.1	0.6	2.6	0.1
1	46.1	42.0	60.4	81.2	29.8
2	33.3	32.2	28.5	14.4	16.5
3	8.3	9.0	5.0	1.1	6.5
4	3.8	4.1	2.2	0.5	8.1
5	3.9	5.0	1.5	0.1	13.6
6	4.5	7.6	1.8	0.1	25.4
forecast horizon	normalized average squared forecast errors				
1	2.63	2.68	2.52	2.37	3.09
2	2.66	2.72	2.51	2.41	3.04
3	2.58	2.67	2.45	2.35	3.05
VAR order	$T=100$				
	frequency distributions of estimated VAR orders in %				
0	0.0	0.0	0.0	0.0	0.0
1	17.6	17.4	42.7	73.1	20.8
2	69.5	69.5	55.5	26.7	53.6
3	8.4	8.4	1.7	0.2	5.3
4	2.8	2.8	0.1	0.0	6.2
5	1.0	1.0	0.0	0.0	5.4
6	0.7	0.9	0.0	0.0	8.7
forecast horizon	normalized average squared forecast errors				
1	2.15	2.15	2.15	2.17	2.22
2	2.20	2.20	2.20	2.22	2.25
3	2.12	2.12	2.13	2.12	2.17

It must be emphasized, however, that these results are very special and hold for the single bivariate VAR(2) process used in the simulations. Different results may be obtained for other processes. To substantiate this statement, we have also simulated 1000 time series based on the VAR(1) process (4.2.2). Some results are given in Table 4.7. While for sample size $T = 30$ again none of the criteria and procedures is very successful in detecting the correct VAR order $p = 1$, all four criteria FPE, AIC, HQ, and SC select the correct order in more than 90% of the replications for $T = 100$. The poor approximation of the small sample distribution of the LR statistic by a $\chi^2(9)$-distribution is evident. Note that we have used the critical values for 5% level individual tests from the χ^2-distribution. As in the VAR(2) example, the prediction performance of the SC models is best for $T = 30$, although the criterion underestimates the true order in more than 80% of the replications. For both sample sizes, the worst forecasts are obtained with the sequential testing procedure which overestimates the true order quite often.

Table 4.7. Simulation results based on 1000 realizations of the three-dimensional VAR(1) process (4.2.2)

VAR order	FPE	AIC	HQ	SC	LR
	$T = 30$				
	frequency distributions of estimated VAR orders in %				
0	24.3	17.5	44.5	81.3	0.7
1	50.7	35.3	39.4	18.0	2.2
2	7.5	4.7	3.0	0.3	1.3
3	3.0	2.2	0.9	0.2	1.5
4	1.8	1.7	0.4	0.0	3.7
5	2.9	4.2	1.5	0.0	14.9
6	9.8	34.4	10.3	0.2	75.7
forecast horizon	normalized average squared forecast errors				
1	4.60	6.06	4.43	3.94	8.35
2	4.12	5.42	3.98	3.33	7.87
3	3.87	5.11	3.75	3.19	7.49
VAR order	$T = 100$				
	frequency distributions of estimated VAR orders in %				
0	0.0	0.0	0.3	8.1	0.0
1	94.1	93.8	99.6	91.9	61.2
2	5.0	5.1	0.1	0.0	5.4
3	0.7	0.7	0.0	0.0	4.3
4	0.2	0.3	0.0	0.0	7.3
5	0.0	0.0	0.0	0.0	9.3
6	0.0	0.1	0.0	0.0	12.5
forecast horizon	normalized average squared forecast errors				
1	3.08	3.08	3.06	3.12	3.24
2	3.12	3.12	3.11	3.12	3.24
3	3.11	3.11	3.10	3.10	3.20

After these two simulation experiments, we still do not have a clear answer to the question which criterion to use in small sample situations. One conclusion that emerges from the two examples is that, in very small samples, slight underestimation of the true order is not necessarily harmful to the forecast precision. Moreover, both examples clearly demonstrate that the χ^2-approximation to the small sample distribution of the LR statistics is a poor one. In a simulation study based on many other processes, Lütkepohl (1985) obtained similar results. In that study, for low order VAR processes, the most parsimonious SC criterion was found to do quite well in terms of choosing the correct VAR order and providing good forecasting models. Unfortunately, in practice we often don't even know whether the underlying data generation law is of finite order VAR type. Sometimes we may just approximate an infinite order VAR process by a finite order model. In that case, for moderate sample sizes, some less parsimonious criterion like AIC may give superior results in terms of forecast precision. Therefore, it may be a good strategy to compare the order estimates obtained with different criteria and possibly perform analyses with different VAR orders.

4.4 Checking the Whiteness of the Residuals

In the previous sections, we have considered procedures for choosing the order of a VAR model for the generation process of a given multiple time series. These procedures may be interpreted as methods for determining a filter that transforms the given data into a white noise series. In this context, the criteria for model choice may be regarded as criteria for deciding whether the residuals are close enough to white noise to satisfy the investigator. Of course, if, for example, forecasting is the objective, it may not be of prime importance whether the residuals are really white noise as long as the model forecasts well. There are, however, situations where checking the white noise (whiteness) assumption for the residuals of a particular model is of interest. For instance, if the model order is chosen by nonstatistical methods (for example, on the basis of some economic theory) it may be useful to have statistical tools available for investigating the properties of the residuals. Moreover, because different criteria emphasize different aspects of the data generation process and may therefore all provide useful information for the analyst, it is common not to rely on just one procedure or criterion for model choice but use a number of different statistical tools. Therefore, in this section, we shall discuss statistical tools for checking the autocorrelation properties of the residuals of a given VAR model.

In Sections 4.4.1 and 4.4.2, the asymptotic distributions of the residual autocovariances and autocorrelations are given under the assumption that the model residuals are indeed white noise. In Sections 4.4.3 and 4.4.4, two popular statistics for checking the overall significance of the residual autocorrelations are discussed. The results of this section are adapted from Chitturi (1974), Hosking (1980, 1981a), Li & McLeod (1981), and Ahn (1988).

4.4.1 The Asymptotic Distributions of the Autocovariances and Autocorrelations of a White Noise Process

It is assumed that u_t is a K-dimensional white noise process with nonsingular covariance matrix Σ_u. For instance, u_t may represent the residuals of a

158   4 VAR Order Selection and Checking the Model Adequacy

VAR(p) process. Let $U := (u_1, \ldots, u_T)$. The autocovariance matrices of u_t are estimated as

$$C_i := \widehat{\Gamma}_u(i) := \frac{1}{T} \sum_{t=i+1}^{T} u_t u'_{t-i} = \frac{1}{T} U F_i U', \qquad i = 0, 1, \ldots, h < T. \quad (4.4.1)$$

The $(T \times T)$ matrix F_i is defined in the obvious way. For instance, for $i = 2$,

$$F_i := \begin{bmatrix} 0 & 0 & \cdots & 0 & 0 & 0 \\ 0 & 0 & \cdots & 0 & 0 & 0 \\ 1 & 0 & \cdots & 0 & 0 & 0 \\ 0 & 1 & & 0 & 0 & 0 \\ \vdots & & \ddots & \vdots & \vdots & \vdots \\ 0 & 0 & \cdots & 1 & 0 & 0 \end{bmatrix}$$

$$= \begin{bmatrix} 0 & 0 & \cdots & 0 \\ 0 & 0 & \cdots & 0 \\ 1 & 0 & \cdots & 0 \\ 0 & 1 & & 0 \\ \vdots & & \ddots & \vdots \\ 0 & 0 & \cdots & 1 \end{bmatrix} \begin{bmatrix} 1 & 0 & \cdots & 0 \\ 0 & 1 & & 0 \\ \vdots & & \ddots & \vdots \\ 0 & 0 & \cdots & 1 \\ 0 & 0 & \cdots & 0 \\ 0 & 0 & \cdots & 0 \end{bmatrix}'.$$

Of course, for $i = 0$, $F_0 = I_T$. In the following, the precise form of F_i is not important. It is useful, though, to remember that F_i is defined such that

$$U F_i U' = \sum_{t=i+1}^{T} u_t u'_{t-i}.$$

Let

$$\mathbf{C}_h := (C_1, \ldots, C_h) = U F (I_h \otimes U'), \quad (4.4.2)$$

where $F := (F_1, \ldots, F_h)$ is a $(T \times hT)$ matrix that is understood to depend on h and T without this being indicated explicitly. Furthermore, let

$$\mathbf{c}_h := \text{vec}(\mathbf{C}_h). \quad (4.4.3)$$

The estimated autocorrelation matrices of the u_t are denoted by R_i, that is,

$$R_i := D^{-1} C_i D^{-1}, \quad i = 0, 1, \ldots, h, \quad (4.4.4)$$

where D is a $(K \times K)$ diagonal matrix, the diagonal elements being the square roots of the diagonal elements of C_0. In other words, a typical element of R_i is

$$r_{mn,i} = \frac{c_{mn,i}}{\sqrt{c_{mm,0}} \sqrt{c_{nn,0}}},$$

where $c_{mn,i}$ is the mn-th element of C_i. The matrix R_i in (4.4.4) is an estimator of the true autocorrelation matrix $R_u(i) = 0$ for $i \neq 0$. We use the notation

$$\mathbf{R}_h := (R_1, \ldots, R_h) \quad \text{and} \quad \mathbf{r}_h := \text{vec}(\mathbf{R}_h) \qquad (4.4.5)$$

and we denote by R_u the *true* correlation matrix corresponding to Σ_u. Now we can give the asymptotic distributions of $\mathbf{r}_h$ and $\mathbf{c}_h$.

Proposition 4.4 (*Asymptotic Distributions of White Noise Autocovariances and Autocorrelations*)
Let u_t be a K-dimensional identically distributed standard white noise process, that is, u_t and u_s have the same multivariate distribution with nonsingular covariance matrix Σ_u and corresponding correlation matrix R_u. Then, for $h \geq 1$,

$$\sqrt{T}\mathbf{c}_h \xrightarrow{d} \mathcal{N}(0, I_h \otimes \Sigma_u \otimes \Sigma_u) \qquad (4.4.6)$$

and

$$\sqrt{T}\mathbf{r}_h \xrightarrow{d} \mathcal{N}(0, I_h \otimes R_u \otimes R_u). \qquad (4.4.7)$$

∎

Proof: The result (4.4.6) follows from an appropriate central limit theorem. The i.i.d. assumption for the u_t implies that

$$w_t = \text{vec}(u_t u'_{t-1}, \ldots, u_t u'_{t-h})$$

is a stationary white noise process with covariance matrix $E(w_t w'_t) = I_h \otimes \Sigma_u \otimes \Sigma_u$ so that the result (4.4.6) may, e.g., be obtained from the central limit theorem for stationary processes given in Proposition C.13 of Appendix C. Proofs can also be found in Fuller (1976, Chapter 6) and Hannan (1970, Chapter IV, Section 4) among others.

The result in (4.4.7) is a quite easy consequence of (4.4.6). From Proposition 3.2, we know that C_0 is a consistent estimator of Σ_u. Hence,

$$\sqrt{T}\,\text{vec}(R_i) = \sqrt{T}(D^{-1} \otimes D^{-1})\,\text{vec}(C_i) \xrightarrow{d} \mathcal{N}(0, R_u \otimes R_u)$$

by Proposition C.15(1) of Appendix C and (4.4.6), because

$$\text{plim}(D^{-1} \otimes D^{-1})(\Sigma_u \otimes \Sigma_u)(D^{-1} \otimes D^{-1})$$
$$= \text{plim}(D^{-1}\Sigma_u D^{-1} \otimes D^{-1}\Sigma_u D^{-1}) = R_u \otimes R_u.$$

∎

The result in (4.4.6) means that $\sqrt{T}\,\text{vec}(C_i)$ has the same asymptotic distribution as $\sqrt{T}\,\text{vec}(C_j)$, namely,

160   4 VAR Order Selection and Checking the Model Adequacy

$$\sqrt{T} \text{ vec}(C_i), \sqrt{T} \text{ vec}(C_j) \xrightarrow{d} \mathcal{N}(0, \Sigma_u \otimes \Sigma_u).$$

Moreover, for $i \neq j$, the two estimators are asymptotically independent. By (4.4.7), the same holds for $\sqrt{T} \text{ vec}(R_i)$ and $\sqrt{T} \text{ vec}(R_j)$.

In practice, the u_t and hence U will usually be unknown and the reader may wonder about the relevance of Proposition 4.4. The result is not only useful in proving other propositions but can also be used to check whether a given time series is white noise. Before we explain that procedure, we mention that Proposition 4.4 remains valid if the considered white noise process is allowed to have nonzero mean and the mean vector is estimated by the sample mean vector. That is, we consider covariance matrices

$$C_i = \frac{1}{T} \sum_{t=i+1}^{T} (u_t - \overline{u})(u_{t-i} - \overline{u})',$$

where

$$\overline{u} = \frac{1}{T} \sum_{t=1}^{T} u_t.$$

Next we observe that the diagonal elements of $R_u \otimes R_u$ are all ones. Consequently, the variances of the asymptotic distributions of the elements of $\sqrt{T}\mathbf{r}_h$ are all unity. Hence, in large samples the $\sqrt{T}r_{mn,i}$ for $i > 0$ have approximate standard normal distributions. Denoting by $\rho_{mn}(i)$ the true correlation coefficients corresponding to the $r_{mn,i}$, a test, with level approximately 5%, of the null hypothesis

$$H_0 : \rho_{mn}(i) = 0 \quad \text{against} \quad H_1 \colon \rho_{mn}(i) \neq 0$$

rejects H_0 if $|\sqrt{T}r_{mn,i}| > 2$ or, equivalently, $|r_{mn,i}| > 2/\sqrt{T}$.

Now we have a test for checking the null hypothesis that a given multiple time series is generated by a white noise process. We simply compute the correlations of the original data (possibly after some stationarity transformation) and compare their absolute values with $2/\sqrt{T}$. In Section 4.3.2, we found that the SC and HQ estimate of the order for the generation process of the investment/income/consumption example data is $\hat{p} = 0$. Therefore, one may want to check the white noise hypothesis for this example. The first two correlation matrices for the data from 1960.4 to 1978.4 are

$$R_1 = \begin{bmatrix} -.197 & .103 & .128 \\ .190 & .020 & .228 \\ -.047 & .150 & -.089 \end{bmatrix} \text{ and } R_2 = \begin{bmatrix} -.045 & .067 & .097 \\ .119 & .079 & .009 \\ .255 & .355 & .279 \end{bmatrix}.$$

(4.4.8)

Comparing these quantities with $2/\sqrt{T} = 2/\sqrt{73} = .234$, we find that some are significantly different from zero and, hence, we reject the white noise hypothesis on the basis of this test.

In applied work, the estimated autocorrelations are sometimes plotted and $\pm 2/\sqrt{T}$-bounds around zero are indicated. The white noise hypothesis is then rejected if any of the estimated correlation coefficients reach out of the area between the $\pm 2/\sqrt{T}$-bounds. In Figure 4.1, plots of some autocorrelations are provided for the example data. Some autocorrelations at lags 2, 4, 8, 11, and 12 are seen to be significant under the aforementioned criterion.

There are several points that must be kept in mind in such a procedure. First, in an exact 5% level test, on average the test will reject one out of twenty times it is performed independently, even if the null hypothesis is correct. Thus, one would expect that one out of twenty autocorrelation estimates exceeds $2/\sqrt{T}$ in absolute value even if the underlying process is indeed white noise. Note, however, that although R_i and R_j are asymptotically independent for $i \neq j$, the same is not necessarily true for the elements of R_i. Thus, considering the individual correlation coefficients may provide a misleading picture of their significance as a group. Tests for overall significance of groups of autocorrelations are discussed in Sections 4.4.3 and 4.4.4.

Second, the tests we have considered here are just asymptotic tests. In other words, the actual sizes of the tests may differ from their nominal sizes. In fact, it has been shown by Dufour & Roy (1985) and others that in small samples the variances of the correlation coefficients may differ considerably from $1/T$. They will often be smaller so that the tests are conservative in that they reject the null hypothesis less often than is indicated by the significance level chosen.

Despite this criticism, this check for whiteness of a time series enjoys much popularity as it is very easy to carry out. It is a good idea, however, not to rely on this criterion exclusively.

4.4.2 The Asymptotic Distributions of the Residual Autocovariances and Autocorrelations of an Estimated VAR Process

Theoretical Results

If a VAR(p) model has been fitted to the data, a procedure similar to that described in the previous subsection is often used to check the whiteness of the residuals. Instead of the actual u_t's, the estimation residuals are used, however. We will now consider the consequences of that approach. For that purpose, we assume that the model has been estimated by LS and, using the notation of Section 3.2, the coefficient estimator is denoted by $\widehat{B}$ and the corresponding residuals are $\widehat{U} = (\widehat{u}_1, \ldots, \widehat{u}_T) := Y - \widehat{B}Z$. Furthermore,

$$\widehat{C}_i := \frac{1}{T}\widehat{U}F_i\widehat{U}', \quad i = 0, 1, \ldots h,$$

$$\widehat{\mathbf{C}}_h := (\widehat{C}_1, \ldots, \widehat{C}_h) = \frac{1}{T}\widehat{U}F(I_h \otimes \widehat{U}'), \qquad (4.4.9)$$

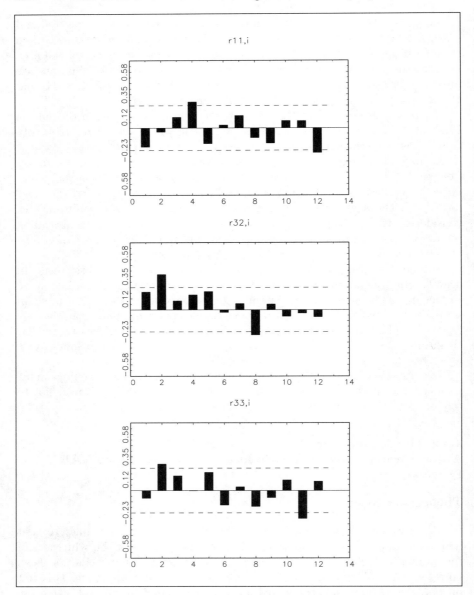

Fig. 4.1. Some estimated autocorrelations of the investment/income/consumption system.

$$\widehat{\mathbf{c}}_h := \text{vec}(\widehat{\mathbf{C}}_h),$$

and, correspondingly,

$$\widehat{R}_i := \widehat{D}^{-1}\widehat{C}_i\widehat{D}^{-1}, \quad \widehat{\mathbf{R}}_h := (\widehat{R}_1, \ldots, \widehat{R}_h), \quad \widehat{\mathbf{r}}_h := \text{vec}(\widehat{\mathbf{R}}_h), \tag{4.4.10}$$

where $\widehat{D}$ is a diagonal matrix with the square roots of the diagonal elements of $\widehat{C}_0$ on the main diagonal. We will consider the asymptotic distribution of $\sqrt{T}\widehat{\mathbf{c}}_h$ first. For that purpose the following lemma is helpful.

Lemma 4.2

Let y_t be a stationary, stable VAR(p) process as in (4.1.1) with identically distributed standard white noise u_t and let $\widehat{B}$ be a consistent estimator of $B = [\nu, A_1, \ldots, A_p]$ such that $\sqrt{T}\,\text{vec}(\widehat{B} - B)$ has an asymptotic normal distribution. Then $\sqrt{T}\widehat{\mathbf{c}}_h$ has the same asymptotic distribution as

$$\sqrt{T}\mathbf{c}_h - \sqrt{T}G\,\text{vec}(\widehat{B} - B), \tag{4.4.11}$$

where $G := \widetilde{G}' \otimes I_K$ with

$$\widetilde{G} := \begin{bmatrix} 0 & 0 & \cdots & 0 \\ \Sigma_u & \Phi_1\Sigma_u & \cdots & \Phi_{h-1}\Sigma_u \\ 0 & \Sigma_u & \cdots & \Phi_{h-2}\Sigma_u \\ \vdots & & & \vdots \\ 0 & 0 & \cdots & \Phi_{h-p}\Sigma_u \end{bmatrix} \quad ((Kp+1) \times Kh). \tag{4.4.12}$$

■

Proof: Using the notation $Y = BZ + U$,

$$\widehat{U} = Y - \widehat{B}Z = BZ + U - \widehat{B}Z = U - (\widehat{B} - B)Z.$$

Hence,

$$\widehat{U}F(I_h \otimes \widehat{U}')$$
$$= UF(I_h \otimes U') - UF\left[I_h \otimes Z'(\widehat{B} - B)'\right] \tag{4.4.13}$$
$$- (\widehat{B} - B)ZF(I_h \otimes U') + (\widehat{B} - B)ZF\left[I_h \otimes Z'(\widehat{B} - B)'\right].$$

Dividing by T and applying the vec operator, this expression becomes $\widehat{\mathbf{c}}_h$. In order to obtain the expression in (4.4.11), we consider the terms on the right-hand side of (4.4.13) in turn. The first term becomes $\sqrt{T}\mathbf{c}_h$ upon division by $\sqrt{T}$ and application of the vec operator.

Dividing the second and last terms by $\sqrt{T}$ they can be shown to converge to zero in probability, that is,

$$\text{plim}\sqrt{T}UF\left[I_h \otimes Z'(\widehat{B} - B)'\right]/T = 0 \tag{4.4.14}$$

and

$$\text{plim}\sqrt{T}(\widehat{B} - B)ZF\left[I_h \otimes Z'(\widehat{B} - B)'\right]/T = 0 \tag{4.4.15}$$

(see Problem 4.3). Thus, it remains to show that dividing the third term in (4.4.13) by $\sqrt{T}$ and applying the vec operator yields an expression which is asymptotically equivalent to the last term in (4.4.11). To see this, consider

$$ZF(I_h \otimes U') = (ZF_1U', \ldots, ZF_hU')$$

and

$$ZF_iU' = \sum_{t=i+1}^{T} Z_{t-1}u'_{t-i} = \sum_{t=i+1}^{T} \begin{bmatrix} 1 \\ y_{t-1} \\ \vdots \\ y_{t-p} \end{bmatrix} u'_{t-i}$$

$$= \sum_{t} \begin{bmatrix} 1 \\ \sum_{j=0}^{\infty} \Phi_j u_{t-1-j} \\ \vdots \\ \sum_{j=0}^{\infty} \Phi_j u_{t-p-j} \end{bmatrix} u'_{t-i},$$

where the Φ_i are the coefficient matrices of the canonical MA representation of y_t (see (2.1.17)). Upon division by T and application of the plim we get

$$\plim \frac{1}{T} ZF_iU' = \begin{bmatrix} 0 \\ \Phi_{i-1}\Sigma_u \\ \vdots \\ \Phi_{i-p}\Sigma_u \end{bmatrix},$$

where $\Phi_j = 0$ for $j < 0$. Hence,

$$\plim \frac{1}{T} ZF(I_h \otimes U') = \begin{bmatrix} 0 & 0 & \cdots & 0 \\ \Sigma_u & \Phi_1\Sigma_u & \cdots & \Phi_{h-1}\Sigma_u \\ 0 & \Sigma_u & \cdots & \Phi_{h-2}\Sigma_u \\ \vdots & & & \vdots \\ 0 & 0 & \cdots & \Phi_{h-p}\Sigma_u \end{bmatrix}_{((Kp+1) \times Kh)} = \widetilde{G}.$$

The lemma follows by noting that

$$\vec\left[(\widehat{B} - B)ZF(I_h \otimes U')\right] = ([ZF(I_h \otimes U')]' \otimes I_K) \vec(\widehat{B} - B).$$

∎

The next lemma is also helpful later.

Lemma 4.3

If y_t is a stable VAR(p) process as in (4.1.1) with identically distributed standard white noise, then

$$\begin{bmatrix} \frac{1}{\sqrt{T}} \operatorname{vec}(UZ') \\ \sqrt{T}\mathbf{c}_h \end{bmatrix} \xrightarrow{d} \mathcal{N}\left(0, \begin{bmatrix} \Gamma & \widetilde{G} \\ \widetilde{G}' & I_h \otimes \Sigma_u \end{bmatrix} \otimes \Sigma_u \right), \qquad (4.4.16)$$

where $\Gamma := \operatorname{plim} ZZ'/T$ and $\widetilde{G}$ is as defined in (4.4.12). ∎

For the two terms $\operatorname{vec}(UZ')/\sqrt{T}$ and $\sqrt{T}\mathbf{c}_h$ separately, the asymptotic distributions are already known from Lemma 3.1 and Proposition 4.4, respectively. So the joint asymptotic distribution is the new result here. The reader is referred to Ahn (1988) for a proof. Now the asymptotic distribution of the residual autocovariances is easily obtained.

Proposition 4.5 (*Asymptotic Distributions of Residual Autocovariances*)
Let y_t be a stationary, stable, K-dimensional VAR(p) process as in (4.1.1) with identically distributed standard white noise process u_t and let the coefficients be estimated by multivariate LS or an asymptotically equivalent procedure. Then

$$\sqrt{T}\,\widehat{\mathbf{c}}_h \xrightarrow{d} \mathcal{N}(0, \Sigma_\mathbf{c}(h)),$$

where

$$\begin{aligned} \Sigma_\mathbf{c}(h) &= (I_h \otimes \Sigma_u - \widetilde{G}'\Gamma^{-1}\widetilde{G}) \otimes \Sigma_u \\ &= (I_h \otimes \Sigma_u \otimes \Sigma_u) - \bar{G}[\Gamma_Y(0)^{-1} \otimes \Sigma_u]\bar{G}'. \end{aligned} \qquad (4.4.17)$$

Here $\widetilde{G}$ and Γ are the same matrices as in Lemma 4.3, $\Gamma_Y(0)$ is the covariance matrix of $Y_t = (y'_t, \ldots, y'_{t-p+1})'$ and $\bar{G} := \check{G}' \otimes I_K$, where $\check{G}$ is a $(Kp \times Kh)$ matrix which has the same form as $\widetilde{G}$ except that the first row of zeros is eliminated. ∎

Proof: Using Lemma 4.2, $\sqrt{T}\,\widehat{\mathbf{c}}_h$ is known to have the same asymptotic distribution as

$$\sqrt{T}\mathbf{c}_h - \sqrt{T}G\,\operatorname{vec}(\widehat{B} - B)$$

$$= \begin{bmatrix} -\widetilde{G}' \otimes I_K : I \end{bmatrix} \begin{bmatrix} \sqrt{T}\operatorname{vec}(\widehat{B} - B) \\ \sqrt{T}\mathbf{c}_h \end{bmatrix}$$

$$= \begin{bmatrix} -\widetilde{G}' \otimes I_K : I \end{bmatrix} \begin{bmatrix} \left(\frac{ZZ'}{T}\right)^{-1} \otimes I_K & 0 \\ 0 & I \end{bmatrix} \begin{bmatrix} \frac{1}{\sqrt{T}}\operatorname{vec}(UZ') \\ \sqrt{T}\mathbf{c}_h \end{bmatrix}.$$

Noting that $\operatorname{plim}(ZZ'/T)^{-1} = \Gamma^{-1}$, the desired result follows from Lemma 4.3 and Proposition C.15(1) of Appendix C because

$$\begin{bmatrix} -\widetilde{G}'\Gamma^{-1} \otimes I_K : I \end{bmatrix} \left(\begin{bmatrix} \Gamma & \widetilde{G} \\ \widetilde{G}' & I_h \otimes \Sigma_u \end{bmatrix} \otimes \Sigma_u \right) \begin{bmatrix} -\Gamma^{-1}\widetilde{G} \otimes I_K \\ I \end{bmatrix}$$

$$= (I_h \otimes \Sigma_u - \widetilde{G}'\Gamma^{-1}\widetilde{G}) \otimes \Sigma_u$$

$$= I_h \otimes \Sigma_u \otimes \Sigma_u - (\widetilde{G}' \otimes I_K)(\Gamma^{-1} \otimes \Sigma_u)(\widetilde{G} \otimes I_K)$$
$$= I_h \otimes \Sigma_u \otimes \Sigma_u - \bar{G}[\Gamma_Y(0)^{-1} \otimes \Sigma_u]\bar{G}'.$$

∎

The form (4.4.17) shows that the variances are smaller than (not greater than) the diagonal elements of $I_h \otimes \Sigma_u \otimes \Sigma_u$. In other words, the variances of the asymptotic distribution of the white noise autocovariances are greater than or equal to the corresponding quantities of the estimated residuals. A similar result can also be shown for the autocorrelations of the estimated residuals.

Proposition 4.6 (*Asymptotic Distributions of Residual Autocorrelations*)
Let D be the $(K \times K)$ diagonal matrix with the square roots of Σ_u on the diagonal and define $G_0 := \widetilde{G}(I_h \otimes D^{-1})$. Then, under the conditions of Proposition 4.5,

$$\sqrt{T}\,\widehat{\mathbf{r}}_h \xrightarrow{d} \mathcal{N}(0, \Sigma_{\mathbf{r}}(h)),$$

where

$$\Sigma_{\mathbf{r}}(h) = [(I_h \otimes R_u) - G_0' \Gamma^{-1} G_0] \otimes R_u. \tag{4.4.18}$$

Specifically,

$$\sqrt{T}\,\text{vec}(\widehat{R}_j) \xrightarrow{d} \mathcal{N}(0, \Sigma_R(j)), \quad j = 1, 2, \ldots,$$

where

$$\Sigma_R(j) = \left(R_u - D^{-1}\Sigma_u \left[0 : \Phi'_{j-1} : \cdots : \Phi'_{j-p} \right] \Gamma^{-1} \begin{bmatrix} 0 \\ \Phi_{j-1} \\ \vdots \\ \Phi_{j-p} \end{bmatrix} \Sigma_u D^{-1} \right) \otimes R_u \tag{4.4.19}$$

with $\Phi_i = 0$ for $i < 0$. ∎

Proof: Noting that

$$\widehat{\mathbf{r}}_h = \text{vec}(\widehat{\mathbf{R}}_h) = \text{vec}\left[\widehat{D}^{-1}\widehat{\mathbf{C}}_h(I_h \otimes \widehat{D}^{-1})\right]$$
$$= (I_h \otimes \widehat{D}^{-1} \otimes \widehat{D}^{-1})\widehat{\mathbf{c}}_h$$

and $\widehat{D}^{-1}$ is a consistent estimator of D^{-1}, we get from Proposition 4.5 that $\sqrt{T}\,\widehat{\mathbf{r}}_h$ has an asymptotic normal distribution with mean zero and covariance matrix

$$(I_h \otimes D^{-1} \otimes D^{-1})\{(I_h \otimes \Sigma_u - \widetilde{G}'\Gamma^{-1}\widetilde{G}) \otimes \Sigma_u\}(I_h \otimes D^{-1} \otimes D^{-1})$$
$$= [(I_h \otimes R_u) - G_0'\Gamma^{-1}G_0] \otimes R_u,$$

where $D^{-1}\Sigma_u D^{-1} = R_u$ has been used. ∎

4.4 Checking the Whiteness of the Residuals

From (4.4.19), it is obvious that the diagonal elements of the asymptotic covariance matrix are not greater that 1 because a positive semidefinite matrix is subtracted from R_u. Hence, if estimated residual autocorrelations are used in a white noise test in a similar fashion as the autocorrelations of the original data, we will get a conservative test that rejects the null hypothesis less often than is indicated by the significance level, provided the asymptotic distribution is a correct indicator of the small sample behavior of the test. In particular, for autocorrelations at small lags the variances will be less than 1, while the asymptotic variances approach one for elements of $\sqrt{T}\widehat{R}_j$ with large j. This conclusion follows because $\Phi_{j-1}, \ldots, \Phi_{j-p}$ approach zero as $j \to \infty$. As a consequence, the matrix subtracted from R_u goes to zero as $j \to \infty$.

In practice, all unknown quantities are replaced by estimates in order to obtain standard errors of the residual autocorrelations and tests of specific hypotheses regarding the autocorrelations. It is perhaps worth noting, though, that if Γ is estimated by ZZ'/T, we have to use the ML estimator $\widetilde{\Sigma}_u$ for Σ_u to ensure positive variances.

An Illustrative Example

As an example, we consider the VAR(2) model for the investment/income/consumption system estimated in Section 3.2.3. For $j = 1$, we get

$$\widehat{R}_1 = \begin{bmatrix} .015 & -.011 & -.010 \\ (.026) & (.033) & (.049) \\ -.007 & -.002 & -.068 \\ (.026) & (.033) & (.049) \\ -.024 & -.045 & -.096 \\ (.026) & (.033) & (.049) \end{bmatrix},$$

where the estimated standard errors are given in parentheses. Obviously, the standard errors of the elements of $\widehat{R}_1$ are much smaller than $1/\sqrt{T} = .117$ which would be obtained if the variances of the elements of $\sqrt{T}\widehat{R}_1$ were 1. In contrast, for $j = 6$, we get

$$\widehat{R}_6 = \begin{bmatrix} .053 & -.008 & -.062 \\ (.117) & (.116) & (.117) \\ .165 & .030 & -.051 \\ (.117) & (.116) & (.117) \\ .068 & .026 & .020 \\ (.117) & (.116) & (.117) \end{bmatrix},$$

where the standard errors are very close to .117.

In Figure 4.2, we have plotted the residual autocorrelations and *twice* their asymptotic standard errors (approximate 95% confidence bounds) around

168   4 VAR Order Selection and Checking the Model Adequacy

zero. It is apparent that the confidence bounds grow with increasing lag length. For a rough check of 5% level significance of autocorrelations at higher lags, we may use the $\pm 2/\sqrt{T}$-bounds in practice, which is convenient from a computational viewpoint.

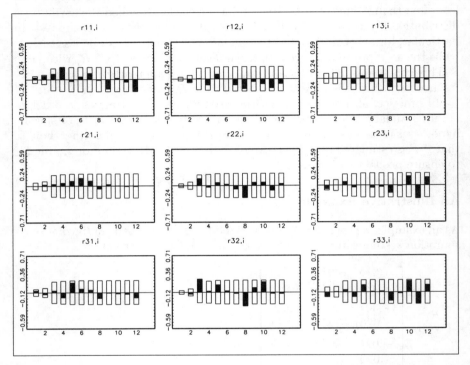

Fig. 4.2. Estimated residual autocorrelations with two-standard error bounds for the investment/income/consumption VAR(2) model.

There are significant residual autocorrelations at lags 3, 4, 8, and 11. While the significant values at lags 3 and 4 may be a reason for concern, one may not worry too much about the higher order lags because one may not be willing to fit a high order model if forecasting is the objective. As we have seen in Section 4.3.4, slight underfitting may even improve the forecast performance. In order to remove the significant residual autocorrelations at low lags, it may help to fit a VAR(3) or VAR(4) model. Of course, this conflicts with choosing the model order on the basis of the model selection criteria. Thus, it has to be decided which criterion is given priority.

It may be worth noting that a plot like that in Figure 4.2 may give a misleading picture of the overall significance of the residual autocorrelations because they are not asymptotically independent. In particular, at low lags there will not only be nonzero correlation between the elements of a specific $\widetilde{R}_j$

but also between $\widehat{R}_j$ and $\widehat{R}_i$ for $i \neq j$. Therefore, it is desirable to have tests for overall significance of the residual autocorrelations of a VAR(p) model. Such tests are discussed in the next subsections.

4.4.3 Portmanteau Tests

The foregoing results may also be used to construct a popular test for the overall significance of the residual autocorrelations up to lag h. This test is commonly called *portmanteau test*. It is designed for testing

$$H_0 : \mathbf{R}_h = (R_1, \ldots, R_h) = 0 \quad \text{against} \quad H_1 : \mathbf{R}_h \neq 0. \quad (4.4.20)$$

The test statistic is

$$\begin{aligned}
Q_h &:= T \sum_{i=1}^{h} \operatorname{tr}(\widehat{R}'_i \widehat{R}_u^{-1} \widehat{R}_i \widehat{R}_u^{-1}) \\
&= T \sum_{i=1}^{h} \operatorname{tr}(\widehat{R}'_i \widehat{R}_u^{-1} \widehat{R}_i \widehat{R}_u^{-1} \widehat{D}^{-1} \widehat{D}) \\
&= T \sum_{i=1}^{h} \operatorname{tr}(\widehat{D} \widehat{R}'_i \widehat{D} \widehat{D}^{-1} \widehat{R}_u^{-1} \widehat{D}^{-1} \widehat{D} \widehat{R}_i \widehat{D} \widehat{D}^{-1} \widehat{R}_u^{-1} \widehat{D}^{-1}) \\
&= T \sum_{i=1}^{h} \operatorname{tr}(\widehat{C}'_i \widehat{C}_0^{-1} \widehat{C}_i \widehat{C}_0^{-1}). \quad (4.4.21)
\end{aligned}$$

Obviously, this statistic is very easy to compute from the estimated residuals. By Proposition 4.5, it has an approximate asymptotic χ^2-distribution.

Proposition 4.7 (*Approximate Distribution of the Portmanteau Statistic*)
Under the conditions of Proposition 4.5, we have, approximately, for large T and h,

$$\begin{aligned}
Q_h &= T \sum_{i=1}^{h} \operatorname{tr}(\widehat{C}'_i \widehat{C}_0^{-1} \widehat{C}_i \widehat{C}_0^{-1}) \\
&= T \operatorname{vec}(\widehat{\mathbf{C}}_h)'(I_h \otimes \widehat{C}_0^{-1} \otimes \widehat{C}_0^{-1}) \operatorname{vec}(\widehat{\mathbf{C}}_h) \approx \chi^2(K^2(h-p)). \quad (4.4.22)
\end{aligned}$$
∎

Sketch of the proof: By Proposition C.15(5) of Appendix C, Q_h has the same asymptotic distribution as

$$T\widehat{\mathbf{c}}'_h(I_h \otimes \Sigma_u^{-1} \otimes \Sigma_u^{-1})\widehat{\mathbf{c}}_h.$$

Defining the ($K \times K$) matrix P such that $PP' = \Sigma_u$ and

$$\widetilde{\mathbf{c}}_h := (I_h \otimes P \otimes P)^{-1}\widehat{\mathbf{c}}_h,$$

it is easily seen that Q_h has the same asymptotic distribution as $T\tilde{c}_h'\tilde{c}_h$. Hence, by Proposition C.15(6), it suffices to show that $\sqrt{T}\tilde{c}_h \xrightarrow{d} \mathcal{N}(0,\Omega)$, where Ω is an idempotent matrix of rank $K^2 h - K^2 p$. Because an approximate limiting χ^2-distribution of Q_h is claimed only, we just show that Ω is approximately equal to an idempotent matrix with rank $K^2(h-p)$.

Using Proposition 4.5, we get

$$\Omega = (I_h \otimes P^{-1} \otimes P^{-1})\Sigma_{\mathbf{c}}(h)(I_h \otimes P'^{-1} \otimes P'^{-1})$$
$$= I_{hK^2} - \mathbf{P}\bar{G}[\Gamma_Y(0)^{-1} \otimes \Sigma_u]\bar{G}'\mathbf{P}',$$

where $\mathbf{P} = I_h \otimes P^{-1} \otimes P^{-1}$ and $\bar{G}$ is defined in Proposition 4.5. Noting that the ij-th block of $\Gamma_Y(0)$ is

$$\text{Cov}(y_{t-i}, y_{t-j}) = \Gamma_y(j-i) = \sum_{n=0}^{\infty} \Phi_{n-i}\Sigma_u \Phi'_{n-j},$$

with $\Phi_k = 0$ for $k < 0$, we get approximately,

$$\Gamma_y(0) \otimes \Sigma_u^{-1} \approx \left[\sum_{n=1}^{h} \Phi_{n-i}\Sigma_u \Phi'_{n-j}\right]_{i,j=1,\ldots,p} \otimes \Sigma_u^{-1}$$
$$= \left[\sum_{n=1}^{h} \Phi_{n-i}\Sigma_u P'^{-1} P^{-1} \Sigma_u \Phi'_{n-j}\right]_{i,j} \otimes \Sigma_u^{-1}$$
$$= \bar{G}'\mathbf{P}'\mathbf{P}\bar{G}.$$

Hence, if h is such that $\Phi_i \approx 0$ for $i > h - p$,

$$\Omega \approx I_{hK^2} - \mathbf{P}\bar{G}(\bar{G}'\mathbf{P}'\mathbf{P}\bar{G})^{-1}\bar{G}'\mathbf{P}'.$$

Thus, Ω is approximately equal to an idempotent matrix with rank

$$\text{tr}(I_{hK^2} - \mathbf{P}\bar{G}(\bar{G}'\mathbf{P}'\mathbf{P}\bar{G})^{-1}\bar{G}'\mathbf{P}') = hK^2 - pK^2,$$

as was to be shown. ∎

Of course, these arguments do not fully prove Proposition 4.7 because we have not shown that an approximately idempotent matrix Ω leads to an approximate χ^2-distribution. To actually obtain the limiting χ^2-distribution, we have to assume that h goes to infinity with the sample size. Because the sketch of the proof should suffice to show in what sense the result is approximate, we do not pursue this issue further and refer the reader to Ahn (1988) for details. For practical purposes, it is important to remember that the χ^2-approximation to the distribution of the test statistic may be misleading for small values of h.

Like in previous sections we have discussed *asymptotic* distributions in this section. Not knowing the small sample distribution is clearly a shortcoming

4.4 Checking the Whiteness of the Residuals

because, in practice, infinite samples are not available. Using Monte Carlo techniques, it was found by some researchers that in small samples the nominal size of the portmanteau test tends to be lower than the significance level chosen (Davies, Triggs & Newbold (1977), Ljung & Box (1978), Hosking (1980)). As a consequence the test has low power against many alternatives. Therefore it has been suggested to use the modified test statistic

$$\bar{Q}_h := T^2 \sum_{i=1}^{h} (T-i)^{-1} \operatorname{tr}(\widehat{C}_i' \widehat{C}_0^{-1} \widehat{C}_i \widehat{C}_0^{-1}). \tag{4.4.23}$$

The modification may be regarded as an adjustment for the number of terms in the sum in

$$\widehat{C}_i = \frac{1}{T} \sum_{t=i+1}^{T} \widehat{u}_t \widehat{u}_{t-i}'.$$

For $T \to \infty$, $T/[T^2(T-i)^{-1}] \to 1$ and, thus, $\bar{Q}_h$ has the same asymptotic distribution as Q_h, that is, approximately in large samples and for large h,

$$\bar{Q}_h \approx \chi^2(K^2(h-p)). \tag{4.4.24}$$

For our example model, we obtained $\bar{Q}_{12} = 81.9$. Comparing this value with $\chi^2(K^2(h-p))_{.95} = \chi^2(90)_{.95} \approx 113$ shows that we cannot reject the white noise hypothesis for the residuals at a 5% level.

As mentioned in the introduction to this section, these tests can also be used in a model selection/order estimation procedure. A sequence of hypotheses as in (4.2.15) is tested in such a procedure by checking whether the residuals are white noise. In the following, Lagrange multiplier tests for residual autocorrelation will be presented.

4.4.4 Lagrange Multiplier Tests

Another way of testing a VAR model for residual autocorrelation is to assume a VAR model for the error vector, $u_t = D_1 u_{t-1} + \cdots + D_h u_{t-h} + v_t$, where v_t is white noise. It is equal to u_t if there is no residual autocorrelation. Therefore, we wish to test the pair of hypotheses

$$\begin{aligned} &H_0 : D_1 = \cdots = D_h = 0 \quad \text{against} \\ &H_1 : D_j \neq 0 \text{ for at least one } j \in \{1, \ldots, h\}. \end{aligned} \tag{4.4.25}$$

In this case, it is convenient to use the LM principle for constructing a test because we then only need to estimate the restricted model where $u_t = v_t$. We determine the test statistic with the help of the auxiliary regression model (see also Appendix C.7)

$$\widehat{u}_t = \nu + A_1 y_{t-1} + \cdots + A_p y_{t-p} + D_1 \widehat{u}_{t-1} + \cdots + D_h \widehat{u}_{t-h} + \varepsilon_t$$

or, for $t = 1, \ldots, T$,

$$\widehat{\mathcal{U}} = BZ + D\widehat{\mathcal{U}} + \mathcal{E},$$

where $D := [D_1 : \cdots : D_h]$ is $(K \times Kh)$, $\widehat{\mathcal{U}} := (I_h \otimes \widehat{U})F'$ with F as in (4.4.2), $\mathcal{E} := [\varepsilon_1, \ldots, \varepsilon_T]$ is a $(K \times T)$ matrix and the other symbols are defined as before. In particular, the $\widehat{u}_t$ are the residuals from LS estimation of the original VAR(p) model and $\widehat{u}_t = 0$ for $t \leq 0$. The LS estimator of $[B : D]$ from the auxiliary model is

$$\begin{aligned}
[\widehat{B} : \widehat{D}] &= \widehat{U}[Z' : \widehat{\mathcal{U}}'] \left(\begin{bmatrix} Z \\ \widehat{\mathcal{U}} \end{bmatrix} [Z' : \widehat{\mathcal{U}}'] \right)^{-1} \\
&= [\widehat{U}Z' : \widehat{U}\widehat{\mathcal{U}}'] \begin{bmatrix} ZZ' & Z\widehat{\mathcal{U}}' \\ \widehat{\mathcal{U}}Z' & \widehat{\mathcal{U}}\widehat{\mathcal{U}}' \end{bmatrix}^{-1} \\
&= [0 : \widehat{U}\widehat{\mathcal{U}}'] \begin{bmatrix} ZZ' & Z\widehat{\mathcal{U}}' \\ \widehat{\mathcal{U}}Z' & \widehat{\mathcal{U}}\widehat{\mathcal{U}}' \end{bmatrix}^{-1},
\end{aligned}$$

where $\widehat{U}Z' = 0$ from the first order conditions for computing the LS estimator has been used. Thus, applying the rules for the partitioned inverse (Appendix A.10, Rule (2)) gives

$$\widehat{D} = \widehat{U}\widehat{\mathcal{U}}'[\widehat{\mathcal{U}}\widehat{\mathcal{U}}' - \widehat{\mathcal{U}}Z'(ZZ')^{-1}Z\widehat{\mathcal{U}}']^{-1}. \tag{4.4.26}$$

The standard χ^2-statistic for testing $D = 0$ then becomes

$$\begin{aligned}
\lambda_{LM}(h) &= \text{vec}(\widehat{D})' \left([\widehat{\mathcal{U}}\widehat{\mathcal{U}}' - \widehat{\mathcal{U}}Z'(ZZ')^{-1}Z\widehat{\mathcal{U}}'] \otimes \widehat{\Sigma}_u^{-1} \right) \text{vec}(\widehat{D}) \\
&= \text{vec}(\widehat{U}\widehat{\mathcal{U}}')' \left([\widehat{\mathcal{U}}\widehat{\mathcal{U}}' - \widehat{\mathcal{U}}Z'(ZZ')^{-1}Z\widehat{\mathcal{U}}']^{-1} \otimes \widehat{\Sigma}_u^{-1} \right) \text{vec}(\widehat{U}\widehat{\mathcal{U}}'),
\end{aligned}$$

where

$$\text{vec}(\widehat{D}) = \left([\widehat{\mathcal{U}}\widehat{\mathcal{U}}' - \widehat{\mathcal{U}}Z'(ZZ')^{-1}Z\widehat{\mathcal{U}}']^{-1} \otimes I_K \right) \text{vec}(\widehat{U}\widehat{\mathcal{U}}')$$

has been used. Noting that $\widehat{U}\widehat{\mathcal{U}}' = \widehat{U}F(I_h \otimes \widehat{U}')$ shows that $T^{-1} \text{vec}(\widehat{U}\widehat{\mathcal{U}}') = \widehat{\mathbf{c}}_h$. Moreover, from results in Section 4.4.2 we get

$$\text{plim} \frac{1}{T}\widehat{\mathcal{U}}\widehat{\mathcal{U}}' = \text{plim} \frac{1}{T}(I_h \otimes \widehat{U})F'F(I_h \otimes \widehat{U}') = I_h \otimes \Sigma_u$$

and

$$\text{plim} \frac{1}{T}\widehat{\mathcal{U}}Z'(ZZ')^{-1}Z\widehat{\mathcal{U}}' = \widetilde{G}\Gamma^{-1}\widetilde{G}$$

(see the proof of Lemma 4.2). Hence,

$$\widehat{\Sigma}_\mathbf{c}(h) = \frac{1}{T}[\widehat{\mathcal{U}}\widehat{\mathcal{U}}' - \widehat{\mathcal{U}}Z'(ZZ')^{-1}Z\widehat{\mathcal{U}}'] \otimes \widehat{\Sigma}_u$$

is a consistent estimator of $\Sigma_{\mathbf{c}}(h)$ and, because the foregoing results imply that

$$\lambda_{LM}(h) = T\widehat{\mathbf{c}}_h' \widehat{\Sigma}_{\mathbf{c}}(h)^{-1}\widehat{\mathbf{c}}_h,$$

the asymptotic χ^2-distribution of this statistic follows from Propositions 4.5 and C.15(5).

Proposition 4.8 (*Asymptotic Distribution of the LM Statistic for Residual Autocorrelation*)
Under the conditions of Proposition 4.5,

$$\lambda_{LM}(h) \xrightarrow{d} \chi^2(hK^2).$$

■

The LM test for residual autocorrelation is sometimes called Breusch-Godfrey test because it was proposed by Breusch (1978) and Godfrey (1978) (see also Godfrey (1988)). Unfortunately, the χ^2-distribution was found to be a poor approximation of the actual null distribution of $\lambda_{LM}(h)$ in many situations (Edgerton & Shukur (1999) and Doornik (1996)). Even a standard F-approximation is unsatisfactory. However, Doornik (1996) finds that the following statistic derived from considerations in Rao (1973, §8c.5) provides satisfactory results in small samples, if it is used with critical values from an $F(hK^2, Ns - \frac{1}{2}K^2h + 1)$-distribution:

$$F_{Rao}(h) = \left[\left(\frac{\det(\widetilde{\Sigma}_u)}{\det(\widetilde{\Sigma}_\varepsilon)} \right)^{1/s} - 1 \right] \frac{Ns - \frac{1}{2}K^2h + 1}{K^2h}.$$

Here

$$s = \left(\frac{K^4 h^2 - 4}{K^2 + K^2 h^2 - 5} \right)^{1/2}, \quad N = T - Kp - 1 - Kh - \frac{1}{2}(K - Kh + 1),$$

and $\widetilde{\Sigma}_\varepsilon$ is the residual covariance estimator from an unrestricted LS estimation of the auxiliary model $\widehat{U} = BZ + D\widehat{\mathcal{U}} + \mathcal{E}$.

We have also applied these tests to our example data and give some results in Table 4.4.4. It turns out that neither of the tests finds strong evidence for remaining residual autocorrelation. All p-values exceed 10%. Recall that a p-value represents the probability of getting a test value greater than the observed one, if the null hypothesis is true. Therefore, even at a significance level of 10%, the null hypothesis of no residual autocorrelation cannot be rejected.

In contrast to the portmanteau tests which should be used for reasonably large h only, the LM tests are more suitable for small values of h. For large h, the degrees of freedom in the auxiliary regression model will be exhausted

Table 4.8. Autocorrelation tests for investment/income/consumption example VAR(2) model, estimation period 1960.4-1978.4

test	h	test value	approximate distribution	p-value
$\lambda_{LM}(h)$	1	6.37	$\chi^2(9)$	0.70
	2	15.52	$\chi^2(18)$	0.62
	3	32.81	$\chi^2(27)$	0.20
	4	46.60	$\chi^2(36)$	0.11
$F_{Rao}(h)$	1	0.62	$F(9, 148)$	0.78
	2	0.76	$F(18, 164)$	0.75
	3	1.14	$F(27, 161)$	0.30
	4	1.26	$F(36, 154)$	0.17

and the statistic cannot be computed in the way described in the foregoing. An LM test for higher order residual autocorrelation may be based on the auxiliary model

$$\widehat{u}_t = \nu + A_1 y_{t-1} + \cdots + A_p y_{t-p} + D_h \widehat{u}_{t-h} + \varepsilon_t$$

and on a test

$$H_0 : D_h = 0 \quad \text{versus} \quad H_1 : D_h \neq 0.$$

The relevant LM statistic can be shown to have an asymptotic $\chi^2(K^2)$-distribution under H_0.

4.5 Testing for Nonnormality

Normality of the underlying data generating process is needed, for instance, in setting up forecast intervals. Nonnormal residuals can also indicate more generally that the model is not a good representation of the data generation process (see Chapter 16 for models for nonnormal data). Therefore, testing this distributional assumption is desirable. We will present tests for multivariate normality of a white noise process first. In Subsection 4.5.2, it is then demonstrated that the tests remain valid if the true residuals are replaced by the residuals of an estimated VAR(p) process.

4.5.1 Tests for Nonnormality of a Vector White Noise Process

The tests developed in the following are based on the third and fourth central moments (skewness and kurtosis) of the normal distribution. If x is a univariate random variable with standard normal distribution, i.e., $x \sim \mathcal{N}(0, 1)$, its third and fourth moments are known to be $E(x^3) = 0$ and $E(x^4) = 3$. Let u_t

be a K-dimensional Gaussian white noise process with $u_t \sim \mathcal{N}(\mu_u, \Sigma_u)$ and let P be a matrix satisfying $PP' = \Sigma_u$. For example, P may be obtained by a Choleski decomposition of Σ_u. Then

$$w_t = (w_{1t}, \ldots, w_{Kt})' := P^{-1}(u_t - \mu_u) \sim \mathcal{N}(0, I_K).$$

In other words, the components of w_t are independent standard normal random variables. Hence,

$$E\begin{bmatrix} w_{1t}^3 \\ \vdots \\ w_{Kt}^3 \end{bmatrix} = 0 \quad \text{and} \quad E\begin{bmatrix} w_{1t}^4 \\ \vdots \\ w_{Kt}^4 \end{bmatrix} = \begin{bmatrix} 3 \\ \vdots \\ 3 \end{bmatrix} =: \mathbf{3}_K. \tag{4.5.1}$$

This result will be utilized in checking the normality of the white noise process u_t. The idea is to compare the third and fourth moments of the transformed process with the theoretical values in (4.5.1) obtained for a Gaussian process. For the univariate case, the corresponding test is known as the *Jarque-Bera* or *Lomnicki-Jarque-Bera test* (see Jarque & Bera (1987) and Lomnicki (1961)).

For constructing the test, we assume to have observations $u_1, \ldots, u_T$ and define

$$\bar{u} := \frac{1}{T} \sum_{t=1}^{T} u_t, \quad S_u := \frac{1}{T-1} \sum_t (u_t - \bar{u})(u_t - \bar{u})',$$

and P_s is a matrix for which $P_s P_s' = S_u$ and such that $\text{plim}(P_s - P) = 0$. Moreover,

$$v_t := (v_{1t}, \ldots, v_{Kt})' = P_s^{-1}(u_t - \bar{u}), \quad t = 1, \ldots, T,$$

$$b_1 := (b_{11}, \ldots, b_{K1})' \quad \text{with} \quad b_{k1} = \frac{1}{T} \sum_t v_{kt}^3, \quad k = 1, \ldots, K, \tag{4.5.2}$$

and

$$b_2 := (b_{12}, \ldots, b_{K2})' \quad \text{with} \quad b_{k2} = \frac{1}{T} \sum_t v_{kt}^4, \quad k = 1, \ldots, K. \tag{4.5.3}$$

Thus, b_1 and b_2 are estimators of the vectors in (4.5.1). In the next proposition, the asymptotic distribution of b_1 and b_2 is given.

Proposition 4.9 (*Asymptotic Distribution of Skewness and Kurtosis*)
If u_t is Gaussian white noise with nonsingular covariance matrix Σ_u and expectation $\mu_u, u_t \sim \mathcal{N}(\mu_u, \Sigma_u)$, then

$$\sqrt{T} \begin{bmatrix} b_1 \\ b_2 - \mathbf{3}_K \end{bmatrix} \xrightarrow{d} \mathcal{N}\left(0, \begin{bmatrix} 6I_K & 0 \\ 0 & 24I_K \end{bmatrix}\right).$$

∎

In other words, b_1 and b_2 are asymptotically independent and normally distributed. The proposition implies that

$$\lambda_s := T b_1' b_1 / 6 \xrightarrow{d} \chi^2(K) \tag{4.5.4}$$

and

$$\lambda_k := T(b_2 - \mathbf{3}_K)'(b_2 - \mathbf{3}_K)/24 \xrightarrow{d} \chi^2(K). \tag{4.5.5}$$

The first statistic can be used to test

$$H_0 : E \begin{bmatrix} w_{1t}^3 \\ \vdots \\ w_{Kt}^3 \end{bmatrix} = 0 \quad \text{against} \quad H_1 : E \begin{bmatrix} w_{1t}^3 \\ \vdots \\ w_{Kt}^3 \end{bmatrix} \neq 0 \tag{4.5.6}$$

and λ_k may be used to test

$$H_0 : E \begin{bmatrix} w_{1t}^4 \\ \vdots \\ w_{Kt}^4 \end{bmatrix} = \mathbf{3}_K \quad \text{against} \quad H_1 : E \begin{bmatrix} w_{1t}^4 \\ \vdots \\ w_{Kt}^4 \end{bmatrix} \neq \mathbf{3}_K. \tag{4.5.7}$$

Furthermore,

$$\lambda_{sk} := \lambda_s + \lambda_k \xrightarrow{d} \chi^2(2K), \tag{4.5.8}$$

which may be used for a joint test of the null hypotheses in (4.5.6) and (4.5.7).

Proof of Proposition 4.9
We state a helpful lemma first.

Lemma 4.4
Let $z_t = (z_{1t}, \ldots, z_{Kt})'$ be a Gaussian white noise process with mean μ_z and covariance matrix I_K, i.e., $z_t \sim \mathcal{N}(\mu_z, I_K)$. Furthermore, let

$$\bar{z} = (\bar{z}_1, \ldots, \bar{z}_K)' := \frac{1}{T} \sum_{t=1}^{T} z_t,$$

$b_{1,z}$ a $(K \times 1)$ vector with k-th component $b_{k1,z} := \frac{1}{T} \sum_{t=1}^{T} (z_{kt} - \bar{z}_k)^3$,

and

$b_{2,z}$ a $(K \times 1)$ vector with k-th component $b_{k2,z} := \frac{1}{T} \sum_{t=1}^{T} (z_{kt} - \bar{z}_k)^4$.

Then

$$\sqrt{T} \begin{bmatrix} b_{1,z} \\ b_{2,z} - \mathbf{3}_K \end{bmatrix} \xrightarrow{d} \mathcal{N}\left(0, \begin{bmatrix} 6I_K & 0 \\ 0 & 24I_K \end{bmatrix}\right). \tag{4.5.9}$$

∎

The proof of this lemma is easily obtained, for instance, from results of Gasser (1975). Proposition 4.9 follows by noting that P_s is a consistent estimator of P (defined such that $PP' = \Sigma_u$) and by defining $z_t = P^{-1}u_t$. Hence,

$$\sqrt{T}(P_s^{-1} \otimes P_s^{-1} \otimes P_s^{-1})\frac{1}{T}\sum_t (u_t - \overline{u}) \otimes (u_t - \overline{u}) \otimes (u_t - \overline{u})$$

$$- \sqrt{T}\left[\frac{1}{T}\sum_t (z_t - \overline{z}) \otimes (z_t - \overline{z}) \otimes (z_t - \overline{z})\right]$$

$$= (P_s^{-1} \otimes P_s^{-1} \otimes P_s^{-1} - P^{-1} \otimes P^{-1} \otimes P^{-1})$$

$$\times \frac{1}{\sqrt{T}}\sum_t (u_t - \overline{u}) \otimes (u_t - \overline{u}) \otimes (u_t - \overline{u}) \xrightarrow{p} 0.$$

An analogous result is obtained for the fourth moments. Consequently,

$$\sqrt{T}\begin{bmatrix} b_1 - b_{1,z} \\ b_2 - b_{2,z} \end{bmatrix} \xrightarrow{p} 0$$

and the proposition follows from Proposition C.2(2) of Appendix C. ∎

Remark 1 In Proposition 4.9, the white noise process is not required to have zero mean. Thus, tests based on λ_s, λ_k, or λ_{sk} may be applied if the original observations are generated by a VAR(0) process. ∎

Remark 2 It is known that in the univariate case tests based on the skewness and kurtosis (third and fourth moments) have small sample distributions that differ substantially from their asymptotic counterparts (see, e.g., White & MacDonald (1980), Jarque & Bera (1987) and the references given there). Therefore, tests based on λ_s, λ_k, and λ_{sk}, in conjunction with the asymptotic χ^2-distributions in (4.5.4), (4.5.5), and (4.5.8), must be interpreted cautiously. They should be regarded as rough checks of normality only. ∎

Remark 3 Tests based on λ_s, λ_k, and λ_{sk} cannot be expected to possess power against distributions having the same first four moments as the normal distribution. Thus, if higher order moment characteristics are of interest, these tests cannot be recommended. Other tests for multivariate normality are described by Mardia (1980), Baringhaus & Henze (1988), and others. ∎

4.5.2 Tests for Nonnormality of a VAR Process

A stationary, stable VAR(p) process, say

$$y_t - \mu = A_1(y_{t-1} - \mu) + \cdots + A_p(y_{t-p} - \mu) + u_t, \tag{4.5.10}$$

is Gaussian (normally distributed) if and only if the white noise process u_t is Gaussian. Therefore, the normality of the y_t's may be checked via the u_t's. In practice, the u_t's are replaced by estimation residuals. In the following we will demonstrate that this is of no consequence for the *asymptotic* distributions of the λ statistics considered in the previous subsection.

The reader may wonder why normality tests are based on the residuals rather than the original observations y_t. The reason is that tests based on the latter may be less powerful than those based on the estimation residuals. For the univariate case this point was demonstrated by Lütkepohl & Schneider (1989). It is also worth recalling that the forecast errors used in the construction of forecast intervals are weighted sums of the u_t's. Therefore, checking the normality of these quantities makes sense if the aim is to establish interval forecasts. The next result states that Proposition 4.9 remains valid if the true white noise innovations u_t are replaced by estimation residuals.

Proposition 4.10 (*Asymptotic Distribution of Residual Skewness and Kurtosis*)
Let y_t be a K-dimensional stationary, stable Gaussian VAR(p) process as in (4.5.10), where u_t is zero mean white noise with nonsingular covariance matrix Σ_u and let $\widehat{A}_1, \ldots, \widehat{A}_p$ be consistent and asymptotically normally distributed estimators of the coefficients based on a sample $y_1, \ldots, y_T$ and possibly some presample values. Define

$$\widehat{u}_t := (y_t - \overline{y}) - \widehat{A}_1(y_{t-1} - \overline{y}) - \cdots - \widehat{A}_p(y_{t-p} - \overline{y}), \quad t = 1, \ldots, T,$$

$$\widehat{\Sigma}_u := \frac{1}{T - Kp - 1} \sum_{t=1}^{T} \widehat{u}_t \widehat{u}_t',$$

and let $\widehat{P}$ be a matrix satisfying $\widehat{P}\widehat{P}' = \widehat{\Sigma}_u$ such that $\text{plim}(\widehat{P} - P) = 0$. Furthermore, define

$$\widehat{w}_t = (\widehat{w}_{1t}, \ldots, \widehat{w}_{Kt})' := \widehat{P}^{-1} \widehat{u}_t,$$

$$\widehat{b}_1 = (\widehat{b}_{11}, \ldots, \widehat{b}_{K1})' \quad \text{with} \quad \widehat{b}_{k1} := \frac{1}{T} \sum_{t=1}^{T} \widehat{w}_{kt}^3, \quad k = 1, \ldots, K,$$

and

$$\widehat{b}_2 = (\widehat{b}_{12}, \ldots, \widehat{b}_{K2})' \quad \text{with} \quad \widehat{b}_{k2} := \frac{1}{T} \sum_{t=1}^{T} \widehat{w}_{kt}^4, \quad k = 1, \ldots, K.$$

Then

$$\sqrt{T} \begin{bmatrix} \widehat{b}_1 \\ \widehat{b}_2 - \mathbf{3}_K \end{bmatrix} \xrightarrow{d} \mathcal{N}\left(0, \begin{bmatrix} 6I_K & 0 \\ 0 & 24I_K \end{bmatrix}\right).$$

∎

Although the proposition is formulated in terms of the mean-adjusted form (4.5.10) of the process, it also holds if estimation residuals from the standard intercept form are used instead. The parameter estimators may be unconstrained ML or LS estimators. However, the proposition does not require this. In other words, the proposition remains valid if, for instance, restricted LS or generalized LS estimators are used, as discussed in the next chapter. The following lemma will be helpful in proving Proposition 4.10.

Lemma 4.5

Under the conditions of Proposition 4.10,

$$\text{plim} \left[\frac{1}{\sqrt{T}} \sum_{t=1}^{T} \widehat{u}_t \otimes \widehat{u}_t \otimes \widehat{u}_t - \frac{1}{\sqrt{T}} \sum_{t=1}^{T} (u_t - \overline{u}) \otimes (u_t - \overline{u}) \otimes (u_t - \overline{u}) \right] = 0$$
(4.5.11)

and

$$\text{plim} \left[\frac{1}{\sqrt{T}} \sum_{t=1}^{T} \widehat{u}_t \otimes \widehat{u}_t \otimes \widehat{u}_t \otimes \widehat{u}_t \right.$$

$$\left. - \frac{1}{\sqrt{T}} \sum_{t=1}^{T} (u_t - \overline{u}) \otimes (u_t - \overline{u}) \otimes (u_t - \overline{u}) \otimes (u_t - \overline{u}) \right] = 0 \quad (4.5.12)$$

■

Proof: A proof for the special case of a VAR(1) process y_t is given and the generalization is left to the reader. Also, we just show the first result. The second one follows with analogous arguments. For the special VAR(1) case,

$$\widehat{u}_t = (y_t - \overline{y}) - \widehat{A}_1(y_{t-1} - \overline{y})$$
$$= (u_t - \overline{u}) + (A_1 - \widehat{A}_1)(y_{t-1} - \overline{y}) + a_T,$$

where $a_T = A_1(y_T - y_0)/T$. Hence,

$$\frac{1}{\sqrt{T}} \sum_t [\widehat{u}_t \otimes \widehat{u}_t \otimes \widehat{u}_t] = \frac{1}{\sqrt{T}} \sum_t [(u_t - \overline{u}) \otimes (u_t - \overline{u}) \otimes (u_t - \overline{u})] + d_T,$$

where d_T is a sum of expressions of the type

$$\frac{1}{\sqrt{T}} \sum_t \left[(A_1 - \widehat{A}_1)(y_{t-1} - \overline{y}) + a_T \right] \otimes (u_t - \overline{u}) \otimes (u_t - \overline{u})$$

$$= \sqrt{T} \left[(A_1 - \widehat{A}_1) \otimes I_{2K} \right] \frac{1}{T} \sum_t [(y_{t-1} - \overline{y}) \otimes (u_t - \overline{u}) \otimes (u_t - \overline{u})]$$

$$+ \sqrt{T} a_T \otimes \frac{1}{T} \sum_t [(u_t - \overline{u}) \otimes (u_t - \overline{u})], \quad (4.5.13)$$

that is, d_T consists of sums of Kronecker products involving $(A_1 - \widehat{A}_1)(y_{t-1} - \overline{y})$, $(u_t - \overline{u})$, and a_T. Therefore, $d_T = o_p(1)$. For instance, (4.5.13) goes to zero in probability because

$$\text{plim} \frac{1}{T} \sum_t (u_t - \overline{u}) \otimes (u_t - \overline{u}) \quad \text{exists and} \quad \sqrt{T} a_T = o_p(1)$$

so that the last term in (4.5.13) vanishes. Moreover, the elements of $\sqrt{T}(A_1 - \widehat{A}_1)$ converge in distribution and

$$\text{plim} \frac{1}{T} \sum_t (y_{t-1} - \overline{y}) \otimes (u_t - \overline{u}) \otimes (u_t - \overline{u}) = 0 \tag{4.5.14}$$

(see Problem 4.4). Hence the first term in (4.5.13) vanishes. ∎

Proof of Proposition 4.10
By Proposition C.2(2) of Appendix C and Proposition 4.9, it suffices to show that

$$(\widehat{P}^{-1} \otimes \widehat{P}^{-1} \otimes \widehat{P}^{-1}) \frac{1}{\sqrt{T}} \sum_t \widehat{u}_t \otimes \widehat{u}_t \otimes \widehat{u}_t$$
$$- (P_s^{-1} \otimes P_s^{-1} \otimes P_s^{-1}) \frac{1}{\sqrt{T}} \sum_t (u_t - \overline{u}) \otimes (u_t - \overline{u}) \otimes (u_t - \overline{u}) \xrightarrow{p} 0 \tag{4.5.15}$$

and the fourth moments possess a similar property. The result (4.5.15) follows from Lemma 4.5 by noting that $\widehat{P}$ and P_s are both consistent estimators of P and, for stochastic vectors h_T, g_T and stochastic matrices H_T, G_T with

$$\text{plim}(h_T - g_T) = 0, \quad h_T \xrightarrow{d} h,$$

and

$$\text{plim } H_T = \text{plim } G_T = H,$$

we get

$$H_T h_T - G_T g_T = (H_T - H) h_T + H(h_T - g_T) + (H - G_T) g_T \xrightarrow{p} 0.$$

∎

Proposition 4.10 implies that

$$\widehat{\lambda}_s := T \widehat{b}_1' \widehat{b}_1 / 6 \xrightarrow{d} \chi^2(K), \tag{4.5.16}$$

$$\widehat{\lambda}_k := T(\widehat{b}_2 - \mathbf{3}_K)'(\widehat{b}_2 - \mathbf{3}_K)/24 \xrightarrow{d} \chi^2(K), \tag{4.5.17}$$

and

$$\widehat{\lambda}_{sk} := \widehat{\lambda}_s + \widehat{\lambda}_k \xrightarrow{d} \chi^2(2K). \quad (4.5.18)$$

Thus, all three statistics may be used for testing nonnormality.

As we have seen, the results hold for any matrix satisfying $\widehat{P}\widehat{P}' = \widehat{\Sigma}_u$. For example, $\widehat{P}$ may be a lower triangular matrix with positive diagonal obtained by a Choleski decomposition of $\widehat{\Sigma}_u$. Clearly, in this case $\widehat{P}$ is a consistent estimator of the corresponding matrix P (see Proposition 3.6). Doornik & Hansen (1994) point out that with this choice the test results will depend on the ordering of the variables. Therefore they suggest using a matrix based on the square root of the correlation matrix corresponding to $\widehat{\Sigma}_u$ instead. In any case, the matrix $\widehat{P}$ is not unique and, hence, the tests will depend to some extent on its choice. Strictly speaking, if one particular $\widehat{P}$ is found for which the null hypothesis can be rejected, this result provides evidence against the normality of the process. Thus, different $\widehat{P}$ matrices could be applied in principle.

For illustrative purposes we consider our standard investment/income/consumption example from Section 3.2.3. Using the least squares residuals from the VAR(2) model with intercepts and a Choleski decomposition of $\widehat{\Sigma}_u$ yields

$$\widehat{\lambda}_s = 3.15 \quad \text{and} \quad \widehat{\lambda}_k = 4.69$$

which are both smaller than $\chi^2(3)_{.90} = 6.25$, the critical value of an asymptotic 10% level test. Also

$$\widehat{\lambda}_{sk} = 7.84 < \chi^2(6)_{.90} = 10.64.$$

Thus, based on these asymptotic tests we cannot reject the null hypothesis of a Gaussian data generation process.

It was pointed out by Kilian & Demiroglu (2000) that the small sample distributions of the test statistics may differ substantially from their asymptotic approximations. Thus, the tests may not be very reliable in practice. Kilian & Demiroglu (2000) proposed bootstrap versions to alleviate the problem.

4.6 Tests for Structural Change

Time invariance or stationarity of the data generation process is an important condition that was used in deriving the properties of estimators and in computing forecasts and forecast intervals. Recall that stationarity is a property that ensures constant means, variances, and autocovariances of the process through time. As we have seen in the investment/income/consumption example, economic time series often have characteristics that do not conform with the assumption of stationarity of the underlying data generation process. For instance, economic time series often have trends or pronounced seasonal components and time varying variances. While these components can sometimes be eliminated by simple transformations, there remains another important source of nonstationarity, namely events that cause turbulence in economic systems in particular time periods. For instance, wars usually change the economic conditions in some areas or countries markedly. Also new tax legislation may have a major impact on some economic variables. Furthermore, the oil price shocks in 1973/74 and 1979/80 are events that have caused drastic changes in some variables (notably the price for gasoline). Such events may be sources of structural change in economic systems.

Because stability and, hence, stationarity is an important assumption in our analysis, it is desirable to have tools for checking this assumed property of the data generation process. In this section, we consider two types of tests that can be used for this purpose. The first set of tests checks whether a change in the parameters has occurred at some point in time by comparing the estimated parameters before and after the possible break date. These tests are known as *Chow tests*. The second set of tests is based on comparing forecasts with actually observed values. More precisely, forecasts are made prior to a period of possible structural change and are compared to the values actually observed during that period. The stability or stationarity hypothesis is rejected if the forecasts differ too much from the actually observed values. These tests are presented in Sections 4.6.1 and 4.6.2. Other tests will be considered in later chapters.

4.6.1 Chow Tests

Suppose a change in the parameters of the VAR(p) process (4.1.1) is suspected after period $T_1 < T$. Given a sample $y_1, \ldots, y_T$ plus the required presample values, the model can be set up as follows for estimation purposes:

$$[Y_{(1)} : Y_{(2)}] = [B_1 : B_2]\mathbf{Z} + [U_{(1)} : U_{(2)}] = \mathbf{BZ} + U,$$

where $Y_{(1)} := [y_1, \ldots, y_{T_1}]$, $Y_{(2)} := [y_{T_1+1}, \ldots, y_T]$, U is partitioned accordingly, $B_1 := [\nu_1, A_{11}, \ldots, A_{p1}]$ and $B_2 := [\nu_2, A_{12}, \ldots, A_{p2}]$ are the $(K \times (pK+1))$ dimensional parameter matrices associated with the first ($t = 1, \ldots, T_1$) and last ($t = T_1+1, \ldots, T$) subperiods, respectively, $\mathbf{B} := [B_1 : B_2]$ is ($K \times 2(Kp+1)$) dimensional and

$$\mathbf{Z} := \begin{bmatrix} Z_{(1)} & 0 \\ 0 & Z_{(2)} \end{bmatrix}.$$

Here $Z_{(1)} := [Z_0, \ldots, Z_{T_1-1}]$ and $Z_{(2)} := [Z_{T_1}, \ldots, Z_{T-1}]$ with $Z_t' := (1, y_t', \ldots, y_{t-p+1}')$, as usual.

In this model setup, a test for parameter constancy checks

$$H_0 : B_1 = B_2 \text{ or } [I : -I]\text{vec}(\mathbf{B}) = 0 \quad \text{versus} \quad H_1 : B_1 \neq B_2.$$

Clearly, this is just a linear hypothesis which can be handled easily within our LS or ML framework under standard assumptions. For example, the LS estimator of $\mathbf{B}$ is

$$\widehat{\mathbf{B}} = [Y_{(1)} : Y_{(2)}] \mathbf{Z}'(\mathbf{ZZ}')^{-1}$$
$$= \left[Y_{(1)} Z'_{(1)} (Z_{(1)} Z'_{(1)})^{-1} : Y_{(2)} Z'_{(2)} (Z_{(2)} Z'_{(2)})^{-1} \right].$$

It has an asymptotic normal distribution under the assumptions of Proposition 3.1. To appeal to that proposition it has to be ensured, however, that $T^{-1}\mathbf{ZZ}'$ converges in probability to a nonsingular matrix. In other words,

$$\operatorname{plim} \frac{T_i}{T} \frac{1}{T_i} Z_{(i)} Z'_{(i)}, \quad i = 1, 2,$$

has to exist and be nonsingular. Hence, T_i/T must not go to zero when T goes to ∞, so that both subperiods before and after the break must be assumed to increase with T. If the assumptions for asymptotic normality can be justified, a Wald test can, for example, be used to test the stability hypothesis. Alternatively, an LR or quasi LR test may be applied. This type of test is often given the label *Chow test* in the literature.

There are some practical matters in applying these tests in the present context that are worth noting. If the possible break date is very close to the sample beginning or the sample end, the LS/ML estimators of B_i may not be available due to lack of degrees of freedom. While at the sample beginning one may be ready to delete a few observations to eliminate the structural break, this option is often undesirable at the end of the sample. For example, if forecasting is the objective of the analysis, a break towards the end of the sample would clearly be problematic. Therefore, the so-called *Chow forecast tests* have been proposed which also work for break dates close to the sample end. In the next subsection, we present a slightly different set of forecast tests which may be applied instead.

Even if the suspected break point is well inside the sample period so that the application of the standard Chow test is unproblematic in principle, in practice, the break may not occur in one period. If there is a longer time phase in which a parameter shift to a new level takes place, it may be useful to eliminate a few observations around the break date and use only the remaining ones in estimating the parameters. One may also argue that using some observations from periods up to T_1 in $Z_{(2)}$ may be problematic and may result in reduced power because observations from both subperiods are mixed in estimating B_2. Under the null hypothesis of parameter constancy, this should be no problem, however, because, under H_0, the same process is in operation before and after T_1. Still, from the point of view of maximizing power, deleting some observations around the possible break point may be a good idea.

Other practical problems may result from multiple structural breaks within the sample period. In principle, it is no problem to test multiple break points simultaneously. Also, to improve power, one may only test some of the parameters or one may wish to test for a changing white noise covariance matrix which is implicitly assumed to be time invariant in the foregoing discussion. Details of such extensions will be discussed in Chapter 17.

184   4 VAR Order Selection and Checking the Model Adequacy

So far we have considered asymptotic results only. Unfortunately, it was found by Candelon & Lütkepohl (2001) that asymptotic theory may be an extremely poor guide for the small sample properties of Chow tests, in particular, if models with many parameters are under consideration. To improve the reliability of the tests, these authors proposed to use bootstrapped versions. Bootstrapped p-values may be obtained as described in Appendix D.3.

For the German investment/income/consumption data we have fitted a VAR(2) model to data up to 1982.4 and we have performed a Chow test for a break in period 1979.1. The test value is 30.5. Comparing that to 29.6, the 10% critical value of a $\chi^2(21)$ distribution, stability is rejected at the 10% level. A bootstrapped p-value based on 2000 bootstrap replications turns out to be 0.21, however. Thus, based on the bootstrapped test, stability is not rejected. It is typical for the test based on the asymptotic χ^2-distribution that it rejects more often in small samples than the specified nominal significance level, even if the model is stable. This distortion is at least partly corrected by the bootstrap.

4.6.2 Forecast Tests for Structural Change

A Test Statistic Based on one Forecast Period

Suppose y_t is a K-dimensional stationary, stable *Gaussian* VAR(p) process as in (4.1.1). The optimal h-step forecast at time T is denoted by $y_T(h)$ and the corresponding forecast error is

$$e_T(h) := y_{T+h} - y_T(h) = \sum_{i=0}^{h-1} \Phi_i u_{T+h-i} = [\Phi_{h-1} : \cdots : \Phi_1 : I_K] \mathbf{u}_{T,h} \tag{4.6.1}$$

where $\mathbf{u}_{T,h} := (u'_{T+1}, \ldots, u'_{T+h})'$, the Φ_i are the coefficient matrices of the canonical MA representation (see Section 2.2.2). Because $\mathbf{u}_{T,h} \sim \mathcal{N}(0, I_h \otimes \Sigma_u)$, the forecast error is a linear transformation of a multivariate normal distribution and, consequently (see Appendix B),

$$e_T(h) \sim \mathcal{N}(0, \Sigma_y(h)), \tag{4.6.2}$$

where

$$\Sigma_y(h) = \sum_{i=0}^{h-1} \Phi_i \Sigma_u \Phi'_i$$

is the forecast MSE matrix (see (2.2.11)). Hence,

$$\tau_h := e_T(h)' \Sigma_y(h)^{-1} e_T(h) \sim \chi^2(K) \tag{4.6.3}$$

by Proposition B.3 of Appendix B.

This derivation assumes that y_{T+h} is generated by the same VAR(p) process that has generated the y_t for $t \leq T$. If this process does not prevail in period $T+h$, the statistic τ_h will, in general, not have a central χ^2-distribution. Hence, τ_h may be used to test the null hypothesis

H_0: (4.6.2) is true, that is, y_{T+h} is generated by the same Gaussian VAR(p) process that has generated $y_1, \ldots, y_T$.

The alternative hypothesis is that y_{T+h} is not generated by the same process as $y_1, \ldots, y_T$. The null hypothesis is rejected if the forecast errors are large so that τ_h exceeds a prespecified critical value from the $\chi^2(K)$-distribution. Such a test may be performed for $h = 1, 2, \ldots$.

It may be worth noting that in these tests we also check the normality assumption for y_t. Even if the same process has generated y_{T+h} and $y_1, \ldots, y_T$, (4.6.2) will not hold if that process is not Gaussian. Thus, the normality assumption for y_t is part of H_0. Other possible deviations from the null hypothesis include changes in the mean and changes in the variance of the process.

In practice, the tests are not feasible in their present form because τ_h involves unknown quantities. The forecast errors $e_T(h)$ and the MSE matrix $\Sigma_y(h)$ are both unknown and must be replaced by estimators. For the forecast errors, we use

$$\widehat{e}_T(h) := y_{T+h} - \widehat{y}_t(h) = \sum_{i=0}^{h-1} \widehat{\Phi}_i \widehat{u}_{T+h-i}, \qquad (4.6.4)$$

where the $\widehat{\Phi}_i$ are obtained from the coefficient estimators $\widehat{A}_i$ in the usual way (see Section 3.5.2) and

$$\widehat{u}_t := y_t - \widehat{\nu} - \widehat{A}_1 y_{t-1} - \cdots - \widehat{A}_p y_{t-p}.$$

The MSE matrix may be estimated by

$$\widehat{\Sigma}_y(h) := \sum_{i=0}^{h-1} \widehat{\Phi}_i \widehat{\Sigma}_u \widehat{\Phi}_i', \qquad (4.6.5)$$

where $\widehat{\Sigma}_u$ is the LS estimator of Σ_u. As usual, we use only data up to period T for estimation and not the data from the forecast period. If the conditions for consistency of the estimators are satisfied, that is,

$$\text{plim } \widehat{\nu} = \nu, \quad \text{plim } \widehat{A}_i = A_i, i = 1, \ldots, p, \quad \text{and} \quad \text{plim } \widehat{\Sigma}_u = \Sigma_u,$$

then plim $\widehat{\Phi}_i = \Phi_i$, plim $\widehat{\Sigma}_y(h) = \Sigma_y(h)$ and

$$\begin{aligned}\text{plim}(\widehat{u}_t - u_t) &= \text{plim}(\nu - \widehat{\nu}) + \text{plim}(A_1 - \widehat{A}_1)y_{t-1} + \cdots \\ &\quad + \text{plim}(A_p - \widehat{A}_p)y_{t-p} \\ &= 0.\end{aligned}$$

Hence, defining

$$\widehat{\tau}_h := \widehat{e}_T(h)' \widehat{\Sigma}_y(h) \widehat{e}_T(h),$$

we get plim $(\widehat{\tau}_h - \tau_h) = 0$ and, thus, by Proposition C.2(2) of Appendix C,

$$\widehat{\tau}_h \xrightarrow{d} \chi^2(K). \tag{4.6.6}$$

In other words, if the unknown coefficients are replaced by consistent estimators, the resulting test statistics $\widehat{\tau}_h$ have the same *asymptotic* distributions as the τ_h.

Of course, it is desirable to know whether the $\chi^2(K)$-distribution is a good approximation to the distribution of $\widehat{\tau}_h$ in small samples. This, however, is not likely because in Section 3.5.1,

$$\Sigma_{\widehat{y}}(h) = \Sigma_y(h) + \frac{1}{T}\Omega(h) \tag{4.6.7}$$

was found to be a better approximation to the MSE matrix than $\Sigma_y(h)$, if the forecasts are based on an estimated process. While asymptotically, as $T \to \infty$, the term $\Omega(h)/T$ vanishes, it seems plausible to include this term in small samples. For univariate processes, it was confirmed in a simulation study by Lütkepohl (1988b) that inclusion of the term results in a better agreement between the small sample and asymptotic distributions. For multivariate vector processes, the simulation results of Section 3.5.4 point in the same direction. Thus, in small samples a statistic of the type

$$\widehat{e}_T(h)' \widehat{\Sigma}_{\widehat{y}}(h)^{-1} \widehat{e}_T(h)$$

is more plausible than $\widehat{\tau}_h$. Here $\widehat{\Sigma}_{\widehat{y}}(h)$ is the estimator given in Section 3.5.2. In addition to this adjustment, it is useful to adjust the statistic for using an estimated rather than known forecast error covariance matrix. Such an adjustment is often done by dividing by the degrees of freedom and using the statistic in conjunction with critical values from an F-distribution. That is, we may use

$$\overline{\tau}_h := \widehat{e}_T(h)' \widehat{\Sigma}_{\widehat{y}}(h)^{-1} \widehat{e}_T(h)/K \approx F(K, T - Kp - 1). \tag{4.6.8}$$

The approximate F-distribution follows from Proposition C.3(2) of Appendix C and the denominator degrees of freedom are chosen by analogy with a result due to Hotelling (e.g., Anderson (1984)). Other choices are possible. Proposition C.3(2) requires, however, that the denominator degrees of freedom go to infinity with the sample size T.

A Test Based on Several Forecast Periods

Another set of stationarity tests is obtained by observing that the errors of forecasts 1- to h-steps ahead are also jointly normally distributed under the null hypothesis of structural stability,

4.6 Tests for Structural Change

$$\mathbf{e}_T(h) := \begin{bmatrix} e_T(1) \\ \vdots \\ e_T(h) \end{bmatrix} = \mathbf{\Phi}_h \mathbf{u}_{T,h} \sim \mathcal{N}(0, \mathbf{\Sigma}_y(h)), \qquad (4.6.9)$$

where

$$\mathbf{\Phi}_h := \begin{bmatrix} I_K & 0 & \cdots & 0 \\ \Phi_1 & I_K & & 0 \\ \vdots & \vdots & \ddots & \vdots \\ \Phi_{h-1} & \Phi_{h-2} & \cdots & I_K \end{bmatrix} \qquad (4.6.10)$$

so that

$$\mathbf{\Sigma}_y(h) := \mathbf{\Phi}_h (I_h \otimes \Sigma_u) \mathbf{\Phi}_h'. \qquad (4.6.11)$$

Using again Proposition B.3 of Appendix B,

$$\begin{aligned} \lambda_h &:= \mathbf{e}_T(h)' \mathbf{\Sigma}_y(h)^{-1} \mathbf{e}_T(h) = \mathbf{u}_{T,h}'(I_h \otimes \Sigma_u^{-1}) \mathbf{u}_{T,h} \\ &= \sum_{i=1}^h u_{T+i}' \Sigma_u^{-1} u_{T+i} = \lambda_{h-1} + u_{T+h}' \Sigma_u^{-1} u_{T+h} \sim \chi^2(hK). \end{aligned} \qquad (4.6.12)$$

Thus, λ_h may be used to check whether a structural change has occurred during the periods $T+1, \ldots, T+h$.

To make this test feasible, it is necessary to replace unknown quantities by estimators just as in the case of the τ-tests. Denoting the test statistics based on estimated VAR processes by $\widehat{\lambda}_h$,

$$\widehat{\lambda}_h \xrightarrow{d} \chi^2(hK) \qquad (4.6.13)$$

follows with the same arguments used for $\widehat{\tau}_h$, provided consistent parameter estimators are used.

Again it seems plausible to make small sample adjustments to the statistics to take into account the fact that estimated quantities are used. The last expression in (4.6.12) suggests that a closer look at the terms

$$u_{T+i}' \Sigma_u^{-1} u_{T+i} \qquad (4.6.14)$$

is useful in searching for a small sample adjustment. This expression involves the 1-step ahead forecast errors $u_{T+i} = y_{T+i} - y_{T+i-1}(1)$. If estimated coefficients are used in the 1-step ahead forecast, the MSE or forecast error covariance matrix is approximately inflated by a factor $(T + Kp + 1)/T$ (see (3.5.13)). Because λ_h is the sum of terms of the form (4.6.14), it may be suitable to replace Σ_u by $(T + Kp + 1)\widehat{\Sigma}_u/T$ when estimated quantities are used. Note, however, that such an adjustment ignores possible dependencies between the estimated $\widehat{u}_{T+i}$ and $\widehat{u}_{T+j}$. Nevertheless, it leads to a computationally extremely simple form and was therefore proposed in the literature

(Lütkepohl (1989b)). Furthermore, it was suggested to divide by the degrees of freedom of the asymptotic χ^2-distribution and, by appeal to Proposition C.3(2) of Appendix C, use the resulting statistic, $\overline{\lambda}_h$ say, in conjunction with critical values from F-distributions to adjust for the fact that Σ_u is replaced by an estimator. In other words,

$$\overline{\lambda}_h := T \sum_{i=1}^{h} \widehat{u}'_{T+i} \widehat{\Sigma}_u^{-1} \widehat{u}_{T+i} / [(T+Kp+1)Kh] \approx F(Kh, T-Kp-1). \quad (4.6.15)$$

The denominator degrees of freedom are chosen by the same arguments used in (4.6.8). Obviously, $\overline{\lambda}_1 = \overline{\tau}_1$.

Now we have different sets of stationarity tests and the question arises which ones to use in practice. To answer this question, it would be useful to know the power characteristics of the tests because it is desirable to use the most powerful test available. For some alternatives the τ- and λ-statistics have noncentral χ^2-distributions (Lütkepohl (1988b, 1989)). In these cases it is possible to investigate and compare their powers. It turns out that for some alternatives the τ-tests are more powerful than the λ-tests and for other alternatives the opposite is true. Because we usually do not know the exact form of the alternative (the exact form of the structural change) it may be a good idea to apply both tests in practice. In addition, a Chow test may be used.

An Example

To illustrate the use of the two tests for stationarity, we use the first differences of logarithms of the West German investment, income, and consumption data and test for a possible structural change caused by the oil price shocks in 1973/74 and 1979/80. Because the first drastic price increase occurred in late 1973, we have estimated a VAR(2) model using the sample period 1960.4–1973.2 and presample values from 1960.2 and 1960.3. Thus $T = 51$. It is important to note that the data from the forecast period are not used for estimation. We have used the estimated process to compute the $\overline{\tau}_h$ and $\overline{\lambda}_h$ for $h = 1, \ldots, 8$. The results are given in Table 4.9 together with the p-values of the tests. The p-value is the probability that the test statistic assumes a value greater than the observed test value, if the null hypothesis is true. Thus, p-values smaller than .10 or .05 would be of concern. Obviously, in this case none of the test values is statistically significant at the 10% level. Thus, the tests do not give rise to concern about the stationarity of the underlying data generation process during the period in question. Although we have given the $\overline{\tau}_h$ and $\overline{\lambda}_h$ values for various forecast horizons h in Table 4.9, we emphasize that the tests are not independent for different h. Thus, the evidence from the set of tests should not lead to overrating the confidence we may have in this result.

Table 4.9. Stability tests for the investment/income/consumption system for 1973–1975

quarter	forecast horizon h	$\overline{\tau}_h$	p-value	$\overline{\lambda}_h$	p-value
1973.3	1	.872	.46	.872	.46
4	2	.271	.85	.717	.64
1974.1	3	.206	.89	.517	.85
2	4	.836	.48	.627	.81
3	5	.581	.63	.785	.69
4	6	.172	.91	.832	.65
1975.1	7	.126	.94	.863	.63
2	8	1.450	.24	1.041	.44

To check the possibility of a structural instability due to the 1979/80 oil price increases, we used the VAR(2) model of Section 3.2.3 which is based on data up to the fourth quarter of 1978. The resulting values of the test statistics for $h = 1, \ldots, 8$ are presented in Table 4.10. Again none of the values is significant at the 10% level. However, in Section 3.5.2, we found that the observed consumption values in 1979 fall outside a 95% forecast interval. Hence, looking at the three series individually, a possible nonstationarity would be detected by a prediction test. This possible instability in 1979 was a reason for using only data up to 1978 in the examples of previous chapters and sections. The example indicates what can also be demonstrated theoretically, namely that the power of a test based on joint forecasts of various variables may be lower than the power of a test based on forecasts for individual variables (see Lütkepohl (1989b)).

4.7 Exercises

4.7.1 Algebraic Problems

Problem 4.1
Show that the restricted ML estimator $\widetilde{\beta}_r$ can be written in the form (4.2.10).

Problem 4.2
Prove Lemma 4.1.
[Hint: Suppose $k < n$. Then

Table 4.10. Stability tests for the investment/income/consumption system for 1979–1980

quarter	forecast horizon h	$\overline{\tau}_h$	p-value	$\overline{\lambda}_h$	p-value
1979.1	1	.277	.84	.277	.84
2	2	2.003	.12	1.077	.38
3	3	2.045	.12	1.464	.18
4	4	.203	.89	1.245	.27
1980.1	5	.630	.60	1.339	.20
2	6	1.898	.86	1.374	.17
3	7	.188	.90	1.204	.28
4	8	.535	.66	1.124	.34

$$\begin{aligned} c_n + a_n &= c_n + (a_n - a_{n-1}) + \cdots + (a_{k+1} - a_k) + a_k \\ &> c_n + (b_n - b_{n-1}) + \cdots + (b_{k+1} - b_k) + a_k \\ &\geq c_k + b_k - b_k + a_k \\ &= c_k + a_k \end{aligned}$$

which contradicts (4.3.13b).[2]]

Problem 4.3
Show (4.4.14) and (4.4.15).
[Hint:

$$\text{vec}(\sqrt{T}UF[I_h \otimes Z'(\widehat{B} - B)']/T)$$
$$= \sqrt{T} \text{ vec}\left[\frac{1}{T}UF(I_h \otimes Z')(I_h \otimes (\widehat{B} - B)')\right]$$
$$= \left[I_K \otimes \frac{1}{T}UF(I_h \otimes Z')\right]\sqrt{T} \text{ vec}(I_h \otimes (\widehat{B} - B)')$$

and

$$\sqrt{T} \text{ vec}\left(\frac{1}{T}(\widehat{B} - B)ZF\left[I_h \otimes Z'(\widehat{B} - B)'\right]\right)$$
$$= \left\{\left[I_h \otimes (\widehat{B} - B)\right]\frac{(I_h \otimes Z)F'Z'}{T} \otimes I_K\right\}\sqrt{T} \text{ vec}(\widehat{B} - B).\right]$$

Problem 4.4
Show (4.5.14).

---
[2] I thank Prof. K. Schürger, Universität Bonn, for pointing out this proof.

[Hint: Note that

$$(y_{t-1} - \bar{y}) \otimes (u_t - \bar{u}) \otimes (u_t - \bar{u}) = (y_{t-1} - \mu) \otimes u_t \otimes u_t$$
$$-(y_{t-1} - \mu) \otimes u_t \otimes \bar{u} + \cdots,$$

define new variables of the type

$$z_t = (y_{t-1} - \mu) \otimes u_t \otimes u_t$$

and use that

$$\text{plim} \frac{1}{T} \sum_t z_t = E(z_t) = 0.]$$

Problem 4.5
Using the notation and assumptions from Proposition 4.1, show that

$$\frac{\partial \ln l(\widetilde{\boldsymbol{\beta}}_r)}{\partial \boldsymbol{\beta}'} \frac{\partial^2 \ln l(\widetilde{\boldsymbol{\beta}}_r)}{\partial \boldsymbol{\beta} \partial \boldsymbol{\beta}'} \frac{\partial \ln l(\widetilde{\boldsymbol{\beta}}_r)}{\partial \boldsymbol{\beta}} = (\widetilde{\boldsymbol{\beta}}_r - \widetilde{\boldsymbol{\beta}})'(ZZ' \otimes (\widetilde{\Sigma}_u^r)^{-1})(\widetilde{\boldsymbol{\beta}}_r - \widetilde{\boldsymbol{\beta}}).$$

4.7.2 Numerical Problems

The following problems require the use of a computer. They refer to the bivariate series $y_t = (y_{1t}, y_{2t})'$ of first differences of the U.S. investment data in File E2, available from the author's webpage.

Problem 4.6
Set up a sequence of tests for the correct VAR order of the data generating process using a maximum order of $M = 4$. Compute the required χ^2 and F likelihood ratio statistics. Which order would you choose?

Problem 4.7
Determine VAR order estimates on the basis of the four criteria FPE, AIC, HQ, and SC. Use a maximum VAR order of $M = 4$ in a first estimation round and $M = 8$ in a second estimation round. Compare the results.

Problem 4.8
Compute the residual autocorrelations $\widehat{R}_1, \ldots, \widehat{R}_{12}$ and estimate their standard errors using the VAR(1) model obtained in Problem 3.12. Interpret your results.

Problem 4.9
Compute LM test values $\lambda_{LM}(1)$, $\lambda_{LM}(2)$, and $\lambda_{LM}(4)$ and portmanteau test values Q_h and $\bar{Q}_h$ for $h = 10$ and 12 for the VAR(1) model of the previous problem. Test the whiteness of the residuals.

Problem 4.10
On the basis of a VAR(1) model, perform a test for nonnormality of the example data.

Problem 4.11
Investigate whether there was a structural change in U.S. investment after 1965 (possibly due to the increasing U.S. engagement in Vietnam).

5
VAR Processes with Parameter Constraints

5.1 Introduction

In Chapter 3, we have discussed estimation of the parameters of a K-dimensional stationary, stable VAR(p) process of the form

$$y_t = \nu + A_1 y_{t-1} + \cdots + A_p y_{t-p} + u_t, \tag{5.1.1}$$

where all the symbols have their usual meanings. In the investment/income/consumption example considered throughout Chapter 3, we found that many of the coefficient estimates were not significantly different from zero. This observation may be interpreted in two ways. First, some of the coefficients may actually be zero and this fact may be reflected in the estimation results. For instance, if some variable is not Granger-causal for the remaining variables, zero coefficients are encountered. Second, insignificant coefficient estimates are found if the information in the data is not rich enough to provide sufficiently precise estimates with confidence intervals that do not contain zero.

In the latter case, one may want to think about better ways to extract the information from the data because, as we have seen in Chapter 3, a large estimation uncertainty for the VAR coefficients leads to poor forecasts (large forecast intervals) and imprecise estimates of the impulse responses and forecast error variance components. Getting imprecise parameter estimates in a VAR analysis is a common practical problem because the number of parameters is often quite substantial relative to the available sample size or time series length. Various cures for this problem have been proposed in the literature. They all amount to putting constraints on the coefficients.

For instance, in the previous chapter, choosing the VAR order p has been discussed. Selecting an order that is less than the maximum order amounts to placing zero constraints on VAR coefficient matrices. This way complete coefficient matrices are eliminated. In the present chapter, we will discuss putting zero constraints on individual coefficients. Such constraints are but one form

5.2 Linear Constraints

In this section, the consequences of estimating the VAR coefficients subject to linear constraints will be considered. Different estimation procedures are treated in Subsections 5.2.2–5.2.5; forecasting and impulse response analysis are discussed in Subsections 5.2.6 and 5.2.7, respectively; strategies for model selection or the choice of constraints are dealt with in Subsection 5.2.8; model checking follows in Subsection 5.2.9; and, finally, an example is discussed in Subsection 5.2.10.

5.2.1 The Model and the Constraints

We consider the model (5.1.1) for $t = 1, \ldots, T$, written in compact form

$$Y = BZ + U, \qquad (5.2.1)$$

where

$$Y := [y_1, \ldots, y_T], \quad Z := [Z_0, \ldots, Z_{T-1}] \text{ with } Z_t := \begin{bmatrix} 1 \\ y_t \\ \vdots \\ y_{t-p+1} \end{bmatrix},$$

$$B := [\nu, A_1, \ldots, A_p], \quad U := [u_1, \ldots, u_T].$$

Suppose that linear constraints for B are given in the form

$$\beta := \text{vec}(B) = R\gamma + r, \qquad (5.2.2)$$

where $\beta = \text{vec}(B)$ is a $(K(Kp+1) \times 1)$ vector, R is a known $(K(Kp+1) \times M)$ matrix of rank M, γ is an unrestricted $(M \times 1)$ vector of unknown parameters, and r is a $K(Kp+1)$-dimensional vector of known constants. All the linear restrictions of interest can be expressed in this form. For instance, the restriction $A_p = 0$ can be written as in (5.2.2) by choosing $M = K^2(p-1)+K$,

$$R = \begin{bmatrix} I_M \\ 0 \end{bmatrix}, \quad \gamma = \text{vec}(\nu, A_1, \ldots, A_{p-1}),$$

and $r = 0$.

Although (5.2.2) is not the most conventional form of representing linear constraints, it is used here because it is particularly useful for our purposes. Often the constraints are expressed as

$$C\boldsymbol{\beta} = c, \tag{5.2.3}$$

where C is a known $(N \times (K^2 p + K))$ matrix of rank N and c is a known $(N \times 1)$ vector (see Chapter 4, Section 4.2.2). Because $\text{rk}(C) = N$, the matrix C has N linearly independent columns. For simplicity we assume that the first N columns are linearly independent and partition C as $C = [C_1 : C_2]$, where C_1 is $(N \times N)$ nonsingular and C_2 is $(N \times (K^2 p + K - N))$. Partitioning $\boldsymbol{\beta}$ conformably gives

$$[C_1 : C_2] \begin{bmatrix} \boldsymbol{\beta}_1 \\ \boldsymbol{\beta}_2 \end{bmatrix} = C_1 \boldsymbol{\beta}_1 + C_2 \boldsymbol{\beta}_2 = c$$

or

$$\boldsymbol{\beta}_1 = -C_1^{-1} C_2 \boldsymbol{\beta}_2 + C_1^{-1} c.$$

Therefore, choosing

$$R = \begin{bmatrix} -C_1^{-1} C_2 \\ I_{pK^2 + K - N} \end{bmatrix}, \qquad \boldsymbol{\gamma} = \boldsymbol{\beta}_2, \qquad \text{and} \qquad r = \begin{bmatrix} C_1^{-1} c \\ 0 \end{bmatrix},$$

the constraints (5.2.3) can be written in the form (5.2.2). Also, it is not difficult to see that restrictions written as in (5.2.2) can be expressed in the form $C\boldsymbol{\beta} = c$ for suitable C and c. Thus, the two forms are equivalent.

The representation (5.2.2) permits to impose the constraints by a simple reparameterization of the original model. Vectorizing (5.2.1) and replacing $\boldsymbol{\beta}$ by $R\boldsymbol{\gamma} + r$ gives

$$\begin{aligned} \mathbf{y} &:= \text{vec}(Y) = (Z' \otimes I_K) \text{vec}(B) + \text{vec}(U) \\ &= (Z' \otimes I_K)(R\boldsymbol{\gamma} + r) + \mathbf{u} \end{aligned}$$

or

$$\mathbf{z} = (Z' \otimes I_K) R\boldsymbol{\gamma} + \mathbf{u}, \tag{5.2.4}$$

where $\mathbf{z} := \mathbf{y} - (Z' \otimes I_K) r$ and $\mathbf{u} := \text{vec}(U)$. This form of the model allows us to derive the estimators and their properties just like in the original unconstrained model. Estimation of $\boldsymbol{\gamma}$ and $\boldsymbol{\beta}$ will be discussed in the following subsections.

5.2.2 LS, GLS, and EGLS Estimation

Asymptotic Properties

Denoting by Σ_u the covariance matrix of u_t, the vector $\widehat{\boldsymbol{\gamma}}$ minimizing

$$\begin{aligned} S(\boldsymbol{\gamma}) &= \mathbf{u}'(I_T \otimes \Sigma_u^{-1}) \mathbf{u} \\ &= [\mathbf{z} - (Z' \otimes I_K) R\boldsymbol{\gamma}]'(I_T \otimes \Sigma_u^{-1})[\mathbf{z} - (Z' \otimes I_K) R\boldsymbol{\gamma}] \end{aligned} \tag{5.2.5}$$

with respect to γ is easily seen to be

$$\begin{aligned}\widehat{\gamma} &= [R'(ZZ' \otimes \Sigma_u^{-1})R]^{-1}R'(Z \otimes \Sigma_u^{-1})\mathbf{z} \\ &= [R'(ZZ' \otimes \Sigma_u^{-1})R]^{-1}R'(Z \otimes \Sigma_u^{-1})[(Z' \otimes I_K)R\gamma + \mathbf{u}] \\ &= \gamma + [R'(ZZ' \otimes \Sigma_u^{-1})R]^{-1}R'(I_{Kp+1} \otimes \Sigma_u^{-1})\operatorname{vec}(UZ')\end{aligned} \quad (5.2.6)$$

(see Chapter 3, Section 3.2.1). This estimator is commonly called a *generalized LS (GLS)* estimator because it minimizes the generalized sum of squared errors $S(\gamma)$ rather than the sum of squared errors $\mathbf{u}'\mathbf{u}$. We will see shortly that in contrast to the unrestricted case considered in Chapter 3, it may make a difference here whether $S(\gamma)$ or $\mathbf{u}'\mathbf{u}$ is used as the objective function. The GLS estimator is in general asymptotically more efficient than the multivariate LS estimator and is therefore preferred here. We will see in Section 5.2.3 that, under Gaussian assumptions, the GLS estimator is equivalent to the ML estimator. From (5.2.6),

$$\sqrt{T}(\widehat{\gamma} - \gamma) = \left[R'\left(\frac{ZZ'}{T} \otimes \Sigma_u^{-1}\right)R\right]^{-1} R'(I_{Kp+1} \otimes \Sigma_u^{-1})\frac{1}{\sqrt{T}}\operatorname{vec}(UZ') \quad (5.2.7)$$

and the asymptotic properties of $\widehat{\gamma}$ are obtained as in Proposition 3.1.

Proposition 5.1 (*Asymptotic Properties of the GLS Estimator*)
Suppose the conditions of Proposition 3.1 are satisfied, that is, y_t is a K-dimensional stable, stationary VAR(p) process and u_t is independent white noise with bounded fourth moments. If $\beta = R\gamma + r$ as in (5.2.2) with $\operatorname{rk}(R) = M$, then $\widehat{\gamma}$ given in (5.2.6) is a consistent estimator of γ and

$$\sqrt{T}(\widehat{\gamma} - \gamma) \xrightarrow{d} \mathcal{N}(0, [R'(\Gamma \otimes \Sigma_u^{-1})R]^{-1}), \quad (5.2.8)$$

where $\Gamma := E(Z_t Z_t') = \operatorname{plim} ZZ'/T$. ∎

Proof: Under the conditions of the proposition, $\operatorname{plim}(ZZ'/T) = \Gamma$ and

$$\frac{1}{T}\operatorname{vec}(UZ') \xrightarrow{d} \mathcal{N}(0, \Gamma \otimes \Sigma_u)$$

(see Lemma 3.1). Hence, by results stated in Appendix C, Proposition C.15(1), using (5.2.7), $\sqrt{T}(\widehat{\gamma} - \gamma)$ has an asymptotic normal distribution with covariance matrix

$$[R'(\Gamma \otimes \Sigma_u^{-1})R]^{-1}R'(I \otimes \Sigma_u^{-1})(\Gamma \otimes \Sigma_u)(I \otimes \Sigma_u^{-1})R[R'(\Gamma \otimes \Sigma_u^{-1})R]^{-1}$$
$$= [R'(\Gamma \otimes \Sigma_u^{-1})R]^{-1}.$$

∎

Unfortunately, the estimator $\widehat{\gamma}$ is of limited value in practice because its computation requires knowledge of Σ_u. Since this matrix is usually unknown, it has to be replaced by an estimator. Using any consistent estimator $\bar{\Sigma}_u$ instead of Σ_u in (5.2.6), we get an *EGLS (estimated GLS)* estimator

$$\widehat{\widehat{\gamma}} = [R'(ZZ' \otimes \bar{\Sigma}_u^{-1})R]^{-1} R'(Z \otimes \bar{\Sigma}_u^{-1})\mathbf{z} \tag{5.2.9}$$

which has the same asymptotic properties as the GLS estimator $\widehat{\gamma}$. This result is an easy consequence of the representation (5.2.7) and Proposition C.15(1) of Appendix C.

Proposition 5.2 (*Asymptotic Properties of the EGLS Estimator*)
Under the conditions of Proposition 5.1, if plim $\bar{\Sigma}_u = \Sigma_u$, the EGLS estimator $\widehat{\widehat{\gamma}}$ in (5.2.9) is asymptotically equivalent to the GLS estimator $\widehat{\gamma}$ in (5.2.6), that is, plim $\widehat{\widehat{\gamma}} = \gamma$ and

$$\sqrt{T}(\widehat{\widehat{\gamma}} - \gamma) \xrightarrow{d} \mathcal{N}(0, [R'(\Gamma \otimes \Sigma_u^{-1})R]^{-1}). \tag{5.2.10}$$

∎

Once an estimator for γ is available, an estimator for β is obtained by substituting in (5.2.2), that is,

$$\widehat{\widehat{\beta}} = R\widehat{\widehat{\gamma}} + r. \tag{5.2.11}$$

The asymptotic properties of this estimator follow immediately from Appendix C, Proposition C.15(2).

Proposition 5.3 (*Asymptotic Properties of the Implied Restricted EGLS Estimator*)

Under the conditions of Proposition 5.2, the estimator $\widehat{\widehat{\beta}} = R\widehat{\widehat{\gamma}} + r$ is consistent and asymptotically normally distributed,

$$\sqrt{T}(\widehat{\widehat{\beta}} - \beta) \xrightarrow{d} \mathcal{N}(0, R[R'(\Gamma \otimes \Sigma_u^{-1})R]^{-1}R'). \tag{5.2.12}$$

∎

To make these EGLS estimators operational, we need a consistent estimator of Σ_u. From Chapter 3, Corollary 3.2.1, we know that, under the conditions of Proposition 5.1,

$$\begin{aligned}\widehat{\Sigma}_u &= \frac{1}{T - Kp - 1}(Y - \widehat{B}Z)(Y - \widehat{B}Z)' \\ &= \frac{1}{T - Kp - 1} Y(I_T - Z'(ZZ')^{-1}Z)Y'\end{aligned} \tag{5.2.13}$$

is a consistent estimator of Σ_u which may thus be used in place of $\bar{\Sigma}_u$. Here $\widehat{B} = YZ'(ZZ')^{-1}$ is the unconstrained multivariate LS estimator of the coefficient matrix B.

Alternatively, the restricted LS estimator minimizing $\mathbf{u}'\mathbf{u}$ with respect to γ may be determined in a first step. The minimizing γ-vector is easily seen to be

$$\check{\gamma} = [R'(ZZ' \otimes I_K)R]^{-1} R'(Z \otimes I_K)\mathbf{z} \tag{5.2.14}$$

(see Problem 5.1). As this LS estimator does not involve the white noise covariance matrix Σ_u, it is generally different from the GLS estimator. We denote the corresponding β-vector by $\check{\beta}$, that is, $\check{\beta} = R\check{\gamma} + r$. Furthermore, $\check{B}$ is the corresponding coefficient matrix, that is, $\text{vec}(\check{B}) = \check{\beta}$. Then we may choose

$$\check{\Sigma}_u = \frac{1}{T}(Y - \check{B}Z)(Y - \check{B}Z)' \tag{5.2.15}$$

as an estimator for Σ_u. The consistency of this estimator is a consequence of Proposition 3.2 and the fact that $\check{B}$ is a consistent estimator of B with asymptotic normal distribution. This result follows from the asymptotic normality of $\check{\gamma}$ which in turn follows by replacing Σ_u with I_K in (5.2.6) and (5.2.7). Thus, $\check{\beta} = R\check{\gamma} + r$ is asymptotically normal. Consequently, we get the following result from Proposition 3.2 and Corollary 3.2.1.

Proposition 5.4 (*Asymptotic Properties of the White Noise Covariance Estimator*)
Under the conditions of Proposition 5.1, $\check{\Sigma}_u$ is consistent and

$$\text{plim}\sqrt{T}(\check{\Sigma}_u - UU'/T) = 0.$$

∎

In (5.2.15), T may be replaced by $T - Kp - 1$ without affecting the consistency of the covariance matrix estimator. However, there is little justification for subtracting $Kp + 1$ from T in the present situation because, due to zero restrictions, some or all of the K equations of the system may contain fewer than $Kp + 1$ parameters.

Of course, in practice one would like to know which one of the possible covariance estimators leads to an EGLS estimator $\widehat{\widehat{\gamma}}$ with best small sample properties. Although we cannot give a general answer to this question, it seems plausible to use an estimator that takes into account the nonsample information concerning the VAR coefficients, provided the restrictions are correct. Thus, if one is confident about the validity of the restrictions, the covariance matrix estimator $\check{\Sigma}_u$ may be used.

As an alternative to the EGLS estimator described in the foregoing, an iterated EGLS estimator may be used. It is obtained by computing a new covariance matrix estimator from the EGLS residuals. This estimator is then used in place of $\bar{\Sigma}_u$ in (5.2.9) and again a new covariance matrix estimator is computed from the corresponding residuals and so on. The procedure is continued until convergence. We will not pursue it here. From Propositions 5.2 and 3.2 it follows that the asymptotic properties of the resulting iterated EGLS estimator are the same as those of the EGLS estimator wherever the iteration is terminated.

Comparison of LS and Restricted EGLS Estimators

A question of interest in this context is how the covariance matrix in (5.2.12) compares with the asymptotic covariance matrix $\Gamma^{-1} \otimes \Sigma_u$ of the unrestricted multivariate LS estimator $\widehat{\boldsymbol{\beta}}$. To see that the restricted estimator has smaller or at least not greater asymptotic variances than the unrestricted estimator, it is helpful to write the restrictions in the form (5.2.3). In that case, the restricted EGLS estimator of $\boldsymbol{\beta}$ turns out to be

$$\widehat{\widehat{\boldsymbol{\beta}}} = \widehat{\boldsymbol{\beta}} + [(ZZ')^{-1} \otimes \bar{\Sigma}_u]C'[C((ZZ')^{-1} \otimes \bar{\Sigma}_u)C']^{-1}(c - C\widehat{\boldsymbol{\beta}}) \tag{5.2.16}$$

(see Chapter 4, Section 4.2.2, and Problem 5.2). Noting that $C\boldsymbol{\beta} - c = 0$, subtracting $\boldsymbol{\beta}$ from both sides of (5.2.16), and multiplying by $\sqrt{T}$ gives

$$\sqrt{T}(\widehat{\widehat{\boldsymbol{\beta}}} - \boldsymbol{\beta}) = \sqrt{T}(\widehat{\boldsymbol{\beta}} - \boldsymbol{\beta}) - F_T\sqrt{T}(\widehat{\boldsymbol{\beta}} - \boldsymbol{\beta}) = (I_{K^2p+K} - F_T)\sqrt{T}(\widehat{\boldsymbol{\beta}} - \boldsymbol{\beta}),$$

where

$$F_T := \left[\left(\frac{ZZ'}{T}\right)^{-1} \otimes \bar{\Sigma}_u\right]C'\left[C\left(\left(\frac{ZZ'}{T}\right)^{-1} \otimes \bar{\Sigma}_u\right)C'\right]^{-1}C$$

so that

$$F := \text{plim } F_T = (\Gamma^{-1} \otimes \Sigma_u)C'[C(\Gamma^{-1} \otimes \Sigma_u)C']^{-1}C.$$

Thus, the covariance matrix of the asymptotic distribution of $\sqrt{T}(\widehat{\widehat{\boldsymbol{\beta}}} - \boldsymbol{\beta})$ is

$$(I - F)(\Gamma^{-1} \otimes \Sigma_u)(I - F)'$$
$$= \Gamma^{-1} \otimes \Sigma_u - (\Gamma^{-1} \otimes \Sigma_u)F' - F(\Gamma^{-1} \otimes \Sigma_u) + F(\Gamma^{-1} \otimes \Sigma_u)F'$$
$$= \Gamma^{-1} \otimes \Sigma_u - (\Gamma^{-1} \otimes \Sigma_u)C'[C(\Gamma^{-1} \otimes \Sigma_u)C']^{-1}C(\Gamma^{-1} \otimes \Sigma_u).$$

In other words, a positive semidefinite matrix is subtracted from the covariance matrix $\Gamma^{-1} \otimes \Sigma_u$ to obtain the asymptotic covariance matrix of the restricted estimator. Hence, the asymptotic variances of the latter will be smaller than or at most equal to those of the unrestricted multivariate LS estimator. Because the two ways of writing the restrictions in (5.2.3) and (5.2.2) are equivalent, the EGLS estimator of $\boldsymbol{\beta}$ subject to restrictions $\boldsymbol{\beta} = R\boldsymbol{\gamma} + r$ must also be asymptotically superior to the unconstrained estimator. In other words,

$$\Gamma^{-1} \otimes \Sigma_u - R[R'(\Gamma \otimes \Sigma_u^{-1})R]^{-1}R'$$

is positive semidefinite. This result shows that imposing restrictions is advantageous in terms of asymptotic efficiency. It must be kept in mind, however, that the restrictions are assumed to be valid in the foregoing derivations. In practice, there is usually some uncertainty with respect to the validity of the constraints.

5.2.3 Maximum Likelihood Estimation

So far in this chapter, no specific distribution of the process y_t is assumed. If the precise distribution of the process is known, ML estimation of the VAR coefficients is possible. In the following, we assume that y_t is Gaussian (normally distributed). The ML estimators of γ and Σ_u are found by equating to zero the first order partial derivatives of the log-likelihood function and solving for γ and Σ_u. The partial derivatives are found as in Section 3.4 of Chapter 3. Note that

$$\frac{\partial \ln l}{\partial \gamma} = \frac{\partial \beta'}{\partial \gamma} \frac{\partial \ln l}{\partial \beta} = R' \frac{\partial \ln l}{\partial \beta},$$

by the chain rule for vector differentiation (Appendix A.13). Proceeding as in Section 3.4, the ML estimator of γ is seen to be

$$\widetilde{\gamma} = [R'(ZZ' \otimes \widetilde{\Sigma}_u^{-1})R]^{-1} R'(Z \otimes \widetilde{\Sigma}_u^{-1})\mathbf{z}, \qquad (5.2.17)$$

where $\widetilde{\Sigma}_u$ is the ML estimator of Σ_u (see Problem 5.3). The resulting ML estimator of β is

$$\widetilde{\beta} = R\widetilde{\gamma} + r. \qquad (5.2.18)$$

Furthermore, the ML estimator of Σ_u is seen to be

$$\widetilde{\Sigma}_u = \frac{1}{T}(Y - \widetilde{B}Z)(Y - \widetilde{B}Z)', \qquad (5.2.19)$$

where $\widetilde{B}$ is the $(K \times (Kp+1))$ matrix satisfying $\text{vec}(\widetilde{B}) = \widetilde{\beta}$.

An immediate consequence of the consistency of the ML estimator $\widetilde{\Sigma}_u$ and of Proposition 5.2 is that the EGLS estimator $\widehat{\widetilde{\gamma}}$ and the ML estimator $\widetilde{\gamma}$ are asymptotically equivalent. In addition, it follows as in Section 3.2.2, Chapter 3, that $\widetilde{\Sigma}_u$ has the same asymptotic properties as in the unrestricted case (see Proposition 3.2) and $\widetilde{\beta}$ and $\widetilde{\Sigma}_u$ are asymptotically independent. In summary, we get the following result.

Proposition 5.5 (*Asymptotic Properties of the Restricted ML Estimators*)
Let y_t be a Gaussian stable K-dimensional VAR(p) process as in (5.1.1) and $\beta = \text{vec}(B) = R\gamma + r$ as in (5.2.2). Then the ML estimators $\widetilde{\beta}$ and $\widetilde{\sigma} = \text{vech}(\widetilde{\Sigma}_u)$ are consistent and asymptotically normally distributed,

$$\sqrt{T}\begin{bmatrix} \tilde{\boldsymbol{\beta}} - \boldsymbol{\beta} \\ \tilde{\boldsymbol{\sigma}} - \boldsymbol{\sigma} \end{bmatrix} \xrightarrow{d} \mathcal{N}\left(0, \begin{bmatrix} R[R'(\Gamma \otimes \Sigma_u^{-1})R]^{-1}R' & 0 \\ 0 & 2\mathbf{D}_K^+(\Sigma_u \otimes \Sigma_u)\mathbf{D}_K^{+\prime} \end{bmatrix}\right),$$

where $\mathbf{D}_K^+ = (\mathbf{D}_K' \mathbf{D}_K)^{-1}\mathbf{D}_K'$ is, as usual, the Moore-Penrose inverse of the $(K^2 \times K(K+1)/2)$ duplication matrix $\mathbf{D}_K$. ∎

Of course, we could have stated the proposition in terms of the joint distribution of $\hat{\boldsymbol{\gamma}}$ and $\tilde{\boldsymbol{\sigma}}$ instead. In the following, the distribution given in the proposition will turn out to be more useful, though.

Both EGLS and ML estimation can be discussed in terms of the mean-adjusted model considered in Section 3.3. However, the present discussion includes restrictions for the intercept terms in a convenient way. If the restrictions are equivalent in the different versions of the model, the asymptotic properties of the estimators of $\boldsymbol{\alpha} := \text{vec}(A_1, \ldots, A_p)$ will not be affected. For instance, the asymptotic covariance matrix of $\sqrt{T}(\tilde{\boldsymbol{\alpha}} - \boldsymbol{\alpha})$, where $\tilde{\boldsymbol{\alpha}}$ is the ML estimator, is just the lower right-hand ($K^2p \times K^2p$) block of $R[R'(\Gamma \otimes \Sigma_u^{-1})R]^{-1}R'$ from Proposition 5.5. If the sample means are subtracted from all variables and the constraints are given in the form $\boldsymbol{\alpha} = R\boldsymbol{\gamma} + r$ for a suitable matrix R and vectors $\boldsymbol{\gamma}$ and r, the covariance matrix of the asymptotic distribution of $\sqrt{T}(\tilde{\boldsymbol{\alpha}} - \boldsymbol{\alpha})$ can be written as

$$R[R'(\Gamma_Y(0) \otimes \Sigma_u^{-1})R]^{-1}R', \tag{5.2.20}$$

where $\Gamma_Y(0) := \Sigma_Y = \text{Cov}(Y_t)$ with $Y_t := (y_t', \ldots, y_{t-p+1}')'$.

5.2.4 Constraints for Individual Equations

In practice, parameter restrictions are often formulated for the K equations of the system (5.1.1) separately. In that case, it may be easier to write the restrictions in terms of the vector $\mathbf{b} := \text{vec}(B')$ which contains the parameters of the first equation in the first $Kp + 1$ positions and those of the second equation in the second $Kp + 1$ positions etc. If the constraints are expressed as

$$\mathbf{b} = \bar{R}\mathbf{c} + \bar{r}, \tag{5.2.21}$$

where $\bar{R}$ is a known $((K^2p + K) \times M)$ matrix of rank M, $\mathbf{c}$ is an unknown ($M \times 1$) parameter vector, and $\bar{r}$ is a known ($K^2p + K$)-dimensional vector, the restricted EGLS and ML estimators of $\mathbf{b}$ and their properties are easily derived. We get the following proposition:

Proposition 5.6 (*EGLS Estimator of Parameters Arranged Equationwise*)
Under the conditions of Proposition 5.2, if $\mathbf{b} = \text{vec}(B')$ satisfies (5.2.21), the EGLS estimator of $\mathbf{c}$ is

$$\widehat{\hat{\mathbf{c}}} = [\bar{R}'(\bar{\Sigma}_u^{-1} \otimes ZZ')\bar{R}]^{-1}\bar{R}'(\bar{\Sigma}_u^{-1} \otimes Z)[\text{vec}(Y') - (Z \otimes I_K)\bar{r}], \tag{5.2.22}$$

where $\bar{\Sigma}_u$ is a consistent estimator of Σ_u. The corresponding estimator of **b** is

$$\widehat{\widehat{\mathbf{b}}} = \bar{R}\widehat{\widehat{\mathbf{c}}} + \bar{r}, \tag{5.2.23}$$

which is consistent and asymptotically normally distributed,

$$\sqrt{T}(\widehat{\widehat{\mathbf{b}}} - \mathbf{b}) \xrightarrow{d} \mathcal{N}(0, \bar{R}[\bar{R}'(\bar{\Sigma}_u^{-1} \otimes \Gamma)\bar{R}]^{-1}\bar{R}'). \tag{5.2.24}$$

∎

The proof is left as an exercise (see Problem 5.4). An estimator of $\boldsymbol{\beta}$ is obtained from $\widehat{\widehat{\mathbf{b}}}$ by premultiplying with the commutation matrix $\mathbf{K}_{Kp+1,K}$. If the restrictions in (5.2.21) are equivalent to those in (5.2.2), the estimator for $\boldsymbol{\beta}$ obtained in this way is identical to $\widehat{\widehat{\boldsymbol{\beta}}}$ given in (5.2.11).

5.2.5 Restrictions for the White Noise Covariance Matrix

Occasionally restrictions for the white noise covariance matrix Σ_u are available. For instance, in Chapter 2, Section 2.3.1, we have seen that instantaneous noncausality is equivalent to Σ_u being block-diagonal. Thus, in that case there are zero off-diagonal elements. Zero constraints are, in fact, the most common constraints for the off-diagonal elements of Σ_u. Therefore, we will focus on such restrictions in the following.

Estimation under zero restrictions for Σ_u is often most easily performed in the context of the recursive model introduced in Chapter 2, Section 2.3.2. In order to obtain the recursive form corresponding to the standard VAR model

$$y_t = \nu + A_1 y_{t-1} + \cdots + A_p y_{t-p} + u_t,$$

Σ_u is decomposed as $\Sigma_u = W\Sigma_\varepsilon W'$, where W is lower triangular with unit main diagonal and Σ_ε is a diagonal matrix. Then, premultiplying with W^{-1} gives the recursive system

$$y_t = \eta + A_0^* y_t + A_1^* y_{t-1} + \cdots + A_p^* y_{t-p} + \varepsilon_t,$$

where $\eta := W^{-1}\nu$, $A_0^* := I_K - W^{-1}$ is a lower triangular matrix with zero diagonal, $A_i^* := W^{-1}A_i$, $i = 1, \ldots, p$, and $\varepsilon_t = (\varepsilon_{1t}, \ldots, \varepsilon_{Kt})' := W^{-1}u_t$ has diagonal covariance matrix, $\Sigma_\varepsilon := E(\varepsilon_t \varepsilon_t')$. The characteristic feature of the recursive representation of our process is that the k-th equation may involve $y_{1,t}, \ldots, y_{k-1,t}$ (current values of $y_1, \ldots, y_{k-1}$) on the right-hand side and the components of the white noise process ε_t are uncorrelated.

Many zero restrictions for the off-diagonal elements of Σ_u are equivalent to simple zero restrictions on A_0^* which are easy to impose in equationwise LS estimation. For instance, if Σ_u is block-diagonal, say

$$\Sigma_u = \begin{bmatrix} \Sigma_{11} & 0 \\ 0 & \Sigma_{22} \end{bmatrix},$$

then Σ_{11} and Σ_{22} can be decomposed in the form

$$\Sigma_{ii} = W_i \Sigma_{\varepsilon i} W_i', \quad i = 1, 2,$$

where W_i is lower triangular with unit diagonal and $\Sigma_{\varepsilon i}$ is a diagonal matrix. Hence,

$$A_0^* = I_K - \begin{bmatrix} W_1^{-1} & 0 \\ 0 & W_2^{-1} \end{bmatrix} =: \begin{bmatrix} A_{01}^* & 0 \\ 0 & A_{02}^* \end{bmatrix},$$

where the

$$A_{0i}^* = \begin{bmatrix} 0 & \cdots & & \cdots & 0 \\ * & \ddots & & & \vdots \\ \vdots & \ddots & \ddots & & \vdots \\ * & \cdots & & * & 0 \end{bmatrix}, \quad i = 1, 2,$$

are lower triangular with zero main diagonal. In summary, if Σ_u is block-diagonal with an $(m \times n)$ block of zeros in its lower left-hand corner, the same holds for A_0^*.

Because the error terms of the K equations of the recursive system are uncorrelated it can be shown that estimating each equation separately does not result in a loss of asymptotic efficiency (see Problem 5.6). Using the notation

$$y_{(k)} := \begin{bmatrix} y_{k1} \\ \vdots \\ y_{kT} \end{bmatrix}, \quad \varepsilon_{(k)} := \begin{bmatrix} \varepsilon_{k1} \\ \vdots \\ \varepsilon_{kT} \end{bmatrix}$$

and denoting by $b_{(k)}$ the vector of all nonzero coefficients and by $Z_{(k)}$ the corresponding matrix of regressors in the k-th equation of the recursive form of the system, we may write the k-th equation as

$$y_{(k)} = Z_{(k)} b_{(k)} + \varepsilon_{(k)}.$$

The LS estimator of $b_{(k)}$ is

$$\widehat{b}_{(k)} = (Z_{(k)}' Z_{(k)})^{-1} Z_{(k)}' y_{(k)}.$$

Under Gaussian assumptions, it is equivalent to the ML estimator and is thus asymptotically efficient. Obviously, this framework makes it easy to take into account zero restrictions by just eliminating regressors.

Generally, restrictions on Σ_u imply restrictions for A_0^* and vice versa. Unfortunately, zero restrictions on Σ_u do not always imply zero restrictions for A_0^*. Consider, for instance, the covariance matrix

$$\Sigma_u = \begin{bmatrix} 1 & 1 & -1 \\ 1 & 2 & 0 \\ -1 & 0 & 3 \end{bmatrix} = \begin{bmatrix} 1 & 0 & 0 \\ 1 & 1 & 0 \\ -1 & 1 & 1 \end{bmatrix} I_3 \begin{bmatrix} 1 & 1 & -1 \\ 0 & 1 & 1 \\ 0 & 0 & 1 \end{bmatrix} (= W\Sigma_\varepsilon W').$$

Hence,

$$A_0^* = I_3 - W^{-1} = I_3 - \begin{bmatrix} 1 & 0 & 0 \\ -1 & 1 & 0 \\ 2 & -1 & 1 \end{bmatrix} = \begin{bmatrix} 0 & 0 & 0 \\ 1 & 0 & 0 \\ -2 & 1 & 0 \end{bmatrix}.$$

Thus, although Σ_u has a zero off-diagonal element, all elements of A_0^* below the main diagonal are nonzero.

In practice, subject matter theory is often more likely to provide restrictions for the A_0^* matrix than for Σ_u because, as we have seen in Section 2.3.2, the elements of A_0^* can sometimes be interpreted as impact multipliers which represent the instantaneous effects of impulses in the variables. For this reason, the recursive form of the system has considerable appeal.

Note, however, that if restrictions are available on the coefficients A_i of the standard VAR form, the implied constraints for the A_i^* matrices should be taken into account in the estimation. Such restrictions may be cross-equation restrictions that involve coefficients from different equations. Taking them into account may require simultaneous estimation of all or some equations of the system rather than single equation LS estimation. In the following sections, we return to the standard form of the VAR model. Further discussion of covariance restrictions will be provided in the context of structural VAR models in Chapter 9.

5.2.6 Forecasting

Forecasting with estimated processes was discussed in Section 3.5 of Chapter 3. The general results of that section remain valid even if parameter restrictions are imposed in the estimation procedure. Some differences in details will be pointed out in the following.

We focus on the standard form (5.1.1) of the VAR model and denote the parameter estimators by $\widehat{\nu}, \widehat{A}_1, \ldots, \widehat{A}_p$, and $\widehat{\boldsymbol{\beta}}$. These estimators may be EGLS or ML estimators. The resulting h-step forecast at origin t is

$$\widehat{y}_t(h) = \widehat{\nu} + \widehat{A}_1 \widehat{y}_t(h-1) + \cdots + \widehat{A}_p \widehat{y}_t(h-p) \tag{5.2.25}$$

with $\widehat{y}_t(j) := y_{t+j}$ for $j \leq 0$, as in Section 3.5. In line with that section, we assume that forecasting and parameter estimation are based on independent processes with identical stochastic structure. Then we get the approximate MSE matrix

$$\Sigma_{\widehat{y}}(h) = \Sigma_y(h) + \frac{1}{T}\Omega(h), \tag{5.2.26}$$

where

$$\Sigma_y(h) := E\{[y_{t+h} - y_t(h)][y_{t+h} - y_t(h)]'\} = \sum_{i=0}^{h-1} \Phi_i \Sigma_u \Phi_i',$$

Φ_i being, as usual, the i-th coefficient matrix of the canonical MA representation of y_t, and

$$\Omega(h) := E\left[\frac{\partial y_t(h)}{\partial \beta'} \Sigma_{\widehat{\beta}} \frac{\partial y_t(h)'}{\partial \beta}\right],$$

where $\Sigma_{\widehat{\beta}}$ is the covariance matrix of the asymptotic distribution of $\sqrt{T}(\widehat{\beta} - \beta)$.

In Chapter 3, the matrix $\Omega(h)$ has a particularly simple form because in that chapter $\Sigma_{\widehat{\beta}} = \Gamma^{-1} \otimes \Sigma_u$. In the present situation, where parameter restrictions are imposed,

$$\Sigma_{\widehat{\beta}} = R[R'(\Gamma \otimes \Sigma_u^{-1})R]^{-1}R',$$

and the form of $\Omega(h)$ is not quite so simple. Because the covariance matrix $\Sigma_{\widehat{\beta}}$ is now smaller than in Chapter 3, $\Omega(h)$ will also become smaller (not greater). Using

$$\frac{\partial y_t(h)}{\partial \beta'} = \sum_{i=0}^{h-1} Z_t'(\mathbf{B}')^{h-1-i} \otimes \Phi_i$$

from Chapter 3, (3.5.11), we may now estimate $\Omega(h)$ by

$$\widehat{\Omega}(h) = \frac{1}{T}\sum_{t=1}^{T}\left[\sum_{i=0}^{h-1} Z_t'(\mathbf{B}')^{h-1-i} \otimes \Phi_i\right] \Sigma_{\widehat{\beta}} \left[\sum_{i=0}^{h-1} Z_t'(\mathbf{B}')^{h-1-i} \otimes \Phi_i\right]'.$$

(5.2.27)

Here

$$\mathbf{B} := \begin{bmatrix} 1 & 0 & \cdots & 0 \\ & B & & \\ 0 & I_{K(p-1)} & & 0 \end{bmatrix} \quad ((Kp+1) \times (Kp+1))$$

(see Section 3.5.2). In practice, the unknown matrices $\mathbf{B}$, Φ_i, and $\Sigma_{\widehat{\beta}}$ are replaced by consistent estimators. Of course, if T is large we may simply ignore the term $\Omega(h)/T$ in (5.2.26) because it approaches zero as $T \to \infty$. An estimator of $\Sigma_{\widehat{y}}(h)$ is then obtained by simply replacing the unknown quantities in $\Sigma_y(h)$ by estimators. Assuming that y_t is Gaussian, forecast intervals and regions can be determined exactly as in Section 3.5.

5.2.7 Impulse Response Analysis and Forecast Error Variance Decomposition

Impulse response analysis and forecast error variance decomposition with restricted VAR models can be done as described in Section 3.7. Proposition 3.6

is formulated in sufficiently general form to accommodate the case of restricted estimation. The impulse responses are then estimated from the restricted estimators of $A_1, \ldots, A_p$. As mentioned earlier, the covariance matrix of the restricted estimator of $\boldsymbol{\alpha} := \text{vec}(A_1, \ldots, A_p)$ is obtained by considering the lower right-hand $(K^2p \times K^2p)$ block of

$$\Sigma_{\widehat{\boldsymbol{\beta}}} = R[R'(\Gamma \otimes \Sigma_u^{-1})R]^{-1}R'.$$

As we have seen in Subsection 5.2.3, Proposition 5.5, the asymptotic covariance matrix $\Sigma_{\widetilde{\boldsymbol{\sigma}}}$ of $\sqrt{T}(\widetilde{\boldsymbol{\sigma}} - \boldsymbol{\sigma})$ is not affected by the restrictions for $\boldsymbol{\beta}$. However, the estimator of

$$\Sigma_{\widetilde{\boldsymbol{\sigma}}} = 2\mathbf{D}_K^+(\Sigma_u \otimes \Sigma_u)\mathbf{D}_K^{+\prime}$$

may be affected. As discussed in Section 5.2.2, we have the choice of different consistent estimators for Σ_u which may or may not take into account the parameter constraints. In other words, we may estimate Σ_u from the residuals of an unrestricted estimation or we may use the residuals of the restricted LS or EGLS estimation. The lower triangular matrix P that is used in estimating the impulse responses for orthogonal innovations is estimated accordingly. In the examples considered below, we will usually base the estimators of Σ_u and P on the residuals of the restricted EGLS estimation. In contrast, Γ will usually be estimated by ZZ'/T, as in the unrestricted case. Of course, instead of the intercept version of the process we may use the mean-adjusted form for estimation, as mentioned in Section 5.2.3.

5.2.8 Specification of Subset VAR Models

A VAR model with zero constraints on the coefficients is called a *subset VAR model*. Formally zero restrictions can be written as in (5.2.2) or (5.2.21) with $r = \bar{r} = 0$. We have encountered such models in previous chapters. For instance, when Granger-causality restrictions are imposed, we get subset VAR models. This example suggests possibilities how to obtain such restrictions, namely, from prior nonsample information and/or from tests of particular hypotheses. Subject matter theory sometimes implies a set of restrictions on a VAR model that can be taken into account, using the estimation procedures outlined in the foregoing. However, in many cases generally accepted a priori restrictions are not available. In that situation, statistical procedures may be used to detect or confirm possible zero constraints. In the following, we will discuss such procedures.

If little or no a priori knowledge of possible zero constraints is available, one may want to compare various different processes or models and choose the one which is optimal under a specific criterion. Using hypothesis tests in such a situation may create problems because the different possible models may not be nested. In that case, statistical tests may not lead to a unique answer as to which model to use. Therefore, in subset VAR modelling it is not uncommon to base the model choice on model selection criteria. For instance, appropriately modified versions of AIC, SC, or HQ may be employed. Generally speaking, in such an approach the subset VAR model is chosen that optimizes some prespecified criterion.

Suppose it is just known that the order of the process is not greater than some number p and otherwise no prior knowledge of possible zero constraints is available. In that situation, one would ideally fit all possible subset VAR models and select the one that optimizes the criterion chosen. The practicability of such a procedure is limited by its computational expense. Note that for a K-dimensional VAR(p) process, even if we do not take into account the intercept terms for the moment, there exist $K^2 p$ coefficients from which

$$\binom{K^2 p}{j}$$

subsets with j elements can be chosen. Thus, there is a total of

$$\sum_{j=0}^{K^2 p - 1} \binom{K^2 p}{j} = 2^{K^2 p} - 1$$

subset VAR models, not counting the full VAR(p) model which is also a possible candidate. For instance, for a bivariate VAR(4) process, there are as many as $2^{16} - 1 = 65{,}535$ subset models plus the full VAR(4) model. Of course, in practice the dimension and order of the process will often be greater than in this example and there may be many more subset VAR models. Therefore, specific strategies for subset VAR modelling have been proposed which avoid fitting all potential candidates. Some possibilities will be described briefly in the following.

Elimination of Complete Matrices

Penm & Terrell (1982) considered subset models where complete coefficient matrices A_j rather than individual coefficients are set to zero. Such a strategy reduces the models to be compared to

$$\sum_{j=0}^{p} \binom{p}{j} = 2^p.$$

For instance, for a VAR(4) process, only 16 models need to be compared.

An obvious advantage of the procedure is its relatively small computational expense. Deleting complete coefficient matrices may be reasonable if seasonal data with strong seasonal components are considered for which only coefficients at seasonal lags are different from zero. On the other hand, there may still be potential for further parameter restrictions. Moreover, some of the deleted coefficient matrices may contain elements that would not have been deleted had they been checked individually. Therefore, the following strategy may be more useful.

Top-Down Strategy

The top-down strategy starts from the full VAR(p) model and coefficients are deleted in the K equations separately. The k-th equation may be written as

$$y_{kt} = \nu_k + \alpha_{k1,1}y_{1,t-1} + \cdots + \alpha_{kK,1}y_{K,t-1}+ \\ \vdots \\ + \alpha_{k1,p}y_{1,t-p} + \cdots + \alpha_{kK,p}y_{K,t-p} + u_{kt}. \quad (5.2.28)$$

The goal is to find the zero restrictions for the coefficients of this equation that lead to the *minimum* value of a prespecified criterion. For this purpose, the equation is estimated by LS and the corresponding value of the criterion is evaluated. Then the last coefficient $\alpha_{kK,p}$ is set to zero (i.e., $y_{K,t-p}$ is deleted from the equation) and the equation is estimated again with this restriction. If the value of the criterion for the restricted model is greater than for the unrestricted model, $y_{K,t-p}$ is kept in the equation. Otherwise it is eliminated. Then the same procedure is repeated for the second last coefficient, $\alpha_{k,K-1,p}$, or variable $y_{K-1,t-p}$ and so on up to ν_k. In each step a lag of a variable is deleted if the criterion does not increase by that additional constraint compared to the smallest value obtained in the previous steps.

Criteria that may be used in this procedure are

$$\text{AIC} = \ln \tilde{\sigma}^2 + \frac{2}{T}(\text{number of estimated parameters}), \quad (5.2.29)$$

$$\text{HQ} = \ln \tilde{\sigma}^2 + \frac{2 \ln \ln T}{T}(\text{number of estimated parameters}), \quad (5.2.30)$$

or

$$\text{SC} = \ln \tilde{\sigma}^2 + \frac{\ln T}{T}(\text{number of estimated parameters}). \quad (5.2.31)$$

Here $\tilde{\sigma}^2$ stands for the sum of squared estimation residuals divided by the sample size T. For instance, the AIC value for a model with or without zero restrictions is computed by estimating the k-th equation, computing the residual sum of squares and dividing by T to obtain $\tilde{\sigma}^2$. Then two times the number of parameters contained in the estimated equation is divided by T and added to the natural logarithm of $\tilde{\sigma}^2$. In the final equation, only those variables and coefficients are retained that lead to the minimum AIC value.

In a more formal manner, this procedure can be described as follows. The k-th equation of the system may be written as

$$y_{(k)} = \begin{bmatrix} y_{k1} \\ \vdots \\ y_{kT} \end{bmatrix} = Z'b_k + u_{(k)} = Z'\bar{R}_k c_k + u_{(k)},$$

where $b_k = \bar{R}_k c_k$ reflects the zero restrictions imposed on the parameters b_k of the k-th equation. $\bar{R}_k$ is the restriction matrix. The LS estimator of c_k is

$$\widehat{c}_k = (\bar{R}'_k ZZ' \bar{R}_k)^{-1} \bar{R}'_k Z y_{(k)}$$

and the implied restricted LS estimator for b_k is

$$\widehat{b}_k = \bar{R}_k \widehat{c}_k.$$

Furthermore, a corresponding estimator of the residual variance is

$$\widetilde{\sigma}^2(\bar{R}_k) = (y_{(k)} - Z'\widehat{b}_k)'(y_{(k)} - Z'\widehat{b}_k)/T.$$

Thus, the AIC value for a model with these restrictions is

$$\text{AIC}(\bar{R}_k) = \ln \widetilde{\sigma}^2(\bar{R}_k) + \frac{2}{T}\text{rk}(\bar{R}_k).$$

The other criteria are determined in a similar way.

In the foregoing subset procedure based on AIC, the unrestricted model with $\bar{R}_k = I_{Kp+1}$ is estimated first and the corresponding value $\text{AIC}(I_{Kp+1})$ is determined. Then the last column of I_{Kp+1} is eliminated. Let us denote the resulting restriction matrix by $\bar{R}_k^{(1)}$. If

$$\text{AIC}(\bar{R}_k^{(1)}) \leq \text{AIC}(I_{Kp+1}),$$

the next restriction matrix $\bar{R}_k^{(2)}$, say, is obtained by deleting the last column of $\bar{R}_k^{(1)}$ and $\text{AIC}(\bar{R}_k^{(2)})$ is compared to $\text{AIC}(\bar{R}_k^{(1)})$. If, however,

$$\text{AIC}(\bar{R}_k^{(1)}) > \text{AIC}(I_{Kp+1}),$$

the restriction matrix $\bar{R}_k^{(2)}$ is obtained by deleting the second last column of I_{Kp+1} and the next restriction matrix is decided upon by comparing $\text{AIC}(\bar{R}_k^{(2)})$ to $\text{AIC}(I_{Kp+1})$. In each step, a column of the restriction matrix is deleted if that leads to a reduction or at least not to an increase of the AIC criterion. Otherwise the column is retained.

The procedure is repeated for each of the K equations of the K-dimensional system, that is, a restriction matrix, $\bar{R}_k$ say, is determined for each equation separately. Once all zero restrictions have been determined by this strategy, the K equations of the restricted model with overall restriction matrix

$$\bar{R} = \begin{bmatrix} \bar{R}_1 & & 0 \\ & \ddots & \\ 0 & & \bar{R}_K \end{bmatrix}$$

can be estimated simultaneously by EGLS, as described in Sections 5.2.2 and 5.2.4. Note that SC tends to choose the most parsimonious models with the fewest coefficients whereas AIC has a tendency to select the most lavish models.

The advantage of this top-down procedure, starting from the top (largest model) and then working down gradually, is that it permits to check all individual coefficients. Also, the computational expense is very reasonable. The disadvantage of the method is that it requires estimation of each full equation in the initial selection step. This may exhaust the available degrees of freedom if a model with large order is deemed necessary for some high-dimensional system. Therefore, a slightly more elaborate bottom-up strategy may be preferred occasionally.

Bottom-Up Strategy

Again the restrictions are chosen for each equation separately. In the k-th equation, only lags of the first variable are considered initially and an optimal lag length p_1, say, for that variable is selected. That is, we select the optimal model of the form

$$y_{kt} = \nu_k + \alpha_{k1,1} y_{1,t-1} + \cdots + \alpha_{k1,p_1} y_{1,t-p_1} + u_{kt}$$

by fitting models

$$y_{kt} = \nu_k + \alpha_{k1,1} y_{1,t-1} + \cdots + \alpha_{k1,n} y_{1,t-n} + u_{kt},$$

where n ranges from zero to some prespecified upper bound p for the order. p_1 is that order for which the selection criterion, e.g., AIC, HQ, or SC, is minimized.

In the next step, p_1 is held fixed and lags of y_2 are added into the equation. Denoting the optimal lag length for y_2 by p_2 gives

$$\begin{aligned}y_{kt} &= \nu_k + \alpha_{k1,1} y_{1,t-1} + \cdots + \alpha_{k1,p_1} y_{1,t-p_1} + \alpha_{k2,1} y_{2,t-1} + \cdots \\ &\quad + \alpha_{k2,p_2} y_{2,t-p_2} + u_{kt}.\end{aligned}$$

Note that p_2 may, of course, be zero in which case y_2 does not enter the equation.

In the third step, p_1 and p_2 are both held fixed and the third variable, y_3, is absorbed into the equation in the same way. This procedure is continued until an optimal lag length for each of the K variables is obtained, conditional on the "optimal" lags of the previous variables.

Due to omitted variables effects, some of the lag lengths may be overstated in the final equation. For instance, when none of the other variables enters the equation, lags of y_1 may be useful in explaining y_{kt} and in reducing the selection criterion. In contrast, lags of y_1 may not contribute to explaining y_k when lags of all the other variables are present too. Therefore, once $p_1, \ldots, p_K$ are chosen, a top-down run, as described in the previous subsection, may complete the search for zero restrictions for the k-th equation.

After zero constraints have been obtained for each equation in this fashion, the K restricted equations may be estimated as one system, using EGLS or ML procedures.

Obviously, it is possible in this bottom-up approach that the largest model, where all K variables enter with p lags in each equation is never fitted. Thereby considerable savings of degrees of freedom may be possible, especially if the maximum order p is substantial. A drawback of the procedure is that the final set of restrictions may depend on the order of the variables.

Sequential Elimination of Regressors

Individual zero coefficients can also be chosen on the basis of the t-ratios of the parameter estimators. A possible strategy is to sequentially delete those regressors with the smallest absolute values of t-ratios until all t-ratios (in absolute value) are greater than some threshold value, say η. In this procedure one regressor is eliminated at a time. Then new t-ratios are computed for the reduced model. Brüggemann & Lütkepohl (2001) showed that this strategy is equivalent to the sequential elimination based on model selection criteria if the threshold value η is chosen accordingly. More precisely, they considered a regression equation

$$y_{kt} = \beta_1 x_{1t} + \cdots + \beta_N x_{Nt} + u_{kt},$$

where all regressors are denoted by x_{jt}, that is, x_{jt} may represent an intercept term or lags of the variables involved in our analysis. Brüggemann & Lütkepohl (2001) studied a procedure where those regressors are deleted sequentially, one at a time, which lead to the largest reduction of the given selection criterion until no further reduction is possible. For a model selection criterion of the type

$$\text{Cr}(i_1, \ldots, i_n) = \ln(SSE(i_1, \ldots, i_n)/T) + c_T n/T,$$

where $SSE(i_1, \ldots, i_n)$ is the sum of squared errors obtained by including $x_{i_1 t}, \ldots, x_{i_n t}$ in the regression model and c_T is a sequence indexed by the sample size. Brüggemann & Lütkepohl (2001) showed that choosing $\eta = \{[\exp(c_T/T) - 1](T - N + j - 1)\}^{1/2}$ in the j-th step of the elimination procedure based on t-ratios results in the same final model that is also obtained by sequentially minimizing the selection criterion defined by the penalty term c_T. Hence, the threshold value depends on the selection criterion via c_T, the sample size, and the number of regressors in the model. The threshold values for the t-ratios correspond to the critical values of the tests. For the criteria AIC, HQ, and SC, the c_T sequences are $c_T(\text{AIC}) = 2$, $c_T(\text{HQ}) = 2 \ln \ln T$, and $c_T(\text{SC}) = \ln T$, respectively. Using these criteria in the procedure, for an equation with 20 regressors and a sample size of $T = 100$, roughly corresponds to eliminating all regressors with t-values that are not significant at the 15–20%, 10% or 2–3% levels, respectively (see Brüggemann & Lütkepohl (2001)).

Procedures similar to those discussed here were, for instance, applied by Hsiao (1979, 1982) and Lütkepohl (1987). Other subset VAR strategies were proposed by Penm & Terrell (1984, 1986), Penm, Brailsford & Terrell (2000), and Brüggemann (2004). Moreover, more elaborate, computer-automated model specification and subset selection strategies based on a mixture of testing and model selection criteria were recently implemented in the software package PcGets (see Hendry & Krolzig (2001)). The alternative subset modelling procedures all have their advantages and drawbacks. Therefore, at this stage, none of them can be recommended as a universally best choice in practice.

5.2.9 Model Checking

After a subset VAR model has been fitted, some checks of the model adequacy are in order. Of course, one check is incorporated in the model selection procedure if some criterion is optimized. By definition, the best model is the one that leads to the optimum criterion value. In practice, the choice of the criterion is often ad hoc or even arbitrary and, in fact, several competing criteria are often employed. It is then left to the applied researcher to decide on the final model to be used for forecasting or economic analysis. In some cases, statistical tests of restrictions may aid in that decision. For example, F-tests, as described in Section 4.2.2, may be helpful for that purpose. In the following, we will discuss tests for residual autocorrelation.

Residual Autocovariances and Autocorrelations

The autocorrelation tests considered in Chapter 4 can also be used to check the white noise assumption for the u_t process in a subset VAR model, if suitable adjustments are made. For that purpose, we will first consider the residual autocovariances and autocorrelations. In analogy with Section 4.4 of Chapter 4, we use the following notation:

$$C_i := \frac{1}{T} \sum_{t=i+1}^{T} u_t u'_{t-i}, \qquad i = 0, 1, \ldots, h,$$

$$\mathbf{C}_h := (C_1, \ldots, C_h),$$

$$\mathbf{c}_h := \mathrm{vec}(\mathbf{C}_h),$$

$\widehat{u}_t$ is the t-th estimation residual of a restricted estimation,

$$\widehat{C}_i := \frac{1}{T} \sum_{t=i+1}^{T} \widehat{u}_t \widehat{u}'_{t-i}, \qquad i = 0, 1, \ldots, h,$$

$$\widehat{\mathbf{C}}_h := (\widehat{C}_1, \ldots, \widehat{C}_h),$$

$\widehat{\mathbf{c}}_h := \text{vec}(\widehat{\mathbf{C}}_h),$

$\widehat{D}$ is the diagonal matrix with the square roots of the diagonal elements of $\widehat{C}_0$ on the diagonal,

$\widehat{R}_i := \widehat{D}^{-1}\widehat{C}_i\widehat{D}^{-1}, \qquad i = 0, 1, \ldots, h,$

$\widehat{\mathbf{R}}_h := (\widehat{R}_1, \ldots, \widehat{R}_h),$

$\widehat{\mathbf{r}}_h := \text{vec}(\widehat{\mathbf{R}}_h).$

In the following proposition, the asymptotic distributions of $\widehat{\mathbf{c}}_h$ and $\widehat{\mathbf{r}}_h$ are given under the assumption of a correctly specified model.

Proposition 5.7 (*Asymptotic Distributions of Residual Autocovariances and Autocorrelations*)

Suppose y_t is a stable, stationary, K-dimensional VAR(p) process with identically distributed standard white noise u_t and the parameter vector $\boldsymbol{\beta}$ satisfies the restrictions $\boldsymbol{\beta} = R\boldsymbol{\gamma} + r$ with R being a known $(K(Kp+1) \times M)$ matrix of rank M. Furthermore suppose that $\boldsymbol{\beta}$ is estimated by EGLS such that $\widehat{\widehat{\boldsymbol{\beta}}} = R\widehat{\widehat{\boldsymbol{\gamma}}} + r$. Then

$$\sqrt{T}\,\widehat{\mathbf{c}}_h \xrightarrow{d} \mathcal{N}(0, \Sigma^r_{\mathbf{c}}(h)), \tag{5.2.32}$$

where

$$\Sigma^r_{\mathbf{c}}(h) = I_h \otimes \Sigma_u \otimes \Sigma_u - GR[R'(\Gamma \otimes \Sigma_u^{-1})R]^{-1}R'G'$$

and $G := \widetilde{G}' \otimes I_K$ is the matrix defined in Chapter 4, Lemma 4.2. Furthermore,

$$\sqrt{T}\,\widehat{\mathbf{r}}_h \xrightarrow{d} \mathcal{N}(0, \Sigma^r_{\mathbf{r}}(h)), \tag{5.2.33}$$

where

$$\Sigma^r_{\mathbf{r}}(h) = I_h \otimes R_u \otimes R_u - (G'_0 \otimes D^{-1})R[R'(\Gamma \otimes \Sigma_u^{-1})R]^{-1}R'(G_0 \otimes D^{-1})$$

and $G_0 := \widetilde{G}(I_h \otimes D^{-1})$ is defined in Proposition 4.6, D is the diagonal matrix with the square roots of the diagonal elements of Σ_u on the diagonal, and $R_u := D^{-1}\Sigma_u D^{-1}$ is the correlation matrix corresponding to Σ_u. ∎

Proof: The proof is similar to that of Propositions 4.5 and 4.6. Defining $\widetilde{G}$ as in Lemma 4.2, the lemma implies that $\sqrt{T}\,\widehat{\mathbf{c}}_h$ is known to have the same asymptotic distribution as

$$\sqrt{T}\mathbf{c}_h - \sqrt{T}G\,\text{vec}(\widehat{\widehat{B}} - B)$$

$$= [-\widetilde{G}' \otimes I_K : I] \begin{bmatrix} \sqrt{T}\,\text{vec}(\widehat{\widehat{B}} - B) \\ \sqrt{T}\mathbf{c}_h \end{bmatrix}$$

$$= [-(\widetilde{G}' \otimes I_K)R : I] \begin{bmatrix} \sqrt{T}(\widehat{\overline{\gamma}} - \gamma) \\ \sqrt{T}\mathbf{c}_h \end{bmatrix}$$

$$= [-(\widetilde{G}' \otimes I_K)R : I] \begin{bmatrix} \left[R' \left(\frac{ZZ'}{T} \otimes \bar{\Sigma}_u^{-1} \right) R \right]^{-1} & R'(I_{Kp+1} \otimes \bar{\Sigma}_u^{-1}) & 0 \\ 0 & & I \end{bmatrix}$$

$$\times \begin{bmatrix} \frac{1}{\sqrt{T}} \text{vec}(UZ') \\ \sqrt{T}\mathbf{c}_h \end{bmatrix}$$

(see (5.2.7)). The asymptotic distribution in (5.2.32) then follows from Lemma 4.3 and Proposition C.2 by noting that $\Gamma = \text{plim}(ZZ'/T)$ and $\Sigma_u = \text{plim}\,\bar{\Sigma}_u$. The limiting distribution of $\sqrt{T}\,\widehat{\mathbf{r}}_h$ follows as in the proof of Proposition 4.6. ∎

The results in Proposition 5.7 can be used to check the white noise assumption for the u_t's. As in Section 4.4, residual autocorrelations are often plotted and evaluated on the basis of two-standard error bounds about zero. Estimators of the standard errors are obtained by replacing all unknown quantities in $\Sigma_{\mathbf{r}}^r(h)$ by consistent estimators. Specifically Σ_u may be estimated by $\widehat{C}_0$. We will illustrate the resulting white noise test in Section 5.2.10 with an example.

Portmanteau Tests

For the portmanteau statistic

$$\begin{aligned} Q_h &:= T \sum_{i=1}^{h} \text{tr}(\widehat{C}_i' \widehat{C}_0^{-1} \widehat{C}_i \widehat{C}_0^{-1}) \\ &= T \widehat{\mathbf{c}}_h'(I_h \otimes \widehat{C}_0^{-1} \otimes \widehat{C}_0^{-1})\widehat{\mathbf{c}}_h \end{aligned} \quad (5.2.34)$$

we get the following result.

Proposition 5.8 (*Approximate Distribution of the Portmanteau Statistic*)
Suppose the conditions of Proposition 5.7 are satisfied and there are no restrictions linking the intercept terms to the $A_1, \ldots, A_p$ coefficients, that is,

$$R = \begin{bmatrix} R_{(1)} & 0 \\ 0 & R_{(2)} \end{bmatrix}$$

is block-diagonal with $R_{(1)}$ and $R_{(2)}$ having row-dimensions K and K^2p, respectively. Then Q_h has an approximate limiting χ^2-distribution with $K^2h - \text{rk}(R_{(2)})$ degrees of freedom for large T and h. ∎

Proof: Under the conditions of the proposition, the covariance matrix of the asymptotic distribution in (5.2.32) is

$$\Sigma_{\bar{\mathbf{c}}}^r(h) = I_h \otimes \Sigma_u \otimes \Sigma_u - GR_{(2)}\{R'_{(2)}[\Gamma_Y(0) \otimes \Sigma_u^{-1}]R_{(2)}\}^{-1}R'_{(2)}G',$$

where G is the matrix defined in Lemma 4.2. Using this fact, Proposition 5.8 can be proven just as Proposition 4.7 by replacing $\bar{G}$ in that proof by $GR_{(2)}$ (see Section 4.4.3). ∎

The degrees of freedom in Proposition 5.8 are obtained by subtracting the number of unconstrained A_i coefficients from K^2h. As in Section 4.4.3, the modified portmanteau statistic

$$\bar{Q}_h := T^2 \sum_{i=1}^{h}(T-i)^{-1}\mathrm{tr}(\widehat{C}'_i\widehat{C}_0^{-1}\widehat{C}_i\widehat{C}_0^{-1}) \qquad (5.2.35)$$

may be preferable for testing the white noise assumption in small samples. In other words, under the white noise hypothesis, the small sample distribution of $\bar{Q}_h$ may be closer to the approximate χ^2-distribution than that of Q_h.

LM Test for Residual Autocorrelation

As for unrestricted models, an LM test for residual autocorrelation can also be constructed if parameter restrictions are imposed on a VAR model. For simplicity of exposition, we assume now that the restrictions can be written in the form $\beta = \mathrm{vec}(B) = R\gamma$. For example, there may be zero restrictions. In that case, a possible test statistic may be obtained by considering the auxiliary model

$$\widehat{\Sigma}_u^{-1/2}\widehat{U} = \widehat{\Sigma}_u^{-1/2}BZ + \widehat{\Sigma}_u^{-1/2}D\widehat{\mathcal{U}} + \mathcal{E}, \qquad (5.2.36)$$

where $D = [D_1 : \cdots : D_h]$ is $(K \times Kh)$, $\widehat{\mathcal{U}} = (I_h \otimes \widehat{U})F'$ with F as in (4.4.2), $\mathcal{E} = [\varepsilon_1, \ldots, \varepsilon_T]$ is a $(K \times T)$ error matrix, $\widehat{\Sigma}_u$ is some consistent estimator of Σ_u which has been used in EGLS estimation, $\widehat{\Sigma}_u^{-1/2}$ is a symmetric matrix such that $\widehat{\Sigma}_u^{-1/2}\widehat{\Sigma}_u^{-1/2} = \widehat{\Sigma}_u^{-1}$, and the other symbols are defined as before. Note, however, that the $\widehat{u}_t$ are now the residuals from EGLS estimation of the original restricted VAR(p) model and $\widehat{u}_t = 0$ for $t \leq 0$. The vectorized version of the auxiliary model (5.2.36) is

$$(I_T \otimes \widehat{\Sigma}_u^{-1/2})\mathrm{vec}(\widehat{U}) = (Z' \otimes \widehat{\Sigma}_u^{-1/2})R\gamma + (\widehat{\mathcal{U}}' \otimes \widehat{\Sigma}_u^{-1/2})\mathrm{vec}(D) + \mathrm{vec}(\mathcal{E}).$$

Defining $\boldsymbol{\delta} = \mathrm{vec}(D)$, the (EG)LS estimator from this auxiliary model is

$$\begin{bmatrix} \widehat{\gamma} \\ \widehat{\boldsymbol{\delta}} \end{bmatrix} = \begin{bmatrix} R'(ZZ' \otimes \widehat{\Sigma}_u^{-1})R & R'(Z\widehat{\mathcal{U}}' \otimes \widehat{\Sigma}_u^{-1}) \\ (\widehat{\mathcal{U}}Z' \otimes \widehat{\Sigma}_u^{-1})R & \widehat{\mathcal{U}}\widehat{\mathcal{U}}' \otimes \widehat{\Sigma}_u^{-1} \end{bmatrix}^{-1} \begin{bmatrix} R'(Z \otimes \widehat{\Sigma}_u^{-1}) \\ \widehat{\mathcal{U}} \otimes \widehat{\Sigma}_u^{-1} \end{bmatrix} \mathrm{vec}(\widehat{U}).$$

The first order conditions for a minimum of the EGLS objective function for the original restricted VAR model are

$$\frac{\partial [\mathbf{y} - (Z' \otimes I_K)R\gamma]'(I_K \otimes \widehat{\Sigma}_u^{-1})[\mathbf{y} - (Z' \otimes I_K)R\gamma]}{\partial \gamma}\bigg|_{\widehat{\gamma}}$$

$$= -2R'(Z \otimes I_K)(I_K \otimes \widehat{\Sigma}_u^{-1})[\mathbf{y} - (Z' \otimes I_K)R\widehat{\gamma}] = 0.$$

Hence, $R'(Z \otimes \widehat{\Sigma}_u^{-1}) \text{vec}(\widehat{U}) = 0$. Applying the rules for the partitioned inverse (see Appendix A.10) thus gives

$$\widehat{\boldsymbol{\delta}} = \Big(\widehat{U}\widehat{U}' \otimes \widehat{\Sigma}_u^{-1} - (\widehat{U}Z' \otimes \widehat{\Sigma}_u^{-1})R[R'(ZZ' \otimes \widehat{\Sigma}_u^{-1})R]^{-1}R'(Z\widehat{U}' \otimes \widehat{\Sigma}_u^{-1})\Big)^{-1}$$

$$\times \text{vec}(\widehat{\Sigma}_u^{-1}\widehat{U}\widehat{U}').$$

The usual χ^2-statistic for testing $\boldsymbol{\delta} = 0$ is

$$\lambda_{LM}(h) = \widehat{\boldsymbol{\delta}}'\Big(\widehat{U}\widehat{U}' \otimes \widehat{\Sigma}_u^{-1} - (\widehat{U}Z' \otimes \widehat{\Sigma}_u^{-1})R[R'(ZZ' \otimes \widehat{\Sigma}_u^{-1})R]^{-1}R'(Z\widehat{U}' \otimes \widehat{\Sigma}_u^{-1})\Big)\widehat{\boldsymbol{\delta}}.$$

Substituting the expression for $\widehat{\boldsymbol{\delta}}$, it can be seen that

$$\lambda_{LM}(h) = T\widehat{\mathbf{c}}_h' \widehat{\Sigma}_{\mathbf{c}}^r(h)^{-1} \widehat{\mathbf{c}}_h,$$

where

$$\widehat{\Sigma}_{\mathbf{c}}^r(h) = \frac{1}{T}\Big(\widehat{U}\widehat{U}' \otimes \widehat{\Sigma}_u^{-1} - (\widehat{U}Z' \otimes \widehat{\Sigma}_u^{-1})R[R'(ZZ' \otimes \widehat{\Sigma}_u^{-1})R]^{-1}R'(Z\widehat{U}' \otimes \widehat{\Sigma}_u^{-1})\Big)$$

is a consistent estimator of $\Sigma_{\mathbf{c}}^r(h)$. Thus, the situation is completely analogous to the case of an unrestricted model treated in Section 4.4.4 and we get the following result from Propositions 5.7 and C.15(5).

Proposition 5.9 (*Asymptotic Distribution of LM Statistic for Residual Autocorrelation of Restricted VAR*)
Under the conditions of Proposition 5.7,

$$\lambda_{LM}(h) \xrightarrow{d} \chi^2(hK^2).$$

■

Notice that unlike for the portmanteau test, the asymptotic distribution of the LM statistic is identical to that obtained for unrestricted VARs in Proposition 4.8. However, $\lambda_{LM}(h)$ is in general not exactly an LM statistic because the restricted estimator $\widehat{\gamma}$ is not identical to the ML estimator. Clearly, this does not affect the asymptotic properties of the test statistic.

Other Checks of Restricted Models

It must also be kept in mind that our discussion has been based on a number of further assumptions that should be checked. Prominent among them are stationarity, stability, and normality. The latter is used in setting up forecast intervals and regions and the former properties are basic conditions underlying much of our analysis (see, for instance, Propositions 5.1–5.6). The stability tests based on predictions and described in Section 4.6 of Chapter 4 may be applied in the same way as for full unrestricted VAR processes. Of course, now the forecasts and MSE matrix estimators should be based on the restricted coefficient estimators as discussed in Section 5.2.6. Also, it is easy to see from Section 4.5 that the tests for nonnormality remain valid when true restrictions are placed on the VAR coefficient matrices.

5.2.10 An Example

As an example, we use again the same data as in Section 3.2.3 and some other previous sections. That is, y_{1t}, y_{2t}, and y_{3t} are first differences of logarithms of investment, income, and consumption, respectively. We keep four presample values and use sample values from the period 1961.2–1978.4. Hence, the time series length is $T = 71$. We have applied the top-down strategy with selection criteria AIC, HQ, and SC and a VAR order of $p = 4$. In other words, we use the same maximum order as in the order selection procedure for full VAR models in Chapter 4. Because HQ and SC choose the same model, we get two different models only which are shown in Table 5.1. As usual, the HQ-SC model is more parsimonious than the AIC model.

In Table 5.1, modified portmanteau statistics with corresponding p-values are also given for both models. Obviously, none of the test values gives rise to concern about the models. In Figure 5.1, residual autocorrelations of the HQ-SC model with estimated two-standard error bounds about zero are depicted. The rather unusual looking estimated two-standard error bounds for some low lags are a consequence of the zero elements in the estimated VAR coefficient matrices. Recall that the asymptotic standard errors are bounded from above by $1/\sqrt{T}$. For low lags, they can be substantially smaller, however, and this property is clearly reflected in Figure 5.1. Although some individual autocorrelations fall outside the two-standard error bounds about zero, this is not necessarily a reason for modifying the model. As in Chapter 4, such a decision depends on which criterion is given priority.

We have also produced forecasts with the HQ-SC model and give them in Table 5.2 together with forecasts from a full VAR(4) model. In this example, the forecasts from the two models are quite close and the estimated forecast intervals from the subset model are all smaller than those of a full VAR(4) model. Although theoretically the more parsimonious subset model produces more precise forecasts if the restrictions are correct, it must be kept in mind that in the present case the restrictions, the forecasts and forecast intervals

Table 5.1. EGLS estimates of subset VAR models for the investment/income/consumption data

	model selection criterion	
	AIC	HQ-SC
$\widehat{\widehat{\nu}}$	$\begin{bmatrix} .015^* \\ _{(.006)} \\ .015 \\ _{(.003)} \\ .013 \\ _{(.003)} \end{bmatrix}$	$\begin{bmatrix} .015 \\ _{(.006)} \\ .020 \\ _{(.001)} \\ .016 \\ _{(.003)} \end{bmatrix}$
$\widehat{\widehat{A}}_1$	$\begin{bmatrix} -.219 & 0 & 0 \\ _{(.104)} & & \\ 0 & 0 & .235 \\ & & _{(.133)} \\ 0 & .274 & -.391 \\ & _{(.082)} & _{(.116)} \end{bmatrix}$	$\begin{bmatrix} -.225 & 0 & 0 \\ _{(.104)} & & \\ 0 & 0 & 0 \\ 0 & .261 & -.439 \\ & _{(.081)} & _{(.095)} \end{bmatrix}$
$\widehat{\widehat{A}}_2$	$\begin{bmatrix} 0 & 0 & 0 \\ .010 & 0 & 0 \\ _{(.024)} & & \\ 0 & .335 & 0 \\ & _{(.073)} & \end{bmatrix}$	$\begin{bmatrix} 0 & 0 & 0 \\ 0 & 0 & 0 \\ 0 & .329 & 0 \\ & _{(.074)} & \end{bmatrix}$
$\widehat{\widehat{A}}_3$	$\begin{bmatrix} 0 & 0 & 0 \\ 0 & 0 & 0 \\ 0 & .095 & 0 \\ & _{(.076)} & \end{bmatrix}$	$\begin{bmatrix} 0 & 0 & 0 \\ 0 & 0 & 0 \\ 0 & 0 & 0 \end{bmatrix}$
$\widehat{\widehat{A}}_4$	$\begin{bmatrix} .340 & 0 & 0 \\ _{(.103)} & & \\ 0 & 0 & 0 \\ 0 & 0 & 0 \end{bmatrix}$	$\begin{bmatrix} .331 & 0 & 0 \\ _{(.103)} & & \\ 0 & 0 & 0 \\ 0 & 0 & 0 \end{bmatrix}$
	$\bar{Q}_{12} = 79.3\ [.937]^{**}$	$\bar{Q}_{12} = 85.5\ [.893]$
	$\bar{Q}_{20} = 144\ [.943]$	$\bar{Q}_{20} = 152\ [.898]$

*Estimated standard errors in parentheses.
**p-value.

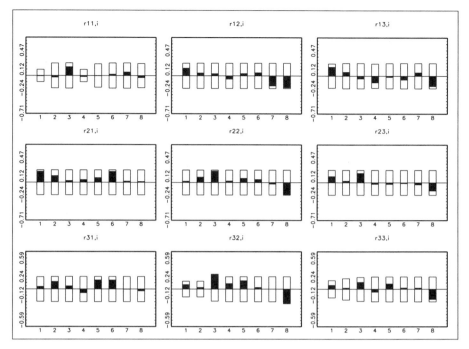

Fig. 5.1. Estimated residual autocorrelations of the investment/income/consumption HQ-SC subset VAR model with estimated asymptotic two-standard error bounds.

are estimated on the basis of a single realization of an unknown data generation process. Under these circumstances, a subset model may produce less precise forecasts than a heavily parameterized full VAR model. Note that in the present subset model, the income forecasts are the same for all forecast horizons because income is generated by a white noise process in the HQ-SC model.

We have also computed impulse responses from the HQ-SC subset VAR model. The Θ_i responses of consumption to an impulse in income based on orthogonalized residuals are depicted in Figure 5.2. Comparing them with Figure 3.8 shows that they are qualitatively similar to the impulse responses from the full VAR(2) model. Considering the responses of investment to a consumption innovation reveals that they are all zero in the subset VAR model. A closer look at Table 5.1 shows that income/consumption are not Granger-causal for investment in both subset models. This result was also obtained in the full VAR model (see Section 3.6.2). However, now it is directly seen in the model without further causality testing. In other words, the causality testing is built into the model selection procedure.

Table 5.2. Point and interval forecasts from full and subset VAR(4) models for the investment/income/consumption example

variable	forecast horizon	full VAR(4)		HQ-SC subset VAR(4)	
		point forecast	95% interval forecast	point forecast	95% interval forecast
investment	1	.006	[−.091, .103]	.015	[−.074, .105]
	2	.025	[−.075, .125]	.023	[−.068, .115]
	3	.028	[−.071, .126]	.018	[−.073, .110]
	4	.026	[−.074, .125]	.023	[−.069, .115]
income	1	.021	[−.005, .047]	.020	[−.004, .044]
	2	.022	[−.004, .049]	.020	[−.004, .044]
	3	.017	[−.009, .043]	.020	[−.004, .044]
	4	.022	[−.004, .049]	.020	[−.004, .044]
consumption	1	.022	[.001, .042]	.023	[.004, .042]
	2	.015	[−.006, .036]	.013	[−.007, .033]
	3	.020	[−.004, .043]	.022	[.001, .044]
	4	.019	[−.004, .042]	.018	[−.004, .040]

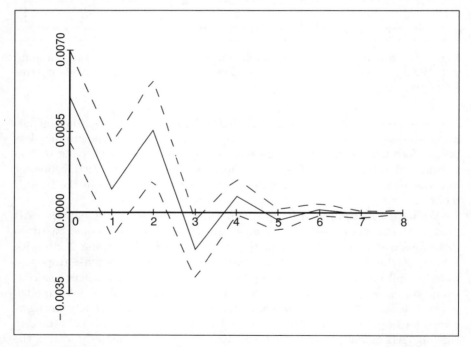

Fig. 5.2. Estimated responses of consumption to an orthogonalized impulse in income with two-standard error bounds based on the HQ-SC subset VAR model.

5.3 VAR Processes with Nonlinear Parameter Restrictions

Some authors have suggested nonlinear constraints for the coefficients of a VAR model. For instance, multiplicative models with VAR operator

$$\begin{aligned} A(L) &= I_K - A_1 L - \cdots - A_p L^p \\ &= (I_K - B_1 L^s - \cdots - B_Q L^{sQ})(I_K - C_1 L - \cdots - C_q L^q) \\ &= B(L^s) C(L) \end{aligned}$$

have been considered. Here L is the lag operator defined in Chapter 2, Section 2.1.2, the B_i's and C_j's are $(K \times K)$ coefficient matrices and $B(L^s)$ contains "seasonal" powers of L only. Such models may be useful for seasonal data. For instance, for quarterly data, a multiplicative seasonal operator may have the form

$$(I_K - B_1 L^4)(I_K - C_1 L - C_2 L^2).$$

The corresponding VAR operator is

$$\begin{aligned} A(L) &= I_K - A_1 L - \cdots - A_6 L^6 \\ &= I_K - C_1 L - C_2 L^2 - B_1 L^4 + B_1 C_1 L^5 + B_1 C_2 L^6, \end{aligned}$$

so that $A_1 = C_1$, $A_2 = C_2$, $A_3 = 0$, $A_4 = B_1$, $A_5 = -B_1 C_1$, $A_6 = -B_1 C_2$. Hence, the coefficients $\boldsymbol{\alpha} := \text{vec}[A_1, \ldots, A_p]$ are determined by $\boldsymbol{\gamma} := \text{vec}[B_1, C_1, C_2]$, that is,

$$\boldsymbol{\alpha} = g(\boldsymbol{\gamma}). \tag{5.3.1}$$

There are also other types of nonlinear constraints that may be written in this way. For example, the VAR operator may have the form $A(L) = B(L)C(L)$, where

$$C(L) = \begin{bmatrix} c_1(L) & & 0 \\ & \ddots & \\ 0 & & c_K(L) \end{bmatrix}$$

is a diagonal operator with $c_k(L) = 1 + c_{k1} L + \cdots + c_{kq} L^q$, which represents the individual dynamics of the variables and $B(L) = I_K - B_1 L - \cdots - B_n L^n$ takes care of joint relations. Again the implied restrictions for $\boldsymbol{\alpha}$ can easily be cast in the form (5.3.1).

In principle, under general conditions, if restrictions are given in the form (5.3.1), the analysis can proceed analogously to the linear restriction case. That is, we need to find an estimator $\widehat{\boldsymbol{\gamma}}$ of $\boldsymbol{\gamma}$, for instance, by minimizing

$$S(\boldsymbol{\gamma}) = [\mathbf{y} - (Z' \otimes I_K) g(\boldsymbol{\gamma})]'(I_T \otimes \Sigma_u^{-1})[\mathbf{y} - (Z' \otimes I_K) g(\boldsymbol{\gamma})],$$

where $\mathbf{y}$, Z, and Σ_u are as defined in Section 5.2. The minimization may require an iterative algorithm. Such algorithms are described in Section 12.3.2 in the context of estimating VARMA models. Once we have an estimator $\widehat{\gamma}$, we may estimate $\boldsymbol{\alpha}$ as $\widehat{\boldsymbol{\alpha}} = g(\widehat{\gamma})$. Under similar conditions as for the linear case, the estimators will be consistent and asymptotically normally distributed, e.g.,

$$\sqrt{T}(\widehat{\boldsymbol{\alpha}} - \boldsymbol{\alpha}) \xrightarrow{d} \mathcal{N}(0, \Sigma_{\widehat{\boldsymbol{\alpha}}}). \tag{5.3.2}$$

The corresponding estimators $\widehat{A}_1, \ldots, \widehat{A}_p$ may be used in computing forecasts and impulse responses etc. The asymptotic properties of these quantities then follow exactly as in the previous sections (see in particular Sections 5.2.6 and 5.2.7).

Another type of "multiplicative" VAR operator has the form

$$A(L) = I_K - B(L)C(L), \tag{5.3.3}$$

where

$$B(L) = B_0 + B_1 L + \cdots + B_q L^q$$

is of dimension $(K \times r)$, that is, the B_i's have dimension $(K \times r)$, and

$$C(L) = C_1 L + \cdots + C_p L^p$$

is of dimension $(r \times K)$, with $r < K$. For $p = 1$, neglecting the intercept terms, the process becomes

$$y_t = B_0 C_1 y_{t-1} + \cdots + B_q C_1 y_{t-q-1} + u_t$$

which is sometimes called an *index model* because y_t is represented in terms of lagged values of the "index" $C_1 y_t$. In the extreme case where $r = 1$, $C_1 y_t$ is simply a weighted sum or index of the components of y_t which justifies the name of the model. Such models have been investigated by Reinsel (1983) in some detail. Alternatively, if $q = 0$, the process is called reduced rank (RR)VAR process which has been analyzed by Velu, Reinsel & Wichern (1986), Tso (1981), Ahn & Reinsel (1988), Reinsel (1993, Chapter 6), Reinsel & Velu (1998) and Anderson (1999, 2002) among others. Models with a reduced rank structure in the coefficients will be of considerable importance in Part II, where VAR processes with cointegrated variables are considered. We will therefore not discuss them here.

5.4 Bayesian Estimation

5.4.1 Basic Terms and Notation

Although the reader is assumed to be familiar with Bayesian estimation, we summarize some basics here. In the Bayesian approach, it is assumed that the nonsample or prior information is available in the form of a density. Denoting the parameters of interest by α, let us assume that the prior information is summarized in the *prior probability density function (p.d.f.)* $g(\alpha)$. The sample information is summarized in the sample p.d.f., say $f(\mathbf{y}|\alpha)$, which is algebraically identical to the likelihood function $l(\alpha|\mathbf{y})$. The two types of information are combined via Bayes' theorem which states that

$$g(\alpha|\mathbf{y}) = \frac{f(\mathbf{y}|\alpha)g(\alpha)}{f(\mathbf{y})},$$

where $f(\mathbf{y})$ denotes the unconditional sample density which, for a given sample, is just a normalizing constant. In other words, the distribution of α, given the sample information contained in $\mathbf{y}$, can be summarized by $g(\alpha|\mathbf{y})$. This function is proportional to the likelihood function times the prior density $g(\alpha)$,

$$g(\alpha|\mathbf{y}) \propto f(\mathbf{y}|\alpha)g(\alpha) = l(\alpha|\mathbf{y})g(\alpha). \tag{5.4.1}$$

The conditional density $g(\alpha|\mathbf{y})$ is the *posterior p.d.f.*. It contains all the information available on the parameter vector α. Point estimators of α may be derived from the posterior distribution. For instance, the mean of that distribution, called the *posterior mean*, is often used as a point estimator for α. In the next subsection this general framework is specialized to VAR models.

5.4.2 Normal Priors for the Parameters of a Gaussian VAR Process

Suppose y_t is a zero mean, stable, stationary Gaussian VAR(p) process of the form

$$y_t = A_1 y_{t-1} + \cdots + A_p y_{t-p} + u_t$$

and the prior distribution for $\alpha := \text{vec}(A) = \text{vec}(A_1, \ldots, A_p)$ is a multivariate normal with known mean α^* and covariance matrix V_α,

$$g(\alpha) = \left(\frac{1}{2\pi}\right)^{K^2 p/2} |V_\alpha|^{-1/2} \exp\left[-\frac{1}{2}(\alpha - \alpha^*)' V_\alpha^{-1} (\alpha - \alpha^*)\right]. \tag{5.4.2}$$

Combining this information with the sample information summarized in the Gaussian likelihood function,

$$l(\alpha|\mathbf{y}) = \left(\frac{1}{2\pi}\right)^{KT/2} |I_T \otimes \Sigma_u|^{-1/2}$$
$$\times \exp\left[-\frac{1}{2}(\mathbf{y} - (X' \otimes I_K)\alpha)'(I_T \otimes \Sigma_u^{-1})(\mathbf{y} - (X' \otimes I_K)\alpha)\right]$$

(see Chapter 3, Section 3.4, for the definitions), gives the posterior density

$$g(\boldsymbol{\alpha}|\mathbf{y}) \propto g(\boldsymbol{\alpha})l(\boldsymbol{\alpha}|\mathbf{y})$$
$$\propto \exp\Big\{-\frac{1}{2}\Big[(V_{\boldsymbol{\alpha}}^{-1/2}(\boldsymbol{\alpha}-\boldsymbol{\alpha}^*))'(V_{\boldsymbol{\alpha}}^{-1/2}(\boldsymbol{\alpha}-\boldsymbol{\alpha}^*))$$
$$+ \{(I_T \otimes \Sigma_u^{-1/2})\mathbf{y} - (X' \otimes \Sigma_u^{-1/2})\boldsymbol{\alpha}\}'$$
$$\times\{(I_T \otimes \Sigma_u^{-1/2})\mathbf{y} - (X' \otimes \Sigma_u^{-1/2})\boldsymbol{\alpha}\}\Big]\Big\}. \quad (5.4.3)$$

Here $V_{\boldsymbol{\alpha}}^{-1/2}$ and $\Sigma_u^{-1/2}$ denote the symmetric square root matrices of $V_{\boldsymbol{\alpha}}^{-1}$ and Σ_u^{-1}, respectively (see Appendix A.9.2). The white noise covariance matrix Σ_u is assumed to be known for the moment. Defining

$$w := \begin{bmatrix} V_{\boldsymbol{\alpha}}^{-1/2}\boldsymbol{\alpha}^* \\ (I_T \otimes \Sigma_u^{-1/2})\mathbf{y} \end{bmatrix} \text{ and } W := \begin{bmatrix} V_{\boldsymbol{\alpha}}^{-1/2} \\ X' \otimes \Sigma_u^{-1/2} \end{bmatrix},$$

the exponent in (5.4.3) can be rewritten as

$$-\tfrac{1}{2}(w - W\boldsymbol{\alpha})'(w - W\boldsymbol{\alpha})$$
$$= -\tfrac{1}{2}[(\boldsymbol{\alpha} - \bar{\boldsymbol{\alpha}})'W'W(\boldsymbol{\alpha} - \bar{\boldsymbol{\alpha}}) + (w - W\bar{\boldsymbol{\alpha}})'(w - W\bar{\boldsymbol{\alpha}})], \quad (5.4.4)$$

where

$$\bar{\boldsymbol{\alpha}} := (W'W)^{-1}W'w = [V_{\boldsymbol{\alpha}}^{-1} + (XX' \otimes \Sigma_u^{-1})]^{-1}[V_{\boldsymbol{\alpha}}^{-1}\boldsymbol{\alpha}^* + (X \otimes \Sigma_u^{-1})\mathbf{y}]. \quad (5.4.5)$$

Because the second term on the right-hand side of (5.4.4) does not contain $\boldsymbol{\alpha}$, it may be absorbed into the constant of proportionality. Hence,

$$g(\boldsymbol{\alpha}|\mathbf{y}) \propto \exp\left[-\frac{1}{2}(\boldsymbol{\alpha} - \bar{\boldsymbol{\alpha}})'\bar{\Sigma}_{\boldsymbol{\alpha}}^{-1}(\boldsymbol{\alpha} - \bar{\boldsymbol{\alpha}})\right],$$

where $\bar{\boldsymbol{\alpha}}$ is given in (5.4.5) and

$$\bar{\Sigma}_{\boldsymbol{\alpha}} := (W'W)^{-1} = [V_{\boldsymbol{\alpha}}^{-1} + (XX' \otimes \Sigma_u^{-1})]^{-1}. \quad (5.4.6)$$

Thus, the posterior density is easily recognizable as the density of a multivariate normal distribution with mean $\bar{\boldsymbol{\alpha}}$ and covariance matrix $\bar{\Sigma}_{\boldsymbol{\alpha}}$, that is, the posterior distribution of $\boldsymbol{\alpha}$ is $\mathcal{N}(\bar{\boldsymbol{\alpha}}, \bar{\Sigma}_{\boldsymbol{\alpha}})$. This distribution may be used for inference regarding $\boldsymbol{\alpha}$.

Sometimes one would like to leave some of the coefficients without any restrictions because no prior information is available. In the above framework, this case can be handled by setting the corresponding prior variance to infinity. Unfortunately, such a choice is inconvenient here because algebraic operations have to be performed with the elements of $V_{\boldsymbol{\alpha}}$ in order to compute $\bar{\boldsymbol{\alpha}}$ and $\bar{\Sigma}_{\boldsymbol{\alpha}}$. Therefore, in such cases it is preferable to write the prior information in the form

$$C\boldsymbol{\alpha} = c + e \quad \text{with} \quad e \sim \mathcal{N}(0, I). \quad (5.4.7)$$

5.4 Bayesian Estimation

Here C is a fixed matrix and c is a fixed vector. If C is a $(K^2p \times K^2p)$ nonsingular matrix,

$$\boldsymbol{\alpha} \sim \mathcal{N}(C^{-1}c, C^{-1}C^{-1\prime}).$$

That is, the prior information is given in the form of a multivariate normal distribution with mean $C^{-1}c$ and covariance matrix $(C'C)^{-1}$. From (5.4.5), under Gaussian assumptions, the resulting posterior mean is

$$\bar{\boldsymbol{\alpha}} = [C'C + (XX' \otimes \Sigma_u^{-1})]^{-1}[C'c + (X \otimes \Sigma_u^{-1})\mathbf{y}]. \qquad (5.4.8)$$

A practical advantage of this representation of the posterior mean is that it does not require the inversion of $V_{\boldsymbol{\alpha}}$. Moreover, this form can also be used if no prior information is available for some of the coefficients. For instance, if no prior information on the first coefficient is available, we may simply eliminate one row from C and put zeros in the first column. Although the prior information cannot be represented in the form of a proper multivariate normal distribution in this case, the estimator $\bar{\boldsymbol{\alpha}}$ in (5.4.8) can still be used.

In order to make these concepts useful, the prior mean $\boldsymbol{\alpha}^*$ and covariance matrix $V_{\boldsymbol{\alpha}}$ or C and c must be specified. In the next subsection possible choices are considered.

5.4.3 The Minnesota or Litterman Prior

In Litterman (1986) and Doan, Litterman & Sims (1984), a specific prior, often referred to as Minnesota prior or Litterman prior, for the parameters of a VAR model is described. A similar prior will be considered here as an example. The so-called Minnesota prior was suggested for certain nonstationary processes. We will adapt it for the stationary case because we are still dealing with stationary, stable processes. The nonstationary version of the Minnesota prior will be presented in Chapter 7.

If the intertemporal dependence of the variables is believed to be weak, one way to describe this is to set the prior mean of the VAR coefficients to zero with nonzero prior variances. In other words, $\boldsymbol{\alpha}^* = 0$ and $V_{\boldsymbol{\alpha}} \neq 0$. With this choice of $\boldsymbol{\alpha}^*$ the posterior mean in (5.4.5) reduces to

$$\bar{\boldsymbol{\alpha}} = [V_{\boldsymbol{\alpha}}^{-1} + (XX' \otimes \Sigma_u^{-1})]^{-1}(X \otimes \Sigma_u^{-1})\mathbf{y}. \qquad (5.4.9)$$

This estimator for $\boldsymbol{\alpha}$ looks like the multivariate LS estimator except for the inverse covariance matrix $V_{\boldsymbol{\alpha}}^{-1}$.

In the spirit of Litterman (1986), the prior covariance matrix $V_{\boldsymbol{\alpha}}$ may be specified as a diagonal matrix with diagonal elements

$$v_{ij,l} = \begin{cases} (\lambda/l)^2 & \text{if } i = j, \\ (\lambda \theta \sigma_i / l \sigma_j)^2 & \text{if } i \neq j, \end{cases} \qquad (5.4.10)$$

where $v_{ij,l}$ is the prior variance of $\alpha_{ij,l}$, λ is the prior standard deviation of the coefficients $\alpha_{kk,1}$, $k = 1, \ldots, K$, $0 < \theta < 1$, and σ_i^2 is the i-th diagonal element of Σ_u. For each equation, λ controls how tightly the coefficient of the first lag of the dependent variable is believed to be concentrated around zero. For instance, in the k-th equation of the system it is the prior standard deviation of $\alpha_{kk,1}$. In practice, different values of λ are sometimes tried. Using different λ's in different equations may also be considered.

Because it is believed that coefficients of high order lags are likely to be close to zero, the prior variance decreases with increasing lag length l. Furthermore, it is believed that most of the variation in each of the variables is accounted for by own lags. Therefore coefficients of variables other than the dependent variable are assigned a smaller variance in relative terms by choosing θ between 0 and 1, for instance, $\theta = .2$. The ratio σ_i^2/σ_j^2 is included to take care of the differences in the variability of the different variables. Here the residual variances are preferred over the y_k variances because it is assumed that the response of one variable to another is largely determined by the unexpected movements reflected in the residual variance. Finally, the assumption of a diagonal V_α matrix means that independent prior distributions of the different coefficients are specified. This specification mainly reflects our inability to model dependencies between the coefficients.

As an example consider a bivariate VAR(2) system consisting of the two equations

$$y_{1t} = \alpha_{11,1}y_{1,t-1} + \alpha_{12,1}y_{2,t-1} + \alpha_{11,2}y_{1,t-2} + \alpha_{12,2}y_{2,t-2} + u_{1t},$$
$$\quad\quad (\lambda) \quad\quad (\lambda\theta\sigma_1/\sigma_2) \quad (\lambda/2) \quad\quad (\lambda\theta\sigma_1/2\sigma_2)$$

$$y_{2t} = \alpha_{21,1}y_{1,t-1} + \alpha_{22,1}y_{2,t-1} + \alpha_{21,2}y_{1,t-2} + \alpha_{22,2}y_{2,t-2} + u_{2t},$$
$$\quad\quad (\lambda\theta\sigma_2/\sigma_1) \quad (\lambda) \quad\quad (\lambda\theta\sigma_2/2\sigma_1) \quad (\lambda/2)$$

(5.4.11)

where the prior standard deviations are given in parentheses. The prior covariance matrix of the eight coefficients of this system is

$$V_\alpha = \begin{bmatrix} \lambda^2 & & & & & & & \\ & \left(\frac{\lambda\theta\sigma_2}{\sigma_1}\right)^2 & & & & 0 & & \\ & & \left(\frac{\lambda\theta\sigma_1}{\sigma_2}\right)^2 & & & & & \\ & & & \lambda^2 & & & & \\ & & & & \left(\frac{\lambda}{2}\right)^2 & & & \\ & & & & & \left(\frac{\lambda\theta\sigma_2}{2\sigma_1}\right)^2 & & \\ & 0 & & & & & \left(\frac{\lambda\theta\sigma_1}{2\sigma_2}\right)^2 & \\ & & & & & & & \left(\frac{\lambda}{2}\right)^2 \end{bmatrix}.$$

In terms of (5.4.7), this prior may be specified by choosing $c = 0$ and C an (8×8) diagonal matrix with the square roots of the reciprocals of the diagonal elements of V_α on the main diagonal.

5.4.4 Practical Considerations

In specifying the Minnesota priors, even if λ and θ are chosen appropriately, there remain some practical problems. The first results from the fact that Σ_u is usually unknown. In a strict Bayesian approach, a prior p.d.f. for the elements of Σ_u would be chosen. However, that would lead to a more difficult posterior distribution for α. Therefore, a more practical approach is to replace the σ_i by the square roots of the diagonal elements of the LS or ML estimator of Σ_u, e.g.,

$$\widetilde{\Sigma}_u = Y(I_T - X'(XX')^{-1}X)Y'/T.$$

A second problem is the computational expense that may result from the inversion of the matrix $V_\alpha^{-1} + (XX' \otimes \Sigma_u^{-1})$ or $C'C + (XX' \otimes \Sigma_u^{-1})$ in the posterior mean $\bar{\alpha}$ which is usually used as an estimator for α. This matrix has dimension $(K^2 p \times K^2 p)$. Because in a Bayesian analysis sometimes one may want to choose a large order p and put tight zero priors on the coefficients of large lags rather than make them zero with probability 1, like in an order selection approach, the dimension of the matrix to be inverted in computing $\bar{\alpha}$ may be quite substantial, although this may not be a concern with modern computing technology. Still, Bayesian estimation is sometimes applied to each of the K equations of the system individually. For instance, for the k-th equation,

$$\bar{a}_k := [V_k^{-1} + \sigma_k^{-2}XX']^{-1}[V_k^{-1}a_k^* + \sigma_k^{-2}Xy_{(k)}] \tag{5.4.12}$$

may be used as an estimator of the parameters a_k (the transpose of the k-th row of $A = [A_1, \ldots, A_p]$). Here a_k^* is the prior mean and V_k is the prior covariance matrix of a_k and $y'_{(k)}$ is the k-th row of Y. Using (5.4.12) instead of (5.4.5) reduces the computational expense a bit.

A further problem is related to the zero mean assumption made in the foregoing for the process y_t. In practice, one may simply subtract the sample mean from each variable and then perform a Bayesian analysis for the mean-adjusted data. This amounts to assuming that no prior information exists for the mean terms. Alternatively, intercept terms may be included in the analysis. If the prior information is specified in terms of (5.4.7), it is easy to leave the intercept terms unrestricted, if desired.

5.4.5 An Example

To illustrate the Bayesian approach, we have computed estimates $\bar{a}_k$ as in (5.4.12) for the investment/income/consumption example data with different values of λ and θ. Again we use first differences of logarithms of the data for the years 1960–1978. In Table 5.3, we give estimates for the investment equation of a VAR(2) model. In a Bayesian analysis, one would usually choose a larger VAR order. For illustrative purposes, the VAR(2) model is helpful, however.

Table 5.3. Bayesian estimates of the investment equation from the investment/income/consumption system

λ	θ	ν_1	$\alpha_{11,1}$	$\alpha_{12,1}$	$\alpha_{13,1}$	$\alpha_{11,2}$	$\alpha_{12,2}$	$\alpha_{13,2}$
∞	1	−.017	−.320	.146	.961	−.161	.115	.934
1	.99	−.015	−.309	.159	.921	−.147	.135	.854
.1	.99	.008	−.096	.150	.297	−.011	.062	.100
.01	.99	.018	−.001	.003	.005	−.000	.000	.001
1	.50	−.013	−.301	.194	.847	−.141	.165	.718
1	.10	.009	−.245	.190	.369	−.099	.074	.137
1	.01	.023	−.208	.004	.007	−.078	.001	.002

In the investment equation, the parameter λ controls the overall prior variance of all VAR coefficients while θ controls the tightness of the variances of the coefficients of lagged income and consumption. Roughly speaking, θ specifies the fraction of the prior standard deviation λ attached to the coefficients of lagged income and consumption. Thus, a value of θ close to one means that all coefficients of lag 1 have about the same prior variance except for a scaling factor that takes care of the different variability of different variables. Note that the intercept terms are not restricted (prior variance set to ∞).

We assume a prior mean of zero for all coefficients, $a_k^* = 0$, and thus shrink towards zero by tightening the prior standard deviation λ. The effect is clearly reflected in Table 5.3. For $\theta = .99$ and $\lambda = 1$ we get coefficient estimates which are quite similar to unrestricted LS estimates ($\lambda = \infty, \theta = 1$). Decreasing λ to zero tightens the prior variance and shrinks all VAR coefficients to zero. For $\lambda = .01$, they are quite close to zero already. On the other hand, moving the variance fraction θ towards zero shrinks the consumption and income coefficients ($\alpha_{12,i}, \alpha_{13,i}$) towards zero. In Table 5.3, for $\lambda = 1$ and $\theta = .01$ they are seen to be almost zero. This, of course, has some impact on the investment coefficients ($\alpha_{11,i}$) too.

5.4.6 Classical versus Bayesian Interpretation of $\bar{\alpha}$ in Forecasting and Structural Analysis

If the coefficients of a VAR process are estimated by a Bayesian procedure, the estimated process may be used for prediction and economic analysis, as described in the previous sections. Again one question of interest concerns the statistical properties of the resulting forecasts and impulse responses. It is possible to interpret $\bar{\alpha}$ in (5.4.5) or (5.4.8) as an estimator in the classical sense and to answer this question in terms of asymptotic theory, as in the previous sections. In the classical context, $\bar{\alpha}$ may be interpreted as a shrinkage estimator or under the heading of estimation with stochastic restrictions (e.g., Judge, Griffiths, Hill, Lütkepohl & Lee (1985, Chapter 3)). In regression models with nonstochastic regressors, such estimators, under suitable conditions, have smaller mean squared errors than ML estimators in small samples. In the present framework, the small sample properties are unknown in general.

To derive asymptotic properties, let us consider the representation (5.4.8). It is easily seen that, under our standard conditions,

$$\begin{aligned}
\text{plim } \bar{\alpha} &= \text{plim} \left(\frac{C'C}{T} + \frac{XX'}{T} \otimes \Sigma_u^{-1} \right)^{-1} \\
&\quad \times \text{plim} \left[\frac{C'c}{T} + \text{vec}\left(\frac{\Sigma_u^{-1} Y X'}{T} \right) \right] \\
&= \left[\text{plim} \left(\frac{XX'}{T} \right)^{-1} \otimes \Sigma_u \right] \text{plim vec}\left(\frac{\Sigma_u^{-1} Y X'}{T} \right) \\
&= \alpha.
\end{aligned}$$

Here $\text{plim } C'C/T = \lim C'C/T = 0$ and $\text{plim } C'c/T = 0$ has been used. Moreover, viewing $\bar{\alpha}$ as an estimator in the classical sense, it has the same asymptotic distribution as the unconstrained multivariate LS estimator,

$$\hat{\alpha} = \text{vec}(YX'(XX')^{-1}),$$

because

$$\begin{aligned}
\sqrt{T}(\bar{\alpha} - \hat{\alpha}) &= \left[\frac{C'C}{T} + \frac{XX'}{T} \otimes \Sigma_u^{-1} \right]^{-1} \left[\frac{C'c}{\sqrt{T}} + \frac{1}{\sqrt{T}} \text{vec}(\Sigma_u^{-1} Y X') \right] \\
&\quad - \left[\left(\frac{XX'}{T} \right)^{-1} \otimes \Sigma_u \right] \frac{1}{\sqrt{T}} \text{vec}(\Sigma_u^{-1} Y X') \xrightarrow{p} 0.
\end{aligned}$$

Thus, $\bar{\alpha}$ and $\hat{\alpha}$ have the same asymptotic distribution by Proposition C.2(2) of Appendix C. This result is intuitively appealing because it shows that the contribution of the prior information becomes negligible when the sample size approaches infinity and the sample information becomes exhaustive. Yet the result is not very helpful when a small sample is given in a practical situation.

Consequently, it may be preferable to base the analysis on the posterior distribution of α. In general, it will be difficult to derive the distribution of, say, the impulse responses from the posterior distribution of α analytically. In that case, one may obtain, for instance, confidence intervals of these quantities from a simulation. That is, a large number of samples is drawn from the posterior distribution of α and the corresponding impulse response coefficients are computed. The required percentage points are then estimated from the empirical distributions of the estimated impulse responses (see Appendix D).

5.5 Exercises

In the following exercises, the notation of the previous sections of this chapter is used.

5.5.1 Algebraic Exercises

Problem 5.1
Show that $\check{\gamma}$ given in (5.2.14) minimizes

$$(\mathbf{z} - (Z' \otimes I_K)R\gamma)'(\mathbf{z} - (Z' \otimes I_K)R\gamma)$$

with respect to γ.

Problem 5.2
Prove that $\widehat{\widehat{\beta}}$ given in (5.2.16) minimizes

$$[\mathbf{y} - (Z' \otimes I_K)\beta]'(I_T \otimes \bar{\Sigma}_u^{-1})[\mathbf{y} - (Z' \otimes I_K)\beta]$$

subject to the restriction $C\beta = c$, where C is $(N \times K(Kp+1))$ of rank N and c is $(N \times 1)$. (*Hint:* Specify the appropriate Lagrange function and find its stationary point as described in Appendix A.14.)

Problem 5.3
Show that $\tilde{\tilde{\gamma}}$ given in (5.2.17) is the ML estimator of γ. (*Hint:* Use the partial derivatives from Section 3.4.)

Problem 5.4
Prove Proposition 5.6.

Problem 5.5
Derive the asymptotic distribution of the EGLS estimator of the parameter vector $\alpha := \text{vec}(A_1, \ldots, A_p)$, based on mean-adjusted data, subject to restrictions $\alpha = R\gamma + r$, where R, γ, and r have suitable dimensions.

Problem 5.6
Consider the recursive system of Section 5.2.5,

$$y_t = \eta + A_0^* y_t + \cdots + A_p^* y_{t-p} + \varepsilon_t,$$

where ε_t has a diagonal covariance matrix Σ_ε. Show that $\sum_t \varepsilon_t' \Sigma_\varepsilon^{-1} \varepsilon_t$ and $\sum_t \varepsilon_t' \varepsilon_t$ assume their minima with respect to the unknown parameters for the same values of $\eta, A_0^*, \ldots, A_p^*$.
(*Hint:* Note that

$$\sum_{t=1}^T \varepsilon_t' \Sigma_\varepsilon^{-1} \varepsilon_t = \sum_{k=1}^K \sum_{t=1}^T \varepsilon_{kt}^2 / \sigma_{\varepsilon_k}^2$$

and consider the partial derivatives with respect to the coefficients of the k-th equation. Here ε_{kt} is the k-th element of ε_t and $\sigma_{\varepsilon_k}^2$ is the k-th diagonal element of Σ_ε.)

5.5.2 Numerical Problems

The following problems require the use of a computer. They are based on the bivariate time series $y_t = (y_{1t}, y_{2t})'$ of first differences of the U.S. investment data provided in File E2.

Problem 5.7
Fit a VAR(2) model to the first differences of the data from File E2 subject to the restrictions $\alpha_{12,i} = 0, i = 1, 2$. Determine the EGLS parameter estimates and estimates of their asymptotic standard errors. Perform an F-test to check the restrictions.

Problem 5.8
Based on the result of the previous problem, perform an impulse response analysis for y_1 and y_2.

Problem 5.9
Use a maximum order of 4 and the AIC criterion to determine an optimal subset VAR model for y_t with the top-down strategy described in Section 5.2.8. Repeat the exercise with the HQ criterion. Compare the two models and interpret.

Problem 5.10
Based on the results of Problem 5.9, perform an impulse response analysis for y_1 and y_2 and compare the result to those of Problem 5.8.

Problem 5.11
Use the Minnesota prior with $\lambda = 1$ and $\theta = .2$ and compute the posterior mean of the coefficients of a VAR(4) model for the mean-adjusted y_t. Compare this estimator to the unconstrained multivariate LS estimator of a VAR(4) model for the mean-adjusted data. Repeat the exercise with a VAR(4) model that contains intercept terms.

Part II

Cointegrated Processes

In Part I, stationary, stable VAR processes have been considered. Recall that a process is stationary if it has time invariant first and second moments. This property implies that there are no trends (trending means) or shifts in the mean or in the covariances. Moreover, there are no deterministic seasonal patterns. In this part, nonstationary processes of a very specific type will be considered. In particular, the processes will be allowed to have stochastic trends. They are then called integrated. If some of the variables move together in the long-run although they have stochastic trends, they are driven by a common stochastic trend and they are called cointegrated. VAR processes with integrated and cointegrated variables are analyzed in this part. In Chapter 6, some important theoretical properties of cointegrated processes are discussed and it is shown that they can be conveniently summarized in a vector error correction model (VECM). Estimation of such models is treated in Chapter 7. Specification of VECMs and model checking are considered in Chapter 8.

6
Vector Error Correction Models

As defined in Chapter 2, a process is stationary if it has time invariant first and second moments. In particular, it does not have trends or changing variances. A VAR process has this property if the determinantal polynomial of its VAR operator has all its roots outside the complex unit circle. Clearly, stationary processes cannot capture some main features of many economic time series. For example, trends (trending means) are quite common in practice. For instance, the *original* investment, income, and consumption data used in many previous examples have trends (see Figure 3.1). Thus, if interest centers on analyzing the original variables (or their logarithms) rather than the rates of change, it is necessary to have models that accommodate the nonstationary features of the data. It turns out that a VAR process can generate stochastic and deterministic trends if the determinantal polynomial of the VAR operator has roots on the unit circle. In fact, it is even sufficient to allow for unit roots (roots for $z = 1$) to obtain a trending behavior of the variables. We will consider this case in some detail in this chapter. In the next section, the effect of unit roots in the AR operator of a univariate process will be analyzed. Variables generated by such processes are called *integrated variables* and the underlying generating processes are *integrated processes*. Vector processes with unit roots are considered in Section 6.2. In these processes, some of the variables can have common trends so that they move together to some extent. They are then called *cointegrated*. This feature is considered in detail in Section 6.3 and it is shown that *vector error correction models* (*VECMs*) offer a convenient way to parameterize and specify them. In Section 6.3, the processes are assumed to be purely stochastic and do not have deterministic terms. How to incorporate these terms is the subject of Section 6.4. Once we have a suitable model setup, it can be used for forecasting, causality analysis, and impulse response analysis. These issues are treated in Sections 6.5–6.7.

6.1 Integrated Processes

Recall that a VAR(p) process,

$$y_t = A_1 y_{t-1} + \cdots + A_p y_{t-p} + u_t, \qquad (6.1.1)$$

is stable if the polynomial defined by

$$\det(I_K - A_1 z - \cdots - A_p z^p)$$

has no roots in and on the complex unit circle. For a univariate AR(1) process, $y_t = \alpha y_{t-1} + u_t$, this property means that

$$1 - \alpha z \neq 0 \quad \text{for } |z| \leq 1$$

or, equivalently, $|\alpha| < 1$.

Consider the borderline case, where $\alpha = 1$. The resulting process $y_t = y_{t-1} + u_t$ is called a *random walk*. Starting the process at $t = 0$ with some fixed y_0, it is easy to see by successive substitution for lagged y_t's, that

$$y_t = y_{t-1} + u_t = y_{t-2} + u_{t-1} + u_t = \cdots = y_0 + \sum_{i=1}^{t} u_i. \qquad (6.1.2)$$

Thus, y_t consists of the sum of all disturbances or innovations of the previous periods so that each disturbance has a lasting impact on the process. If u_t is white noise with variance σ_u^2,

$$E(y_t) = y_0$$

and

$$\text{Var}(y_t) = t\text{Var}(u_t) = t\sigma_u^2.$$

Hence, the variance of a random walk tends to infinity. Furthermore, the correlation

$$\text{Corr}(y_t, y_{t+h}) = \frac{E\left[\left(\sum_{i=1}^{t} u_i\right)\left(\sum_{i=1}^{t+h} u_i\right)\right]}{[t\sigma_u^2(t+h)\sigma_u^2]^{1/2}}$$

$$= \frac{t}{(t^2 + th)^{1/2}} \xrightarrow[t \to \infty]{} 1$$

for any integer h. This latter property of a random walk means that y_t and y_s are strongly correlated even if they are far apart in time. It can also be shown that the expected time between two crossings of zero is infinite. These properties are often reflected in trending behavior. Examples are depicted in Figure 6.1. This kind of trend is, of course, not a deterministic one but a stochastic trend.

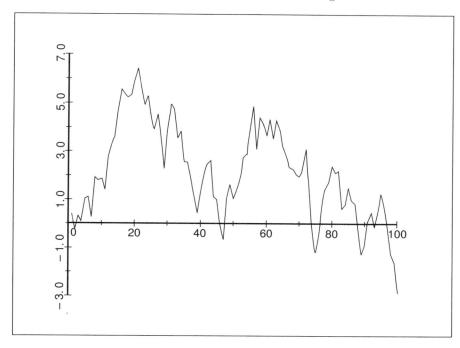

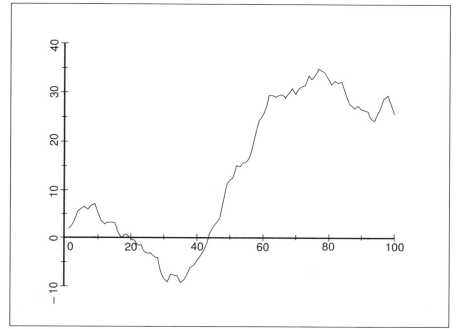

Fig. 6.1. Artificially generated random walks.

If the process has a nonzero constant term ν, $y_t = \nu + y_{t-1} + u_t$ is called a *random walk with drift* and it has a deterministic linear trend in the mean. To see this property, suppose again that the process is started at $t = 0$ with a fixed y_0. Then

$$y_t = y_0 + t\nu + \sum_{i=1}^{t} u_i$$

and $E(y_t) = y_0 + t\nu$. An example of a time series generated by a random walk with drift is shown in Figure 6.2.

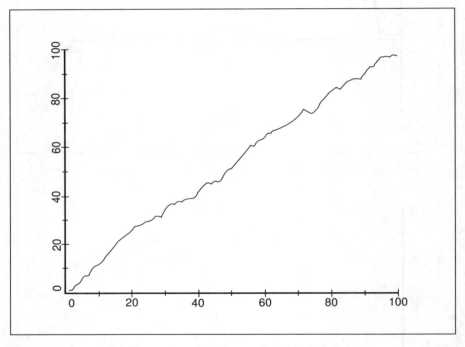

Fig. 6.2. An artificially generated random walk with drift.

The previous discussion suggests that starting unstable processes at some finite time t_0 is useful to obtain processes with finite moments. On the other hand, if an AR process starts at some finite time, it is strictly speaking not necessarily stationary, even if it is stable. To see this property, let $y_t = \nu + \alpha y_{t-1} + u_t$ be a univariate stable AR(1) process with $|\alpha| < 1$. Starting with a random variable y_0 at $t = 0$, gives

$$y_t = \nu \sum_{i=0}^{t-1} \alpha^i + \alpha^t y_0 + \sum_{i=0}^{t-1} \alpha^i u_{t-i}.$$

Hence,
$$E(y_t) = \nu \sum_{i=0}^{t-1} \alpha^i + \alpha^t E(y_0)$$

is generally not time invariant if α and $\nu \neq 0$. A similar result is obtained for the second moments,
$$\text{Var}(y_t) = \alpha^{2t} \text{Var}(y_0) + \sigma_u^2 \sum_{i=0}^{t-1} \alpha^{2i}.$$

However, the first and second moments approach limit values as $t \to \infty$ and one might call such a process *asymptotically stationary*. To simplify matters, the term "asymptotically" is sometimes dropped and such processes are then simply called stationary. Moreover, if we consider purely stochastic processes without deterministic terms ($\nu = 0$), the initial variable can be chosen such that y_t is stationary if the process is stable. In particular, if we choose

$$y_0 = \sum_{i=0}^{\infty} \alpha^i u_{-i}$$

we get, for $\nu = 0$,
$$y_t = \alpha^t \sum_{i=0}^{\infty} \alpha^i u_{-i} + \sum_{i=0}^{t-1} \alpha^i u_{t-i} = \sum_{i=0}^{\infty} \alpha^i u_{t-i}, \quad t = 1, 2, \ldots,$$

and, hence, for $t = 1, 2, \ldots$,

$E(y_t) = 0$,

$\text{Var}(y_t) = \sigma_u^2 / (1 - \alpha^2)$,

and also the autocovariances are time invariant. Thus, for a stable process we may in fact choose the initial variable such that y_t is stationary even if the process is started in some given period. This result can also be used as a justification for simply calling stable processes stationary in this situation. We may implicitly assume that the starting value is chosen to justify the terminology. For our purposes, this point is of limited importance because in later chapters we will be interested in the parameters of the processes considered and possibly in their asymptotic moments. Without further warning, nonstationary, unstable processes will be assumed to begin at some given finite time period.

A behavior similar to that of a random walk is also observed for higher order AR processes such as

$$y_t = \nu + \alpha_1 y_{t-1} + \cdots + \alpha_p y_{t-p} + u_t,$$

if $1 - \alpha_1 z - \cdots - \alpha_p z^p$ has a root for $z = 1$. Note that

$$1 - \alpha_1 z - \cdots - \alpha_p z^p = (1 - \lambda_1 z) \cdots (1 - \lambda_p z),$$

where $\lambda_1, \ldots, \lambda_p$ are the reciprocals of the roots of the polynomial. If the process has just one *unit root* (a root equal to 1) and all other roots are outside the complex unit circle, its behavior is similar to that of a random walk, that is, its variances increase linearly, the correlation between variables h periods apart tends to 1 and the process has a linear trend in mean if $\nu \neq 0$. In case one of the roots is strictly inside the unit circle, the process becomes explosive, that is, its variances go to infinity at an exponential rate. Many researchers feel that such processes are unrealistic models for most economic data. Although processes with roots on the unit circle other than one are often useful, we shall concentrate on the case of unit roots and all other roots outside the unit circle. This situation is of considerable practical interest.

Univariate processes with d unit roots (d roots equal to 1) in their AR operators are called *integrated of order d* ($I(d)$). If there is just one unit root, i.e., the process is $I(1)$, it is quite easy to see how a stable and possibly stationary process can be obtained: simply by taking first differences, $\Delta y_t := (1 - L)y_t = y_t - y_{t-1}$, of the original process. More generally, if the process is $I(d)$ it can be made stable by differencing d times, that is, $\Delta^d y_t = (1 - L)^d y_t$ is stable and, again, initial values can be chosen such that it is stationary. In the following, it will often be convenient to extend this terminology also to stable, stationary processes and to call them $I(0)$.

More generally, y_t may be defined to be an $I(1)$ process, if $\Delta y_t = w_t$ is a stationary process with infinite MA representation, $w_t = \sum_{j=0}^{\infty} \theta_j u_{t-j} = \theta(L)u_t$, where the MA coefficients satisfy the condition $\sum_{j=0}^{\infty} j|\theta_j| < \infty$, $\theta(1) = \sum_{j=0}^{\infty} \theta_j \neq 0$, and $u_t \sim (0, \sigma_u^2)$ is white noise. In that case, $y_t = y_{t-1} + w_t$ can be rewritten as

$$y_t = y_0 + w_1 + \cdots + w_t = y_0 + \theta(1)(u_1 + \cdots + u_t) + \sum_{j=0}^{\infty} \theta_j^* u_{t-j} - w_0^*, \quad (6.1.3)$$

where $\theta_j^* = -\sum_{i=j+1}^{\infty} \theta_i$, $j = 0, 1, \ldots$, and $w_0^* = \sum_{j=0}^{\infty} \theta_j^* u_{-j}$ contains initial values. Thus, y_t can be represented as the sum of a random walk $[\theta(1)(u_1 + \cdots + u_t)]$, a stationary process $[\sum_{j=0}^{\infty} \theta_j^* u_{t-j}]$, and initial values $[y_0 - w_0^*]$. Notice that the condition $\sum_{j=0}^{\infty} j|\theta_j| < \infty$ ensures that $\sum_{j=0}^{\infty} |\theta_j^*| < \infty$, so that $\sum_{j=0}^{\infty} \theta_j^* u_{t-j}$ is indeed well-defined by Proposition C.7 of Appendix C.3. Although the condition for the θ_j is stronger than absolute summability, it is satisfied for many processes of practical interest. The decomposition of y_t in (6.1.3) is known as the *Beveridge-Nelson decomposition* (see also Appendix C.8). A similar decomposition for multivariate processes is helpful in some of the subsequent analysis. It will be discussed in Section 6.3.

6.2 VAR Processes with Integrated Variables

Consider now a K-dimensional VAR(p) process without a deterministic term as in (6.1.1). It can be written as

$$A(L)y_t = u_t, \qquad (6.2.1)$$

where $A(L) := I_K - A_1 L - \cdots - A_p L^p$ and L is the lag operator. Multiplying from the left by the adjoint $A(L)^{adj}$ of $A(L)$ gives

$$|A(L)|y_t = A(L)^{adj} u_t \qquad (6.2.2)$$

(see Appendix A.4.1 for the definition of the adjoint of a matrix). Thus, the VAR(p) process in (6.2.1) can be written as a process with univariate AR operator, that is, all components have the same AR operator $|A(L)|$. The right-hand side of (6.2.2), $A(L)^{adj} u_t$, is a finite order MA process (see Chapter 11 for further discussion of such processes). If $|A(L)|$ has d unit roots and otherwise all roots are outside the unit circle, the AR operator can be written as

$$|A(L)| = \alpha(L)(1-L)^d = \alpha(L)\Delta^d,$$

where $\alpha(L)$ is an invertible operator. Consequently, $\Delta^d y_t$ is a stable process. Hence, each component becomes stable upon differencing.

Because we are considering processes which are started at some specific time t_0, we should perhaps think for a moment about the treatment of initial values when multiplying by an operator such as $A(L)^{adj}$ in (6.2.2). One possible assumption is that the new representation is valid for all t for which the y_t's are defined in (6.2.1).

The foregoing discussion shows that if a VAR(p) process is unstable because of unit roots only, it can be made stable by differencing its components. Note, however, that, due to cancellations, it may not be necessary to difference each component as many times as there are unit roots in $|A(L)|$. To illustrate this point, consider the bivariate VAR(1) process

$$\left(\begin{bmatrix} 1 & 0 \\ 0 & 1 \end{bmatrix} - \begin{bmatrix} 1 & 0 \\ 0 & 1 \end{bmatrix} L\right) \begin{bmatrix} y_{1t} \\ y_{2t} \end{bmatrix} = \begin{bmatrix} (1-L)y_{1t} \\ (1-L)y_{2t} \end{bmatrix} = u_t.$$

Obviously, each component is stationary after differencing once, i.e., each component is $I(1)$, although

$$|A(L)| = \left|\begin{bmatrix} 1-L & 0 \\ 0 & 1-L \end{bmatrix}\right| = (1-L)^2$$

has two unit roots. It is also possible that some components are stable and stationary as univariate processes whereas others need differencing. Examples are easy to construct.

If the VAR(p) process has a nonzero intercept term so that

$$A(L)y_t = \nu + u_t$$

and $|A(z)|$ has one or more unit roots, then some of the components of y_t may have deterministic trends in their mean values. Unlike the univariate case, it is also possible, however, that none of the components of y_t has a deterministic trend in mean. This occurs if $A(L)^{adj}\nu = 0$. For instance, if

$$A(L) = \begin{bmatrix} 1-L & \eta L \\ 0 & 1 \end{bmatrix},$$

$|A(z)|$ has a unit root and

$$A(L)^{adj} = \begin{bmatrix} 1 & -\eta L \\ 0 & 1-L \end{bmatrix}.$$

Hence,

$$A(L)^{adj}\begin{bmatrix} \nu_1 \\ \nu_2 \end{bmatrix} = \begin{bmatrix} \nu_1 - \eta\nu_2 \\ \nu_2 - \nu_2 \end{bmatrix}$$

which is zero if $\nu_1 = \eta\nu_2$. Thus, in a VAR analysis an intercept term cannot be excluded a priori if there are unit roots and none of the component series has a deterministic trend.

The following question comes to mind in this context. Suppose each component of a VAR(p) process is $I(d)$, is it possible that differencing each component individually distorts interesting features of the relationship between the original variables? If the latter were not the case, a VAR analysis could be performed as described in previous chapters after differencing the individual components. It turns out, however, that differencing may indeed distort the relationship between the original variables. Systems with cointegrated variables are examples, where fitting VAR models upon differencing may be inadequate. Such systems are introduced next.

6.3 Cointegrated Processes, Common Stochastic Trends, and Vector Error Correction Models

Equilibrium relationships are suspected between many economic variables such as household income and expenditures or prices of the same commodity in different markets. Suppose the variables of interest are collected in the vector $y_t = (y_{1t}, \ldots, y_{Kt})'$ and their long-run equilibrium relation is $\boldsymbol{\beta}'y_t = \beta_1 y_{1t} + \cdots + \beta_K y_{Kt} = 0$, where $\boldsymbol{\beta} = (\beta_1, \ldots, \beta_K)'$. In any particular period, this relation may not be satisfied exactly but we may have $\boldsymbol{\beta}'y_t = z_t$, where z_t is a stochastic variable representing the deviations from the equilibrium. If there really is an equilibrium, it seems plausible to assume that the y_t variables move together and that z_t is stable. This setup, however, does not exclude the possibility that the y_t variables wander extensively as a group. Thus, they may be driven by a *common stochastic trend*. In other words, it is not excluded that each variable is integrated, yet there exists a linear combination of the variables which is stationary. Integrated variables with this property are called *cointegrated*. In Figure 6.3, two artificially generated cointegrated time series are depicted.

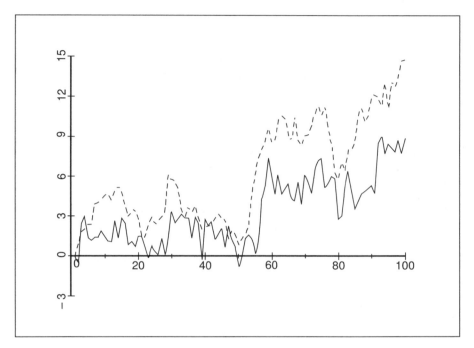

Fig. 6.3. A bivariate cointegrated time series.

Generally, the variables in a K-dimensional process y_t are called *cointegrated of order* (d, b), briefly, $y_t \sim CI(d, b)$, if all components of y_t are $I(d)$ and there exists a linear combination $z_t := \beta' y_t$ with $\beta = (\beta_1, \ldots, \beta_K)' \neq 0$ such that z_t is $I(d - b)$. For instance, if all components of y_t are $I(1)$ and $\beta' y_t$ is stationary ($I(0)$), then $y_t \sim CI(1, 1)$. The vector β is called a *cointegrating vector* or a *cointegration vector*. A process consisting of cointegrated variables is called a *cointegrated process*. These processes were introduced by Granger (1981) and Engle & Granger (1987). Since then they have become popular in theoretical and applied econometric work.

In the following, a slightly different definition of cointegration will be used in order to simplify the terminology. We call a K-dimensional process y_t integrated of order d, briefly, $y_t \sim I(d)$, if $\Delta^d y_t$ is stable and $\Delta^{d-1} y_t$ is not stable. The $I(d)$ process y_t is called cointegrated if there is a linear combination $\beta' y_t$ with $\beta \neq 0$ which is integrated of order less than d. This definition differs from the one given by Engle & Granger (1987) in that we do not exclude components of y_t with order of integration less than d. If there is just one $I(d)$ component in y_t and all other components are stable ($I(0)$), then the vector y_t is $I(d)$ according to our definition because $\Delta^d y_t$ is stable and $\Delta^{d-1} y_t$ is not. In such a case a relation $\beta' y_t$ that involves the stationary components only is a cointegration relation in our terms. Clearly, this aspect of our definition is not in line with the original idea of cointegration as a special relation between integrated variables with common stochastic trends. In the following, our definition is still useful because it simplifies the terminology as it avoids distinguishing between variables with different orders of integration. The reader should keep in mind the basic ideas of cointegration when it comes to interpreting specific relationships, however.

Obviously, a cointegrating vector is not unique. Multiplying by a nonzero constant yields a further cointegrating vector. Also, there may be various linearly independent cointegrating vectors. For instance, if there are four variables in a system, the first two may be connected by a long-run equilibrium relation and also the last two. Thus, there may be a cointegrating vector with zeros in the last two positions and one with zeros in the first two positions. In addition, there may be a cointegration relation involving all four variables.

Before the concept of cointegration was introduced, the closely related *error correction models* were discussed in the econometrics literature (see, e.g., Davidson, Hendry, Srba & Yeo (1978), Hendry & von Ungern-Sternberg (1981), Salmon (1982)). In an error correction model, the changes in a variable depend on the deviations from some equilibrium relation. Suppose, for instance, that y_{1t} represents the price of a commodity in a particular market and y_{2t} is the corresponding price of the same commodity in another market. Assume furthermore that the equilibrium relation between the two variables is given by $y_{1t} = \beta_1 y_{2t}$ and that the changes in y_{1t} depend on the deviations from this equilibrium in period $t - 1$,

$$\Delta y_{1t} = \alpha_1 (y_{1,t-1} - \beta_1 y_{2,t-1}) + u_{1t}.$$

A similar relation may hold for y_{2t},

$$\Delta y_{2t} = \alpha_2 (y_{1,t-1} - \beta_1 y_{2,t-1}) + u_{2t}.$$

In a more general error correction model, the Δy_{it} may in addition depend on previous changes in both variables as, for instance, in the following model:

$$\begin{aligned}\Delta y_{1t} &= \alpha_1 (y_{1,t-1} - \beta_1 y_{2,t-1}) + \gamma_{11,1} \Delta y_{1,t-1} + \gamma_{12,1} \Delta y_{2,t-1} + u_{1t}, \\ \Delta y_{2t} &= \alpha_2 (y_{1,t-1} - \beta_1 y_{2,t-1}) + \gamma_{21,1} \Delta y_{1,t-1} + \gamma_{22,1} \Delta y_{2,t-1} + u_{2t}.\end{aligned}$$

(6.3.1)

Further lags of the Δy_{it}'s may also be included.

6.3 Cointegrated Processes and VECMs

To see the close relationship between error correction models and the concept of cointegration, suppose that y_{1t} and y_{2t} are both $I(1)$ variables. In that case all terms in (6.3.1) involving the Δy_{it} are stable. In addition, u_{1t} and u_{2t} are white noise errors which are also stable. Because an unstable term cannot equal a stable process,

$$\alpha_i(y_{1,t-1} - \beta_1 y_{2,t-1}) = \Delta y_{it} - \gamma_{i1,1}\Delta y_{1,t-1} - \gamma_{i2,1}\Delta y_{2,t-1} - u_{it}$$

must be stable too. Hence, if $\alpha_1 \neq 0$ or $\alpha_2 \neq 0$, $y_{1t} - \beta_1 y_{2t}$ is stable and, thus, represents a cointegration relation.

In vector and matrix notation the model (6.3.1) can be written as

$$\Delta y_t = \alpha\beta' y_{t-1} + \Gamma_1 \Delta y_{t-1} + u_t,$$

or

$$y_t - y_{t-1} = \alpha\beta' y_{t-1} + \Gamma_1(y_{t-1} - y_{t-2}) + u_t, \qquad (6.3.2)$$

where $y_t := (y_{1t}, y_{2t})'$, $u_t := (u_{1t}, u_{2t})'$,

$$\alpha := \begin{bmatrix} \alpha_1 \\ \alpha_2 \end{bmatrix}, \quad \beta' := (1, -\beta_1), \quad \text{and} \quad \Gamma_1 := \begin{bmatrix} \gamma_{11,1} & \gamma_{12,1} \\ \gamma_{21,1} & \gamma_{22,1} \end{bmatrix}.$$

Rearranging terms in (6.3.2) gives the VAR(2) representation

$$y_t = (I_K + \Gamma_1 + \alpha\beta') y_{t-1} - \Gamma_1 y_{t-2} + u_t.$$

Hence, cointegrated variables may be generated by a VAR process.

To see how cointegration can arise more generally in K-dimensional VAR models, consider the VAR(2) process

$$y_t = A_1 y_{t-1} + A_2 y_{t-2} + u_t \qquad (6.3.3)$$

with $y_t = (y_{1t}, \ldots, y_{Kt})'$. Suppose the process is unstable with

$$|I_K - A_1 z - A_2 z^2| = (1 - \lambda_1 z) \cdots (1 - \lambda_n z) = 0 \quad \text{for } z = 1.$$

Because the λ_i are the reciprocals of the roots of the determinantal polynomial, one or more of them must be equal to 1. All other roots are assumed to lie outside the unit circle, that is, all λ_i which are not 1 are inside the complex unit circle. Because $|I_K - A_1 - A_2| = 0$, the matrix

$$\Pi := -(I_K - A_1 - A_2)$$

is singular. Suppose $\text{rk}(\Pi) = r < K$. Then Π can be decomposed as $\Pi = \alpha\beta'$, where α and β are $(K \times r)$ matrices. From the discussion in the previous section, we know that each variable becomes stationary upon differencing. Let us assume that differencing once is sufficient, subtract y_{t-1} on both sides of (6.3.3) and rearrange terms as

$$y_t - y_{t-1} = -(I_K - A_1 - A_2)y_{t-1} - A_2 y_{t-1} + A_2 y_{t-2} + u_t$$

or

$$\Delta y_t = \Pi y_{t-1} + \Gamma_1 \Delta y_{t-1} + u_t, \tag{6.3.4}$$

where $\Gamma_1 := -A_2$, or

$$\alpha\beta' y_{t-1} = \Delta y_t - \Gamma_1 \Delta y_{t-1} - u_t.$$

Because the right-hand side involves stationary terms only, $\alpha\beta' y_{t-1}$ must also be stationary and it remains stationary upon multiplication by $(\alpha'\alpha)^{-1}\alpha'$. In other words, $\beta' y_t$ is stationary and, hence, each element of $\beta' y_t$ represents a cointegrating relation. Note that simply taking first differences of all variables in (6.3.3) eliminates the cointegration term which may well contain relations of great importance for a particular analysis. Moreover, in general, a VAR process with cointegrated variables does not admit a pure VAR representation in first differences.

It may also be worth emphasizing that here we have worked under the assumption that all variables are stationary after differencing once. In general, variables with higher integration orders may also be present. In that case, $\beta' y_t$ may not be stationary even if $\text{rk}(\Pi) = r < K$. The components of y_t may still be cointegrated of a higher order if linear combinations exist which have a reduced order of integration.

In the following, we will be interested in the specific case where all individual variables are $I(1)$ or $I(0)$. The K-dimensional VAR(p) process

$$y_t = A_1 y_{t-1} + \cdots + A_p y_{t-p} + u_t, \tag{6.3.5}$$

is called *cointegrated of rank r* if

$$\Pi := -(I_K - A_1 - \cdots - A_p)$$

has rank r and, thus, Π can be written as a matrix product $\alpha\beta'$ with α and β being of dimension $(K \times r)$ and of rank r. The matrix β is called a *cointegrating* or *cointegration matrix* or a *matrix of cointegrating* or *cointegration vectors* and α is sometimes called the *loading matrix*. If $r = 0$, Δy_t has a stable VAR($p-1$) representation and, for $r = K$, $|I_K - A_1 - \cdots - A_p| = |-\Pi| \neq 0$ and, hence, the VAR operator has no unit roots so that y_t is a stable VAR(p) process.

Rewriting (6.3.5) as in (6.3.4) it has a *vector error correction model* (*VECM*) representation

$$\begin{aligned}\Delta y_t &= \Pi y_{t-1} + \Gamma_1 \Delta y_{t-1} + \cdots + \Gamma_{p-1} \Delta y_{t-p+1} + u_t \\ &= \alpha\beta' y_{t-1} + \Gamma_1 \Delta y_{t-1} + \cdots + \Gamma_{p-1} \Delta y_{t-p+1} + u_t,\end{aligned} \tag{6.3.6}$$

where

$$\Gamma_i := -(A_{i+1} + \cdots + A_p), \quad i = 1, \ldots, p-1.$$

If this representation of a cointegrated process is given, it is easy to recover the corresponding VAR form (6.3.5) by noting that

$$\begin{aligned} A_1 &= \Pi + I_K + \Gamma_1 \\ A_i &= \Gamma_i - \Gamma_{i-1}, \quad i = 2, \ldots, p-1, \\ A_p &= -\Gamma_{p-1}. \end{aligned} \tag{6.3.7}$$

It may be worth pointing out that we can also rearrange the terms in a different way and obtain a representation

$$\Delta y_t = D_1 \Delta y_{t-1} + \cdots + D_{p-1} \Delta y_{t-p+1} + \Pi y_{t-p} + u_t, \tag{6.3.8}$$

where the error correction term appears at lag p and

$$D_i = -(I_K - A_1 - \cdots - A_i), \quad i = 1, \ldots, p-1.$$

In the following sections, we will usually work with (6.3.5) or (6.3.6). Of course, thereby we work within a much more narrow framework than that allowed for in the general definition of cointegration. First, we consider $I(1)$ processes only and, second, the discussion is limited to finite order VAR processes or VECMs.

It is important to note that the decomposition of the $(K \times K)$ matrix Π as the product of two $(K \times r)$ matrices, $\Pi = \alpha\beta'$, is not unique. In fact, for every nonsingular $(r \times r)$ matrix Q, we can define $\alpha^* = \alpha Q'$ and $\beta^* = \beta Q^{-1}$ and get $\Pi = \alpha^* \beta^{*'}$. This nonuniqueness of the decomposition of Π shows again that the cointegration relations are not unique. It is possible, however, to impose restrictions on β and/or α to get unique relations. Such restrictions may be implied by subject matter considerations or they may be imposed for convenience, using the algebraic properties of the associated matrices.

As an example, consider a system of three interest rates, $y_t = (y_{1t}, y_{2t}, y_{3t})'$, where y_{1t} is a short-term rate, y_{2t} is a medium-term rate, and y_{3t} is a long-term rate. Suppose all three interest rates are $I(1)$ variables whereas the interest rate spreads, $y_{it} - y_{jt}$ ($i \neq j$) are stationary ($I(0)$). Then we have two linearly independent cointegrating relations which can, for example, be written as

$$\beta' y_t = \begin{bmatrix} 1 & 0 & -1 \\ 0 & 1 & -1 \end{bmatrix} y_t$$

or, alternatively, as

$$\beta^{*'} y_t = \begin{bmatrix} 1 & -1 & 0 \\ 0 & 1 & -1 \end{bmatrix} y_t.$$

Using the fact that $\text{rk}(\beta) = r$, there must be r linearly independent rows. Thus, by a suitable rearrangement of the variables it can always be ensured that the first r rows of β are linearly independent. Hence, the upper $(r \times r)$

submatrix consisting of the first r rows of β is nonsingular. Choosing Q then equal to this matrix gives a cointegration matrix

$$\beta^* = \begin{bmatrix} I_r \\ \beta_{(K-r)} \end{bmatrix}, \qquad (6.3.9)$$

where $\beta_{(K-r)}$ is $((K-r) \times r)$. This normalization will occasionally be used in the following because it is quite convenient to ensure a unique cointegration matrix. It does not imply a loss of generality except that it is assumed that the variables are arranged in the right way so that the normalization is feasible. If the system is known, as implicitly assumed here, rearranging the variables in a suitable way is no problem, of course. In fact, we just need to know the cointegration properties between all subsets of variables in order to find a suitable arrangement of the variables.

To see this, consider again a three-dimensional system, $y_t = (y_{1t}, y_{2t}, y_{3t})'$, with cointegrating rank 1 so that there is just one cointegration vector β. In that case, the normalization in (6.3.9) amounts to setting the first component of the cointegration vector to one. Hence, $\beta^{*\prime} y_t = [1, \beta'_{(K-1)}] y_t = y_{1t} + \beta_2 y_{2t} + \beta_3 y_{3t}$. Clearly, this normalization is only feasible if the first component of y_t actually belongs to the cointegration relation and has nonzero coefficient. If we know that y_{2t} and y_{3t} are not cointegrated while y_{1t}, y_{2t}, and y_{3t} together are cointegrated, then we know already that y_{1t} is part of the cointegration relation and, thus, has a nonzero coefficient in β.

As another example, suppose y_t has cointegrating rank 2. In that case the normalized cointegrating relations are

$$\begin{bmatrix} 1 & 0 & \beta_1 \\ 0 & 1 & \beta_2 \end{bmatrix} y_t = \begin{bmatrix} y_{1t} + \beta_1 y_{3t} \\ y_{2t} + \beta_2 y_{3t} \end{bmatrix}.$$

Thus, a cointegration relation must exist in the bivariate systems $(y_{1t}, y_{3t})'$ and $(y_{2t}, y_{3t})'$. By checking these subsystems separately, a possible ordering of the variables is easy to find. It may be worth mentioning, however, that given our general definition of cointegration, it is possible that in this example y_{1t} or/and y_{2t} are in fact stationary $I(0)$ variables. For instance, if both are $I(0)$, $\beta_1 = \beta_2 = 0$. Recall that a process y_t is called $I(1)$ even if only a single component is $I(1)$ and the other components are $I(0)$.

Generally, any stationary variables in the system must be placed in the upper r-dimensional subvector of y_t. If y_{kt}, the k-th component of y_t, is stationary, there is a 'cointegrating relation' $\beta'_k y_t$ with β_k being a vector with a one as the k-th component and zeros elsewhere so that $\beta'_k y_t = y_{kt}$. Thus, there is a cointegrating relation for each of the stationary components of y_t. Because the associated cointegrating vectors are linearly independent, the cointegrating rank must be at least as great as the number of $I(0)$ variables in the system.

The important result to remember from this discussion is that the normalization of the cointegration matrix given in (6.3.9) is always possible if the variables are arranged in a suitable way. Finding the proper ordering is easy if the cointegration properties of all subsystems are known, including the univariate subsystems. In other words, we also need to know the order of integration of the individual variables in the system. In practice, the order of integration and the cointegrating rank of a given system and its subsystems will not be known. Statistical procedures for determining the cointegrating rank which can help to overcome this practical problem are discussed in Chapter 8.

If the normalization in (6.3.9) is made, the system may also be set up as

$$y_t^{(1)} = -\beta'_{(K-r)} y_t^{(2)} + z_t^{(1)},$$
$$\Delta y_t^{(2)} = z_t^{(2)}, \tag{6.3.10}$$

where $y_t^{(1)}$ and $z_t^{(1)}$ are $(r \times 1)$, $y_t^{(2)}$ and $z_t^{(2)}$ are $((K-r) \times 1)$ and $z_t = (z_t^{(1)\prime}, z_t^{(2)\prime})'$ is a stationary process. There cannot be any cointegrating relations between the components of the subsystem $y_t^{(2)}$, because otherwise there would be more than r linearly independent cointegrating relations and the cointegrating rank would be larger than r. Thus, the variables in $y_t^{(2)}$ represent stochastic trends in the system. The representation (6.3.10) is known as the *triangular representation* of a cointegrated system. It has been used extensively in some of the literature related to cointegration analysis (see, e.g., Park & Phillips (1988, 1989)).

Yet another useful representation of a cointegrated system is given by Johansen (1995, Theorem 4.2). The underlying result is often referred to as *Granger representation theorem*. To state this representation, we use the following notation. For $m \geq n$, we denote by $M_\perp$ an orthogonal complement of the $(m \times n)$ matrix M with $\text{rk}(M) = n$ (see also Appendix A.8.2). In other words, $M_\perp$ is any $(m \times (m-n))$ matrix with $\text{rk}(M_\perp) = m-n$ and $M'M_\perp = 0$. If M is a nonsingular square matrix $(m = n)$, then $M_\perp = 0$ and if $n = 0$, we define $M_\perp = I_m$. This latter convention is sometimes useful to avoid clumsy notation and looking at different cases separately. We assume that y_t is a K-dimensional cointegrated $I(1)$ process as in (6.3.6) with cointegration rank r, $0 \leq r < K$. Then the following proposition holds.

Proposition 6.1 (*Granger Representation Theorem*)
Suppose

$$\Delta y_t = \alpha\beta' y_{t-1} + \Gamma_1 \Delta y_{t-1} + \cdots + \Gamma_{p-1} \Delta y_{t-p+1} + u_t, \quad t = 1, 2, \ldots,$$

where $y_t = 0$ for $t \leq 0$, u_t is white noise for $t = 1, 2, \ldots$, and $u_t = 0$ for $t \leq 0$. Moreover, define

$$C(z) := (1-z)I_K - \alpha\beta' z - \sum_{i=1}^{p-1} \Gamma_i (1-z) z^i$$

and let the following conditions hold for the parameters:

(a) $\det C(z) = 0 \Rightarrow |z| > 1$ or $z = 1$.
(b) The number of unit roots, $z = 1$, is exactly $K - r$.
(c) α and β are $(K \times r)$ matrices with $\operatorname{rk}(\alpha) = \operatorname{rk}(\beta) = r$.

Then y_t has the representation

$$y_t = \Xi \sum_{i=1}^{t} u_i + \Xi^*(L)u_t + y_0^*, \qquad (6.3.11)$$

where

$$\Xi = \beta_\perp \left[\alpha_\perp' \left(I_K - \sum_{i=1}^{p-1} \Gamma_i \right) \beta_\perp \right]^{-1} \alpha_\perp', \qquad (6.3.12)$$

$\Xi^*(L)u_t = \sum_{j=0}^{\infty} \Xi_j^* u_{t-j}$ is an $I(0)$ process and y_0^* contains initial values. ∎

Remark 1 The proposition is of fundamental importance because it decomposes the process y_t into $I(1)$ and $I(0)$ components which have to be treated accordingly, for example, when asymptotic properties of parameter estimators are derived (see Chapter 7). It makes precise under what conditions the process y_t is driven by $K - r$ $I(1)$ components and r $I(0)$ components. The representation in (6.3.11) is a multivariate version of the *Beveridge-Nelson decomposition* of y_t. The first term on the right-hand side of (6.3.11) consists of K random walks $\sum_{i=1}^{t} u_i$ which are multiplied by a matrix of rank $K - r$, denoted by Ξ. Thus, there are actually $K - r$ stochastic trends driving the system. They determine to a large extent the development of y_t. Therefore one may call y_t an $I(1)$ process if there are actually $I(1)$ trends (random walks) in the representation (6.3.11). In other words, y_t is $I(1)$ if it has the representation (6.3.11) with $\Xi \neq 0$. Clearly, for Ξ to have the form given in (6.3.12), the $((K - r) \times (K - r))$ matrix

$$\alpha_\perp' \left(I_K - \sum_{i=1}^{p-1} \Gamma_i \right) \beta_\perp$$

must be invertible. Only under that condition, $\operatorname{rk}(\Xi) = K - r$. Therefore the latter condition ensures that y_t is actually driven by $K - r$ random walk components. ∎

Remark 2 The parameter matrices Ξ_j^* in (6.3.11) are determined by the model parameters. To state the precise relation, we define

$$\bar{\beta} := \beta(\beta'\beta)^{-1} \quad (K \times r),$$

$$Q := \begin{bmatrix} \beta' \\ \beta_\perp' \end{bmatrix}_{(K \times K)} \quad \text{so that} \quad Q^{-1} = [\bar{\beta} : \beta_\perp],$$

$$\boldsymbol{\Gamma}(z) := I_K - \sum_{i=1}^{p-1} \boldsymbol{\Gamma}_i z^i,$$

$$B_*(z) := Q[\boldsymbol{\Gamma}(z)\bar{\boldsymbol{\beta}}(1-z) - \alpha z : \boldsymbol{\Gamma}(z)\boldsymbol{\beta}_\perp],$$

$$B(z) = I_K - \sum_{i=1}^{p} B_i z^i := Q^{-1} B_*(z) Q, \qquad (6.3.13)$$

and

$$\boldsymbol{\Theta}(z) := B(z)^{-1} = \sum_{j=0}^{\infty} \boldsymbol{\Theta}_j z^j.$$

Notice that $B(0) = Q^{-1} B_*(0) Q = [\bar{\boldsymbol{\beta}} : \boldsymbol{\beta}_\perp] Q = I_K$. Hence, $B(z)$ has the representation $I_K - \sum_{i=1}^{p} B_i z^i$ stated in (6.3.13). Moreover, the matrix operator $\boldsymbol{\Theta}(z)$ can be decomposed as

$$\boldsymbol{\Theta}(z) = \boldsymbol{\Theta}(1) + (1-z)\boldsymbol{\Theta}^*(z),$$

where expressions for the $\boldsymbol{\Theta}_j^*$'s can be found by comparing coefficients in $\boldsymbol{\Theta}(z) = \sum_{j=0}^{\infty} \boldsymbol{\Theta}_j z^j$ and

$$\begin{aligned}
\boldsymbol{\Theta}(1) + (1-z)\boldsymbol{\Theta}^*(z) &= \boldsymbol{\Theta}(1) + \sum_{j=0}^{\infty} \boldsymbol{\Theta}_j^* z^j (1-z) \\
&= (\boldsymbol{\Theta}(1) + \boldsymbol{\Theta}_0^*) + \sum_{j=1}^{\infty} (\boldsymbol{\Theta}_j^* - \boldsymbol{\Theta}_{j-1}^*) z^j.
\end{aligned}$$

Hence,

$$\boldsymbol{\Theta}_0 = \boldsymbol{\Theta}(1) + \boldsymbol{\Theta}_0^*$$

and

$$\boldsymbol{\Theta}_i = \boldsymbol{\Theta}_i^* - \boldsymbol{\Theta}_{i-1}^*, \quad i = 1, 2, \ldots.$$

Using the last expression, we get by successive substitution,

$$\begin{aligned}
\boldsymbol{\Theta}_i^* &= \boldsymbol{\Theta}_i + \boldsymbol{\Theta}_{i-1}^* = \sum_{j=1}^{i} \boldsymbol{\Theta}_{i-j} + \boldsymbol{\Theta}_0^* \\
&= \sum_{j=1}^{i} \boldsymbol{\Theta}_{i-j} + \boldsymbol{\Theta}_0 - \boldsymbol{\Theta}(1) = -\sum_{j=i+1}^{\infty} \boldsymbol{\Theta}_j, \quad i = 1, 2, \ldots. \qquad (6.3.14)
\end{aligned}$$

From these quantities the operator $\boldsymbol{\Xi}^*(z)$ in (6.3.11) can be obtained as

$$\boldsymbol{\Xi}^*(z) = [\boldsymbol{\Theta}^*(z) + \bar{\boldsymbol{\beta}} \boldsymbol{\beta}' B(z)^{-1}] \qquad (6.3.15)$$

(see the proof of Proposition 6.1). The representation (6.3.11) will turn out to be useful, for example, in Chapter 9, where structural VECMs are discussed. The coefficient matrices $\boldsymbol{\Xi}_j^*$ of the operator $\boldsymbol{\Xi}^*(z)$ will then play an important role as specific impulse response coefficients. ∎

Proof of Proposition 6.1
The proof is adapted from Saikkonen (2005). We use the notation from Remark 2 and first show that under the conditions of Proposition 6.1,

$$C(z) = Q^{-1}B_*(z)P(z), \qquad (6.3.16)$$

where

$$P(z) := \begin{bmatrix} \beta' \\ (1-z)\bar{\beta}'_\perp \end{bmatrix} = \begin{bmatrix} I_r & 0 \\ 0 & (1-z)I_{K-r} \end{bmatrix} Q.$$

This representation is obtained by noting that

$$\begin{aligned} C(z) &= [\mathbf{\Gamma}(z)(1-z) - \alpha\beta'z]Q^{-1}Q \\ &= [\mathbf{\Gamma}(z)\bar{\beta}(1-z) - \alpha\beta'\bar{\beta}z : \mathbf{\Gamma}(z)\beta_\perp(1-z) - \alpha\beta'\beta_\perp z]Q \\ &= [\mathbf{\Gamma}(z)\bar{\beta}(1-z) - \alpha z : \mathbf{\Gamma}(z)\beta_\perp(1-z)]\begin{bmatrix} \beta' \\ \bar{\beta}'_\perp \end{bmatrix} \\ &= Q^{-1}Q[\mathbf{\Gamma}(z)\bar{\beta}(1-z) - \alpha z : \mathbf{\Gamma}(z)\beta_\perp]\begin{bmatrix} \beta' \\ (1-z)\bar{\beta}'_\perp \end{bmatrix}. \end{aligned}$$

Clearly, $\det P(z)$ has exactly $K - r$ unit roots and, thus, $\det B_*(z)$ cannot have any such roots so that $\det B_*(z) \neq 0$ for $|z| \leq 1$ must hold. In other words, $B_*(L)$ is an invertible operator.

Now define

$$z_t := Q^{-1}P(L)y_t = \bar{\beta}\beta' y_t + \beta_\perp \bar{\beta}'_\perp \Delta y_t \qquad (6.3.17)$$

and note that

$$\beta' z_t = \beta' y_t. \qquad (6.3.18)$$

For the operator $B(z) = Q^{-1}B_*(z)Q$, we have $B(0) = Q^{-1}B_*(0)Q = I_K$ and $\det B(z) \neq 0$ for $|z| \leq 1$ because $\det B_*(z)$ has no roots inside or on the complex unit circle. Moreover,

$$B(L)z_t = Q^{-1}B_*(L)QQ^{-1}P(L)y_t = C(L)y_t = u_t.$$

Thus,

$$z_t = \sum_{i=1}^p B_i z_{t-i} + u_t \qquad (6.3.19)$$

is a stable VAR(p) process with the same residual process u_t as y_t. We know from Chapter 2 that it has an MA representation

$$z_t = B(L)^{-1} u_t = \Theta(L) u_t = \sum_{j=0}^\infty \Theta_j u_{t-j}. \qquad (6.3.20)$$

As we have seen in Remark 2, the matrix operator $\Theta(z)$ can be decomposed as

$$\Theta(z) = \Theta(1) + (1-z)\Theta^*(z).$$

Hence, we get from (6.3.20),

$$z_t = \Theta(1)u_t + \Theta^*(L)\Delta u_t = B(1)^{-1}u_t + \Theta^*(L)\Delta u_t. \qquad (6.3.21)$$

Using

$$y_t = Q^{-1}Qy_t = [\bar{\beta} : \beta_\perp]\begin{bmatrix}\beta' y_t \\ \bar{\beta}'_\perp y_t\end{bmatrix} = \bar{\beta}\beta' y_t + \beta_\perp \bar{\beta}'_\perp y_t$$

and, hence,

$$\Delta y_t = \bar{\beta}\beta' \Delta y_t + \beta_\perp \bar{\beta}'_\perp \Delta y_t,$$

it follows from (6.3.17) and (6.3.18) that $\Delta y_t = z_t - \bar{\beta}\beta' z_{t-1}$. Thus,

$$\beta_\perp \bar{\beta}'_\perp \Delta y_t = \beta_\perp \bar{\beta}'_\perp z_t.$$

Substituting the expression from (6.3.21) for z_t gives

$$\begin{aligned}\Delta y_t &= \beta_\perp \bar{\beta}'_\perp z_t + \bar{\beta}\beta' \Delta y_t \\ &= \beta_\perp \bar{\beta}'_\perp B(1)^{-1}u_t + \Theta^*(L)\Delta u_t + \bar{\beta}\beta' \Delta y_t := w_t.\end{aligned}$$

Solving for $y_t = y_{t-1} + w_t$ results in

$$\begin{aligned}y_t &= y_0 + \sum_{i=1}^{t} w_i \\ &= y_0 + \beta_\perp \bar{\beta}'_\perp B(1)^{-1}\sum_{i=1}^{t} u_i + \Theta^*(L)\sum_{i=1}^{t}\Delta u_t + \bar{\beta}\beta'\sum_{i=1}^{t}\Delta y_t \\ &= y_0 + \beta_\perp \bar{\beta}'_\perp B(1)^{-1}\sum_{i=1}^{t} u_i + \Theta^*(L)(u_t - u_0) + \bar{\beta}\beta'(y_t - y_0) \\ &= \beta_\perp \bar{\beta}'_\perp B(1)^{-1}\sum_{i=1}^{t} u_i + \Theta^*(L)u_t + \bar{\beta}\beta' y_t + y_0^*, \qquad (6.3.22)\end{aligned}$$

where $y_0^* := y_0 - \Theta^*(L)u_0 - \bar{\beta}\beta' y_0$. Using $\beta' y_t = \beta' z_t$, the term $\bar{\beta}\beta' y_t = \bar{\beta}\beta' z_t$ is seen to have a representation

$$\bar{\beta}\beta' z_t = \bar{\beta}\beta' \Theta(L)u_t$$

and, thus, $\Theta^*(L)u_t + \bar{\beta}\beta' y_t$ has an MA representation

$$\Xi^*(L)u_t = [\Theta^*(L) + \bar{\beta}\beta'\Theta(L)]u_t.$$

For the first term on the right-hand side of (6.3.22) we have

$$\begin{aligned}\beta_\perp \bar{\beta}'_\perp B(1)^{-1} &= \beta_\perp \bar{\beta}'_\perp Q^{-1} B_*(1)^{-1} Q \\ &= \beta_\perp \bar{\beta}'_\perp [\bar{\beta} : \beta_\perp][-\alpha : \Gamma(1)\beta_\perp]^{-1} \\ &= \beta_\perp [0 : I_{K-r}][-\alpha : \Gamma(1)\beta_\perp]^{-1} \\ &= \beta_\perp [\alpha'_\perp \Gamma(1)\beta_\perp]^{-1} \alpha'_\perp, \end{aligned}$$

because

$$[-\alpha : \Gamma(1)\beta_\perp]^{-1} = \begin{bmatrix} (\alpha'\alpha)^{-1}\alpha'\{\Gamma(1)\beta_\perp [\alpha'_\perp \Gamma(1)\beta_\perp]^{-1}\alpha'_\perp - I_K\} \\ [\alpha'_\perp \Gamma(1)\beta_\perp]^{-1}\alpha'_\perp \end{bmatrix}.$$

Hence, $\Xi = \beta_\perp \bar{\beta}'_\perp B(1)^{-1}$ is as stated in the proposition. Notice that the invertibility of $\alpha'_\perp \Gamma(1)\beta_\perp$ follows from the invertibility of $B(1)$ which in turn is implied by $\det B(z) \neq 0$ for $|z| \leq 1$. ∎

6.4 Deterministic Terms in Cointegrated Processes

In the previous section, we have ignored deterministic terms in the DGP. Clearly, deterministic terms may also be present in cointegrated processes and VECMs. Actually, from the discussion of the random walk with drift it should be clear that deterministic terms in a VAR process with unit roots may have a different impact than in a stable VAR. For example, an intercept term in a random walk generates a linear trend in the mean of the process, whereas an intercept term in a stable AR process just implies a constant mean value. To explore the implications of the deterministic term, the following model is assumed:

$$y_t = \mu_t + x_t, \qquad (6.4.1)$$

where x_t is a zero mean VAR(p) process with possibly cointegrated variables and μ_t stands for the deterministic term. For example, the deterministic term may just be a constant, $\mu_t = \mu_0$, or it may be a linear trend term, $\mu_t = \mu_0 + \mu_1 t$, where μ_0 and μ_1 are fixed K-dimensional parameter vectors. Other possible deterministic terms that may be included are seasonal dummy variables or other dummies to account for special events. The advantage of setting up the process in the form (6.4.1) by adding the deterministic part to the zero mean stochastic part is that the mean of the y_t variables is clearly specified by the deterministic term and does not need to be derived from quantities that involve the parameters of the stochastic part in addition. The disadvantage is that the stochastic part x_t is not directly observable in general. Therefore, for estimation purposes, for instance, we have to rewrite the process in terms of the observable y_t's. We will do so in the following for some cases of specific interest.

6.4 Deterministic Terms in Cointegrated Processes   257

It is assumed that the DGP of x_t can be represented as a VECM such as (6.3.6),

$$\begin{aligned}\Delta x_t &= \alpha\beta' x_{t-1} + \Gamma_1 \Delta x_{t-1} + \cdots + \Gamma_{p-1} \Delta x_{t-p+1} + u_t \\ &= \Pi x_{t-1} + \Gamma_1 \Delta x_{t-1} + \cdots + \Gamma_{p-1} \Delta x_{t-p+1} + u_t.\end{aligned} \quad (6.4.2)$$

Considering now the case of a constant deterministic term, $\mu_t = \mu_0$, we have $x_t = y_t - \mu_0$ so that $\Delta y_t = \Delta x_t$ and from (6.4.2) we get

$$\begin{aligned}\Delta y_t &= \alpha\beta'(y_{t-1} - \mu_0) + \Gamma_1 \Delta y_{t-1} + \cdots + \Gamma_{p-1} \Delta y_{t-p+1} + u_t \\ &= \alpha\beta^{o\prime} \begin{bmatrix} y_{t-1} \\ 1 \end{bmatrix} + \Gamma_1 \Delta y_{t-1} + \cdots + \Gamma_{p-1} \Delta y_{t-p+1} + u_t \\ &= \Pi^o y^o_{t-1} + \Gamma_1 \Delta y_{t-1} + \cdots + \Gamma_{p-1} \Delta y_{t-p+1} + u_t,\end{aligned} \quad (6.4.3)$$

where $\beta^{o\prime} := [\beta' : \tau']$ with $\tau' := -\beta'\mu_0$ an $(r \times 1)$ vector,

$$y^o_{t-1} := \begin{bmatrix} y_{t-1} \\ 1 \end{bmatrix}$$

and $\Pi^o := [\Pi : \nu_0]$ is $(K \times (K+1))$ with $\nu_0 := -\Pi\mu_0 = \alpha\tau'$. Hence, if there is just a constant mean, it can be absorbed into the cointegration relations. In other words, the constant mean becomes an intercept term in the cointegration relations. Of course, the model can also be written with an overall intercept term as

$$\begin{aligned}\Delta y_t &= \nu_0 + \alpha\beta' y_{t-1} + \Gamma_1 \Delta y_{t-1} + \cdots + \Gamma_{p-1} \Delta y_{t-p+1} + u_t \\ &= \nu_0 + \Pi y_{t-1} + \Gamma_1 \Delta y_{t-1} + \cdots + \Gamma_{p-1} \Delta y_{t-p+1} + u_t.\end{aligned} \quad (6.4.4)$$

Here ν_0 cannot be an arbitrary $(K \times 1)$ vector but has to satisfy the indicated restrictions ($\nu_0 = \alpha\tau'$) in order to ensure that the intercept term in this model does not generate a linear trend in the mean of the y_t variables. By specifying the deterministic term in additive form as in (6.4.1), the properties of the mean of y_t are easy to see.

A process with a linear trend in the mean, $\mu_t = \mu_0 + \mu_1 t$, is another case of practical importance. Using $x_t = y_t - \mu_0 - \mu_1 t$, $\Delta x_t = \Delta y_t - \mu_1$, and (6.4.2), gives

$$\begin{aligned}\Delta y_t - \mu_1 &= \alpha\beta'(y_{t-1} - \mu_0 - \mu_1(t-1)) + \Gamma_1(\Delta y_{t-1} - \mu_1) + \cdots \\ &\quad + \Gamma_{p-1}(\Delta y_{t-p+1} - \mu_1) + u_t\end{aligned} \quad (6.4.5)$$

or, collecting deterministic terms,

$$\begin{aligned}\Delta y_t &= \nu + \alpha[\beta' : \eta'] \begin{bmatrix} y_{t-1} \\ t-1 \end{bmatrix} + \Gamma_1 \Delta y_{t-1} + \cdots + \Gamma_{p-1} \Delta y_{t-p+1} + u_t \\ &= \nu + \Pi^+ y^+_{t-1} + \Gamma_1 \Delta y_{t-1} + \cdots + \Gamma_{p-1} \Delta y_{t-p+1} + u_t,\end{aligned} \quad (6.4.6)$$

where $\nu := -\Pi\mu_0 + (I_K - \Gamma_1 - \cdots - \Gamma_{p-1})\mu_1$, $\eta' := -\beta'\mu_1$, $\Pi^+ := \alpha[\beta' : \eta']$ is a $(K \times (K+1))$ matrix and

$$y_t^+ := \begin{bmatrix} y_t \\ t \end{bmatrix}.$$

Now the general intercept term ν is in fact unrestricted and can take on any value from $\mathbb{R}^K$, depending of course on μ_0, μ_1, and the other parameters. In contrast, the trend term can be absorbed into the cointegration relations. Writing the model with unrestricted linear trend term in the form

$$\Delta y_t = \nu_0 + \nu_1 t + \mathbf{\Pi} y_{t-1} + \mathbf{\Gamma}_1 \Delta y_{t-1} + \cdots + \mathbf{\Gamma}_{p-1} \Delta y_{t-p+1} + u_t,$$

the model is actually in principle capable of generating quadratic trends in the means of the variables.

It is also possible, that the trend slope parameter μ_1 is orthogonal to the cointegration matrix so that $\boldsymbol{\beta}'\mu_1 = 0$ and, hence, $\eta = 0$ and the trend term disappears from the cointegration relations. This situation can also occur if $\mu_1 \neq 0$ and the variables actually have linear trends in their means. The linear trends will then be generated via the intercept term ν. The resulting model,

$$\begin{aligned} \Delta y_t &= \nu + \boldsymbol{\alpha\beta}' y_{t-1} + \mathbf{\Gamma}_1 \Delta y_{t-1} + \cdots + \mathbf{\Gamma}_{p-1} \Delta y_{t-p+1} + u_t \\ &= \nu + \mathbf{\Pi} y_{t-1} + \mathbf{\Gamma}_1 \Delta y_{t-1} + \cdots + \mathbf{\Gamma}_{p-1} \Delta y_{t-p+1} + u_t, \end{aligned} \qquad (6.4.7)$$

with unrestricted intercept term ν will be of some importance later on. It represents a situation where a linear trend appears in the variables but not in the cointegration relations. Notice, however, that in this situation the cointegration rank must be smaller than K. If the process has cointegrating rank K, it is stable and, hence, it cannot generate a linear trend when just an intercept is included in the model. Formally, a "cointegrating matrix" $\boldsymbol{\beta}$ of rank K is nonsingular so that $\boldsymbol{\beta}'\mu_1$ cannot be zero if μ_1 is nonzero.

It may also be worth noting that the specification of the deterministic component in additive form as in (6.4.1) has the additional advantage that the Beveridge-Nelson representation of y_t is obtained by adding the deterministic term to the Beveridge-Nelson representation of x_t. Thus, a suitable generalization of the Granger representation theorem (Proposition 6.1) is readily available.

6.5 Forecasting Integrated and Cointegrated Variables

If forecasting is the objective, the VAR form of a process is quite convenient. Because forecasting the deterministic part is trivial, a purely stochastic process will be considered initially. For a VAR(p) process,

$$y_t = A_1 y_{t-1} + \cdots + A_p y_{t-p} + u_t, \qquad (6.5.1)$$

the optimal h-step forecast with minimal MSE is given by the conditional expectation, provided that expectation exists, even if $\det(I_K - A_1 z - \cdots - A_p z^p)$ has roots on the unit circle. In the proof of the optimality of the conditional

6.5 Forecasting Integrated and Cointegrated Variables

expectation in Section 2.2.2, we have not used the stationarity and stability of the system. Thus, assuming that u_t is independent white noise, the optimal h-step forecast at origin t is

$$y_t(h) = A_1 y_t(h-1) + \cdots + A_p y_t(h-p), \tag{6.5.2}$$

where $y_t(j) := y_{t+j}$ for $j \leq 0$, just as in the stationary, stable case.

Also the forecast errors are of the same form as in the stable case. To see this, we write the process (6.5.1) in VAR(1) form as

$$Y_t = \mathbf{A} Y_{t-1} + U_t, \tag{6.5.3}$$

where

$$Y_t := \begin{bmatrix} y_t \\ \vdots \\ y_{t-p+1} \end{bmatrix}_{(Kp \times 1)}, \quad \mathbf{A} := \begin{bmatrix} A_1 & A_2 & \cdots & A_{p-1} & A_p \\ I_K & 0 & \cdots & 0 & 0 \\ 0 & I_K & & 0 & 0 \\ \vdots & & \ddots & \vdots & \vdots \\ 0 & 0 & \cdots & I_K & 0 \end{bmatrix}_{(Kp \times Kp)}, \text{ and } U_t := \begin{bmatrix} u_t \\ 0 \\ \vdots \\ 0 \end{bmatrix}_{(Kp \times 1)}.$$

If u_t is independent white noise, the optimal h-step forecast of Y_t is

$$Y_t(h) = \mathbf{A} Y_t(h-1) = \mathbf{A}^h Y_t.$$

Moreover,

$$\begin{aligned} Y_{t+h} &= \mathbf{A} Y_{t+h-1} + U_{t+h} \\ &= \mathbf{A}^h Y_t + U_{t+h} + \mathbf{A} U_{t+h-1} + \cdots + \mathbf{A}^{h-1} U_{t+1}. \end{aligned}$$

Hence, the forecast error for the process Y_t is

$$Y_{t+h} - Y_t(h) = U_{t+h} + \mathbf{A} U_{t+h-1} + \cdots + \mathbf{A}^{h-1} U_{t+1}.$$

Premultiplying by the $(K \times Kp)$ matrix $J := [I_K : 0 : \cdots : 0]$ gives

$$\begin{aligned} y_{t+h} - y_t(h) &= J U_{t+h} + J \mathbf{A} J' J U_{t+h-1} + \cdots + J \mathbf{A}^{h-1} J' J U_{t+1} \\ &= u_{t+h} + \Phi_1 u_{t+h-1} + \cdots + \Phi_{h-1} u_{t+1}, \end{aligned} \tag{6.5.4}$$

where $J' J U_t = U_t$ and $\Phi_i = J \mathbf{A}^i J'$ have been used. Thus, the form of the forecast error is exactly the same as in the stable case and the forecast is easily seen to be unbiased, that is,

$$E[y_{t+h} - y_t(h)] = 0.$$

Furthermore, the Φ_i's may be obtained from the A_i's by the recursions

$$\Phi_i = \sum_{j=1}^{i} \Phi_{i-j} A_j, \quad i = 1, 2, \ldots, \tag{6.5.5}$$

with $\Phi_0 = I_K$, just as in Chapter 2. Also the forecast MSE matrix becomes

$$\Sigma_y(h) = \sum_{i=0}^{h-1} \Phi_i \Sigma_u \Phi_i', \qquad (6.5.6)$$

as in the stable case. Yet there is a very important difference. In the stable case, the Φ_i's converge to zero as $i \to \infty$ and $\Sigma_y(h)$ converges to the covariance matrix of y_t as $h \to \infty$. This result was obtained because the eigenvalues of $\mathbf{A}$ have modulus less than one in the stable case. Hence, $\Phi_i = J\mathbf{A}^i J' \to 0$ as $i \to \infty$. Because the eigenvalues of $\mathbf{A}$ are just the reciprocals of the roots of the determinantal polynomial $\det(I_K - A_1 z - \cdots - A_p z^p)$, the Φ_i's do not converge to zero in the presently considered unstable case where one or more of the eigenvalues of $\mathbf{A}$ are 1. Consequently, some elements of the forecast MSE matrix $\Sigma_y(h)$ will approach infinity as $h \to \infty$. In other words, the forecast MSEs will be unbounded and the forecast uncertainty may become extremely large as we make forecasts for the distant future, even if the structure of the process does not change.

To illustrate this point, consider the following bivariate VAR(1) example process with cointegrating rank 1:

$$\begin{bmatrix} y_{1t} \\ y_{2t} \end{bmatrix} = \begin{bmatrix} 0 & 1 \\ 0 & 1 \end{bmatrix} \begin{bmatrix} y_{1,t-1} \\ y_{2,t-1} \end{bmatrix} + \begin{bmatrix} u_{1t} \\ u_{2t} \end{bmatrix}. \qquad (6.5.7)$$

The corresponding VECM representation is

$$\Delta y_t = -\begin{bmatrix} 1 & -1 \\ 0 & 0 \end{bmatrix} y_{t-1} + u_t = \begin{bmatrix} -1 \\ 0 \end{bmatrix} [1, -1] y_{t-1} + u_t,$$

that is,

$$\alpha = \begin{bmatrix} -1 \\ 0 \end{bmatrix}, \qquad \beta' = [1, -1].$$

For this process, it is easily seen that $\Phi_0 = I_2$ and

$$\Phi_j = A_1^j = \begin{bmatrix} 0 & 1 \\ 0 & 1 \end{bmatrix}, \qquad j = 1, 2, \ldots,$$

which implies

$$\Sigma_y(h) = \sum_{j=0}^{h-1} \Phi_j \Sigma_u \Phi_j' = \Sigma_u + (h-1) \begin{bmatrix} \sigma_2^2 & \sigma_2^2 \\ \sigma_2^2 & \sigma_2^2 \end{bmatrix}, \qquad h = 1, 2, \ldots,$$

where σ_2^2 is the variance of u_{2t}. Moreover, the conditional expectations are $y_{k,t}(h) = y_{2,t}$ ($k = 1, 2$). Hence, the forecast intervals are

$$\left[y_{2,t} - z_{(\alpha/2)} \sqrt{\sigma_k^2 + (h-1)\sigma_2^2},\; y_{2,t} + z_{(\alpha/2)} \sqrt{\sigma_k^2 + (h-1)\sigma_2^2} \right], \quad k = 1, 2,$$

where $z_{(\alpha/2)}$ is the $(1-\frac{\alpha}{2})100$ percentage point of the standard normal distribution. It is easy to see that the length of this interval is unbounded for $h \to \infty$.

If there are cointegrated variables, some linear combinations can be forecasted with bounded forecast error variance, however. To see this, multiply (6.5.7) by

$$\begin{bmatrix} 1 & -1 \\ 0 & 1 \end{bmatrix}.$$

Thereby we get

$$\begin{bmatrix} 1 & -1 \\ 0 & 1 \end{bmatrix} y_t = \begin{bmatrix} 0 & 0 \\ 0 & 1 \end{bmatrix} y_{t-1} + \begin{bmatrix} 1 & -1 \\ 0 & 1 \end{bmatrix} u_t,$$

which implies that the cointegration relation $z_t := y_{1t} - y_{2t} = u_{1t} - u_{2t}$ is zero mean white noise. Thus, the forecast intervals for z_t for any forecast horizon h are of constant length,

$$[z_t(h) - z_{(\alpha/2)}\sigma_z(h),\ z_t(h) + z_{(\alpha/2)}\sigma_z(h)] = [-z_{(\alpha/2)}\sigma_z,\ z_{(\alpha/2)}\sigma_z],$$

where $\sigma_z^2 := \text{Var}(u_{1t}) + \text{Var}(u_{2t}) - 2\text{Cov}(u_{1t}, u_{2t})$ is the variance of z_t and $z_t(h) = 0$ for $h \geq 1$ has been used.

If deterministic terms are present, we may use the foregoing formulas for the mean-adjusted variables and then add the deterministic terms for the forecast period to the mean-adjusted forecasts. More precisely, if $y_t = \mu_t + x_t$, where μ_t is the deterministic term and x_t is the stochastic part, a forecast for y_{t+h} is obtained from a forecast $x_t(h)$ for x_{t+h} by simply adding μ_{t+h}, $y_t(h) = \mu_{t+h} + x_t(h)$. By the very nature of a deterministic term, μ_{t+h} is known, of course.

In practice, the parameters $A_1, \ldots, A_p$, Σ_u, and and those of the deterministic part are usually unknown. The consequences of replacing them by estimators will be discussed in Chapter 7.

6.6 Causality Analysis

From the discussion in the previous subsection, it follows easily that the restrictions characterizing Granger-noncausality are exactly the same as in the stable case. More precisely, suppose that the vector y_t in (6.5.1) is partitioned in M- and $(K-M)$-dimensional subvectors z_t and x_t,

$$y_t = \begin{bmatrix} z_t \\ x_t \end{bmatrix} \quad \text{and} \quad A_i = \begin{bmatrix} A_{11,i} & A_{12,i} \\ A_{21,i} & A_{22,i} \end{bmatrix}, \quad i = 1, \ldots, p,$$

where the A_i are partitioned in accordance with the partitioning of y_t. Then x_t does not Granger-cause z_t if and only if

$$A_{12,i} = 0, \quad i = 1, \ldots, p. \tag{6.6.1}$$

In turn, z_t does not Granger-cause x_t if and only if $A_{21,i} = 0$ for $i = 1, \ldots, p$. It is also easy to derive the corresponding restrictions for the VECM,

$$\begin{bmatrix} \Delta z_t \\ \Delta x_t \end{bmatrix} = \begin{bmatrix} \Pi_{11} & \Pi_{12} \\ \Pi_{21} & \Pi_{22} \end{bmatrix} \begin{bmatrix} z_{t-1} \\ x_{t-1} \end{bmatrix} + \sum_{i=1}^{p-1} \begin{bmatrix} \Gamma_{11,i} & \Gamma_{12,i} \\ \Gamma_{21,i} & \Gamma_{22,i} \end{bmatrix} \begin{bmatrix} \Delta z_{t-i} \\ \Delta x_{t-i} \end{bmatrix} + u_t,$$

where all matrices are partitioned in line with y_t. From (6.3.6) it follows immediately, that the restrictions in (6.6.1) can be written equivalently as

$$\Pi_{12} = 0 \quad \text{and} \quad \Gamma_{12,i} = 0 \quad \text{for } i = 1, \ldots, p-1. \tag{6.6.2}$$

In other words, in order to check Granger-causality, we just have to test a set of linear hypotheses. It will be seen in the next chapter that in the case of cointegrated processes, testing these restrictions is not as straightforward as for stationary processes.

Also restrictions for multi-step causality and instantaneous causality can be placed on the VAR coefficients and the residual covariance matrix in the same way as in Chapter 2. Especially for the former restrictions, constructing valid asymptotic tests is not straightforward, however.

6.7 Impulse Response Analysis

Integrated and cointegrated systems must be interpreted cautiously. As mentioned in Section 6.3, in cointegrated systems the term $\beta' y_t$ is usually thought of as representing the long-run equilibrium relations between the variables. Suppose there is just one such relation, say

$$\beta_1 y_{1t} + \cdots + \beta_K y_{Kt} = 0,$$

or, if $\beta_1 \neq 0$,

$$y_{1t} = -\frac{\beta_2}{\beta_1} y_{2t} - \cdots - \frac{\beta_K}{\beta_1} y_{Kt}.$$

It is tempting to argue that the long-run effect of a unit increase in y_2 will be a change of size β_2/β_1 in y_1. This, however, ignores all the other relations between the variables which are summarized in a VAR(p) model or the corresponding VECM. A one-time unit innovation in y_2 may affect various other variables which also have an impact on y_1. Therefore, the long-run effect of a y_2-innovation on y_1 may be quite different from $-\beta_2/\beta_1$. The impulse responses may give a better picture of the relations between the variables.

In Chapter 2, Section 2.3.2, the impulse responses of stationary, stable VAR(p) processes were shown to be the coefficients of specific MA representations. An unstable, integrated or cointegrated VAR(p) process does not possess valid MA representations of the types discussed in Chapter 2. Yet the Φ_i and Θ_i matrices can be computed as in Section 2.3.2. For the Φ_i's we have seen this in Section 6.5 and, from the discussion in that section, it is easy to see that the elements of the $\Phi_i = (\phi_{jk,i})$ matrices may represent impulse responses just as in the stable case. More precisely, $\phi_{jk,i}$ represents the response of variable j to a unit forecast error in variable k, i periods ago, if the system reflects the actual responses to forecast errors. Recall that in stable processes the responses taper off to zero as $i \to \infty$. This property does not necessarily hold in unstable systems where the effect of a one-time impulse may not die out asymptotically.

In Section 2.3, we have also considered accumulated impulse responses, responses to orthogonalized residuals and forecast error variance decompositions. These tools for structural analysis are all available for unstable systems as well, using precisely the same formulas as in Chapter 2. The only quantities that cannot be computed in general are the total "long-run effects" or total multipliers Ψ_∞ and Ξ_∞ because they may not be finite.

To illustrate impulse response analysis of cointegrated systems, we consider the following VECM:

$$\begin{bmatrix} \Delta R_t \\ \Delta Dp_t \end{bmatrix} = \begin{bmatrix} -0.07 \\ 0.17 \end{bmatrix} (R_{t-1} - 4Dp_{t-1}) + \begin{bmatrix} 0.24 & -0.08 \\ 0 & -0.31 \end{bmatrix} \begin{bmatrix} \Delta R_{t-1} \\ \Delta Dp_{t-1} \end{bmatrix}$$

$$+ \begin{bmatrix} 0 & -0.13 \\ 0 & -0.37 \end{bmatrix} \begin{bmatrix} \Delta R_{t-2} \\ \Delta Dp_{t-2} \end{bmatrix} + \begin{bmatrix} 0.20 & -0.06 \\ 0 & -0.34 \end{bmatrix} \begin{bmatrix} \Delta R_{t-3} \\ \Delta Dp_{t-3} \end{bmatrix} + \begin{bmatrix} u_{1,t} \\ u_{2,t} \end{bmatrix}, \tag{6.7.1}$$

$$\Sigma_u = \begin{bmatrix} 2.61 & -0.15 \\ -0.15 & 2.31 \end{bmatrix} \times 10^{-5}$$

and the corresponding correlation matrix is

$$R_u = \begin{bmatrix} 1 & -0.06 \\ -0.06 & 1 \end{bmatrix}.$$

This model is from Lütkepohl (2004, Eq. (3.41)). The variables are a long-term interest rate (R_t) and the quarterly inflation rate (Dp_t). The coefficients are estimated from quarterly German data. Deterministic terms have been deleted because they are not important for the present analysis.

In contrast to the inflation/interest rate example system considered in Chapter 2, the two variables in the present system are $I(1)$. The cointegration relation, $R_t - 4Dp_t$, is just the real interest rate because $4Dp_t$ is the annual inflation rate and R_t is an annual nominal interest rate. Thus, in the present model the real interest rate is stationary. This relation is sometimes called the Fisher effect. The zero restrictions have been determined by a subset modelling algorithm. The residual covariance matrix is almost diagonal.

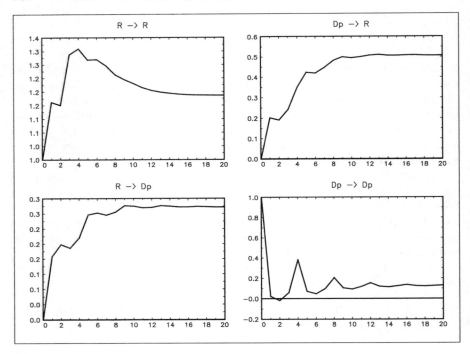

Fig. 6.4. Forecast error impulse responses of VECM (6.7.1).

Therefore, forecast error impulse responses should be similar to orthogonalized impulse responses, except for the scaling. The two types of impulse responses are shown in Figures 6.4 and 6.5, respectively. Indeed, the shape of corresponding impulse responses in the two figures is quite similar. A remarkable feature of the impulse responses is that they do not die out to zero when the time span after the impulse increases but approach some nonzero value. Clearly, this reflects the nonstationarity of the system where a one-time impulse can have permanent effects.

Using the orthogonalized impulse responses, it is also possible to compute forecast error variance decompositions based on the same formulas as in Chapter 2, Section 2.3.3. For the example system, they are shown in Figure 6.6. They look similar to forecast error variance decompositions from a stationary VAR process. Of course, there is no reason why they should look differently than in the stationary case.

As discussed in Chapter 2, interpreting the forecast error and orthogonalized impulse responses used here is often problematic if there is significant correlation between the components of the residuals u_t. It will be discussed in Chapter 9 how identifying restrictions for impulse responses can be imposed in the VECM framework.

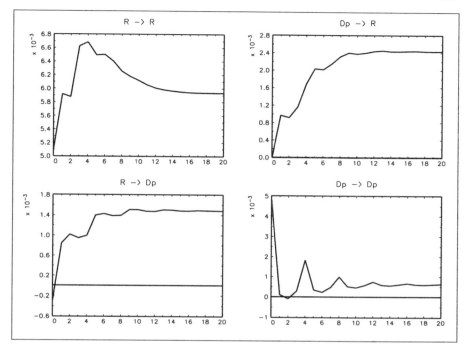

Fig. 6.5. Orthogonalized impulse responses of VECM (6.7.1).

6.8 Exercises

Problem 6.1
Consider the process

$$y_t = \begin{bmatrix} 1 & 0 \\ 0 & \psi \end{bmatrix} y_{t-1} + u_t$$

with residual covariance matrix

$$\Sigma_u = \begin{bmatrix} 1 & \rho \\ \rho & 1 \end{bmatrix}.$$

(a) What is the cointegrating rank of the process?
(b) Write the process in VECM form.

Problem 6.2
Determine the roots of the reverse characteristic polynomial and, if applicable, the cointegrating rank of the process

$$y_t = \begin{bmatrix} 1.1 & -0.2 \\ -0.2 & 1.4 \end{bmatrix} y_{t-1} + u_t.$$

Can you write the process in VECM form?

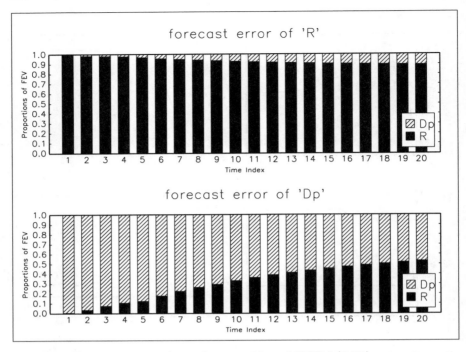

Fig. 6.6. Forecast error variance decomposition of VECM (6.7.1).

Problem 6.3
What is the maximum possible cointegrating rank of a three-dimensional process $y_t = (y_{1t}, y_{2t}, y_{3t})'$,

(a) if y_{1t}, y_{2t} are $I(0)$ and y_{3t} is $I(1)$?
(b) if y_{1t}, y_{2t}, and y_{3t} are $I(1)$ and y_{1t} and y_{2t} are not cointegrated in a bivariate system?
(c) if y_{1t}, y_{2t}, and y_{3t} are $I(1)$ and $(y_{1t}, y_{2t})'$ and $(y_{2t}, y_{3t})'$ are not cointegrated as bivariate systems?

Problem 6.4
Find the Beveridge-Nelson decomposition associated with the VECM

$$\Delta y_t = \alpha\beta' y_{t-1} + u_t,$$

(a) if all initial values are zero ($y_t = u_t = 0$ for $t \leq 0$),
(b) if y_0 is nonzero.

Problem 6.5
Derive the VECM form of y_t if the deterministic term is $\mu_t = \mu_0 + \delta I_{(t>T_B)}$, where $I_{(t>T_B)}$ is a shift dummy variable which is zero up to time T_B and then jumps to one and δ is the associated ($K \times 1$) parameter vector.

Problem 6.6
Consider the quarterly process $y_t = \mu_t + x_t$, where x_t has a VECM representation as in (6.4.2) and

$$\mu_t = \mu_0 + \mu_1 t + \delta_1 s_{1t} + \delta_2 s_{2t} + \delta_3 s_{3t}.$$

Here μ_0, μ_1, δ_1, δ_2, and δ_3 are K-dimensional parameter vectors and the s_{it}'s ($i = 1, 2, 3$) are seasonal dummy variables. Determine the VECM representation of y_t.

Problem 6.7
Consider the VECM

$$\Delta y_t = \begin{bmatrix} -0.1 \\ 0.1 \end{bmatrix} (1, -1) y_{t-1} + u_t.$$

(a) Rewrite the process in VAR form.
(b) Determine the roots of the reverse characteristic polynomial.
(c) Determine forecast intervals for the two variables for forecast horizon h.
(d) Has a forecast error impulse in y_{1t} a permanent impact on y_{2t}? Has a forecast error impulse in y_{2t} a permanent impact on y_{1t}?

7

Estimation of Vector Error Correction Models

In this chapter, estimation of VECMs is discussed. The asymptotic properties of estimators for nonstationary models differ in important ways from those of stationary processes. Therefore, in the first section, a simple special case model with no lagged differences and no deterministic terms is considered and different estimation methods for the parameters of the error correction term are treated. For this simple case, the asymptotic properties can be derived with a reasonable amount of effort and the difference to estimation in stationary models can be seen fairly easily. Therefore it is useful to treat this case in some detail. The results can then be extended to more general VECMs which are considered in Section 7.2. In Section 7.3, Bayesian estimation including the Minnesota or Litterman prior for integrated processes is discussed and forecasting and structural analysis based on estimated processes are considered in Sections 7.4–7.6.

7.1 Estimation of a Simple Special Case VECM

In this section, a simple VECM without lagged differences and deterministic terms is considered. More precisely, the model of interest is

$$\Delta y_t = \Pi y_{t-1} + u_t = \alpha \beta' y_{t-1} + u_t, \quad t = 1, 2, \ldots, \tag{7.1.1}$$

where y_t is K-dimensional, Π is a $(K \times K)$ matrix of rank r, $0 < r < K$, α and β are $(K \times r)$ with rank r, and u_t is K-dimensional white noise with mean zero and nonsingular covariance matrix Σ_u. For simplicity, we assume that u_t is standard white noise so that certain limiting results hold which will be discussed and used in the following. For the time being, the initial vector y_0 is arbitrary with some fixed distribution. We also assume that y_t is an $I(1)$ vector so that we know from Section 6.3 that the $((K-r) \times (K-r))$ matrix

$$\alpha'_\perp \beta_\perp$$

is invertible (see Eq. (6.3.12)). Here $\alpha_\perp$ and $\beta_\perp$ are, as usual, orthogonal complements of α and β, respectively.

The cointegration rank r is assumed to be known and it is strictly between 0 and K. For $r = 0$, Δy_t is stable and for $r = K$, y_t is stable. For the present purposes, these two boundary cases are of limited interest because they can be treated in the stationary framework considered in Part I. If r is not known, however, it may be of interest to consider the case $r = 0$. The matrix Π is then zero, of course. We will comment on this case at the end of this section.

We will discuss different estimators of the matrix Π, assuming that a sample $y_1, \ldots, y_T$ and a presample vector y_0 are available. Our first estimator is the unrestricted LS estimator,

$$\widehat{\Pi} = \left(\sum_{t=1}^{T} \Delta y_t y_{t-1}'\right) \left(\sum_{t=1}^{T} y_{t-1} y_{t-1}'\right)^{-1}. \tag{7.1.2}$$

Substituting $\Pi y_{t-1} + u_t$ for Δy_t gives

$$\widehat{\Pi} - \Pi = \left(\sum_{t=1}^{T} u_t y_{t-1}'\right) \left(\sum_{t=1}^{T} y_{t-1} y_{t-1}'\right)^{-1}. \tag{7.1.3}$$

To derive the asymptotic distribution of this quantity, we multiply from the left with the $(K \times K)$ matrix

$$Q := \begin{bmatrix} \beta' \\ \alpha_\perp' \end{bmatrix}$$

and from the right by

$$Q^{-1} = [\alpha(\beta'\alpha)^{-1} : \beta_\perp(\alpha_\perp'\beta_\perp)^{-1}]$$

which yields

$$\begin{aligned} Q(\widehat{\Pi} - \Pi)Q^{-1} &= Q\left(\sum_{t=1}^{T} u_t y_{t-1}'\right) Q' Q^{-1'} \left(\sum_{t=1}^{T} y_{t-1} y_{t-1}'\right)^{-1} Q^{-1} \\ &= \left(\sum_{t=1}^{T} v_t z_{t-1}'\right) \left(\sum_{t=1}^{T} z_{t-1} z_{t-1}'\right)^{-1}, \end{aligned} \tag{7.1.4}$$

where $v_t := Q u_t$ and $z_t := Q y_t$. Notice that invertibility of $\alpha_\perp' \beta_\perp$ follows from our assumption of an $I(1)$ system, as mentioned earlier, and it implies that the inverse of Q exists because

$$\begin{bmatrix} \beta' \\ \alpha_\perp' \end{bmatrix} [\beta : \beta_\perp] = \begin{bmatrix} \beta'\beta & 0 \\ \alpha_\perp'\beta & \alpha_\perp'\beta_\perp \end{bmatrix}$$

is invertible if $\alpha_\perp'\beta_\perp$ is nonsingular. Hence, Q must be invertible and, thus, $\beta'\alpha$ is also nonsingular.

7.1 Estimation of a Simple Special Case VECM

Premultiplying the VECM (7.1.1) by Q shows that

$$\Delta z_t = Q\Pi Q^{-1} z_{t-1} + v_t = \begin{bmatrix} \beta'\alpha & 0 \\ 0 & 0 \end{bmatrix} z_{t-1} + v_t.$$

Hence, denoting the first r components of z_t by $z_t^{(1)}$, we know that $z_t^{(1)} = \beta' y_t$ consists of the cointegrating relations and is therefore stationary while the last $K - r$ components of z_t, denoted by $z_t^{(2)}$, constitute a $(K - r)$-dimensional random walk because $\Delta z_t^{(2)}$ is white noise. Thus, stationary and nonstationary components are separated in z_t. To derive the asymptotic properties of the LS estimator, it is useful to write

$$Q(\widehat{\Pi} - \Pi)Q^{-1}$$
$$= \left[\sum_{t=1}^{T} v_t z_{t-1}^{(1)\prime} : \sum_{t=1}^{T} v_t z_{t-1}^{(2)\prime} \right] \begin{bmatrix} \sum_t z_{t-1}^{(1)} z_{t-1}^{(1)\prime} & \sum_t z_{t-1}^{(1)} z_{t-1}^{(2)\prime} \\ \sum_t z_{t-1}^{(2)} z_{t-1}^{(1)\prime} & \sum_t z_{t-1}^{(2)} z_{t-1}^{(2)\prime} \end{bmatrix}^{-1} . \quad (7.1.5)$$

For the cross product terms in this relation, we have the following special case results from Ahn & Reinsel (1990).

Lemma 7.1

(1) $T^{-1} \sum_{t=1}^{T} z_{t-1}^{(1)} z_{t-1}^{(1)\prime} = T^{-1} \sum_{t=1}^{T} \beta' y_{t-1} y_{t-1}' \beta \xrightarrow{p} \Gamma_z^{(1)}$.

(2) $T^{-1/2} \text{vec} \left(\sum_{t=1}^{T} v_t z_{t-1}^{(1)\prime} \right) \xrightarrow{d} \mathcal{N}(0, \Gamma_z^{(1)} \otimes \Sigma_v)$,
where $\Sigma_v := Q \Sigma_u Q'$ is the covariance matrix of v_t.

(3) $T^{-1} \sum_{t=1}^{T} v_t z_{t-1}^{(2)\prime} \xrightarrow{d} \Sigma_v^{1/2} \left(\int_0^1 \mathbf{W}_K d\mathbf{W}_K' \right)' \Sigma_v^{1/2} \begin{bmatrix} 0 \\ I_{K-r} \end{bmatrix}$,
where $\mathbf{W}_K$ abbreviates a standard Wiener process $\mathbf{W}_K(s)$ of dimension K (see Appendix C.8.2).

(4) $T^{-3/2} \sum_{t=1}^{T} z_{t-1}^{(1)} z_{t-1}^{(2)\prime} \xrightarrow{p} 0$.

(5) $T^{-2} \sum_{t=1}^{T} z_{t-1}^{(2)} z_{t-1}^{(2)\prime} \xrightarrow{d} [0 : I_{K-r}] \Sigma_v^{1/2} \left(\int_0^1 \mathbf{W}_K \mathbf{W}_K' ds \right) \Sigma_v^{1/2} \begin{bmatrix} 0 \\ I_{K-r} \end{bmatrix}$.

The quantities in (2), (3), and (5) converge jointly. ■

In this lemma we encounter asymptotic distributions of random matrices. As in Appendix C.8.2, these are understood as the limits in distribution of the vectorized quantities. Because the asymptotic distributions are also conveniently stated in matrix form, not using vectorization here is a useful simplification. Moreover, in the lemma as well as in the following analysis we denote the square root of a positive definite matrix Σ by $\Sigma^{1/2}$, that is, $\Sigma^{1/2}$ is the positive definite symmetric matrix for which $\Sigma^{1/2}\Sigma^{1/2} = \Sigma$ (see Appendix A.9.2).

Proof: The proof follows Ahn & Reinsel (1990). Lemma 7.1(1) is implied by a standard weak law of large numbers (see, e.g., Proposition C.12(7)) because $z_{t-1}^{(1)}$ contains stationary components only.

The second result also involves stationary processes only. Therefore it follows from a martingale difference central limit theorem for stationary processes. Notice that $\text{vec}(v_t z_{t-1}^{(1)\prime})$ is a martingale difference sequence and, hence, a martingale difference array which satisfies the conditions of Proposition C.13(2). Thus, the result follows from that proposition.

To show Lemma 7.1(3), we define a random walk

$$z_t^* = \begin{bmatrix} z_t^{*(1)} \\ z_t^{(2)} \end{bmatrix} = z_{t-1}^* + v_t, \quad t = 1, 2, \ldots,$$

with $z_0^{*(1)} = 0$ and notice that the second part of z_t^* is identical to the last $K - r$ components of z_t. Hence, it follows from Proposition C.18(6) that

$$T^{-1} \sum_{t=1}^{T} v_t z_{t-1}^{*\prime} \xrightarrow{d} \Sigma_v^{1/2} \left(\int_0^1 \mathbf{W}_K d\mathbf{W}_K' \right)' \Sigma_v^{1/2}.$$

Considering the last $K - r$ columns only gives the desired result.

Part (4) of the lemma can be shown by defining

$$z_t^+ = \begin{bmatrix} z_t^{+(1)} \\ z_t^{(2)} \end{bmatrix} = z_{t-1}^+ + v_t^+, \quad t = 1, 2, \ldots,$$

with $z_0^{+(1)} = 0$ and

$$v_t^+ = \begin{bmatrix} z_t^{(1)} \\ v_t^{(2)} \end{bmatrix}.$$

Thus, v_t^+ is an $I(0)$ process. By Proposition C.18(5), we have

$$\sum_{t=1}^{T} z_{t-1}^+ v_t^{+\prime} = \begin{bmatrix} \sum_t z_{t-1}^{+(1)} z_t^{(1)\prime} & \sum_t z_{t-1}^{+(1)} v_t^{(2)\prime} \\ \sum_t z_{t-1}^{(2)} z_t^{(1)\prime} & \sum_t z_{t-1}^{(2)} v_t^{(2)\prime} \end{bmatrix} = O_p(T),$$

which implies the desired result.

Lemma 7.1(5) is just a special case of Proposition C.18(9) because z_t^* is a random walk and the last $K - r$ components of z_t^* are just $z_t^{(2)}$.

Finally, the joint convergence of the quantities in Lemma 7.1(2), (3), and (5) follows because all quantities are eventually made up of the same u_t's. ∎

7.1 Estimation of a Simple Special Case VECM

The lemma implies the following limiting result for the LS estimator $\widehat{\Pi}$.

Result 1

Let
$$D = \begin{bmatrix} T^{1/2} & 0 \\ 0 & T \end{bmatrix}.$$

Then

$$\text{vec}[Q(\widehat{\Pi} - \Pi)Q^{-1}D]$$

$$\xrightarrow{d} \begin{bmatrix} \mathcal{N}(0, (\Gamma_z^{(1)})^{-1} \otimes \Sigma_v) \\ \text{vec}\left\{ \Sigma_v^{1/2} \left(\int_0^1 \mathbf{W}_K d\mathbf{W}'_K \right)' \Sigma_v^{1/2} \begin{bmatrix} 0 \\ I_{K-r} \end{bmatrix} \\ \times \left([0 : I_{K-r}] \Sigma_v^{1/2} \left(\int_0^1 \mathbf{W}_K \mathbf{W}'_K ds \right) \Sigma_v^{1/2} \begin{bmatrix} 0 \\ I_{K-r} \end{bmatrix} \right)^{-1} \right\} \end{bmatrix}.$$

(7.1.6)

∎

Proof:

$$Q(\widehat{\Pi} - \Pi)Q^{-1}D$$

$$= \left[T^{-1/2} \sum_{t=1}^T v_t z_{t-1}^{(1)\prime} : T^{-1} \sum_{t=1}^T v_t z_{t-1}^{(2)\prime} \right]$$

$$\times D \begin{bmatrix} \sum_t z_{t-1}^{(1)} z_{t-1}^{(1)\prime} & \sum_t z_{t-1}^{(1)} z_{t-1}^{(2)\prime} \\ \sum_t z_{t-1}^{(2)} z_{t-1}^{(1)\prime} & \sum_t z_{t-1}^{(2)} z_{t-1}^{(2)\prime} \end{bmatrix}^{-1} D$$

$$= \left[\left(T^{-1/2} \sum_{t=1}^T v_t z_{t-1}^{(1)\prime} \right) \left(T^{-1} \sum_{t=1}^T z_{t-1}^{(1)} z_{t-1}^{(1)\prime} \right)^{-1} \right.$$

$$\left. : \left(T^{-1} \sum_{t=1}^T v_t z_{t-1}^{(2)\prime} \right) \left(T^{-2} \sum_{t=1}^T z_{t-1}^{(2)} z_{t-1}^{(2)\prime} \right)^{-1} \right] + o_p(1).$$

The last equality follows from Lemma 7.1(4). The result in (7.1.6) is obtained by vectorizing this matrix and applying Lemma 7.1(2), (3), and (5) and the continuous mapping theorem (see Appendix C.8). ∎

An immediate implication of Result 1 follows.

Result 2
The estimator $\widehat{\mathbf{\Pi}}$ is asymptotically normal,

$$\sqrt{T}\text{vec}(\widehat{\mathbf{\Pi}} - \mathbf{\Pi}) \xrightarrow{d} \mathcal{N}\left(0, \beta(\Gamma_z^{(1)})^{-1}\beta' \otimes \Sigma_u\right), \tag{7.1.7}$$

and $\beta(\Gamma_z^{(1)})^{-1}\beta'$ can be estimated consistently by

$$\left(T^{-1}\sum_{t=1}^{T} y_{t-1}y'_{t-1}\right)^{-1}.$$

∎

Proof:

$$\sqrt{T}Q(\widehat{\mathbf{\Pi}} - \mathbf{\Pi})Q^{-1}$$
$$= Q(\widehat{\mathbf{\Pi}} - \mathbf{\Pi})Q^{-1}D\begin{bmatrix} 1 & 0 \\ 0 & T^{-1/2} \end{bmatrix}$$
$$= \left[\left(T^{-1/2}\sum_{t=1}^{T} v_t z_{t-1}^{(1)\prime}\right)\left(T^{-1}\sum_{t=1}^{T} z_{t-1}^{(1)} z_{t-1}^{(1)\prime}\right)^{-1}\right.$$
$$\left. : T^{-1/2}\left(T^{-1}\sum_{t=1}^{T} v_t z_{t-1}^{(2)\prime}\right)\left(T^{-2}\sum_{t=1}^{T} z_{t-1}^{(2)} z_{t-1}^{(2)\prime}\right)^{-1}\right] + o_p(1)$$

from the proof of Result 1 and, hence,

$$\sqrt{T}\text{vec}[Q(\widehat{\mathbf{\Pi}} - \mathbf{\Pi})Q^{-1}] = (Q^{-1\prime} \otimes Q)\sqrt{T}\text{vec}(\widehat{\mathbf{\Pi}} - \mathbf{\Pi})$$
$$\xrightarrow{d} \begin{bmatrix} \mathcal{N}\left(0, (\Gamma_z^{(1)})^{-1} \otimes \Sigma_v\right) \\ 0 \end{bmatrix}.$$

Premultiplying by $Q' \otimes Q^{-1}$ and recalling the definition of Q, gives a multivariate normal limiting distribution with covariance matrix

$$(Q' \otimes Q^{-1})\left(\begin{bmatrix} (\Gamma_z^{(1)})^{-1} & 0 \\ 0 & 0 \end{bmatrix} \otimes \Sigma_v\right)(Q \otimes Q^{-1\prime})$$

or

$$[\beta : \alpha_\perp]\begin{bmatrix} (\Gamma_z^{(1)})^{-1} & 0 \\ 0 & 0 \end{bmatrix}\begin{bmatrix} \beta' \\ \alpha'_\perp \end{bmatrix} \otimes Q^{-1}\Sigma_v Q^{-1\prime}$$

which implies (7.1.7) because $\Sigma_v = Q\Sigma_u Q'$.

Now consider

$$\left(T^{-1}\sum_{t=1}^{T} y_{t-1}y'_{t-1}\right)^{-1} = Q'\left(T^{-1}\sum_{t=1}^{T} z_{t-1}z'_{t-1}\right)^{-1} Q$$

7.1 Estimation of a Simple Special Case VECM

$$= Q' \begin{bmatrix} T^{-1}\sum_t z_{t-1}^{(1)} z_{t-1}^{(1)\prime} & T^{-1}\sum_t z_{t-1}^{(1)} z_{t-1}^{(2)\prime} \\ T^{-1}\sum_t z_{t-1}^{(2)} z_{t-1}^{(1)\prime} & T^{-1}\sum_t z_{t-1}^{(2)} z_{t-1}^{(2)\prime} \end{bmatrix}^{-1} Q$$

$$= Q' \begin{bmatrix} S_{11}^{-1} + S_{11}^{-1} S_{12} S^* S_{21} S_{11}^{-1} & -S_{11}^{-1} S_{12} S^* \\ -S^* S_{21} S_{11}^{-1} & S^* \end{bmatrix} Q,$$

where the rules for the partitioned inverse from Appendix A.10 have been used and $S^* := (S_{22}^{-1} - S_{21} S_{11}^{-1} S_{12})^{-1}$. Moreover,

$$S_{11} := T^{-1} \sum_t z_{t-1}^{(1)} z_{t-1}^{(1)\prime} \xrightarrow{p} \Gamma_z^{(1)}$$

by Lemma 7.1(1),

$$S_{12} = S_{21}' := T^{-1} \sum_t z_{t-1}^{(1)} z_{t-1}^{(2)\prime} = o_p(T^{1/2})$$

by Lemma 7.1(4), and

$$S_{22} := T^{-1} \sum_t z_{t-1}^{(2)} z_{t-1}^{(2)\prime}.$$

By Lemma 7.1(5) and the continuous mapping theorem, $S_{22}^{-1} = O_p(T^{-1})$. Using again the rules for the partitioned inverse from Appendix A.10,

$$\begin{aligned} S^* &= S_{22}^{-1} + S_{22}^{-1} S_{21} (S_{11} - S_{12} S_{22}^{-1} S_{21})^{-1} S_{12} S_{22}^{-1} \\ &= O_p(T^{-1}) + O_p(T^{-1}) o_p(T^{1/2}) O_p(1) o_p(T^{1/2}) O_p(T^{-1}) \\ &= O_p(T^{-1}), \end{aligned}$$

because

$$S_{11} - S_{12} S_{22}^{-1} S_{21} = S_{11} - o_p(T^{1/2}) O_p(T^{-1}) o_p(T^{1/2}) = S_{11} + o_p(1)$$

so that

$$(S_{11} - S_{12} S_{22}^{-1} S_{21})^{-1} = O_p(1).$$

Hence, we get

$$\begin{aligned} S_{11}^{-1} + S_{11}^{-1} S_{12} S^* S_{21} S_{11}^{-1} &= (\Gamma_z^{(1)})^{-1} \\ &\quad + O_p(1) o_p(T^{1/2}) O_p(T^{-1}) o_p(T^{1/2}) O_p(1) \\ &= (\Gamma_z^{(1)})^{-1} + o_p(1) \end{aligned}$$

and

$$-S_{11}^{-1} S_{12} S^* = O_p(1) o_p(T^{1/2}) O_p(T^{-1}) = o_p(1).$$

Thus,

$$\left(T^{-1}\sum_{t=1}^{T} y_{t-1}y'_{t-1}\right)^{-1} = Q'\begin{bmatrix} (\Gamma_z^{(1)})^{-1} + o_p(1) & o_p(1) \\ o_p(1) & o_p(1) \end{bmatrix} Q$$
$$= \beta(\Gamma_z^{(1)})^{-1}\beta' + o_p(1),$$

which proves Result 2. ∎

Thus, the limiting distribution of $\sqrt{T}\mathrm{vec}(\widehat{\Pi} - \Pi)$ is singular because $\Gamma_z^{(1)}$ is an $(r \times r)$ matrix. Still, we can use the usual estimator of the covariance matrix based on the regressor matrix. Thus, t-ratios can be set up in the standard way and have their usual asymptotic standard normal distributions, if a consistent estimator of Σ_u is used. In Result 8, we will see that the usual residual covariance matrix is in fact a consistent estimator for Σ_u, as in the stationary case. On the other hand, it is not difficult to see that the covariance matrix in the limiting distribution (7.1.7) has rank rK. Therefore, setting up a Wald test for more general restrictions may be problematic. As explained in Appendix C.7, a nonsingular weighting matrix is needed for the Wald test to have its usual limiting χ^2-distribution under the null hypothesis. Thus, if we want to test, for example,

$$H_0: \Pi = 0 \quad \text{versus} \quad H_1: \Pi \neq 0,$$

the corresponding Wald statistic is

$$\lambda_W = T\mathrm{vec}(\widehat{\Pi})'\left(\left(T^{-1}\sum_{t=1}^{T} y_{t-1}y'_{t-1}\right) \otimes \widehat{\Sigma}_u^{-1}\right)\mathrm{vec}(\widehat{\Pi}).$$

Under H_0, the arguments in the proof of Result 2 can be used to show that $T^{-1}\sum_{t=1}^{T} y_{t-1}y'_{t-1}$ converges to zero in probability and, hence, the limit of the weighting matrix in the Wald statistic is singular. Thus, λ_W will not have an asymptotic $\chi^2(K^2)$-distribution. Therefore, caution is necessary in setting up F-tests, for example. In the nonstationary case, they may not have an asymptotic justification. We will provide more discussion of this problem in Section 7.6 in the context of testing for Granger-causality.

It is interesting to note that the asymptotic distribution in (7.1.7) is the same one that is obtained if the cointegration matrix β is known and only α is estimated by LS. To see this result, we consider the LS estimator

$$\widehat{\alpha} = \left(\sum_{t=1}^{T} \Delta y_t y'_{t-1}\beta\right)\left(\sum_{t=1}^{T} \beta' y_{t-1}y'_{t-1}\beta\right)^{-1}. \quad (7.1.8)$$

This estimator has the following properties.

Result 3

$$\sqrt{T}\text{vec}(\widehat{\alpha} - \alpha) \xrightarrow{d} \mathcal{N}(0, (\Gamma_z^{(1)})^{-1} \otimes \Sigma_u) \tag{7.1.9}$$

and, thus,

$$\sqrt{T}\text{vec}(\widehat{\alpha}\beta' - \Pi) \xrightarrow{d} \mathcal{N}(0, \beta(\Gamma_z^{(1)})^{-1}\beta' \otimes \Sigma_u).$$

■

Proof: Substituting $\alpha\beta' y_{t-1} + u_t$ for Δy_t in (7.1.8) and rearranging terms gives

$$\widehat{\alpha} - \alpha = \left(\sum_{t=1}^{T} u_t y_{t-1}' \beta\right)\left(\sum_{t=1}^{T} \beta' y_{t-1} y_{t-1}' \beta\right)^{-1}$$

from which we get (7.1.9) by similar arguments as in the proof of Lemma 7.1. Noting that $\text{vec}(\widehat{\alpha}\beta' - \Pi) = (\beta \otimes I_K)\text{vec}(\widehat{\alpha} - \alpha)$, gives the stated asymptotic distribution of $\sqrt{T}\text{vec}(\widehat{\alpha}\beta - \Pi)$.

■

Clearly, this result may seem a bit surprising because it means that knowledge of β does not improve our estimator for Π, at least asymptotically. In turn, not knowing β does not lead to a reduction in asymptotic precision of our estimator. This is a consequence of the fact that β can be estimated with a better convergence rate than $\sqrt{T}$. To see this fact, suppose for the moment that α is known and that β is normalized as in (6.3.9) such that

$$\beta = \begin{bmatrix} I_r \\ \beta_{(K-r)} \end{bmatrix}. \tag{7.1.10}$$

We know from the discussion in Section 6.3 that this normalization is always possible if the variables are arranged appropriately. Thus, upon normalization, the only unknown elements of β are in the $((K-r) \times r)$ matrix $\beta_{(K-r)}$. This matrix can be estimated from

$$\Delta y_t - \alpha y_{t-1}^{(1)} = \alpha \beta_{(K-r)}' y_{t-1}^{(2)} + u_t = (y_{t-1}^{(2)\prime} \otimes \alpha)\text{vec}(\beta_{(K-r)}') + u_t, \tag{7.1.11}$$

where $y_{t-1}^{(1)}$ and $y_{t-1}^{(2)}$ consist of the first r and the last $K-r$ elements of y_{t-1}, respectively. Because this is a multivariate regression model where the regressors are not identical in the different equations, we assume for the moment that Σ_u is also known and consider the GLS estimator

$$\text{vec}(\widehat{\beta}_{(K-r)}') = \left[\left(\sum_{t=1}^{T} y_{t-1}^{(2)} y_{t-1}^{(2)\prime}\right)^{-1} \otimes (\alpha' \Sigma_u^{-1} \alpha)^{-1}\right]$$

$$\times (I_T \otimes \alpha' \Sigma_u^{-1})\text{vec}\left(\sum_{t=1}^{T}(\Delta y_t - \alpha y_{t-1}^{(1)})y_{t-1}^{(2)\prime}\right).$$

7 Estimation of Vector Error Correction Models

or

$$\widehat{\boldsymbol{\beta}}'_{(K-r)} = (\boldsymbol{\alpha}' \Sigma_u^{-1} \boldsymbol{\alpha})^{-1} \boldsymbol{\alpha}' \Sigma_u^{-1}$$
$$\times \left(\sum_{t=1}^{T} (\Delta y_t - \boldsymbol{\alpha} y_{t-1}^{(1)}) y_{t-1}^{(2)\prime} \right) \left(\sum_{t=1}^{T} y_{t-1}^{(2)} y_{t-1}^{(2)\prime} \right)^{-1}. \quad (7.1.12)$$

This estimator has the following asymptotic distribution.

Result 4

$$T(\widehat{\boldsymbol{\beta}}'_{(K-r)} - \boldsymbol{\beta}'_{(K-r)}) \xrightarrow{d} \left(\int_0^1 \mathbf{W}^*_{K-r} d\mathbf{W}^{*\prime}_r \right)' \left(\int_0^1 \mathbf{W}^*_{K-r} \mathbf{W}^{*\prime}_{K-r} ds \right)^{-1}, \quad (7.1.13)$$

where

$$\mathbf{W}^*_{K-r} := Q^{22}[0 : I_{K-r}] \Sigma_v^{1/2} \mathbf{W}_K,$$

Q^{22} denotes the lower right-hand $((K-r) \times (K-r))$ block of Q^{-1} and

$$\mathbf{W}^*_r := (\boldsymbol{\alpha}' \Sigma_u^{-1} \boldsymbol{\alpha})^{-1} \boldsymbol{\alpha}' \Sigma_u^{-1} Q^{-1} \Sigma_v^{1/2} \mathbf{W}_K.$$

Thus, the asymptotic distribution depends on functionals of a standard Wiener process. ∎

Proof: Replacing $\Delta y_t - \boldsymbol{\alpha} y_{t-1}^{(1)}$ in (7.1.12) with $\boldsymbol{\alpha} \boldsymbol{\beta}'_{(K-r)} y_{t-1}^{(2)} + u_t$ and rearranging terms gives

$$\widehat{\boldsymbol{\beta}}'_{(K-r)} - \boldsymbol{\beta}'_{(K-r)} = (\boldsymbol{\alpha}' \Sigma_u^{-1} \boldsymbol{\alpha})^{-1} \boldsymbol{\alpha}' \Sigma_u^{-1} \left(\sum_{t=1}^{T} u_t y_{t-1}^{(2)\prime} \right) \left(\sum_{t=1}^{T} y_{t-1}^{(2)} y_{t-1}^{(2)\prime} \right)^{-1}. \quad (7.1.14)$$

Thus, we have to consider the quantity

$$T \left(\sum_{t=1}^{T} u_t y_{t-1}^{(2)\prime} \right) \left(\sum_{t=1}^{T} y_{t-1}^{(2)} y_{t-1}^{(2)\prime} \right)^{-1}$$

$$= \left(T^{-1} \sum_{t=1}^{T} u_t y_{t-1}^{(2)\prime} \right) \left(T^{-2} \sum_{t=1}^{T} y_{t-1}^{(2)} y_{t-1}^{(2)\prime} \right)^{-1}.$$

For the first matrix on the right-hand side we have

$$T^{-1} \sum_{t=1}^{T} u_t y_{t-1}^{(2)\prime}$$

$$= \left(T^{-1}\sum_{t=1}^{T} u_t y'_{t-1}\right)\begin{bmatrix} 0 \\ I_{K-r} \end{bmatrix}$$

$$= Q^{-1}\left(T^{-1}\sum_{t=1}^{T} v_t z'_{t-1}\right) Q^{-1'} \begin{bmatrix} 0 \\ I_{K-r} \end{bmatrix}$$

$$= Q^{-1}\left[o_p(1) : T^{-1}\sum_{t=1}^{T} v_t z_{t-1}^{(2)'}\right] Q^{-1'} \begin{bmatrix} 0 \\ I_{K-r} \end{bmatrix}$$

$$\stackrel{d}{\to} Q^{-1}\Sigma_v^{1/2}\left(\int_0^1 \mathbf{W}_K d\mathbf{W}'_K\right)' \Sigma_v^{1/2} \begin{bmatrix} 0 \\ I_{K-r} \end{bmatrix} Q^{22'},$$

where Lemma 7.1(2) and (3) have been used for the last equality and the limiting result, respectively. Thus,

$$T^{-1}\sum_{t=1}^{T} y_{t-1}^{(2)} u'_t \Sigma_u^{-1} \alpha (\alpha' \Sigma_u^{-1} \alpha)^{-1} \stackrel{d}{\to} \int_0^1 \mathbf{W}^*_{K-r} d\mathbf{W}^{*'}_r. \tag{7.1.15}$$

The matrix

$$T^{-2}\sum_{t=1}^{T} y_{t-1}^{(2)} y_{t-1}^{(2)'}$$

$$= [0 : I_{K-r}]\left(T^{-2}\sum_{t=1}^{T} y_{t-1} y'_{t-1}\right)\begin{bmatrix} 0 \\ I_{K-r} \end{bmatrix}$$

$$= [0 : I_{K-r}] Q^{-1}\left(T^{-2}\sum_{t=1}^{T} z_{t-1} z'_{t-1}\right) Q^{-1'}\begin{bmatrix} 0 \\ I_{K-r} \end{bmatrix}$$

$$= [0 : I_{K-r}] Q^{-1}\begin{bmatrix} o_p(1) & o_p(1) \\ o_p(1) & T^{-2}\sum_{t=1}^{T} z_{t-1}^{(2)} z_{t-1}^{(2)'} \end{bmatrix} Q^{-1'}\begin{bmatrix} 0 \\ I_{K-r} \end{bmatrix}$$

$$= Q^{22}\left(T^{-2}\sum_{t=1}^{T} z_{t-1}^{(2)} z_{t-1}^{(2)'}\right) Q^{22'} + o_p(1)$$

$$\stackrel{d}{\to} Q^{22}[0 : I_{K-r}]\Sigma_v^{1/2}\left(\int_0^1 \mathbf{W}_K \mathbf{W}'_K ds\right) \Sigma_v^{1/2}\begin{bmatrix} 0 \\ I_{K-r} \end{bmatrix} Q^{22'}$$

$$= \int_0^1 \mathbf{W}^*_{K-r} \mathbf{W}^{*'}_{K-r} ds, \tag{7.1.16}$$

where Lemma 7.1(5) has been applied. Using (7.1.14) and combining (7.1.15) and (7.1.16), gives the result in (7.1.13). ∎

Clearly, in the present model setup, the GLS estimator of $\boldsymbol{\beta}'_{(K-r)}$ does not have the usual normal limiting distribution. In fact, it converges with rate T rather than the usual rate $\sqrt{T}$, at least under our present rather restrictive assumptions. The asymptotic distribution consists of functionals of a standard Wiener process. It is also interesting to note that the two Wiener processes $\mathbf{W}_r^*$ and $\mathbf{W}_{K-r}^*$ are independent because their cross-covariance matrix is

$$Q^{22}[0:I_{K-r}]\Sigma_v Q^{-1'}\Sigma_u^{-1}\boldsymbol{\alpha}(\boldsymbol{\alpha}'\Sigma_u^{-1}\boldsymbol{\alpha})^{-1}$$
$$= Q^{22}[0:I_{K-r}]Q\boldsymbol{\alpha}(\boldsymbol{\alpha}'\Sigma_u^{-1}\boldsymbol{\alpha})^{-1}$$
$$= Q^{22}\boldsymbol{\alpha}'_\perp\boldsymbol{\alpha}(\boldsymbol{\alpha}'\Sigma_u^{-1}\boldsymbol{\alpha})^{-1}$$
$$= 0,$$

where $\Sigma_v = Q\Sigma_u Q'$ has been used to obtain the first equality. The independence of the two Wiener processes implies that the conditional distribution of

$$\operatorname{vec}\left(\int_0^1 \mathbf{W}_{K-r}^* d\mathbf{W}_r^{*'}\right)'$$

given $\mathbf{W}_{K-r}^*$ is

$$\mathcal{N}\left(0, \int_0^1 \mathbf{W}_{K-r}^* \mathbf{W}_{K-r}^{*'} ds \otimes (\boldsymbol{\alpha}'\Sigma_u^{-1}\boldsymbol{\alpha})^{-1}\right)$$

(see Ahn & Reinsel (1990), Phillips & Park (1988) or Johansen (1995)). This reasoning leads to the following interesting result.

Result 5

$$\operatorname{vec}\left[(\widehat{\boldsymbol{\beta}}'_{(K-r)} - \boldsymbol{\beta}'_{(K-r)})\left(\sum_{t=1}^T y_{t-1}^{(2)}y_{t-1}^{(2)'}\right)^{1/2}\right]$$
$$\xrightarrow{d} \mathcal{N}\left(0, I_{K-r} \otimes (\boldsymbol{\alpha}'\Sigma_u^{-1}\boldsymbol{\alpha})^{-1}\right). \tag{7.1.17}$$

■

Proof: From (7.1.16) we have

$$T^{-2}\sum_{t=1}^T y_{t-1}^{(2)}y_{t-1}^{(2)'} \xrightarrow{d} \int_0^1 \mathbf{W}_{K-r}^* \mathbf{W}_{K-r}^{*'} ds.$$

Hence, Result 5 follows because

$$\operatorname{vec}\left[(\widehat{\boldsymbol{\beta}}'_{(K-r)} - \boldsymbol{\beta}'_{(K-r)})\left(\sum_{t=1}^T y_{t-1}^{(2)}y_{t-1}^{(2)'}\right)^{1/2}\right]$$
$$= \left(\left(\sum_{t=1}^T y_{t-1}^{(2)}y_{t-1}^{(2)'}\right)^{1/2} \otimes I_K\right)\operatorname{vec}(\widehat{\boldsymbol{\beta}}'_{(K-r)} - \boldsymbol{\beta}'_{(K-r)}).$$

■

Result 5 means that, although the GLS estimator $\widehat{\beta}'_{(K-r)}$ has a nonstandard limiting distribution, a transformation is asymptotically normal and can, for example, be used to construct hypothesis tests with standard limiting distributions. For example, t-ratios can be constructed in the usual way by considering an element of $\widehat{\beta}'_{(K-r)}$ and dividing by its asymptotic standard deviation obtained from

$$\left(\sum_{t=1}^{T} y_{t-1}^{(2)} y_{t-1}^{(2)\prime}\right)^{-1} \otimes (\alpha' \Sigma_u^{-1} \alpha)^{-1}.$$

Also Wald tests can be constructed as usual (see Appendix C.7).

Of course, the GLS estimator is only available under the very restrictive assumption that both α and Σ_u are known. It turns out, however, that the same asymptotic distribution is obtained for the corresponding EGLS estimator,

$$\widehat{\widehat{\beta}}'_{(K-r)} = (\widehat{\alpha}' \widehat{\Sigma}_u^{-1} \widehat{\alpha})^{-1} \widehat{\alpha}' \widehat{\Sigma}_u^{-1} \left(\sum_{t=1}^{T} (\Delta y_t - \widehat{\alpha} y_{t-1}^{(1)}) y_{t-1}^{(2)\prime}\right) \left(\sum_{t=1}^{T} y_{t-1}^{(2)} y_{t-1}^{(2)\prime}\right)^{-1}, \tag{7.1.18}$$

where $\widehat{\alpha}$ and $\widehat{\Sigma}_u$ are consistent estimators of α and Σ_u, respectively. Fortunately, such estimators are available in the present case. A consistent estimator $\widehat{\alpha}$ follows from Result 2. If β is normalized as in (7.1.10), the first r columns of Π are equal to α. Hence, the first r columns of $\widehat{\Pi}$ are a consistent estimator of α and the usual white noise covariance matrix estimator from the unrestricted LS estimation can be shown to be a consistent estimator of Σ_u, as we will demonstrate later (see Result 8). The following result can be established.

Result 6

$$T(\widehat{\widehat{\beta}}'_{(K-r)} - \widehat{\beta}'_{(K-r)}) = o_p(1). \tag{7.1.19}$$

∎

Proof: Defining $u_t^* = \Delta y_t - \widehat{\alpha} \beta' y_{t-1}$ and substituting $\widehat{\alpha} \beta'_{(K-r)} y_{t-1}^{(2)} + u_t^*$ for $\Delta y_t - \widehat{\alpha} y_{t-1}^{(1)}$ in (7.1.18) gives, after rearrangement of terms,

$$\widehat{\widehat{\beta}}'_{(K-r)} - \beta'_{(K-r)} = (\widehat{\alpha}' \widehat{\Sigma}_u^{-1} \widehat{\alpha})^{-1} \widehat{\alpha}' \widehat{\Sigma}_u^{-1} \left(\sum_{t=1}^{T} u_t^* y_{t-1}^{(2)\prime}\right) \left(\sum_{t=1}^{T} y_{t-1}^{(2)} y_{t-1}^{(2)\prime}\right)^{-1}.$$

Hence,

$$T(\widehat{\widehat{\beta}}'_{(K-r)} - \beta'_{(K-r)})$$

$$= \left[(\widehat{\alpha}'\widehat{\Sigma}_u^{-1}\widehat{\alpha})^{-1}\widehat{\alpha}'\widehat{\Sigma}_u^{-1} - (\alpha'\Sigma_u^{-1}\alpha)^{-1}\alpha'\Sigma_u^{-1}\right]$$

$$\times \left(T^{-1}\sum_{t=1}^{T} u_t y_{t-1}^{(2)\prime}\right) \left(T^{-2}\sum_{t=1}^{T} y_{t-1}^{(2)} y_{t-1}^{(2)\prime}\right)^{-1}$$

$$+ (\widehat{\alpha}'\widehat{\Sigma}_u^{-1}\widehat{\alpha})^{-1}\widehat{\alpha}'\widehat{\Sigma}_u^{-1}$$

$$\times \left(T^{-1}\sum_{t=1}^{T}(u_t^* - u_t) y_{t-1}^{(2)\prime}\right) \left(T^{-2}\sum_{t=1}^{T} y_{t-1}^{(2)} y_{t-1}^{(2)\prime}\right)^{-1}.$$

The term in brackets is $o_p(1)$ because $\widehat{\alpha}$ and $\widehat{\Sigma}_u$ are consistent estimators by assumption. Moreover, $T^{-1}\sum_{t=1}^{T}(u_t^* - u_t) y_{t-1}^{(2)\prime} = o_p(1)$ (see Problem 7.1). Thus, the desired result follows because all other terms converge as established previously. ∎

If the process is assumed to be Gaussian, ML estimation may be used alternatively. In case α and Σ_u are known, the ML estimator is identical to the GLS estimator for $\beta'_{(K-r)}$ and, hence, $\widehat{\beta}'_{(K-r)}$ is also the ML estimator. If α and Σ_u are unknown, ML estimation under the constraint $\text{rk}(\Pi) = r$ may be used. The log-likelihood function is

$$\ln l = -\frac{KT}{2}\ln 2\pi - \frac{T}{2}\ln|\Sigma_u| - \frac{1}{2}\sum_{t=1}^{T}(\Delta y_t - \Pi y_{t-1})'\Sigma_u^{-1}(\Delta y_t - \Pi y_{t-1}).$$

(7.1.20)

From Chapter 3, we know that maximizing this function is equivalent to minimizing the determinant

$$\left|T^{-1}\sum_{t=1}^{T}(\Delta y_t - \Pi y_{t-1})(\Delta y_t - \Pi y_{t-1})'\right|.$$

To impose the rank restriction $\text{rk}(\Pi) = r$, we write $\Pi = \alpha\beta'$, where α and β are $(K \times r)$ matrices with rank r. For the moment we do not impose any normalization restrictions and consider minimization of the determinant

$$\left|T^{-1}\sum_{t=1}^{T}(\Delta y_t - \alpha\beta' y_{t-1})(\Delta y_t - \alpha\beta' y_{t-1})'\right|$$

with respect to α and β. This minimization problem is solved in Proposition A.7 in Appendix A.14 and the solution is obtained by considering the eigenvalues $\lambda_1 \geq \cdots \geq \lambda_K$ and the associated orthonormal eigenvectors $v_1, \ldots, v_K$ of the matrix

$$\left(\sum_{t=1}^{T} y_{t-1} y_{t-1}'\right)^{-1/2} \left(\sum_{t=1}^{T} y_{t-1}\Delta y_t'\right) \left(\sum_{t=1}^{T} \Delta y_t \Delta y_t'\right)$$

$$\times \left(\sum_{t=1}^{T} \Delta y_t y'_{t-1}\right) \left(\sum_{t=1}^{T} y_{t-1} y'_{t-1}\right)^{-1/2}.$$

The minimum of the determinant is attained for

$$\widetilde{\beta} = [\mathbf{v}_1, \ldots, \mathbf{v}_r]' \left(\sum_{t=1}^{T} y_{t-1} y'_{t-1}\right)^{-1/2} \tag{7.1.21}$$

and

$$\widetilde{\alpha} = \left(\sum_{t=1}^{T} \Delta y_t y'_{t-1} \widetilde{\beta}\right) \left(\sum_{t=1}^{T} \widetilde{\beta}' y_{t-1} y'_{t-1} \widetilde{\beta}\right)^{-1}. \tag{7.1.22}$$

Clearly, the resulting ML estimator $\widetilde{\Pi} = \widetilde{\alpha}\widetilde{\beta}'$ for Π must have the same asymptotic properties as the unrestricted LS estimator of Π because even the estimator in Result 3, which is based on a known β does not have better properties. Notice that, for a Gaussian model, the LS estimator based on a known β is equal to the ML estimator because the same regressors appear in all equations. Thus, we can draw the following conclusion.

Result 7

$$\sqrt{T}\mathrm{vec}(\widetilde{\alpha}\widetilde{\beta}' - \Pi) \xrightarrow{d} \mathcal{N}(0, \beta(\Gamma_z^{(1)})^{-1}\beta' \otimes \Sigma_u). \tag{7.1.23}$$

■

This result was derived by Johansen (1995) and other authors for more general models. It is also interesting to note that we can, of course, normalize the ML estimator for β as in (7.1.10), that is, we postmultiply the estimator in (7.1.21) by the inverse of the upper $(r \times r)$ submatrix. Denoting the normalized estimator by $\check{\beta}$ and using the corresponding estimator for α from (7.1.22),

$$\check{\alpha} = \left(\sum_{t=1}^{T} \Delta y_t y'_{t-1} \check{\beta}\right) \left(\sum_{t=1}^{T} \check{\beta}' y_{t-1} y'_{t-1} \check{\beta}\right)^{-1},$$

gives an estimator $\check{\alpha}\check{\beta}'$ of Π which is identical to $\widetilde{\alpha}\widetilde{\beta}'$. Thus, the asymptotic properties must also be identical. It follows that $\check{\alpha}$ has the same asymptotic distribution as the LS estimator in (7.1.9). Moreover, the asymptotic distribution of the lower $((K-r) \times r)$ part of $\check{\beta}$ is the same as that of the GLS estimator in Result 4 because

$$\check{\beta}'_{(K-r)} =$$
$$(\check{\alpha}'\widetilde{\Sigma}_u^{-1}\check{\alpha})^{-1}\check{\alpha}'\widetilde{\Sigma}_u^{-1}\left(\sum_{t=1}^{T}(\Delta y_t - \check{\alpha} y_{t-1}^{(1)}) y_{t-1}^{(2)\prime}\right)\left(\sum_{t=1}^{T} y_{t-1}^{(2)} y_{t-1}^{(2)\prime}\right)^{-1},$$

where the ML estimator $\widetilde{\Sigma}_u$ is substituted for Σ_u. Thus, the asymptotic distribution of $\breve{\beta}_{(K-r)}$ follows from Result 6 and the consistency of the ML estimators $\breve{\alpha}$ and $\widetilde{\Sigma}_u$.

In fact, any of the estimators for Π which we have considered so far, leads to a consistent estimator of the white noise covariance matrix of the form

$$\widetilde{\Sigma}_u = T^{-1} \sum_{t=1}^{T} (\Delta y_t - \widehat{\Pi} y_{t-1})(\Delta y_t - \widehat{\Pi} y_{t-1})'. \tag{7.1.24}$$

Here $\widehat{\Pi}$ can be any of the estimators for Π considered so far, because they are all asymptotically equivalent. The following result can be established.

Result 8

$$\text{plim } \widetilde{\Sigma}_u = \Sigma_u. \tag{7.1.25}$$

∎

Proof: Notice that

$$\begin{aligned}
\widetilde{\Sigma}_u &= T^{-1} \sum_{t=1}^{T} (\Pi y_{t-1} - \widehat{\Pi} y_{t-1} + u_t)(\Pi y_{t-1} - \widehat{\Pi} y_{t-1} + u_t)' \\
&= T^{-1} \sum_{t=1}^{T} u_t u_t' + (\Pi - \widehat{\Pi}) \left(T^{-1} \sum_{t=1}^{T} y_{t-1} y_{t-1}' \right) (\Pi - \widehat{\Pi})' \\
&\quad + \left(T^{-1} \sum_{t=1}^{T} u_t y_{t-1}' \right) (\Pi - \widehat{\Pi})' \\
&\quad + (\Pi - \widehat{\Pi}) \left(T^{-1} \sum_{t=1}^{T} y_{t-1} u_t' \right).
\end{aligned} \tag{7.1.26}$$

Using a standard law of large numbers,

$$\text{plim } T^{-1} \sum_{t=1}^{T} u_t u_t' = \Sigma_u.$$

Thus, it suffices to show that all other terms are $o_p(1)$. This property follows because from Lemma 7.1 we have

$$T^{-1} \sum_{t=1}^{T} y_{t-1} u_t' = O_p(1)$$

and

$$T^{-1} \sum_{t=1}^{T} \beta' y_{t-1} y_{t-1}' \beta = O_p(1).$$

Using the estimator $\widehat{\alpha}\widehat{\beta}'$ for Π, it is easily seen that all terms but the first on the right-hand side of the last equality sign in (7.1.26) converge to zero in probability. The argument is easily extended to the other estimators by noting that their difference to the previously treated estimator is $o_p(T^{-1/2})$. ∎

So far we have assumed that $r \neq 0$ and, hence, $\Pi \neq 0$. This assumption is of obvious importance for some of the results to hold and some of the proofs to work. If $\Pi = 0$, the analysis becomes even simpler in some respects. In that case, y_t is a multivariate random walk and we can apply Proposition C.18 directly to evaluate the asymptotic properties of the term

$$T(\widehat{\Pi} - \Pi) = \left(T^{-1} \sum_{t=1}^{T} u_t y'_{t-1} \right) \left(T^{-2} \sum_{t=1}^{T} y_{t-1} y'_{t-1} \right)^{-1},$$

where $\widehat{\Pi}$ is again the LS estimator. Using Proposition C.18(6) and (9) gives the following result.

Result 9
If the cointegrating rank $r = 0$,

$$T(\widehat{\Pi} - \Pi) \xrightarrow{d} \Sigma_u^{1/2} \left(\int_0^1 \mathbf{W}_K d\mathbf{W}'_K \right)' \left(\int_0^1 \mathbf{W}_K \mathbf{W}'_K ds \right)^{-1} \Sigma_u^{-1/2}. \quad (7.1.27)$$

∎

The LS estimator is again identical to the ML estimator and, hence, the same result is obtained for the latter. On the other hand, the GLS estimator is not applicable here. Now we cannot even use the usual t-ratios anymore in a standard way because they do not have a limiting standard normal distribution in this case. For the special case of a univariate model this can be seen from Appendix C.8.1. Notice that for $K = 1$, $\widehat{\Pi} = \widehat{\rho} - 1$ in Proposition C.17 and, thus, the asymptotic distribution of $T\widehat{\Pi} = T(\widehat{\rho} - 1)$ is clearly different from the standard normal in this case.

The results for the estimator of the VECM imply analogous results for the parameters of the corresponding levels VAR form $y_t = A_1 y_{t-1} + u_t$. Notice that $A_1 = \Pi + I_K$. Consequently, we have for the LS estimator, for example,

$$\widehat{A}_1 - A_1 = \widehat{\Pi} - \Pi. \quad (7.1.28)$$

Hence, the asymptotic properties of $\widehat{A}_1$ follow immediately from those of $\widehat{\Pi}$.

The simple model we have discussed in this section shows the main differences to the stationary case. All the results can be extended to richer models with short-term dynamics and deterministic terms. Estimation of such models will be considered in the next section.

7.2 Estimation of General VECMs

We first consider a model without deterministic terms,

$$\Delta y_t = \Pi y_{t-1} + \Gamma_1 \Delta y_{t-1} + \cdots + \Gamma_{p-1} \Delta y_{t-p+1} + u_t, \qquad (7.2.1)$$

where y_t is a process of dimension K, $\text{rk}(\Pi) = r$ with $0 < r < K$ so that $\Pi = \alpha \beta'$, where α and β are $(K \times r)$ matrices with $\text{rk}(\alpha) = \text{rk}(\beta) = r$. All other symbols have their conventional meanings, that is, the Γ_j $(j = 1, \ldots, p-1)$ are $(K \times K)$ parameter matrices and $u_t \sim (0, \Sigma_u)$ is standard white noise. Also, y_t is assumed to be an $I(1)$ process so that

$$\alpha'_\perp \left(I_K - \sum_{i=1}^{p-1} \Gamma_i \right) \beta_\perp \qquad (7.2.2)$$

is nonsingular (see Section 6.3, Eq. (6.3.12)). These conditions are always assumed to hold without further notice when the VECM (7.2.1) is considered in this chapter.

For estimation purposes, we assume that a sample $y_1, \ldots, y_T$ and the needed presample values are available. It is then often convenient to write the VECM (7.2.1), for $t = 1, \ldots, T$, in matrix notation as

$$\Delta Y = \Pi Y_{-1} + \Gamma \Delta X + U, \qquad (7.2.3)$$

where

$$\Delta Y := [\Delta y_1, \ldots, \Delta y_T],$$

$$Y_{-1} := [y_0, \ldots, y_{T-1}],$$

$$\Gamma := [\Gamma_1, \ldots, \Gamma_{p-1}],$$

$$\Delta X := [\Delta X_0, \ldots, \Delta X_{T-1}] \quad \text{with} \quad \Delta X_{t-1} := \begin{bmatrix} \Delta y_{t-1} \\ \vdots \\ \Delta y_{t-p+1} \end{bmatrix}$$

and

$$U := [u_1, \ldots, u_T].$$

We will now consider LS, EGLS, and ML estimation of the parameters of this model. Estimation of the parameters of the corresponding levels VAR form will also be discussed and, moreover, we comment on the implications of including deterministic terms.

7.2.1 LS Estimation

From the matrix version (7.2.3) of our VECM, the LS estimator is seen to be

$$[\widehat{\Pi} : \widehat{\Gamma}] = [\Delta Y Y'_{-1} : \Delta Y \Delta X'] \begin{bmatrix} Y_{-1} Y'_{-1} & Y_{-1} \Delta X' \\ \Delta X Y'_{-1} & \Delta X \Delta X' \end{bmatrix}^{-1}, \qquad (7.2.4)$$

using the usual formulas from Chapter 3. The corresponding white noise covariance matrix estimator is

$$\widehat{\Sigma}_u := (T - Kp)^{-1}(\Delta Y - \widehat{\Pi} Y_{-1} - \widehat{\Gamma} \Delta X)(\Delta Y - \widehat{\Pi} Y_{-1} - \widehat{\Gamma} \Delta X)'. \qquad (7.2.5)$$

The asymptotic properties of these estimators are given in the next proposition.

Proposition 7.1 (*Asymptotic Properties of the LS Estimator for a VECM*)
Consider the VECM (7.2.1). The LS estimator given in (7.2.4) is consistent and

$$\sqrt{T} \operatorname{vec}([\widehat{\Pi} : \widehat{\Gamma}] - [\Pi : \Gamma]) \xrightarrow{d} \mathcal{N}(0, \Sigma_{\mathrm{co}}), \qquad (7.2.6)$$

where

$$\Sigma_{\mathrm{co}} = \left(\begin{bmatrix} \beta & 0 \\ 0 & I_{Kp-K} \end{bmatrix} \Omega^{-1} \begin{bmatrix} \beta' & 0 \\ 0 & I_{Kp-K} \end{bmatrix} \right) \otimes \Sigma_u$$

and

$$\Omega = \operatorname{plim} \frac{1}{T} \begin{bmatrix} \beta' Y_{-1} Y'_{-1} \beta & \beta' Y_{-1} \Delta X' \\ \Delta X Y'_{-1} \beta & \Delta X \Delta X' \end{bmatrix}.$$

The matrix

$$\begin{bmatrix} \beta & 0 \\ 0 & I_{Kp-K} \end{bmatrix} \Omega^{-1} \begin{bmatrix} \beta' & 0 \\ 0 & I_{Kp-K} \end{bmatrix}$$

is consistently estimated by

$$T \begin{bmatrix} Y_{-1} Y'_{-1} & Y_{-1} \Delta X' \\ \Delta X Y'_{-1} & \Delta X \Delta X' \end{bmatrix}^{-1}$$

and $\widehat{\Sigma}_u$ is a consistent estimator for Σ_u. ∎

This proposition generalizes Result 2 of Section 7.1. Therefore similar remarks can be made.

Remark 1 The covariance matrix Σ_{co} is singular. This property is easily seen by noting that Ω is a $[(Kp - K + r) \times (Kp - K + r)]$ matrix. Thus, the rank of the $(K^2 p \times K^2 p)$ matrix Σ_{co} cannot be greater than $K(Kp - K + r)$ which is smaller than $K^2 p$ under our assumption that $r < K$. Still, t-ratios can be set up and interpreted in the usual way because they have standard normal limiting distributions under our assumptions. In contrast, Wald tests and the corresponding F-tests of linear restrictions on the parameters may not have the usual asymptotic χ^2- or approximate F-distributions that are obtained for stationary processes. A more detailed discussion of this issue will be given in Section 7.6. ∎

Remark 2 If β is known, the LS estimator

$$[\widehat{\alpha}:\widehat{\Gamma}] = [\Delta Y Y'_{-1}\beta : \Delta Y \Delta X'] \begin{bmatrix} \beta' Y_{-1} Y'_{-1} \beta & \beta' Y_{-1} \Delta X' \\ \Delta X Y'_{-1} \beta & \Delta X \Delta X' \end{bmatrix}^{-1} \quad (7.2.7)$$

of $[\alpha : \Gamma]$ may be considered. Using standard arguments for stationary processes, its asymptotic distribution is seen to be

$$\sqrt{T}\, \text{vec}([\widehat{\alpha}:\widehat{\Gamma}] - [\alpha:\Gamma]) \xrightarrow{d} \mathcal{N}(0, \Sigma_{\alpha,\Gamma}), \quad (7.2.8)$$

where

$$\Sigma_{\alpha,\Gamma} = \Omega^{-1} \otimes \Sigma_u = \text{plim}\, T \begin{bmatrix} \beta' Y_{-1} Y'_{-1} \beta & \beta' Y_{-1} \Delta X' \\ \Delta X Y'_{-1} \beta & \Delta X \Delta X' \end{bmatrix}^{-1} \otimes \Sigma_u.$$

The asymptotic distribution in (7.2.8) is nonsingular so that, for given β, asymptotic inference for α and Γ is standard. Noting that

$$[\widehat{\alpha}\beta' : \widehat{\Gamma}] - [\Pi : \Gamma] = ([\widehat{\alpha}:\widehat{\Gamma}] - [\alpha:\Gamma]) \begin{bmatrix} \beta' & 0 \\ 0 & I_{Kp-K} \end{bmatrix},$$

it is easy to see that

$$\text{vec}([\widehat{\alpha}\beta' : \widehat{\Gamma}] - [\Pi : \Gamma])$$

has the same asymptotic distribution as the LS estimator in Proposition 7.1. This finding corresponds to Result 3 in Section 7.1. It means that, whether the cointegrating matrix β is known or estimated is of no consequence for the asymptotic distribution of the LS estimators of Π and Γ. The reason is that β is estimated "superconsistently" even if LS estimation is used. This point will be discussed further in Section 7.2.2. ∎

Remark 3 If the cointegrating rank $r = 0$ and, thus, $\Pi = 0$,

$$\sqrt{T}[\widehat{\Pi} - \Pi] = o_p(1),$$

that is, the LS estimator of Π converges faster than with the usual rate $\sqrt{T}$. Therefore, Proposition 7.1 remains valid in the sense that all parts of the asymptotic covariance matrix in (7.2.6) related to Π have to be set to zero. In other words, the first K^2 rows and columns of Σ_{co} are zero. ∎

Remark 4 From Proposition 7.1 it is also easy to derive the asymptotic distribution of the LS estimator for the parameters of the levels VAR form corresponding to our VECM,

$$y_t = A_1 y_{t-1} + \cdots + A_p y_{t-p} + u_t. \tag{7.2.9}$$

The A_i's are related to the VECM parameters by

$$\begin{aligned} A_1 &= \mathbf{\Pi} + I_K + \mathbf{\Gamma}_1 \\ A_i &= \mathbf{\Gamma}_i - \mathbf{\Gamma}_{i-1}, \quad i = 2, \ldots, p-1, \\ A_p &= -\mathbf{\Gamma}_{p-1} \end{aligned} \tag{7.2.10}$$

(see also (6.3.7)). Hence, they are obtained by a linear transformation,

$$A := [A_1 : \cdots : A_p] = [\mathbf{\Pi} : \mathbf{\Gamma}]W + J, \tag{7.2.11}$$

where

$$J := [I_K : 0 : \cdots : 0] \quad (K \times Kp)$$

and

$$W := \begin{bmatrix} I_K & 0 & 0 & \cdots & 0 & 0 \\ I_K & -I_K & 0 & \cdots & 0 & 0 \\ 0 & I_K & -I_K & & 0 & 0 \\ \vdots & & \ddots & \ddots & & \vdots \\ \vdots & & & \ddots & \ddots & \vdots \\ 0 & 0 & \cdots & \cdots & I_K & -I_K \end{bmatrix} \quad (Kp \times Kp).$$

Consequently, using

$$\text{vec}([\mathbf{\Pi} : \mathbf{\Gamma}]W) = (W' \otimes I_K)\text{vec}[\mathbf{\Pi} : \mathbf{\Gamma}],$$

we get the following implication of Proposition 7.1 (see also Sims, Stock & Watson (1990)).

Corollary 7.1.1
Under the conditions of Proposition 7.1,

$$\sqrt{T}\,\text{vec}(\widehat{A} - A) \xrightarrow{d} \mathcal{N}(0, \Sigma_{\boldsymbol{\alpha}}^{\text{co}}),$$

where $\widehat{A}$ is the LS estimator of A and

$$\begin{aligned} \Sigma_{\boldsymbol{\alpha}}^{\text{co}} &:= \left(W' \begin{bmatrix} \boldsymbol{\beta} & 0 \\ 0 & I_{Kp-K} \end{bmatrix} \Omega^{-1} \begin{bmatrix} \boldsymbol{\beta}' & 0 \\ 0 & I_{Kp-K} \end{bmatrix} W \right) \otimes \Sigma_u \\ &= (W' \otimes I_K)\Sigma_{\text{co}}(W \otimes I_K). \end{aligned}$$

Furthermore,

$$\widehat{\Sigma}_{\alpha}^{co} = (XX')^{-1} \otimes [(Y - \widehat{A}X)(Y - \widehat{A}X)']$$

is a consistent estimator of Σ_{α}^{co}. Here $Y := [y_1, \ldots, y_T]$ and

$$X := [Y_0, \ldots, Y_{T-1}] \quad \text{with} \quad Y_{t-1} := \begin{bmatrix} y_{t-1} \\ \vdots \\ y_{t-p} \end{bmatrix}.$$

■

Because Σ_{α}^{co} is singular, $\widehat{A}$ also has a singular asymptotic distribution. The distribution in Corollary 7.1.1 remains, in fact, valid if $r = 0$. ■

Discussion of the Proof of Proposition 7.1

The proof of Proposition 7.1 is a generalization of that of Result 2 in Section 7.1. Multiplying

$$\begin{bmatrix} y_t \\ \Delta X_t \end{bmatrix}$$

by

$$Q^* := \begin{bmatrix} \beta' & 0 \\ 0 & I_{K(p-1)} \\ \alpha'_{\perp} & 0 \end{bmatrix}$$

gives a process

$$z_t = \begin{bmatrix} z_t^{(1)} \\ z_t^{(2)} \end{bmatrix} := Q^* \begin{bmatrix} y_t \\ \Delta X_t \end{bmatrix}, \tag{7.2.12}$$

where

$$z_t^{(1)} := \begin{bmatrix} \beta' y_t \\ \Delta X_t \end{bmatrix}$$

contains $I(0)$ components only and $z_t^{(2)} := \alpha'_{\perp} y_t$ consists of $I(1)$ components (see Proposition 6.1). Therefore, a lemma analogous to Lemma 7.1 can be established and used to prove Proposition 7.1. We leave the details as an exercise (see Problem 7.2).

In fact, via the process z_t, we can get the following useful lemma from standard weak laws of large numbers and central limit theorems for stationary processes (see Appendix C.4) as well as Proposition C.18 of Appendix C. It summarizes a number of convergence results for variables generated by the VECM (7.2.1). Some of these or similar results were derived by different authors including Phillips & Durlauf (1986), Johansen (1988), Ahn & Reinsel (1990) and Park & Phillips (1989).

Lemma 7.2

(1) $\Delta X \Delta X' = O_p(T)$ and $(T^{-1}\Delta X \Delta X')^{-1} = O_p(1)$.
(2) $\beta' Y_{-1} \Delta X' = O_p(T)$.
(3) $\beta' Y_{-1} Y'_{-1} \beta = O_p(T)$ and $(T^{-1}\beta' Y_{-1} Y'_{-1} \beta)^{-1} = O_p(1)$.
(4) $\beta' Y_{-1} U' = O_p(T^{1/2})$.
(5) $\beta' Y_{-1} \Delta Y' = O_p(T^{1/2})$.
(6) $Y_{-1} U' = O_p(T)$.
(7) $Y_{-1} \Delta X' = O_p(T)$.
(8) $\beta' Y_{-1} Y'_{-1} = O_p(T)$.
(9) $Y_{-1} Y'_{-1} = O_p(T^2)$.

■

Some of these results are helpful in deriving Proposition 7.1 and they are also useful in proving the next propositions. Because ΔY, $\beta' Y_{-1}$, and ΔX contain $I(0)$ variables only, essentially the same results as in the stable case hold for these quantities. This is reflected in Lemma 7.2(1)–(5). On the other hand, Y_{-1} contains $I(1)$ variables that behave differently from $I(0)$ variables. For instance, for a stable process, $Y_{-1} Y'_{-1}/T$ has a fixed probability limit (see Chapter 3). Now the corresponding quantity $Y_{-1} Y'_{-1}$ is $O_p(T^2)$. Intuitively, the reason is that integrated variables do not fluctuate around a constant mean but are trending. Thus, the sums of products and cross-products go to infinity (or minus infinity) more rapidly than for stable processes.

7.2.2 EGLS Estimation of the Cointegration Parameters

For GLS estimation we assume that β is normalized as in (7.1.10),

$$\beta = \begin{bmatrix} I_r \\ \beta_{(K-r)} \end{bmatrix}.$$

Because we are primarily interested in estimating $\beta_{(K-r)}$, we concentrate on the error correction term and replace the short-run parameters Γ by their LS estimators for a given matrix Π,

$$\widehat{\Gamma}(\Pi) = (\Delta Y - \Pi Y_{-1})\Delta X'(\Delta X \Delta X')^{-1}.$$

Hence,

$$\Delta Y = \Pi Y_{-1} + (\Delta Y - \Pi Y_{-1})\Delta X'(\Delta X \Delta X')^{-1}\Delta X + U^*.$$

Rearranging terms and defining the $(T \times T)$ matrix

$$M := I_T - \Delta X'(\Delta X \Delta X')^{-1}\Delta X,$$

gives

$$R_0 = \Pi R_1 + U^* = \alpha\beta' R_1 + U^*, \tag{7.2.13}$$

where

$$R_0 := \Delta Y M \quad \text{and} \quad R_1 := Y_{-1} M.$$

Notice that R_0 is just the residual matrix from a (multivariate) regression of Δy_t on ΔX_{t-1} and R_1 is the matrix of residuals from a regression of y_{t-1} on ΔX_{t-1}. Denoting the first r and last $K-r$ rows of R_1 by $R_1^{(1)}$ and $R_1^{(2)}$, respectively, and using the normalization of β, (7.2.13) can be rewritten as

$$R_0 - \alpha R_1^{(1)} = \alpha\beta'_{(K-r)} R_1^{(2)} + U^*. \tag{7.2.14}$$

Based on this "concentrated model" the GLS estimator of $\beta'_{(K-r)}$ is

$$\widehat{\beta}'_{(K-r)} = (\alpha' \Sigma_u^{-1} \alpha)^{-1} \alpha' \Sigma_u^{-1} (R_0 - \alpha R_1^{(1)}) R_1^{(2)'} \left(R_1^{(2)} R_1^{(2)'}\right)^{-1} \tag{7.2.15}$$

(see Eq. (7.1.12)). Note that the same estimator is obtained if the short-run parameters are not concentrated out first because Γ has been replaced by the optimal matrix for any given matrix Π. As in the simple special case model considered in Section 7.1, it is now obvious how to obtain a feasible GLS estimator. In a first estimation round we determine the LS estimator of $[\Pi : \Gamma]$ as in (7.2.4) and Σ_u as in (7.2.5). Using the first r columns of $\widehat{\Pi}$ as an estimator $\widehat{\alpha}$, we get the EGLS estimator

$$\widehat{\widehat{\beta}}'_{(K-r)} = (\widehat{\alpha}' \widehat{\Sigma}_u^{-1} \widehat{\alpha})^{-1} \widehat{\alpha}' \widehat{\Sigma}_u^{-1} (R_0 - \widehat{\alpha} R_1^{(1)}) R_1^{(2)'} \left(R_1^{(2)} R_1^{(2)'}\right)^{-1}. \tag{7.2.16}$$

This estimator was proposed by Ahn & Reinsel (1990) and Saikkonen (1992) (see also Reinsel (1993, p. 171)). Its asymptotic properties are analogous to those of the EGLS estimator for the simple model considered in Section 7.1. They are summarized in the following proposition which was proven by Ahn & Reinsel (1990).

Proposition 7.2 (*Asymptotic Properties of the EGLS Estimator for the Cointegration Matrix*)
Consider the VECM (7.2.1) with cointegration matrix β normalized as in (7.1.10). Suppose $\widehat{\alpha}$ and $\widehat{\Sigma}_u$ are consistent estimators of α and Σ_u, respectively. Then the EGLS estimator of $\beta'_{(K-r)}$ given in (7.2.16) has the following asymptotic distribution:

$$T(\widehat{\widehat{\beta}}'_{(K-r)} - \beta'_{(K-r)}) \xrightarrow{d} \left(\int_0^1 \mathbf{W}^\#_{K-r} d\mathbf{W}^{\#'}_r\right)' \left(\int_0^1 \mathbf{W}^\#_{K-r} \mathbf{W}^{\#'}_{K-r} ds\right)^{-1}, \tag{7.2.17}$$

where $\mathbf{W}^\#_{K-r}$ and $\mathbf{W}^\#_r$ are suitable independent $(K-r)$- and r-dimensional Wiener processes, respectively, whose parameters depend on those of the VECM. Furthermore,

$$\text{vec}\left[(\widehat{\widehat{\beta}}'_{(K-r)} - \beta'_{(K-r)})\left(R_1^{(2)}R_1^{(2)\prime}\right)^{1/2}\right] \xrightarrow{d} \mathcal{N}\left(0, I_{K-r} \otimes (\alpha'\Sigma_u^{-1}\alpha)^{-1}\right).$$
(7.2.18)

■

Remark 1 The EGLS estimator has the same asymptotic distribution as the GLS estimator. Moreover, it has the same asymptotic distribution one would obtain if all parameters (α, Γ, and Σ_u) except $\beta_{(K-r)}$ were known. It converges at rate T. Hence, $\widehat{\widehat{\beta}}_{(K-r)}$ is a superconsistent estimator of $\beta_{(K-r)}$ and, thus,

$$\widehat{\widehat{\beta}} = \begin{bmatrix} I_r \\ \widehat{\widehat{\beta}}_{(K-r)} \end{bmatrix}$$

is a superconsistent estimator of β. The precise form of the Wiener processes $\mathbf{W}^{\#}_{K-r}$ and $\mathbf{W}^{\#}_r$ depends on the short-run dynamics of the process y_t. It is given, for example, in Ahn & Reinsel (1990). ■

Remark 2 The matrix

$$\begin{aligned} T^{-2}R_1R_1' &= T^{-2}Y_{-1}MY_{-1}' \\ &= T^{-2}Y_{-1}Y_{-1}' - T^{-2}Y_{-1}\Delta X'(T^{-1}\Delta X\Delta X')^{-1}T^{-1}\Delta XY_{-1}' \\ &= T^{-2}Y_{-1}Y_{-1}' + o_p(1)O_p(1)O_p(1) \\ &= T^{-2}Y_{-1}Y_{-1}' + o_p(1), \end{aligned}$$

where Lemma 7.2(1) and (7) have been used. This result implies that (7.2.18) could be stated alternatively as

$$\text{vec}\left[(\widehat{\widehat{\beta}}'_{(K-r)} - \beta'_{(K-r)})\left(Y_{-1}^{(2)}Y_{-1}^{(2)\prime}\right)^{1/2}\right] \xrightarrow{d} \mathcal{N}\left(0, I_K \otimes (\alpha'\Sigma_u^{-1}\alpha)^{-1}\right),$$

where $Y_{-1}^{(2)}$ contains the last $K-r$ rows of Y_{-1}. For practical purposes, the result as stated in (7.2.18) is more useful because it can be used directly for setting up meaningful t-ratios and Wald or F-tests for hypotheses about the coefficients of $\beta_{(K-r)}$. These quantities have the usual asymptotic or approximate distributions. Of course, the same is true if $(R_1^{(2)}R_1^{(2)\prime})^{1/2}$ is replaced by $(Y_{-1}^{(2)}Y_{-1}^{(2)\prime})^{1/2}$. Still, in small samples it is advantageous to take the short-run dynamics into account as in $(R_1^{(2)}R_1^{(2)\prime})^{1/2}$. ■

Remark 3 It is also possible to replace β in $\Pi = \alpha\beta'$ in (7.2.3) by the EGLS estimator and estimate the other parameters by LS from the model

$$\Delta Y = \alpha\widehat{\widehat{\beta}}'Y_{-1} + \Gamma\Delta X + \widehat{U}^*.$$

The resulting estimator $[\widehat{\widehat{\alpha}} : \widehat{\widehat{\Gamma}}]$ has the same asymptotic properties as $[\widehat{\alpha} : \widehat{\Gamma}]$ in (7.2.7) which is based on a known β. As a consequence, $[\widehat{\widehat{\alpha}}\widehat{\widehat{\beta}}' : \widehat{\widehat{\Gamma}}]$ also has the same asymptotic properties as $[\widehat{\alpha}\beta' : \widehat{\Gamma}]$. ∎

Remark 4 The EGLS estimator was actually presented in a slightly different form by Ahn & Reinsel (1990) and Saikkonen (1992). These authors use the representation

$$\widehat{\widehat{\beta}}'_{(K-r)} = (\widehat{\alpha}'\widehat{\Sigma}_u^{-1}\widehat{\alpha})^{-1}\widehat{\alpha}'\widehat{\Sigma}_u^{-1}\widehat{\Pi}_2,$$

where $\widehat{\Pi}_2$ is the $(K \times (K-r))$ matrix of the last $K-r$ columns of the LS estimator $\widehat{\Pi}$ of Π (see Reinsel (1993, p. 171) for a discussion of the equivalence of this estimator and the EGLS estimator (7.2.16)). ∎

7.2.3 ML Estimation

If the process y_t is Gaussian or, equivalently, $u_t \sim \mathcal{N}(0, \Sigma_u)$, the VECM (7.2.1) can be estimated by maximum likelihood (ML) taking also the rank restriction for $\Pi = \alpha\beta'$ into account (see Johansen (1988, 1995)). The log-likelihood function for a sample of size T is

$$\ln l = -\frac{KT}{2}\ln 2\pi - \frac{T}{2}\ln|\Sigma_u| \\ -\frac{1}{2}\text{tr}\left[(\Delta Y - \alpha\beta' Y_{-1} - \Gamma\Delta X)'\Sigma_u^{-1}(\Delta Y - \alpha\beta' Y_{-1} - \Gamma\Delta X)\right].$$

(7.2.19)

In the following, we will first discuss the computation of the estimators and then consider their asymptotic properties.

The Estimator

For ML estimation we do not assume that β is normalized. We only make the assumption $\text{rk}(\Pi) = r$ which implies that the matrix can be represented as $\Pi = \alpha\beta'$, where α and β are $(K \times r)$ with $\text{rk}(\alpha) = \text{rk}(\beta) = r$. In the next proposition the ML estimators are given. The proposition generalizes the special case estimators given in (7.1.21) and (7.1.22).

Proposition 7.3 (*ML Estimators of a VECM*)
Let $M := I_T - \Delta X'(\Delta X \Delta X')^{-1}\Delta X$, $R_0 := \Delta Y M$ and $R_1 := Y_{-1}M$, as before, and define

$$S_{ij} := R_i R_j'/T, \quad i = 0, 1,$$

$\lambda_1 \geq \cdots \geq \lambda_K$ are the eigenvalues of $S_{11}^{-1/2} S_{10} S_{00}^{-1} S_{01} S_{11}^{-1/2}$,

and

$v_1, \ldots, v_K$ are the corresponding orthonormal eigenvectors.

The log-likelihood function in (7.2.19) is maximized for

$$\begin{aligned}
\beta &= \widetilde{\beta} := [v_1, \ldots, v_r]' S_{11}^{-1/2}, \\
\alpha &= \widetilde{\alpha} := \Delta Y M Y'_{-p} \widetilde{\beta} \left(\widetilde{\beta}' Y_{-1} M Y'_{-1} \widetilde{\beta} \right)^{-1} = S_{01} \widetilde{\beta} (\widetilde{\beta}' S_{11} \widetilde{\beta})^{-1}, \\
\Gamma &= \widetilde{\Gamma} := (\Delta Y - \widetilde{\alpha} \widetilde{\beta}' Y_{-1}) \Delta X' (\Delta X \Delta X')^{-1}, \\
\Sigma_u &= \widetilde{\Sigma}_u := (\Delta Y - \widetilde{\alpha} \widetilde{\beta}' Y_{-1} - \widetilde{\Gamma} \Delta X)(\Delta Y - \widetilde{\alpha} \widetilde{\beta}' Y_{-1} - \widetilde{\Gamma} \Delta X)'/T.
\end{aligned}$$

The maximum is

$$\max \ln l = -\frac{KT}{2} \ln 2\pi - \frac{T}{2} \left[\ln |S_{00}| + \sum_{i=1}^{r} \ln(1 - \lambda_i) \right] - \frac{KT}{2}. \quad (7.2.20)$$

∎

Proof: From Chapter 3, Section 3.4, it is known that for any fixed α and β the maximum of $\ln l$ is attained for

$$\widetilde{\Gamma}(\alpha \beta') = (\Delta Y - \alpha \beta' Y_{-1}) \Delta X' (\Delta X \Delta X')^{-1}.$$

Thus, we replace Γ in (7.2.19) by $\widetilde{\Gamma}(\alpha \beta')$ and get the concentrated log-likelihood

$$-\frac{KT}{2} \ln 2\pi - \frac{T}{2} \ln |\Sigma_u|$$

$$-\frac{1}{2} \text{tr} \left[(\Delta Y M - \alpha \beta' Y_{-1} M)' \Sigma_u^{-1} (\Delta Y M - \alpha \beta' Y_{-1} M) \right].$$

Hence, we just have to maximize this expression with respect to α, β, and Σ_u. We also know from Chapter 3 that, for given α and β, the maximum is attained if

$$\widetilde{\Sigma}(\alpha \beta') = (\Delta Y M - \alpha \beta' Y_{-1} M)(\Delta Y M - \alpha \beta' Y_{-1} M)'/T$$

is substituted for Σ_u. Consequently, we have to maximize

$$-\frac{T}{2} \ln |(\Delta Y M - \alpha \beta' Y_{-1} M)(\Delta Y M - \alpha \beta' Y_{-1} M)'/T|$$

or, equivalently, minimize the determinant with respect to α and β. Thus, all results of Proposition 7.3 follow from Proposition A.7 of Appendix A.14. ∎

The solutions $\widetilde{\beta}$ and $\widetilde{\alpha}$ of the optimization problem given in the proposition are not unique because, for any nonsingular $(r \times r)$ matrix Q, $\widetilde{\alpha} Q^{-1}$

and $\widetilde{\beta}Q'$ represent another set of ML estimators for α and β. However, the proposition shows that explicit expressions for ML estimators are available. If $r = K$, the proposition still remains valid. Also, ML estimators for the levels VAR representation corresponding to the VECM (7.2.1) observing the rank restriction are readily available via the relations in (7.2.10).

The next question concerns the properties of the ML estimators of a cointegrated system. They are discussed in the following.

Asymptotic Properties of the ML Estimator

The following proposition generalizes Result 7 of Section 7.1.

Proposition 7.4 (*Asymptotic Properties of the ML Estimators of a VECM*) The ML estimators for the VECM (7.2.1) given in Proposition 7.3 have the following asymptotic properties:

$$\sqrt{T}\,\mathrm{vec}([\widetilde{\alpha}\widetilde{\beta}' : \widetilde{\Gamma}] - [\Pi : \Gamma]) \xrightarrow{d} \mathcal{N}(0, \Sigma_{\mathrm{co}}), \qquad (7.2.21)$$

where Σ_{co} is as defined in Proposition 7.1, and

$$\sqrt{T}\,\mathrm{vech}(\widetilde{\Sigma}_u - \Sigma_u) \xrightarrow{d} \mathcal{N}(0, 2\mathbf{D}_K^+(\Sigma_u \otimes \Sigma_u)\mathbf{D}_K^{+\prime}). \qquad (7.2.22)$$

Furthermore, $\widetilde{\Sigma}_u$ is asymptotically independent of $\widetilde{\alpha}\widetilde{\beta}'$ and $\widetilde{\Gamma}$. Here, as usual, $\mathbf{D}_K^+ = (\mathbf{D}_K'\mathbf{D}_K)^{-1}\mathbf{D}_K'$ and $\mathbf{D}_K$ is the $(K^2 \times \frac{1}{2}K(K+1))$ duplication matrix. ∎

Remark 1 It is clear that the ML estimator of $[\Pi : \Gamma]$ must have the same asymptotic distribution as the LS estimator in Proposition 7.1 because the ML estimator with known or given cointegration matrix β also has the same asymptotic distribution. The ML estimator $\widetilde{\alpha}\widetilde{\beta}'$ of Π in Proposition 7.3 may be viewed as a restricted LS estimator which is not as much restricted as the one with known β. Thus, the asymptotic result in (7.2.21) is not surprising. A rigorous proof of the result is given in Johansen (1995). ∎

Remark 2 The covariance matrix Σ_{co} is singular, as noted in Remark 1 for Proposition 7.1. The rank of the $(K^2p \times K^2p)$ matrix Σ_{co} cannot be greater than $K(Kp - K + r)$ which is smaller than K^2p if $r < K$. ∎

Remark 3 Individually, the matrices α and β cannot be estimated consistently without further constraints. Under the assumptions of Proposition 7.4, these matrices are not identified (not unique). If we make specific identifying assumptions in order to obtain unique parameter values and estimators, consistent estimation is possible. For instance, we may use

$$\beta = \begin{bmatrix} I_r \\ \beta_{(K-r)} \end{bmatrix}.$$

The ML estimator of $\beta_{(K-r)}$ may be obtained from the ML estimator of β given in Proposition 7.3 by denoting the first r rows of $\widetilde{\beta}$ by $\widetilde{\beta}_{(r)}$ and letting $\widecheck{\beta}_{(K-r)}$ consist of the last $K-r$ rows of $\widetilde{\beta}\widetilde{\beta}_{(r)}^{-1}$. This ML estimator has the same asymptotic properties as the EGLS estimator in Proposition 7.2 (see Ahn & Reinsel (1990)). In other words, inference procedures based on the ML estimator can be derived from the result

$$\text{vec}\left[(\widecheck{\beta}'_{(K-r)} - \beta'_{(K-r)})\left(R_1^{(2)} R_1^{(2)'}\right)^{1/2}\right] \xrightarrow{d} \mathcal{N}\left(0, I_{K-r} \otimes (\alpha' \Sigma_u^{-1} \alpha)^{-1}\right).$$

It was found in a number of studies that the ML estimator $\widecheck{\beta}_{(K-r)}$ may have some undesirable properties in small samples and, in particular, it may produce occasional outlying estimates which are far away from the true parameter values (e.g., Phillips (1994), Hansen, Kim & Mittnik (1998)). This behavior of the estimator is due to the lack of finite sample moments. Brüggemann & Lütkepohl (2004) compared the EGLS and ML estimators in a small Monte Carlo study and found that the EGLS estimator is more robust in this respect. ■

Remark 4 If β is identified, the corresponding ML estimator of α is asymptotically normal, i.e., $\sqrt{T}\,\text{vec}(\widetilde{\alpha} - \alpha)$ converges to the same asymptotic distribution as in Remark 2 for Proposition 7.1. ■

Remark 5 The normality of the process is not essential for the asymptotic properties of the estimators $\widetilde{\Gamma}$ and $\widetilde{\Pi} = \widetilde{\alpha}\widetilde{\beta}'$. Much of Proposition 7.4 holds under weaker conditions when quasi ML estimators based on the Gaussian likelihood function are considered. We have chosen the normality assumption for convenience. ■

Remark 6 The asymptotic distribution of $\widetilde{\Sigma}_u$ may be different if u_t is not Gaussian. The limiting distribution in (7.2.22) is obtained from the following lemma. ■

Lemma 7.3

$$\text{plim } \sqrt{T}(\widetilde{\Sigma}_u - UU'/T) = 0.$$

■

This lemma not only implies consistency of $\widetilde{\Sigma}_u$ but also shows that the asymptotic distribution of

$$\sqrt{T}\,\text{vech}(\widetilde{\Sigma}_u - \Sigma_u)$$

is the same as that of

$$\sqrt{T}\,\text{vech}(T^{-1}UU' - \Sigma_u).$$

In other words, it is independent of the other coefficients of the system and has the form given in (7.2.22) (see also Section 3.4, Proposition 3.4).

Proof of Lemma 7.3:

$$\begin{aligned}
\widetilde{\Sigma}_u &= T^{-1}(\Delta Y - \widetilde{\alpha}\widetilde{\beta}'Y_{-1} - \widetilde{\Gamma}\Delta X)(\Delta Y - \widetilde{\alpha}\widetilde{\beta}'Y_{-1} - \widetilde{\Gamma}\Delta X)' \\
&= T^{-1}[U + (\Pi - \widetilde{\alpha}\widetilde{\beta}')Y_{-1} + (\Gamma - \widetilde{\Gamma})\Delta X] \\
&\quad \times [U + (\Pi - \widetilde{\alpha}\widetilde{\beta}')Y_{-1} + (\Gamma - \widetilde{\Gamma})\Delta X]' \\
&= \frac{UU'}{T} + (\Pi - \widetilde{\alpha}\widetilde{\beta}')\frac{Y_{-1}U'}{T} + \frac{UY'_{-1}}{T}(\Pi - \widetilde{\alpha}\widetilde{\beta}')' \\
&\quad + (\Pi - \widetilde{\alpha}\widetilde{\beta}')\frac{Y_{-1}Y'_{-1}}{T}(\Pi - \widetilde{\alpha}\widetilde{\beta}')' \\
&\quad + (\Pi - \widetilde{\alpha}\widetilde{\beta}')\frac{Y_{-1}\Delta X'}{T}(\Gamma - \widetilde{\Gamma})' + (\Gamma - \widetilde{\Gamma})\frac{\Delta X Y'_{-1}}{T}(\Pi - \widetilde{\alpha}\widetilde{\beta}')' \\
&\quad + (\Gamma - \widetilde{\Gamma})\frac{\Delta X U'}{T} + \frac{U \Delta X'}{T}(\Gamma - \widetilde{\Gamma})' \\
&\quad + (\Gamma - \widetilde{\Gamma})\frac{\Delta X \Delta X'}{T}(\Gamma - \widetilde{\Gamma})'.
\end{aligned}$$

Using $\widetilde{\alpha}\widetilde{\beta}' - \Pi = O_p(T^{-1/2})$, $\widetilde{\Gamma} - \Gamma = O_p(T^{-1/2})$ and the results in Lemma 7.2, we get

$$\sqrt{T}(\Gamma - \widetilde{\Gamma})\frac{\Delta X U'}{T} = o_p(1),$$

$$\sqrt{T}(\Gamma - \widetilde{\Gamma})\frac{\Delta X \Delta X'}{T}(\Gamma - \widetilde{\Gamma})' = o_p(1),$$

$$\sqrt{T}(\Pi - \widetilde{\alpha}\widetilde{\beta}')\frac{Y_{-1}\Delta X'}{T}(\Gamma - \widetilde{\Gamma})' = o_p(1),$$

and

$$\sqrt{T}(\Pi - \widetilde{\alpha}\widetilde{\beta}')\frac{Y_{-1}Y'_{-1}}{T}(\Pi - \widetilde{\alpha}\widetilde{\beta}')' = o_p(1).$$

Thus, Lemma 7.3 is proven if we can show that

$$\sqrt{T}(\widetilde{\alpha}\widetilde{\beta}' - \Pi)\frac{Y_{-1}U'}{T} = o_p(1). \tag{7.2.23}$$

To prove this result, we define $\widetilde{\alpha}(\beta)$ to be the ML estimator of α given β and note that

$$\begin{aligned}
\sqrt{T}(\widetilde{\alpha}\widetilde{\beta}' - \Pi)\frac{Y_{-1}U'}{T} &= \sqrt{T}[\widetilde{\alpha}\widetilde{\beta}' - \widetilde{\alpha}(\beta)\beta']\frac{Y_{-1}U'}{T} \\
&\quad + \sqrt{T}[\widetilde{\alpha}(\beta) - \alpha]\frac{\beta'Y_{-1}U'}{T}.
\end{aligned}$$

This quantity converges to zero in probability by Lemma 7.2(4), the fact that $\sqrt{T}[\widetilde{\alpha}(\beta) - \alpha] = O_p(1)$ (see (7.2.8)) and because $\sqrt{T}[\widetilde{\alpha}\widetilde{\beta}' - \widetilde{\alpha}(\beta)\beta'] = o_p(1)$. We leave the latter result as an exercise (see Problem 7.3). ∎

7.2.4 Including Deterministic Terms

So far we have assumed that there are no deterministic terms in the data generation process, to simplify the exposition. In practice, such terms are typically needed for a proper representation of the data generation process. It turns out, however, that they can be easily accommodated in the estimation procedures for VECMs discussed so far, if the setup of Section 6.4 is used. Suppose the observed process y_t can be represented as

$$y_t = \mu_t + x_t, \tag{7.2.24}$$

where x_t is a zero mean process with VECM representation as in (7.2.1) and μ_t stands for the deterministic term. In general, the latter term may consist of polynomial trends, seasonal and other dummy variables as well as constant means. As in Section 6.4, we can then set up the VECM for the observed y_t variables as

$$\begin{aligned}\Delta y_t &= \alpha[\beta' : \eta']\begin{bmatrix} y_{t-1} \\ D_{t-1}^{co} \end{bmatrix} + \Gamma_1\Delta y_{t-1} + \cdots + \Gamma_{p-1}\Delta y_{t-p+1} + CD_t + u_t \\ &= \Pi^+ y_{t-1}^+ + \Gamma_1\Delta y_{t-1} + \cdots + \Gamma_{p-1}\Delta y_{t-p+1} + CD_t + u_t, \quad (7.2.25)\end{aligned}$$

where D_t^{co} contains all the deterministic terms which are present in the cointegration relations, D_t contains all remaining deterministics, and η' and C are the corresponding parameter matrices. Moreover, $\Pi^+ := \alpha[\beta' : \eta'] = \alpha\beta^{+\prime}$ and

$$y_t^+ := \begin{bmatrix} y_t \\ D_t^{co} \end{bmatrix}.$$

Notice that we assume that a specific deterministic term appears only once, either in D_t^{co} or in D_t.

Now we can simply modify the matrices used for representing the estimators in the previous subsections and then use basically the same formulas as before for computing the estimators. For example, defining

$$Y_{-1}^+ := [y_0^+, \ldots, y_{T-1}^+],$$

$$\Gamma^+ := [\Gamma_1, \ldots, \Gamma_{p-1}, C],$$

and

$$\Delta X^+ := [\Delta X_0^+, \ldots, \Delta X_{T-1}^+] \quad \text{with} \quad \Delta X_{t-1}^+ := \begin{bmatrix} \Delta y_{t-1} \\ \vdots \\ \Delta y_{t-p+1} \\ D_t \end{bmatrix}$$

gives the LS estimator

$$[\widehat{\boldsymbol{\Pi}}^+ : \widehat{\boldsymbol{\Gamma}}^+] = [\Delta Y Y_{-1}^{+\prime} : \Delta Y \Delta X^{+\prime}] \begin{bmatrix} Y_{-1}^+ Y_{-1}^{+\prime} & Y_{-1}^+ \Delta X^{+\prime} \\ \Delta X^+ Y_{-1}^{+\prime} & \Delta X^+ \Delta X^{+\prime} \end{bmatrix}^{-1}.$$

The EGLS or ML estimators may be obtained analogously.

Hence, the computation of the estimators is equally easy as in the case without deterministic terms. Also, the asymptotic properties of the parameter estimators are essentially unchanged. The asymptotic theory for the deterministic terms requires some care, however, because their convergence rates depend on the specific terms included. For instance, if linear trends are included, the convergence rates of the associated slope parameters are different from $\sqrt{T}$. Generally, if the VECM is specified properly, including the cointegrating rank r, and if EGLS or ML methods are used, the usual inference methods are available. In particular, likelihood ratio tests for parameter restrictions related to the deterministic terms permit standard χ^2 asymptotics (see, e.g., Johansen (1995)).

A question of interest in this context is, for example, whether a particular deterministic term can indeed be constrained to the cointegration relations or needs to be maintained in unrestricted form in the model. The i-th component of D_t can be absorbed in the error correction term if the i-th column of the coefficient matrix C, denoted by C_i, satisfies $C_i = \alpha \eta_i$ for some r-dimensional vector η_i. Thus, the relevant null hypothesis is

$$\alpha'_\perp C_i = 0.$$

In other words, there are $K-r$ restrictions for each component that is confined to the cointegration relations. They are easy to test by a likelihood ratio test because the ML estimators and, hence, the likelihood maxima are easy to obtain for both the restricted and unrestricted model by just specifying the terms in D_t^{co} and D_t accordingly. If m deterministic components are restricted to the cointegration relations, the LR statistic has an asymptotic $\chi^2(m(K-r))$-distribution under our usual assumptions.

7.2.5 Other Estimation Methods for Cointegrated Systems

Some other estimation methods for cointegration relations and VECMs have been proposed in the literature. For example, other systems methods for estimating the cointegrating parameters were considered by Phillips (1991) who discussed nonparametric estimation of the short-run parameters. Stock & Watson (1988) proposed an estimator based on principal components and Bossaerts (1988) used canonical correlations. The latter two estimators were shown to be inferior to the ML estimators in a small sample comparison by Gonzalo (1994) and are therefore not considered here.

If there is just a single cointegration relation, it may also be estimated by single equation LS. Suppose that β is normalized as in (7.1.10) such that $\beta = (1, \beta_2, \ldots, \beta_K)'$ and $\beta' y_t = y_{1t} + \beta_2 y_{2t} + \cdots + \beta_K y_{Kt}$. Hence,

$$y_{1t} = \gamma_2 y_{2t} + \cdots + \gamma_K y_{Kt} + ec_t,$$

where $\gamma_i := -\beta_i$ and ec_t is a stable, stationary process. Defining

$$y_{(1)} := \begin{bmatrix} y_{11} \\ \vdots \\ y_{1T} \end{bmatrix} \quad \text{and} \quad Y_{(2)} := \begin{bmatrix} y_{21} & \cdots & y_{K1} \\ \vdots & & \vdots \\ y_{2T} & \cdots & y_{KT} \end{bmatrix},$$

the LS estimator for $\gamma' := (\gamma_2, \ldots, \gamma_K)$ is

$$\widehat{\gamma}' = y_{(1)}' Y_{(2)} (Y_{(2)}' Y_{(2)})^{-1}.$$

Stock (1987) showed that $\widehat{\gamma}$ is superconsistent and, more precisely, $T(\widehat{\gamma} - \gamma)$ converges in distribution. Thus, $\widehat{\gamma} - \gamma = O_p(T^{-1})$. However, there is some evidence that $\widehat{\gamma}$ is biased in small samples (Phillips & Hansen (1990)). Therefore, using LS estimation of the cointegration parameters without any correction for further dynamics in the model is not recommended.

A large number of single equation estimators for cointegration relations were reviewed and compared by Caporale & Pittis (2004). In addition to the simple LS estimator presented in the foregoing, they also considered estimators which are corrected for short-run dynamics. For example, this may be accomplished by including leads and lags of the differenced regressor variables in the estimation equation (e.g., Stock & Watson (1993)) or by adding also lagged differences of the dependent variable (e.g., Banerjee, Dolado, Galbraith & Hendry (1993), Wickens & Breusch (1988)). Another possible choice in this context is the fully modified estimator of Phillips & Hansen (1990) which takes care of the short-run dynamics nonparametrically and a semiparametric variant of this estimator proposed by Inder (1993). In addition, Caporale & Pittis (2004) presented a large number of modifications. Some of these estimators have rather undesirable small sample properties compared to the systems ML estimator presented in Section 7.2.3. Even those modifications that lead to small sample improvements were only shown to work in a rather limited framework. Also, of course, some of these estimators are only designed for situations where only one cointegration relation exists.

Two-Stage Estimation

Generally, if a superconsistent estimator $\widehat{\beta}$ of the cointegration matrix β is available, this estimator may be substituted for the true β and all the other parameters may be estimated in a second stage from

$$\Delta y_t = \alpha \widehat{\beta}' y_{t-1} + \Gamma_1 \Delta y_{t-1} + \cdots + \Gamma_{p-1} \Delta y_{t-p+1} + u_t^*, \qquad (7.2.26)$$

where deterministic terms are again ignored for simplicity. If no restrictions are imposed on α and the Γ_i's ($i = 1, \ldots, p-1$), LS estimation can be used without loss of asymptotic efficiency. Denoting the two-stage estimators of α and Γ by $\widehat{\alpha}_{2s}$ and $\widehat{\Gamma}_{2s}$, respectively, we have

$$\widehat{\alpha}_{2s} = \Delta Y M Y'_{-1}\widehat{\beta}\left(\widehat{\beta}'Y_{-1}MY'_{-1}\widehat{\beta}\right)^{-1} \tag{7.2.27}$$

and

$$\widehat{\Gamma}_{2s} = (\Delta Y - \widehat{\alpha}_{2s}\widehat{\beta}'Y_{-1})\Delta X'(\Delta X \Delta X')^{-1}, \tag{7.2.28}$$

where the notation from the previous subsections has been used. For these estimators the following proposition holds, which is stated without proof.

Proposition 7.5 (*Asymptotic Properties of the Two-Stage LS Estimator*)
Let y_t be a K-dimensional, cointegrated process with VECM representation (7.2.1). Then the two-stage estimator is consistent and

$$\sqrt{T}\operatorname{vec}([\widehat{\alpha}_{2s} : \widehat{\Gamma}_{2s}] - [\alpha : \Gamma]) \xrightarrow{d} \mathcal{N}(0, \Sigma_{\alpha,\Gamma}), \tag{7.2.29}$$

where $\Sigma_{\alpha,\Gamma}$ is the same covariance matrix as in (7.2.8). ∎

The proposition implies that if a superconsistent estimator of the cointegration matrix β is available, the loading coefficients and short-run parameters of the VECM can be estimated by LS and these estimators have the same asymptotic properties we would obtain by using the true β. Thus, standard inference procedures can be used for the short-run parameters. An analogous result is also available for VECMs with parameter restrictions (see Section 7.3 for the extension).

The second stage in the procedure may be modified. For instance, one may just be interested in the first equation of the system. In this case, the first equation may be estimated separately without taking into account the remaining ones. Thus, the two-stage procedure may be applied in a single equation modelling context.

Results similar to those in Proposition 7.5 were derived by many authors (see, e.g., Stock (1987), Phillips & Durlauf (1986), Park & Phillips (1989), and Johansen (1991)). Generally there has been a considerable amount of research on estimation and hypothesis testing in systems with integrated and cointegrated variables. For instance, Johansen (1991), Johansen & Juselius (1990), and Lütkepohl & Reimers (1992b) considered estimation with restrictions on the cointegration and loading matrices; Park & Phillips (1988, 1989) and Phillips (1988) provided general results on estimating systems with integrated and cointegrated exogenous variables; Stock (1987) considered a so-called nonlinear LS estimator, and Phillips & Hansen (1990) discussed instrumental variables estimation of models containing integrated variables.

7.2.6 An Example

As an example, we use the bivariate system of quarterly, seasonally unadjusted German long-term interest rate ($R_t = y_{1t}$) and inflation rate ($Dp_t = y_{2t}$)

which was also analyzed in Lütkepohl (2004). The sample period is the second quarter of 1972 to the end of 1998. Thus we have $T = 107$ observations. The data are available in File E6 and the two time series are plotted in Figure 7.1. Preliminary tests indicated that both series have a unit root and there are also theoretical reasons for a cointegration relation between them. The so-called Fisher effect implies that the real interest rate is stationary. Because R_t is a nominal yearly interest rate while Dp_t is a quarterly inflation rate, one would therefore expect $R_t - 4Dp_t$ to be stationary, that is, this relation is expected to be a cointegration relation.

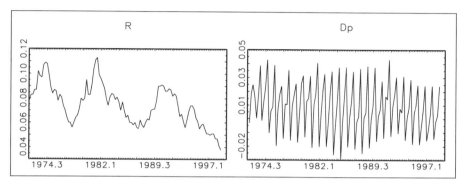

Fig. 7.1. Seasonally unadjusted, quarterly German interest rate (left) and inflation rate (right), 1972.2–1998.4.

We have fitted a VECM with a constant, seasonal dummy variables, and three lagged differences and the pre-specified cointegration relation $R_t - 4Dp_t$ to the data. The results are shown in Table 7.1. Notice that three lagged differences in the VECM imply a model with four lags in the levels. Including at least lags of one year seems plausible because the inflation series has a strong seasonal pattern (see Figure 7.1). Formal statistical procedures for determining the lag length will be discussed in the next chapter. The seasonal movement in Dp_t is also the reason for including seasonal dummy variables in addition to a constant. The deterministic term, $D_t = (1, s_{1t}, s_{2t}, s_{3t})'$, where the s_{it} are seasonal dummy variables, is placed outside the cointegration relation. We have also estimated a VECM with cointegrating rank $r = 1$ using the reduced rank ML procedure and the EGLS method. The estimates are also shown in Table 7.1.

The two estimated cointegration relations are

$$R_t - \underset{(0.63)}{3.96} \; Dp_t = ec_t^{ML} \tag{7.2.30}$$

and

$$R_t - \underset{(0.61)}{3.63} \; Dp_t = ec_t^{EGLS}, \tag{7.2.31}$$

Table 7.1. Estimated VECMs for interest rate/inflation example system

	known β	ML estimator	EGLS estimator
α	$\begin{bmatrix} -0.10 \\ (-2.3) \\ 0.16 \\ (3.8) \end{bmatrix}$	$\begin{bmatrix} -0.10 \\ (-2.3) \\ 0.16 \\ (3.8) \end{bmatrix}$	$\begin{bmatrix} -0.14 \\ (-2.8) \\ 0.14 \\ (2.9) \end{bmatrix}$
β'	$[1:-4]$	$\begin{bmatrix} 1.00 : & -3.96 \\ & (-6.3) \end{bmatrix}$	$\begin{bmatrix} 1.00 : & -3.63 \\ & (-6.0) \end{bmatrix}$
Γ_1	$\begin{bmatrix} 0.27 & -0.21 \\ (2.7) & (-1.4) \\ 0.07 & -0.34 \\ (0.7) & (-2.4) \end{bmatrix}$	$\begin{bmatrix} 0.27 & -0.21 \\ (2.7) & (-1.4) \\ 0.07 & -0.34 \\ (0.7) & (-2.4) \end{bmatrix}$	$\begin{bmatrix} 0.29 & -0.16 \\ (2.9) & (-1.1) \\ 0.08 & -0.31 \\ (0.8) & (-2.2) \end{bmatrix}$
Γ_2	$\begin{bmatrix} -0.02 & -0.22 \\ (-0.2) & (-1.8) \\ -0.00 & -0.39 \\ (-0.0) & (-3.4) \end{bmatrix}$	$\begin{bmatrix} -0.02 & -0.22 \\ (-0.2) & (-1.8) \\ -0.00 & -0.39 \\ (-0.0) & (-3.4) \end{bmatrix}$	$\begin{bmatrix} 0.01 & -0.19 \\ (0.1) & (-1.6) \\ 0.01 & -0.37 \\ (0.1) & (-3.2) \end{bmatrix}$
Γ_3	$\begin{bmatrix} 0.22 & -0.11 \\ (2.3) & (-1.3) \\ 0.02 & -0.35 \\ (0.2) & (-4.5) \end{bmatrix}$	$\begin{bmatrix} 0.22 & -0.11 \\ (2.3) & (-1.3) \\ 0.02 & -0.35 \\ (0.2) & (-4.5) \end{bmatrix}$	$\begin{bmatrix} 0.26 & -0.09 \\ (2.6) & (-1.1) \\ 0.04 & -0.34 \\ (0.4) & (-4.4) \end{bmatrix}$
C'	$\begin{bmatrix} 0.001 & 0.010 \\ (0.4) & (3.0) \\ 0.001 & -0.034 \\ (0.3) & (-7.5) \\ 0.009 & -0.018 \\ (1.8) & (-3.8) \\ -0.000 & -0.016 \\ (-0.1) & (-3.6) \end{bmatrix}$	$\begin{bmatrix} 0.002 & 0.010 \\ (0.4) & (3.0) \\ 0.001 & -0.034 \\ (0.3) & (-7.5) \\ 0.009 & -0.018 \\ (1.8) & (-3.8) \\ -0.000 & -0.016 \\ (-0.1) & (-3.6) \end{bmatrix}$	$\begin{bmatrix} 0.005 & 0.012 \\ (1.2) & (3.1) \\ 0.001 & -0.034 \\ (0.3) & (-7.5) \\ 0.009 & -0.018 \\ (1.8) & (-3.8) \\ -0.000 & -0.016 \\ (-0.1) & (-3.6) \end{bmatrix}$

Note: t-values in parentheses underneath parameter estimates; deterministic terms: constant and seasonal dummies ($D_t = (1, s_{1t}, s_{2t}, s_{3t})'$).

where estimated standard errors are given in parentheses. The first coefficient is normalized to be 1. Thereby the t-ratios and the standard errors of the inflation coefficient can be interpreted in the usual way. Clearly, -4 is well within a two-standard error interval around both estimates. Therefore one could argue that restricting the inflation coefficient to 4 is in line with the data. Using the result in Proposition 7.2, a formal test of the null hypothesis $H_0: \beta_2 = -4$, where β_2 denotes the second component of β, can be based on the t-statistic

$$\frac{-3.96 - (-4)}{0.63} = 0.06$$

for the ML estimator or on

$$\frac{-3.63 - (-4)}{0.61} = 0.61$$

for the EGLS estimator. Both t-values are small compared to critical values from the standard normal distribution corresponding to usual significance levels. Hence, the null hypothesis cannot be rejected for either of the two estimators.

Comparing the other estimates of the three models in Table 7.1, it is obvious that corresponding estimates do not differ much, especially when the sampling uncertainty reflected in the t-ratios is taken into account. In particular, the ML estimates are very close to those of the model with fixed cointegration vector. Thus, imposing the theoretically expected cointegration vector does not appear to be a problematic constraint.

Another observation that can be made in Table 7.1 is that there are some insignificant coefficients in the short-run matrices Γ_i and the estimated deterministic terms (C). Because some of the parameters in $\widehat{\Gamma}_3$ have rather large t-ratios, it is clear that simply reducing the lag order is not likely to be a good strategy for reducing the number of parameters in the model. It makes sense, however, to consider restricting some of the parameter values to zero. This issue is discussed in the next section.

7.3 Estimating VECMs with Parameter Restrictions

As for other models, restrictions may be imposed on the parameters of VECMs to increase the estimation precision. We will first discuss restrictions for the cointegration relations and then turn to restrictions on the loading coefficients and short-run parameters.

7.3.1 Linear Restrictions for the Cointegration Matrix

In case just-identifying restrictions for the cointegration relations are available, estimation may proceed as described in Section 7.2 and then the identified estimator of β may be obtained by a suitable transformation of the estimator $\widehat{\beta}$. For example, if β is just a single vector and ML estimation is used, a normalization of the first component may be obtained by dividing the vector $\widetilde{\beta}$ by its first component, as discussed earlier.

Sometimes over-identifying restrictions are available for the cointegration matrix. In general, if the restrictions can be expressed in the form

$$\text{vec}(\beta'_{(K-r)}) = \mathbf{R}\gamma + \mathbf{r}, \tag{7.3.1}$$

where $\mathbf{R}$ is a fixed $(r(K-r) \times m)$ matrix of rank m, $\mathbf{r}$ is a fixed $r(K-r)$-dimensional vector, and γ is a vector of free parameters, the EGLS estimator is still available. The GLS estimator may be obtained from the vectorized "concentrated model" (7.2.14),

$$\begin{aligned}\text{vec}(R_0 - \alpha R_1^{(1)}) &= (R_1^{(2)\prime} \otimes \alpha)\text{vec}(\beta'_{(K-r)}) + \text{vec}(U^*) \\ &= (R_1^{(2)\prime} \otimes \alpha)(\mathbf{R}\gamma + \mathbf{r}) + \text{vec}(U^*),\end{aligned}$$

so that

$$\text{vec}(R_0 - \alpha R_1^{(1)}) - (R_1^{(2)\prime} \otimes \alpha)\mathbf{r} = (R_1^{(2)\prime} \otimes \alpha)\mathbf{R}\gamma + \text{vec}(U^*). \quad (7.3.2)$$

Thus, the GLS estimator for γ is

$$\begin{aligned}
\widehat{\gamma} &= \left[\mathbf{R}'(R_1^{(2)} R_1^{(2)\prime} \otimes \alpha' \Sigma_u^{-1} \alpha)\mathbf{R}\right]^{-1} \\
&\quad \times \mathbf{R}'(R_1^{(2)} \otimes \alpha' \Sigma_u^{-1}) \left[\text{vec}(R_0 - \alpha R_1^{(1)}) - (R_1^{(2)\prime} \otimes \alpha)\mathbf{r}\right].
\end{aligned}$$

Substituting consistent estimators $\widehat{\alpha}$ and $\widehat{\Sigma}_u$ for α and Σ_u, respectively, gives the EGLS estimator

$$\begin{aligned}
\widehat{\widehat{\gamma}} &= \left[\mathbf{R}'(R_1^{(2)} R_1^{(2)\prime} \otimes \widehat{\alpha}' \widehat{\Sigma}_u^{-1} \widehat{\alpha})\mathbf{R}\right]^{-1} \\
&\quad \times \mathbf{R}'(R_1^{(2)} \otimes \widehat{\alpha}' \widehat{\Sigma}_u^{-1}) \left[\text{vec}(R_0 - \widehat{\alpha} R_1^{(1)}) - (R_1^{(2)\prime} \otimes \widehat{\alpha})\mathbf{r}\right].
\end{aligned} \quad (7.3.3)$$

Extending the arguments used for proving Proposition 7.2, the following asymptotic properties of the EGLS estimator can be shown.

Proposition 7.6 (*Asymptotic Properties of the Restricted EGLS Estimator*) Suppose y_t is generated by the VECM (7.2.1) and β satisfies the restrictions in (7.3.1). Then

$$\left[\mathbf{R}'(R_1^{(2)} R_1^{(2)\prime} \otimes \widehat{\alpha}' \widehat{\Sigma}_u^{-1} \widehat{\alpha})\mathbf{R}\right]^{1/2} (\widehat{\widehat{\gamma}} - \gamma) \xrightarrow{d} \mathcal{N}(0, I_m). \quad (7.3.4)$$

∎

Thus, standard inference procedures can be based on the transformed estimator. It can also be shown that $\widehat{\widehat{\gamma}} - \gamma = O_p(T^{-1})$. In other words, the estimator is superconsistent. Clearly, consistent estimators of α and Σ_u are readily available from unrestricted LS estimation as in Section 7.2.2.

Defining $\widehat{\widehat{\beta}}_{(K-r)}^R$ such that $\text{vec}\,\widehat{\widehat{\beta}}_{(K-r)}^R = \mathbf{R}\widehat{\widehat{\gamma}} + \mathbf{r}$,

$$\widehat{\widehat{\beta}}^R := \begin{bmatrix} I_r \\ \widehat{\widehat{\beta}}_{(K-r)}^R \end{bmatrix}$$

is a restricted estimator of the cointegration matrix. It can, for example, be used in the two-stage procedure described in Section 7.2.5.

If the restrictions for the cointegration matrix can be written in the form $\beta = H\varphi$, where H is some known, fixed $(K \times s)$ matrix and φ is $(s \times r)$ with $s \geq r$, ML estimation is also straightforward. For example, in a system with three variables and one cointegration relation, if $\beta_{31} = -\beta_{21}$, we have

$$\beta = \begin{bmatrix} \beta_{11} \\ \beta_{21} \\ -\beta_{21} \end{bmatrix} = \begin{bmatrix} 1 & 0 \\ 0 & 1 \\ 0 & -1 \end{bmatrix} \begin{bmatrix} \beta_{11} \\ \beta_{21} \end{bmatrix} = H\varphi,$$

where $\varphi := (\beta_{11}, \beta_{21})'$ and H is defined in the obvious way. If the restrictions can be represented in this form, Y_{-1} is simply replaced by $H'Y_{-1}$ in the quantities entering the eigenvalue problem in Proposition 7.3. Denoting the resulting estimator by $\widetilde{\varphi}$ gives a restricted estimator $\widetilde{\beta} = H\widetilde{\varphi}$ for β and corresponding estimators of α and Γ as in Proposition 7.3. If the restrictions in (7.3.1) can be written in this form, the EGLS and the ML estimators have again identical asymptotic properties.

However, the restrictions in (7.3.1) can in general not be written in the form $\beta = H\varphi$. For instance, if there are three variables ($K = 3$) and two cointegrating relations ($r = 2$), a single zero restriction on the second cointegration vector cannot be expressed in the form $\beta = H\varphi$, whereas it may still be written in the form (7.3.1). Moreover, it may be expressed in the form $\beta = [H_1\varphi_1, H_2\varphi_2]$ with suitable matrices H_1 and H_2 and vectors φ_1 and φ_2. For example, if a zero restriction is placed on the last element of the second cointegrating vector, we get

$$\beta = \begin{bmatrix} \beta_{11} & \beta_{12} \\ \beta_{21} & \beta_{22} \\ \beta_{31} & 0 \end{bmatrix} = [H_1\varphi_1, H_2\varphi_2]$$

with $H_1 := I_3$, $\varphi_1 := (\beta_{11}, \beta_{21}, \beta_{31})'$,

$$H_2 := \begin{bmatrix} 1 & 0 \\ 0 & 1 \\ 0 & 0 \end{bmatrix}$$

and $\varphi_2 := (\beta_{12}, \beta_{22})'$. In that case, restricted ML estimation is still not difficult but requires an iterative optimization (see Boswijk & Doornik (2002)).

7.3.2 Linear Restrictions for the Short-Run and Loading Parameters

If a superconsistent estimator of the cointegration matrix $\widehat{\beta}$ is available, the two-stage procedure described in Section 7.2.5 can be used for estimating the loading and short-run parameters of a VECM. The method can be readily extended to models with parameter restrictions. Suppose linear restrictions of the form

$$\text{vec}[\alpha : \Gamma] = \Re\varphi, \tag{7.3.5}$$

where $\Re$ is a fixed $(K(r + K(p-1)) \times n)$ matrix and φ is an n-dimensional vector. Then we can write the model in matrix form as

$$\Delta Y = [\alpha : \Gamma]\begin{bmatrix} \widehat{\beta}'Y_{-1} \\ \Delta X \end{bmatrix} + U^*$$

and in vectorized form we get

$$\text{vec}(\Delta Y) = \left([Y'_{-1}\widehat{\beta} : \Delta X'] \otimes I_K\right) \text{vec}[\alpha : \Gamma] + \text{vec}(U^*)$$
$$= \left([Y'_{-1}\widehat{\beta} : \Delta X'] \otimes I_K\right) \Re\varphi + \text{vec}(U^*).$$

Hence, the GLS estimator of φ is

$$\widehat{\varphi} = \left[\Re'\left(\begin{bmatrix} \widehat{\beta}'Y_{-1}Y'_{-1}\widehat{\beta} & \widehat{\beta}'Y_{-1}\Delta X' \\ \Delta XY'_{-1}\widehat{\beta} & \Delta X\Delta X' \end{bmatrix} \otimes \Sigma_u^{-1}\right)\Re\right]^{-1}$$
$$\times \Re'\left(\begin{bmatrix} \widehat{\beta}'Y_{-1} \\ \Delta X \end{bmatrix} \otimes \Sigma_u^{-1}\right)\text{vec}(\Delta Y), \qquad (7.3.6)$$

from which an EGLS estimator $\widehat{\widehat{\varphi}}$ is obtained by replacing the residual covariance matrix Σ_u by a consistent estimator. The latter estimator may, for example, be obtained from an unrestricted estimation of the model. The resulting EGLS estimator has the following asymptotic properties.

Proposition 7.7 (*Asymptotic Properties of the Restricted EGLS Estimator of the Short-Run Parameters*)
Suppose y_t is generated by the VECM (7.2.1), $\widehat{\beta}$ is a superconsistent estimator of β, $\widehat{\Sigma}_u$ is a consistent estimator of Σ_u, and the short-run and loading parameters satisfy (7.3.5). Then

$$\sqrt{T}(\widehat{\widehat{\varphi}} - \varphi)$$
$$\xrightarrow{d} \mathcal{N}\left(0, \text{plim } T\left[\Re'\left(\begin{bmatrix} \widehat{\beta}'Y_{-1}Y'_{-1}\widehat{\beta} & \widehat{\beta}'Y_{-1}\Delta X' \\ \Delta XY'_{-1}\widehat{\beta} & \Delta X\Delta X' \end{bmatrix} \otimes \Sigma_u^{-1}\right)\Re\right]^{-1}\right).$$
$$(7.3.7)$$

∎

We do not prove the proposition but just note that it follows from the fact that only stationary variables are involved if $\widehat{\beta}$ is replaced by the true cointegration matrix β and the resulting estimator for φ differs from $\widehat{\widehat{\varphi}}$ by a quantity which is $o_p(T^{-1/2})$. Moreover, the asymptotic normal distribution of $\text{vec}[\widehat{\widehat{\alpha}} : \widehat{\widehat{\Gamma}}] = \Re\widehat{\widehat{\varphi}}$ follows in the usual way.

It is straightforward to extend these result to the case where the restrictions are of the form

$$\text{vec}[\alpha : \Gamma] = \Re\varphi + \mathbf{r}, \qquad (7.3.8)$$

where $\mathbf{r}$ is now a fixed $(K(r + K(p-1)) \times 1$ vector (see Problem 7.6). The more special restrictions in (7.3.5) are considered here for convenience and because they cover most cases of practical importance.

7.3.3 An Example

In Section 7.2.6, we have seen that in the short-run dynamics of the German interest rate/inflation example models a number of coefficients have quite low t-ratios (see Table 7.1). Therefore it makes sense to restrict some of the coefficients to zero. The following model from Lütkepohl (2004, Equation (3.41)) for our data set is an example of a restricted (subset) VECM:

$$\begin{bmatrix} \Delta R_t \\ \Delta Dp_t \end{bmatrix} = \begin{bmatrix} -0.07 \\ \scriptstyle(-3.1) \\ 0.17 \\ \scriptstyle(4.5) \end{bmatrix} (R_{t-1} - 4Dp_{t-1})$$

$$+ \begin{bmatrix} 0.24 & -0.08 \\ \scriptstyle(2.5) & \scriptstyle(-1.9) \\ 0 & -0.31 \\ & \scriptstyle(-2.5) \end{bmatrix} \begin{bmatrix} \Delta R_{t-1} \\ \Delta Dp_{t-1} \end{bmatrix} + \begin{bmatrix} 0 & -0.13 \\ & \scriptstyle(-2.5) \\ 0 & -0.37 \\ & \scriptstyle(-3.6) \end{bmatrix} \begin{bmatrix} \Delta R_{t-2} \\ \Delta Dp_{t-2} \end{bmatrix}$$

$$+ \begin{bmatrix} 0.20 & -0.06 \\ \scriptstyle(2.1) & \scriptstyle(-1.6) \\ 0 & -0.34 \\ & \scriptstyle(-4.7) \end{bmatrix} \begin{bmatrix} \Delta R_{t-3} \\ \Delta Dp_{t-3} \end{bmatrix}$$

$$+ \begin{bmatrix} 0 & 0 & 0.010 & 0 \\ & & \scriptstyle(2.8) & \\ 0.010 & -0.034 & -0.018 & -0.016 \\ \scriptstyle(3.0) & \scriptstyle(-7.6) & \scriptstyle(-3.8) & \scriptstyle(-3.6) \end{bmatrix} \begin{bmatrix} c \\ s_{1,t} \\ s_{2,t} \\ s_{3,t} \end{bmatrix} + \begin{bmatrix} \widehat{u}_{1,t} \\ \widehat{u}_{2,t} \end{bmatrix},$$

$$\tag{7.3.9}$$

$$\widetilde{\Sigma}_u = \begin{bmatrix} 2.61 & -0.15 \\ -0.15 & 2.31 \end{bmatrix} \times 10^{-5}.$$

Here we have used the fixed cointegration vector that was found in Section 7.2.6 and EGLS estimation of the loading coefficients and short-term parameters is used. t-ratios are again given in parentheses underneath the parameter estimates. They are all relatively large. In fact, with two exceptions they are all larger than two. Recall that t-ratios can be interpreted in the usual way as asymptotically standard normally distributed by Proposition 7.7. Comparing the model (7.3.9) to those in Table 7.1, it turns out that the parameters with very small t-ratios in the unrestricted models are just the ones restricted to zero in (7.3.9). The model was actually found by a sequential model selection procedure which will be discussed in the next chapter.

7.4 Bayesian Estimation of Integrated Systems

It is also possible to place Bayesian restrictions on VECMs. A very important constraint in these models is the cointegrating rank, however. In Bayesian analysis, a basic idea is to allow the data to revise the prior restrictions imposed by the analyst. Using this principle also for the unit roots and, hence, for the cointegration relations, setting up the system in VECM form may not be the most plausible approach anymore. Therefore, Bayesian restrictions have often been imposed on the levels VAR form, even if the variables are possibly integrated. A popular prior in this context is the Minnesota or Litterman prior which ignores possible cointegration between the variables altogether. We will present this prior in the following after the general setting has been discussed.

7.4.1 The Model Setup

In Chapter 5, Section 5.4, we have discussed Bayesian estimation of stationary, stable VAR(p) processes. For a Gaussian process with integrated variables and a normal prior, the posterior distribution of the VAR coefficients can be derived in a similar manner. We now consider a levels VAR(p) model of the form

$$y_t = \nu + A_1 y_{t-1} + \cdots + A_p y_{t-p} + u_t.$$

As usual, $\boldsymbol{\beta} := \text{vec}[\nu, A_1, \ldots, A_p]$ is the vector of VAR coefficients including an intercept vector and we assume a prior

$$\boldsymbol{\beta} \sim \mathcal{N}(\boldsymbol{\beta}^*, V_{\boldsymbol{\beta}}). \tag{7.4.1}$$

Then, using the same line of reasoning as in Section 5.4, the posterior mean is

$$\bar{\boldsymbol{\beta}} = [V_{\boldsymbol{\beta}}^{-1} + (ZZ' \otimes \Sigma_u^{-1})]^{-1}[V_{\boldsymbol{\beta}}^{-1}\boldsymbol{\beta}^* + (Z \otimes \Sigma_u^{-1})\mathbf{y}]$$

and the posterior covariance matrix is

$$\bar{\Sigma}_{\boldsymbol{\beta}} = [V_{\boldsymbol{\beta}}^{-1} + (ZZ' \otimes \Sigma_u^{-1})]^{-1},$$

where

$$\mathbf{y} := \text{vec}[y_1, \ldots, y_T] \quad \text{and} \quad Z := [Z_0, \ldots, Z_{T-1}] \quad \text{with } Z_t := \begin{bmatrix} 1 \\ y_t \\ \vdots \\ y_{t-p+1} \end{bmatrix}.$$

7.4.2 The Minnesota or Litterman Prior

A possible choice of $\boldsymbol{\beta}^*$ and $V_{\boldsymbol{\beta}}$ for stable processes was discussed in Section 5.4.3. If the variables are believed to be integrated, the following prior discussed by Doan et al. (1984) and Litterman (1986), sometimes known as Minnesota prior, could be used: (1) Set the prior mean of the first lag of each variable equal to one in its own equation and set all other coefficients at zero. In other words, if the prior means were the true parameter values each variable were a random walk. (2) Choose the prior variances of the coefficients as in Section 5.4.3. In other words, the prior variances of the intercept terms are infinite and the prior variance of $\alpha_{ij,l}$, the ij-th element of A_l, is

$$v_{ij,l} = \begin{cases} (\lambda/l)^2 & \text{if } i = j, \\ (\lambda\theta\sigma_i/l\sigma_j)^2 & \text{if } i \neq j, \end{cases}$$

where λ is the prior standard deviation of $\alpha_{ii,1}, 0 < \theta < 1$, and σ_i^2 is the i-th diagonal element of Σ_u. Thus, we get, for instance, for a bivariate VAR(2) system,

$$y_{1t} = \underset{(\infty)}{0} + \underset{(\lambda)}{1 \cdot y_{1,t-1}} + \underset{(\lambda\theta\sigma_1/\sigma_2)}{0 \cdot y_{2,t-1}} + \underset{(\lambda/2)}{0 \cdot y_{1,t-2}} + \underset{(\lambda\theta\sigma_1/2\sigma_2)}{0 \cdot y_{2,t-2}} + u_{1t},$$

$$y_{2t} = \underset{(\infty)}{0} + \underset{(\lambda\theta\sigma_2/\sigma_1)}{0 \cdot y_{1,t-1}} + \underset{(\lambda)}{1 \cdot y_{2,t-1}} + \underset{(\lambda\theta\sigma_2/2\sigma_1)}{0 \cdot y_{1,t-2}} + \underset{(\lambda/2)}{0 \cdot y_{2,t-2}} + u_{2t},$$

where all coefficients are set to their prior means and the numbers in parentheses are their prior standard deviations. Forgetting about the latter numbers for the moment, each of these two equations is seen to specify a random walk for one of the variables. The nonzero prior standard deviations indicate that we are not sure about such a simple model. The standard deviations decline with increasing lag length because more recent lags are assumed to be more likely to belong into the model. The infinite standard deviations for the intercept terms simply reflect that we do not have any prior guess for these coefficients. Also, we do not impose covariance priors and, hence, choose V_β to be a diagonal matrix. Its inverse is

$$V_\beta^{-1} = \begin{bmatrix} 0 & & & & & & & & & \\ & 0 & & & & & & & & \\ & & \frac{1}{\lambda^2} & & & & 0 & & & \\ & & & \frac{\sigma_1^2}{(\lambda\theta\sigma_2)^2} & & & & & & \\ & & & & \frac{\sigma_2^2}{(\lambda\theta\sigma_1)^2} & & & & & \\ & & & & & \frac{1}{\lambda^2} & & & & \\ & & 0 & & & & \frac{2^2}{\lambda^2} & & & \\ & & & & & & & \frac{2^2\sigma_1^2}{(\lambda\theta\sigma_2)^2} & & \\ & & & & & & & & \frac{2^2\sigma_2^2}{(\lambda\theta\sigma_1)^2} & \\ & & & & & & & & & \frac{2^2}{\lambda^2} \end{bmatrix},$$

where 0 is also substituted for the inverse (infinite) standard deviation of the intercept terms.

To compute $\bar{\boldsymbol{\beta}}$ requires the inversion of $V_{\boldsymbol{\beta}}^{-1} + (ZZ' \otimes \Sigma_u^{-1})$. Because this matrix is usually quite large, in the past, Bayesian estimation has often been performed separately for each of the K equations of the system. In that case,

$$\bar{b}_k = [V_k^{-1} + \sigma_k^{-2} ZZ']^{-1}(V_k^{-1} b_k^* + \sigma_k^{-2} Z y_{(k)})$$

is used as an estimator for the parameters b_k of the k-th equation, that is, b_k' is the k-th row of $B := [\nu, A_1, \ldots, A_p]$. Here V_k is the prior covariance matrix of b_k, b_k^* is its prior mean, and $y_{(k)} := (y_{k1}, \ldots, y_{kT})'$. As in Chapter 5, σ_k^2 is replaced by the k-th diagonal element of the ML estimator

$$\widetilde{\Sigma}_u = Y(I_T - Z'(ZZ')^{-1}Z)Y'/T$$

of the white noise covariance matrix.

Clearly, in this prior, possible cointegration between the variables is not taken into account. Given the growing importance of the concept of cointegration in the recent literature, it is perhaps not surprising that the Minnesota prior has lately lost some of its appeal. Bayesians have responded to the success of the concept of cointegration and of VECMs in classical econometrics. Some recent contributions to Bayesian analysis of VECMs include Kleibergen & van Dijk (1994), Kleibergen & Paap (2002), Strachan (2003), and Strachan & Inder (2004). A survey with many more references was given by Koop, Strachan, van Dijk & Villani (2005).

7.4.3 An Example

As an example illustrating Bayesian estimation based on the Minnesota prior, we consider the following four-dimensional system of U.S. economic variables:

y_1 - logarithm of the real money stock M1 (ln M1),

y_2 - logarithm of GNP in billions of 1982 dollars (ln GNP),

y_3 - discount interest rate on new issues of 91-day Treasury bills (r^s),

y_4 - yield on long-term (20 years) Treasury bonds (r^l).

Quarterly data for the years 1954 to 1987 are used. The data are available in File E3. They are plotted in Figure 7.2. The GNP and M1 data are seasonally adjusted. The variables r^s and r^l are regarded as short- and long-term interest rates, respectively. The plots in Figure 7.2 show that the series are trending. Thus, they may be integrated and, given that this is a small monetary system, there may in fact be cointegration. For example, there may be a long-run money demand relation and perhaps the interest rate spread $r^l - r^s$ may be a stationary variable. Although the system may be cointegrated, we will consider the Minnesota prior in the following.

We have first fitted an unrestricted VAR(2) model to the data and present the results in Table 7.2. It can be seen that at least the last three of the four diagonal elements of A_1 are estimated to be close to 1. The first diagonal

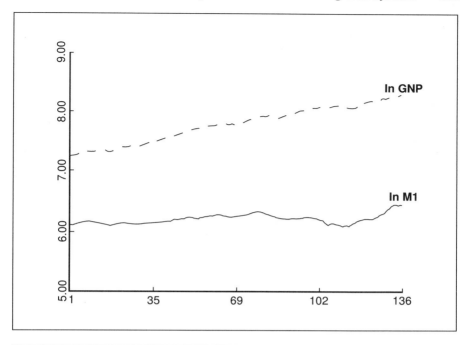

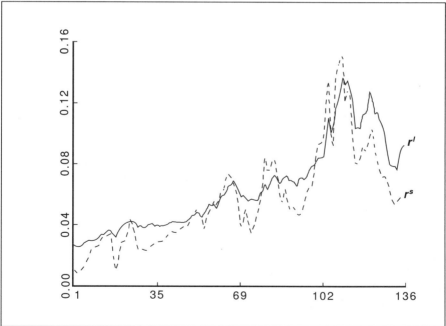

Fig. 7.2. U.S. ln M1, ln GNP, and interest rate time series.

314   7 Estimation of Vector Error Correction Models

element is also not drastically different from 1, although 1 is not within a two-standard error interval around the estimate. On the basis of the unrestricted estimates, a prior with mean 1 for the diagonal elements of A_1 does not appear to be unreasonable for this example. Of course, in a Bayesian analysis the prior is usually not chosen on the basis of an unrestricted estimation.

Table 7.2. VAR(2) coefficient estimates for the U.S. example system with estimated standard errors in parentheses

estimation method	ν	A_1				A_2			
	.028	1.307	.106	−.554	−.814	−.318	−.101	.318	1.022
		(.070)	(.075)	(.107)	(.224)	(.070)	(.076)	(.115)	(.221)
	.129	.080	1.045	−.177	.473	−.135	−.014	−.197	−.416
unrestricted		(.083)	(.088)	(.126)	(.265)	(.083)	(.090)	(.136)	(.261)
LS	.096	.193	.068	.978	.284	−.248	−.035	.053	−.644
		(.077)	(.081)	(.116)	(.245)	(.077)	(.083)	(.125)	(.240)
	.030	.042	.042	.034	1.065	−.064	−.027	.070	−.308
		(.038)	(.041)	(.058)	(.122)	(.038)	(.041)	(.063)	(.120)
	.041	1.332	.098	−.556	−.838	−.346	−.091	.354	.969
		(.067)	(.073)	(.104)	(.216)	(.064)	(.073)	(.110)	(.207)
	.086	.071	1.052	−.169	.549	−.099	−.039	−.239	−.286
ML		(.079)	(.086)	(.123)	(.256)	(.076)	(.087)	(.131)	(.245)
($r=1$)	.005	.179	.080	.991	.425	−.181	−.079	−.022	−.405
		(.076)	(.082)	(.118)	(.245)	(.073)	(.083)	(.125)	(.235)
	−.014	.037	.047	.041	1.138	−.032	−.050	.033	−.186
		(.038)	(.041)	(.059)	(.122)	(.036)	(.042)	(.062)	(.117)

We have estimated the system with the Minnesota prior and different values of λ and θ. Some results for a VAR(2) process are given in Table 7.3 to illustrate the effect of the choice of the prior variance parameters λ and θ. For this particular data set, a combination $\lambda = 1$ and $\theta = .25$ leads to mild changes in the estimates only relative to unrestricted estimates ($\lambda = \infty, \theta = 1$). Decreasing θ has the effect of shrinking the off-diagonal elements towards zero. Thus, a small θ is reasonable if the variables are expected to be unrelated. The effect of a small θ is seen in Table 7.3 in the panel corresponding to $\lambda = 1$ and $\theta = .01$. On the other hand, lowering λ shrinks the diagonal elements of A_1 towards 1 and all other coefficients (except the intercept terms) towards zero. This effect is clearly observed for $\lambda = .01$, $\theta = .25$. Hence, if the analyst has a strong prior in favor of unrelated random walks, a small λ is appropriate.

In practice, one would usually choose a higher VAR order than 2 in a Bayesian analysis because chopping off the process at $p = 2$ implies a very strong prior with mean zero and variances zero for $A_3, A_4, \ldots$, which is a

Table 7.3. Bayesian estimates of the U.S. example system

prior	ν	A_1				A_2			
$\lambda = \infty$	.028	1.307	.106	−.554	−.814	−.318	−.101	.318	1.022
$\theta = 1$	.129	.080	1.045	−.177	.473	−.135	−.014	−.197	−.416
(unrestricted)	.096	.193	.068	.978	.284	−.248	−.035	.053	−.644
	.030	.042	.042	.034	1.065	−.064	−.027	.070	−.308
	.061	1.307	.021	−.514	−.465	−.331	−.009	.212	.679
$\lambda = 1$	.110	.060	1.088	−.173	.283	−.108	−.060	−.162	−.238
$\theta = .25$	.078	.119	.064	1.060	.025	−.167	−.034	−.069	−.316
	.029	.021	.029	.050	1.044	−.043	−.014	.031	−.265
	.083	1.550	.004	−.012	−.007	−.570	−.002	−.000	.004
$\lambda = 1$	−.015	.005	1.270	−.011	−.011	−.001	−.271	−.003	−.002
$\theta = .01$	−.032	−.003	.008	1.095	−.001	−.002	.002	−.216	−.001
	−.016	−.003	.004	.002	1.187	−.001	.001	.000	−.252
	−.045	1.009	.002	−.001	−.000	−.003	.000	−.000	.000
$\lambda = .01$	.018	.001	.999	−.001	−.002	.000	−.002	−.000	−.000
$\theta = .25$	−.004	.001	−.000	.993	−.001	.000	−.000	−.002	−.000
	−.003	.000	.000	.000	.994	.000	.000	−.000	−.002

bit unrealistic. The above analysis is just meant to illustrate the effect of the parameters that determine the prior variances. Also, if the variables are believed to be cointegrated, the Minnesota prior is not a good choice. It is more suited for a process which has a VAR representation in first differences because the basic idea underlying this prior is that the variables are roughly unrelated random walks. Notice, however, that for the present system, if a VECM with cointegration rank $r = 1$ and one lagged difference is fitted by ML and the corresponding levels VAR coefficients are determined via (7.2.10), the estimates in the lower part of Table 7.2 are obtained. If the system is actually cointegrated, the rank restriction should not lead to major distortions in the estimates. Therefore, it should not be surprising that the diagonal elements of the ML estimator of A_1 are again not far from 1. Thus, even if the variables are cointegrated, the Minnesota prior may not lead to substantial distortions. This property may explain why the prior has been used successfully in many applications, in particular, for forecasting (see Litterman (1986)).

7.5 Forecasting Estimated Integrated and Cointegrated Systems

As seen in Chapter 6, Section 6.5, forecasting integrated and cointegrated variables is conveniently discussed in the framework of the levels VAR representation of the data generation process. Therefore we consider a VAR(p) model,

$$y_t = A_1 y_{t-1} + \cdots + A_p y_{t-p} + u_t, \qquad (7.5.1)$$

with integrated and possibly cointegrated variables. All symbols have their usual meanings (see Section 6.5). Deterministic terms are left out for convenience. Adding them is a straightforward exercise which is left to the reader.

Replacing the coefficients $A_1, \ldots, A_p$, and the white noise covariance matrix Σ_u by estimators in the forecasting formulas of Section 6.5 creates similar problems as in the stationary, stable case considered in Chapter 3, Section 3.5. Denoting the h-step forecast based on estimated coefficients by $\widehat{y}_t(h)$ and indicating estimators by hats gives

$$\widehat{y}_t(h) = \widehat{A}_1 \widehat{y}_t(h-1) + \cdots + \widehat{A}_p \widehat{y}_t(h-p), \qquad (7.5.2)$$

where $\widehat{y}_t(j) := y_{t+j}$ for $j \leq 0$. For this predictor, the forecast error becomes

$$\begin{aligned} y_{t+h} - \widehat{y}_t(h) &= [y_{t+h} - y_t(h)] + [y_t(h) - \widehat{y}_t(h)] \\ &= \sum_{i=0}^{h-1} \Phi_i u_{t+h-i} + [y_t(h) - \widehat{y}_t(h)], \end{aligned} \qquad (7.5.3)$$

where the last equality sign follows from Eq. (6.5.4) in Chapter 6. The last two terms in (7.5.3) are uncorrelated if parameter estimation is based on data up to period t only. In fact, under standard assumptions, the last term has zero probability limit, $y_t(h) - \widehat{y}_t(h) = o_p(1)$, as in the stationary case (see Problem 7.7). Thus, the forecast errors from estimated processes and processes with known coefficients are asymptotically equivalent. However, in the present case, the MSE correction for estimated processes derived in Section 3.5 is difficult to justify (see Problem 7.8 and Basu & Sen Roy (1987)). This problem must be kept in mind when forecast intervals are constructed. One possible MSE estimator is

$$\widehat{\Sigma}_y(h) = \sum_{i=0}^{h-1} \widehat{\Phi}_i \widehat{\Sigma}_u \widehat{\Phi}_i', \qquad (7.5.4)$$

where the $\widehat{\Phi}_i$'s are obtained from the estimated A_i's by the recursions in (6.5.5) in Section 6.5. This estimator is likely to underestimate the true forecast uncertainty on average in small samples. Therefore, there is some danger that the confidence level of corresponding forecast intervals is overstated. Reimers (1991) derived a small sample correction especially for models with cointegrated variables and Engle & Yoo (1987) and Reinsel & Ahn (1992) reported on simulation studies in which imposing the cointegration restriction in the estimation gave better long-range forecasts than the use of unrestricted multivariate LS estimators.

7.6 Testing for Granger-Causality

7.6.1 The Noncausality Restrictions

In Section 6.6, we have seen that the restrictions characterizing Granger-noncausality are the same as in the stable case. If the levels VAR(p) representation (7.5.1) of the data generation process is considered again and the vector y_t is partitioned in M- and $(K-M)$-dimensional subvectors z_t and x_t,

$$y_t = \begin{bmatrix} z_t \\ x_t \end{bmatrix} \quad \text{and} \quad A_i = \begin{bmatrix} A_{11,i} & A_{12,i} \\ A_{21,i} & A_{22,i} \end{bmatrix}, \quad i = 1, \ldots, p,$$

where the A_i are partitioned in accordance with the partitioning of y_t, then x_t does not Granger-cause z_t if and only if the hypothesis

$$H_0 : A_{12,i} = 0 \quad \text{for} \quad i = 1, \ldots, p, \tag{7.6.1}$$

is true. Hence, we just have to test a set of linear restrictions. A Wald test is a standard choice for this purpose. In the present case, it may be problematic, however. We will discuss the potential problem next and then present a modification that has a limiting χ^2-distribution, as usual, and, hence, resolves the problem.

7.6.2 Problems Related to Standard Wald Tests

If the process is estimated by one of the procedures described in Section 7.2 such that the estimator $\widehat{\alpha}$ of $\alpha := \text{vec}[A_1, \ldots, A_p]$ has the asymptotic distribution given in Corollary 7.1.1, then a Wald test can be conducted for the pair of hypotheses

$$H_0 : C\alpha = 0 \quad \text{against} \quad H_1 : C\alpha \neq 0. \tag{7.6.2}$$

Here C is an $(N \times pK^2)$ matrix of rank N. The relevant Wald statistic is

$$\lambda_W = T\widehat{\alpha}' C' (C\widehat{\Sigma}_\alpha^{\text{co}} C')^{-1} C\widehat{\alpha}, \tag{7.6.3}$$

where $\widehat{\Sigma}_\alpha^{\text{co}}$ is a consistent estimator of $\Sigma_\alpha^{\text{co}}$. The statistic λ_W has an asymptotic $\chi^2(N)$-distribution, provided the null hypothesis is true and

$$\text{rk}(C\widehat{\Sigma}_\alpha^{\text{co}} C') = \text{rk}(C\Sigma_\alpha^{\text{co}} C') = N. \tag{7.6.4}$$

This result follows from standard asymptotic theory (see Appendix C.7). We have chosen to state it here again because the rank condition (7.6.4) now becomes important. It is automatically satisfied for stable, full VAR processes as discussed in Chapter 3, because in that case the asymptotic covariance matrix of the coefficient estimator is nonsingular. Now, however, $\Sigma_\alpha^{\text{co}}$ is singular if the cointegration rank r is less than K (see Corollary 7.1.1). Therefore, it is possible in principle that $\text{rk}(C\Sigma_\alpha^{\text{co}} C') < N$, even if C has full row rank N.

A limiting χ^2-distribution of λ_W can also be obtained if the inverse of $C\widehat{\Sigma}_\alpha^{\text{co}} C'$ in (7.6.3) is replaced by a generalized inverse. In that case, the asymptotic distribution of λ_W is $\chi^2(\text{rk}(C\Sigma_\alpha^{\text{co}} C'))$ if

$$\text{rk}(C\widehat{\Sigma}_\alpha^{\text{co}} C') = \text{rk}(C\Sigma_\alpha^{\text{co}} C') \tag{7.6.5}$$

with probability one (see Andrews (1987)). Unfortunately, the latter condition will not hold in general. In particular, if a cointegrated system is estimated in unconstrained form by multivariate LS and if Σ_α^{co} is estimated as in Corollary 7.1.1, $C\widehat{\Sigma}_\alpha^{co}C'$ has rank N with probability 1, while $\mathrm{rk}(C\Sigma_\alpha^{co}C')$ may be less than N. Andrews (1987) showed that in such a case the asymptotic distribution of λ_W may not even be χ^2. A detailed analysis of the problem for the particular case of testing for Granger-causality in cointegrated systems was provided by Toda & Phillips (1993). In this context, it is perhaps worth pointing out that the equality in (7.6.5) may not hold, even if the cointegration rank has been specified correctly and the corresponding restrictions have been imposed in the estimation procedure (see Problem 7.9). For the hypothesis of interest here, a possible solution to the problem was proposed by Dolado & Lütkepohl (1996) and Toda & Yamamoto (1995). It will be presented next. Our discussion follows the former article.

Another possible approach to overcome inference problems in levels VARs with integrated variables was described by Phillips (1995). It is known as *fully modified VAR estimation* and is based on nonparametric corrections. Some of its drawbacks are pointed out by Kauppi (2004).

7.6.3 A Wald Test Based on a Lag Augmented VAR

As discussed in Section 7.2 (see in particular Section 7.2.1), the estimators of coefficients attached to stationary regressors converge at the usual $T^{1/2}$ rate to a nonsingular normal distribution. Therefore, the problem of the previous subsection can be solved if the model can be rewritten in such a way that all parameters under test are attached to stationary regressors. To this end, the following reparameterization is helpful:

$$y_t = \sum_{j=1, j\neq i}^{p} A_j y_{t-j} + A_i y_{t-i} + u_t$$

$$= \sum_{j=1, j\neq i}^{p} A_j (y_{t-j} - y_{t-i}) + \left(\sum_{j=1}^{p} A_j\right) y_{t-i} + u_t.$$

Defining a differencing operator Δ_k such that $\Delta_k y_t = y_t - y_{t-k}$ for $k = \pm 1, \pm 2, \ldots$, the model can be written as

$$\Delta_i y_t = \sum_{j=1, j\neq i}^{p} A_j \Delta_{i-j} y_{t-j} + \mathbf{\Pi} y_{t-i} + u_t, \qquad (7.6.6)$$

where $\mathbf{\Pi} = -(I_K - A_1 - \cdots - A_p)$, as usual. For $k > 0$, $\Delta_k y_t = (y_t - y_{t-1}) + (y_{t-1} - y_{t-2}) + \cdots + (y_{t-k+1} - y_{t-k})$ is stationary as the sum of stationary processes and the same is easily seen to hold for $k < 0$. Therefore, it follows from the previously mentioned results in Section 7.2 that the LS estimators of the A_j, $j \neq i$, have a nonsingular joint asymptotic normal distribution.

Notice that these estimators are, of course, identical to those based on the levels VAR model (7.5.1) because we have just reparameterized the model. Hence, the following proposition from Dolado & Lütkepohl (1996, Theorem 1) is obtained.

Proposition 7.8 (*Asymptotic Distribution of the Wald Statistic*)
Let y_t be a K-dimensional $I(1)$ process generated by the VAR(p) process in (7.5.1) and denote the LS estimator of A_i by $\widehat{A}_i$ ($i = 1, \ldots, p$). Moreover, let $\boldsymbol{\alpha}_{(-i)}$ be a $K^2(p-1)$-dimensional vector obtained by deleting A_i from $[A_1, \ldots, A_p]$ and vectorizing the remaining matrix. Analogously, let $\widehat{\boldsymbol{\alpha}}_{(-i)}$ be a $K^2(p-1)$-dimensional vector obtained by deleting $\widehat{A}_i$ from $[\widehat{A}_1, \ldots, \widehat{A}_p]$ and vectorizing the remainder. Then

$$\sqrt{T}(\widehat{\boldsymbol{\alpha}}_{(-i)} - \boldsymbol{\alpha}_{(-i)}) \xrightarrow{d} \mathcal{N}(0, \Sigma_{\boldsymbol{\alpha}_{(-i)}}), \qquad (7.6.7)$$

where the $(K^2(p-1) \times K^2(p-1))$ covariance matrix $\Sigma_{\boldsymbol{\alpha}_{(-i)}}$ is nonsingular and the Wald statistic λ_W for testing $H_0 : C\boldsymbol{\alpha}_{(-i)} = 0$ has a limiting $\chi^2(N)$-distribution, that is,

$$\lambda_W = T\widehat{\boldsymbol{\alpha}}'_{(-i)}C'(C\widehat{\Sigma}_{\boldsymbol{\alpha}_{(-i)}}C')^{-1}C\widehat{\boldsymbol{\alpha}}_{(-i)} \xrightarrow{d} \chi^2(N)$$

under H_0. Here C is an $(N \times K^2(p-1))$ matrix with $\text{rk}(C) = N$ and $\widehat{\Sigma}_{\boldsymbol{\alpha}_{(-i)}}$ is a consistent estimator of $\Sigma_{\boldsymbol{\alpha}_{(-i)}}$. ∎

Note that

$$\Sigma_{\boldsymbol{\alpha}_{(-i)}} = \text{plim } T(X_{(-i)}X'_{(-i)})^{11} \otimes \Sigma_u,$$

where $X_{(-i)} = [X_0^{(-i)}, \ldots, X_{T-1}^{(-i)}]$ with

$$X_{t-1}^{(-i)} = \begin{bmatrix} \Delta_{i-1}y_{t-1} \\ \vdots \\ \Delta_{i-p}y_{t-p} \\ y_{t-i} \end{bmatrix} \quad (K^2 p \times 1)$$

and $(X_{(-i)}X'_{(-i)})^{11}$ denotes the upper left-hand $(K^2(p-1) \times K^2(p-1))$ dimensional submatrix of $(X_{(-i)}X'_{(-i)})^{-1}$. Thereby a consistent estimator of $\Sigma_{\boldsymbol{\alpha}_{(-i)}}$ is obtained as

$$\widehat{\Sigma}_{\boldsymbol{\alpha}_{(-i)}} = T(X_{(-i)}X'_{(-i)})^{11} \otimes \widehat{\Sigma}_u,$$

where $\widehat{\Sigma}_u$ is the residual covariance matrix obtained from the LS residuals.

Proposition 7.8 shows that, whenever the elements in at least one of the complete coefficient matrices A_i are not restricted under H_0, the Wald statistic has its usual asymptotic χ^2-distribution. In other words, if restrictions are placed on all A_i's, $i = 1, \ldots, p$, as in the noncausality hypothesis (7.6.1), we can get a χ^2 Wald test by adding an extra lag in estimating the parameters of the process. If the true data generation process is a VAR(p), then a VAR($p+1$) with $A_{p+1} = 0$ is also a correct model. Because we know that $A_{p+1} = 0$, the causality test can be based on the estimator $\widehat{\boldsymbol{\alpha}}_{(-(p+1))}$, that is, an estimator of the first $K^2 p$ elements of $\text{vec}[\widehat{A}_1, \ldots, \widehat{A}_{p+1}]$. Notice that LS estimation may be applied to the levels VAR($p+1$) model. To carry out the causality test, it is not necessary to actually perform the reparameterization of the process in (7.6.6) because the LS estimators of the A_j matrices do not change due to the reparameterization. Also, the covariance matrix of the asymptotic distribution may be estimated as usual from the levels VAR($p+1$).

We do not have to know the cointegration properties of the system to use this lag augmentation test procedure. Of course, there may be a loss of power due to over-specifying the lag length. The loss in power may not be substantial if the true order p is large and the dimension K is small or moderate, because, in this case, the relative reduction in the estimation precision due to one extra VAR coefficient matrix may be small. On the other hand, if the true order is small and K is large, an extra lag of all variables may lead to a sizeable decline in overall estimation precision and, hence, in the power of the modified Wald test. There are in fact cases, where the extra lag is not necessary to obtain the asymptotic χ^2-distribution of the Wald test for Granger-causality. For example, for bivariate processes with cointegrating rank 1, no extra lag is needed, if both variables are $I(1)$ (e.g., Lütkepohl & Reimers (1992a)).

Proposition 7.8 remains valid if deterministic terms are included in the VAR model. This result follows from the discussion in Section 7.2 because including such terms leaves the asymptotic properties of the VAR coefficients unaffected. It may also be of interest that a similar result can be obtained for VAR systems with $I(d)$ variables where $d > 1$. In that case, d coefficient matrices A_i must be unrestricted under H_0 (see Dolado & Lütkepohl (1996)). Alternatively, d lags must be added if all parameter matrices of the original process are restricted. This result can also be obtained from Sims et al. (1990).

7.6.4 An Example

We follow again Lütkepohl (2004) and use the German interest rate/inflation example to illustrate causality testing for cointegrated variables. The data generation process is assumed to be a VAR(4). The model is augmented by one lag and, hence, a VAR(5) is fitted and used in the actual tests for Granger-causality, while a VAR(4) is used for testing instantaneous causality. The results are given in Table 7.4, where F-versions of the Granger-causality test statistics are reported. The asymptotic χ^2-distribution is often a poor approximation to the small sample distribution of the causality test statistics. Therefore, an F-version is preferred which is obtained in the usual way by dividing the χ^2-statistic by its degrees of freedom parameter (see Section 3.6). As in Section 3.6, the test for instantaneous causality is based on the residual covariance matrix. This approach is justified by Lemma 7.3 which shows that the asymptotic distribution of the usual residual covariance matrix estimator is the same as in the stationary case. Hence, the same test for instantaneous causality can be used under normality assumptions.

Table 7.4. Tests for causality between German interest rate and inflation

causality hypothesis	test value	distribution	p-value
R Granger-causal for Dp	2.24	$F(4, 152)$	0.07
Dp Granger-causal for R	0.31	$F(4, 152)$	0.87
R and Dp instantaneously causal	0.61	$\chi^2(1)$	0.44

None of the p-values in Table 7.4 is smaller than 0.05. Therefore, none of the noncausality hypotheses can be rejected at the 5% significance level. Given the subset model (7.3.9), this outcome is somewhat surprising because there are clearly significant estimates in that model. Of course, using the present tests is a different way of looking at the data than considering the individual coefficients in the subset model. The relatively large number of parameters in the presently considered unrestricted model which even includes an extra lag, makes it difficult for the sample information to clearly distinguish the sets of parameters from their values specified in the null hypothesis.

The insignificant value of the test for instantaneous causality is not surprising, however. The correlation matrix corresponding to the covariance matrix in (7.3.9) is

$$\begin{bmatrix} 1 & -0.01 \\ -0.01 & 1 \end{bmatrix}.$$

Thus, the instantaneous correlation between the two residual series is very small. This property is reflected in the test result in Table 7.4.

7.7 Impulse Response Analysis

In Section 6.7, we have seen that, in principle, impulse response analysis in cointegrated systems can be conducted in the same way as for stationary systems. If estimated processes are used, the asymptotic properties of the impulse response coefficients and forecast error variance components follow from Proposition 3.6 in conjunction with Corollary 7.1.1. In other words, the relevant covariance matrices $\Sigma_{\widehat{\alpha}}$ and $\Sigma_{\widehat{\sigma}}$ have to be used in Proposition 3.6. Of course, the remarks on Proposition 3.6 regarding the estimation of standard errors etc. apply for the present case too. In practice, confidence intervals for impulse responses are typically computed with bootstrap methods.

To illustrate the impulse response analysis we use again our German interest rate/inflation example system. We have performed an impulse response analysis on the basis of the subset VECM (7.3.9) and show forecast error impulse responses with bootstrap confidence intervals determined by Hall's percentile method (see Appendix D.3) in Figure 7.3. Using forecast error impulse responses is unproblematic here because no instantaneous causality and no significant instantaneous correlation between the two residual series was diagnosed in Section 7.6.4. The point estimates of the impulse responses look very much like those in Figure 6.4 in Chapter 6. This similarity is not surprising because the model assumed in that chapter is very similar to the present one. Because the variables are integrated of order one, the impulses have permanent effects. This conclusion can be defended even if the estimation uncertainty is taken into account.

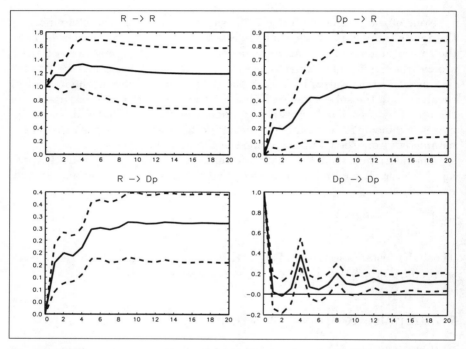

Fig. 7.3. Forecast error impulse responses for model (7.3.9) with 95% Hall percentile bootstrap confidence intervals based on 2000 bootstrap replications.

We emphasize again that an uncritical impulse response analysis is problematic. In particular, different sets of impulse responses exist and it is not clear which one properly reflects the actual reactions of the variables. The caveats of impulse response analysis are discussed in Sections 2.3 and 3.7. They are therefore not repeated here. We will return to impulse response analysis in Chapter 9, when structural restrictions are discussed for identifying meaningful shocks.

7.8 Exercises

7.8.1 Algebraic Exercises

Problem 7.1
Show that, in the proof of Result 6 of Section 7.1,

$$T^{-1}\sum_{t=1}^{T}(u_t^* - u_t)y_{t-1}^{(2)\prime} = o_p(1).$$

$\Big($Hint: Use

$$T^{-1}\sum_{t=1}^{T}(u_t^* - u_t)y_{t-1}^{(2)\prime} = (\widehat{\alpha} - \alpha)T^{-1}\sum_{t=1}^{T}\beta' y_{t-1}y_{t-1}^{(2)\prime}.\Big)$$

Problem 7.2
Prove Proposition 7.1 based on the ideas presented in Section 7.2.1. (Hint: See Ahn & Reinsel (1990).)

Problem 7.3
Prove that $\sqrt{T}[\widetilde{\alpha}\widetilde{\beta}' - \widetilde{\alpha}(\beta)\beta'] = o_p(1)$ holds in the proof of Lemma 7.3. $\Big($Hint: note that

$$\widetilde{\alpha}\widetilde{\beta}' - \widetilde{\alpha}(\beta)\beta' = \widetilde{\alpha}[\widetilde{\beta}' - \beta'] + [\widetilde{\alpha} - \widetilde{\alpha}(\beta)]\beta'.\Big)$$

Problem 7.4
Determine the ML estimators in a cointegrated VAR(p) process with cointegration rank r, under the assumption that the cointegration matrix satisfies restrictions $\beta = H\varphi$, where H and φ are $(K \times s)$ and $(s \times r)$ matrices, respectively, with $r < s < K$. (Hint: Proceed as in the proof of Proposition 7.3.)

Problem 7.5
Show that the expressions in (7.2.27) and (7.2.28) are the LS estimators of α and Γ, respectively, conditional on $\beta = \widehat{\beta}$.

Problem 7.6
Derive the EGLS estimator for restrictions of the form $\text{vec}[\alpha : \Gamma] = \Re\varphi + \mathbf{r}$ on the short-run parameters of the VECM (7.2.1) and state its asymptotic distribution (see (7.3.8) for the definition of the notation).

Problem 7.7
Consider a cointegrated VAR(1) process without intercept, $y_t = A_1 y_{t-1} + u_t$, and show that

$$\text{plim}\,[y_T(1) - \widehat{y}_T(1)] = \text{plim}\,(A_1 - \widetilde{A}_1)y_T = 0.$$

Assume that y_t is Gaussian with initial vector $y_0 = 0$ and the ML estimator $\widetilde{A}_1$ is based on $y_1, \ldots, y_T$. (Hint: Use Lemma 7.2 and plim $y_T/T = 0$ from Phillips & Durlauf (1986).)

Problem 7.8
Consider the matrix $\Omega(h)$ used in the MSE correction in Section 3.5 and argue why it is problematic for unstable processes. Analyze in particular the derivation in (3.5.12).

Problem 7.9
Consider a three-dimensional VAR(1) process with cointegration rank 1 and suppose the cointegrating matrix has the form $\beta = (\beta_1, \beta_2, 0)'$. Use Corollary 7.1.1 to demonstrate that the elements in the last column of A_1 have zero asymptotic variances. Formulate a linear hypothesis for the coefficients of A_1 for which the rank condition (7.6.4) is likely to be violated if the covariance estimator of Corollary 7.1.1 is used.

7.8.2 Numerical Exercises

The following problems are based on the U.S. data given in File E3 and described in Section 7.4.3. The variables are defined as in that subsection.

Problem 7.10
Apply the ML procedure described in Section 7.2.3 to estimate a VAR(3) process with cointegration rank $r = 1$ and intercept vector. Determine the estimates $\tilde{\nu}, \tilde{A}_1, \tilde{A}_2$, and $\tilde{A}_3$ and compare them to unrestricted LS estimates of a VAR(3) process.

Problem 7.11
Compute forecasts up to 10 periods ahead using both the unrestricted VAR(3) model and the VAR(3) model with cointegration rank 1. Compare the forecasts.

Problem 7.12
Compare the impulse responses obtained from an unrestricted and restricted VAR(3) model with cointegration rank 1.

8
Specification of VECMs

In specifying VECMs, the lag order, the cointegration rank and possibly further restrictions have to be determined. The lag order and the cointegration rank are typically determined before further restrictions are imposed on the parameter matrices. Moreover, the specification of a VECM usually starts by determining a suitable lag length because, in choosing the lag order, the cointegration rank does not have to be known, whereas many procedures for specifying the cointegration rank require knowledge of the lag order. Therefore, in the following, we will first discuss the lag order choice (Section 8.1) and then consider procedures for determining the cointegration rank (Section 8.2). We will comment on subset modelling in a VECM framework in Section 8.3 and, in Section 8.4, we will discuss checking the adequacy of such models. More precisely, residual autocorrelation analysis, testing for nonnormality and structural change are dealt with.

8.1 Lag Order Selection

It was mentioned in Section 7.2.1 that Wald tests for zero restrictions on coefficient matrices of the lagged differences can be constructed. Hence, the number of lagged differences in a VECM can be chosen by a sequence of tests similar to that in Section 4.2. Because the procedure and its problems are discussed in some detail in that section, we will not repeat the discussion here but focus on order selection criteria such as AIC, HQ, and SC in this section.

In Section 4.3, the FPE criterion was introduced for stationary, stable processes as a criterion that minimizes the forecast MSE and therefore has a justification if forecasting is the objective. We have seen in Section 7.5 that the forecast MSE correction used for estimated stationary processes is difficult to justify in the cointegrated case and, hence, the FPE criterion cannot be based on the same footing in the latter case. This argument does not mean, however, that the criterion is not a useful one in some other sense for nonstationary processes. For instance, it is possible that it still provides models with excellent small sample forecasting properties. It was also shown in Section 4.3 that Akaike's AIC is asymptotically equivalent to the FPE criterion. Therefore, similar comments apply for AIC.

The criteria HQ and SC were justified by their ability to choose the order "correctly in large samples", that is, they are consistent criteria. It was shown by Paulsen (1984) and Tsay (1984) that the consistency property of these criteria is maintained for integrated processes. To make that statement precise, we give the following result from Paulsen (1984) without proof.

Proposition 8.1 (*Consistent VAR Order Estimation*)
Let

$$y_t = \nu + A_1 y_{t-1} + \cdots + A_p y_{t-p} + u_t$$

be a K-dimensional VAR(p) process with $A_p \neq 0$ and standard white noise u_t and suppose that $\det(I_K - A_1 z - \cdots - A_p z^p)$ has s roots equal to one, that is, $z = 1$ is a root with multiplicity s, and all other roots are outside the complex unit circle. Furthermore, let

$$\mathrm{Cr}(m) = \ln |\widetilde{\Sigma}_u(m)| + m c_T/T, \qquad (8.1.1)$$

where $\widetilde{\Sigma}_u(m)$ is the Gaussian ML or quasi ML estimator of Σ_u for a VAR(m) model based on a sample of size T and m fixed presample values as in Proposition 4.2, and c_T is a nondecreasing sequence indexed by T. Let $\widehat{p}$ be such that

$$\mathrm{Cr}(\widehat{p}) = \min\{\mathrm{Cr}(m) | m = 0, 1, \ldots, M\}$$

and suppose $M \geq p$. Then $\widehat{p}$ is a consistent estimator of p if and only if $c_T \to \infty$ and $c_T/T \to 0$ as $T \to \infty$. ∎

This proposition extends Proposition 4.2 to processes with integrated variables. It implies that AIC is not a consistent criterion while HQ and SC are both consistent. Thus, if consistent estimation is the objective, we may apply HQ and SC for stationary and integrated processes.

Denoting the orders chosen by AIC, HQ, and SC by $\widehat{p}(\mathrm{AIC})$, $\widehat{p}(\mathrm{HQ})$, and $\widehat{p}(\mathrm{SC})$, respectively, we also get from Proposition 4.3 that

$$\widehat{p}(\mathrm{SC}) \leq \widehat{p}(\mathrm{HQ}) \leq \widehat{p}(\mathrm{AIC}) \quad \text{for} \quad T \geq 16.$$

This result is obtained because Proposition 4.3 does not require any stationarity or stability assumptions. It follows as in Chapter 4 that AIC asymptotically overestimates the true order with positive probability (see Corollary 4.3.1).

Although these results are nice because they generalize the stationary case in an easy way, they do not mean that AIC or FPE are order selection criteria inferior to HQ and SC. Recall that consistent order estimation may not be a relevant objective in small sample situations. In fact, the true data generating process may not admit a finite order VAR representation.

Notice also that, while we have considered specifying the VAR order p, the criteria are also applicable for choosing the number of lagged differences in a VECM because $p-1$ lagged differences in a VECM correspond to a VAR order p. Thus, once we know p, we know the number of lagged differences. If some of the variables are known to be integrated, the VAR order must be at least 1. This information can be taken into account in model selection by searching only over orders $1,\dots,M$ rather then $0,1,\dots,M$.

We have applied the three criteria AIC, HQ, and SC to our German interest rate/inflation example data from Section 7.2.6 with a maximum order of $M = 8$ and a constant and seasonal dummies in the model. The values of the criteria are shown in Table 8.1. SC and HQ both recommend the order $\hat{p} = 1$ while $\hat{p}(\text{AIC}) = 4$. Thus, in a VECM based on SC and HQ, no lagged differences appear, whereas three lagged differences have to be included according to AIC. We have chosen to go with the AIC estimate in the example in Section 7.2.6.

Table 8.1. VAR order estimation for interest rate/inflation system

VAR order m	AIC(m)	HQ(m)	SC(m)
0	−18.75	−18.75	−18.75
1	−20.98	−20.94*	−20.88*
2	−20.97	−20.89	−20.76
3	−20.89	−20.77	−20.58
4	−20.99*	−20.82	−20.57
5	−20.93	−20.72	−20.41
6	−20.89	−20.63	−20.26
7	−20.85	−20.55	−20.12
8	−20.80	−20.46	−19.96

*Minimum.

In Chapter 4, we have mentioned that model selection may be based on the residual autocorrelations or portmanteau tests. These statistics can also be used for VECMs. They are discussed in Section 8.4.1.

8.2 Testing for the Rank of Cointegration

Although model selection criteria have also been used in specifying the cointegrating rank of a VECM (e.g., Lütkepohl & Poskitt (1998)), it is more common in practice to use statistical tests for this purpose. Many different tests have been proposed in the literature and the properties of most of them depend on the deterministic terms included in the model. In the following, we will therefore discuss models with different deterministic terms separately. The general model is assumed to be of the form

$$y_t = \mu_t + x_t,$$

where x_t is the stochastic part which is assumed to have a VECM representation without deterministic terms and μ_t is the deterministic term, as in Chapter 6, Section 6.4. We will start with the easiest although most unrealistic case where no deterministic term is present and, thus, $\mu_t = 0$. Most of the discussion will focus on likelihood ratio (LR) tests and close relatives of them because they are very common in applied work and they also fit well into the present framework. Some comments on other procedures will be provided in Section 8.2.9.

8.2.1 A VECM without Deterministic Terms

Based on Proposition 7.3, it is easy to derive the likelihood ratio statistic for testing a specific cointegration rank $r = r_0$ of a VECM against a larger rank of cointegration, say $r = r_1$. Consider the VECM without determinist terms,

$$\Delta y_t = \mathbf{\Pi} y_{t-1} + \mathbf{\Gamma}_1 \Delta y_{t-1} + \cdots + \mathbf{\Gamma}_{p-1} \Delta y_{t-p+1} + u_t, \qquad (8.2.1)$$

where y_t is a process of dimension K, $\mathrm{rk}(\mathbf{\Pi}) = r$ with $0 \leq r \leq K$, the $\mathbf{\Gamma}_j$'s $(j = 1, \ldots, p-1)$ are $(K \times K)$ parameter matrices and $u_t \sim \mathcal{N}(0, \Sigma_u)$ is Gaussian white noise, as in Chapter 7, Section 7.2.3. For simplicity we assume that the process starts at time $t = 1$ with zero initial values (i.e., $y_t = 0$ for $t \leq 0$). Alternatively, the initial values may be any fixed values.

Suppose we wish to test

$$H_0 : \mathrm{rk}(\mathbf{\Pi}) = r_0 \quad \text{against} \quad H_1 : r_0 < \mathrm{rk}(\mathbf{\Pi}) \leq r_1. \qquad (8.2.2)$$

Under normality assumptions, the maximum of the likelihood function for a model with cointegration rank r is given in Proposition 7.3. From that result, the LR statistic for testing (8.2.2) is seen to be

$$\begin{aligned}
\lambda_{LR}(r_0, r_1) &= 2[\ln l(r_1) - \ln l(r_0)] \\
&= T \left[-\sum_{i=1}^{r_1} \ln(1 - \lambda_i) + \sum_{i=1}^{r_0} \ln(1 - \lambda_i) \right] \\
&= -T \sum_{i=r_0+1}^{r_1} \ln(1 - \lambda_i), \qquad (8.2.3)
\end{aligned}$$

where $l(r_i)$ denotes the maximum of the Gaussian likelihood function for cointegration rank r_i. Obviously, the test value is quite easy to compute, using the eigenvalues from Proposition 7.3.

It turns out, however, that the asymptotic distribution of the LR statistic under the null hypothesis for given r_0 and r_1 is nonstandard. In particular, it is not a χ^2-distribution. It depends on the number of common trends $K - r_0$ under H_0 and on the alternative hypothesis. Two different pairs of hypotheses have received prime attention in the related literature:

$$H_0 : \mathrm{rk}(\mathbf{\Pi}) = r_0 \quad \text{versus} \quad H_1 : r_0 < \mathrm{rk}(\mathbf{\Pi}) \leq K \qquad (8.2.4)$$

and

$$H_0 : \text{rk}(\boldsymbol{\Pi}) = r_0 \quad \text{versus} \quad H_1 : \text{rk}(\boldsymbol{\Pi}) = r_0 + 1. \tag{8.2.5}$$

The LR statistic $\lambda_{LR}(r_0, K)$ for checking (8.2.4) is often referred to as the *trace statistic* for testing the cointegrating rank and $\lambda_{LR}(r_0, r_0+1)$ is called the *maximum eigenvalue statistic*. Johansen (1988, 1995) shows that the asymptotic distributions of these LR statistics under the null hypothesis are

$$\lambda_{LR}(r_0, K) \xrightarrow{d} \text{tr}(\mathcal{D}) \tag{8.2.6}$$

and

$$\lambda_{LR}(r_0, r_0 + 1) \xrightarrow{d} \lambda_{\max}(\mathcal{D}), \tag{8.2.7}$$

where $\lambda_{\max}(\mathcal{D})$ denotes the maximum eigenvalue of the matrix $\mathcal{D}$ and

$$\mathcal{D} := \left(\int_0^1 \mathbf{W} d\mathbf{W}' \right)' \left(\int_0^1 \mathbf{W}\mathbf{W}' ds \right)^{-1} \left(\int_0^1 \mathbf{W} d\mathbf{W}' \right). \tag{8.2.8}$$

Here $\mathbf{W} := \mathbf{W}_{K-r_0}(s)$ stands for a $(K - r_0)$-dimensional standard Wiener process. In other words, the limiting null distributions are functionals of a $(K - r_0)$-dimensional standard Wiener process. Percentage points of the asymptotic distributions and, thus, critical values for the LR tests can be generated easily. Tables are, for example, available in Johansen (1995). Hence, a LR test is available under Gaussian assumptions and, as usual, the test statistics have the same limiting distributions even if the underlying process is not normally distributed but satisfies the more general assumptions used in Section 7.2, for example.

The strategy for determining the cointegrating rank of a given system of K variables is to test a sequence of null hypotheses,

$$H_0 : \text{rk}(\boldsymbol{\Pi}) = 0, \ H_0 : \text{rk}(\boldsymbol{\Pi}) = 1, \ldots, H_0 : \text{rk}(\boldsymbol{\Pi}) = K - 1, \tag{8.2.9}$$

and terminate the tests when the null hypothesis cannot be rejected for the first time. The cointegrating rank is then chosen accordingly. Both the maximum eigenvalue and the trace tests may be used here. For example, if there are three variables ($K = 3$), we first test $\text{rk}(\boldsymbol{\Pi}) = 0$. If this null hypothesis cannot be rejected, the analysis proceeds with a cointegration rank of $r = 0$ and, hence, a model in first differences is considered in the subsequent analysis. If, however, $\text{rk}(\boldsymbol{\Pi}) = 0$ is rejected, we test $\text{rk}(\boldsymbol{\Pi}) = 1$. Should the test not reject this hypothesis, the analysis may proceed with a VECM with cointegrating rank $r = 1$. Otherwise $\text{rk}(\boldsymbol{\Pi}) = 2$ is tested and $r = 2$ is chosen as the cointegrating rank if this hypothesis cannot be rejected. If $\text{rk}(\boldsymbol{\Pi}) = 2$ is also rejected, one may consider working with a stationary VAR model for the levels of the variables.

Clearly, in these tests the lag order has to be known. In practice, it is often chosen by one of the model selection criteria discussed in the previous section, based on the levels VAR model, before the cointegrating rank is tested.

As mentioned previously, the model framework in (8.2.1) is too simple for practical purposes because deterministic terms are usually needed to describe the generation process of a given set of time series properly. Therefore, we will now consider processes with deterministic terms.

8.2.2 A Nonzero Mean Term

We now assume that the deterministic term consists of a simple constant mean term only,

$$\mu_t = \mu_0. \tag{8.2.10}$$

Although we typically think of μ_0 as a fixed nonzero $(K \times 1)$ vector, the case $\mu_0 = 0$ is not explicitly excluded. In other words, the user of the test is not sure that the process mean is zero and therefore allows for the possibility of a nonzero mean term. In Section 6.4, we have seen that in this case the VECM for the observable variables y_t can be written as

$$\Delta y_t = \mathbf{\Pi}^o y_{t-1}^o + \mathbf{\Gamma}_1 \Delta y_{t-1} + \cdots + \mathbf{\Gamma}_{p-1} \Delta y_{t-p+1} + u_t, \tag{8.2.11}$$

where

$$y_{t-1}^o := \begin{bmatrix} y_{t-1} \\ 1 \end{bmatrix}$$

and $\mathbf{\Pi}^o := [\mathbf{\Pi} : \nu_0]$ is $(K \times (K+1))$ with $\nu_0 := -\mathbf{\Pi}\mu_0$. Thus, the LR statistic for testing the cointegration rank can be determined exactly as in the zero mean case considered in the previous subsection, except that y_{t-1} has to be replaced by y_{t-1}^o in the relevant formulas from which the eigenvalues are computed in Proposition 7.3. In this case, the LR statistics have asymptotic null distributions as in (8.2.6) and (8.2.7), where now

$$\mathcal{D} := \left(\int_0^1 \mathbf{W}^o d\mathbf{W}' \right)' \left(\int_0^1 \mathbf{W}^o \mathbf{W}^{o\prime} ds \right)^{-1} \left(\int_0^1 \mathbf{W}^o d\mathbf{W}' \right) \tag{8.2.12}$$

with

$$\mathbf{W}^o := \mathbf{W}^o(s) := \begin{bmatrix} \mathbf{W}_{K-r_0}(s) \\ 1 \end{bmatrix}$$

(see Johansen (1991)). Again, critical values may be found in Johansen (1995).

8.2.3 A Linear Trend

A process with a linear trend is also of interest from a practical point of view. Hence, let

$$\mu_t = \mu_0 + \mu_1 t, \tag{8.2.13}$$

where μ_0 and μ_1 are arbitrary $(K \times 1)$ vectors. In Section 6.4, we have seen that in this case the VECM for the observable y_t can be represented as

$$\Delta y_t = \nu + \mathbf{\Pi}^+ y_{t-1}^+ + \mathbf{\Gamma}_1 \Delta y_{t-1} + \cdots + \mathbf{\Gamma}_{p-1} \Delta y_{t-p+1} + u_t, \tag{8.2.14}$$

where $\nu := -\mathbf{\Pi}\mu_0 + (I_K - \mathbf{\Gamma}_1 - \cdots - \mathbf{\Gamma}_{p-1})\mu_1$, $\mathbf{\Pi}^+ := [\mathbf{\Pi} : \nu_1]$ is a $(K \times (K+1))$ matrix with $\nu_1 := -\mathbf{\Pi}\mu_1$, and

$$y_{t-1}^+ := \begin{bmatrix} y_{t-1} \\ t-1 \end{bmatrix}.$$

Thus, the LR statistics of interest can again be determined exactly as in the zero mean case of Section 8.2.1 by replacing y_{t-1} with y_{t-1}^+ and accounting for the intercept term by adding a row of ones in ΔX in the relevant formulas in Proposition 7.3 (see Section 7.2.4). For the present case, the LR statistics have asymptotic null distributions as in (8.2.6) and (8.2.7) with

$$\mathcal{D} := \left(\int_0^1 \mathbf{W}^+ d\mathbf{W}' \right)' \left(\int_0^1 \mathbf{W}^+ \mathbf{W}^{+\prime} ds \right)^{-1} \left(\int_0^1 \mathbf{W}^+ d\mathbf{W}' \right). \tag{8.2.15}$$

Here $\mathbf{W}^+$ abbreviates the $(K - r_0 + 1)$-dimensional stochastic process $\mathbf{W}^+(s) := [\overline{\mathbf{W}}(s)', s - \tfrac{1}{2}]'$ with $\overline{\mathbf{W}}(s) := \mathbf{W}_{K-r_0}(s) - \int_0^1 \mathbf{W}_{K-r_0}(u) du$ being a demeaned standard Wiener process, as shown by Johansen (1994, 1995). Critical values may also be found in the latter reference.

8.2.4 A Linear Trend in the Variables and Not in the Cointegration Relations

In the model (8.2.14), the linear trend term is unrestricted and therefore may also be part of the cointegration relations. Even if the variables have a linear trend, it is possible that there is no such term in the cointegration relations. In other words, the cointegration relations are drifting along a common linear trend. This situation can arise if the trend slope is the same for all variables which have a linear trend. Formally this case occurs if $\mu_1 \neq 0$ and $\mathbf{\Pi}\mu_1 = \alpha\beta'\mu_1 = 0$ or, equivalently, if $\beta'\mu_1 = 0$. In other words, this situation is present if the trend parameter μ_1 is nonzero and it is orthogonal to the cointegration relations. In this case, (8.2.14) reduces to

$$\Delta y_t = \nu + \mathbf{\Pi} y_{t-1} + \mathbf{\Gamma}_1 \Delta y_{t-1} + \cdots + \mathbf{\Gamma}_{p-1} \Delta y_{t-p+1} + u_t. \tag{8.2.16}$$

Thus, in this situation we have a model just like (8.2.1), except that there is an intercept term in addition. Again, the LR statistics for testing (8.2.4) or (8.2.5) can be determined easily as in the zero mean case of Section 8.2.1 by adding a row of ones in ΔX in the relevant formulas in Proposition 7.3 (see Section 7.2.4). The limiting distributions of the LR statistics under the null hypothesis are also as in (8.2.6) and (8.2.7), where now

$$\mathcal{D} := \left(\int_0^1 \check{\mathbf{W}} d\mathbf{W}' \right)' \left(\int_0^1 \check{\mathbf{W}}\check{\mathbf{W}}' ds \right)^{-1} \left(\int_0^1 \check{\mathbf{W}} d\mathbf{W}' \right). \tag{8.2.17}$$

Here $\check{\mathbf{W}} := \check{\mathbf{W}}(s) := \mathbf{W}^c(s) - \int_0^1 \mathbf{W}^c(u)du$, where $\mathbf{W}^c(s) := [\mathbf{W}_{K-r_0-1}(s)', s]'$ is a $(K - r_0)$-dimensional stochastic process. This result and corresponding critical values for the tests may also be found in Johansen (1995).

Notice that the condition $\mu_1 \neq 0$ and $\Pi \mu_1 = 0$ rules out the situation where $\text{rk}(\Pi) = K$ because, for a nonsingular matrix Π, the relation $\Pi \mu_1 = 0$ cannot hold for a nonzero μ_1. Thus, the assumptions made for deriving the limiting distributions of the test statistics make a test of

$$H_0 : \text{rk}(\Pi) = K - 1 \quad \text{versus} \quad H_1 : \text{rk}(\Pi) = K$$

meaningless. Intuitively, this result is obtained because, if Π has full rank, the data generation process is stationary and, in that case, a VAR process with an intercept does not generate a linear trend. Thus, if a linear trend is known to be present in the variables, Π cannot have full rank in a model where an intercept is the only deterministic term.

8.2.5 Summary of Results and Other Deterministic Terms

The results of the previous subsections are summarized in the following proposition.

Proposition 8.2 (*Limiting Distributions of LR Tests for the Cointegrating Rank*)
Suppose $y_t = \mu_t + x_t$, where μ_t is a deterministic term and x_t is a purely stochastic Gaussian process defined by

$$\Delta x_t = \Pi x_{t-1} + \Gamma_1 \Delta x_{t-1} + \cdots + \Gamma_{p-1} \Delta x_{t-p+1} + u_t, \quad t = 1, 2, \ldots,$$

where all symbols are defined as in (8.2.1) and $x_t = 0$ for $t \leq 0$. Then the LR statistics for testing (8.2.4) and (8.2.5) have limiting null distributions

$$\lambda_{LR}(r_0, K) \xrightarrow{d} \text{tr}(\mathcal{D})$$

and

$$\lambda_{LR}(r_0, r_0 + 1) \xrightarrow{d} \lambda_{\max}(\mathcal{D}),$$

respectively, where

$$\mathcal{D} = \left(\int_0^1 \mathbf{F} d\mathbf{W}'_{K-r_0}\right)' \left(\int_0^1 \mathbf{F}\mathbf{F}' ds\right)^{-1} \left(\int_0^1 \mathbf{F} d\mathbf{W}'_{K-r_0}\right)$$

with

(1) $\mathbf{F}(s) = \mathbf{W}_{K-r_0}(s)$, if $\mu_t = 0$ a priori,
(2) $\mathbf{F}(s) = \mathbf{W}^\circ(s) = [\mathbf{W}_{K-r_0}(s)' : 1]'$, if $\mu_t = \mu_0$ is a constant,
(3) $\mathbf{F}(s) = [\overline{\mathbf{W}}(s)', s - \frac{1}{2}]'$ as in (8.2.15), if $\mu_t = \mu_0 + \mu_1 t$ is a linear trend,
(4) $\mathbf{F}(s) = \check{\mathbf{W}}(s)$ as in (8.2.17), if $\mu_t = \mu_0 + \mu_1 t$ is a linear trend with $\mu_1 \neq 0$ and $\boldsymbol{\beta}'\mu_1 = 0$, that is, the trend is orthogonal to the cointegration relations. ∎

Several remarks are worthwhile with respect to this result.

Remark 1 Percentage points of the asymptotic distributions in Proposition 8.2 are easy to simulate by considering multivariate random walks of the form

$$x_t = x_{t-1} + u_t, \quad t = 1, 2, \ldots, T,$$

where $x_0 = 0$ and $u_t \sim \mathcal{N}(0, I_K)$ is Gaussian white noise, that is,

$$x_t = \sum_{i=1}^t u_i.$$

Noting that

$$T^{-2} \sum_{t=1}^T x_{t-1} x'_{t-1} \xrightarrow{d} \int_0^1 \mathbf{W}\mathbf{W}' ds,$$

$$T^{-1} \sum_{t=1}^T x_{t-1} u'_t \xrightarrow{d} \int_0^1 \mathbf{W} d\mathbf{W}',$$

and so on (see Appendix C.8, Proposition C.18), we can, for example, approximate

$$\operatorname{tr}\left[\left(\int_0^1 \mathbf{W} d\mathbf{W}'\right)' \left(\int_0^1 \mathbf{W}\mathbf{W}' ds\right)^{-1} \left(\int_0^1 \mathbf{W} d\mathbf{W}'\right)\right]$$

by

$$\operatorname{tr}\left[\left(\sum_{t=1}^T x_{t-1} u'_t\right)' \left(\sum_{t=1}^T x_{t-1} x'_{t-1}\right)^{-1} \left(\sum_{t=1}^T x_{t-1} u'_t\right)\right]$$

for a large sample size T. Similar approximations can be used for the other asymptotic distributions (see also Problem 8.2). ∎

Remark 2 Although we only give the limiting distributions of the LR statistics under the null hypothesis in the proposition, the asymptotic distributions under local alternatives of the form

$$\Pi = \alpha\beta' + \frac{1}{T}\alpha_1\beta_1'$$

were also derived (see Johansen (1995) and Saikkonen & Lütkepohl (1999, 2000a)). Here α and β are fixed $(K \times r_0)$ matrices of rank r_0 and α_1 and β_1 are fixed $(K \times (r - r_0))$ matrices of rank $r - r_0$ and such that the matrices $[\alpha : \alpha_1]$ and $[\beta : \beta_1]$ have full column rank r. Thus, in this setup, the matrix Π is assumed to depend on the sample size. Local power studies have been performed to shed light on the power properties of the LR tests when the alternative is true but the corresponding parameter values are close to the region where the null hypothesis holds. ∎

Remark 3 Power comparisons between the alternative test versions can help in deciding whether to use trace or maximum eigenvalue tests. Lütkepohl, Saikkonen & Trenkler (2001) performed a detailed small sample and local power comparison of several test versions and concluded that trace and maximum eigenvalue tests have very similar local power in many situations, whereas each test version has its relative advantages in small samples, depending on the criterion for comparison. Thus, neither of the tests is generally preferable in practice. ∎

Remark 4 It is also possible to derive the asymptotic properties of the LR tests for other deterministic terms. For example, higher order polynomial trends may be considered. Such terms lead to changes in the null distributions of the test statistics. We do not consider them here because they seem to be of lesser importance from a practical point of view. ∎

Remark 5 Seasonal dummy variables are another type of deterministic terms which are of practical importance. They are often used to account for seasonal fluctuations in the variables (see, e.g., the example in Section 7.2.6). If seasonal dummies are added in addition to an unrestricted intercept term, they do not affect the asymptotic distributions of the LR statistics for the cointegration rank. We have considered two models, however, where no unrestricted intercept term was included. The first one was the model of Section 8.2.1 without any deterministic terms at all. As this model is of limited practical use anyway, we do not consider the implications of adding seasonal dummy variables. The other model without an unrestricted intercept term was the one with a nonzero mean discussed in Section 8.2.2. It is of more use in practice and it is therefore of interest to consider the possibility of adding seasonal dummies.

Suppose there are q seasons and the deterministic term is of the form

$$\mu_t = \mu_0 + \sum_{i=1}^{q-1} \delta_i s_{it},$$

where μ_0 and δ_i ($i = 1, \dots, q-1$) are ($K \times 1$) parameter vectors and the seasonal dummies are denoted by s_{it}. Suppose that they are defined such that they are orthogonal to the intercept term, that is,

$$s_{it} = \begin{cases} 1 & \text{if } t \text{ is associated with season } i, \\ \frac{-1}{q-1} & \text{otherwise,} \end{cases}$$

for $i = 1, \dots, q$. In that case, using the same line of reasoning as in Section 6.4, the corresponding VECM for y_t is

$$\Delta y_t = \Pi^o y_{t-1}^o + \Gamma_1 \Delta y_{t-1} + \cdots + \Gamma_{p-1} \Delta y_{t-p+1} + \sum_{i=1}^{q-1} \delta_i^* s_{it} + u_t,$$

where the δ_i^*'s are ($K \times 1$) parameter vectors. Notice that $Ls_{it} = s_{i,t-1} = s_{i-1,t}$ for $i = 2, \dots, q$ and $Ls_{1t} = s_{qt}$ and, for any t, $\sum_{i=1}^{q} s_{it} = 0$ so that $s_{qt} = -\sum_{i=1}^{q-1} s_{it}$. Hence, the latter sum can be substituted for s_{qt} (see also Problem 8.1). In this model, the seasonal dummies have no impact on the asymptotic distribution of the LR statistic for the cointegrating rank (Johansen (1991)). ∎

Remark 6 A different situation arises if the deterministic term includes a shift dummy variable $I_{(t>T_B)}$ which is zero up to time T_B and then jumps to one. Such a variable affects the asymptotic distributions of the LR test statistics for the cointegrating rank. In fact, Johansen, Mosconi & Nielsen (2000) showed that in this case the asymptotic distributions depend on where the shift occurs in the sample. More precisely, it depends on the fraction of the sample before the break. In contrast, impulse dummy variables which are always zero except in one specific period, do not affect the asymptotic properties of the LR tests. ∎

8.2.6 An Example

We have applied LR trace tests for the cointegrating rank to the German interest rate/inflation example data from Section 7.2.6 and give results for different lag orders in Table 8.2. Notice that, although we report the results for the trace tests, the maximum eigenvalue variant is equivalent if $H_0 : \text{rk}(\Pi) = 1$ is tested in a bivariate system. In that case, the alternative hypotheses in (8.2.4) and (8.2.5) coincide. Because the inflation rate has a strong seasonal pattern, we have included seasonal dummy variables in the deterministic term. Given the theoretical considerations in Section 7.2.6, one may not see the need for a general trend in the model. Clearly, one would not expect the cointegration relation to include a linear trend. In fact, one may wonder about the need to consider a deterministic linear trend at all in the model because one could argue that neither interest rates nor inflation rates are likely to have such components in Germany. Even if there is a strong case for excluding the possibility of a linear trend term in a long-run analysis of these two variables, it may still be useful to include such a trend for a particular sample period. Recall that any model is just an approximation to the data generation process for a specific period of time. In Table 8.2, we therefore report results for different deterministic terms.

Table 8.2. LR trace tests for the cointegration rank of the German interest rate/inflation system

deterministic term	no. of lagged differences	null hypothesis	test value	critical values 10%	5%
constant, seasonal dummies	0	$rk(\Pi) = 0$	89.72	17.79	19.99
		$rk(\Pi) = 1$	1.54	7.50	9.13
	3	$rk(\Pi) = 0$	21.78	17.79	19.99
		$rk(\Pi) = 1$	4.77	7.50	9.13
orthogonal linear trend, seasonal dummies	0	$rk(\Pi) = 0$	89.10	13.31	15.34
	3	$rk(\Pi) = 0$	20.80	13.31	15.34
linear trend, seasonal dummies	0	$rk(\Pi) = 0$	97.21	22.95	25.47
		$rk(\Pi) = 1$	4.45	10.56	12.39
	3	$rk(\Pi) = 0$	24.78	22.95	25.47
		$rk(\Pi) = 1$	7.72	10.56	12.39

Notes: Sample period: 1972.2 − 1998.4 (including presample values). Critical values from Johansen (1995, Tables 15.2, 15.3, and 15.4).

For all deterministic terms and all lag orders, the tests reject a cointegrating rank of zero. The only possible exception is the case, where a fully general linear trend and three lagged differences are included in the model. In that case, the cointegration rank zero can only be rejected at the 10% level and not at the 5% level, whereas in all other cases the tests reject at a 5% level. Of course, the model with three lagged differences and a linear deterministic trend is the least restricted model considered in Table 8.2. Thus, if any one of the other models describes the DGP well, the same is true for the latter model. Therefore, one may argue that the tests based on this model should be the most reliable. Unfortunately, such an argument is valid for the size of the test at best. In small sample studies, some evidence was found that redundant lags or deterministic terms can have a negative effect on the powers of the LR tests (see Hubrich, Lütkepohl & Saikkonen (2001) for an overview of small sample studies). Thus, taking the small sample properties of the tests into account, there is substantial evidence that the cointegrating rank is larger than zero.

For the models with a constant and a linear trend, none of the tests can reject a cointegration rank of $r = 1$. If a deterministic linear trend is assumed to be present in at least one of the variables and not in the cointegration relations, that is, the trend is orthogonal to the cointegration relations, then testing the null hypothesis $\mathrm{rk}(\Pi) = 1$ does not make sense for a bivariate system, as explained in Section 8.2.4. Therefore, no results are reported for that null hypothesis in Table 8.2. Thus, the evidence in favor of a single cointegration relation in our example system is overall quite strong. Therefore, we have used this rank in previous models for the two series.

The discussion of which deterministic terms to include in the model for our example data shows that there is a need for statistical procedures to assist in the decision. There are indeed appropriate tests available, as discussed in Section 7.2.4. We will return to some such tests for specific hypotheses of interest in the present context in Section 8.2.8. Before we do so, we will discuss some other ideas for testing the cointegrating rank of a VECM. In the next subsection, we consider the possibility of subtracting the deterministic part first and then applying LR type tests to the adjusted series.

8.2.7 Prior Adjustment for Deterministic Terms

LR tests for the cointegrating rank were found to have low power, in particular in large models (large dimension and/or long lag order). Therefore, other tests and test variants have been proposed which have advantages at least in some situations. One variant was, for instance, proposed by Saikkonen & Lütkepohl (2000d). They suggested a two-step procedure in which the deterministic part is estimated first. Then the observed series are adjusted for the deterministic terms and an 'LR test' is applied to the adjusted system. We will discuss their approach for the case of a model with a linear trend term. The other cases of interest can be handled with straightforward modifications.

Thus, we consider a data generation process of the form

$$y_t = \mu_0 + \mu_1 t + x_t, \qquad (8.2.18)$$

where μ_0 and μ_1 are fully general ($K \times 1$) vectors and x_t has a VECM representation of the form (8.2.1). Hence, the data generation process of y_t has the VECM representation (8.2.14). Suppose we want to test the pair of hypotheses

$$H_0 : \mathrm{rk}(\Pi) = r_0 \quad \text{versus} \quad H_1 : \mathrm{rk}(\Pi) > r_0.$$

Then the model (8.2.14) is estimated by ML with a cointegration rank r_0 and estimators $\widetilde{\alpha}$, $\widetilde{\beta}$, $\widetilde{\Gamma}_j$ ($j = 1, \ldots, p-1$) as well as estimators of the other parameters are obtained. From these estimators we can get estimators of the levels VAR parameter matrices as follows (see Section 6.3, Eq. (6.3.7)):

$$\begin{aligned}
\widetilde{A}_1 &= I_K + \widetilde{\alpha}\widetilde{\beta}' + \widetilde{\Gamma}_1, \\
\widetilde{A}_i &= \widetilde{\Gamma}_i - \widetilde{\Gamma}_{i-1}, \quad i = 2, \ldots, p-1, \\
\widetilde{A}_p &= -\widetilde{\Gamma}_{p-1}.
\end{aligned}$$

These estimators are used to estimate the parameters μ_0 and μ_1 in (8.2.18) by an EGLS procedure. To present the estimator, we define $\widetilde{A}(L) := I_K - \widetilde{A}_1 L - \cdots - \widetilde{A}_p L^p$, $\widetilde{G}_t := \widetilde{A}(L)a_t$, and $\widetilde{H}_t := \widetilde{A}(L)b_t$, with

$$a_t := \begin{cases} 1 & \text{for } t \geq 1, \\ 0 & \text{for } t \leq 0, \end{cases} \qquad b_t := \begin{cases} t & \text{for } t \geq 1, \\ 0 & \text{for } t \leq 0. \end{cases}$$

Moreover, we define

$$\widetilde{Q} := \begin{bmatrix} (\widetilde{\alpha}' \widetilde{\Sigma}_u^{-1} \widetilde{\alpha})^{-1/2} \widetilde{\alpha}' \widetilde{\Sigma}_u^{-1} \\ (\widetilde{\alpha}'_\perp \widetilde{\Sigma}_u \widetilde{\alpha}_\perp)^{-1/2} \widetilde{\alpha}'_\perp \end{bmatrix}.$$

Premultiplying (8.2.18) by $\widetilde{Q}\widetilde{A}(L)$ gives

$$\widetilde{Q}\widetilde{A}(L)y_t = \widetilde{Q}\widetilde{G}_t \mu_0 + \widetilde{Q}\widetilde{H}_t \mu_1 + \eta_t^*, \quad t = p+1, \ldots, T, \tag{8.2.19}$$

where the transformation ensures that the error term has roughly a unit covariance matrix because $\widetilde{Q}'\widetilde{Q} = \widetilde{\Sigma}_u^{-1}$. Thus, estimating the transformed model (8.2.19) by LS amounts to EGLS estimation of μ_0 and μ_1 in the untransformed model $y_t = \mu_0 + \mu_1 t + x_t$. The resulting estimators of μ_0 and μ_1 will be denoted by $\widetilde{\mu}_0^{GLS}$ and $\widetilde{\mu}_1^{GLS}$, respectively.

Using these estimators, y_t can now be trend-adjusted as $\widetilde{x}_t := y_t - \widetilde{\mu}_0^{GLS} - \widetilde{\mu}_1^{GLS} t$ and an 'LR test' can be applied to $\widetilde{x}_t$, as described in Section 8.2.1. Of course, although the test statistics are computed in the same way as described in that section except that y_t is replaced by $\widetilde{x}_t$, the tests are now not really LR tests anymore because they are applied to adjusted data rather than the original ones. To distinguish the resulting tests from the actual LR tests, we will refer to them as GLS-LR tests and we denote the trace and maximum eigenvalue test statistics as $\lambda_{LR}^{GLS}(r_0, K)$ and $\lambda_{LR}^{GLS}(r_0, r_0+1)$, respectively, in the following. Given that these tests are not actual LR tests, it may also not be surprising that the limiting distributions of the test statistics are different from those of the actual LR statistics. They also depend on the deterministic terms that are included in the model. To state the asymptotic distributions formally, we use the following conventions. A *Brownian bridge* of dimension $K - r_0$ is defined as

$$\mathbf{W}^B(s) = \mathbf{W}_{K-r_0}(s) - s\mathbf{W}_{K-r_0}(1)$$

and an integral of a stochastic process $\mathbf{F}$ with respect to a Brownian bridge is defined as

$$\int_0^1 \mathbf{F} d\mathbf{W}^B := \int_0^1 \mathbf{F} d\mathbf{W}_{K-r_0} - \int_0^1 \mathbf{F} ds \mathbf{W}_{K-r_0}(1).$$

Now we can state the limiting null distributions of the λ_{LR}^{GLS} statistics for the different deterministic terms of interest.

Proposition 8.3 (*Limiting Distributions of GLS-LR Tests for the Cointegrating Rank*)
Under the conditions of Proposition 8.2, the GLS-LR test statistics have the following limiting null distributions:

$$\lambda_{LR}^{GLS}(r_0, K) \xrightarrow{d} \text{tr}(\mathcal{D})$$

and

$$\lambda_{LR}^{GLS}(r_0, r_0 + 1) \xrightarrow{d} \lambda_{\max}(\mathcal{D}),$$

where $\mathcal{D}$ depends on the deterministic terms included in the model as follows:
(1) If $\mu_t = \mu_0$ is a constant,

$$\mathcal{D} = \left(\int_0^1 \mathbf{W} d\mathbf{W}'\right)' \left(\int_0^1 \mathbf{W}\mathbf{W}' ds\right)^{-1} \left(\int_0^1 \mathbf{W} d\mathbf{W}'\right)$$

with $\mathbf{W} := \mathbf{W}_{K-r_0}(s)$.
(2) If $\mu_t = \mu_0 + \mu_1 t$ is a linear trend,

$$\mathcal{D} = \left(\int_0^1 \mathbf{W}^B d\mathbf{W}^{B\prime}\right)' \left(\int_0^1 \mathbf{W}^B \mathbf{W}^{B\prime} ds\right)^{-1} \left(\int_0^1 \mathbf{W}^B d\mathbf{W}^{B\prime}\right).$$

(3) If $\mu_t = \mu_0 + \mu_1 t$ is a linear trend with $\mu_1 \neq 0$ and $\beta'\mu_1 = 0$,

$$\mathcal{D} = \left(\int_0^1 \mathbf{\check{W}} d\mathbf{W}'\right)' \left(\int_0^1 \mathbf{W}^c \mathbf{W}^{c\prime} ds\right)^{-1} \left(\int_0^1 \mathbf{\check{W}} d\mathbf{W}'\right)$$

with $\mathbf{W} := \mathbf{W}_{K-r_0}(s)$, $\mathbf{W}^c(s) := [\mathbf{W}_{K-r_0-1}(s)', s]'$, and $\mathbf{\check{W}}(s)$ as in (8.2.17).
■

Proofs of these results can be found in Saikkonen & Lütkepohl (2000b, d) and Lütkepohl et al. (2001). The following remarks may be of interest.

Remark 1 The adjustment for deterministic terms may appear to be complicated at first sight. One may, for instance, wonder why the deterministic terms are not directly estimated by LS and then subtracted from the observed y_t. Unfortunately, in the present case, the LS estimators do not have the same asymptotic properties as the EGLS estimators described here and also the resulting cointegration tests will have different properties. The present procedure is useful because it results in tests with attractive asymptotic properties, as we will argue in the next remark. ■

Remark 2 Comparing the asymptotic distributions in Propositions 8.2 and 8.3, it turns out that the λ_{LR}^{GLS} statistics for the case of a constant deterministic term ($\mu_t = \mu_0$) have the same asymptotic distributions as the corresponding LR statistics for the case without any deterministic term. Thus, estimation of the constant mean term does not affect the asymptotic distributions of the λ_{LR}^{GLS} statistics, while it has an impact on the LR statistics in Proposition 8.2. This observation suggests that the GLS-LR tests may have better power properties, at least asymptotically. This conjecture was actually confirmed in a local power comparison by Saikkonen & Lütkepohl (1999). The situation is not as clear for the other situations. In other words, if there is a linear trend term in the model, a local power comparison does not lead to a unique ranking of the tests. In some situations the LR tests are preferable and in other situations the GLS-LR variants may be preferable, depending on the properties of the data generation process. Also, local power is an asymptotic concept which allows to investigate the power properties of tests in regions close to the null hypothesis when the sample size goes to infinity. Because asymptotic theory is not always a good guide for small sample properties, these results do not guarantee superior performance of the GLS-LR tests, even when only a constant mean term is included in the model. In particular, the latter tests may have size distortions in small samples. ■

Remark 3 Although the asymptotic distributions in Proposition 8.3 look a little more complicated than those in Proposition 8.2, critical values can again be simulated easily because the asymptotic distributions are still functionals of Wiener processes. Percentage points for all three asymptotic distributions are tabulated in the literature (see Johansen (1995), Lütkepohl & Saikkonen (2000) and Saikkonen & Lütkepohl (2000b)). ■

Remark 4 The GLS-LR tests can also be adopted for other deterministic terms such as higher order polynomials and seasonal dummy variables. For the former case, different asymptotic distributions will result, whereas seasonal dummies can be added to all three deterministic terms considered in Proposition 8.3 without affecting the limiting distributions of the test statistics. An advantage of the GLS-LR tests is that these asymptotic distributions are also not affected by including shift dummies in the deterministic term. This property is in contrast to the LR tests and means that the same critical values can be used as for the corresponding tests without shift dummies (see Saikkonen & Lütkepohl (2000c)). In particular, there is no need to compute new critical values for each break point. Given the computing power which is available today, this may not seem as a great advantage over the LR tests at first sight. It makes it possible, however, to also consider cases where the actual break date is unknown and has to be estimated in addition to the other parameters of the process. Lütkepohl, Saikkonen & Trenkler (2004) consider that case and show that a number of different estimators of the break date can be used without affecting the asymptotic distributions of the λ_{LR}^{GLS} statistics under the null hypothesis. ■

Example

We have also applied the GLS-LR tests to the German interest rate/inflation example series and present the results in Table 8.3. Although the evidence is again clearly in favor of a cointegrating rank of $r = 1$, all tests have more trouble rejecting $r_0 = 0$ if the larger lag order is used. In that case, the hypothesis $\text{rk}(\mathbf{\Pi}) = 0$ cannot even be rejected at the 10% level if only a constant and seasonal dummies are included in the model. Thus, although the GLS-LR tests have good local power properties especially for this case, superior small sample power is not guaranteed. Of course, it must also be kept in mind that a test with higher power does not necessarily reject a specific null hypothesis for a particular data set more easily than a test with lower power. Moreover, our theoretical models underlying the asymptotic analysis may not fully capture all features of the actual data generation process.

Table 8.3. GLS-LR trace tests for the cointegration rank of the German interest rate/inflation system

deterministic term	no. of lagged differences	null hypothesis	test value	critical values 10%	5%
constant, seasonal dummies	0	$\text{rk}(\mathbf{\Pi}) = 0$	28.21	10.35	12.21
		$\text{rk}(\mathbf{\Pi}) = 1$	0.41	2.98	4.14
	3	$\text{rk}(\mathbf{\Pi}) = 0$	10.13	10.35	12.21
		$\text{rk}(\mathbf{\Pi}) = 1$	2.42	2.98	4.14
orthogonal linear trend, seasonal dummies	0	$\text{rk}(\mathbf{\Pi}) = 0$	28.16	8.03	9.79
	3	$\text{rk}(\mathbf{\Pi}) = 0$	9.75	8.03	9.79
linear trend, seasonal dummies	0	$\text{rk}(\mathbf{\Pi}) = 0$	49.42	13.89	15.92
		$\text{rk}(\mathbf{\Pi}) = 1$	1.83	5.43	6.83
	3	$\text{rk}(\mathbf{\Pi}) = 0$	14.43	13.89	15.92
		$\text{rk}(\mathbf{\Pi}) = 1$	4.71	5.43	6.83

Notes: Sample period: 1972.2 − 1998.4 (including presample values). Critical values from Johansen (1995, Tables 15.1), Saikkonen & Lütkepohl (2000b, Table 1) and Lütkepohl & Saikkonen (2000, Table 1) for the case of a constant, an orthogonal trend, and a general linear trend, respectively.

8.2.8 Choice of Deterministic Terms

As mentioned earlier, including redundant deterministic terms in the models on which cointegration rank tests are based, may result in a substantial loss of power (see also Doornik, Hendry & Nielsen (1998) and Hubrich et al. (2001)). Therefore, it is helpful that statistical procedures are available for investigating which terms to include. Johansen (1994, 1995) proposed LR tests for hypotheses regarding the deterministic terms. These tests are obvious choices because the ML estimators and, hence, the corresponding maxima of the likelihood functions are easy to compute for various different deterministic terms (see Section 7.2.4).

Apart from dummy variables, a linear trend

$$\mu_0 + \mu_1 t \tag{8.2.20}$$

is the most general deterministic term considered in the foregoing. A possible pair of hypotheses of interest related to this term when $\mu_1 \neq 0$ is

$$H_0 : \beta' \mu_1 = 0 \quad \text{versus} \quad H_1 : \beta' \mu_1 \neq 0. \tag{8.2.21}$$

Hence, there is a deterministic linear trend in the variables and the test checks whether the trend is orthogonal to the cointegration relations. In other words, the test checks the model (8.2.14) against (8.2.16). The corresponding LR test has a standard χ^2 limiting distribution under the null hypothesis, as we have seen in Section 7.2.4. If the underlying VECM has cointegrating rank r and, thus, β is a $(K \times r)$ matrix, r zero restrictions are specified in H_0. Therefore we have r degrees of freedom, that is, the LR test statistic has an asymptotic $\chi^2(r)$-distribution.

Another pair of hypotheses of interest is

$$H_0 : \mu_1 = 0 \quad \text{versus} \quad H_1 : \mu_1 \neq 0, \beta' \mu_1 = 0. \tag{8.2.22}$$

In this case, a model with an unrestricted intercept, (8.2.16), is tested against one where no linear trend is present and, thus, the constant can be absorbed into the cointegration relations as in (8.2.11). Again, the LR test has standard asymptotic properties, that is, for a VECM of dimension K and with cointegration rank r, it has a $\chi^2(K-r)$ limiting distribution.

If these tests are used for deciding on the deterministic term in a VECM, it may be worth keeping in mind that they introduce additional uncertainty into the modelling procedure. The tests are performed for a model with a specific cointegrating rank. Thus, ideally the cointegrating rank has to be determined before the deterministic terms are tested, whereas one motivation for them was that cointegrating rank tests may have better power if the deterministic term is specified properly. Thus, the tests present only a partial solution to the problem. Proceeding as in the example and checking the robustness of the rank tests with respect to different specifications of the deterministic terms is a useful strategy.

8.2.9 Other Approaches to Testing for the Cointegrating Rank

The literature on cointegration rank tests has grown rapidly in recent years. Many related issues have been discussed and investigated. Examples are nonnormal processes (Lucas (1997, 1998), Boswijk & Lucas (2002), Caner (1998)), the presence of higher order integration and long memory (Gonzalo & Lee (1998), Breitung & Hassler (2002)), the impact of the dimension of the data generation process (Ho & Sørensen (1996)) and using a reversed sequence of null hypotheses in testing for the cointegrating rank (Snell (1999)). Also, a number of studies considered the small sample properties of the tests. A recent review of the related literature with many more references was provided by Hubrich et al. (2001).

Moreover, a number of other test procedures were proposed. For instance, Lütkepohl & Saikkonen (1999a) used the idea underlying the causality test which was presented in Section 7.6.3 and augmented the number of lags to obtain a χ^2-test for the cointegrating rank. Bewley & Yang (1995) and Yang & Bewley (1996) constructed a test based on canonical correlations of the levels variables. Stock & Watson (1988) considered the use of principal component analysis and Bierens (1997) presented a fully nonparametric approach to cointegration rank testing. These and many other proposals were also reviewed in Hubrich et al. (2001), including the possibility of choosing the cointegrating rank by model selection criteria. A range of cointegration tests was also proposed and investigated in a single equation framework (e.g., Engle & Granger (1987), Phillips & Ouliaris (1990), Banerjee et al. (1993), Choi (1994), Shin (1994), Haug (1996)). They are of limited usefulness for the situation we have considered here, where several cointegrating relations may be present in a system of variables. Therefore, no details are presented.

8.3 Subset VECMs

When the lag order and the cointegration rank of a VECM have been determined, specifying further restrictions may be useful to reduce the dimensionality of the parameter space and thereby improve the estimation precision. As we have seen in Sections 7.2 and 7.3, the standard t-ratios and F-tests retain their usual asymptotic properties if they are applied to the short-run and loading parameters of a VECM. Therefore, subset modelling for cointegrated systems may be based on statistical tests. Instead of using testing procedures, restrictions for individual parameters or groups of parameters may also be based on model selection criteria in a similar way as in Chapter 5. In particular, the strategies applied to individual equations of the system may be used. Consider, for instance, the j-th equation of a VECM,

$$y_{jt} = x_{1t}\theta_1 + \cdots + x_{Nt}\theta_N + u_{jt}, \quad t = 1, \ldots, T. \tag{8.3.1}$$

Here all right-hand side regressor variables are denoted by x_{kt}, including deterministic terms and the cointegration relations. Thus, $x_{kt} = \beta'_i y_{t-1}$, where β_i is the i-th column of the cointegration matrix β, is a possible regressor. If β is unknown, it may be replaced by a superconsistent estimator $\widehat{\beta}$, which may be based on the unrestricted model and variables $x_{kt} = \widehat{\beta}'_i y_{t-1}$ may be added as regressors in (8.3.1). Using this setup, all the standard procedures described in Section 5.2.8 are available, including the full search procedure, sequential elimination of regressors as well as top-down and bottom-up strategies.

For the German interest rate/inflation example with cointegration relation $\beta' y_t = R_t - 4Dp_t$, we have used the sequential elimination of regressors procedure in conjunction with the AIC criterion based on a search for restrictions on individual equations and found the following model, using the sample period 1973.2–1998.4 plus the required presample values:

$$\begin{bmatrix} \Delta R_t \\ \Delta Dp_t \end{bmatrix} = \begin{bmatrix} -0.07 \\ {\scriptstyle (-3.1)} \\ 0.17 \\ {\scriptstyle (4.5)} \end{bmatrix} (R_{t-1} - 4Dp_{t-1})$$

$$+ \begin{bmatrix} 0.24 & -0.08 \\ {\scriptstyle (2.5)} & {\scriptstyle (-1.9)} \\ 0 & -0.31 \\ & {\scriptstyle (-2.5)} \end{bmatrix} \begin{bmatrix} \Delta R_{t-1} \\ \Delta Dp_{t-1} \end{bmatrix} + \begin{bmatrix} 0 & -0.13 \\ & {\scriptstyle (-2.5)} \\ 0 & -0.37 \\ & {\scriptstyle (-3.6)} \end{bmatrix} \begin{bmatrix} \Delta R_{t-2} \\ \Delta Dp_{t-2} \end{bmatrix}$$

$$+ \begin{bmatrix} 0.20 & -0.06 \\ {\scriptstyle (2.1)} & {\scriptstyle (-1.6)} \\ 0 & -0.34 \\ & {\scriptstyle (-4.7)} \end{bmatrix} \begin{bmatrix} \Delta R_{t-3} \\ \Delta Dp_{t-3} \end{bmatrix}$$

$$+ \begin{bmatrix} 0 & 0 & 0.010 & 0 \\ & & {\scriptstyle (2.8)} & \\ 0.010 & -0.034 & -0.018 & -0.016 \\ {\scriptstyle (3.0)} & {\scriptstyle (-7.6)} & {\scriptstyle (-3.8)} & {\scriptstyle (-3.6)} \end{bmatrix} \begin{bmatrix} c \\ s_{1,t} \\ s_{2,t} \\ s_{3,t} \end{bmatrix} + \begin{bmatrix} \widehat{u}_{1,t} \\ \widehat{u}_{2,t} \end{bmatrix}.$$

(8.3.2)

Here t-ratios are given in parentheses underneath the parameter estimates. This is precisely the model that was also used in Section 7.3.3, see (7.3.9), to illustrate EGLS estimation and that procedure is used here as well. Notice, however, that the search procedure was based on LS estimation of individual equations. Hence, different t-ratios were the basis for variable selection. Still, generally the coefficients with large absolute t-ratios in the unrestricted model (see Table 7.1) are maintained in the restricted subset VECM.

In the present example, we have pretended that the cointegration relation is known. Such an assumption is not required for the subset procedures to be applicable. The same subset model selection procedure may be applied if the cointegration relations contain estimated parameters. In other words, it may be used as the second stage in a two-stage procedure, where the cointegration matrix β is estimated first and then the estimated β matrix is substituted for the true one in the second stage. The subset restrictions are determined in the second stage.

8.4 Model Diagnostics

Diagnostic checking is also an important stage of the general modelling procedure for VECMs. Many of the tests for model adequacy discussed for stationary VAR processes can be extended to the VECM case. Tests for residual autocorrelation, nonnormality, and structural change will be treated in turn in the following. We will start with a discussion of the properties of residual autocorrelations of an estimated VECM. The underlying model is assumed to be of the simple form

$$\Delta y_t = \alpha\beta' y_{t-1} + \Gamma_1 \Delta y_{t-1} + \cdots + \Gamma_{p-1}\Delta y_{t-p+1} + u_t, \tag{8.4.1}$$

where α and β are $(K \times r)$ matrices of rank r and all other symbols are defined as in (8.2.1). We assume that the model has been estimated by reduced rank ML or the two-stage procedure discussed in Section 7.2.5. If not explicitly stated otherwise, no restrictions are placed on the loading and short-run parameters.

8.4.1 Checking for Residual Autocorrelation

Asymptotic Properties of Residual Autocovariances and Autocorrelations

To study the properties of the autocovariances and autocorrelations of the residuals of a VECM, we denote the estimated residuals by $\widehat{u}_t$ and otherwise use the notation from Section 4.4 of Chapter 4 and Section 5.2.9 of Chapter 5, that is,

$$\widehat{C}_i := \frac{1}{T} \sum_{t=i+1}^{T} \widehat{u}_t \widehat{u}'_{t-i}, \quad i = 0, 1, \ldots, h,$$

$$\widehat{\mathbf{C}}_h := (\widehat{C}_1, \ldots, \widehat{C}_h), \quad \widehat{\mathbf{c}}_h := \text{vec}(\widehat{\mathbf{C}}_h),$$

are the residual autocovariances and $\widehat{R}_i$ $(i = 0, 1, \ldots, h)$,

$$\widehat{\mathbf{R}}_h := (\widehat{R}_1, \ldots, \widehat{R}_h), \quad \text{and} \quad \widehat{\mathbf{r}}_h := \text{vec}(\widehat{\mathbf{R}}_h)$$

denote the corresponding residual autocorrelations.

To derive the asymptotic properties of these quantities, it is convenient to also treat the case of a known cointegration matrix. Suppose the short-run and loading parameters of the VECM (8.4.1) are estimated with the same method as before, except that the true cointegration matrix is used instead of the estimated one. For the resulting estimation residuals we denote the previously defined quantities by tildes instead of hats. In other words, we have $\widetilde{C}_i$, $\widetilde{\mathbf{C}}_h$, and $\widetilde{\mathbf{c}}_h$ instead of $\widehat{C}_i$, $\widehat{\mathbf{C}}_h$, and $\widehat{\mathbf{c}}_h$, respectively, and so on. Brüggemann, Lütkepohl & Saikkonen (2004) showed that $\widetilde{C}_i$ and $\widehat{C}_i$ have the same asymptotic distributions. More precisely they proved the following lemma.

Lemma 8.1

$$\widetilde{C}_i - \widehat{C}_i = O_p(T^{-1}) \quad \text{for} \quad i = 1, 2, \ldots.$$

■

Although Brüggemann et al. (2004) showed this result for full VECMs estimated by reduced rank ML or unrestricted LS, it is clear from their proof that it also applies for other asymptotically equivalent estimation methods. The lemma enables us to get the asymptotic distributions of residual autocovariances, for example, with the same arguments as previously derived results (see, e.g., Proposition 5.7) because, if the cointegration matrix is known, all regressors in the VECM are stationary variables. Therefore, the same arguments apply as in Section 5.2.9 in Chapter 5. From Lemma 8.1 it then follows that

$$\sqrt{T}\widetilde{C}_i - \sqrt{T}\widehat{C}_i = o_p(1)$$

so that $\sqrt{T}\widetilde{\mathbf{c}}$ and $\sqrt{T}\widehat{\mathbf{c}}$ have identical asymptotic distributions. From the asymptotic distributions of the residual autocovariances we also get those of the residual autocorrelations in the familiar way.

Portmanteau and LM Tests for Residual Autocorrelation

Brüggemann et al. (2004) also showed that portmanteau and LM tests for residual autocorrelation can be used in conjunction with VECMs. In this case, the portmanteau statistic

$$Q_h := T \sum_{i=1}^{h} \operatorname{tr}(\widehat{C}_i' \widehat{C}_0^{-1} \widehat{C}_i \widehat{C}_0^{-1}) = T\widehat{\mathbf{c}}_h'(I_h \otimes \widehat{C}_0^{-1} \otimes \widehat{C}_0^{-1})\widehat{\mathbf{c}}_h$$

has an approximate $\chi^2(hK^2 - K^2(p-1) - Kr)$-distribution. Notice that the degrees of freedom are adjusted relative to the stationary full VAR case. Now we subtract from the number of autocovariances included in the statistic (hK^2) the number of estimated parameters not counting the elements of the cointegration matrix. Again this result follows from Lemma 8.1 which allows us to treat the cointegration matrix as known for asymptotic derivations, even if it is estimated.

It may be worth emphasizing that this result also holds if the VECM is estimated by unrestricted LS or, equivalently, the corresponding VAR in levels is estimated by unrestricted LS. In other words, if the integration and cointegration properties of a system of time series are not clear and an analyst therefore decides to use a levels VAR model, the portmanteau test cannot be used because the degrees of freedom of the approximating χ^2-distribution are not known. If one ignores this problem and simply uses the smaller degrees of freedom for the stationary full VAR case ($hK^2 - pK^2$), the test is likely to reject a true null hypothesis far too often. Also, recall that the approximate χ^2-distribution is obtained under the assumption that h goes to infinity with the sample size. Thus, the portmanteau test is not suitable for testing for residual autocorrelation of low order. As in the stationary case, in small samples it may be preferable to use the modified portmanteau statistic

$$\bar{Q}_h := T^2 \sum_{i=1}^{h} (T-i)^{-1} \mathrm{tr}(\widehat{C}'_i \widehat{C}_0^{-1} \widehat{C}_i \widehat{C}_0^{-1}).$$

The asymptotic distribution of the LM statistic for residual autocorrelation is not affected by the presence of integrated variables. We may use the auxiliary regression model

$$\begin{aligned}\widehat{u}_t &= \alpha \beta' y_{t-1} + \Gamma_1 \Delta y_{t-1} + \cdots + \Gamma_{p-1} \Delta y_{t-p+1} \\ &\quad + D_1 \widehat{u}_{t-1} + \cdots + D_h \widehat{u}_{t-h} + \varepsilon_t, \quad t = 1, \ldots, T,\end{aligned} \quad (8.4.2)$$

with $\widehat{u}_s = 0$ for $s < 1$, and compute the LM statistic for the hypotheses

$$H_0 : D_1 = \cdots = D_h = 0 \text{ vs. } H_1 : D_j \neq 0 \text{ for at least one } j \in \{1, \ldots, h\}.$$

The resulting LM statistic has an asymptotic χ^2-distribution,

$$\lambda_{LM}(h) \xrightarrow{d} \chi^2(hK^2),$$

if the null hypothesis of no autocorrelation is true, as in the stationary case (see Section 4.4.4). In contrast to the portmanteau test, the LM test is especially useful for testing for low order residual autocorrelation. For large h, it may in fact not be possible to estimate the parameters in the auxiliary model (8.4.2) because of an insufficient sample size.

Both the portmanteau tests and the LM tests are also applicable for subset VECMs with restrictions on the short-run and loading parameters. In that case, modifications analogous to those described in Section 5.2.9 have to be used. For the portmanteau tests, this means that the degrees of freedom in the approximate distributions have to be adjusted. More precisely, the number of estimated loading and short-term parameters is subtracted from the number of autocovariances included in the statistic. Here restricted parameters are not counted. For the LM tests, the auxiliary model has to be modified. The estimated residuals may now come from a two-stage estimation as described in Section 7.3.2. Moreover, the restrictions should also be accounted for in the auxiliary model as described in Section 5.2.9.

To illustrate these tests, we have applied them to the subset VECM (8.3.2) for the German interest rate/inflation example data. In this case, the cointegration relation is assumed to be known. According to our previous results, the same asymptotic distributions of the autocorrelation test statistics are obtained for an estimated cointegration relation. Moreover, deterministic terms are included in the model (8.3.2). Again, it can be shown that such terms do

Table 8.4. Residual autocorrelation tests for subset VECM (8.3.2)

test	$\lambda_{LM}(1)$	$\lambda_{LM}(2)$	$\lambda_{LM}(3)$	$\lambda_{LM}(4)$	Q_{24}	$\bar{Q}_{24}$	Q_{30}	$\bar{Q}_{30}$
test statistic	3.91	6.62	6.89	10.26	77.2	89.3	93.5	111.5
approximate distribution	$\chi^2(4)$	$\chi^2(8)$	$\chi^2(12)$	$\chi^2(16)$	$\chi^2(86)$	$\chi^2(86)$	$\chi^2(110)$	$\chi^2(110)$
p-value	0.42	0.58	0.86	0.85	0.74	0.38	0.87	0.44

not affect the asymptotic distributions of the portmanteau and LM tests for residual autocorrelation (see Brüggemann et al. (2004) for details).

Both types of tests have been applied with different lag orders h and the results are given in Table 8.4. The LM tests are useful for testing for low order residual autocorrelation. Therefore, only lags one to four are considered. Clearly, for a very long lag length (high order autocorrelation) the degrees of freedom may be exhausted in the auxiliary regression. In contrast, the lag length h has to be large for the approximate χ^2-distribution to be valid for the portmanteau tests. Therefore, only large lag orders are considered for these tests. All asymptotic p-values in Table 8.4 are substantially larger than conventional significance levels for such tests. Hence, there is no apparent residual autocorrelation problem for our example model.

8.4.2 Testing for Nonnormality

The tests for nonnormality considered in Chapter 4, Section 4.5, are based on the estimated residuals from a VAR process. We can use the residuals of a VECM instead without affecting the asymptotic distributions of the test statistics. This result follows again from the previously used superconsistency of the estimator for the cointegration matrix and the properties of the empirical moment matrices of integrated variables (see also Kilian & Demiroglu (2000)).

8.4.3 Tests for Structural Change

Time invariance is an important property of a VECM for valid statistical inference as well as for proper economic analysis and forecasting. Therefore, tests for structural change are also important tools for diagnostic checking of VECMs. The Chow tests and the prediction tests considered in Section 4.6 for stationary VARs can be extended easily to the case of cointegrated systems. We will discuss both types of tests in the following.

Chow Tests

Analogously to Section 4.6.1, in deriving the Chow tests, we assume that a change in the parameters of the VECM (8.4.1) is suspected after period

$T_1 < T$. For a sample $y_1, \ldots, y_T$ plus the required presample values, we can then set up the model in two parts:

$$\Delta Y_{(1)} = \alpha_{(1)} \beta'_{(1)} Y_{-1(1)} + \Gamma_{(1)} \Delta X_{(1)} + U_{(1)} \qquad (8.4.3)$$

and

$$\Delta Y_{(2)} = \alpha_{(2)} \beta'_{(2)} Y_{-1(2)} + \Gamma_{(2)} \Delta X_{(2)} + U_{(2)}, \qquad (8.4.4)$$

where $\Delta Y_{(1)} := [\Delta y_1, \ldots, \Delta y_{T_1}]$, $\Delta Y_{(2)} := [\Delta y_{T_1+1}, \ldots, \Delta y_T]$ and the other data matrices are partitioned accordingly. The parameter matrices $\alpha_{(i)}$, $\beta_{(i)}$ and $\Gamma_{(i)} := [\Gamma_{1(i)}, \ldots, \Gamma_{p-1(i)}]$ contain the values for the i-th subperiod, where $i = 1, 2$. Using similar arguments as in the proof of Proposition 7.3, it follows that the ML estimators of these parameter matrices can be determined by two separate reduced rank regressions applied to each of the two models (see also Problem 8.5). Notice that the presample values used in the second subsample coincide with the last observations of the first subperiod. To avoid this overlap, one may consider starting the second subsample only with observation y_{T_1+p+1}. Such a modification may have advantages in small samples if there is actually a structural break. If the null hypothesis of constant parameters in both subperiods is tested, however, there is no strong case for dropping observations between the two subsamples because, under the null hypothesis, all observations are generated by the same process.

Assuming, as in Section 4.6.1, that both parts of the sample go to infinity at a fixed proportion when T gets large, the asymptotic theory of Section 7.2 can be applied to derive the asymptotic distributions of the estimators. These asymptotic results can then be used to test parameter constancy hypotheses of the type

$$H_0 : \beta_{(1)} = \beta_{(2)}, \alpha_{(1)} = \alpha_{(2)}, \Gamma_{(1)} = \Gamma_{(2)} \qquad (8.4.5)$$

against the alternative that at least one of the equalities is violated. From the results in Section 7.2, it follows that the relevant Wald or LR tests have asymptotic χ^2-distributions. To determine the number of degrees of freedom, it has to be kept in mind, however, that a nonsingular asymptotic distribution for the estimator of β is only obtained upon suitable normalization. Hence, the equalities $\beta_{(1)} = \beta_{(2)}$ account only for $r(K-r)$ restrictions so that the LR statistic corresponding to (8.4.5) has a limiting χ^2-distribution with $r(K-r) + rK + (p-1)K^2$ degrees of freedom. It is also possible to construct similar tests for constancy of only a subset of the parameters (see Hansen (2003)). Moreover, the tests can be extended to models with deterministic terms.

Prediction Tests for Structural Change

In Chapter 4, Section 4.6.2, we have considered two tests for structural change that may be applied with small modifications if the data generation process is integrated or cointegrated. To see this, consider a K-dimensional Gaussian VECM with cointegration rank r, as in (8.4.1). Denoting the optimal h-step forecast at origin T by $y_T(h)$ and its MSE matrix by $\Sigma_y(h)$, as in Section 6.5, the quantity

$$\tau_h = [y_{T+h} - y_T(h)]' \Sigma_y(h)^{-1}[y_{T+h} - y_T(h)] \tag{8.4.6}$$

has a $\chi^2(K)$-distribution (see Section 4.6.2). If the parameters of the process were known, this statistic could be used to test whether y_{T+h} is generated by a Gaussian process of the type (8.4.1).

In practice, the process parameters have to be replaced by estimators and, in Section 4.6.2, we have modified the forecast MSE matrix accordingly. In Section 7.5, we have seen that the MSE approximation used for stationary, stable processes is not appropriate in the present integrated case. Therefore, we propose the statistic

$$\tau_h^\# = [y_{T+h} - \widetilde{y}_T(h)]' \widetilde{\Sigma}_y(h)^{-1}[y_{T+h} - \widetilde{y}_T(h)]/K, \tag{8.4.7}$$

which has an approximate $F(K, T - Kp - 1)$-distribution. Here

$$\widetilde{y}_T(h) = \widetilde{A}_1 \widetilde{y}_T(h-1) + \cdots + \widetilde{A}_p \widetilde{y}_T(h-p),$$

with $\widetilde{y}_T(j) := y_{T+j}$ for $j \le 0$, and the $\widetilde{A}_i$'s are the ML estimators of the A_i's obtained from ML estimation of the VECM and converting to the levels VAR representation. Moreover,

$$\widetilde{\Sigma}_y(h) = \sum_{i=0}^{h-1} \widetilde{\Phi}_i \widetilde{\Sigma}_u \widetilde{\Phi}_i',$$

where $\widetilde{\Sigma}_u$ is the ML estimator of Σ_u (see Proposition 7.3) and the $\widetilde{\Phi}_i$'s are computed from the $\widetilde{A}_i$'s by the recursions in (6.5.5). The F approximation to the distribution of $\tau_h^\#$ follows by noting that

$$\text{plim}(\tau_h - K\tau_h^\#) = 0.$$

Hence $K\tau_h^\#$ has an asymptotic $\chi^2(K)$-distribution and

$$\tau_h^\# \approx \chi^2(K)/K \approx F(K, T - Kp - 1), \tag{8.4.8}$$

where the numerator degrees of freedom are chosen in analogy with the stationary case. The quality of the F approximation in small samples is presently unknown.

A test based on several forecasts, as discussed in Section 4.6.2, may be generalized to integrated processes in a similar way. We may use

$$\overline{\lambda}_h = T \sum_{i=1}^{h} \widetilde{u}_{T+i}' \widetilde{\Sigma}_u^{-1} \widetilde{u}_{T+i} / [(T + Kp + 1)Kh] \tag{8.4.9}$$

as a test statistic with an approximate $F(Kh, T - Kp - 1)$-distribution. Here the $\tilde{u}_{T+i}$'s are the residuals obtained for the postsample period by using the ML estimators. The approximate distribution follows from asymptotic theory as in the stationary, stable case (see Problem 8.7).

A number of other tests for structural change are available for VECMs. For instance, Hansen & Johansen (1999) proposed tests which are based on the eigenvalues from the ML estimation procedure (see Proposition 7.3).

8.5 Exercises

8.5.1 Algebraic Exercises

Problem 8.1
Consider the model $y_t = \mu_t + x_t$, as in Section 8.2, for quarterly series with deterministic term

$$\mu_t = \mu_0 + \sum_{i=1}^{3} \delta_i s_{it},$$

where μ_0 and δ_i ($i = 1, 2, 3$) are ($K \times 1$) parameter vectors and the seasonal dummies are denoted by s_{it}, that is, s_{it} has a value of 1 in season i and $-1/3$ otherwise. Show that the VECM for y_t can be written as

$$\Delta y_t = \mathbf{\Pi}^\circ y_{t-1}^\circ + \mathbf{\Gamma}_1 \Delta y_{t-1} + \cdots + \mathbf{\Gamma}_{p-1} \Delta y_{t-p+1} + \sum_{i=1}^{3} \delta_i^* s_{it} + u_t.$$

Show also that the vector $(s_{1t}, s_{2t}, s_{3t}, s_{4t})'$ is orthogonal to $(1, 1, 1, 1)'$. In other words, the seasonal dummies are orthogonal to the constant term.

Problem 8.2
Use Proposition C.18 from Appendix C.8.2 to construct a mechanism for approximating the distribution

$$\left(\int_0^1 \mathbf{W}^\circ d\mathbf{W}' \right)' \left(\int_0^1 \mathbf{W}^\circ \mathbf{W}^{\circ \prime} ds \right)^{-1} \left(\int_0^1 \mathbf{W}^\circ d\mathbf{W}' \right)$$

in (8.2.12) via simulation.

Problem 8.3
Write down the EGLS estimation problem for the cointegration rank tests described in Section 8.2.7 for a model with $\mu_t = \mu_0$.

Problem 8.4
Consider residual autocorrelation tests for a three-dimensional VECM with two lagged differences ($p = 3$) and a cointegrating rank of $r = 2$. What are the approximate distributions of the Q_{20}, Q_{25}, and Q_{30} portmanteau statistics? What are the asymptotic distributions of $\lambda_{LM}(2)$ and $\lambda_{LM}(5)$?

Problem 8.5
Show that, for a sample $y_1, \ldots, y_T$ with a possible structural break in period T_1, $1 < T_1 < T$, a VECM can be estimated by two separate reduced rank regressions as in Proposition 7.3. (Hint: Use similar arguments as in the proof of Proposition 7.3.)

Problem 8.6
Consider the model

$$[\Delta Y_{(1)} : \Delta Y_{(2)}] = \alpha\beta' Y_{-1} + [\Gamma_{(1)} : \Gamma_{(2)}]\begin{bmatrix} \Delta X_{(1)} & 0 \\ 0 & \Delta X_{(2)} \end{bmatrix} + [U_{(1)} : U_{(2)}],$$

where the symbols are defined as in (8.4.3) and (8.4.4). Derive the ML estimators of the parameters. (Hint: Use similar arguments as in the proof of Proposition 7.3.)

Problem 8.7
Under the conditions of Section 8.4.3, show that

$$(T + Kp + 1)Kh\overline{\lambda}_h/T \overset{d}{\to} \chi^2(hK),$$

where $\overline{\lambda}_h$ is the statistic defined in (8.4.9).

8.5.2 Numerical Exercises

The following problems are based on the U.S. data given in File E3 and described in Section 7.4.3. The variables are defined in the same way as in that section.

Problem 8.8
Use a maximum order of 10 and determine the VAR order of the example system by the three model selection criteria AIC, HQ, and SC.

Problem 8.9
Assume that the data are generated by a VAR(3) process and determine the cointegration rank with the tests described in Section 8.2.

Problem 8.10
Modify the AIC criterion appropriately and choose the order and cointegration rank simultaneously with this criterion. Compare the result with that from Problem 8.9.

Problem 8.11
Apply the ML procedure described in Section 7.2.3 and the EGLS estimator of Section 7.2.2 to estimate the cointegration relation and the other parameters of a VECM with cointegration rank $r = 1$, two lagged differences (i.e., $p = 3$) and an intercept. Compare the estimates.

Problem 8.12
Use diagnostic tests to check the adequacy of the model estimated in Problem 8.11.

Part III

Structural and Conditional Models

In Parts I and II, we have assumed that the time series of interest are generated by stationary or cointegrated reduced form VAR processes. In this part, structural models and systems with unmodelled, exogenous variables are discussed. In Chapter 9, structural VARs and VECMs are considered and, in Chapter 10, conditional or partial models are treated, where we condition on some variables whose generation process is not part of the model. These systems may be stationary if the unmodelled variables are generated by stationary processes. Alternatively, some or all of the unmodelled variables may be nonstochastic fixed quantities. In that case, the mean vectors of the time series variables of interest may be time varying and, hence, the series may not be stationary. They may still be stationary when the deterministic terms are removed, however. Generally, some of the endogenous and unmodelled stochastic variables may be integrated and have stochastic trends. Suitable models for this case will also be considered in Chapter 10.

9
Structural VARs and VECMs

In Chapters 2 and 6, we have seen that, on the one hand, impulse responses are an important tool to uncover the relations between the variables in a VAR or VECM and, on the other hand, there are some obstacles in their interpretation. In particular, impulse responses are generally not unique and it is often not clear which set of impulse responses actually reflects the ongoings in a given system. Because the different sets of impulses can be computed from the same underlying VAR or VECM, it is clear that nonsample information has to be used to decide on the proper set for a particular given model. In econometric terminology, VARs are reduced form models and structural restrictions are required to identify the relevant innovations and impulse responses. In this chapter, different possible restrictions that have been proposed in the literature will be considered. The resulting models are known as *structural VAR* (SVAR) models (see, e.g., Sims (1981, 1986), Bernanke (1986), Shapiro & Watson (1988), Blanchard & Quah (1989)) or *structural VECMs* (SVECMs) (e.g., King, Plosser, Stock & Watson (1991), Jacobson, Vredin & Warne (1997), Gonzalo & Ng (2001), Breitung, Brüggemann & Lütkepohl (2004)).

In the next section, structural restrictions will be discussed for stationary processes. Some of them will also be relevant for VARs with integrated variables. Such variables are explicitly taken into account in VECMs for which structural restrictions will be discussed in Section 9.2. It will be seen that VECMs offer additional possibilities for structural restrictions. The general modelling strategy for both SVARs and SVECMs is to specify and estimate a reduced form model first and then focus on the structural parameters and the resulting structural impulse responses. Estimation of structural VARs and VECMs will be discussed in Section 9.3 and impulse response analysis and forecast error variance decomposition based on such models are considered in Section 9.4. Some extensions of the setup used in this chapter are pointed out in Section 9.5.

9.1 Structural Vector Autoregressions

Our point of departure is a K-dimensional stationary, stable VAR(p) process,

$$y_t = A_1 y_{t-1} + \cdots + A_p y_{t-p} + u_t, \tag{9.1.1}$$

where, as usual, y_t is a ($K \times 1$) vector of observable time series variables, the A_j's ($j = 1, \ldots, p$) are ($K \times K$) coefficient matrices and u_t is K-dimensional white noise with $u_t \sim (0, \Sigma_u)$. Deterministic terms have been excluded for simplicity. In other words, we just consider the stochastic part of a data generation process because it is the part of interest from the point of view of structural modelling and impulse response analysis. From Chapter 2, it is known that the process (9.1.1) has a Wold MA representation

$$y_t = u_t + \Phi_1 u_{t-1} + \Phi_2 u_{t-2} + \cdots, \tag{9.1.2}$$

where

$$\Phi_s = \sum_{j=1}^{s} \Phi_{s-j} A_j, \quad s = 1, 2, \ldots, \tag{9.1.3}$$

with $\Phi_0 = I_K$.

In Chapter 2, we have also seen that the elements of the Φ_j matrices are the forecast error impulse responses. They may not reflect the relations between the variables properly because the components of u_t may be instantaneously correlated, that is, Σ_u may not be a diagonal matrix. Thus, isolated shocks in the components of u_t may not be likely in practice. From Chapter 2, we also know that there are different ways to orthogonalize the impulses. One possibility is based on a Choleski decomposition of the white noise covariance matrix, $\Sigma_u = PP'$, where P is a lower-triangular matrix with positive elements on the main diagonal. Again such an approach is arbitrary and therefore unsatisfactory, unless there are special reasons for a recursive structure. We will now discuss different ways to use nonsample information in specifying unique innovations and, hence, unique impulse responses. The relevant models will be referred to as A-model, B-model and AB-model. The latter label was also used by Amisano & Giannini (1997). The models will be considered in turn in the following.

9.1.1 The A-Model

A conventional approach to finding a model with instantaneously uncorrelated residuals is to model the instantaneous relations between the observable variables directly. That may be done by considering a structural form model,

$$\mathsf{A} y_t = A_1^* y_{t-1} + \cdots + A_p^* y_{t-p} + \varepsilon_t, \tag{9.1.4}$$

where $A_j^* := \mathsf{A}A_j$ $(j = 1, \ldots, p)$ and $\varepsilon_t := \mathsf{A}u_t \sim (0, \Sigma_\varepsilon = \mathsf{A}\Sigma_u \mathsf{A}')$. Thus, for a proper choice of A, ε_t will have a diagonal covariance matrix. An MA representation based on the ε_t is given by

$$y_t = \Theta_0 \varepsilon_t + \Theta_1 \varepsilon_{t-1} + \Theta_2 \varepsilon_{t-2} + \cdots, \tag{9.1.5}$$

where $\Theta_j = \Phi_j \mathsf{A}^{-1}$ $(j = 0, 1, 2, \ldots)$. The elements of the Θ_j matrices represent the responses to ε_t shocks. If an identified structural form (9.1.4) can be found, the corresponding impulse responses will be unique.

It may be worth reflecting a little on the restrictions required for a unique matrix A of instantaneous effects. From the relation

$$\Sigma_\varepsilon = \mathsf{A}\Sigma_u \mathsf{A}'$$

and the assumption of a diagonal Σ_ε matrix, we get $K(K-1)/2$ independent equations, that is, all $K(K-1)/2$ off-diagonal elements of $\mathsf{A}\Sigma_u \mathsf{A}'$ are equal to zero. To solve uniquely for all K^2 elements of A, we need a set of K^2 equations, however. In other words, we need $K(K+1)/2$ additional equations. They may be set up in the form of restrictions for the elements of A. Clearly, we may want to choose the diagonal elements of A to be unity. This normalization enables us to write the k-th equation of (9.1.4) with y_{kt} as the left-hand variable. In addition to this normalization, we still need another $K(K-1)/2$ restrictions. Such restrictions have to come from nonsample sources. For example, if a Wold causal ordering is possible, where y_{1t} may have an instantaneous impact on all the other variables, y_{2t} may have an instantaneous impact on all other variables except y_{1t}, and so on (see Section 2.3.2), then

$$\mathsf{A} = \begin{bmatrix} 1 & 0 & \cdots & 0 \\ a_{21} & 1 & & 0 \\ \vdots & & \ddots & \vdots \\ a_{K1} & a_{K2} & \cdots & 1 \end{bmatrix}$$

is a lower-triangular matrix. Thus, we have just enough restrictions ($K(K-1)/2$ zeros above the main diagonal) so that the innovations and the associated impulse responses are just-identified. The zeros can also appear in a different arrangement as off-diagonal elements of A. There can also be more than $K(K-1)/2$ restrictions, of course. In SVAR modelling it is common, however, that just-identified models are considered. In other words, only as few restrictions are imposed as are necessary for obtaining unique impulse responses. If at some stage of the analysis it turns out that further restrictions are compatible with the data, it is also possible to impose them, of course.

In the presently considered model, the identifying restrictions are imposed on the matrix A such that $\varepsilon_t = \mathsf{A}u_t$ has a diagonal covariance matrix. This model will be called the A-model in the following. Given the way we have introduced the associated restrictions, it is plausible to assume that A has a unit main diagonal. In that case $K(K-1)/2$ restrictions are required for the off-diagonal elements of A to ensure just-identified shocks ε_t and, hence, just-identified impulse responses. If the restrictions are such that A is lower-triangular, the same is true for A^{-1}. Thus, the resulting Θ_j impulse responses are qualitatively the same as the orthogonalized impulse responses based on a Choleski decomposition of Σ_u which were considered in Chapter 2. The only difference is that, for the latter case, the w_t impulses have unit variances which may not be the case for the presently considered ε_t impulses.

Regarding the restrictions for A, it should be understood that they cannot be arbitrary restrictions. Writing them in the form $C_A \text{vec}(A) = c_A$, where C_A is a $(\frac{1}{2}K(K+1) \times K^2)$ selection matrix and c_A is a suitable $(\frac{1}{2}K(K+1) \times 1)$ fixed vector, the restrictions have to be such that the system of equations

$$A^{-1}\Sigma_\varepsilon A'^{-1} = \Sigma_u \quad \text{and} \quad C_A \text{vec}(A) = c_A \tag{9.1.6}$$

has a unique solution, at least locally. Clearly, this system is nonlinear in A. Therefore, we can only hope for local uniqueness or identification in general. The following proposition gives a necessary and sufficient condition for (9.1.6) to have a locally unique solution and, thus, for local identification of the structural parameters.

Proposition 9.1 (*Identification of the A-Model*)
Let Σ_ε be a $(K \times K)$ positive definite diagonal matrix and let A be a $(K \times K)$ nonsingular matrix. Then, for a given symmetric, positive definite $(K \times K)$ matrix Σ_u, an $(N \times K^2)$ matrix C_A and a fixed $(N \times 1)$ vector c_A, the system of equations in (9.1.6) has a locally unique solution for A and the diagonal elements of Σ_ε if and only if

$$\text{rk} \begin{bmatrix} -2\mathbf{D}_K^+(\Sigma_u \otimes A^{-1}) & \mathbf{D}_K^+(A^{-1} \otimes A^{-1})\mathbf{D}_K \\ C_A & 0 \\ 0 & C_\sigma \end{bmatrix} = K^2 + \tfrac{1}{2}K(K+1).$$

Here $\mathbf{D}_K$ is a $(K^2 \times \frac{1}{2}K(K+1))$ duplication matrix, $\mathbf{D}_K^+ := (\mathbf{D}_K'\mathbf{D}_K)^{-1}\mathbf{D}_K'$, and C_σ is a $(\frac{1}{2}K(K-1) \times \frac{1}{2}K(K+1))$ selection matrix which selects the elements of $\text{vech}(\Sigma_\varepsilon)$ below the main diagonal. ∎

Proof: For an n-dimensional function $\varphi(x)$ of the m-dimensional vector x, the system of equations $\varphi(x) = 0$ can be solved locally uniquely for x in a neighborhood of a given vector x_0 if and only if $\text{rk}(\partial \varphi / \partial x'|_{x=x_0}) = m$ (see, e.g., Rothenberg (1971, Theorem 6)). Hence, considering the function

$$\begin{bmatrix} \text{vec}(A) \\ \text{vech}(\Sigma_\varepsilon) \end{bmatrix} \mapsto \begin{bmatrix} \text{vec}(A^{-1}\Sigma_\varepsilon A'^{-1} - \Sigma_u) \\ C_A \text{vec}(A) - c_A \\ C_\sigma \text{vech}(\Sigma_\varepsilon) \end{bmatrix},$$

a locally unique solution for A and $\text{vech}(\Sigma_\varepsilon)$ exists for a given Σ_u if and only if

$$\text{rk}\begin{bmatrix} \dfrac{\partial\,\text{vech}(\mathbf{A}^{-1}\Sigma_\varepsilon\mathbf{A}'^{-1})}{\partial\,\text{vec}(\mathbf{A})'} & \dfrac{\partial\,\text{vech}(\mathbf{A}^{-1}\Sigma_\varepsilon\mathbf{A}'^{-1})}{\partial\,\text{vech}(\Sigma_\varepsilon)'} \\ C_\mathbf{A} & 0 \\ 0 & C_\sigma \end{bmatrix} = K^2 + \tfrac{1}{2}K(K+1).$$

Taking into account that the off-diagonal elements of Σ_ε are uniquely determined by $C_\sigma\,\text{vech}(\Sigma_\varepsilon) = 0$, a locally unique solution for $\mathbf{A}$ and the diagonal elements of Σ_ε exists if and only if the rank condition is satisfied. Thus, the proposition follows by using the rules for matrix and vector differentiation from Appendix A.13 and noting that

$$\begin{aligned}
\dfrac{\partial\,\text{vech}(\mathbf{A}^{-1}\Sigma_\varepsilon\mathbf{A}'^{-1})}{\partial\,\text{vech}(\Sigma_\varepsilon)'} &= \mathbf{D}_K^+ \dfrac{\partial\,\text{vec}(\mathbf{A}^{-1}\Sigma_\varepsilon\mathbf{A}'^{-1})}{\partial\,\text{vech}(\Sigma_\varepsilon)'} \\
&= \mathbf{D}_K^+(\mathbf{A}^{-1}\otimes\mathbf{A}^{-1}) \dfrac{\partial\,\text{vec}(\Sigma_\varepsilon)}{\partial\,\text{vech}(\Sigma_\varepsilon)'} \\
&= \mathbf{D}_K^+(\mathbf{A}^{-1}\otimes\mathbf{A}^{-1})\mathbf{D}_K
\end{aligned}$$

and

$$\begin{aligned}
\dfrac{\partial\,\text{vech}(\mathbf{A}^{-1}\Sigma_\varepsilon\mathbf{A}'^{-1})}{\partial\,\text{vec}(\mathbf{A})'} &= \mathbf{D}_K^+ \dfrac{\partial\,\text{vec}(\mathbf{A}^{-1}\Sigma_\varepsilon\mathbf{A}'^{-1})}{\partial\,\text{vec}(\mathbf{A}^{-1})'} \dfrac{\partial\,\text{vec}(\mathbf{A}^{-1})}{\partial\,\text{vec}(\mathbf{A})'} \\
&= \mathbf{D}_K^+\left[(\mathbf{A}^{-1}\Sigma_\varepsilon\otimes I_K)\dfrac{\partial\,\text{vec}(\mathbf{A}^{-1})}{\partial\,\text{vec}(\mathbf{A}^{-1})'} \right. \\
&\qquad \left. + (I_K\otimes\mathbf{A}^{-1}\Sigma_\varepsilon)\dfrac{\partial\,\text{vec}(\mathbf{A}'^{-1})}{\partial\,\text{vec}(\mathbf{A}^{-1})'}\right]\dfrac{\partial\,\text{vec}(\mathbf{A}^{-1})}{\partial\,\text{vec}(\mathbf{A})'} \\
&= -\mathbf{D}_K^+(I_{K^2}+\mathbf{K}_{KK})(\mathbf{A}^{-1}\Sigma_\varepsilon\otimes I_K)(\mathbf{A}'^{-1}\otimes\mathbf{A}^{-1}) \\
&= -\mathbf{D}_K^+(I_{K^2}+\mathbf{K}_{KK})(\Sigma_u\otimes\mathbf{A}^{-1}) \\
&= -2\mathbf{D}_K^+(\Sigma_u\otimes\mathbf{A}^{-1}),
\end{aligned}$$

where $\mathbf{K}_{KK}$ denotes a $(K^2\times K^2)$ commutation matrix and the last equality sign holds because $\mathbf{D}_K^+\mathbf{K}_{KK} = \mathbf{D}_K^+$ (see Appendix A.12.2). ∎

Although this proposition provides a condition for local identification of the A-model only, a globally unique solution is obtained if the diagonal elements of $\mathbf{A}$ are restricted to 1. A discussion of the nonuniqueness problem resulting from sign changes of some elements will be deferred to Section 9.1.2.

For practical purposes, it is problematic that the identification condition in Proposition 9.1 involves unknown parameters. Therefore, strictly speaking, it can only be checked when the true parameters are known. In practice, the unknown quantities may be replaced by estimates and the condition may be checked using the estimated matrix because it can be shown that the rank of the relevant matrix is either smaller than $K^2 + \tfrac{1}{2}K(K+1)$ everywhere in the parameter space or the rank condition is satisfied almost everywhere. In the latter case, it can fail only on a set of Lebesgue measure zero. Thus, if a randomly drawn vector from the parameter space is considered, it should satisfy the rank condition with probability one, if the model is locally identified. In any case, C_A must have at least $K(K+1)/2$ rows to ensure identification. In other words, having $K(K+1)/2$ restrictions is a necessary condition for identification.

Although we have stated the restrictions for the A matrix in the form $C_\mathsf{A} \text{vec}(\mathsf{A}) = c_\mathsf{A}$ in the foregoing, we note that they can be written alternatively in the form

$$\text{vec}(\mathsf{A}) = R_\mathsf{A} \gamma_\mathsf{A} + r_\mathsf{A},$$

where R_A and r_A are a suitable fixed matrix and a suitable vector, respectively, and γ_A is the vector of unrestricted parameters (see Chapter 5, Section 5.2.1).

9.1.2 The B-Model

Generally, in impulse response analysis the emphasis has shifted from specifying the relations between the observable variables directly to interpreting the unexpected part of their changes or the shocks. Therefore, it is not uncommon to identify the structural innovations ε_t directly from the forecast errors or reduced form residuals u_t. One way to do so is to think of the forecast errors as linear functions of the structural innovations. In that case, we have the relations $u_t = \mathsf{B}\varepsilon_t$. Hence, $\Sigma_u = \mathsf{B}\Sigma_\varepsilon \mathsf{B}'$. Normalizing the variances of the structural innovations to one, i.e., assuming $\varepsilon_t \sim (0, I_K)$, gives

$$\Sigma_u = \mathsf{B}\mathsf{B}'. \tag{9.1.7}$$

Due to the symmetry of the covariance matrix, these relations specify only $K(K+1)/2$ different equations and we need again $K(K-1)/2$ further relations to identify all K^2 elements of B. As in the previous A-model case, choosing B to be lower-triangular, for example, provides sufficiently many restrictions. Hence, choosing B by a Choleski decomposition solves the identification or uniqueness problem, as we have also seen in Chapter 2, Section 2.3.2. Now it is assumed, however, that this recursive structure is chosen only if it has some theoretical justification so that the ε_t's can be regarded as structural innovations. This property makes them potentially different from the w_t innovations in Chapter 2 which were obtained by a mechanical application of the Choleski decomposition. In principle, there could be other zero restrictions for B in the present context. The triangular form is just an example. In practice, it is perhaps the most important case (e.g., Eichenbaum & Evans (1995), Christiano, Eichenbaum & Evans (1996)).

The present model with

$$u_t = \mathsf{B}\varepsilon_t$$

and $\varepsilon_t \sim (0, I_K)$ will be called B-model in the following and it is worth remembering that at least $K(K-1)/2$ restrictions have to be imposed to identify B. If there are just zero restrictions they can be written in the form

$$C_\mathsf{B} \text{vec}(\mathsf{B}) = 0, \qquad (9.1.8)$$

where C_B is an $(N \times K^2)$ selection matrix. A necessary and sufficient rank condition for local identification of the model is given in the next proposition.

Proposition 9.2 (*Local Identification of the* B-*Model*)
Let B be a nonsingular $(K \times K)$ matrix. Then, for a given symmetric, positive definite $(K \times K)$ matrix Σ_u and an $(N \times K^2)$ matrix C_B, the system of equations in (9.1.7)/(9.1.8) has a locally unique solution if and only if

$$\text{rk} \begin{bmatrix} 2\mathbf{D}_K^+ (\mathsf{B} \otimes I_K) \\ C_\mathsf{B} \end{bmatrix} = K^2.$$

■

Proof: Using the same kind of reasoning as in the proof of Proposition 9.1, the result of Proposition 9.2 follows by noting that

$$\frac{\partial \text{vech}(\mathsf{BB}')}{\partial \text{vec}(\mathsf{B})'} = \mathbf{D}_K^+ (I_{K^2} + \mathbf{K}_{KK})(\mathsf{B} \otimes I_K) = 2\mathbf{D}_K^+ (\mathsf{B} \otimes I_K).$$

■

A necessary condition for the $((\frac{1}{2}K(K+1) + N) \times K^2)$ matrix

$$\begin{bmatrix} 2\mathbf{D}_K^+ (\mathsf{B} \otimes I_K) \\ C_\mathsf{B} \end{bmatrix}$$

to have rank K^2 is that $N = \frac{1}{2}K(K-1)$. In other words, we need $\frac{1}{2}K(K-1)$ restrictions for identification, as mentioned earlier.

It is easy to see that the solution of the system (9.1.7)/(9.1.8) will not be globally unique because for any matrix B satisfying the equations, $-\mathsf{B}$ will also be a solution. This result is due to the fact that B enters the equations (9.1.7) in "squared" form. In fact, for any solution B, the matrix $\mathsf{B}\Lambda$ will also be a solution for any diagonal matrix Λ which has only 1 and -1 elements on the main diagonal. Obviously, if B is such that (9.1.7) and (9.1.8) hold, $\Sigma_u = \mathsf{B}\Lambda\Lambda'\mathsf{B}'$ also holds because $\Lambda\Lambda' = I_K$. Moreover,

$$C_\mathsf{B} \text{vec}(\mathsf{B}\Lambda) = C_\mathsf{B}(\Lambda \otimes I_K)\text{vec}(\mathsf{B}) = 0,$$

because for each element $\mathsf{b}_{ij} = 0$ we have $-\mathsf{b}_{ij} = 0$. Thus, each column of B can be replaced by a column with opposite sign. Hence, the restrictions in (9.1.8) identify B only locally in general. Uniqueness can potentially be obtained by fixing the signs of the diagonal elements, however. The signs of the diagonal elements of B determine the signs of shocks. Thus, if we want to study the effect of a positive shock to a particular variable while the corresponding diagonal element of B is negative, we can just reverse the signs of all elements in the relevant column of B or, in other words, we can just reverse the signs of all instantaneous responses to the corresponding shock to find the desired result.

For later purposes, it is also worth noting that the restrictions can be expressed in the alternative form

$$\text{vec}(\mathsf{B}) = R_\mathsf{B}\gamma_\mathsf{B}, \tag{9.1.9}$$

where γ_B contains all the unrestricted coefficients of B and R_B is a fixed matrix of zeros and ones (see Section 5.2.1).

9.1.3 The AB-Model

It is also possible to consider both types of restrictions of the previous subsections simultaneously. That is, we may consider the so-called AB-model,

$$\mathsf{A}u_t = \mathsf{B}\varepsilon_t, \qquad \varepsilon_t \sim (0, I_K). \tag{9.1.10}$$

In this case, a simultaneous equations system is formulated for the errors of the reduced form model rather than the observable variables directly. Thereby the model accounts for the shift from specifying direct relations for the observable variables to formulating relations for the innovations. Applications of this methodology can, for instance, be found in Galí (1992) and Pagan (1995) (see also Breitung et al. (2004) for further discussion and an illustration).

In this model, we get from (9.1.10), $u_t = \mathsf{A}^{-1}\mathsf{B}\varepsilon_t$ and, hence, $\Sigma_u = \mathsf{A}^{-1}\mathsf{B}\mathsf{B}'\mathsf{A}^{-1'}$. Thus, we have $K(K+1)/2$ equations

$$\text{vech}(\Sigma_u) = \text{vech}(\mathsf{A}^{-1}\mathsf{B}\mathsf{B}'\mathsf{A}^{-1'}), \tag{9.1.11}$$

whereas the two matrices A and B have K^2 elements each. Thus, we need additionally $2K^2 - \frac{1}{2}K(K+1)$ restrictions to identify all $2K^2$ elements of A and B at least locally. Even if the diagonal elements of A are set to one, $2K^2 - K - \frac{1}{2}K(K+1)$ further restrictions are needed for identification. Therefore, it is perhaps not surprising that most applications consider special cases with $\mathsf{A} = I_K$ (B-model) or $\mathsf{B} = I_K$ (A-model). Still, the general model is a useful framework for SVAR analysis. The restrictions are typically normalization or zero restrictions which can be written in the form of linear equations,

$$\text{vec}(\mathsf{A}) = R_\mathsf{A}\gamma_\mathsf{A} + r_\mathsf{A} \qquad \text{and} \qquad \text{vec}(\mathsf{B}) = R_\mathsf{B}\gamma_\mathsf{B} + r_\mathsf{B}, \tag{9.1.12}$$

where R_A and R_B are suitable fixed matrices of zeros and ones, γ_A and γ_B are vectors of free parameters and r_A and r_B are vectors of fixed parameters which allow, for instance, to normalize the diagonal elements of A. Although r_B is typically zero, as in (9.1.9), we present the restrictions for B here with a general r_B vector because this additional term will not complicate the analysis.

Multiplying the two sets of equations in (9.1.12) by orthogonal complements of R_A and R_B, $R_{\mathsf{A}\perp}$ and $R_{\mathsf{B}\perp}$, respectively, it is easy to see that they can be written alternatively in the form

$$C_\mathsf{A}\operatorname{vec}(\mathsf{A}) = c_\mathsf{A} \quad \text{and} \quad C_\mathsf{B}\operatorname{vec}(\mathsf{B}) = c_\mathsf{B}, \tag{9.1.13}$$

where $C_\mathsf{A} = R_{\mathsf{A}\perp}$, $C_\mathsf{B} = R_{\mathsf{B}\perp}$, $c_\mathsf{A} = R_{\mathsf{A}\perp} r_\mathsf{A}$ and $c_\mathsf{B} = R_{\mathsf{B}\perp} r_\mathsf{B}$ (see Appendix A.8.2 for the definition of an orthogonal complement of a matrix). The matrices C_A and C_B may be thought of as appropriate selection matrices. Again, in general, the restrictions will ensure only local uniqueness of A and B due to the nonlinear nature of the full set of equations from which to solve for the two matrices. The following proposition states a rank condition for local identification.

Proposition 9.3 (*Local Identification of the* AB-*Model*)
Let A and B be nonsingular $(K \times K)$ matrices. Then, for a given symmetric, positive definite $(K \times K)$ matrix Σ_u, the system of equations in (9.1.11)/(9.1.13) has a locally unique solution if and only if

$$\operatorname{rk}\begin{bmatrix} -2\mathbf{D}_K^+(\Sigma_u \otimes \mathsf{A}^{-1}) & 2\mathbf{D}_K^+(\mathsf{A}^{-1}\mathsf{B} \otimes \mathsf{A}^{-1}) \\ C_\mathsf{A} & 0 \\ 0 & C_\mathsf{B} \end{bmatrix} = 2K^2. \tag{9.1.14}$$

■

Proof: Again, we can use the same reasoning as in the proof of Proposition 9.1. The result of Proposition 9.3 is then obtained by noting that

$$\begin{aligned}
\frac{\partial \operatorname{vech}(\mathsf{A}^{-1}\mathsf{B}\mathsf{B}'\mathsf{A}'^{-1})}{\partial \operatorname{vec}(\mathsf{A})'} &= \frac{\partial \operatorname{vech}(\mathsf{A}^{-1}\mathsf{B}\mathsf{B}'\mathsf{A}'^{-1})}{\partial \operatorname{vec}(\mathsf{A}^{-1}\mathsf{B})'} \frac{\partial \operatorname{vec}(\mathsf{A}^{-1}\mathsf{B})}{\partial \operatorname{vec}(\mathsf{A})'} \\
&= \mathbf{D}_K^+ \frac{\partial \operatorname{vec}(\mathsf{A}^{-1}\mathsf{B}\mathsf{B}'\mathsf{A}'^{-1})}{\partial \operatorname{vec}(\mathsf{A}^{-1}\mathsf{B})'} \frac{\partial \operatorname{vec}(\mathsf{A}^{-1}\mathsf{B})}{\partial \operatorname{vec}(\mathsf{A})'} \\
&= \mathbf{D}_K^+ \left[(\mathsf{A}^{-1}\mathsf{B} \otimes I_K) \frac{\partial \operatorname{vec}(\mathsf{A}^{-1}\mathsf{B})}{\partial \operatorname{vec}(\mathsf{A}^{-1}\mathsf{B})'} \right. \\
&\quad \left. + (I_K \otimes \mathsf{A}^{-1}\mathsf{B}) \frac{\partial \operatorname{vec}(\mathsf{B}'\mathsf{A}'^{-1})}{\partial \operatorname{vec}(\mathsf{A}^{-1}\mathsf{B})'} \right] \frac{\partial \operatorname{vec}(\mathsf{A}^{-1}\mathsf{B})}{\partial \operatorname{vec}(\mathsf{A})'} \\
&= \mathbf{D}_K^+ \left[(\mathsf{A}^{-1}\mathsf{B} \otimes I_K) + (I_K \otimes \mathsf{A}^{-1}\mathsf{B})\mathbf{K}_{KK}\right] \\
&\quad \times (\mathsf{B}' \otimes I_K) \frac{\partial \operatorname{vec}(\mathsf{A}^{-1})}{\partial \operatorname{vec}(\mathsf{A})'} \\
&= -\mathbf{D}_K^+ (I_{K^2} + \mathbf{K}_{KK})(\Sigma_u \otimes \mathsf{A}^{-1}) \\
&= -2\mathbf{D}_K^+ (\Sigma_u \otimes \mathsf{A}^{-1})
\end{aligned}$$

and

$$\begin{aligned}
\frac{\partial \operatorname{vech}(\mathsf{A}^{-1}\mathsf{B}\mathsf{B}'\mathsf{A}'^{-1})}{\partial \operatorname{vec}(\mathsf{B})'} &= \frac{\partial \operatorname{vech}(\mathsf{A}^{-1}\mathsf{B}\mathsf{B}'\mathsf{A}'^{-1})}{\partial \operatorname{vec}(\mathsf{A}^{-1}\mathsf{B})'} \frac{\partial \operatorname{vec}(\mathsf{A}^{-1}\mathsf{B})}{\partial \operatorname{vec}(\mathsf{B})'} \\
&= \mathbf{D}_K^+ (I_{K^2} + \mathbf{K}_{KK})(\mathsf{A}^{-1}\mathsf{B} \otimes \mathsf{A}^{-1}) \\
&= 2\mathbf{D}_K^+ (\mathsf{A}^{-1}\mathsf{B} \otimes \mathsf{A}^{-1}),
\end{aligned}$$

because $\mathbf{D}_K^+ \mathbf{K}_{KK} = \mathbf{D}_K^+$ (see Appendix A.12.2). ■

To illustrate the AB-model, we follow Breitung et al. (2004) and use a small macro system from Pagan (1995) for output q_t, an interest rate i_t, and real money m_t. The residuals of the reduced form VAR model will be denoted by $u_t = (u_t^q, u_t^i, u_t^m)'$. Pagan (1995) uses Keynesian arguments to specify the following relations between the reduced form residuals and the structural innovations:

$$\begin{aligned}
u_t^q &= -a_{12} u_t^i + b_{11} \varepsilon_t^{IS} & \text{(IS curve)}, \\
u_t^i &= -a_{21} u_t^q - a_{23} u_t^m + b_{22} \varepsilon_t^{LM} & \text{(inverse LM curve)}, \\
u_t^m &= b_{33} \varepsilon_t^m & \text{(money supply rule)}.
\end{aligned}$$

Here $\varepsilon_t = (\varepsilon_t^{IS}, \varepsilon_t^{LM}, \varepsilon_t^m)'$ is the vector of structural innovations with $\varepsilon_t \sim (0, I_K)$ (see Breitung et al. (2004) for further discussion of this example system).

For our purposes the three equations can be written in AB-model form as

$$\begin{bmatrix} 1 & a_{12} & 0 \\ a_{21} & 1 & a_{23} \\ 0 & 0 & 1 \end{bmatrix} u_t = \begin{bmatrix} b_{11} & 0 & 0 \\ 0 & b_{22} & 0 \\ 0 & 0 & b_{33} \end{bmatrix} \varepsilon_t.$$

Thus, we have the following set of restrictions:

$$\text{vec}(\mathsf{A}) = \begin{bmatrix} 1 \\ a_{21} \\ 0 \\ a_{12} \\ 1 \\ 0 \\ 0 \\ a_{23} \\ 1 \end{bmatrix} = \begin{bmatrix} 0 & 0 & 0 \\ 1 & 0 & 0 \\ 0 & 0 & 0 \\ 0 & 1 & 0 \\ 0 & 0 & 0 \\ 0 & 0 & 0 \\ 0 & 0 & 0 \\ 0 & 0 & 1 \\ 0 & 0 & 0 \end{bmatrix} \begin{bmatrix} a_{21} \\ a_{12} \\ a_{23} \end{bmatrix} + \begin{bmatrix} 1 \\ 0 \\ 0 \\ 0 \\ 1 \\ 0 \\ 0 \\ 0 \\ 1 \end{bmatrix}$$

and

$$\text{vec}(\mathsf{B}) = \begin{bmatrix} b_{11} \\ 0 \\ 0 \\ 0 \\ b_{22} \\ 0 \\ 0 \\ 0 \\ b_{33} \end{bmatrix} = \begin{bmatrix} 1 & 0 & 0 \\ 0 & 0 & 0 \\ 0 & 0 & 0 \\ 0 & 0 & 0 \\ 0 & 1 & 0 \\ 0 & 0 & 0 \\ 0 & 0 & 0 \\ 0 & 0 & 0 \\ 0 & 0 & 1 \end{bmatrix} \begin{bmatrix} b_{11} \\ b_{22} \\ b_{33} \end{bmatrix}.$$

Because $K = 3$, we need $2K^2 - \frac{1}{2}K(K+1) = 12$ restrictions on A and B for identification in this example model. There are 3 zeros and 3 ones in A. Thus, we have 6 restrictions on this matrix. In addition, there are 6 zero restrictions for B.

Writing the restrictions in the form (9.1.13), we get

$$\begin{bmatrix} 1 & 0 & 0 & 0 & 0 & 0 & 0 & 0 & 0 \\ 0 & 0 & 1 & 0 & 0 & 0 & 0 & 0 & 0 \\ 0 & 0 & 0 & 0 & 1 & 0 & 0 & 0 & 0 \\ 0 & 0 & 0 & 0 & 0 & 1 & 0 & 0 & 0 \\ 0 & 1 & 0 & 0 & 0 & 0 & 1 & 0 & 0 \\ 0 & 0 & 0 & 0 & 0 & 0 & 0 & 0 & 1 \end{bmatrix} \begin{bmatrix} a_{11} \\ a_{21} \\ a_{31} \\ a_{12} \\ a_{22} \\ a_{32} \\ a_{13} \\ a_{23} \\ a_{33} \end{bmatrix} = \begin{bmatrix} 1 \\ 0 \\ 1 \\ 0 \\ 0 \\ 1 \end{bmatrix}$$

and

$$\begin{bmatrix} 0 & 1 & 0 & 0 & 0 & 0 & 0 & 0 & 0 \\ 0 & 0 & 1 & 0 & 0 & 0 & 0 & 0 & 0 \\ 0 & 0 & 0 & 1 & 0 & 0 & 0 & 0 & 0 \\ 0 & 0 & 0 & 0 & 0 & 1 & 0 & 0 & 0 \\ 0 & 1 & 0 & 0 & 0 & 0 & 1 & 0 & 0 \\ 0 & 0 & 0 & 0 & 0 & 0 & 0 & 1 & 0 \end{bmatrix} \begin{bmatrix} b_{11} \\ b_{21} \\ b_{31} \\ b_{12} \\ b_{22} \\ b_{32} \\ b_{13} \\ b_{23} \\ b_{33} \end{bmatrix} = \begin{bmatrix} 0 \\ 0 \\ 0 \\ 0 \\ 0 \\ 0 \end{bmatrix}.$$

Thus, the necessary condition for local identification is satisfied. The necessary and sufficient condition from Proposition 9.3 can be checked by selecting randomly drawn matrices A and B from the restricted parameter space and determining the rank of the corresponding matrix in (9.1.14).

9.1.4 Long-Run Restrictions à la Blanchard-Quah

Clearly, it is not always easy to find suitable and generally acceptable restrictions for the matrices A and B. Imposing the restrictions directly on these matrices is in fact not necessary to identify the structural innovations and impulse responses. Another type of restrictions was discussed by Blanchard & Quah (1989). They considered the accumulated effects of shocks to the system. In terms of the structural impulse responses in (9.1.5) they focussed on the *total impact matrix*,

$$\Xi_\infty = \sum_{i=0}^{\infty} \Theta_i = (I_K - A_1 - \cdots - A_p)^{-1} A^{-1} B, \qquad (9.1.15)$$

and they identified the structural innovations by placing zero restrictions on this matrix. In other words, they assumed that some shocks do not have any total long-run effects. In particular, they considered a bivariate system consisting of output growth q_t and an unemployment rate ur_t (i.e., $y_t =$

$(q_t, ur_t)'$) and they assumed that the structural innovations represent supply and demand shocks. Moreover, they assumed that the demand shocks have only transitory effects on q_t and that the accumulated long-run effect of such shocks on q_t is zero. Placing the supply shocks first and the demand shocks last in the vectors of structural innovations $\varepsilon_t = (\varepsilon_t^s, \varepsilon_t^d)'$, the (1,2)-element of Ξ_∞ is restricted to be zero. In other words, we restrict the upper right-hand corner element of

$$\Xi_\infty = (I_K - A_1 - \cdots - A_p)^{-1}\mathsf{A}^{-1}\mathsf{B}$$

to zero. Given the VAR parameters, this set of equations clearly specifies a restriction for $\mathsf{A}^{-1}\mathsf{B}$. Thereby we have enough restrictions for identification of a bivariate system if we set $\mathsf{A} = I_K$, because, for $K = 2$, we have $K(K-1)/2 = 1$. Notice that $\mathsf{A} = I_K$ may be chosen because the idea is to identify the structural shocks from the reduced form residuals only and no restrictions are placed on the instantaneous effects of the observable variables directly. Thus, we have a B-model with restriction

$$(0, 0, 1, 0)\mathrm{vec}[(I_K - A_1 - \cdots - A_p)^{-1}\mathsf{B}]$$
$$= (0, 0, 1, 0)[I_2 \otimes (I_K - A_1 - \cdots - A_p)^{-1}]\mathrm{vec}(\mathsf{B}) = 0.$$

In summary, the AB-model offers a useful general framework for placing identifying restrictions for the structural innovations and impulse responses on a VAR process. The restrictions can be simple normalization and exclusion (zero) restrictions and may also be more general nonlinear restrictions. Clearly, before we can actually use this framework in practice, it will be necessary to estimate the reduced form and structural parameters. Estimation of the former parameters has been discussed in some detail in previous chapters. Thus, it remains to consider estimation of the A, B matrices. We will do so in Section 9.3. Before turning to inference procedures, we will consider structural restrictions for VECMs in the following section.

9.2 Structural Vector Error Correction Models

If all or some of the variables of interest are integrated, the previously discussed AB-model can still be used together with the levels VAR form of the data generation process. In most of the analysis of Section 9.1, the stationarity of the process was not used. Only in the treatment of the Blanchard-Quah restrictions, stability of the VAR operator is required because otherwise the matrix of total accumulated long-run effects does not exist. This result follows from the fact that the matrix $(I_K - A_1 - \cdots - A_p)$ is singular for cointegrated processes, as we have seen in Chapter 6. In other cases, we may use the AB-model even for integrated variables. In fact, we can even specify and fit a reduced form VECM, convert that model to the levels VAR form and then

9.2 Structural Vector Error Correction Models

use it as a basis for an AB-analysis, as discussed in the previous section. There are, however, advantages in utilizing the cointegration properties of the variables. They provide restrictions which can be taken into account beneficially in identifying the structural shocks. Therefore, it is useful to treat SVECMs separately.

As in the previous chapters, we assume that all variables are at most $I(1)$ and that the data generation process can be represented as a VECM with cointegration rank r of the form

$$\Delta y_t = \alpha\beta' y_{t-1} + \Gamma_1 \Delta y_{t-1} + \cdots + \Gamma_{p-1} \Delta y_{t-p+1} + u_t, \qquad (9.2.1)$$

where all symbols have their usual meanings. In other words, y_t is a K-dimensional vector of observable variables, α is a $(K \times r)$ matrix of loading coefficients, β is the $(K \times r)$ cointegration matrix, Γ_j is a $(K \times K)$ short-run coefficient matrix for $j = 1, \ldots, p-1$, and u_t is a white noise error vector with $u_t \sim (0, \Sigma_u)$.

In Chapter 6, Proposition 6.1, we have seen that the process has the Beveridge-Nelson MA representation

$$y_t = \Xi \sum_{i=1}^{t} u_i + \sum_{j=0}^{\infty} \Xi_j^* u_{t-j} + y_0^*, \qquad (9.2.2)$$

where the Ξ_j^* are absolutely summable so that the infinite sum is well-defined and the term y_0^* contains the initial values. Absolute summability of the Ξ_j^* implies that these matrices converge to zero for $j \to \infty$. Thus, the long-run effects of shocks are captured by the common trends term $\Xi \sum_{i=1}^{t} u_i$. The matrix

$$\Xi = \beta_\perp \left[\alpha'_\perp \left(I_K - \sum_{i=1}^{p-1} \Gamma_i \right) \beta_\perp \right]^{-1} \alpha'_\perp$$

has rank $K - r$. Thus, there are $K - r$ common trends and if the structural innovations embodied in the u_i can be recovered, at most r of them can have transitory effects only because the matrix Ξ or a nonsingular transformation of this matrix cannot have more than r columns of zeros. Thus, by knowing the cointegrating rank of the system, we know already the maximum number of transitory shocks.

In this context, the focus of interest is usually on the residuals and, hence, in order to identify the structural innovations, the B-model setup is typically used. In other words, we are looking for a matrix B such that

$$u_t = B\varepsilon_t \quad \text{with} \quad \varepsilon_t \sim (0, I_K).$$

Substituting this relation in the common trends term gives $\Xi B \sum_{i=1}^{t} \varepsilon_i$. Hence, the long-run effects of the structural innovations are given by

$\Xi\mathsf{B}$.

Because the structural innovations represent a regular random vector with nonsingular covariance matrix, the matrix B has to be nonsingular. Recall that $\Sigma_u = \mathsf{BB}'$. Thus, $\mathrm{rk}(\Xi\mathsf{B}) = K - r$ and there can be at most r zero columns in this matrix. In other words, r of the structural innovations can have transitory effects and $K - r$ of them must have permanent effects. If there are r transitory shocks, we can restrict r columns of $\Xi\mathsf{B}$ to zero. Because the matrix has reduced rank $K - r$, each column of zeros stands for $K - r$ independent restrictions only. Thus, the r transitory shocks represent $r(K-r)$ independent restrictions only. Still, it is useful to note that restrictions can be imposed on the basis of our knowledge of the cointegrating rank of the system which can be determined by statistical means. Further theoretical considerations are required for imposing additional restrictions, however.

For local just-identification of the structural innovations in the B-model, we need a total of $K(K-1)/2$ restrictions. Assuming that there are r shocks with transitory effects only, we have already $r(K - r)$ restrictions from the cointegration structure of the model, this leaves us with $\frac{1}{2}K(K-1) - r(K-r)$ further restrictions for just-identifying the structural innovations. In fact, $r(r-1)/2$ additional contemporaneous restrictions are needed to disentangle the transitory shocks and $(K-r)((K-r)-1)/2$ restrictions identify the permanent shocks (see, e.g., King et al. (1991), Gonzalo & Ng (2001)). Then we have a total of $\frac{1}{2}r(r-1) + \frac{1}{2}(K-r)((K-r)-1) = \frac{1}{2}K(K-1) - r(K-r)$ restrictions, as required. Thus, it is not sufficient to impose arbitrary restrictions on B or $\Xi\mathsf{B}$, but we have to choose them to identify the transitory and permanent shocks at least locally. In fact, the transitory shocks can only be identified through restrictions directly on B because they correspond to zero columns in $\Xi\mathsf{B}$. Thus, $r(r-1)/2$ of the restrictions have to be imposed on B directly. Generally, the restrictions have the form

$$C_{\Xi\mathsf{B}}\mathrm{vec}(\Xi\mathsf{B}) = c_l \text{ or } C_l\mathrm{vec}(\mathsf{B}) = c_l \quad \text{and} \quad C_s\mathrm{vec}(\mathsf{B}) = c_s, \qquad (9.2.3)$$

where $C_l := C_{\Xi\mathsf{B}}(I_K \otimes \Xi)$ is a matrix of long-run restrictions, that is, $C_{\Xi\mathsf{B}}$ is a suitable selection matrix such that $C_{\Xi\mathsf{B}}\mathrm{vec}(\Xi\mathsf{B}) = c_l$, and C_s specifies short-run or instantaneous constraints by restricting elements of B directly. Here c_l and c_s are vectors of suitable dimensions. In applied work, they are typically zero vectors. In other words, zero restrictions are specified in (9.2.3) for $\Xi\mathsf{B}$ and B.

As discussed for the stationary case in Section 9.1.2, the matrix B will only be locally identified. In particular, in general we may reverse the signs of the columns of B to find another valid matrix. Formal necessary and sufficient conditions for local identification are given in the following proposition.

Proposition 9.4 (*Local Identification of a SVECM*)
Suppose the reduced form model (9.2.1) with Beveridge-Nelson MA representation (9.2.2) is given. Let B be a nonsingular $(K \times K)$ matrix. Then, the set of equations

$$\Sigma_u = \mathsf{BB}', \quad C_l \mathrm{vec}(\mathsf{B}) = c_l \quad \text{and} \quad C_s \mathrm{vec}(\mathsf{B}) = c_s,$$

with C_l, c_l, C_s, and c_s as in (9.2.3), has a locally unique solution for B if and only if

$$\mathrm{rk} \begin{bmatrix} 2\mathbf{D}_K^+(\mathsf{B} \otimes I_K) \\ C_l \\ C_s \end{bmatrix} = K^2.$$

∎

Proof: The model underlying Proposition 9.4 is a B-model. Therefore the proposition can be shown using the same arguments as for Proposition 9.2. Details are omitted. ∎

As an example, we consider a small model discussed by King et al. (1991). They specified a model for the logarithms of private output (q_t), consumption (c_t), and investment (i_t). Assuming that all three variables are $I(1)$ with cointegrating rank $r = 2$ and that there are two transitory shocks and one permanent shock, the permanent shock is identified without further assumptions because $K - r = 1$ and, hence, $(K - r)((K - r) - 1)/2 = 0$. Moreover, only $1\ (= r(r-1)/2)$ further restriction is necessary to identify the two transitory shocks. Placing the permanent shock first in the ε_t vector and allowing the first transitory shock to have instantaneous effects on all variables, we may use the following restrictions:

$$\Xi\mathsf{B} = \begin{bmatrix} * & 0 & 0 \\ * & 0 & 0 \\ * & 0 & 0 \end{bmatrix} \quad \text{and} \quad \mathsf{B} = \begin{bmatrix} * & * & * \\ * & * & 0 \\ * & * & * \end{bmatrix}. \quad (9.2.4)$$

Here asterisks denote unrestricted elements. The two zero columns in $\Xi\mathsf{B}$ represent two independent restrictions only because $\Xi\mathsf{B}$ has rank 1. A third restriction is placed on B in such a way that the third shock does not have an instantaneous effect on the second variable. Hence, there are $K(K-1)/2 = 3$ independent restrictions in total and the structural innovations are locally just-identified. Uniqueness can be obtained by fixing the signs of the diagonal elements of B.

In our three-dimensional example with two zero columns in $\Xi\mathsf{B}$, it does not suffice to impose a further restriction on this matrix to ensure local uniqueness of B. For that we need to disentangle the two transitory shocks which cannot be identified by restrictions on the long-run matrix $\Xi\mathsf{B}$. Thus, we have to impose a restriction directly on B. In fact, it is necessary to restrict an element in the last two columns of B (see also Problem 9.1 for further details).

In the standard B-model with three variables, we need to specify at least 3 restrictions for identification. In contrast, in the present VECM case, assuming that $r = 2$ and there are two transitory shocks, only one restriction is needed because two columns of $\Xi\mathsf{B}$ are zero. Thus, taking into account the long-run restrictions from the cointegration properties of the variables may result in substantial simplifications. In fact, for a bivariate system with one cointegrating relation, no further restriction is required to identify the permanent and transitory shocks. It is enough to specify that the first shock is allowed to have permanent effects while the second one can only have transitory effects or vice versa. A more detailed higher-dimensional example may be found in Breitung et al. (2004). Further discussion of partitioning the shocks in permanent and transitory ones is also given in Gonzalo & Ng (2001) and Fisher & Huh (1999).

9.3 Estimation of Structural Parameters

We will first consider estimation of the AB-SVAR model and then discuss SVECMs. The A- and B-models are straightforward special cases which are not treated separately in detail. For both SVARs and SVECMs, ML methods are typically used and they will therefore be presented here.

9.3.1 Estimating SVAR Models

Suppose we wish to estimate the following SVAR model

$$\mathsf{A} y_t = \mathsf{A} A Y_{t-1} + \mathsf{B}\varepsilon_t, \tag{9.3.1}$$

where $Y'_{t-1} := [y'_{t-1}, \ldots, y'_{t-p}]$, $A := [A_1, \ldots, A_p]$, and ε_t is assumed to be Gaussian white noise with covariance matrix I_K, $\varepsilon_t \sim \mathcal{N}(0, I_K)$. The normality assumption is just made for convenience to derive the estimators. The asymptotic properties of the estimators will be the same under more general distributional assumptions, as usual. The reduced form residuals corresponding to (9.3.1) have the form $u_t = \mathsf{A}^{-1}\mathsf{B}\varepsilon_t$.

From Chapter 3, Section 3.4, the log-likelihood function for a sample $y_1, \ldots, y_T$ is seen to be

$$\begin{aligned}\ln l(A, \mathsf{A}, \mathsf{B}) &= -\tfrac{KT}{2}\ln 2\pi - \tfrac{T}{2}\ln|\mathsf{A}^{-1}\mathsf{B}\mathsf{B}'\mathsf{A}'^{-1}| \\ &\quad -\tfrac{1}{2}\mathrm{tr}\{(Y-AX)'[\mathsf{A}^{-1}\mathsf{B}\mathsf{B}'\mathsf{A}'^{-1}]^{-1}(Y-AX)\} \\ &= \text{constant} + \tfrac{T}{2}\ln|\mathsf{A}|^2 - \tfrac{T}{2}\ln|\mathsf{B}|^2 \\ &\quad -\tfrac{1}{2}\mathrm{tr}\{\mathsf{A}'\mathsf{B}'^{-1}\mathsf{B}^{-1}\mathsf{A}(Y-AX)(Y-AX)'\},\end{aligned} \tag{9.3.2}$$

where, as usual, $Y := [y_1, \ldots, y_T]$, $X := [Y_0, \ldots, Y_{T-1}]$, and the matrix rules $|\mathsf{A}^{-1}\mathsf{B}\mathsf{B}'(\mathsf{A}^{-1})'| = |\mathsf{A}^{-1}|^2|\mathsf{B}|^2 = |\mathsf{A}|^{-2}|\mathsf{B}|^2$ and $\mathrm{tr}(VW) = \mathrm{tr}(WV)$ have been used (see Appendix A).

Suppose there are no restrictions on the reduced form parameters A. Then, it follows from Section 3.4 that for any given A and B, the log-likelihood function $\ln l(A, \mathsf{A}, \mathsf{B})$ is maximized with respect to A by $\widehat{A} = YX'(XX')^{-1}$. Thus, replacing A with $\widehat{A}$ in (9.3.2) gives the concentrated log-likelihood

$$\ln l_c(\mathsf{A}, \mathsf{B}) = \text{constant} + \frac{T}{2}\ln|\mathsf{A}|^2 - \frac{T}{2}\ln|\mathsf{B}|^2 - \frac{T}{2}\text{tr}(\mathsf{A}'\mathsf{B}'^{-1}\mathsf{B}^{-1}\mathsf{A}\widetilde{\Sigma}_u), \quad (9.3.3)$$

where $\widetilde{\Sigma}_u = T^{-1}(Y - \widehat{A}X)(Y - \widehat{A}X)'$. Maximization of this function with respect to A and B, subject to the structural restrictions (9.1.12) or (9.1.13), has to be done by numerical methods because a closed form solution is usually not available. If the restrictions are of the form (9.1.12), restricted maximization of the concentrated log-likelihood amounts to maximization with respect to γ_A and γ_B. If these parameters are locally identified, the ML estimators have standard asymptotic properties which are summarized in the following proposition.

Proposition 9.5 (*Properties of the SVAR ML Estimators*)
Suppose y_t is a stationary Gaussian VAR(p) as in (9.1.1) and structural restrictions of the form (9.1.12) are available such that γ_A and γ_B are locally identified. Then the ML estimators $\widetilde{\gamma}_\mathsf{A}$ and $\widetilde{\gamma}_\mathsf{B}$ are consistent and asymptotically normally distributed,

$$\sqrt{T}\left(\begin{bmatrix}\widetilde{\gamma}_\mathsf{A}\\\widetilde{\gamma}_\mathsf{B}\end{bmatrix} - \begin{bmatrix}\gamma_\mathsf{A}\\\gamma_\mathsf{B}\end{bmatrix}\right) \xrightarrow{d} \mathcal{N}\left(0, \mathcal{I}_a\begin{pmatrix}\gamma_\mathsf{A}\\\gamma_\mathsf{B}\end{pmatrix}^{-1}\right),$$

where $\mathcal{I}_a(\cdot)$ is the asymptotic information matrix. It has the form

$$\mathcal{I}_a\begin{pmatrix}\gamma_\mathsf{A}\\\gamma_\mathsf{B}\end{pmatrix} = \begin{bmatrix}R'_\mathsf{A} & 0\\0 & R'_\mathsf{B}\end{bmatrix}\mathcal{I}_a\begin{pmatrix}\text{vec }\mathsf{A}\\\text{vec }\mathsf{B}\end{pmatrix}\begin{bmatrix}R_\mathsf{A} & 0\\0 & R_\mathsf{B}\end{bmatrix}$$

and

$$\begin{aligned}&\mathcal{I}_a\begin{pmatrix}\text{vec }\mathsf{A}\\\text{vec }\mathsf{B}\end{pmatrix}\\&= \begin{bmatrix}\mathsf{A}^{-1}\mathsf{B}\otimes\mathsf{B}'^{-1}\\-(I_K\otimes\mathsf{B}'^{-1})\end{bmatrix}(I_{K^2} + \mathbf{K}_{KK})\\&\quad\times\left[(\mathsf{B}'\mathsf{A}'^{-1}\otimes\mathsf{B}^{-1}) : -(I_K\otimes\mathsf{B}^{-1})\right]\end{aligned} \quad (9.3.4)$$

■

Proof: The proposition follows from the general ML theory (see Appendix C.6). For the derivation of the asymptotic information matrix see Problem 9.4. ■

If γ_A and γ_B are identified, the same is true for A and B. Estimating these matrices such that $\text{vec}(\widetilde{\mathsf{A}}) = R_\mathsf{A}\widetilde{\gamma}_\mathsf{A} + r_\mathsf{A}$ and $\text{vec}(\widetilde{\mathsf{B}}) = R_\mathsf{B}\widetilde{\gamma}_\mathsf{B} + r_\mathsf{B}$, respectively, we get the following immediate implication of Proposition 9.5.

Corollary 9.5.1
Under the conditions of Proposition 9.5,

$$\sqrt{T}\left(\begin{bmatrix} \text{vec } \widetilde{\mathsf{A}} \\ \text{vec } \widetilde{\mathsf{B}} \end{bmatrix} - \begin{bmatrix} \text{vec } \mathsf{A} \\ \text{vec } \mathsf{B} \end{bmatrix}\right) \xrightarrow{d} \mathcal{N}(0, \Sigma_{\mathsf{AB}}),$$

where

$$\Sigma_{\mathsf{AB}} = \begin{bmatrix} R_{\mathsf{A}} & 0 \\ 0 & R_{\mathsf{B}} \end{bmatrix} \mathcal{I}_a \begin{pmatrix} \gamma_{\mathsf{A}} \\ \gamma_{\mathsf{B}} \end{pmatrix}^{-1} \begin{bmatrix} R'_{\mathsf{A}} & 0 \\ 0 & R'_{\mathsf{B}} \end{bmatrix}.$$

∎

If only just-identifying restrictions are imposed on the structural parameters, we have for the ML estimator of Σ_u,

$$\widetilde{\Sigma}_u = T^{-1}(Y - \widehat{A}X)(Y - \widehat{A}X)' = \widetilde{\mathsf{A}}^{-1}\widetilde{\mathsf{B}}\widetilde{\mathsf{B}}'\widetilde{\mathsf{A}}'^{-1}.$$

If, however, over-identifying restrictions have been imposed on A and/or B, the corresponding estimator for Σ_u,

$$\widetilde{\Sigma}_u^r := \widetilde{\mathsf{A}}^{-1}\widetilde{\mathsf{B}}\widetilde{\mathsf{B}}'\widetilde{\mathsf{A}}'^{-1}, \tag{9.3.5}$$

will differ from $\widetilde{\Sigma}_u$. In fact, the LR statistic,

$$\lambda_{LR} = T(\ln|\widetilde{\Sigma}_u^r| - \ln|\widetilde{\Sigma}_u|), \tag{9.3.6}$$

can be used to check the over-identifying restrictions. Under the null hypothesis that the restrictions are valid, it has an asymptotic χ^2-distribution with degrees of freedom equal to the number of over-identifying restrictions. In other words, the number of degrees of freedom is equal to the number of independent constraints imposed on A and B minus $2K^2 - \frac{1}{2}K(K+1)$.

Computation of ML Estimates

Because the structural parameters A and B are nonlinearly related to the reduced form parameters, no closed form of the ML estimates exists in general and an iterative optimization algorithm may be used for actually computing the ML estimates. Amisano & Giannini (1997) proposed to use a scoring algorithm for this purpose. The i-th iteration of this algorithm is of the form

$$\begin{bmatrix} \widetilde{\gamma}_{\mathsf{A}} \\ \widetilde{\gamma}_{\mathsf{B}} \end{bmatrix}_{i+1} = \begin{bmatrix} \widetilde{\gamma}_{\mathsf{A}} \\ \widetilde{\gamma}_{\mathsf{B}} \end{bmatrix}_i + \ell \mathcal{I}\left(\begin{bmatrix} \widetilde{\gamma}_{\mathsf{A}} \\ \widetilde{\gamma}_{\mathsf{B}} \end{bmatrix}_i\right)^{-1} \mathbf{s}\left(\begin{bmatrix} \widetilde{\gamma}_{\mathsf{A}} \\ \widetilde{\gamma}_{\mathsf{B}} \end{bmatrix}_i\right), \tag{9.3.7}$$

where $\mathcal{I}(\cdot)$ denotes the information matrix of the free parameters $\gamma_{\mathsf{A}}, \gamma_{\mathsf{B}}$, that is, in this case $\mathcal{I}(\cdot) = T\mathcal{I}_a(\cdot)$, $\mathbf{s}(\cdot)$ is the score vector and ℓ is the step length (see also Chapter 12, Section 12.3.2, for further discussion of optimization algorithms of this type).

9.3 Estimation of Structural Parameters

The score vector can be obtained using the rules for matrix and vector differentiation (Appendix A.13). Applying the chain rule for vector differentiation, it is seen to be

$$\mathbf{s}\begin{pmatrix} \gamma_A \\ \gamma_B \end{pmatrix} = \frac{\partial \ln l}{\partial (\gamma_A', \gamma_B')'} = \begin{bmatrix} R_A' & 0 \\ 0 & R_B' \end{bmatrix} \mathbf{s}\begin{pmatrix} \text{vec A} \\ \text{vec B} \end{pmatrix}, \quad (9.3.8)$$

and

$$\mathbf{s}\begin{pmatrix} \text{vec A} \\ \text{vec B} \end{pmatrix} = \frac{\partial \ln l}{\partial \begin{pmatrix} \text{vec A} \\ \text{vec B} \end{pmatrix}} = \begin{bmatrix} (I_K \otimes \mathsf{B}'^{-1}) \\ -(\mathsf{B}^{-1}\mathsf{A} \otimes \mathsf{B}'^{-1}) \end{bmatrix} \mathbf{s}(\text{vec}[\mathsf{B}^{-1}\mathsf{A}])$$

with

$$\mathbf{s}(\text{vec}[\mathsf{B}^{-1}\mathsf{A}]) = T\text{vec}([\mathsf{B}^{-1}\mathsf{A}]'^{-1}) - T(\widetilde{\Sigma}_u \otimes I_K)\text{vec}(\mathsf{B}^{-1}\mathsf{A})$$

(see Problem 9.3 for further details). In practice, the iterations of the scoring algorithm terminate if prespecified convergence criteria, such as the relative change in the log-likelihood and the parameters, are satisfied. For this algorithm to work, the inverse of the information matrix has to exist which is guaranteed by the identification of the parameters, at least in a neighborhood of the true parameter values. Giannini (1992) used this property to derive alternative conditions for identification of the models presented in Section 9.1. More precisely, he derived identification conditions from the fact that, for instance, the AB-model is locally identified if and only if the matrix

$$\begin{bmatrix} \mathcal{I}_a\begin{pmatrix} \text{vec A} \\ \text{vec B} \end{pmatrix} \\ \begin{bmatrix} C_A & 0 \\ 0 & C_B \end{bmatrix} \end{bmatrix} \quad (9.3.9)$$

has full column rank when $\mathcal{I}_a(\cdot)$ is evaluated at the true parameter values (see Rothenberg (1971)).

Although we have discussed models without deterministic terms and restrictions on the reduced form parameters, the ML estimation procedure for the structural parameters can be extended easily to more general situations which cover these complications. Again, estimation of the structural parameters can be based on the concentrated likelihood function. If there are restrictions for the reduced form parameters A, for example, if a subset model is considered, one may even use the EGLS estimator instead of the ML estimator for these parameters in estimating the structural parameters. Clearly, in that case, the white noise covariance estimator $\widetilde{\Sigma}_u$ will not be the exact ML estimator and the exact concentrated log-likelihood is obtained only if ML estimators are substituted for the reduced form parameters A. Asymptotically, the corresponding estimators $\widetilde{\mathsf{A}}, \widetilde{\mathsf{B}}$ based on the EGLS estimators will have the same properties as the exact ML estimators, however. Even in small samples, exact ML estimation may not result in substantial gains (see, e.g., Brüggemann (2004)).

Estimation with Long-Run Restrictions à la Blanchard-Quah

If the total impact matrix Ξ_∞ is restricted to be triangular as in Blanchard & Quah (1989) and Galí (1999), estimation becomes particularly easy. Specifying $\mathsf{A} = I_K$, using the relation $\Xi_\infty = (I_K - A_1 - \cdots - A_p)^{-1}\mathsf{B}$ and noting that

$$\Xi_\infty \Xi'_\infty = (I_K - A_1 - \cdots - A_p)^{-1} \Sigma_u (I_K - A'_1 - \cdots - A'_p)^{-1},$$

the matrix B can be estimated by premultiplying a Choleski decomposition of the matrix

$$(I_K - \widehat{A}_1 - \cdots - \widehat{A}_p)^{-1} \widetilde{\Sigma}_u (I_K - \widehat{A}'_1 - \cdots - \widehat{A}'_p)^{-1}$$

by $(I_K - \widehat{A}_1 - \cdots - \widehat{A}_p)$.

This latter procedure works only if the VAR operator is stable and the process is stationary because for integrated processes the inverse of $(I_K - A_1 - \cdots - A_p)$ does not exist, as explained earlier. On the other hand, cointegrated variables do not create problems for the other estimation methods for SVAR models.

9.3.2 Estimating Structural VECMs

Suppose the structural restrictions for a VECM are given in the form of linear restrictions on $\Xi\mathsf{B}$ and B, as in (9.2.3). For computing the parameter estimates, we may replace Ξ by its reduced form ML estimator,

$$\widetilde{\Xi} = \widetilde{\beta}_\perp \left[\widetilde{\alpha}'_\perp \left(I_K - \sum_{i=1}^{p-1} \widetilde{\Gamma}_i \right) \widetilde{\beta}_\perp \right]^{-1} \widetilde{\alpha}'_\perp,$$

where the $\widetilde{\Gamma}_i$'s are the ML estimators of the Γ_i's from Proposition 7.3 and $\widetilde{\alpha}_\perp$ and $\widetilde{\beta}_\perp$ are any orthogonal complements of the ML estimators $\widetilde{\alpha}$ and $\widetilde{\beta}$, respectively. The restricted ML estimator of B can be obtained by setting $\mathsf{A} = I_K$ and optimizing the concentrated log-likelihood function (9.3.3) with respect to B, subject to the restrictions (9.2.3), with C_l replaced by

$$\widetilde{C}_l = C_{\Xi\mathsf{B}}(I_K \otimes \widetilde{\Xi})$$

(see Vlaar (2004)). Although this procedure results in a set of stochastic restrictions, from a numerical point of view we have a standard constrained optimization problem which can be solved by a Lagrange approach (see Appendix A.14) because $\widetilde{\Xi}$ is fixed in computing the estimate of B. Due to the fact that for a just-identified structural model the log-likelihood maximum is the same as for the reduced form, a comparison of the log-likelihood values can serve as a check for a proper convergence of the optimization algorithm used for structural estimation.

The properties of the ML estimator of B follow in principle from Corollary 9.5.1. In other words, $\widetilde{\mathsf{B}}$ is consistent and asymptotically normal under standard conditions,

$$\sqrt{T}\operatorname{vec}(\widetilde{\mathsf{B}} - \mathsf{B}) \overset{d}{\to} \mathcal{N}(0, \Sigma_{\mathsf{B}}).$$

The asymptotic distribution is singular because of the restrictions that have been imposed on B. Thus, although t-ratios can be used for assessing the significance of individual parameters, F-tests based on the Wald principle will in general not be valid and have to be interpreted cautiously. Expressions for the covariance matrices of the asymptotic distributions in terms of the model parameters can be obtained in the usual way by working out the corresponding information matrices (see Vlaar (2004)). For practical purposes, it is common to use bootstrap methods for inference in this context.

In principle, the same approach can be used if there are over-identifying restrictions for B. In that case, $\widetilde{\mathsf{B}}\widetilde{\mathsf{B}}'$ will not be equal to the reduced form white noise covariance estimator $\widehat{\Sigma}_u$, however. Still the estimator of B will be consistent and asymptotically normal under general conditions and also the LR statistic given in (9.3.6) can be used to check the validity of the over-identifying restrictions. It will have the usual asymptotic χ^2-distribution with degrees of freedom equal to the number of over-identifying restrictions.

9.4 Impulse Response Analysis and Forecast Error Variance Decomposition

Impulse response analysis can now be based on structural innovations. In other words, the impulse response coefficients are obtained from the matrices

$$\Theta_j = \Phi_j \mathsf{A}^{-1}\mathsf{B}, \quad j = 0, 1, 2, \ldots.$$

Using the same reasoning as in Chapter 3, Section 3.7, the corresponding estimated quantities are asymptotically normal as nonlinear functions of asymptotically normal parameter estimators,

$$\sqrt{T}\operatorname{vec}(\widehat{\Theta}_j - \Theta_j) \overset{d}{\to} \mathcal{N}(0, \Sigma_{\widehat{\Theta}_j}).$$

In practice, bootstrap methods are routinely employed for inference in this context. However, the same inference problems as in Chapter 3, Section 3.7, prevail for structural impulse responses. More precisely, the asymptotic distribution may be singular in which case confidence intervals based on asymptotic theory or bootstrap methods may not have the desired confidence level even asymptotically.

We use a set of quarterly U.S. data for the period 1947.1–1988.4 from King et al. (1991) for the three variables log private output (q_t), consumption

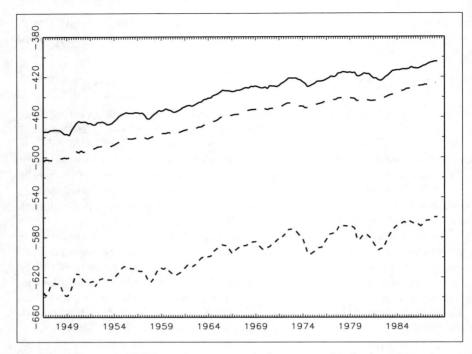

Fig. 9.1. Quarterly U.S. log private output (—), consumption (– –), and investment (- - -).

(c_t), and investment (i_t) (all multiplied by 100) to illustrate structural impulse responses.[1] The three series are plotted in Figure 9.1. They all have a trending behavior and there is some evidence that they are well modelled as $I(1)$ series. Applying LR tests for the cointegrating rank with a trend orthogonal to the cointegration relations to a model with one lagged difference of the variables, provides evidence for two cointegration relations, that is, $r = 2$ (see Section 8.2.4 for the description of the tests). Therefore we proceed from the following estimated reduced form VECM (t-statistics in parentheses):

$$\begin{bmatrix} \Delta q_t \\ \Delta c_t \\ \Delta i_t \end{bmatrix} = \begin{bmatrix} -0.88 \\ {\scriptstyle (-0.2)} \\ -2.83 \\ {\scriptstyle (-1.1)} \\ -30.07 \\ {\scriptstyle (-4.1)} \end{bmatrix}$$

---

[1] The data are available at the website http://www.wws.princeton.edu/mwatson/.

9.4 Impulse Response Analysis and Forecast Error Variance Decomposition

$$+ \begin{bmatrix} -0.23 & 0.20 \\ (-3.6) & (4.6) \\ -0.06 & 0.07 \\ (-1.5) & (2.4) \\ -0.11 & 0.26 \\ (-0.9) & (2.9) \end{bmatrix} \begin{bmatrix} 1 & 0 & -1.02 \\ & & (-27.7) \\ 0 & 1 & -1.10 \\ & & (-24.2) \end{bmatrix} \begin{bmatrix} q_{t-1} \\ c_{t-1} \\ i_{t-1} \end{bmatrix}$$

$$+ \begin{bmatrix} 0.12 & 0.09 & 0.16 \\ (1.2) & (0.7) & (3.4) \\ 0.21 & -0.21 & 0.02 \\ (3.2) & (-2.3) & (0.8) \\ 0.70 & -0.17 & 0.33 \\ (3.6) & (-0.6) & (3.6) \end{bmatrix} \begin{bmatrix} \Delta q_{t-1} \\ \Delta c_{t-1} \\ \Delta i_{t-1} \end{bmatrix} + \begin{bmatrix} u_{1t} \\ u_{2t} \\ u_{3t} \end{bmatrix}. \quad (9.4.1)$$

Before we can proceed with structural estimation, we have to specify identifying restrictions. Using the zero restrictions from (9.2.4), the following estimates are obtained:

$$\widetilde{B} = \begin{bmatrix} 0.08 & 1.03 & -0.45 \\ (0.4) & (3.9) & (-0.8) \\ -0.60 & 0.43 & 0 \\ (-0.7) & (4.1) & \\ 0.26 & 1.96 & 1.00 \\ (0.6) & (5.1) & (1.9) \end{bmatrix} \quad (9.4.2)$$

and

$$\widetilde{\Xi}\widetilde{B} = \begin{bmatrix} -0.71 & 0 & 0 \\ (-0.8) & & \\ -0.76 & 0 & 0 \\ (-0.8) & & \\ -0.69 & 0 & 0 \\ (-0.8) & & \end{bmatrix}.$$

Here bootstrapped t-statistics based on 2000 bootstrap replications are given in parentheses. In other words, the standard deviations of the estimates are obtained with a bootstrap (see Appendix D.3) and then the estimated coefficients are divided by their respective bootstrap standard deviations to get the t-ratios. Clearly, some of the t-ratios are quite small. Thus, it may be possible to impose over-identifying restrictions. In fact, because all t-ratios of the nonzero long-run effects are small, it may be tempting to argue that no significant permanent effect is found. Recall, however, that, based on the unit root and cointegration analysis, there cannot be more shocks with transitory effects. We have used the just-identified model for an impulse response analysis to shed more light on this issue.

There are three structural innovations, one of which must have permanent effects if the cointegration rank is 2. In Figure 9.2, the responses of all three variables to the shock with potentially permanent effects are depicted. The 95% confidence intervals are based on 2000 replications. Considering the confidence intervals determined with Hall's percentile method (see Appendix D.3), it turns out that none of the confidence intervals associated with longer term responses contains zero. Hence, a significant long-run effect may actually

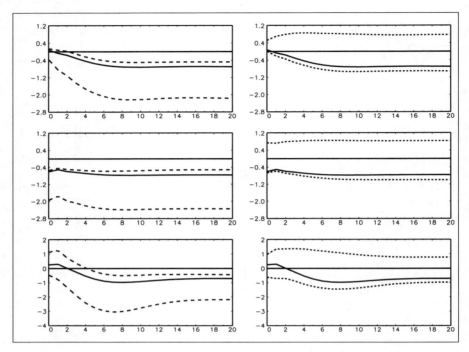

Fig. 9.2. Responses of output, consumption, and investment (top to bottom) to a permanent shock with Hall percentile (left) and standard percentile (right) 95% bootstrap confidence intervals based on 2000 bootstrap replications.

be present for each of the three variables. If, however, the standard percentile bootstrap confidence intervals are used for the impulse responses, the situation is quite different. These confidence intervals are also shown in Figure 9.2 and they all include zero for longer term horizons. Thus, the results are not very robust with respect to the methods used. Clearly, the confidence intervals are quite asymmetric around the point estimates. In such a situation the Hall percentile confidence intervals may be more reliable due to their built-in bias correction.

The estimated responses to the permanent shock are all negative in the long-run. To see the effects of an impulse which leads to positive long-run effects, we can just reverse the signs of the responses. This follows from the unidentified signs of the columns of B discussed in Sections 9.1.2 and 9.2. Generally, the effects of positive and negative shocks of the same size are identical in absolute value because our model is a linear one which does not permit asymmetric reactions to positive and negative shocks.

In Figure 9.3, the responses of the variables to the two transitory shocks are shown. All impulse responses approach zero quickly after some periods and the effects of the shocks after 20 periods are practically negligible. The

9.4 Impulse Response Analysis and Forecast Error Variance Decomposition

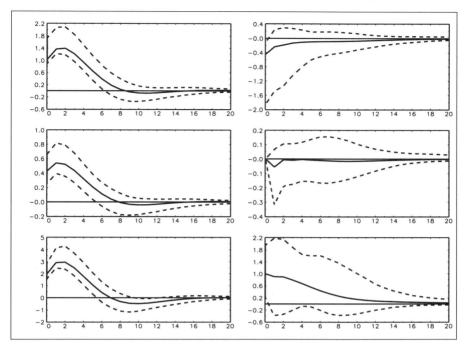

Fig. 9.3. Responses of output, consumption, and investment (top to bottom) to transitory shocks with 95% Hall percentile bootstrap confidence intervals based on 2000 bootstrap replications (identification restriction (9.4.2)).

identifying restriction on the B matrix is clearly seen in the right-hand panel in the middle row of Figure 9.3. Here the instantaneous effect of the second transitory shock on c_t is zero. If a zero restriction is imposed instead on the upper right-hand corner element of B, the estimated matrix becomes

$$\widetilde{\mathsf{B}} = \begin{bmatrix} \underset{(0.4)}{0.08} & \underset{(5.7)}{1.12} & 0 \\ \underset{(-0.7)}{-0.60} & \underset{(2.9)}{0.39} & \underset{(1.4)}{0.17} \\ \underset{(0.6)}{0.26} & \underset{(4.5)}{1.39} & \underset{(11.2)}{1.70} \end{bmatrix} \tag{9.4.3}$$

and the corresponding structural impulse responses are depicted in Figure 9.4. Obviously, the identification restriction determines to some extent the shape of the impulse responses. At least the responses to the second transitory shock are quite different from those based on the identification restriction (9.4.2). Now, of course, q_t reacts only with a delay to the second transitory shock. The first column of $\widetilde{\mathsf{B}}$ in (9.4.3) is unchanged relative to (9.4.2) and, more generally, the responses to the permanent shock (not shown) are unaffected because that shock is identified without additional restrictions.

382   9 Structural VARs and VECMs

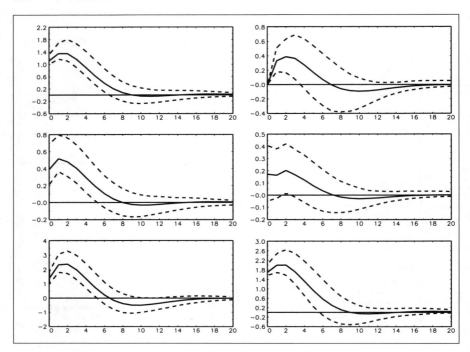

Fig. 9.4. Responses of output, consumption, and investment (top to bottom) to transitory shocks with 95% Hall percentile bootstrap confidence intervals based on 2000 bootstrap replications (identification restriction (9.4.3)).

Forecast error variance decompositions can also be based on the structural innovations. The computations are based on the Θ_j as in Section 2.3.3. The interpretation may be different, however. It may not be possible to associate the structural innovations uniquely with the variables of the system. Therefore, the forecast errors are not decomposed into contributions of the different variables but into the contributions of the structural innovations. For instance, for the example system with identifying restriction on B as in (9.4.2), a forecast error variance decomposition is shown in Figure 9.5. Now we can see that the permanent shocks (the first components of the ε_t's) have a growing importance with increasing forecast horizon, where the estimation uncertainty is ignored, however. In turn, the importance of the transitory shocks (shocks number 2 and 3) declines for all three variables. Actually, the third shock (the second transitory shock) does not contribute much to the forecast errors of any of the three variables.

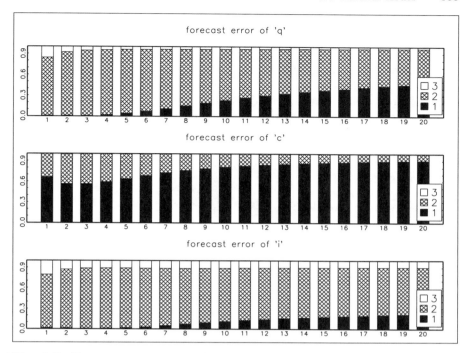

Fig. 9.5. Forecast error variance decomposition of the output, consumption, and investment system based on identification scheme (9.4.2) with relative contributions of the permanent shock (1) and the two transitory shocks (2 and 3).

9.5 Further Issues

Structural VARs and VECMs have not only found widespread use in applied work but there are also numerous further methodological contributions. For example, confidence bands for impulse responses are sometimes constructed with Bayesian methods (e.g., Koop (1992)). In fact, the practice of reporting confidence intervals around individual impulse response coefficients was questioned by Sims & Zha (1999). They proposed likelihood-characterizing error bands as alternatives.

Also other forms of identifying restrictions were considered by some authors. For example, Uhlig (1994) proposed to use inequality constraints for the impulse responses for identifying them. In contrast, Lee, Pesaran & Pierse (1992) and Pesaran & Shin (1996) considered persistence profiles which measure the persistence of certain shocks without imposing structural identification restrictions.

It may be worth remembering, however, that structural impulse responses are not immune to some of the problems discussed in Chapter 2 in the context of impulse response analysis. In particular, omitted variables, filtering and adjusting series prior to using them for a VAR analysis and using aggregated or transformed data can lead to major changes in the dynamic behavior of the model. For instance, if an important variable is omitted from a system of interest, adding it can change in principle all the impulse responses. Similarly, using seasonally adjusted and, hence, filtered data can change the dynamic structure of the variables and, thus, may lead to impulse responses which are quite different from those for unadjusted variables. These problems are not solved by imposing identifying restrictions and are worth keeping in mind also in a structural VAR analysis.

9.6 Exercises

9.6.1 Algebraic Problems

Problem 9.1
Show that for a three-dimensional VECM with cointegration rank $r = 2$, the set of restrictions

$$\Xi \mathsf{B} = \begin{bmatrix} 0 & 0 & 0 \\ * & 0 & 0 \\ * & 0 & 0 \end{bmatrix}$$

is not sufficient for identification. Moreover, show that the restrictions

$$\Xi \mathsf{B} = \begin{bmatrix} * & 0 & 0 \\ * & 0 & 0 \\ * & 0 & 0 \end{bmatrix} \quad \text{and} \quad \mathsf{B} = \begin{bmatrix} 0 & * & * \\ * & * & * \\ * & * & * \end{bmatrix}.$$

do not identify B locally.
(Hint: Choose

$$\mathsf{B} = \begin{bmatrix} b_{11} & 0 \\ 0 & \mathsf{B}_2 \end{bmatrix},$$

where B_2 is a (2×2) matrix. Show that B_2 is not unique.)

Problem 9.2
Suppose a four-dimensional process y_t can be written in VECM form (9.2.1) with cointegrating rank 2. Impose just-identifying restrictions on B and $\Xi \mathsf{B}$.

Problem 9.3
Define $C = \mathsf{B}^{-1}\mathsf{A}$ and write the concentrated log-likelihood (9.3.3) as

$$\ln l_c(C) = \text{constant} + T \ln |C| - \frac{T}{2} \text{tr}(C'C\widetilde{\Sigma}_u).$$

Use the rules for matrix differentiation from Appendix A.13 to show that

$$\frac{\partial \ln l_c}{\partial C} = TC'^{-1} - TC\widetilde{\Sigma}_u.$$

Next show that
$$\frac{\partial \operatorname{vec}(\mathsf{B}^{-1}\mathsf{A})}{\partial \operatorname{vec}(\mathsf{A})'} = I_K \otimes \mathsf{B}^{-1}$$

and
$$\frac{\partial \operatorname{vec}(\mathsf{B}^{-1}\mathsf{A})}{\partial \operatorname{vec}(\mathsf{B})'} = -(\mathsf{A}'\mathsf{B}'^{-1} \otimes \mathsf{B}^{-1}).$$

Use these results to derive an explicit expression for the score vector
$$\mathbf{s}\begin{pmatrix}\gamma_A\\\gamma_B\end{pmatrix} = \frac{\partial \ln l}{\partial(\gamma_A', \gamma_B')'}.$$

Problem 9.4
Define $\mathbf{a} := [\operatorname{vec}(\mathsf{A})', \operatorname{vec}(\mathsf{B})']'$ and $\gamma := (\gamma_A', \gamma_B')'$ and show that, for the setup in Proposition 9.5,
$$-E\left(\frac{\partial^2 \ln l}{\partial\gamma\partial\gamma'}\right) = -\begin{bmatrix} R_A' & 0 \\ 0 & R_B' \end{bmatrix} E\left(\frac{\partial^2 \ln l}{\partial\mathbf{a}\partial\mathbf{a}'}\right)\begin{bmatrix} R_A & 0 \\ 0 & R_B \end{bmatrix}.$$

Moreover, show that (9.3.4) holds by proving that
$$E\left(\frac{\partial^2 \ln l}{\partial\mathbf{a}\partial\mathbf{a}'}\right) = \frac{\partial \operatorname{vec}(\Sigma_u)'}{\partial \mathbf{a}} E\left(\frac{\partial^2 \ln l}{\partial \operatorname{vec}(\Sigma_u)\partial \operatorname{vec}(\Sigma_u)'}\right)\frac{\partial \operatorname{vec}(\Sigma_u)}{\partial \mathbf{a}'}$$

and, for C such that $CC' = \Sigma_u$,
$$\frac{\partial \operatorname{vec}(\Sigma_u)}{\partial \mathbf{a}'} = \frac{\partial \operatorname{vec}(CC')}{\partial \operatorname{vec}(C)'}\frac{\partial \operatorname{vec}(C)}{\partial \mathbf{a}'} = (I_{K^2} + \mathbf{K}_{KK})(C \otimes I_K)\frac{\partial \operatorname{vec}(C)}{\partial \mathbf{a}'}$$

(see also Chapter 3 for related derivations).

9.6.2 Numerical Problems

Problem 9.5
Specify, estimate, and analyze a model for U.S. quarterly log output (q_t) and the unemployment rate (ur_t) for the period 1948.2–1987.4 as given in the *Journal of Applied Econometrics* data archive at
  http://www.econ.queensu.ca/jae/
(see the data for Weber (1995)). Blanchard & Quah (1989) considered this system in their study.

(a) Analyze the integration and cointegration properties of the data.
(b) Fit a suitable VAR model to the bivariate series.
(c) Check the adequacy of the model.
(d) Impose an identifying restriction on the long-run total impact matrix and perform a structural impulse response analysis.

(e) Compare standard and Hall percentile confidence intervals for the impulse responses and interpret possible differences.
(f) Perform a forecast error variance decomposition and comment on the results.

(Hint: See Breitung et al. (2004) for a similar analysis.)

Problem 9.6

Analyze the Canadian labor market data from Breitung et al. (2004) (see www.jmulti.de → datasets
for the data). The variables are:
p_t – ln productivity,
e_t – ln employment,
ur_t – unemployment rate,
w_t – ln real wage index.
Thus, $y_t = (p_t, e_t, ur_t, w_t)'$ is four-dimensional. The data are quarterly for the period 1980.1–2000.4. They are constructed as described in Breitung et al. (2004) based on data from the OECD database. Note that Breitung et al. (2004) use a slightly different notation for the variables.

(a) Analyze the integration and cointegration properties of the data.
(b) Fit a VECM with cointegration rank $r = 1$ for y_t.
(c) Check the adequacy of your model.
(d) Impose identifying restrictions of the form

$$B = \begin{bmatrix} * & * & * & * \\ * & * & * & * \\ * & * & * & * \\ * & 0 & * & * \end{bmatrix} \quad \text{and} \quad \Xi B = \begin{bmatrix} * & 0 & 0 & 0 \\ * & * & * & 0 \\ * & * & * & 0 \\ * & * & * & 0 \end{bmatrix}$$

and perform a structural impulse response analysis.
(e) Compare standard and Hall percentile confidence intervals for the impulse responses and interpret possible differences.
(f) Impose another zero restriction on B and repeat the structural impulse response analysis.
(g) Perform forecast error variance decompositions based on the structural innovations for different identification schemes and comment on the results.

(Hint: See Breitung et al. (2004) for a detailed analysis of the system.)

10
Systems of Dynamic Simultaneous Equations

10.1 Background

This chapter serves to point out some possible extensions of the models considered so far and to draw attention to potential problems related to such extensions. So far, we have assumed that all stochastic variables of a system have essentially the same status in that they are all determined within the system. In other words, the model describes the joint generation process of all the observable variables of interest. In practice, the generation process may be affected by other observable variables which are determined outside the system of interest. Such variables are called *exogenous* or *unmodelled* variables. In contrast, the variables determined within the system are called *endogenous*. Although deterministic terms can be included in the set of unmodelled variables, we often have stochastic variables in mind in this category. For instance, weather related variables such as rainfall or hours of sunshine are usually regarded as stochastic exogenous variables. As another example of the latter type of variables, if a small open economy is being studied, the price level or the output of the rest of the world may be regarded as exogenous. A model which specifies the generation process of some variables conditionally on some other unmodelled variables is sometimes called a *conditional* or *partial model* because it describes the generation process of a subset of the variables only.

A model with unmodelled variables may have the structural form

$$A y_t = A_1^* y_{t-1} + \cdots + A_p^* y_{t-p} + B_0^* x_t + B_1^* x_{t-1} + \cdots + B_s^* x_{t-s} + w_t, \quad (10.1.1)$$

where $y_t = (y_{1t}, \ldots, y_{Kt})'$ is a K-dimensional vector of endogenous variables, $x_t = (x_{1t}, \ldots, x_{Mt})'$ is an M-dimensional vector of unmodelled variables, A is $(K \times K)$ and represents the instantaneous relations between the endogenous variables, the A_i^*'s and B_j^*'s are $(K \times K)$ and $(K \times M)$ coefficient matrices, respectively, and w_t is a K-dimensional error vector. The vector x_t may contain both stochastic and non-stochastic components. For example, it may include intercept terms, seasonal dummies, and the amount of rainfall in a specific region. If the error term w_t is white noise, a model of the type (10.1.1) is sometimes called a VARX(p, s) model in the following. More generally, models of the form (10.1.1) are often called *linear systems* because they are obviously linear in all variables. In the econometrics literature, the label (linear) *dynamic simultaneous equations model* (SEM) is used for such a model. Because we often have systems of economic variables in mind in the following discussion, we will use this name occasionally. We will also consider a vector error correction version of the model which is useful when cointegrated variables are involved.

Other names that are occasionally found in the related literature are *transfer function models* or *distributed lag models*. These terms will become more plausible in the next section, where different representations and some properties of our basic model (10.1.1) will be discussed. Estimation is briefly considered in Section 10.3 and some remarks on model specification and model checking follow in Section 10.4. Possible uses of such models, namely forecasting, multiplier analysis, and control, are treated in Sections 10.5–10.7. Concluding remarks are contained in Section 10.8. It is not the purpose of this chapter to give a detailed and complete account of all these topics. The chapter is just meant to give some guidance to possible extensions of the by now familiar VAR models and VECMs, the related problems and some further reading.

10.2 Systems with Unmodelled Variables

10.2.1 Types of Variables

In the dynamic simultaneous equations model (10.1.1), we have partitioned the observables in two groups, y_t and x_t. The components of y_t are endogenous variables and the components of x_t are the unmodelled or exogenous variables. Although we have given some explanation of the differences between the two groups of variables, we have not given a precise definition of the terms endogenous and exogenous so far. The idea is that the endogenous variables are determined within the system, whereas the unmodelled, exogenous variables are those on which we can condition the analysis without affecting the results of interest. Because there are different possible objectives of an analysis, there are also different notions of exogeneity. For example, if we are interested in estimating a particular parameter vector γ, say, x_t is exogenous if the estimation properties do not suffer from conditioning on x_t rather than using a full model for the data generation process of all the variables involved. In that case, x_t is called *weakly exogenous* for γ. This and other types of exogeneity have been formalized by Engle, Hendry & Richard (1983). They call x_t *strongly exogenous* if we can condition on this set of variables for forecasting purposes without loosing forecast precision and they classify x_t as *super-exogenous* if policy analysis can be made conditional on these variables (see also Geweke (1982), Hendry (1995, Chapter 5), Ericsson (1994) for more discussion of exogeneity).

A simple technical definition is to call x_t exogenous if $x_t, x_{t-1}, \ldots, x_{t-s}$ are independent of the error term w_t. Moreover, x_t is sometimes called *strictly exogenous* if all its leads and lags are independent of all leads and lags of the error process w_t, that is, if x_t and w_t are independent processes. Such assumptions simplify derivations of properties of estimators and are therefore convenient. They may be unnecessarily restrictive, however, for some purposes. In the following, we will implicitly make the assumption that x_t and w_t are independent processes for convenience, although most results can be obtained under less restrictive conditions.

For much of the present discussion, a formal definition of the types of variables involved is not necessary. It suffices to have a partitioning into two groups of variables. The reader should, however, have some intuition of which variables are contained in y_t and which ones are included in x_t. As mentioned previously, roughly speaking, y_t contains the observable *outputs* of the system, that is, the observable variables that are determined by the system. In contrast, the x_t variables may be regarded as *observable input variables* which are determined outside the system. In this setting, the error variables w_t may be viewed as *unobservable inputs* to the system. As we have seen, nonstochastic components may be absorbed into the set of x_t variables. All or some of the components of x_t may be under full or partial control of the government or a decision or policy maker. In a control context, such variables are often referred to as *instruments* or *instrument variables* (see Section 10.7). Sometimes the lagged endogenous variables together with the exogenous variables of a system are called *predetermined variables*. If x_t contains just a constant and $s = 0$, the model (10.1.1) reduces to a VAR model, provided w_t is white noise.

For illustrative purposes, consider the following example system relating investment (x_{1t}), income (y_{1t}), and consumption (y_{2t}) variables:

$$y_{1t} = \nu_1^* + \alpha_{11,1}^* y_{1,t-1} + \alpha_{12,1}^* y_{2,t-1} + \beta_{12,1}^* x_{1,t-1} + w_{1t},$$
$$y_{2t} = \nu_2^* + \alpha_{22,1}^* y_{2,t-1} + a_{21,0} y_{1t} + \alpha_{21,1}^* y_{1,t-1} + w_{2t}. \tag{10.2.1}$$

This model is similar to those obtained for West German data in Chapter 5. An important difference is that current income appears in the consumption equation and there is no equation for investment. Thus, only income and consumption are determined within the system whereas investment is not. The fact that investment is, of course, determined within the economic system as a whole does not necessarily mean that we have to specify its generation mechanism if our main interest is with the generation mechanism of income and consumption. In terms of the representation (10.1.1), the example system can be written as

$$\begin{bmatrix} 1 & 0 \\ -a_{21,0} & 1 \end{bmatrix} \begin{bmatrix} y_{1t} \\ y_{2t} \end{bmatrix} = \begin{bmatrix} \alpha_{11,1}^* & \alpha_{12,1}^* \\ \alpha_{21,1}^* & \alpha_{22,1}^* \end{bmatrix} \begin{bmatrix} y_{1,t-1} \\ y_{2,t-1} \end{bmatrix}$$
$$+ \begin{bmatrix} \nu_1^* & \beta_{12,1}^* \\ \nu_2^* & 0 \end{bmatrix} \begin{bmatrix} 1 \\ x_{1,t-1} \end{bmatrix} + \begin{bmatrix} w_{1t} \\ w_{2t} \end{bmatrix}. \tag{10.2.2}$$

Thus, $y_t = (y_{1t}, y_{2t})'$ and $x_t = (1, x_{1t})'$ are both two-dimensional. The predetermined variables are y_{t-1} and x_{t-1}.

In dynamic SEMs there are sometimes identities or exact relations between some variables. For instance, the same figures may be used for supply and demand of a product. In that case, an identity equating supply and demand may appear as a separate equation of a system. So far we have not excluded this possibility. However, in later sections the covariance matrix of w_t will be assumed to be nonsingular which excludes identities. Then we assume without further notice that they have been eliminated by substitution. For instance, the demand variable may be substituted for the supply variable in all instances where it appears in the system.

10.2.2 Structural Form, Reduced Form, Final Form

The representation (10.1.1) is called the *structural form* of the model if it represents the instantaneous effects of the endogenous variables properly. The instantaneous effects are reflected in the elements of A. The idea is that the instantaneous causal links are derived from theoretical considerations and are used to place restrictions on A. Of course, multiplication of (10.1.1) with any other nonsingular $(K \times K)$ matrix results in an equivalent representation of the process generating y_t. Such a representation is not called a structural form, however, unless it reflects the actual relations of interest.

The *reduced form* of the system is obtained by premultiplying (10.1.1) with A^{-1} which gives

$$y_t = A_1 y_{t-1} + \cdots + A_p y_{t-p} + B_0 x_t + \cdots + B_s x_{t-s} + u_t, \qquad (10.2.3)$$

where $A_i := \mathsf{A}^{-1} A_i^*$ ($i = 1, \ldots, p$), $B_j := \mathsf{A}^{-1} B_j^*$ ($j = 0, 1, \ldots, s$), and $u_t := \mathsf{A}^{-1} w_t$. We always assume without notice that the inverse of A exists. In Sections 10.5–10.7, we will see that the reduced form is useful for forecasting, multiplier analysis, and control purposes.

For the example model given in (10.2.2), we have

$$\mathsf{A}^{-1} = \begin{bmatrix} 1 & 0 \\ \mathsf{a}_{21,0} & 1 \end{bmatrix}$$

and, hence, the reduced form is

$$\begin{bmatrix} y_{1t} \\ y_{2t} \end{bmatrix} = A_1 \begin{bmatrix} y_{1,t-1} \\ y_{2,t-1} \end{bmatrix} + B_1 \begin{bmatrix} 1 \\ x_{1,t-1} \end{bmatrix} + \begin{bmatrix} u_{1t} \\ u_{2t} \end{bmatrix}, \qquad (10.2.4)$$

where

$$A_1 = \begin{bmatrix} \alpha_{11,1} & \alpha_{12,1} \\ \alpha_{21,1} & \alpha_{22,1} \end{bmatrix} = \begin{bmatrix} \alpha_{11,1}^* & \alpha_{12,1}^* \\ \mathsf{a}_{21,0}\alpha_{11,1}^* + \alpha_{21,1}^* & \mathsf{a}_{21,0}\alpha_{12,1}^* + \alpha_{22,1}^* \end{bmatrix},$$

$$(10.2.5)$$

$$B_1 = \begin{bmatrix} \beta_{11,1} & \beta_{12,1} \\ \beta_{21,1} & \beta_{22,1} \end{bmatrix} = \begin{bmatrix} \nu_1^* & \beta_{12,1}^* \\ a_{21,0}\nu_1^* + \nu_2^* & a_{21,0}\beta_{12,1}^* \end{bmatrix}, \tag{10.2.6}$$

and

$$\begin{bmatrix} u_{1t} \\ u_{2t} \end{bmatrix} = \begin{bmatrix} w_{1t} \\ a_{21,0}w_{1t} + w_{2t} \end{bmatrix}.$$

It is important to note that the reduced form parameters are in general non-linear functions of the structural form parameters.

In lag operator notation, the reduced form (10.2.3) can be written as

$$A(L)y_t = B(L)x_t + u_t, \tag{10.2.7}$$

where

$$A(L) := I_K - A_1 L - \cdots - A_p L^p$$

and

$$B(L) := B_0 + B_1 L + \cdots + B_s L^s.$$

If the effect of a change in an exogenous variable on the endogenous variables is of interest, it is useful to solve the system (10.2.7) for the endogenous variables by multiplying with $A(L)^{-1}$. The resulting representation,

$$y_t = D(L)x_t + A(L)^{-1}u_t, \tag{10.2.8}$$

where $D(L) := A(L)^{-1}B(L)$, is sometimes called the *final form* of the system. Of course, using $A(L)^{-1}$ requires invertibility of $A(L)$ which is guaranteed if

$$\det A(z) \neq 0 \quad \text{for } |z| \leq 1. \tag{10.2.9}$$

If y_t contains just one variable, $A(L)$ is a scalar operator and the form (10.2.8) is often called a *distributed lag model* in the econometrics literature because it describes how lagged effects of changes in x_t are distributed over time. Because the lag distribution for each exogenous variable can be written as a ratio of two finite order polynomials in the lag operator $(A(L)^{-1}B(L))$, the model is referred to as a *rational distributed lag model*. In the time series literature, the label *rational transfer function model* is often attached to (10.2.8) in both the scalar and the vector case. The operator $D(L)$ represents the *transfer function* transferring the observable inputs into the outputs of the system.

For the example model with reduced form (10.2.4), we get a final form

$$\begin{bmatrix} y_{1t} \\ y_{2t} \end{bmatrix} = (I_2 - A_1 L)^{-1} B_1 L \begin{bmatrix} 1 \\ x_{1t} \end{bmatrix} + (I_2 - A_1 L)^{-1} \begin{bmatrix} u_{1t} \\ u_{2t} \end{bmatrix}$$

$$= \left(\sum_{i=1}^{\infty} A_1^{i-1} B_1 L^i\right) \begin{bmatrix} 1 \\ x_{1t} \end{bmatrix} + \left(\sum_{i=0}^{\infty} A_1^i L^i\right) \begin{bmatrix} u_{1t} \\ u_{2t} \end{bmatrix}. \tag{10.2.10}$$

Note that $B_0 = 0$ and thus, $D_0 = 0$ and $D_i = A_1^{i-1} B_1$ for $i = 1, 2, \ldots$.

The coefficient matrices $D_i = (d_{kj,i})$ of the transfer function operator

$$D(L) = \sum_{i=0}^{\infty} D_i L^i$$

contain the effects that changes in the exogenous variables have on the endogenous variables. Everything else held constant, a unit change in the j-th exogenous variable in period t induces a marginal change of $d_{kj,i}$ units in the k-th endogenous variable in period $t + i$. The elements of the D_i matrices are therefore called *dynamic multipliers*. The accumulated effects contained in $\sum_{i=0}^{n} D_i$ are the n-th *interim multipliers* and the elements of $\sum_{i=0}^{\infty} D_i$ are the *long-run effects* or *total multipliers*. We will return to multiplier analysis in Section 10.6.

As in the example, the transfer function operator $D(L)$ has infinite order in general. A finite order representation of the system is obtained by noting that $A(L)^{-1} = A(L)^{adj}/|A(L)|$, where $A(L)^{adj}$ denotes, as usual, the adjoint of $A(L)$. Thus, multiplying the reduced form by $A(L)^{adj}$ gives

$$|A(L)| y_t = A(L)^{adj} B(L) x_t + A(L)^{adj} u_t \qquad (10.2.11)$$

which involves finite order operators only. In the econometrics literature these equations are sometimes called *final equations*. Because $|A(L)|$ is a scalar operator, each equation contains only one of the endogenous variables.

Assuming that the unmodelled variables x_t are driven by a VAR(q) process, say

$$x_t = C_1 x_{t-1} + \cdots + C_q x_{t-q} + v_t,$$

where $q \leq p$ and v_t is white noise, then the joint generation process of x_t and y_t is

$$\begin{bmatrix} I_K & -B_0 \\ 0 & I_M \end{bmatrix} \begin{bmatrix} y_t \\ x_t \end{bmatrix} = \begin{bmatrix} A_1 & B_1 \\ 0 & C_1 \end{bmatrix} \begin{bmatrix} y_{t-1} \\ x_{t-1} \end{bmatrix} + \cdots$$

$$+ \begin{bmatrix} A_p & B_p \\ 0 & C_p \end{bmatrix} \begin{bmatrix} y_{t-p} \\ x_{t-p} \end{bmatrix} + \begin{bmatrix} u_t \\ v_t \end{bmatrix},$$

where it is assumed without loss of generality that $s, q \leq p$, $B_i := 0$ for $i > s$ and $C_j := 0$ for $j > q$. If u_t is also white noise, premultiplying by

$$\begin{bmatrix} I_K & -B_0 \\ 0 & I_M \end{bmatrix}^{-1} = \begin{bmatrix} I_K & B_0 \\ 0 & I_M \end{bmatrix}$$

shows that the joint generation process of y_t and x_t is a VAR(p).

10.2.3 Models with Rational Expectations

Sometimes the endogenous variables are assumed to depend not only on other endogenous and exogenous variables but also on expectations on endogenous variables. If only expectations formed in the previous period for the present period are of importance, one could simply add another term involving the expectations variables to the structural form (10.1.1). Denoting the expectations variables by y_t^e may then result in a reduced form

$$y_t = A_1 y_{t-1} + \cdots + A_p y_{t-p} + F y_t^e + B_0 x_t + \cdots + B_s x_{t-s} + u_t \quad (10.2.12)$$

or

$$A(L) y_t = F y_t^e + B(L) x_t + u_t, \quad (10.2.13)$$

where F is a $(K \times K)$ matrix of parameters and $A(L)$ and $B(L)$ are the matrix polynomials in the lag operator from (10.2.7).

Following Muth (1961), the expectations y_t^e formed in period $t-1$ are called *rational* if they are the best possible predictions, given the information in period $t-1$. In other words, y_t^e is the conditional expectation $E_{t-1}(y_t)$, given all information available in period $t-1$. In forming the predictions or expectations, not only the past values of the endogenous and unmodelled variables are assumed to be known but also the model (10.2.12) and the generation process of the unmodelled variables. It is easy to see that, if the unmodelled variables are generated by a VAR process, the expectations variables can be eliminated from (10.2.12)/(10.2.13). The resulting reduced form is of VARX type. To show this result, suppose that u_t is independent white noise and, as before, denote by E_t the conditional expectation, given all information available in period t. Applying E_{t-1} to (10.2.12) then gives

$$\begin{aligned} y_t^e &= E_{t-1}(y_t) \\ &= A_1 y_{t-1} + \cdots + A_p y_{t-p} \\ &\quad + F y_t^e + B_0 E_{t-1}(x_t) + B_1 x_{t-1} + \cdots + B_s x_{t-s} \end{aligned} \quad (10.2.14)$$

or

$$y_t^e = (A(L) - I_K) y_t + F y_t^e + B_0 E_{t-1}(x_t) + (B(L) - B_0) x_t. \quad (10.2.15)$$

Assuming that $I_K - F$ is invertible, this system can be solved for y_t^e:

$$y_t^e = (I_K - F)^{-1} [(A(L) - I_K) y_t + B_0 E_{t-1}(x_t) + (B(L) - B_0) x_t]. \quad (10.2.16)$$

If x_t is generated by a VAR(q) process, say

$$x_t = C_1 x_{t-1} + \cdots + C_q x_{t-q} + v_t,$$

where v_t is independent white noise, then

$$E_{t-1}(x_t) = C_1 x_{t-1} + \cdots + C_q x_{t-q}.$$

Substituting this expression in (10.2.16) shows that y_t^e depends on lagged y_t and x_t only. Thus, substituting for y_t^e in (10.2.12) or (10.2.13), we get a standard VARX form of the model.

Thus, in theory, when the true coefficient matrices are known, we can simply eliminate the term involving expectations variables and work with a standard reduced form without an expectations term. It should be clear, however, that substituting the right-hand side of (10.2.16) for y_t^e in (10.2.12) implies nonlinear restrictions on the coefficient matrices of the reduced form without expectations terms. Taking into account such restrictions may increase the efficiency of parameter estimators. The same is true, of course, for the structural form. Therefore, it is important in practice whether or not the actual relationship between the variables is partly determined by agents' expectations.

For expository purposes we have just treated a very special case where only expectations formed in period $t-1$ for period t enter the model. Extensions can be treated in a similar way. For instance, past expectations for more than one period ahead or expectations formed in various previous periods may be of importance. If x_t is generated by a VAR(q) process, they can be eliminated like in the special case considered in the foregoing.

A complication of the basic model that makes life a bit more difficult is the inclusion of future expectations. It is quite realistic to suppose that, for instance, the expected future price of a commodity may determine the supply in the present period. For example, if bond prices are expected to fall during the next period, an investor may decide to sell now. If future expectations enter the model, the solution for the endogenous variables will in general not be unique. In other words, the process that generates the endogenous variables may not be uniquely determined by the model, even if the generation process of the exogenous variables is uniquely specified. Further extensive discussions of rational expectations models can be found in volumes by Lucas & Sargent (1981) and Pesaran (1987).

10.2.4 Cointegrated Variables

Many of the results discussed so far in this section hold for systems of stationary or integrated variables. More precisely, whenever the VAR operator $A(L)$ is not required to be invertible, integrated variables may be present as endogenous as well as unmodelled variables. If there are cointegrated variables, it may be preferable, however, to separate the short- and long-run dynamics as in a VECM. Assuming that there are r cointegration relations among the endogenous variables and they are not cointegrated with the unmodelled variables, the corresponding form of the model is

$$\begin{aligned} A \Delta y_t &= \alpha^* \beta' y_{t-1} + \Gamma_1^* \Delta y_{t-1} + \cdots + \Gamma_{p-1}^* \Delta y_{t-p+1} \\ &\quad + B_0^* x_t + B_1^* x_{t-1} + \cdots + B_s^* x_{t-s} + w_t, \end{aligned} \quad (10.2.17)$$

where A is a $(K \times K)$ matrix of instantaneous effects, as before, α^* is a $(K \times r)$ matrix of structural loading coefficients, β is the $(K \times r)$ cointegration matrix, Γ_j^* $(j = 1, \ldots, p-1)$ is a $(K \times K)$ matrix of structural short-run coefficients, and all other symbols are defined as in (10.1.1). In many respects, this model can be dealt with in essentially the same way as the VECMs considered in Part II of this volume.

It is also possible, however, that there is cointegration between endogenous and unmodelled variables. In that case, a suitable form of the model is

$$A\Delta y_t = \alpha^* \beta^{+\prime} \begin{bmatrix} y_{t-1} \\ x_{t-1} \end{bmatrix} + \Gamma_1^* \Delta y_{t-1} + \cdots + \Gamma_{p-1}^* \Delta y_{t-p+1}$$
$$+ \Upsilon_0^* \Delta x_t + \Upsilon_1^* \Delta x_{t-1} + \cdots + \Upsilon_{s-1}^* \Delta x_{t-s+1} + w_t, \quad (10.2.18)$$

where now the unmodelled variables appear in levels form in the error correction term only and otherwise enter in differenced form with suitable coefficient matrices Υ_j^* $(j = 0, 1, \ldots, s-1)$. It is easy to see that such a model form can be obtained if the joint generation process of y_t and x_t has a (reduced form) VECM representation

$$\begin{bmatrix} \Delta y_t \\ \Delta x_t \end{bmatrix} = \begin{bmatrix} \alpha \\ \alpha_x \end{bmatrix} \beta^{+\prime} \begin{bmatrix} y_{t-1} \\ x_{t-1} \end{bmatrix} + \begin{bmatrix} \Gamma_1 & \Upsilon_1 \\ 0 & \Gamma_1^x \end{bmatrix} \begin{bmatrix} \Delta y_{t-1} \\ \Delta x_{t-1} \end{bmatrix} + \cdots$$
$$+ \begin{bmatrix} \Gamma_{p-1} & \Upsilon_{p-1} \\ 0 & \Gamma_{p-1}^x \end{bmatrix} \begin{bmatrix} \Delta y_{t-p+1} \\ \Delta x_{t-p+1} \end{bmatrix} + \begin{bmatrix} u_t \\ v_t \end{bmatrix}, \quad (10.2.19)$$

where $p \geq s$ is assumed without loss of generality and all symbols have obvious definitions. Premultiplying this model form with

$$\begin{bmatrix} A & -\Upsilon_0^* \\ 0 & I_M \end{bmatrix}$$

gives a model where the first K equations are just the structural form (10.2.18). Notice, however, that the y_t may enter the x_t equations in (10.2.19) via the cointegration relations if $\alpha_x \neq 0$. It turns out that x_t is weakly exogenous for β^+, if $\alpha_x = 0$. Thus, if the cointegration relations are of primary interest, considering the partial model for Δy_t is justified if $\alpha_x = 0$.

Both models (10.2.17) and (10.2.18) can be rewritten in levels form. The result is then a structural form as in (10.1.1). Moreover, the structural forms can be converted into reduced form by premultiplying with A^{-1}.

10.3 Estimation

Parameter estimation in the presence of unmodelled variables will be discussed separately for stationary and cointegrated variables. We begin with the stationary case.

10.3.1 Stationary Variables

Suppose $(y_t', x_t')'$ is generated by a stationary process and we wish to estimate the parameters of the reduced form (10.2.3) which can be written as

$$y_t = AY_{t-1} + BX_{t-1} + B_0 x_t + u_t, \qquad (10.3.1)$$

where $A := [A_1, \ldots, A_p]$, $B := [B_1, \ldots, B_s]$,

$$Y_t := \begin{bmatrix} y_t \\ \vdots \\ y_{t-p+1} \end{bmatrix}, \quad X_t := \begin{bmatrix} x_t \\ \vdots \\ x_{t-s+1} \end{bmatrix}.$$

Here u_t is assumed to be standard white noise with *nonsingular* covariance matrix Σ_u. Moreover, we allow for parameter restrictions and assume that a matrix R and a vector γ exist such that

$$\boldsymbol{\beta} := \text{vec}[A, B, B_0] = R\gamma. \qquad (10.3.2)$$

With these assumptions, estimation of $\boldsymbol{\beta}$ and, hence, of A, B, and B_0 is straightforward.

For a sample of size T, the system can be written compactly as

$$Y = [A, B, B_0]Z + U, \qquad (10.3.3)$$

where

$$Y := [y_1, \ldots, y_T], \quad Z := \begin{bmatrix} Y_0, \ldots, Y_{T-1} \\ X_0, \ldots, X_{T-1} \\ x_1, \ldots, x_T \end{bmatrix} \quad \text{and} \quad U := [u_1, \ldots, u_T].$$

Vectorizing gives

$$\mathbf{y} = (Z' \otimes I_K) R\gamma + \mathbf{u},$$

where $\mathbf{y} := \text{vec}(Y)$ and $\mathbf{u} := \text{vec}(U)$. From Chapter 5, the GLS estimator is known to be

$$\widehat{\gamma} = [R'(ZZ' \otimes \Sigma_u^{-1})R]^{-1} R'(Z \otimes \Sigma_u^{-1})\mathbf{y}. \qquad (10.3.4)$$

This estimator is not operational because in practice Σ_u is unknown. However, as in Section 5.2.2, Σ_u may be estimated from the LS estimator

$$\check{\gamma} = [R'(ZZ' \otimes I_K)R]^{-1} R'(Z \otimes I_K)\mathbf{y}$$

which gives residuals $\check{\mathbf{u}} = \mathbf{y} - (Z' \otimes I_K)R\check{\gamma}$ and an estimator

$$\check{\Sigma}_u = \check{U}\check{U}'/T \qquad (10.3.5)$$

of Σ_u, where $\breve{U}$ is such that $\text{vec}(\breve{U}) = \breve{\mathbf{u}}$. Using this estimator of the white noise covariance matrix results in the EGLS estimator

$$\widehat{\widehat{\gamma}} = [R'(ZZ' \otimes \breve{\Sigma}_u^{-1})R]^{-1} R'(Z \otimes \breve{\Sigma}_u^{-1})\mathbf{y}. \tag{10.3.6}$$

Under standard assumptions, this estimator is consistent and asymptotically normal,

$$\sqrt{T}(\widehat{\widehat{\gamma}} - \gamma) \xrightarrow{d} \mathcal{N}(0, \Sigma_{\widehat{\widehat{\gamma}}}), \tag{10.3.7}$$

where

$$\Sigma_{\widehat{\widehat{\gamma}}} = (R'[\text{plim}(T^{-1}ZZ') \otimes \Sigma_u^{-1}]R)^{-1}. \tag{10.3.8}$$

One condition for this result to hold is, of course, that both plim $T^{-1}ZZ'$ and the inverse of the matrix in (10.3.8) exist. Further assumptions are required to guarantee the asymptotic normal distribution of the EGLS estimator. The assumptions may include the following ones: (i) u_t is standard white noise, (ii) the VAR part is stable, that is,

$$|A(z)| = |I_K - A_1 z - \cdots - A_p z^p| \neq 0 \quad \text{for } |z| \leq 1,$$

and (iii) x_t is generated by a stationary, stable VAR process which is independent of the white noise process u_t. A precise statement of more general conditions and a proof are given, e.g., by Hannan & Deistler (1988). The latter part of our set of assumptions requires that all the exogenous variables are stochastic. It can be modified so as to include nonstochastic variables as well. In that case, the plim in (10.3.8) reduces to a nonstochastic limit in some or all components (see, e.g., Anderson (1971, Chapter 5), Harvey (1981)).

An estimator for $\beta = R\gamma$ is obtained as $\widehat{\widehat{\beta}} = R\widehat{\widehat{\gamma}}$. If (10.3.7) holds, this estimator also has an asymptotic normal distribution,

$$\sqrt{T}(\widehat{\widehat{\beta}} - \beta) \xrightarrow{d} \mathcal{N}(0, \Sigma_{\widehat{\widehat{\beta}}} = R\Sigma_{\widehat{\widehat{\gamma}}}R'), \tag{10.3.9}$$

Moreover, under general conditions, the corresponding estimator $\widehat{\widehat{\Sigma}}_u$ of the white noise covariance matrix is asymptotically independent of $\widehat{\widehat{\beta}}$ and has the same asymptotic distribution as the estimator UU'/T based on the unobserved true residuals. For instance, for a Gaussian process,

$$\sqrt{T} \text{vech}(\widehat{\widehat{\Sigma}}_u - \Sigma_u) \xrightarrow{d} \mathcal{N}(0, 2\mathbf{D}_K^+(\Sigma_u \otimes \Sigma_u)\mathbf{D}_K^{+\prime}), \tag{10.3.10}$$

where $\mathbf{D}_K^+ = (\mathbf{D}_K'\mathbf{D}_K)^{-1}\mathbf{D}_K'$ is the Moore-Penrose inverse of the $(K^2 \times \frac{1}{2}K(K+1))$ duplication matrix $\mathbf{D}_K$.

In discussing direct reduced form estimation with white noise errors, we have treated a particularly simple case. The following complications are possible.

(1) Usually there will be restrictions on the structural form coefficients A, A_i^*, $i = 1,\dots,p$, and B_j^*, $j = 0,\dots,s$. Such restrictions may imply nonlinear constraints on the reduced form coefficients which are not covered by the above approach. Rational expectations assumptions may be another source of nonlinear restrictions on the reduced form parameters. Theoretically, it is not difficult to handle nonlinear restrictions on the reduced form parameters. In practice, numerical problems may arise in a multivariate LS or GLS estimation with nonlinear restrictions.

(2) Interest may focus on the structural rather than the reduced form. Estimation of the structural form has been discussed extensively in the econometrics literature. For recent surveys and many further references see Judge et al. (1985), Hausman (1983), or textbooks such as Hayashi (2000). A major complication in estimating the structural form of a SEM such as (10.1.1) results from its possible nonuniqueness. Note that we have not assumed a triangular A matrix or a diagonal covariance matrix of w_t. Premultiplication of (10.1.1) by any nonsingular matrix results in an equivalent representation of the process. Thus, for proper estimation there must be restrictions on the structural form coefficients that guarantee uniqueness or identification of the structural form coefficients.

(3) So far we have just discussed models which are linear in the variables. In practice, there may be nonlinear relations between the variables. Estimation of nonlinear dynamic models where the endogenous as well as the unmodelled conditioning variables may enter in a nonlinear way are, for instance, discussed by Bierens (1981), Gallant (1987), and Gallant & White (1988).

In the next section, we will consider models with integrated and cointegrated variables.

10.3.2 Estimation of Models with $I(1)$ Variables

If there are integrated and cointegrated variables in the model and a reduced form VECM corresponding to the structural form (10.2.18),

$$\Delta y_t = \alpha\beta^{+\prime}\begin{bmatrix} y_{t-1} \\ x_{t-1} \end{bmatrix} + \Gamma_1 \Delta y_{t-1} + \cdots + \Gamma_{p-1}\Delta y_{t-p+1}$$
$$+ \Upsilon_0 \Delta x_t + \Upsilon_1 \Delta x_{t-1} + \cdots + \Upsilon_{s-1} \Delta x_{t-s+1} + u_t, \qquad (10.3.11)$$

is set up, estimation can in principle proceed as in Section 7.2. Assuming that a sample of size T and all required presample values are available and defining

$$\Delta Y := [\Delta y_1, \dots, \Delta y_T],$$

$$Y_{-1}^+ := [y_0^+, \dots, y_{T-1}^+], \quad \text{with} \quad y_{t-1}^+ := \begin{bmatrix} y_{t-1} \\ x_{t-1} \end{bmatrix},$$

$$\Delta X^+ := [\Delta X_0^+, \ldots, \Delta X_{T-1}^+] \quad \text{with} \quad \Delta X_{t-1}^+ := \begin{bmatrix} \Delta y_{t-1} \\ \vdots \\ \Delta y_{t-p+1} \\ \Delta x_t \\ \Delta x_{t-1} \\ \vdots \\ \Delta x_{t-s+1} \end{bmatrix},$$

and

$$U := [u_1, \ldots, u_T],$$

we get

$$\Delta Y = \alpha \beta^{+\prime} Y_{-1}^+ + \Gamma^+ \Delta X^+ + U, \tag{10.3.12}$$

where

$$\Gamma^+ := [\Gamma_1 : \cdots : \Gamma_{p-1} : \Upsilon_0 : \Upsilon_1 : \cdots : \Upsilon_{s-1}].$$

Thus, we have precisely the same model form as in Section 7.2 (see, e.g., (7.2.3)) and, in principle, all the estimators of that section are available. Notice, however, that now β^+ is a $((K+M) \times r)$ matrix whereas α is still $(K \times r)$. Because the error correction term now involves all the cointegration relations between the endogenous and unmodelled variables, it is possible that $r > K$. In that case, it is easy to see that most of the estimators of Section 7.2 are not available. Thus, we have to assume that $r \leq K$. In fact, if $r = K$, the matrix $\Pi^+ := \alpha \beta^{+\prime}$ is of full row rank under our usual assumption that $\text{rk}(\alpha) = \text{rk}(\beta^+) = r$. Therefore, if $K = r$, we do not even need reduced rank regression but can simply estimate the matrix $\Pi^+ = \alpha \beta^{+\prime}$ by applying multivariate LS to (10.3.12). An estimator of β^+ can then be obtained by normalizing the cointegration matrix as in Section 7.2 such that

$$\beta^+ = \begin{bmatrix} I_K \\ \beta_{(M)}^+ \end{bmatrix} \tag{10.3.13}$$

and, using

$$\widehat{\beta}^{+\prime} = (\widehat{\Pi}_{(1)}^+)^{-1} \widehat{\Pi}^+,$$

where $\widehat{\Pi}_{(1)}^+$ is the $(K \times K)$ submatrix consisting of the first K columns of the LS estimator $\widehat{\Pi}^+$ of Π^+.

If $r < K$, there is nothing special here relative to the procedures discussed in Section 7.2. Reduced rank ML estimation, as discussed in Section 7.2.3, is available just as the EGLS estimator of the cointegration parameters of Section 7.2.2 and the two-stage estimator described in Section 7.2.5. Moreover, the two-stage procedure can also be used to estimate models with parameter restrictions on α and Γ^+, as in Section 7.3.2. In fact, a similar procedure can even be used for the estimation of structural form models of the type (10.2.18).

In this context, it is, of course, of interest to know the properties of the resulting estimators. They are available under suitable assumptions for the model and the variables (see, e.g., Johansen (1992) or Davidson (2000, Section 16.5)). Under general assumptions, the estimator of the cointegration matrix continues to be superconsistent, that is,

$$T(\widehat{\beta}^+ - \beta^+) = O_p(1),$$

if all variables are at most $I(1)$ and β^+ is identified. If the cointegration relations do not enter the generation process of x_t, that is, $\alpha_x = 0$ in (10.2.19), x_t is weakly exogenous for β^+ and the ML and EGLS estimators of β^+ have mixed normal distributions similar to those discussed in Section 7.2. Therefore standard inference is possible, as discussed in that section. The estimators of the α and Γ^+ parameters have again standard properties which are the same as in the case where the β^+ matrix is known.

10.4 Remarks on Model Specification and Model Checking

The basic principles of model specification and checking the model adequacy have been discussed in some detail in previous chapters. We will therefore make just a few remarks here. With respect to the specification there is, however, a major difference between the models considered previously and the dynamic SEMs of this chapter. While in a reduced form VAR analysis usually relatively little prior knowledge from economic or other subject matter theory is used, such theories may well be the major building block in specifying SEMs. In that case, model checking becomes of central importance in investigating the validity of the theory. Quite often, theories are not available that specify the data generation process completely. For instance, the lag lengths of the endogenous and/or exogenous variables may have to be specified with statistical tools. Also, some researchers may not be prepared to rely on the available theories and therefore prefer to substitute statistical investigations for uncertain prior knowledge. Statistical specification strategies for general dynamic SEMs were, for instance, proposed and discussed by Hannan & Kavalieris (1984), Hannan & Deistler (1988), and Poskitt (1992). These strategies are based on model selection criteria of the type considered in previous chapters. An extensive literature exists on the specification of special models. For instance, distributed lag models are discussed at length in the econometrics literature (for some references see Judge et al. (1985, Chapters 9 and 10)). Specification proposals for transfer function models with one dependent variable y_t go back to the pioneering work of Box & Jenkins (1976). Other suggestions have been made by Haugh & Box (1977), Young, Jakeman & McMurtrie (1980), Liu & Hanssens (1982), Tsay (1985), and Poskitt (1989) to name just a few.

If some of the variables are integrated, one may also want to investigate the number of cointegration relations with statistical tests. From the discussion in Section 10.3.2, it is clear that rank tests can be used for that purpose, as in Section 8.2. These tests may now be based either on a VECM for the full joint generation process of y_t and x_t or on a partial model with some unmodelled variables. The latter approach may be preferable if a large number of variables is involved. Johansen's LR tests for the cointegrating rank may be unreliable in that situation because of size distortions and lack of power. Therefore, testing for the cointegrating rank in a partial model may be advantageous. The asymptotic distributions of the relevant LR test statistics in this case depend on the conditioning variables, however. This result is not surprising, of course, because the conditioning variables can in fact be deterministic terms and we have seen in Section 8.2 that such terms have an impact on the asymptotic properties of the LR tests. The relevant tests for conditional models were derived by Harbo, Johansen, Nielsen & Rahbek (1998) and critical values were given in MacKinnon, Haug & Michelis (1999).

In checking the model adequacy one may want to test various restrictions. These may range from constraints suggested by some kind of theory such as the rational expectations hypothesis, to tests of the significance of extra lags. The three testing principles discussed previously, namely the LR, LM, and Wald principles (see Appendix C.7) can be used in the present context. Their asymptotic properties follow in the usual way from properties of the estimators and the model.

A residual analysis is another tool which is available in the present case. Plots of residuals may help to identify unusual values or patterns that suggest model deficiencies. Plots of residual autocorrelations may aid in checking the white noise assumption. Also a portmanteau test for overall residual autocorrelation may be developed for dynamic models with exogenous variables; see Poskitt & Tremayne (1981) for a discussion of this issue and further references.

10.5 Forecasting

10.5.1 Unconditional and Conditional Forecasts

If the future paths of the unmodelled variables are unknown to the forecaster, then forecasts of these variables are needed in order to predict the future values of the endogenous variables on the basis of a dynamic SEM. For simplicity, suppose that the exogenous variables are generated by a zero mean VAR(q) process as in Section 10.2.3,

$$x_t = C_1 x_{t-1} + \cdots + C_q x_{t-q} + v_t. \tag{10.5.1}$$

Now this process can be used to produce optimal forecasts $x_t(h)$ of x_t in the usual way. If the endogenous variables are generated by the reduced form model (10.2.3) with u_t being independent white noise which is also independent of the x_t process, the optimal h-step forecast of y_{t+h} at origin t is

$$y_t(h) = A_1 y_t(h-1) + \cdots + A_p y_t(h-p) + B_0 x_t(h) + \cdots + B_s x_t(h-s), \quad (10.5.2)$$

where $y_t(j) := y_{t+j}$ and $x_t(j) := x_{t+j}$ for $j \leq 0$. This formula can be used for recursively determining forecasts for $h = 1, 2, \ldots$.

An alternative way for getting these forecasts is obtained by writing the generation processes of the exogenous variables in one overall model together with the reduced form SEM:

$$\begin{bmatrix} I_K & -B_0 \\ 0 & I_M \end{bmatrix} \begin{bmatrix} y_t \\ x_t \end{bmatrix} = \begin{bmatrix} A_1 & B_1 \\ 0 & C_1 \end{bmatrix} \begin{bmatrix} y_{t-1} \\ x_{t-1} \end{bmatrix} + \cdots$$
$$+ \begin{bmatrix} A_p & B_p \\ 0 & C_p \end{bmatrix} \begin{bmatrix} y_{t-p} \\ x_{t-p} \end{bmatrix} + \begin{bmatrix} u_t \\ v_t \end{bmatrix}, \quad (10.5.3)$$

where we assume without loss of generality that $p \geq \max(s, q)$ and set $B_i = 0$ for $i > s$ and $C_j = 0$ for $j > q$. As in Section 10.2.2, premultiplying by

$$\begin{bmatrix} I_K & -B_0 \\ 0 & I_M \end{bmatrix}^{-1} = \begin{bmatrix} I_K & B_0 \\ 0 & I_M \end{bmatrix}$$

gives a standard reduced form VAR(p) model. It is easy to see that the optimal forecasts for y_t and x_t from that model are exactly the same as those obtained by getting forecasts for x_t from (10.5.1) first and using them in the prediction formula for y_t given in (10.5.2) (see Problem 10.5). Thus, under the present assumptions, the discussion of forecasting VAR(p) processes applies. It will not be repeated here. Also, it is not difficult to extend these ideas to sets of unmodelled variables with nonstochastic components such as intercept terms or seasonal dummies.

We will refer to forecasts of y_t obtained in this way as *unconditional forecasts* because they are based on forecasts of the exogenous variables for the forecast period. Occasionally, the forecaster may know some or all of the future values of the exogenous variables, for instance, because they are under the control of some decision maker. In that case he or she may be interested in *forecasts* of y_t *conditional* on a specific future path of x_t. In order to derive the optimal conditional forecasts, we write the reduced form (10.2.3) in VARX(1,0) form,

$$Y_t = \mathbf{A} Y_{t-1} + \mathbf{B} x_t + U_t, \quad (10.5.4)$$

where

$$Y_t := \begin{bmatrix} y_t \\ \vdots \\ y_{t-p+1} \\ x_t \\ \vdots \\ x_{t-s+1} \end{bmatrix}, \quad U_t := \begin{bmatrix} u_t \\ 0 \\ \vdots \\ 0 \end{bmatrix} \quad ((Kp+Ms) \times 1),$$

$$\mathbf{A} := \begin{bmatrix} A_1 & \cdots & A_{p-1} & A_p & | & B_1 & \cdots & B_{s-1} & B_s \\ I_K & & 0 & 0 & | & 0 & \cdots & 0 & 0 \\ & \ddots & \vdots & \vdots & | & \vdots & \ddots & \vdots & \vdots \\ 0 & \cdots & I_K & 0 & | & 0 & \cdots & 0 & 0 \\ \hline & & & & | & 0 & \cdots & 0 & 0 \\ & & 0 & & | & I_M & & 0 & 0 \\ & & & & | & & \ddots & \vdots & \vdots \\ & & & & | & 0 & \cdots & I_M & 0 \end{bmatrix},$$
$$((Kp+Ms) \times (Kp+Ms))$$

and

$$\mathbf{B} := \begin{bmatrix} \left.\begin{matrix} B_0 \\ 0 \\ \vdots \\ 0 \end{matrix}\right\} (Kp \times M) \\ \left.\begin{matrix} I_M \\ 0 \\ \vdots \\ 0 \end{matrix}\right\} (Ms \times M) \end{bmatrix}$$

Successive substitution for lagged Y_t's gives

$$Y_t = \mathbf{A}^h Y_{t-h} + \sum_{i=0}^{h-1} \mathbf{A}^i \mathbf{B} x_{t-i} + \sum_{i=0}^{h-1} \mathbf{A}^i U_{t-i}. \tag{10.5.5}$$

Hence, premultiplying by the $(K \times (Kp+Ms))$ matrix $J := [I_K : 0 : \cdots : 0]$ results in

$$y_{t+h} = J\mathbf{A}^h Y_t + \sum_{i=0}^{h-1} J\mathbf{A}^i \mathbf{B} x_{t+h-i} + \sum_{i=0}^{h-1} J\mathbf{A}^i J' u_{t+h-i}, \tag{10.5.6}$$

where $U_t = J'JU_t = J'u_t$, has been used. Now the optimal h-step forecast of y_t at origin t, given $x_{t+1}, \ldots, x_{t+h}$, and all present and past information, is easily seen to be

$$y_t(h|x) := J\mathbf{A}^h Y_t + \sum_{i=0}^{h-1} J\mathbf{A}^i \mathbf{B} x_{t+h-i} \tag{10.5.7}$$

and the corresponding forecast error is

$$y_{t+h} - y_t(h|x) = \sum_{i=0}^{h-1} J\mathbf{A}^i J' u_{t+h-i}. \tag{10.5.8}$$

Thus, the MSE of the conditional forecast is

$$\Sigma_y(h|x) := \text{MSE}[y_t(h|x)] = \sum_{i=0}^{h-1} J\mathbf{A}^i J' \Sigma_u J(\mathbf{A}^i)' J'. \tag{10.5.9}$$

Although this MSE matrix formally looks like the MSE matrix of the optimal forecast from a VAR model, where $J\mathbf{A}^i J'$ is replaced by Φ_i, the MSE matrix in (10.5.9) is in general different from the one of an unconditional forecast. This fact is easy to see by considering the different definition of the matrix $\mathbf{A}$ used in the pure VAR(p) case.

To illustrate the difference between conditional and unconditional forecasts, we consider the simple reduced form

$$y_t = A_1 y_{t-1} + B_0 x_t + u_t, \tag{10.5.10}$$

where x_t is assumed to be generated by a zero mean VAR(1) process,

$$x_t = C_1 x_{t-1} + v_t.$$

Moreover, we assume that u_t and v_t are independent white noise processes with covariance matrices Σ_u and Σ_v, respectively. The unconditional forecasts are obtained from the VAR process

$$\begin{bmatrix} I_K & -B_0 \\ 0 & I_M \end{bmatrix} \begin{bmatrix} y_t \\ x_t \end{bmatrix} = \begin{bmatrix} A_1 & 0 \\ 0 & C_1 \end{bmatrix} \begin{bmatrix} y_{t-1} \\ x_{t-1} \end{bmatrix} + \begin{bmatrix} u_t \\ v_t \end{bmatrix}$$

which, upon premultiplying with

$$\begin{bmatrix} I_K & -B_0 \\ 0 & I_M \end{bmatrix}^{-1} = \begin{bmatrix} I_K & B_0 \\ 0 & I_M \end{bmatrix},$$

has the standard VAR(1) from

$$\begin{bmatrix} y_t \\ x_t \end{bmatrix} = \begin{bmatrix} A_1 & B_0 C_1 \\ 0 & C_1 \end{bmatrix} \begin{bmatrix} y_{t-1} \\ x_{t-1} \end{bmatrix} + \begin{bmatrix} u_t + B_0 v_t \\ v_t \end{bmatrix}.$$

The optimal 1-step forecast from this model is

$$\begin{bmatrix} y_t(1) \\ x_t(1) \end{bmatrix} = \begin{bmatrix} A_1 & B_0 C_1 \\ 0 & C_1 \end{bmatrix} \begin{bmatrix} y_t \\ x_t \end{bmatrix}.$$

The corresponding MSE matrix is

$$\Sigma_*(1) = E\left(\begin{bmatrix} u_t + B_0 v_t \\ v_t \end{bmatrix} [(u_t + B_0 v_t)', v_t']\right)$$
$$= \begin{bmatrix} \Sigma_u + B_0 \Sigma_v B_0' & B_0 \Sigma_v \\ \Sigma_v B_0' & \Sigma_v \end{bmatrix}.$$

The upper left-hand corner block of this matrix is the MSE matrix of $y_t(1)$, the unconditional forecast of the endogenous variables. Thus,

$$\Sigma_y(1) = \Sigma_u + B_0 \Sigma_v B_0'. \tag{10.5.11}$$

On the other hand, in the VARX(1,0) representation (10.5.4), we have $\mathbf{A} = A_1$ and $\mathbf{B} = B_0$ for the present example. Hence, the conditional 1-step forecast of y_t is

$$y_t(1|x) = A_1 y_t + B_0 x_{t+1}$$

with corresponding MSE matrix

$$\Sigma_y(1|x) = \Sigma_u.$$

Obviously, $\Sigma_y(1) - \Sigma_y(1|x) = B_0 \Sigma_v B_0'$ is positive semidefinite and, thus, the unconditional forecast is inferior to the conditional forecast, if $B_0 \neq 0$. It must be kept in mind, however, that the conditional forecast is only feasible if the future values of the exogenous variables are either known or assumed. If only hypothetical values are used, the conditional forecast may be quite poor if the actual values of the exogenous variables turn out to be different from the hypothetical ones. The smaller MSE of the conditional forecast is simply due to ignoring any uncertainty regarding the future paths of the exogenous variables.

Using the foregoing results, interval forecasts and forecast regions can be set up as usual. It may also be worth pointing out that we have not used the stability of the VAR operator or stationarity of the variables. Hence, the formulas are also valid for systems with integrated and cointegrated variables. So far we have discussed forecasting with known models. The case of estimated models will be considered next.

10.5.2 Forecasting Estimated Dynamic SEMs

In order to evaluate the consequences of using estimated instead of known processes for unconditional forecasts, we can use a joint model for the endogenous and exogenous variables and then draw on results of the previous chapters. Therefore, in this section we will focus on conditional forecasts only. We denote by $\widehat{y}_t(h|x)$ the conditional h-step forecast (10.5.7) based on the estimated reduced form (10.2.3). The forecast error is

$$y_{t+h} - \widehat{y}_t(h|x) = [y_{t+h} - y_t(h|x)] + [y_t(h|x) - \widehat{y}_t(h|x)]. \tag{10.5.12}$$

Conditional on the exogenous variables, the two terms in brackets are uncorrelated. Hence, assuming, as in previous chapters, that the processes used for estimation and forecasting are independent, an MSE approximation

$$\Sigma_{\hat{y}}(h|x) = \Sigma_y(h|x) + \frac{1}{T}\Omega_y(h|x) \tag{10.5.13}$$

is obtained in the by now familiar way. Here

$$\Omega_y(h|x) := E\left[\frac{\partial y_t(h|x)}{\partial \beta'} \Sigma_{\hat{\beta}} \frac{\partial y_t(h|x)'}{\partial \beta}\right], \tag{10.5.14}$$

$\beta := \text{vec}[A_1, \ldots, A_p, B_1, \ldots, B_s, B_0]$ and $\Sigma_{\hat{\beta}}$ is the covariance matrix of the asymptotic distribution of $\sqrt{T}(\hat{\beta} - \beta)$. It is straightforward to show that

$$\begin{aligned}
\frac{\partial y_t(h|x)}{\partial \beta'} &= \frac{\partial(J\mathbf{A}^h Y_t)}{\partial \beta'} + \sum_{i=0}^{h-1} \frac{\partial(J\mathbf{A}^i \mathbf{B} x_{t+h-i})}{\partial \beta'} \\
&= \sum_{i=0}^{h-1} \Big[Y_t'(\mathbf{A}')^{h-1-i} \otimes J\mathbf{A}^i J' \\
&\quad + \sum_{j=0}^{i-1} x'_{t+h-i} \mathbf{B}'(\mathbf{A}')^{i-1-j} \otimes J\mathbf{A}^j J' : x'_{t+h-i} \otimes J\mathbf{A}^i J'\Big].
\end{aligned}$$

$$\tag{10.5.15}$$

For stationary processes, an estimator of $\Omega_y(h|x)$ is obtained in the usual way be replacing all unknown parameters in this expression and in $\Sigma_{\hat{\beta}}$ by estimators and by using the average over $t = 1, \ldots, T$ for the expectation in (10.5.14).

Although we have discussed forecasting with estimated coefficients in terms of a simple VARX(p, s) model with white noise residuals, it is possible to generalize these results to models with autocorrelated error processes. The more general case was treated, for instance, by Yamamoto (1980) and Baillie (1981).

10.6 Multiplier Analysis

In an econometric simultaneous equations analysis, the marginal impact of changes in the exogenous variables is sometimes investigated. For example, if the exogenous variables are instruments for, say, the government or a central bank the consequences of changes in these instruments may be of interest. A government may, for instance, desire to know the effects of a change in a tax rate. In that case, *policy simulation* is of interest. In other cases, the consequences of changes in the exogenous variables that are not under the control of any decision maker may be of interest. For instance, it may be desirable to study the future consequences of the present weather conditions.

Therefore, the dynamic multipliers discussed in Section 10.2.2 are considered. They are contained in the D_i matrices of the final form operator,

$$D(L) = \sum_{i=0}^{\infty} D_i L^i := A(L)^{-1} B(L),$$

where $A(L) := I_K - A_1 L - \cdots - A_p L^p$ and $B(L) := B_0 + B_1 L + \cdots + B_s L^s$ are the reduced form operators, as before. Here stability and, hence, invertibility of the VAR operator $A(L)$ is assumed. The D_i matrices are conveniently obtained from the VARX$(1,0)$ representation (10.5.4) which implies

$$y_t = \sum_{i=0}^{\infty} J\mathbf{A}^i \mathbf{B} x_{t-i} + \sum_{i=0}^{\infty} J\mathbf{A}^i J' u_{t-i}, \tag{10.6.1}$$

because $J\mathbf{A}^h Y_t \to 0$ as $h \to \infty$, if y_t is a stable, stationary process (see (10.5.6)). The D_i's are coefficient matrices of the exogenous variables in the final form representation. Thus,

$$D_i = J\mathbf{A}^i \mathbf{B}, \quad i = 0, 1, \ldots, \tag{10.6.2}$$

the n-th interim multipliers are

$$M_n := D_0 + D_1 + \cdots + D_n = J(I + \mathbf{A} + \cdots + \mathbf{A}^n)\mathbf{B}, \quad n = 0, 1, \ldots, \tag{10.6.3}$$

and the total multipliers are

$$M_\infty := \sum_{i=0}^{\infty} D_i = J(I - \mathbf{A})^{-1}\mathbf{B} = A(1)^{-1} B(1). \tag{10.6.4}$$

If the model contains integrated variables and the generation mechanism is started at time $t = 0$, say, from a set of initial values, then we get from (10.5.5),

$$y_t = J\mathbf{A}^t Y_0 + \sum_{i=0}^{t-1} J\mathbf{A}^i \mathbf{B} x_{t-i} + \sum_{i=0}^{t-1} J\mathbf{A}^i J' u_{t-i}. \tag{10.6.5}$$

Thus, the D_i matrices in (10.6.2) still reflect the marginal impacts of changes in the unmodelled variables and, hence, contain the multipliers. Also the n-th interim multipliers can be computed as in (10.6.3), whereas the total multipliers in (10.6.4) will not exist in general.

Having obtained the foregoing representations of the multipliers, estimation of these quantities is straightforward. Estimators of the dynamic multipliers are obtained by substituting estimators $\widehat{A}_i$ and $\widehat{B}_j$ of the coefficient matrices in $\mathbf{A}$ and $\mathbf{B}$. The asymptotic properties of the estimators then follow in the usual way. For completeness we mention the following result from Schmidt (1973).

In the framework of Section 10.3, suppose $\widehat{\boldsymbol{\beta}}$ is a consistent estimator of $\boldsymbol{\beta} := \operatorname{vec}[A, B, B_0]$ satisfying

$$\sqrt{T}(\widehat{\boldsymbol{\beta}} - \boldsymbol{\beta}) \xrightarrow{d} \mathcal{N}(0, \Sigma_{\widehat{\boldsymbol{\beta}}}).$$

Then

$$\sqrt{T} \operatorname{vec}(\widehat{D}_i - D_i) \xrightarrow{d} \mathcal{N}(0, G_i \Sigma_{\widehat{\boldsymbol{\beta}}} G_i'), \qquad (10.6.6)$$

where $G_0 := [0 : I_{KM}]$ and

$$G_i := \frac{\partial \operatorname{vec}(D_i)}{\partial \boldsymbol{\beta}'} = \left[\sum_{j=0}^{i-1} \mathbf{B}'(\mathbf{A}')^{i-1-j} \otimes J\mathbf{A}^j J' : I_M \otimes J\mathbf{A}^i J' \right],$$
$$i = 1, 2, \ldots,$$

are $[KM \times (K^2 p + KM(s+1))]$ matrices. The proof of this result is left as an exercise. It is also easy to find the asymptotic distribution of the interim multipliers (accumulated multipliers) and the total multipliers if they exist (see Problem 10.8).

10.7 Optimal Control

A policy or decision maker who has control over some of the exogenous variables can use a dynamic simultaneous equations model to assess interventions with a multiplier or simulation analysis, as described in the previous section. However, if the decision maker has specific target values of the endogenous variables in mind, he or she may wish to go a step further and determine which values of the instrument variables will produce the desired values of the endogenous variables.

Usually it will not be possible to actually achieve all targets simultaneously and sometimes the decision maker is not completely free to choose the instruments. For instance, doubling a particular tax rate or increasing the price of specific government services drastically may result in the overthrow of the government or in social unrest and is therefore not a feasible option. Therefore, a loss function is usually set up in which the loss of deviations from the target values is specified. For instance, if the desired paths of the endogenous and instrument variables after period T are $y^0_{T+1}, \ldots, y^0_{T+n}$ and $x^0_{T+1}, \ldots, x^0_{T+n}$, respectively, a *quadratic* loss function has the form

$$\mathcal{L} = \sum_{i=1}^{n} [(y_{T+i} - y^0_{T+i})' K_i (y_{T+i} - y^0_{T+i})$$
$$+ (x_{T+i} - x^0_{T+i})' P_i (x_{T+i} - x^0_{T+i})], \qquad (10.7.1)$$

where the K_i and P_i are symmetric positive semidefinite matrices. Because the variables are assumed to be stochastic, the loss is a random variable too.

Therefore, minimization of the average or expected loss, $E(\mathcal{L})$, is usually the objective.

In a quadratic loss function the same weight is assigned to positive and negative deviations from the target values. For many situations and variables this specification is not quite realistic. For example, if the target is to have an unemployment rate of 2%, then having less than 2% may not be a problem at all while any higher rate may be regarded as a serious problem. Nevertheless, quadratic loss functions are the most common ones in applied and theoretical studies. Therefore, we will also use them in the following. One reason for the popularity of this type of loss function is clearly its tractability.

In order to approach a formal solution of the optimal control problem outlined in the foregoing, we assume that the economic system is described by a model like (10.1.1) with reduced form (10.2.3). However, to be able to distinguish between instrument variables and other exogenous variables, we introduce a new symbol for the latter. Suppose x_t represents an $(M \times 1)$ vector of instrument variables, the $(N \times 1)$ vector z_t contains all other unmodelled variables and the reduced form of the model is

$$y_t = A_1 y_{t-1} + \cdots + A_p y_{t-p} + B_0 x_t + \cdots + B_s x_{t-s} + C z_t + u_t, \qquad (10.7.2)$$

where u_t is white noise. Some of the components of z_t may be lagged variables. To summarize them in a vector indexed by t is just a matter of convenience.

For the present purposes, it is useful to write the model in VARX$(1,0)$ form similar to (10.5.4),

$$Y_t = \mathbf{A} Y_{t-1} + \mathbf{B} x_t + \mathbf{C} z_t + U_t, \qquad (10.7.3)$$

where $Y_t, U_t, \mathbf{A}$, and $\mathbf{B}$ are as defined in (10.5.4) and

$$\mathbf{C} := \begin{bmatrix} C \\ 0 \\ \vdots \\ 0 \end{bmatrix}$$

is a $((Kp + Ms) \times N)$ matrix. Recall that

$$Y_t := \begin{bmatrix} y_t \\ \vdots \\ y_{t-p+1} \\ x_t \\ \vdots \\ x_{t-s+1} \end{bmatrix}$$

contains current and lagged endogenous and instrument variables. Thus, the quadratic loss function specified in (10.7.1) may be rewritten in the form

$$\mathfrak{L} = \sum_{i=1}^{n}(Y_{T+i} - Y_{T+i}^0)'Q_i(Y_{T+i} - Y_{T+i}^0), \tag{10.7.4}$$

where the Q_i are symmetric positive semidefinite matrices involving the K_i's and P_i's.

In this framework, the problem of *optimal control* may be stated as follows: Given the model (10.7.3), given the vector Y_T, given values $z_{T+1}, \ldots, z_{T+n}$ of the uncontrolled variables and given target values $y_{T+1}^0, \ldots, y_{T+n}^0$ and $x_{T+1}^0, \ldots, x_{T+n}^0$, find the values $x_{T+1}^*, \ldots, x_{T+n}^*$ that minimize the expected loss $E(\mathfrak{L})$ specified in (10.7.4). The solution to this dynamic programming problem is well documented in the control theory literature. It turns out to be

$$x_{T+i}^* = G_i Y_{T+i-1} + g_i, \quad i = 1, \ldots, n, \tag{10.7.5}$$

where the Y_{T+i} are assumed to be obtained as

$$Y_{T+i} = \mathbf{A} Y_{T+i-1} + \mathbf{B} x_{T+i}^* + \mathbf{C} z_{T+i} + u_{T+i}.$$

Here the $(M \times (Kp + Ms))$ matrix G_i is defined as

$$G_i := -(\mathbf{B}' H_i \mathbf{B})^{-1} \mathbf{B}' H_i \mathbf{A}$$

and the $(M \times 1)$ vector g_i is defined as

$$g_i := -(\mathbf{B}' H_i \mathbf{B})^{-1} \mathbf{B}'(H_i \mathbf{C} z_{T+i} - h_i)$$

with

$$H_n := Q_n \quad \text{and} \quad H_{i-1} := Q_{i-1} + (\mathbf{A} + \mathbf{B} G_i)' H_i (\mathbf{A} + \mathbf{B} G_i),$$
$$\text{for } i = 1, \ldots, n-1,$$

and

$$h_n := Q_n Y_{T+n}^0 \quad \text{and}$$
$$h_{i-1} := Q_{i-1} Y_{T+i-1}^0 - \mathbf{A}' H_i(\mathbf{C} z_{T+i} + \mathbf{B} g_i) + \mathbf{A}' h_i$$
$$\text{for } i = 1, \ldots, n-1.$$

The actual computation of these quantities proceeds in the order H_n, G_n, h_n, g_n, H_{n-1}, G_{n-1}, h_{n-1}, g_{n-1}, H_{n-2}, ... This solution can be found in various variations in the control theory literature (e.g., Chow (1975, 1981), Murata (1982)). Obviously, because the Y_t are random, the same is true for the optimal decision rule x_{T+i}^*, $i = 1, \ldots, n$.

There are a number of problems that arise in practice in the context of optimal control as presented here. For instance, we have considered a finite planning horizon of n periods. In some situations it is of interest to find the optimal decision rule for an infinite planning period. Moreover, in practice the parameter matrices $\mathbf{A}$, $\mathbf{B}$, and $\mathbf{C}$ are usually unknown and have to be replaced by estimators. More generally, stochastic parameter models may be considered. This, of course, introduces an additional stochastic element into the optimal decision rule. A further complication arises if the relations between the variables cannot be captured adequately by a *linear* model such as (10.7.2) but require a nonlinear specification. It is also possible to consider other types of optimization rules. In this section, we have assumed that the optimal decision rule for period $T + i$ is determined on the basis of all available information in period $T + i - 1$. In particular, the realization Y_{T+i-1} is assumed to be given in setting up the decision rule x^*_{T+i}. Such an approach is often referred to as a *closed-loop strategy*. An alternative approach would be to determine the decision rule at the beginning of the planning period for the entire planning horizon. This approach is called an *open-loop strategy*. Although it is in general inferior to closed-loop optimization, it may be of interest occasionally. These and many other topics are treated in the optimal control literature. Chow (1975, 1981) and Murata (1982) are books on the topic with emphasis on optimal decision making related to economic and econometric models. Friedmann (1981) provided the asymptotic properties of the optimal decision rule when estimators are substituted for the parameters in the control rule.

10.8 Concluding Remarks on Dynamic SEMs

In this chapter, we have summarized some problems related to the estimation, specification, and analysis of dynamic models with unmodelled variables. Major problem areas that were identified without giving details of possible solutions are the distinction between endogenous and exogenous variables, the identification or unique parameterization of dynamic models, the estimation, specification, and checking of structural form models as well as the treatment of nonlinear specifications. Also, we have just scratched the surface of control problems which represent one important area of applications of dynamic SEMs.

Other problems of obvious importance in the context of these models relate to the choice of the data associated with the variables. If a structural form is derived from some economic or other subject matter theory, it is important that the available data represents realizations of the variables related to the theory. In particular, the level of aggregation (temporal and contemporaneous) and seasonal characteristics (seasonally adjusted or unadjusted) may be of importance. The models we have considered do not allow specifically for seasonality, except perhaps for seasonal dummies and other seasonal components among the unmodelled variables. The seasonality aspect in the context of dynamic SEMs and models specifically designed for seasonal data were discussed, for example, by Hylleberg (1986).

So far, we have essentially considered stationary and integrated processes. Mild deviations from the stationarity assumption are possible in dynamic

SEMs where unmodelled variables may cause changes in the mean or conditional mean of the endogenous variables. However, in discussing properties of estimators or long-run multipliers, we have made assumptions that come close to assuming stationarity or cointegration. For instance, if the unmodelled variables are driven by a stationary VAR process, the means and second moments of the endogenous variables may be time invariant. Unfortunately, in practice, changes in the data generation process may occur. Therefore, we will discuss specific types of models with time varying parameters in later chapters (see Chapters 17 and 18).

10.9 Exercises

Problem 10.1
Consider the following structural form

$$Q_t = \alpha_0 + \alpha_1 R_{t-1} + w_{1t},$$
$$P_t = \beta_0 + \beta_1 Q_t + w_{2t},$$

where R_t is a measure for the rainfall in period t, Q_t is the quantity of an agricultural product supplied in period t, and P_t is the price of the product. Derive the reduced form, the final equations, and the final form of the model.

Problem 10.2
Suppose that the rainfall variable R_t in Problem 10.1 is generated by a white noise process with mean μ_R. Determine the unconditional 3-step ahead forecasts for Q_t and P_t based on the model from Problem 10.1. Determine also the conditional 3-step ahead forecasts given $R_{t+i} = \mu_R$, $i = 1, 2, 3$. Compare the two forecasts.

Problem 10.3
Given the model of Problem 10.1, what is the marginal total or long-run effect of an additional unit of rainfall in period t?

Problem 10.4
Suppose the system y_t has the structural form

$$A^*(L)y_t = F^* y_t^e + B^*(L)x_t + w_t,$$

where $A^*(L) := \mathsf{A} - A_1^* L - \cdots - A_p^* L^p$, $B^*(L) := B_0^* + B_1^* L + \cdots + B_s^* L^s$ and x_t is generated by a VAR(q) process

$$C(L)x_t = v_t.$$

Assume that y_t^e represents rational expectations formed in period $t - 1$ and eliminate the expectations variables from the structural form.

Problem 10.5
Show that the 1-step ahead forecast for y_t obtained from the VAR(p) model (10.5.3) is identical to the one determined from (10.5.2) if
$$x_t(1) = C_1 x_t + \cdots + C_q x_{t-q+1}$$
is used as forecast for the exogenous variables.

Problem 10.6
Show that the partial derivatives $\partial y_t(h|x)/\partial \beta'$ have the form given in (10.5.15).

Problem 10.7
Derive a prediction test for structural change on the basis of the conditional forecasts of the endogenous variables of a dynamic SEM.

Problem 10.8
Show that the dynamic multipliers have the asymptotic distributions given in Section 10.6. Show also that the n-th interim multipliers have an asymptotic normal distribution,
$$\sqrt{T} \operatorname{vec}(\widehat{M}_n - M_n) \xrightarrow{d} \mathcal{N}(0, \Sigma_{\widehat{\mathbf{m}}}(n)),$$
where
$$\Sigma_{\widehat{\mathbf{m}}}(n) = (G_0 + \cdots + G_n) \Sigma_{\widehat{\beta}} (G_0 + \cdots + G_n)'$$
and the G_i are the $[KM \times K(Kp + M(s+1))]$ matrices defined in Section 10.6. Furthermore,
$$\sqrt{T} \operatorname{vec}(\widehat{M}_\infty - M_\infty) \xrightarrow{d} \mathcal{N}(0, \Sigma_{\widehat{\mathbf{m}}}(\infty)),$$
where
$$\Sigma_{\widehat{\mathbf{m}}}(\infty) = G_\infty \Sigma_{\widehat{\beta}} G'_\infty$$
with
$$G_\infty := [((I - \mathbf{A})^{-1} \mathbf{B})' : I_M] \otimes J(I - \mathbf{A})^{-1} J'.$$
Here the notation from Section 10.6 is used.

Problem 10.9
Derive the optimal decision rule for the control problem stated in Section 10.7. (Hint: See Chow (1975).)

Part IV

Infinite Order Vector Autoregressive Processes

So far we have considered finite order VAR processes. A more flexible and perhaps more realistic class of processes is obtained by allowing for an infinite VAR order. Of course, having only a finite string of time series data, the infinitely many VAR coefficients cannot be estimated without further assumptions. There are two competing approaches that have been used in practice in order to overcome this problem. In one approach, it is assumed that the infinite number of VAR coefficients depend on finitely many parameters. In Chapter 11, vector autoregressive moving average (VARMA) processes are introduced that may be viewed as finite parameterizations of potentially infinite order VAR processes. Estimation and specification of these processes are discussed in Chapters 12 and 13, respectively. Cointegrated VARMA processes are considered in Chapter 14. In Chapter 15, another approach is pursued. In that approach, the infinite order VAR operator is truncated at some finite lag and the resulting finite order VAR model is estimated. It is assumed, however, that the truncation point depends on the time series length available for estimation. A suitable asymptotic theory for the resulting estimators is discussed both for stationary as well as cointegrated processes.

11
Vector Autoregressive Moving Average Processes

11.1 Introduction

In this chapter, we extend our standard finite order VAR model,

$$y_t = \nu + A_1 y_{t-1} + \cdots + A_p y_{t-p} + \varepsilon_t,$$

by allowing the error terms, here ε_t, to be autocorrelated rather than white noise. The autocorrelation structure is assumed to be of a relatively simple type so that ε_t has a finite order moving average (MA) representation,

$$\varepsilon_t = u_t + M_1 u_{t-1} + \cdots + M_q u_{t-q},$$

where, as usual, u_t is zero mean white noise with nonsingular covariance matrix Σ_u. A finite order VAR process with finite order MA error term is called a VARMA (*vector autoregressive moving average*) process.

Before we study VARMA processes in general, we will discuss some properties of finite order MA processes in Section 11.2. In Section 11.3, we consider the more general stationary VARMA processes with stable VAR part and we will learn that generally they have infinite order pure VAR and MA representations. Their autocovariance and autocorrelation properties are treated in Section 11.4 and forecasting VARMA processes is discussed in Section 11.5. In Section 11.6, transforming and aggregating these processes is considered. In that section, we will see that a linearly transformed finite order VAR(p) process, in general, does not admit a finite order VAR representation but becomes a VARMA process. Because transformations of variables are quite common in practice, this result is a powerful argument in favor of the more general VARMA class. Finally, Section 11.7 contains discussions of causality issues and impulse response analysis in the context of VARMA systems. Throughout this chapter, we consider stationary processes only.

11.2 Finite Order Moving Average Processes

In Chapter 2, we have encountered MA processes of possibly infinite order. Specifically, we have seen that stationary, stable finite order VAR processes can be represented as MA processes. Now we deal explicitly with *finite order* MA processes. Let us begin with the simplest case of a K-dimensional MA process of order 1 (MA(1) process), $y_t = \mu + u_t + M_1 u_{t-1}$, where $y_t = (y_{1t}, \ldots, y_{Kt})'$, u_t is zero mean white noise with nonsingular covariance matrix Σ_u, and $\mu = (\mu_1, \ldots, \mu_K)'$ is the mean vector of y_t, i.e., $E(y_t) = \mu$ for all t. For notational simplicity we will assume in the following that $\mu = 0$, that is, y_t is a zero mean process. Thus, we consider

$$y_t = u_t + M_1 u_{t-1}, \quad t = 0, \pm 1, \pm 2, \ldots, \tag{11.2.1}$$

which may be rewritten as

$$u_t = y_t - M_1 u_{t-1}.$$

By successive substitution we get

$$\begin{aligned} u_t &= y_t - M_1(y_{t-1} - M_1 u_{t-2}) = y_t - M_1 y_{t-1} + M_1^2 u_{t-2} \\ &= \cdots = y_t - M_1 y_{t-1} + \cdots + (-M_1)^n y_{t-n} + (-M_1)^{n+1} u_{t-n-1} \\ &= y_t + \sum_{i=1}^{\infty} (-M_1)^i y_{t-i}, \end{aligned}$$

if $M_1^i \to 0$ as $i \to \infty$. Hence,

$$y_t = -\sum_{i=1}^{\infty} (-M_1)^i y_{t-i} + u_t, \tag{11.2.2}$$

which is the potentially infinite order VAR representation of the process. Because $(-M_1)^i$ may be equal to zero for i greater than some finite number p, the process may in fact be a finite order VAR(p). For instance, we get $p = 1$ for a bivariate process with

$$M_1 = \begin{bmatrix} 0 & m \\ 0 & 0 \end{bmatrix},$$

where m is some nonzero real number.

For the representation (11.2.2) to be meaningful, M_1^i must approach zero as $i \to \infty$, which in turn requires that the eigenvalues of M_1 are all less than 1 in modulus or, equivalently,

$$\det(I_K + M_1 z) \neq 0 \quad \text{for } z \in \mathbb{C}, |z| \leq 1.$$

This condition is analogous to the stability condition for a VAR(1) process. It guarantees that the infinite sum in (11.2.2) exists as a mean square limit.

11.2 Finite Order Moving Average Processes

More generally, it can be shown that a (zero mean) MA(q) process (moving average process of order q),

$$y_t = u_t + M_1 u_{t-1} + \cdots + M_q u_{t-q}, \quad t = 0, \pm 1, \pm 2, \ldots, \tag{11.2.3}$$

has a pure VAR representation

$$y_t = \sum_{i=1}^{\infty} \Pi_i y_{t-i} + u_t, \tag{11.2.4}$$

if

$$\det(I_K + M_1 z + \cdots + M_q z^q) \neq 0 \quad \text{for } z \in \mathbb{C}, |z| \leq 1. \tag{11.2.5}$$

An MA(q) process with this property is called *invertible* in the following because we can invert from the MA to a VAR representation. Writing the process in lag operator notation as

$$y_t = (I_K + M_1 L + \cdots + M_q L^q) u_t = M(L) u_t$$

the MA operator $M(L) := I_K + M_1 L + \cdots + M_q L^q$ is invertible if it satisfies (11.2.5) and we may formally write

$$M(L)^{-1} y_t = u_t.$$

The actual computation of the coefficient matrices Π_i in

$$M(L)^{-1} = \Pi(L) = I_K - \sum_{i=1}^{\infty} \Pi_i L^i$$

can be done recursively using $\Pi_1 = M_1$ and

$$\Pi_i = M_i - \sum_{j=1}^{i-1} \Pi_{i-j} M_j, \quad i = 2, 3, \ldots, \tag{11.2.6}$$

where $M_j := 0$ for $j > q$. These recursions follow immediately from the corresponding recursions used to compute the MA coefficients of a pure VAR process (see Chapter 2, (2.1.22)).

The autocovariances of the MA(q) process (11.2.3) are particularly easy to obtain. They follow directly from those of an infinite order MA process given in Chapter 2, Section 2.1.2, (2.1.18):

$$\Gamma_y(h) = E(y_t y'_{t-h}) = \begin{cases} \sum_{i=0}^{q-h} M_{i+h} \Sigma_u M'_i, & h = 0, 1, \ldots, q, \\ 0, & h = q+1, q+2, \ldots, \end{cases} \tag{11.2.7}$$

with $M_0 := I_K$. As before, $\Gamma_y(-h) = \Gamma_y(h)'$. Thus, the vectors y_t and y_{t-h} are uncorrelated if $h > q$. Obviously, the process (11.2.3) is stationary because the $\Gamma_y(h)$ do not depend on t and the mean $E(y_t) = 0$ for all t.

It can be shown that a noninvertible MA(q) process violating (11.2.5) also has a pure VAR representation if the determinantal polynomial in (11.2.5) has no roots on the complex unit circle, i.e., if

$$\det(I_K + M_1 z + \cdots + M_q z^q) \neq 0 \quad \text{for } |z| = 1. \tag{11.2.8}$$

The VAR representation will, however, not be of the type (11.2.4) in that the white noise process will in general not be the one appearing in (11.2.3). The reason is that for any noninvertible MA(q) process satisfying (11.2.8), there is an equivalent invertible MA(q) satisfying (11.2.5) which has an identical autocovariance structure (see Hannan & Deistler (1988, Chapter 1, Section 3)). For instance, for the univariate MA(1) process

$$y_t = u_t + m u_{t-1}, \tag{11.2.9}$$

the invertibility condition requires that $1 + mz$ has no roots for $|z| \leq 1$ or, equivalently, $|m| < 1$. For any m, the process has autocovariances

$$E(y_t y_{t-h}) = \begin{cases} (1+m^2)\sigma_u^2 & \text{for } h = 0, \\ m\sigma_u^2 & \text{for } h = \pm 1, \\ 0 & \text{otherwise}, \end{cases}$$

where $\sigma_u^2 := \text{Var}(u_t)$. It is easy to check that the process $v_t + \frac{1}{m}v_{t-1}$, where v_t is a white noise process with $\sigma_v^2 := \text{Var}(v_t) = m^2 \sigma_u^2$, has the very same autocovariance structure. Thus, if $|m| > 1$, we may choose the invertible MA(1) representation

$$y_t = v_t + \frac{1}{m}v_{t-1} \tag{11.2.10}$$

with

$$v_t = \left(1 + \frac{1}{m}L\right)^{-1} y_t = \sum_{i=0}^{\infty} \left(\frac{-1}{m}\right)^i y_{t-i}$$

$$= \left(1 + \frac{1}{m}L\right)^{-1} (1 + mL)u_t.$$

The reader is invited to check that v_t is indeed a white noise process with $\sigma_v^2 = m^2 \sigma_u^2$ (see Problem 11.10). Only if $|m| = 1$ and, hence, $1 + mz = 0$ for some z on the unit circle ($z = 1$ or -1), an invertible representation does not exist.

Although for higher order and higher-dimensional processes, where roots inside and outside the unit circle may exist, it is more complicated to find the invertible representation, it can be done whenever (11.2.8) is satisfied. In the remainder of this chapter, we will therefore assume without notice that all MA processes are invertible unless stated otherwise. It should be understood that this assumption implies a slight loss of generality because MA processes with roots on the complex unit circle are excluded.

11.3 VARMA Processes

11.3.1 The Pure MA and Pure VAR Representations of a VARMA Process

As mentioned in the introduction to this chapter, allowing finite order VAR processes to have finite order MA instead of white noise error terms, results in the broad and flexible class of vector autoregressive moving average (VARMA) processes. The general form of a process from this class with VAR order p and MA order q is

$$y_t = \nu + A_1 y_{t-1} + \cdots + A_p y_{t-p} + u_t + M_1 u_{t-1} + \cdots + M_q u_{t-q},$$
$$t = 0, \pm 1, \pm 2, \ldots. \qquad (11.3.1)$$

Such a process is briefly called a VARMA(p,q) process. As before, u_t is zero mean white noise with nonsingular covariance matrix Σ_u.

It may be worth elaborating a bit on this specification. What kind of process y_t is defined by the VARMA(p,q) model (11.3.1)? To look into this question, let us denote the MA part by ε_t, that is, $\varepsilon_t = u_t + M_1 u_{t-1} + \cdots + M_q u_{t-q}$ and

$$y_t = \nu + A_1 y_{t-1} + \cdots + A_p y_{t-p} + \varepsilon_t.$$

If this process is *stable*, that is, if

$$\det(I_K - A_1 z - \cdots - A_p z^p) \neq 0 \quad \text{for } |z| \leq 1, \qquad (11.3.2)$$

then, by the same arguments used in Chapter 2, Section 2.1.2, and by Proposition C.9 of Appendix C.3,

$$\begin{aligned}
y_t &= \mu + \sum_{i=0}^{\infty} D_i \varepsilon_{t-i} \\
&= \mu + \sum_{i=0}^{\infty} D_i (u_{t-i} + M_1 u_{t-i-1} + \cdots + M_q u_{t-i-q}) \\
&= \mu + \sum_{i=0}^{\infty} \Phi_i u_{t-i} \qquad (11.3.3)
\end{aligned}$$

is well-defined as a limit in mean square, given a well-defined white noise process u_t. Here

424   11 Vector Autoregressive Moving Average Processes

$$\mu := (I_K - A_1 - \cdots - A_p)^{-1}\nu,$$

the D_i are $(K \times K)$ matrices satisfying

$$\sum_{i=0}^{\infty} D_i z^i = (I_K - A_1 z - \cdots - A_p z^p)^{-1},$$

and the Φ_i are $(K \times K)$ matrices satisfying

$$\sum_{i=0}^{\infty} \Phi_i z^i = \left(\sum_{i=0}^{\infty} D_i z^i\right)(I_K + M_1 z + \cdots + M_q z^q).$$

In the following, when we call y_t a stable VARMA(p,q) process, we mean the well-defined process given in (11.3.3). For instance, if u_t is Gaussian white noise, it can be shown that y_t is a Gaussian process with all finite subcollections of vectors $y_t, \ldots, y_{t+h}$ having joint multivariate normal distributions. The representation (11.3.3) is a pure MA or simply MA representation of y_t.

To make the derivation of the MA representation more transparent, let us write the process (11.3.1) in lag operator notation,

$$A(L)y_t = \nu + M(L)u_t, \tag{11.3.4}$$

where $A(L) := I_K - A_1 L - \cdots - A_p L^p$ and $M(L) := I_K + M_1 L + \cdots + M_q L^q$. A pure MA representation of y_t is obtained by premultiplying with $A(L)^{-1}$,

$$y_t = A(1)^{-1}\nu + A(L)^{-1}M(L)u_t = \mu + \sum_{i=0}^{\infty} \Phi_i u_{t-i}.$$

Hence, multiplying from the left by $A(L)$ gives

$$(I_K - A_1 L - \cdots - A_p L^p)\left(\sum_{i=0}^{\infty} \Phi_i L^i\right)$$
$$= I_K + \sum_{i=1}^{\infty}\left(\Phi_i - \sum_{j=1}^{i} A_j \Phi_{i-j}\right)L^i$$
$$= I_K + M_1 L + \cdots + M_q L^q$$

and, thus, comparing coefficients results in

$$M_i = \Phi_i - \sum_{j=1}^{i} A_j \Phi_{i-j}, \quad i = 1, 2, \ldots,$$

with $\Phi_0 := I_K$, $A_j := 0$ for $j > p$, and $M_i := 0$ for $i > q$. Rearranging terms gives

$$\Phi_i = M_i + \sum_{j=1}^{i} A_j \Phi_{i-j}, \quad i = 1, 2, \ldots. \tag{11.3.5}$$

If the MA operator $M(L)$ satisfies the invertibility condition (11.2.5), then the VARMA process (11.3.4) is called *invertible*. In that case, it has a pure VAR representation,

$$y_t - \sum_{i=1}^{\infty} \Pi_i y_{t-i} = M(L)^{-1} A(L) y_t = M(1)^{-1} \nu + u_t,$$

and the Π_i matrices are obtained by comparing coefficients in

$$I_K - \sum_{i=1}^{\infty} \Pi_i L^i = M(L)^{-1} A(L).$$

Alternatively, multiplying this expression from the left by $M(L)$ gives

$$(I_K + M_1 L + \cdots + M_q L^q) \left(I_K - \sum_{i=1}^{\infty} \Pi_i L^i \right)$$

$$= I_K + \sum_{i=1}^{\infty} \left(M_i - \sum_{j=1}^{i} M_{i-j} \Pi_j \right) L^i$$

$$= I_K - A_1 L - \cdots - A_p L^p,$$

where $M_0 := I_K$ and $M_i := 0$ for $i > q$. Setting $A_i := 0$ for $i > p$ and comparing coefficients gives

$$-A_i = M_i - \sum_{j=1}^{i-1} M_{i-j} \Pi_j - \Pi_i$$

or

$$\Pi_i = A_i + M_i - \sum_{j=1}^{i-1} M_{i-j} \Pi_j \quad \text{for } i = 1, 2, \ldots. \tag{11.3.6}$$

As usual, the sum is defined to be zero if the lower bound for the summation index exceeds its upper bound.

For instance, for the zero mean VARMA(1, 1) process

$$y_t = A_1 y_{t-1} + u_t + M_1 u_{t-1}, \tag{11.3.7}$$

we get

$$\Pi_1 = A_1 + M_1$$
$$\Pi_2 = A_2 + M_2 - M_1 \Pi_1 = -M_1 A_1 - M_1^2$$
$$\vdots$$
$$\Pi_i = (-1)^{i-1} (M_1^i + M_1^{i-1} A_1), \quad i = 1, 2, \ldots,$$

and the coefficients of the pure MA representation are

$$\Phi_0 = I_K$$
$$\Phi_1 = M_1 + A_1$$
$$\Phi_2 = M_2 + A_1\Phi_1 + A_2\Phi_0 = A_1(M_1 + A_1)$$
$$\vdots$$
$$\Phi_i = A_1^{i-1}M_1 + A_1^i, \quad i = 1, 2, \ldots.$$

If y_t is a stable and invertible VARMA process, then the pure MA representation (11.3.3) is called the *canonical* or *prediction error MA representation*, in accordance with the terminology used in the finite order VAR case. In addition to the pure MA and VAR representations considered in this section, a VARMA process also has VAR(1) representations. One such representation is introduced next.

11.3.2 A VAR(1) Representation of a VARMA Process

Suppose y_t has the VARMA(p, q) representation (11.3.1). For simplicity, we assume that its mean is zero and, hence, $\nu = 0$. Let

$$Y_t := \begin{bmatrix} y_t \\ \vdots \\ y_{t-p+1} \\ u_t \\ \vdots \\ u_{t-q+1} \end{bmatrix}_{(K(p+q)\times 1)}, \quad U_t := \begin{bmatrix} u_t \\ 0 \\ \vdots \\ 0 \\ u_t \\ 0 \\ \vdots \\ 0 \end{bmatrix} \begin{matrix} \} (Kp \times 1) \\ \\ \} (Kq \times 1) \end{matrix}$$

and

$$\mathbf{A} := \begin{bmatrix} \mathbf{A}_{11} & \mathbf{A}_{12} \\ \mathbf{A}_{21} & \mathbf{A}_{22} \end{bmatrix} \quad [K(p+q) \times K(p+q)],$$

where

$$\mathbf{A}_{11} := \begin{bmatrix} A_1 & \cdots & A_{p-1} & A_p \\ I_K & & 0 & 0 \\ & \ddots & & \vdots \\ 0 & \cdots & I_K & 0 \end{bmatrix},$$
$$(Kp \times Kp)$$

$$\mathbf{A}_{12} := \begin{bmatrix} M_1 & \cdots & M_{q-1} & M_q \\ 0 & \cdots & 0 & 0 \\ \vdots & & \vdots & \vdots \\ 0 & \cdots & 0 & 0 \end{bmatrix},$$
$$(Kp \times Kq)$$

$$\mathbf{A}_{21} := \underset{(Kq \times Kp)}{0}, \quad \mathbf{A}_{22} := \underset{(Kq \times Kq)}{\begin{bmatrix} 0 & \cdots & 0 & 0 \\ I_K & & 0 & 0 \\ & \ddots & & \vdots \\ 0 & \cdots & I_K & 0 \end{bmatrix}}.$$

With this notation, we get the VAR(1) representation of Y_t,

$$Y_t = \mathbf{A} Y_{t-1} + U_t. \tag{11.3.8}$$

If the VAR order is zero ($p = 0$), we choose $p = 1$ and set $A_1 = 0$ in this representation.

The $K(p+q)$-dimensional VAR(1) process in (11.3.8) is stable if and only if y_t is stable. This result follows because

$$\begin{aligned} \det(I_{K(p+q)} - \mathbf{A}z) &= \det(I_{Kp} - \mathbf{A}_{11}z)\det(I_{Kq} - \mathbf{A}_{22}z) \\ &= \det(I_K - A_1 z - \cdots - A_p z^p). \end{aligned} \tag{11.3.9}$$

Here the rules for the determinant of a partitioned matrix from Appendix A.10 have been used and we have also used that $I_{Kq} - \mathbf{A}_{22}z$ is a lower triangular matrix with ones on the main diagonal which has determinant 1. Furthermore, $\det(I_{Kp} - \mathbf{A}_{11}z) = \det(I_K - A_1 z - \cdots - A_p z^p)$ follows as in Section 2.1.1.

From Chapter 2, we know that if y_t and, hence, Y_t is stable, the latter process has an MA representation

$$Y_t = \sum_{i=0}^{\infty} \mathbf{A}^i U_{t-i}.$$

Premultiplying by the $(K \times K(p+q))$ matrix $J := [I_K : 0 : \cdots : 0]$ gives

$$y_t = \sum_{i=0}^{\infty} J \mathbf{A}^i U_{t-i} = \sum_{i=0}^{\infty} J \mathbf{A}^i H J U_{t-i} = \sum_{i=0}^{\infty} J \mathbf{A}^i H u_{t-i} = \sum_{i=0}^{\infty} \Phi_i u_{t-i},$$

where

$$H = \begin{bmatrix} I_K \\ 0 \\ \vdots \\ 0 \\ I_K \\ 0 \\ \vdots \\ 0 \end{bmatrix} \begin{matrix} \left.\vphantom{\begin{matrix}I_K\\0\\\vdots\\0\end{matrix}}\right\} (Kp \times K) \\ \left.\vphantom{\begin{matrix}I_K\\0\\\vdots\\0\end{matrix}}\right\} (Kq \times K) \end{matrix}.$$

Thus,

$$\Phi_i = J \mathbf{A}^i H. \tag{11.3.10}$$

As an example, consider the zero mean VARMA(1, 1) process from (11.3.7),

$$y_t = A_1 y_{t-1} + u_t + M_1 u_{t-1}.$$

For this process

$$Y_t = \begin{bmatrix} y_t \\ u_t \end{bmatrix}, \quad \mathbf{A} = \begin{bmatrix} A_1 & M_1 \\ 0 & 0 \end{bmatrix}, \quad U_t = \begin{bmatrix} u_t \\ u_t \end{bmatrix},$$

$$J = [I_K : 0] \quad (K \times 2K),$$

and

$$H = \begin{bmatrix} I_K \\ I_K \end{bmatrix} \quad (2K \times K).$$

Hence,

$$\begin{aligned}
\Phi_0 &= JH = I_K, \\
\Phi_1 &= J\mathbf{A}H = [A_1 : M_1]H = A_1 + M_1, \\
\Phi_2 &= J\mathbf{A}^2 H = J \begin{bmatrix} A_1^2 & A_1 M_1 \\ 0 & 0 \end{bmatrix} H = A_1^2 + A_1 M_1, \\
&\vdots \\
\Phi_i &= J\mathbf{A}^i H = J \begin{bmatrix} A_1^i & A_1^{i-1} M_1 \\ 0 & 0 \end{bmatrix} H = A_1^i + A_1^{i-1} M_1, \quad i = 1, 2, \ldots.
\end{aligned} \quad (11.3.11)$$

This, of course, is precisely the same formula obtained from the recursions in (11.3.5).

The foregoing method of computing the MA matrices is just another way of computing the coefficient matrices of the power series

$$I_K + \sum_{i=1}^{\infty} \Phi_i L^i = (I_K - A_1 L - \cdots - A_p L^p)^{-1}(I_K + M_1 L + \cdots + M_q L^q).$$

Therefore, it can just as well be used to compute the Π_i coefficient matrices of the pure VAR representation of a VARMA process. Recall that

$$I_K - \sum_{i=1}^{\infty} \Pi_i L^i = (I_K + M_1 L + \cdots + M_q L^q)^{-1}(I_K - A_1 L - \cdots - A_p L^p).$$

Hence, if we define

$$\mathbf{M} := \begin{bmatrix} \mathbf{M}_{11} & \mathbf{M}_{12} \\ \mathbf{M}_{21} & \mathbf{M}_{22} \end{bmatrix}, \quad (11.3.12)$$

where

11.4 The Autocovariances and Autocorrelations of a VARMA(p, q) Process

$$\mathbf{M}_{11} := \begin{bmatrix} -M_1 & \cdots & -M_{q-1} & -M_q \\ I_K & & 0 & 0 \\ & \ddots & \vdots & \vdots \\ 0 & \cdots & I_K & 0 \end{bmatrix}_{(Kq \times Kq)},$$

$$\mathbf{M}_{12} := \begin{bmatrix} -A_1 & \cdots & -A_{p-1} & -A_p \\ 0 & \cdots & 0 & 0 \\ \vdots & & \vdots & \vdots \\ 0 & \cdots & 0 & 0 \end{bmatrix}_{(Kq \times Kp)},$$

$$\mathbf{M}_{21} := 0, \quad \mathbf{M}_{22} := \begin{bmatrix} 0 & \cdots & 0 & 0 \\ I_K & & 0 & 0 \\ & \ddots & \vdots & \vdots \\ 0 & \cdots & I_K & 0 \end{bmatrix}_{(Kp \times Kp)},$$

we get

$$-\Pi_i = J\mathbf{M}^i H \tag{11.3.13}$$

with

$$H := \begin{bmatrix} I_K \\ 0 \\ \vdots \\ 0 \\ I_K \\ 0 \\ \vdots \\ 0 \end{bmatrix} \begin{matrix} \left.\vphantom{\begin{matrix}I_K\\0\\\vdots\\0\end{matrix}}\right\} (Kq \times K) \\ \\ \left.\vphantom{\begin{matrix}I_K\\0\\\vdots\\0\end{matrix}}\right\} (Kp \times K) \end{matrix}.$$

11.4 The Autocovariances and Autocorrelations of a VARMA(p, q) Process

For the K-dimensional, zero mean, stable VARMA(p, q) process

$$y_t = A_1 y_{t-1} + \cdots + A_p y_{t-p} + u_t + M_1 u_{t-1} + \cdots + M_q u_{t-q}, \tag{11.4.1}$$

the autocovariances can be obtained formally from its pure MA representation as in Section 2.1.2. For instance, if y_t has the canonical MA representation

$$y_t = \sum_{i=0}^{\infty} \Phi_i u_{t-i},$$

the autocovariance matrices are

$$\Gamma_y(h) := E(y_t y'_{t-h}) = \sum_{i=0}^{\infty} \Phi_{h+i} \Sigma_u \Phi'_i.$$

For the actual computation of the autocovariance matrices, the following approach is more convenient. Postmultiplying (11.4.1) by y'_{t-h} and taking expectations gives

$$E(y_t y'_{t-h}) = A_1 E(y_{t-1} y'_{t-h}) + \cdots + A_p E(y_{t-p} y'_{t-h}) + E(u_t y'_{t-h}) + \cdots \\ + M_q E(u_{t-q} y'_{t-h}).$$

From the pure MA representation of the process, it can be seen that $E(u_t y'_s) = 0$ for $s < t$. Hence, we get for $h > q$,

$$\Gamma_y(h) = A_1 \Gamma_y(h-1) + \cdots + A_p \Gamma_y(h-p). \quad (11.4.2)$$

If $p > q$ and $\Gamma_y(0), \ldots, \Gamma_y(p-1)$ are available, this relation can be used to compute the autocovariances recursively for $h = p, p+1, \ldots$.

The initial matrices can be obtained from the VAR(1) representation (11.3.8), just as in Chapter 2, Section 2.1.4. In that section, we obtained the relation

$$\Gamma_Y(0) = \mathbf{A} \Gamma_Y(0) \mathbf{A}' + \Sigma_U \quad (11.4.3)$$

for the covariance matrix of the VAR(1) process Y_t. Here $\Sigma_U = E(U_t U'_t)$ is the covariance matrix of the white noise process in (11.3.8). Applying the vec operator to (11.4.3) and rearranging terms gives

$$\text{vec } \Gamma_Y(0) = (I_{K^2(p+q)^2} - \mathbf{A} \otimes \mathbf{A})^{-1} \text{vec}(\Sigma_U), \quad (11.4.4)$$

where the existence of the inverse follows again from the stability of the process, as in Section 2.1.4, by appealing to the determinantal relation (11.3.9).

Having computed $\Gamma_Y(0)$ as in (11.4.4), we may collect $\Gamma_y(0), \ldots, \Gamma_y(p-1)$ from

$$\Gamma_Y(0) = \begin{bmatrix} \mathbf{\Gamma}_{11}(0) & \mathbf{\Gamma}_{12}(0) \\ \mathbf{\Gamma}_{12}(0)' & \mathbf{\Gamma}_{22}(0) \end{bmatrix},$$

where

$$\mathbf{\Gamma}_{11}(0) = \begin{bmatrix} \Gamma_y(0) & \Gamma_y(1) & \cdots & \Gamma_y(p-1) \\ \Gamma_y(-1) & \Gamma_y(0) & \cdots & \Gamma_y(p-2) \\ \vdots & \vdots & \ddots & \vdots \\ \Gamma_y(-p+1) & \Gamma_y(-p+2) & \cdots & \Gamma_y(0) \end{bmatrix},$$

11.4 The Autocovariances and Autocorrelations of a VARMA(p,q) Process

$$\Gamma_{12}(0) = \begin{bmatrix} E(y_t u_t') & E(y_t u_{t-1}') & \cdots & E(y_t u_{t-q+1}') \\ 0 & E(y_{t-1} u_{t-1}') & \cdots & E(y_{t-1} u_{t-q+1}') \\ \vdots & & \ddots & \vdots \\ 0 & 0 & \cdots & E(y_{t-p+1} u_{t-q+1}') \end{bmatrix},$$

and

$$\Gamma_{22}(0) = \begin{bmatrix} \Sigma_u & 0 & \cdots & 0 \\ 0 & \Sigma_u & & 0 \\ \vdots & & \ddots & \vdots \\ 0 & 0 & \cdots & \Sigma_u \end{bmatrix}.$$

As mentioned previously, the recursions (11.4.2) are valid for $h > q$ only. Thus, this way of computing the autocovariances requires that $p > q$. If the VAR order is less than q, then it may be increased artificially by adding lags of y_t with zero coefficient matrices until the VAR order p exceeds the MA order q. Then the aforementioned procedure can be applied. A computationally more efficient method of computing the autocovariances of a VARMA process is described by Mittnik (1990).

The autocorrelations of a VARMA(p,q) process are obtained from its autocovariances as in Chapter 2, Section 2.1.4. That is,

$$R_y(h) = D^{-1} \Gamma_y(h) D^{-1}, \tag{11.4.5}$$

where D is a diagonal matrix with the square roots of the diagonal elements of $\Gamma_y(0)$ on the main diagonal.

To illustrate the computation of the covariance matrices, we consider the VARMA(1,1) process (11.3.7). Because $p = q$, we add a second lag of y_t so that

$$y_t = A_1 y_{t-1} + A_2 y_{t-2} + u_t + M_1 u_{t-1}$$

with $A_2 := 0$. Thus, in this case,

$$Y_t = \begin{bmatrix} y_t \\ y_{t-1} \\ u_t \end{bmatrix}, \quad \mathbf{A} = \begin{bmatrix} A_1 & 0 & M_1 \\ I_K & 0 & 0 \\ 0 & 0 & 0 \end{bmatrix},$$

$$U_t = \begin{bmatrix} u_t \\ 0 \\ u_t \end{bmatrix}, \quad \Sigma_U = \begin{bmatrix} \Sigma_u & 0 & \Sigma_u \\ 0 & 0 & 0 \\ \Sigma_u & 0 & \Sigma_u \end{bmatrix}.$$

With this notation, we get from (11.4.4),

$$\text{vec} \begin{bmatrix} \Gamma_y(0) & \Gamma_y(1) & \Sigma_u \\ \Gamma_y(-1) & \Gamma_y(0) & 0 \\ \Sigma_u & 0 & \Sigma_u \end{bmatrix} = (I_{9K^2} - \mathbf{A} \otimes \mathbf{A})^{-1} \text{vec}(\Sigma_U).$$

Now, because we have the starting-up matrices $\Gamma_y(0)$ and $\Gamma_y(1)$, the recursions (11.4.2) may be applied, giving

$$\Gamma_y(h) = A_1 \Gamma_y(h-1) \quad \text{for } h = 2, 3, \ldots.$$

In stating the assumptions for the VARMA(p, q) process at the beginning of this section, invertibility has not been mentioned. This is no accident because this condition is actually not required for computing the autocovariances of a VARMA(p, q) process. The same formulas may be used for invertible and noninvertible processes. On the other hand, the stability condition is essential here, because it ensures invertibility of the matrix $I - \mathbf{A} \otimes \mathbf{A}$.

11.5 Forecasting VARMA Processes

Suppose the K-dimensional zero mean VARMA(p, q) process

$$y_t = A_1 y_{t-1} + \cdots + A_p y_{t-p} + u_t + M_1 u_{t-1} + \cdots + M_q u_{t-q} \tag{11.5.1}$$

is stable and invertible. As we have seen in Section 11.3.1, it has a pure VAR representation,

$$y_t = \sum_{i=1}^{\infty} \Pi_i y_{t-i} + u_t, \tag{11.5.2}$$

and a pure MA representation,

$$y_t = \sum_{i=0}^{\infty} \Phi_i u_{t-i}. \tag{11.5.3}$$

Formulas for optimal forecasts can be given in terms of each of these representations.

Assuming that u_t is *independent* white noise and applying the conditional expectation operator E_t, given information up to time t, to (11.5.1) gives an optimal h-step forecast

$$y_t(h) = \begin{cases} A_1 y_t(h-1) + \cdots + A_p y_t(h-p) \\ \quad + M_h u_t + \cdots + M_q u_{t+h-q} & \text{for } h \leq q, \\ A_1 y_t(h-1) + \cdots + A_p y_t(h-p) & \text{for } h > q, \end{cases} \tag{11.5.4}$$

where, as usual, $y_t(j) := y_{t+j}$ for $j \leq 0$. Analogously, we get from (11.5.2),

$$y_t(h) = \sum_{i=1}^{\infty} \Pi_i y_t(h-i), \tag{11.5.5}$$

and, in Chapter 2, Section 2.2.2, we have seen that the optimal forecast in terms of the infinite order MA representation is

$$y_t(h) = \sum_{i=h}^{\infty} \Phi_i u_{t+h-i} = \sum_{i=0}^{\infty} \Phi_{h+i} u_{t-i} \tag{11.5.6}$$

(see (2.2.10)). Although in Chapter 2 this result was derived in the slightly more special setting of finite order VAR processes, it is not difficult to see that it carries over to the present situation. All three formulas (11.5.4)–(11.5.6) result, of course, in equivalent predictors or forecasts. They are different representations of the *linear* minimum MSE predictors if u_t is uncorrelated but not necessarily independent white noise.

A forecasting formula can also be obtained from the VAR(1) representation (11.3.8) of the VARMA(p,q) process. From Section 2.2.2, the optimal h-step forecast of a VAR(1) process at origin t is known to be

$$Y_t(h) = \mathbf{A}^h Y_t = \mathbf{A} Y_t(h-1). \tag{11.5.7}$$

Premultiplying with the $(K \times K(p+q))$ matrix $J := [I_K : 0 : \cdots : 0]$ results precisely in the recursive relation (11.5.4) (see Problem 11.4).

The forecasts at origin t are based on the information set

$$\Omega_t = \{y_s | s \leq t\}.$$

This information set has the drawback of being unavailable in practice. Usually a finite sample of y_t data is given only and, hence, the u_t cannot be determined exactly. Thus, even if the parameters of the process are known, the prediction formulas (11.5.4)–(11.5.6) cannot be used. However, the invertibility of the process implies that the Π_i coefficient matrices go to zero exponentially with increasing i and we have the approximation

$$\sum_{i=1}^{\infty} \Pi_i y_t(h-i) \approx \sum_{i=1}^{n} \Pi_i y_t(h-i)$$

for large n. Consequently, in practice, if the information set is

$$\{y_1, \ldots, y_T\} \tag{11.5.8}$$

and T is large, then the forecast

$$\breve{y}_T(h) = \sum_{i=1}^{T+h-1} \Pi_i \breve{y}_T(h-i), \tag{11.5.9}$$

where $\breve{y}_T(j) := y_{T+j}$ for $j \leq 0$, will be almost identical to the optimal forecast. For a low order process, as it is commonly used in practice, for which the roots of

$$\det(I_K + M_1 z + \cdots + M_q z^q)$$

are not close to the unit circle, $T > 50$ will usually result in forecasts that cannot be distinguished from the optimal forecasts. It is worth noting, however, that the optimal forecasts based on the finite information set (11.5.8) can be determined. The resulting forecast formulas are, for instance, given by Brockwell & Davis (1987, Chapter 11, §11.4). A similar problem is not encountered in forecasting finite order VAR processes because there the optimal forecast depends on a finite string of past variables only.

In the presently considered theoretical setting, the forecast MSE matrices are most easily obtained from the representation (11.5.6). The forecast error is

$$y_{t+h} - y_t(h) = \sum_{i=0}^{h-1} \Phi_i u_{t+h-i}$$

and, hence, the forecast MSE matrix turns out to be

$$\begin{aligned} \Sigma_y(h) &:= E[(y_{t+h} - y_t(h))(y_{t+h} - y_t(h))'] \\ &= \sum_{i=0}^{h-1} \Phi_i \Sigma_u \Phi_i', \end{aligned} \quad (11.5.10)$$

as in the finite order VAR case. Note, however, that, in the present case, the M_i coefficient matrices enter in computing the Φ_i matrices. Because the forecasts are unbiased, that is, the forecast errors have mean zero, the MSE matrix is the forecast error covariance matrix. Consequently, if the process is Gaussian, i.e., for all t and $h, y_t, \ldots, y_{t+h}$ have a multivariate normal distribution and also the u_t's are normally distributed, then the forecast errors are normally distributed,

$$y_{t+h} - y_t(h) \sim \mathcal{N}(0, \Sigma_y(h)). \quad (11.5.11)$$

This result may be used in the usual fashion in setting up forecast intervals.

If a process with nonzero mean vector μ is considered, the mean vector may simply be added to the prediction formula for the mean-adjusted process. For example, if y_t has zero mean and $x_t = y_t + \mu$, then the optimal h-step forecast of x_t is

$$x_t(h) = y_t(h) + \mu.$$

The forecast MSE matrix is not affected, that is, $\Sigma_x(h) = \Sigma_y(h)$.

11.6 Transforming and Aggregating VARMA Processes

In practice, the original variables of interest are often transformed before their generation process is modelled. For example, data are often seasonally adjusted prior to an analysis. Also, sometimes they are temporally aggregated. For instance, quarterly data may have been obtained by adding up the corresponding monthly values or by taking their averages. Moreover, contemporaneous aggregation over a number of households, regions or sectors of the

11.6 Transforming and Aggregating VARMA Processes

economy is quite common. For example, the GNP (gross national product) value for some period is the sum of private consumption, investment expenditures, net exports, and government spending for that period. It is often of interest to see what these transformations do to the generation processes of the variables in order to assess the consequences of transformations for forecasting and structural analysis. In the following, we assume that the original data are generated by a VARMA process and we study the consequences of linear transformations. These results are of importance because many temporal as well as contemporaneous aggregation procedures can be represented as linear transformations.

11.6.1 Linear Transformations of VARMA Processes

We shall begin with the result that a linear transformation of a process possessing an MA(q) representation gives a process that also has a finite order MA representation with order not greater than q.

Proposition 11.1 (*Linear Transformation of an MA(q) Process*)
Let u_t be a K-dimensional white noise process with nonsingular covariance matrix Σ_u and let

$$y_t = \mu + u_t + M_1 u_{t-1} + \cdots + M_q u_{t-q}$$

be a K-dimensional invertible MA(q) process. Furthermore, let F be an ($M \times K$) matrix of rank M. Then the M-dimensional process $z_t = F y_t$ has an invertible MA($\check{q}$) representation,

$$z_t = F\mu + v_t + N_1 v_{t-1} + \cdots + N_{\check{q}} v_{t-\check{q}},$$

where v_t is M-dimensional white noise with nonsingular covariance matrix Σ_v, the N_i are ($M \times M$) coefficient matrices and $\check{q} \leq q$. ∎

We will not give a proof of this result here but refer the reader to Lütkepohl (1984) or Lütkepohl (1987, Chapter 4). The proposition is certainly not surprising because considering the autocovariance matrices of z_t, it is seen that

$$\Gamma_z(h) = E[(Fy_t - F\mu)(Fy_{t-h} - F\mu)'] = F\Gamma_y(h)F'$$
$$= \begin{cases} \sum_{i=0}^{q-h} FM_{i+h}\Sigma_u M_i' F', & h = 0, 1, \ldots, q, \\ 0, & h = q+1, q+2, \ldots, \end{cases}$$

by (11.2.7). Thus, the autocovariances of z_t for lags greater than q are all zero. This result is a necessary requirement for the proposition to be true. It also helps to understand that the MA order of z_t may be lower than that of y_t because $\Gamma_z(h) = F\Gamma_y(h)F'$ may be zero even if $\Gamma_y(h)$ is nonzero.

The proposition has some interesting implications. As we will see in the following (Corollary 11.1.1), it implies that a linearly transformed VARMA(p, q) process has again a finite order VARMA representation. Thus, the VARMA class is closed with respect to linear transformations. The same is not true for the class of finite order VAR processes because, as we will see shortly, a linearly transformed VAR(p) process may not admit a finite order VAR representation. This, of course, is an argument in favor of considering the VARMA class rather than restricting the analysis to finite order VAR processes.

Corollary 11.1.1

Let y_t be a K-dimensional, stable, invertible VARMA(p, q) process and let F be an $(M \times K)$ matrix of rank M. Then the process $z_t = Fy_t$ has a VARMA$(\check{p}, \check{q})$ representation with

$$\check{p} \leq Kp$$

and

$$\check{q} \leq (K-1)p + q.$$

∎

Proof: We write the process y_t in lag operator notation as

$$A(L)y_t = M(L)u_t, \tag{11.6.1}$$

where the mean is set to zero without loss of generality as y_t may represent deviations from the mean. Premultiplying by the adjoint $A(L)^{adj}$ of $A(L)$ gives

$$|A(L)|y_t = A(L)^{adj}M(L)u_t, \tag{11.6.2}$$

where $A(L)^{adj}A(L) = |A(L)|$ has been used. It is easy to check that $|A(z)^{adj}| \neq 0$ for $|z| \leq 1$. Thus, (11.6.2) is a stable and invertible VARMA representation of y_t. Premultiplying (11.6.2) with F results in

$$|A(L)|z_t = FA(L)^{adj}M(L)u_t. \tag{11.6.3}$$

The operator $A(L)^{adj}M(L)$ is easily seen to have degree at most $p(K-1) + q$ and, thus, the right-hand side of (11.6.3) is just a linearly transformed finite order MA process which, by Proposition 11.1, has an MA$(\check{q})$ representation with

$$\check{q} \leq p(K-1) + q.$$

The degree of the AR operator $|A(L)|$ is at most Kp because the determinant is just a sum of products involving one operator from each row and each column of $A(L)$. This proves the corollary. ∎

11.6 Transforming and Aggregating VARMA Processes

The corollary gives upper bounds for the VARMA orders of a linearly transformed VARMA process. For instance, if y_t is a VAR(p)=VARMA($p, 0$) process, a linear transformation $z_t = Fy_t$ has a VARMA($\check{p}, \check{q}$) representation with $\check{p} \leq Kp$ and $\check{q} \leq (K-1)p$. For some linear transformations, $\check{q}$ will be zero. We will see in the following, however, that generally there are transformations for which the upper bounds for the orders are attained and a representation with lower orders does not exist. This result implies that a linear transformation of a finite order VAR(p) process may not admit a finite order VAR representation. Specifically, the subprocesses or marginal processes of a K-dimensional process y_t are obtained by using transformation matrices such as $F = [I_M : 0]$. Hence, a subprocess of a VAR(p) process may not have a finite order VAR but just a mixed VARMA representation.

For some transformations the result in Corollary 11.1.1 can, in fact, be tightened. Generally, tighter bounds for the VARMA orders are available if $M > 1$, as is seen in the following corollary.

Corollary 11.1.2

Let y_t be a K-dimensional, stable, invertible VARMA(p, q) process and let F be an $(M \times K)$ matrix of rank M. Then the process $z_t = Fy_t$ has a VARMA($\check{p}, \check{q}$) representation with

$$\check{p} \leq (K - M + 1)p$$

and

$$\check{q} \leq (K - M)p + q.$$

■

Proof: We first consider the case where z_t is a subprocess of y_t consisting of the first M components. To treat this case, we denote the first M and last $K - M$ components of the process y_t by y_{1t} and y_{2t}, respectively, and we partition the VAR and MA operators as well as the white noise process u_t accordingly. Thus, we can write the process as

$$A_{11}(L)y_{1t} + A_{12}(L)y_{2t} = M_{11}(L)u_{1t} + M_{12}(L)u_{2t}, \quad (11.6.4)$$

$$A_{21}(L)y_{1t} + A_{22}(L)y_{2t} = M_{21}(L)u_{1t} + M_{22}(L)u_{2t}. \quad (11.6.5)$$

Premultiplying (11.6.5) by the adjoint of $A_{22}(L)$ gives

$$\begin{aligned}|A_{22}(L)|y_{2t} &= -A_{22}(L)^{adj}A_{21}(L)y_{1t} + A_{22}(L)^{adj}M_{21}(L)u_{1t} \\ &\quad + A_{22}(L)^{adj}M_{22}(L)u_{2t}.\end{aligned} \quad (11.6.6)$$

Moreover, premultiplying (11.6.4) by $|A_{22}(L)|$, replacing $|A_{22}(L)|y_{2t}$ by the right-hand side of (11.6.6) and rearranging terms, we get

$$[|A_{22}(L)|A_{11}(L) - A_{12}(L)A_{22}(L)^{adj}A_{21}(L)]y_{1t}$$
$$= [|A_{22}(L)|M_{11}(L) - A_{12}(L)A_{22}(L)^{adj}M_{21}(L)]u_{1t} \qquad (11.6.7)$$
$$+ [|A_{22}(L)|M_{12}(L) - A_{12}(L)A_{22}(L)^{adj}M_{22}(L)]u_{2t}.$$

The VAR part of this representation has order

$$\check{p} \leq \max\{(K-M)p+p, (K-M-1)p+p+p\} = (K-M+1)p$$

and, by Proposition 11.1, the right-hand side of (11.6.7) has an MA representation with order

$$\check{q} \leq \max\{(K-M)p+q, p+(K-M-1)p+q\} = (K-M)p+q.$$

Hence, we have established the corollary for transformations $F = [I_M : 0]$.

For a general $(M \times K)$ transformation matrix F with $\text{rk}(F) = M$, we choose a $((K-M) \times K)$ matrix C such that the $(K \times K)$ matrix

$$\mathfrak{F} = \begin{bmatrix} F \\ C \end{bmatrix}$$

is nonsingular and we consider the process $x_t = \mathfrak{F}y_t$. Because nonsingular transformations do not increase the orders of a VARMA process, x_t also has a VARMA(p,q) representation. Now we get the result of the corollary by considering the transformation $z_t = Fy_t = [I_M : 0]x_t$. ∎

Other bounds for the VARMA orders than those provided in Corollaries 11.1.1 and 11.1.2 for linearly transformed VARMA processes and bounds for special linear transformations are given in various articles in the literature. For further results and references see Lütkepohl (1987, Chapter 4; 1986, Kapitel 2).

To illustrate Corollaries 11.1.1 and 11.1.2, we consider the bivariate VAR(1) process

$$\begin{bmatrix} 1-0.5L & 0.66L \\ 0.5L & 1+0.3L \end{bmatrix} \begin{bmatrix} y_{1t} \\ y_{2t} \end{bmatrix} = \begin{bmatrix} u_{1t} \\ u_{2t} \end{bmatrix} \quad \text{with } \Sigma_u = I_2. \qquad (11.6.8)$$

Here $K = 2$, $p = 1$, and $q = 0$. Thus, $z_t = [1,0]y_t = y_{1t}$ as a univariate $(M = 1)$ marginal process has an ARMA representation with orders not greater than $(2, 1)$. The precise form of the process can be determined with the help of the representation (11.6.3). Using that representation gives

$$[(1+0.3L)(1-0.5L) - 0.66 \cdot 0.5L^2]z_t$$
$$= [1,0] \begin{bmatrix} 1+0.3L & -0.66L \\ -0.5L & 1-0.5L \end{bmatrix} \begin{bmatrix} u_{1t} \\ u_{2t} \end{bmatrix} \qquad (11.6.9)$$
$$= (1+0.3L)u_{1t} - 0.66Lu_{2t}.$$

11.6 Transforming and Aggregating VARMA Processes

The right-hand side, say w_{1t}, is the sum of an MA(1) process and a white noise process. Thus, by Proposition 11.1, it is known to have an MA(1) representation, say $w_{1t} = v_{1t} + \gamma v_{1,t-1}$. To determine γ and $\sigma_1^2 = \text{Var}(v_{1t})$, we use

$$\begin{aligned} E(w_{1t}^2) &= E(v_{1t} + \gamma v_{1,t-1})^2 = (1+\gamma^2)\sigma_1^2 \\ &= E[(1+0.3L)u_{1t} - 0.66Lu_{2t}]^2 = 1.53 \end{aligned}$$

and

$$\begin{aligned} E(w_t w_{t-1}) &= E[(v_{1t} + \gamma v_{1,t-1})(v_{1,t-1} + \gamma v_{1,t-2})] = \gamma \sigma_1^2 \\ &= E[((1+0.3L)u_{1t} - 0.66u_{2,t-1}) \\ &\quad \times ((1+0.3L)u_{1,t-1} - 0.66u_{2,t-2})] \\ &= 0.3. \end{aligned}$$

Solving this nonlinear system of two equations for γ and σ_1^2 gives

$$\gamma = 0.204 \quad \text{and} \quad \sigma_1^2 = 1.47.$$

Note that we have picked the invertible solution with $|\gamma| < 1$. Thus, from (11.6.9), we get a marginal process

$$(1 - 0.2L - 0.48L^2)y_{1t} = (1 + 0.204L)v_{1t} \quad \text{with} \quad \sigma_1^2 = 1.47.$$

In other words, y_{1t} has indeed an ARMA(2, 1) representation and it is easy to check that cancellation of the AR and MA operators is not possible. Hence, the ARMA orders are minimal in this case.

As another example, consider again the bivariate VAR(1) process (11.6.8) and suppose we are interested in the process $z_t := y_{1t} + y_{2t}$. Thus, $F = [1, 1]$ is again a (1×2) vector. Multiplying (11.6.8) by the adjoint of the VAR operator gives

$$(1 - 0.2L - 0.48L^2)\begin{bmatrix} y_{1t} \\ y_{2t} \end{bmatrix} = \begin{bmatrix} 1+0.3L & -0.66L \\ -0.5L & 1-0.5L \end{bmatrix}\begin{bmatrix} u_{1t} \\ u_{2t} \end{bmatrix}.$$

Hence, multiplying by F gives

$$(1 - 0.2L - 0.48L^2)(y_{1t} + y_{2t}) = (1 - 0.2L)u_{1t} + (1 - 1.16L)u_{2t}.$$

Using similar arguments as for (11.6.9), it can be shown that the right-hand side of this expression is a process with MA(1) representation $v_t - 0.504v_{t-1}$, where $\sigma_v^2 := \text{Var}(v_t) = 2.70$. Consequently, the process of interest has the ARMA(2, 1) representation

$$(1 - 0.2L - 0.48L^2)z_t = (1 - 0.504L)v_t \quad \text{with} \quad \sigma_v^2 = 2.70. \qquad (11.6.10)$$

The following result is of interest if forecasting is the objective of the analysis.

Proposition 11.2 (*Forecast Efficiency of Linearly Transformed VARMA Processes*)
Let y_t be a stable, invertible, K-dimensional VARMA(p,q) process, let F be an $(M \times K)$ matrix of rank M, and let $z_t = Fy_t$. Furthermore, denote the MSE matrices of the optimal h-step predictors of y_t and z_t by $\Sigma_y(h)$ and $\Sigma_z(h)$, respectively. Then

$$\Sigma_z(h) - F\Sigma_y(h)F'$$

is positive semidefinite. ∎

This result means that $Fy_t(h)$ is generally a better predictor of z_{t+h} with smaller (at least not greater) MSEs than $z_t(h)$. In other words, forecasting the original process y_t and transforming the forecasts is generally better than forecasting the transformed process directly. A proof and references for related results were given by Lütkepohl (1987, Chapter 4). To see the point more clearly, consider again the example process (11.6.8) and suppose we are interested in the sum of its components $z_t = y_{1t} + y_{2t}$. Forecasting the bivariate process one step ahead results in a forecast MSE matrix $\Sigma_y(1) = \Sigma_u = I_2$. Thus, the corresponding 1-step ahead forecast of z_t has MSE

$$[1,1]\Sigma_y(1)\begin{bmatrix} 1 \\ 1 \end{bmatrix} = 2.$$

In contrast, if a univariate forecast is obtained on the basis of the ARMA$(2,1)$ representation (11.6.10), the 1-step ahead forecast MSE becomes $\sigma_v^2 = 2.70$. Clearly, the latter forecast is inferior in terms of MSE.

Of course, these results hold for VARMA processes for which all the parameters are known. They do not necessarily carry over to estimated processes, a case which was also investigated and reviewed by Lütkepohl (1987).

11.6.2 Aggregation of VARMA Processes

There is little to be added to the foregoing results for the case of contemporaneous aggregation. Suppose $y_t = (y_{1t}, \ldots, y_{Kt})'$ consists of K variables. If all or some of them are contemporaneously aggregated by taking their sum or average, this just means that y_t is transformed linearly and the foregoing results apply directly. In particular, the aggregated process has a finite order VARMA representation if the original process does. Moreover, if forecasts for the aggregated variables are desired it is generally preferable to forecast the disaggregated process and aggregate the forecasts rather than forecast the aggregated process directly.

The foregoing results are also helpful in studying the consequences of temporal aggregation. Suppose we wish to aggregate the variables y_t generated by

$$y_t = A_1 y_{t-1} + A_2 y_{t-2} + u_t + M_1 u_{t-1}$$

11.6 Transforming and Aggregating VARMA Processes

over, say, $m = 3$ subsequent periods. To be able to use the previous framework, we construct a process

$$\begin{bmatrix} I_K & 0 & 0 \\ -A_1 & I_K & 0 \\ -A_2 & -A_1 & I_K \end{bmatrix} \begin{bmatrix} y_{m(\tau-1)+1} \\ y_{m(\tau-1)+2} \\ y_{m\tau} \end{bmatrix}$$

$$= \begin{bmatrix} 0 & A_2 & A_1 \\ 0 & 0 & A_2 \\ 0 & 0 & 0 \end{bmatrix} \begin{bmatrix} y_{m(\tau-2)+1} \\ y_{m(\tau-2)+2} \\ y_{m(\tau-1)} \end{bmatrix} + \begin{bmatrix} I_K & 0 & 0 \\ M_1 & I_K & 0 \\ 0 & M_1 & I_K \end{bmatrix} \begin{bmatrix} u_{m(\tau-1)+1} \\ u_{m(\tau-1)+2} \\ u_{m\tau} \end{bmatrix}$$

$$+ \begin{bmatrix} 0 & 0 & M_1 \\ 0 & 0 & 0 \\ 0 & 0 & 0 \end{bmatrix} \begin{bmatrix} u_{m(\tau-2)+1} \\ u_{m(\tau-2)+2} \\ u_{m(\tau-1)} \end{bmatrix}.$$

Defining

$$\mathfrak{y}_\tau := \begin{bmatrix} y_{m(\tau-1)+1} \\ y_{m(\tau-1)+2} \\ y_{m\tau} \end{bmatrix} \quad \text{and} \quad \mathfrak{u}_\tau := \begin{bmatrix} u_{m(\tau-1)+1} \\ u_{m(\tau-1)+2} \\ u_{m\tau} \end{bmatrix},$$

we get

$$\mathfrak{A}_0 \mathfrak{y}_\tau = \mathfrak{A}_1 \mathfrak{y}_{\tau-1} + \mathfrak{M}_0 \mathfrak{u}_\tau + \mathfrak{M}_1 \mathfrak{u}_{\tau-1}, \tag{11.6.11}$$

where $\mathfrak{A}_0$, $\mathfrak{A}_1$, $\mathfrak{M}_0$, and $\mathfrak{M}_1$ have the obvious definitions. This form is a VARMA$(1, 1)$ representation of the $3K$-dimensional process $\mathfrak{y}_\tau$. Our standard form of a VARMA$(1, 1)$ process can be obtained from this form by premultiplying with $\mathfrak{A}_0^{-1}$ and defining $\mathfrak{v}_\tau = \mathfrak{A}_0^{-1} \mathfrak{M}_0 \mathfrak{u}_\tau$ which gives

$$\mathfrak{y}_\tau = \mathfrak{A}_0^{-1} \mathfrak{A}_1 \mathfrak{y}_{\tau-1} + \mathfrak{v}_\tau + \mathfrak{A}_0^{-1} \mathfrak{M}_1 \mathfrak{M}_0^{-1} \mathfrak{A}_0 \mathfrak{v}_{\tau-1}.$$

Now temporal aggregation over $m = 3$ periods can be represented as a linear transformation of the process $\mathfrak{y}_\tau$. Clearly, it is not difficult to see that this method generalizes for higher order processes and temporal aggregation over more than three periods. Moreover, different types of temporal aggregation can be handled. For instance, the aggregate may be the sum of subsequent values or it may be their average. Furthermore, temporal and contemporaneous aggregation can be dealt with simultaneously. In all of these cases, the aggregate has a VARMA representation if the original variables are generated by a finite order VARMA process and its structure can be studied using the foregoing framework. Moreover, by Proposition 11.2, if forecasts of the aggregate are of interest, it is in general preferable to forecast the original disaggregated process and aggregate the forecasts rather than forecast the aggregate directly. A detailed discussion of these issues and also of forecasting with estimated processes can be found in Lütkepohl (1987).

11.7 Interpretation of VARMA Models

The same tools and concepts that we have used for interpreting VAR models may also be applied in the VARMA case. We will consider Granger-causality and impulse response analysis in turn.

11.7.1 Granger-Causality

To study Granger-causality in the context of VARMA processes, we partition y_t in two groups of variables, z_t and x_t, and we partition the VAR and MA operators as well as the white noise process u_t accordingly. Hence, we get

$$\begin{bmatrix} A_{11}(L) & A_{12}(L) \\ A_{21}(L) & A_{22}(L) \end{bmatrix} \begin{bmatrix} z_t \\ x_t \end{bmatrix} = \begin{bmatrix} M_{11}(L) & M_{12}(L) \\ M_{21}(L) & M_{22}(L) \end{bmatrix} \begin{bmatrix} u_{1t} \\ u_{2t} \end{bmatrix}, \qquad (11.7.1)$$

where again a zero mean is assumed for simplicity and without loss of generality. The results derived in the following are not affected by a nonzero mean term. The process (11.7.1) is assumed to be stable and invertible and its pure, canonical MA representation is

$$\begin{bmatrix} z_t \\ x_t \end{bmatrix} = \begin{bmatrix} \Phi_{11}(L) & \Phi_{12}(L) \\ \Phi_{21}(L) & \Phi_{22}(L) \end{bmatrix} \begin{bmatrix} u_{1t} \\ u_{2t} \end{bmatrix}.$$

From Proposition 2.2, we know that x_t is not Granger-causal for z_t if and only if $\Phi_{12}(L) \equiv 0$. Although the proposition is stated for VAR processes, it is easy to see that it remains correct for the presently considered VARMA case. We also know that

$$\begin{bmatrix} \Phi_{11}(L) & \Phi_{12}(L) \\ \Phi_{21}(L) & \Phi_{22}(L) \end{bmatrix}$$
$$= \begin{bmatrix} A_{11}(L) & A_{12}(L) \\ A_{21}(L) & A_{22}(L) \end{bmatrix}^{-1} \begin{bmatrix} M_{11}(L) & M_{12}(L) \\ M_{21}(L) & M_{22}(L) \end{bmatrix}$$
$$= \begin{bmatrix} D(L) & -D(L)A_{12}(L)A_{22}(L)^{-1} \\ -A_{22}(L)^{-1}A_{21}(L)D(L) & A_{22}(L)^{-1} + A_{22}(L)^{-1}A_{21}(L)D(L)A_{12}(L)A_{22}(L)^{-1} \end{bmatrix}$$
$$\times \begin{bmatrix} M_{11}(L) & M_{12}(L) \\ M_{21}(L) & M_{22}(L) \end{bmatrix},$$

where

$$D(L) := [A_{11}(L) - A_{12}(L)A_{22}(L)^{-1}A_{21}(L)]^{-1}$$

and the rules for the partitioned inverse have been used (see Appendix A.10). Consequently, x_t is not Granger-causal for z_t if and only if

$$0 \equiv D(L)M_{12}(L) - D(L)A_{12}(L)A_{22}(L)^{-1}M_{22}(L)$$

or, equivalently,

$$M_{12}(L) - A_{12}(L)A_{22}(L)^{-1}M_{22}(L) \equiv 0.$$

Moreover, it follows as in Proposition 2.3 that there is no instantaneous causality between x_t and z_t if and only if $E(u_{1t}u'_{2t}) = 0$. We state these results as a proposition.

Proposition 11.3 (*Characterization of Noncausality*)
Let

$$y_t = \begin{bmatrix} z_t \\ x_t \end{bmatrix}$$

be a stable and invertible VARMA(p,q) process as in (11.7.1) with possibly nonzero mean. Then x_t is not Granger-causal for z_t if and only if

$$M_{12}(L) \equiv A_{12}(L)A_{22}(L)^{-1}M_{22}(L). \tag{11.7.2}$$

There is no instantaneous causality between z_t and x_t if and only if

$$E(u_{1t}u'_{2t}) = 0.$$

∎

Remark 1 Obviously, the restrictions characterizing Granger-noncausality are not quite so easy here as in the VAR(p) case. Consider, for instance, a bivariate VARMA$(1,1)$ process

$$\begin{bmatrix} z_t \\ x_t \end{bmatrix} = \begin{bmatrix} \alpha_{11,1} & \alpha_{12,1} \\ \alpha_{21,1} & \alpha_{22,1} \end{bmatrix} \begin{bmatrix} z_{t-1} \\ x_{t-1} \end{bmatrix} + \begin{bmatrix} u_{1t} \\ u_{2t} \end{bmatrix} + \begin{bmatrix} m_{11,1} & m_{12,1} \\ m_{21,1} & m_{22,1} \end{bmatrix} \begin{bmatrix} u_{1,t-1} \\ u_{2,t-1} \end{bmatrix}.$$

For this process, the restrictions (11.7.2) reduce to

$$m_{12,1}L = (-\alpha_{12,1}L)(1 - \alpha_{22,1}L)^{-1}(1 + m_{22,1}L)$$

or

$$(1 - \alpha_{22,1}L)m_{12,1}L = -(1 + m_{22,1}L)\alpha_{12,1}L$$

or

$$m_{12,1} = -\alpha_{12,1} \quad \text{and} \quad \alpha_{22,1}m_{12,1} = \alpha_{12,1}m_{22,1}.$$

This, of course, is a set of nonlinear restrictions whereas only linear constraints were required to characterize Granger-noncausality in the corresponding pure VAR(p) case. However, a sufficient condition for (11.7.2) to hold is

$$M_{12}(L) \equiv A_{12}(L) \equiv 0, \tag{11.7.3}$$

which is again a set of linear constraints. Occasionally, these sufficient conditions may be easier to test than (11.7.2). ∎

Remark 2 To turn the arguments put forward prior to Proposition 11.3 into a formal proof requires that we convince ourselves that all the operations performed with the matrices of lag polynomials are feasible and correct. Because we have not proven these results, the arguments should just be taken as an indication of how a proof may proceed. ∎

11.7.2 Impulse Response Analysis

The impulse responses and forecast error variance decompositions of a VARMA model are obtained from its pure MA representation, as in the finite order VAR case. Thus, the discussion of Sections 2.3.2 and 2.3.3 carries over to the present case, except that the Φ_i's are computed with different formulas. Also, Propositions 2.4 and 2.5 need modification. We will not give the details here but refer the reader to the exercises (see Problem 11.9).

It may be worth reiterating some caveats of impulse response analysis which may be more apparent now after the discussion of transformations in Section 11.6. In particular, we have seen there that dropping variables (considering subprocesses) or aggregating the components of a VARMA process temporally and/or contemporaneously results in possibly quite different VARMA structures. They will in general have quite different coefficients in their pure MA representations. In other words, the impulse responses may change drastically if important variables are excluded from a system or if the level of aggregation is altered, for instance, if quarterly instead of monthly data are considered. Again, this does not necessarily render impulse response analysis useless. It should caution the reader against over interpreting the evidence from VARMA models, though. Some thought must be given to the choice of variables, the level of aggregation, and other transformations of the variables.

11.8 Exercises

Problem 11.1
Write the MA(1) process $y_t = u_t + M_1 u_{t-1}$ in VAR(1) form, $Y_t = \mathbf{A} Y_{t-1} + U_t$, and determine $\mathbf{A}^i$ for $i = 1, 2$.

Problem 11.2
Suppose $y_t = A_1 y_{t-1} + u_t + M_1 u_{t-1} + M_2 u_{t-2}$ is a stable and invertible VARMA(1, 2) process. Determine the coefficient matrices Π_i, $i = 1, 2, 3, 4$, of its pure VAR representation and the coefficient matrices Φ_i, $i = 1, 2, 3, 4$, of its pure MA representation.

Problem 11.3
Evaluate the autocovariances $\Gamma_y(h)$, $h = 1, 2, 3$, of the bivariate VARMA(2, 1) process

11.8 Exercises   445

$$y_t = \begin{bmatrix} .3 \\ .5 \end{bmatrix} + \begin{bmatrix} .5 & .1 \\ .4 & .5 \end{bmatrix} y_{t-1} + \begin{bmatrix} 0 & 0 \\ .25 & 0 \end{bmatrix} y_{t-2} + u_t + \begin{bmatrix} .6 & .2 \\ 0 & .3 \end{bmatrix} u_{t-1}. \quad (11.8.1)$$

(Hint: The use of a computer will greatly simplify this problem.)

Problem 11.4
Write the VARMA(1, 1) process $y_t = A_1 y_{t-1} + u_t + M_1 u_{t-1}$ in VAR(1) form, $Y_t = \mathbf{A} Y_{t-1} + U_t$. Determine forecasts $Y_t(h) = \mathbf{A}^h Y_t$ for $h = 1, 2, 3$, and compare them to forecasts obtained from the recursive formula (11.5.4).

Problem 11.5
Derive a univariate ARMA representation of the second component, y_{2t}, of the process given in (11.6.8).

Problem 11.6
Provide upper bounds for the ARMA orders of the process $z_t = y_{1t} + y_{2t} + y_{3t}$, where $y_t = (y_{1t}, y_{2t}, y_{3t}, y_{4t})'$ is a 4-dimensional VARMA(3, 3) process.

Problem 11.7
Write the VARMA(1, 1) process y_t from Problem 11.4 in a form such as (11.6.11) that permits to analyze temporal aggregation over four periods in the framework of Section 11.6.2. Give upper bounds for the orders of a VARMA representation of the process obtained by temporally aggregating y_t over four periods.

Problem 11.8
Write down explicitly the restrictions characterizing Granger-noncausality for a bivariate VARMA(2, 1) process. Is y_{1t} Granger-causal for y_{2t} in the process (11.8.1)?

Problem 11.9
Generalize Propositions 2.4 and 2.5 to the VARMA(p, q) case.
(Hint: Show that for a K-dimensional VARMA(p, q) process,

$$\phi_{jk,i} = 0, \quad \text{for } i = 1, 2, \ldots,$$

is equivalent to

$$\phi_{jk,i} = 0, \quad \text{for } i = 1, 2, \ldots, p(K-1) + q;$$

and

$$\theta_{jk,i} = 0, \quad \text{for } i = 0, 1, 2, \ldots,$$

is equivalent to

$$\theta_{jk,i} = 0, \quad \text{for } i = 0, 1, \ldots, p(K-1) + q.)$$

Problem 11.10

Suppose that m is a real number with $|m| > 1$ and u_t is a white noise process. Show that the process

$$v_t = \left(1 + \frac{1}{m}L\right)^{-1}(1 + mL)u_t$$

is also white noise with $\text{Var}(v_t) = m^2 \text{Var}(u_t)$.

12
Estimation of VARMA Models

In this chapter, maximum likelihood estimation of the coefficients of a VARMA model is considered. Before we can proceed to the actual estimation, a unique set of parameters must be specified. In this context, the problem of nonuniqueness of a VARMA representation becomes important. This identification problem, that is, the problem of identifying a unique structure among many equivalent ones, is treated in Section 12.1. In Section 12.2, the Gaussian likelihood function of a VARMA model is considered. A numerical algorithm for maximizing it and, thus, for computing the actual estimates is discussed in Section 12.3. The asymptotic properties of the ML estimators are the subject of Section 12.4. Forecasting with estimated processes and impulse response analysis are dealt with in Sections 12.5 and 12.6, respectively.

12.1 The Identification Problem

12.1.1 Nonuniqueness of VARMA Representations

In the previous chapter, we have considered K-dimensional, stationary processes y_t with VARMA(p, q) representations

$$y_t = A_1 y_{t-1} + \cdots + A_p y_{t-p} + u_t + M_1 u_{t-1} + \cdots + M_q u_{t-q}. \qquad (12.1.1)$$

Because the mean term is of no importance for the presently considered problem, we have set it to zero. Therefore, no intercept term appears in (12.1.1). This model can be written in lag operator notation as

$$A(L)y_t = M(L)u_t, \qquad (12.1.2)$$

where $A(L) := I_K - A_1 L - \cdots - A_p L^p$ and $M(L) := I_K + M_1 L + \cdots + M_q L^q$. Assuming that the VARMA representation is stable and invertible, the well-defined process described by the model (12.1.1) or (12.1.2) is given by

$$y_t = \sum_{i=0}^{\infty} \Phi_i u_{t-i} = \Phi(L) u_t = A(L)^{-1} M(L) u_t.$$

In practice, it is sometimes useful to consider a slightly more general type of VARMA model by attaching nonidentity coefficient matrices to y_t and u_t, that is, one may want to consider representations of the type

$$A_0 y_t = A_1 y_{t-1} + \cdots + A_p y_{t-p} + M_0 v_t + M_1 v_{t-1} + \cdots + M_q v_{t-q}, \quad (12.1.3)$$

where v_t is a suitable white noise process. Such a form may be suggested by subject matter theory which may imply instantaneous effects of some variables on other variables. It will also turn out to be useful in finding unique structures for VARMA models. By the specification (12.1.3) we mean the well-defined process

$$y_t = (A_0 - A_1 L - \cdots - A_p L^p)^{-1} (M_0 + M_1 L + \cdots + M_q L^q) v_t.$$

Such a process has a standard VARMA(p, q) representation with identity coefficient matrices attached to the instantaneous y_t and u_t if A_0 and M_0 are nonsingular. To see this, we premultiply (12.1.3) by A_0^{-1} and define $u_t = A_0^{-1} M_0 v_t$ which gives

$$\begin{aligned} y_t = & A_0^{-1} A_1 y_{t-1} + \cdots + A_0^{-1} A_p y_{t-p} + u_t + A_0^{-1} M_1 M_0^{-1} A_0 u_{t-1} + \cdots \\ & + A_0^{-1} M_q M_0^{-1} A_0 u_{t-q}. \end{aligned}$$

Redefining the matrices appropriately, this, of course, is a representation of the type (12.1.1) with identity coefficient matrices at lag zero which describes the same process as (12.1.3). The assumption that both A_0 and M_0 are nonsingular does not entail any loss of generality, as long as none of the components of y_t can be written as a linear combination of the other components. We call a stable and invertible representation as in (12.1.1) a *VARMA representation in standard form* or a *standard VARMA representation* to distinguish it from representations with nonidentity matrices at lag zero as in (12.1.3). This discussion shows that VARMA representations are not unique, that is, a given process y_t can be written in standard form or in nonstandard form by premultiplying by any nonsingular $(K \times K)$ matrix. We have encountered a similar problem in dealing with finite order structural VAR processes in Chapter 9. However, once we consider standard reduced form VAR models only, we have unique representations. This property is in sharp contrast to the presently considered VARMA case, where, in general, a standard form is not a unique representation, as we will see shortly.

It may be useful at this stage to emphasize what we mean by equivalent representations of a process. Generally, two representations of a process y_t are equivalent if they give rise to the same realizations (except on a set of measure zero) and, thus, to the same multivariate distributions of any finite subcollection of variables $y_t, y_{t+1}, \ldots, y_{t+h}$, for arbitrary integers t and h. Of course, this specification just says that equivalent representations really represent the same process. If y_t is a zero mean process with canonical MA representation

$$y_t = \sum_{i=0}^{\infty} \Phi_i u_{t-i}, \quad \Phi_0 = I_K,$$
$$= \Phi(L)u_t, \qquad (12.1.4)$$

where $\Phi(L) := \sum_{i=0}^{\infty} \Phi_i L^i$, then any VARMA model $A(L)y_t = M(L)u_t$ for which

$$A(L)^{-1}M(L) = \Phi(L) \qquad (12.1.5)$$

is an equivalent representation of the process y_t. In other words, all VARMA models are equivalent for which $A(L)^{-1}M(L)$ results in the same operator $\Phi(L)$. Thus, in order to ensure uniqueness of a VARMA representation, we must impose restrictions on the VAR and MA operators such that there is precisely one feasible pair of operators $A(L)$ and $M(L)$ satisfying (12.1.5) for a given $\Phi(L)$.

Obviously, given some stable, invertible VARMA representation $A(L)y_t = M(L)u_t$, an equivalent representation results if we premultiply by any nonsingular matrix A_0. Therefore, to remove this source of nonuniqueness, let us for the moment focus on VARMA representations in standard form. As mentioned earlier, even then uniqueness is not ensured. To see this problem more clearly, let us consider a bivariate VARMA(1, 1) process in standard form,

$$y_t = A_1 y_{t-1} + u_t + M_1 u_{t-1}. \qquad (12.1.6)$$

From Section 11.3.1, we know that this process has the canonical MA representation

$$y_t = \sum_{i=0}^{\infty} \Phi_i u_{t-i} = u_t + \sum_{i=1}^{\infty} (A_1^i + A_1^{i-1} M_1) u_{t-i}. \qquad (12.1.7)$$

Thus, for example, any VARMA(1, 1) representation with $M_1 = -A_1$ will result in the same canonical MA representation. In other words, if it turns out that y_t is such that $M_1 = -A_1$ for some set of coefficients, then any choice of A_1 matrix that gives rise to a stable VAR operator can be matched by an M_1 matrix that leads to an equivalent VARMA(1, 1) representation of y_t. Of course, in this case, the MA coefficient matrices in (12.1.7) are in fact all zero and $y_t = u_t$ is really white noise, that is, y_t actually has a VARMA(0, 0) structure. This fact is also quite easy to see from the lag operator representation of (12.1.6),

$$(I_2 - A_1 L)y_t = (I_2 + M_1 L)u_t.$$

Of course, if $M_1 = -A_1$, the MA operator cancels against the VAR operator. This type of parameter indeterminacy is also known from univariate ARMA processes. It is usually ruled out by the assumption that the AR and MA operators have no common factors. Let us make a similar assumption in the presently considered multivariate case by requiring that y_t is not white noise, i.e., $M_1 \neq -A_1$.

Unfortunately, in the multivariate case, the nonuniqueness problem is not solved by this assumption. To see this, suppose that

$$A_1 = \begin{bmatrix} 0 & \alpha \\ 0 & 0 \end{bmatrix} \quad \text{and} \quad M_1 = 0,$$

where $\alpha \neq 0$. In this case, the canonical MA representation (12.1.4) has coefficient matrices

$$\Phi_1 = A_1, \quad \Phi_2 = \Phi_3 = \cdots = 0, \qquad (12.1.8)$$

because $A_1^i = 0$ for $i > 1$. The same MA representation results if

$$A_1 = 0 \quad \text{and} \quad M_1 = \begin{bmatrix} 0 & \alpha \\ 0 & 0 \end{bmatrix}.$$

More generally, a canonical MA representation with coefficient matrices as in (12.1.8) is obtained if

$$A_1 = \begin{bmatrix} 0 & \alpha+m \\ 0 & 0 \end{bmatrix} \quad \text{and} \quad M_1 = \begin{bmatrix} 0 & -m \\ 0 & 0 \end{bmatrix},$$

whatever the value of m. Note also that the VARMA representation will be stable and invertible for any value of m.

To understand where the parameter indeterminacy comes from, consider the VAR operator

$$I_2 - \begin{bmatrix} 0 & \alpha \\ 0 & 0 \end{bmatrix} L. \qquad (12.1.9)$$

The inverse of this operator is

$$I_2 + \begin{bmatrix} 0 & \alpha \\ 0 & 0 \end{bmatrix} L, \qquad (12.1.10)$$

which is easily checked by multiplying the two operators together. Thus, the operator (12.1.9) has a finite order inverse. Operators of this type are precisely the ones that cause trouble in setting up a uniquely parameterized VARMA representation of a given process because multiplying by such an operator may cancel part of one operator (VAR or MA) while at the same time the finite order of the other operator is maintained.

To get a better sense for this problem, let us look at the following VARMA(1, 1) process:

$$A(L)y_t = M(L)u_t,$$

where

$$A(L) := \begin{bmatrix} 1 - \alpha_{11}L & -\alpha_{12}L \\ 0 & 1 \end{bmatrix} \quad \text{and} \quad M(L) := \begin{bmatrix} 1 + m_{11}L & m_{12}L \\ 0 & 1 \end{bmatrix}.$$

The two operators do not cancel if $\alpha_{11} \neq -m_{11}$ and $\alpha_{12} \neq -m_{12}$. Still we can factor an operator

$$D(L) := \begin{bmatrix} 1 & 0 \\ 0 & 1 \end{bmatrix} + \begin{bmatrix} 0 & \gamma \\ 0 & 0 \end{bmatrix} L = \begin{bmatrix} 1 & \gamma L \\ 0 & 1 \end{bmatrix}$$

from both operators without changing their general structure:

$$A(L) = D(L) \begin{bmatrix} 1 - \alpha_{11}L & -(\gamma + \alpha_{12})L \\ 0 & 1 \end{bmatrix},$$

$$M(L) = D(L) \begin{bmatrix} 1 + m_{11}L & (m_{12} - \gamma)L \\ 0 & 1 \end{bmatrix}.$$

Cancelling $D(L)$ gives operators

$$\begin{bmatrix} 1 - \alpha_{11}L & -(\gamma + \alpha_{12})L \\ 0 & 1 \end{bmatrix} = D(L) \begin{bmatrix} 1 + \alpha_{11}L & -(2\gamma + \alpha_{12})L \\ 0 & 1 \end{bmatrix}$$

and

$$\begin{bmatrix} 1 + m_{11}L & (m_{12} - \gamma)L \\ 0 & 1 \end{bmatrix} = D(L) \begin{bmatrix} 1 + m_{11}L & (m_{12} - 2\gamma)L \\ 0 & 1 \end{bmatrix}.$$

Thus, we can again factor and cancel $D(L)$. In fact, we can cancel $D(L)$ as often as we like without changing the general structure of the process. Hence, even if the orders of both operators cannot be reduced simultaneously by cancellation, it may still be possible to factor some operator from both $A(L)$ and $M(L)$ without changing their general structure. Note that the troubling operator $D(L)$ is again one with finite order inverse,

$$D(L)^{-1} = \begin{bmatrix} 1 & -\gamma L \\ 0 & 1 \end{bmatrix}.$$

Finite order operators that have a finite order inverse are characterized by the property that their determinant is a nonzero constant, that is, it does not involve L or powers of L. Operators with this property are called *unimodular*. For instance, the operator (12.1.9) has determinant,

$$\left| I_2 - \begin{bmatrix} 0 & \alpha \\ 0 & 0 \end{bmatrix} L \right| = \left| \begin{bmatrix} 1 & -\alpha L \\ 0 & 1 \end{bmatrix} \right| = 1$$

and, hence, it is unimodular. The property of a unimodular operator to have a finite order inverse follows because the inverse of an operator $A(L)$ is its adjoint divided by its determinant,

$$A(L)^{-1} = A(L)^{adj}/|A(L)| = |A(L)|^{-1}A(L)^{adj}.$$

The determinant is a univariate operator. A finite order invertible univariate operator, however, has an infinite order inverse, unless its degree is zero, that is, unless it is a constant.

In order to state uniqueness conditions for a VARMA representation, we will first of all require that a representation is chosen for which further cancellation is not possible in the sense that there are no common factors in the VAR and MA parts, except for unimodular operators. Operators $A(L)$ and $M(L)$ with this property are *left-coprime*. This property may be defined by calling the matrix operator $[A(L) : M(L)]$ left-coprime, if the existence of operators $D(L)$, $\bar{A}(L)$, and $\bar{M}(L)$ satisfying

$$D(L)[\bar{A}(L) : \bar{M}(L)] = [A(L) : M(L)] \tag{12.1.11}$$

implies that $D(L)$ is unimodular, that is, $|D(L)|$ is a nonzero constant. From the foregoing examples, it should be understood that in general factoring unimodular operators from $A(L)$ and $M(L)$ is unavoidable if no further constraints are imposed. Thus, to obtain uniqueness of left-coprime operators we have to impose restrictions ensuring that the only feasible unimodular operator $D(L)$ in (12.1.11) is $D(L) = I_K$. We will now give two sets of conditions that ensure uniqueness of a VARMA representation.

12.1.2 Final Equations Form and Echelon Form

Suppose y_t is a stationary zero mean process that has a stable, invertible VARMA representation,

$$A(L)y_t = M(L)u_t, \tag{12.1.12}$$

where $A(L) := A_0 - A_1 L - \cdots - A_p L^p$ and $M(L) := M_0 + M_1 L + \cdots + M_q L^q$. Further suppose that $A(L)$ and $M(L)$ are left-coprime and the white noise covariance matrix Σ_u is nonsingular.

Definition 12.1 (*Final Equations Form*)
The VARMA representation (12.1.12) is said to be in *final equations form* if $M_0 = I_K$ and $A(L) = \alpha(L)I_K$, where $\alpha(L) := 1 - \alpha_1 L - \cdots - \alpha_p L^p$ is a scalar (one-dimensional) operator with $\alpha_p \neq 0$. ∎

For instance, the bivariate VARMA(3, 1) model

$$(1 - \alpha_1 L - \alpha_2 L^2 - \alpha_3 L^3) \begin{bmatrix} y_{1t} \\ y_{2t} \end{bmatrix} = \begin{bmatrix} 1 + m_{11,1}L & m_{12,1}L \\ m_{21,1}L & 1 + m_{22,1}L \end{bmatrix} \begin{bmatrix} u_{1t} \\ u_{2t} \end{bmatrix} \tag{12.1.13}$$

with $\alpha_3 \neq 0$, is in final equations form. The label "final equations form" for this type of VARMA representation is in line with the terminology used in Chapter 10, Section 10.2.2.

Uniqueness of the final equations form

$$\alpha(L)y_t = M(L)u_t$$

is seen by noting that $D(L) = I_K$ is the only operator that retains the scalar AR part upon multiplication. For the operator $D(L)\alpha(L)I_K$ to maintain the order p, the operator $D(L)$ must have degree zero, that is, $D(L) = D$. However, the only possible matrix D that guarantees a zero order matrix I_K for the VAR operator is $D = I_K$.

Definition 12.2 (*Echelon Form*)
The VARMA representation (12.1.12) is said to be in *echelon form* or ARMA$_E$ form if the VAR and MA operators $A(L) = [\alpha_{ki}(L)]_{k,i=1,\ldots,K}$ and $M(L) = [m_{ki}(L)]$ are left-coprime and satisfy the following conditions: The operators $\alpha_{ki}(L)$ $(i = 1, \ldots, K)$ and $m_{kj}(L)$ $(j = 1, \ldots, K)$ in the k-th row of $A(L)$ and $M(L)$ have degree p_k and they have the form

$$\alpha_{kk}(L) = 1 - \sum_{j=1}^{p_k} \alpha_{kk,j} L^j, \quad \text{for } k = 1, \ldots, K,$$

$$\alpha_{ki}(L) = - \sum_{j=p_k - p_{ki}+1}^{p_k} \alpha_{ki,j} L^j, \quad \text{for } k \neq i,$$

and

$$m_{ki}(L) = \sum_{j=0}^{p_k} m_{ki,j} L^j, \quad \text{for } k, i = 1, \ldots, K, \quad \text{with } M_0 = A_0.$$

In the VAR operators $\alpha_{ki}(L)$,

$$p_{ki} := \begin{cases} \min(p_k + 1, p_i) & \text{for } k \geq i, \\ \min(p_k, p_i) & \text{for } k < i, \end{cases} \quad k, i = 1, \ldots, K. \quad (12.1.14)$$

That is, p_{ki} specifies the number of free coefficients in the operator $\alpha_{ki}(L)$ for $i \neq k$. The row degrees $(p_1, \ldots, p_K)$ are called the *Kronecker indices* and their sum $\sum_{k=1}^{K} p_i$ is the *McMillan degree*. Obviously, for the VARMA orders we have, in general, $p = q = \max(p_1, \ldots, p_K)$. ∎

We will sometimes denote an echelon form VARMA model with Kronecker indices $(p_1, \ldots, p_K)$ by ARMA$_E(p_1, \ldots, p_K)$. The following model is an example of a bivariate VARMA process in echelon form or, more precisely, an ARMA$_E(2, 1)$:

$$\begin{bmatrix} 1 - \alpha_{11,1}L - \alpha_{11,2}L^2 & -\alpha_{12,2}L^2 \\ -\alpha_{21,0} - \alpha_{21,1}L & 1 - \alpha_{22,1}L \end{bmatrix} \begin{bmatrix} y_{1t} \\ y_{2t} \end{bmatrix}$$
$$= \begin{bmatrix} 1 + m_{11,1}L + m_{11,2}L^2 & m_{12,1}L + m_{12,2}L^2 \\ -\alpha_{21,0} + m_{21,1}L & 1 + m_{22,1}L \end{bmatrix} \begin{bmatrix} u_{1t} \\ u_{2t} \end{bmatrix} \quad (12.1.15)$$

or

$$\begin{bmatrix} 1 & 0 \\ -\alpha_{21,0} & 1 \end{bmatrix} \begin{bmatrix} y_{1,t} \\ y_{2,t} \end{bmatrix}$$
$$= \begin{bmatrix} \alpha_{11,1} & 0 \\ \alpha_{21,1} & \alpha_{22,1} \end{bmatrix} \begin{bmatrix} y_{1,t-1} \\ y_{2,t-1} \end{bmatrix} + \begin{bmatrix} \alpha_{11,2} & \alpha_{12,2} \\ 0 & 0 \end{bmatrix} \begin{bmatrix} y_{1,t-2} \\ y_{2,t-2} \end{bmatrix}$$
$$+ \begin{bmatrix} 1 & 0 \\ -\alpha_{21,0} & 1 \end{bmatrix} \begin{bmatrix} u_{1,t} \\ u_{2,t} \end{bmatrix} + \begin{bmatrix} m_{11,1} & m_{12,1} \\ m_{21,1} & m_{22,1} \end{bmatrix} \begin{bmatrix} u_{1,t-1} \\ u_{2,t-1} \end{bmatrix}$$
$$+ \begin{bmatrix} m_{11,2} & m_{12,2} \\ 0 & 0 \end{bmatrix} \begin{bmatrix} u_{1,t-2} \\ u_{2,t-2} \end{bmatrix}.$$

In this model, the Kronecker indices (row degrees) are $p_1 = 2$ and $p_2 = 1$. Thus, the McMillan degree is 3. The p_{ki} numbers are

$$\begin{bmatrix} p_{11} & p_{12} \\ p_{21} & p_{22} \end{bmatrix} = \begin{bmatrix} 2 & 1 \\ 2 & 1 \end{bmatrix}$$

(see (12.1.14)). The off-diagonal elements p_{12} and p_{21} of this matrix indicate the numbers of parameters contained in the operators $\alpha_{12}(L)$ and $\alpha_{21}(L)$, respectively. Because $\alpha_{12}(L)$ belongs to the first row or first equation of the system, it has degree $p_1 = 2$. Hence, because it has just one free coefficient ($p_{12} = 1$), it has the form $\alpha_{12}(L) = -\alpha_{12,2}L^2$. Similarly, $\alpha_{21}(L)$ belongs to the second row of the system and, thus, it has degree $p_2 = 1$. Because it has $p_{21} = 2$ free coefficients, it must be of the form $\alpha_{21}(L) = -\alpha_{21,0} - \alpha_{21,1}L$. Another characteristic feature of the echelon form is that A_0 is lower-triangular and has ones on the main diagonal. Moreover, the zero order MA coefficient matrix is identical to the zero order VAR matrix, $M_0 = A_0$.

Some free coefficients of the echelon form of a VARMA model may be zero and, hence, p or q may be less than $\max(p_1, \ldots, p_K)$. For instance, in the example process (12.1.15), $m_{11,2}$ and $m_{12,2}$ may be zero. In that case, $q = 1 < \max(p_1, p_2) = 2$. In order for a representation to be an echelon form with Kronecker indices $(p_1, \ldots, p_K)$, at least one operator in the k-th row of $[A(L) : M(L)]$ must have degree p_k, with nonzero coefficient at lag p_k.

An echelon is a certain positioning of an army in the form of steps. Similarly, the nonzero parameters in an echelon VARMA representation are positioned in a specific way. In particular, the positioning of freely varying parameters in the k-th equation depends only on Kronecker indices $p_i \leq p_k$ and not on Kronecker indices $p_j > p_k$. More precisely, as long as $p_j > p_k$, the positioning of the free parameters in the k-th equation will be the same for any value p_j. For the example process (12.1.15), it is easy to check that the positions of the free parameters in the second equation will remain the same if the row degree of the first equation is increased to $p_1 = 3$. In other words, p_{21} does not change due to an increase in p_1.

It can be shown that the echelon form, just like the final equations form, guarantees uniqueness of the VARMA representation. In other words, if a VARMA representation is in echelon form, then the representation is unique within the class of all echelon representations. A similar statement applies for the final equations form. Also, for any stable, invertible VARMA(p,q) representation, there exists an equivalent echelon form and an equivalent final equations form.

The reader may wonder why we consider the complicated looking echelon representation although the final equations form serves the same purpose. The reason is that the echelon form is usually preferable in practice because it often involves fewer free parameters than the equivalent final equations form. We will see an example of this phenomenon shortly. Having as few free parameters as possible is important to ease the numerical problems in maximizing the likelihood function and to gain efficiency of the parameter estimators.

There are a number of other unique or *identified* parameterizations of VARMA models. We have chosen to present the final equations form and the echelon form because these two forms will play a role when we discuss the issue of specifying VARMA models in Chapter 13. For proofs of the uniqueness of the echelon form and for other identification conditions we refer to Hannan (1969, 1970, 1976, 1979), Deistler & Hannan (1981), and Hannan & Deistler (1988). We now proceed with illustrations of the final equations form and the echelon form.

12.1.3 Illustrations

Starting from some VARMA(p,q) representation $A(L)y_t = M(L)u_t$, one strategy for finding the corresponding final equations form results from premultiplying with the adjoint $A(L)^{adj}$ of the VAR operator $A(L)$ which gives

$$|A(L)|y_t = A(L)^{adj} M(L) u_t, \qquad (12.1.16)$$

where $A(L)^{adj} A(L) = |A(L)|$ has been used. Obviously, (12.1.16) has a scalar VAR operator and, hence, is in final equations form if all superfluous terms are cancelled.

To find the echelon form corresponding to a given VARMA model, we have to cancel as much as possible so as to make the VAR and MA operators left-coprime. Then a unimodular matrix operator has to be determined which, upon premultiplication, transforms the given model into an echelon form. It usually helps to determine the Kronecker indices (row degrees) and the corresponding numbers p_{ki} first. We will now consider examples.

Let us begin with the simple bivariate process

$$\left(I_2 - \begin{bmatrix} 0 & \alpha \\ 0 & 1 \end{bmatrix} L \right) y_t = u_t \qquad (12.1.17)$$

with $\alpha \neq 0$. Noting that

$$|A(L)| = \left| \begin{bmatrix} 1 & -\alpha L \\ 0 & 1 \end{bmatrix} \right| = 1 \quad \text{and} \quad A(L)^{adj} = \begin{bmatrix} 1 & \alpha L \\ 0 & 1 \end{bmatrix},$$

456   12 Estimation of VARMA Models

the final equations form is seen to be

$$y_t = \left(I_2 + \begin{bmatrix} 0 & \alpha \\ 0 & 0 \end{bmatrix} L \right) u_t. \tag{12.1.18}$$

To find the echelon representation, we first determine the Kronecker indices or row degrees and the implied p_{ki} from Definition 12.2. The first row of (12.1.17) has degree $p_1 = 1$ and the second row has degree $p_2 = 0$. Hence,

$$p_{11} = 1, \quad p_{12} = 0, \quad p_{21} = 1, \quad p_{22} = 0,$$

so that

$$\alpha_{11}(L) = 1 - \alpha_{11,1}L, \ \alpha_{12}(L) = 0, \ \alpha_{21}(L) = -\alpha_{21,0}, \text{ and } \alpha_{22}(L) = 1.$$

Thus, the echelon form is

$$\begin{bmatrix} 1 - \alpha_{11,1}L & 0 \\ -\alpha_{21,0} & 1 \end{bmatrix} y_t = \begin{bmatrix} 1 + m_{11,1}L & m_{12,1}L \\ -\alpha_{21,0} & 1 \end{bmatrix} u_t. \tag{12.1.19}$$

The unique parameter values in this representation corresponding to the specific process (12.1.17) are easily seen to be

$$\alpha_{11,1} = \alpha_{21,0} = m_{11,1} = 0 \quad \text{and} \quad m_{12,1} = \alpha.$$

Thus, in this particular case, the final equations form and the echelon form coincide.

As another example, we consider a 3-dimensional process with VARMA(2, 1) representation

$$\begin{bmatrix} 1 - \theta_1 L & -\theta_2 L & 0 \\ 0 & 1 - \theta_3 L - \theta_4 L^2 & -\theta_5 L \\ 0 & 0 & 1 \end{bmatrix} y_t$$

$$= \begin{bmatrix} 1 - \eta_1 L & 0 & 0 \\ 0 & 1 - \eta_2 L & 0 \\ 0 & 0 & 1 - \eta_3 L \end{bmatrix} u_t. \tag{12.1.20}$$

Using (12.1.16), its final equations form is seen to be

$$(1 - \theta_1 L)(1 - \theta_3 L - \theta_4 L^2) y_t$$

$$= \begin{bmatrix} 1 - \theta_3 L - \theta_4 L^2 & \theta_2 L & \theta_2 \theta_5 L^2 \\ 0 & 1 - \theta_1 L & \theta_5 L - \theta_1 \theta_5 L^2 \\ 0 & 0 & (1 - \theta_1 L)(1 - \theta_3 L - \theta_4 L^2) \end{bmatrix}$$

$$\times \begin{bmatrix} 1 - \eta_1 L & 0 & 0 \\ 0 & 1 - \eta_2 L & 0 \\ 0 & 0 & 1 - \eta_3 L \end{bmatrix} u_t$$

12.1 The Identification Problem   457

which is easily recognizable as a VARMA(3, 4) structure with scalar VAR operator.

The Kronecker indices, that is, the row degrees of (12.1.20) are $(p_1, p_2, p_3) = (1, 2, 1)$ and the implied p_{ki}-numbers from (12.1.14) are collected in the following matrix:

$$[p_{ki}]_{k,i=1,2,3} = \begin{bmatrix} 1 & 1 & 1 \\ 1 & 2 & 1 \\ 1 & 2 & 1 \end{bmatrix}.$$

Consequently, the VAR operator of the echelon form becomes

$$\begin{bmatrix} 1 - \alpha_{11,1}L & -\alpha_{12,1}L & -\alpha_{13,1}L \\ -\alpha_{21,2}L^2 & 1 - \alpha_{22,1}L - \alpha_{22,2}L^2 & -\alpha_{23,2}L^2 \\ -\alpha_{31,1}L & -\alpha_{32,0} - \alpha_{32,1}L & 1 - \alpha_{33,1}L \end{bmatrix}$$

or

$$\begin{bmatrix} 1 & 0 & 0 \\ 0 & 1 & 0 \\ 0 & -\alpha_{32,0} & 1 \end{bmatrix} - \begin{bmatrix} \alpha_{11,1} & \alpha_{12,1} & \alpha_{13,1} \\ 0 & \alpha_{22,1} & 0 \\ \alpha_{31,1} & \alpha_{32,1} & \alpha_{33,1} \end{bmatrix} L - \begin{bmatrix} 0 & 0 & 0 \\ \alpha_{21,2} & \alpha_{22,2} & \alpha_{23,2} \\ 0 & 0 & 0 \end{bmatrix} L^2. \tag{12.1.21}$$

Hence, in the echelon representation,

$$A_0 = \begin{bmatrix} 1 & 0 & 0 \\ 0 & 1 & 0 \\ 0 & -\alpha_{32,0} & 1 \end{bmatrix}$$

is different from I_3, if $\alpha_{32,0} \neq 0$, and, thus, $M_0 = A_0$ is also not the identity matrix. The MA operator is

$$\begin{bmatrix} 1 + m_{11,1}L & m_{12,1}L & m_{13,1}L \\ m_{21,1}L + m_{21,2}L^2 & 1 + m_{22,1}L + m_{22,2}L^2 & m_{23,1}L + m_{23,2}L^2 \\ m_{31,1}L & -\alpha_{32,0} + m_{32,1}L & 1 + m_{33,1}L \end{bmatrix}$$

or

$$\begin{bmatrix} 1 & 0 & 0 \\ 0 & 1 & 0 \\ 0 & -\alpha_{32,0} & 1 \end{bmatrix} + \begin{bmatrix} m_{11,1} & m_{12,1} & m_{13,1} \\ m_{21,1} & m_{22,1} & m_{23,1} \\ m_{31,1} & m_{32,1} & m_{33,1} \end{bmatrix} L + \begin{bmatrix} 0 & 0 & 0 \\ m_{21,2} & m_{22,2} & m_{23,2} \\ 0 & 0 & 0 \end{bmatrix} L^2. \tag{12.1.22}$$

The reader may be puzzled by the fact that the last element in the second row of (12.1.21) does not involve a term with first power of L while such a term appears in (12.1.20). This model form shows that there is a VARMA representation equivalent to (12.1.20) with the second but not the first power of L in the last operator in the second row of $A(L)$. The fact, that there always exists an equivalent echelon representation does not mean that there is always an immediately obvious relation between the coefficients of any given VARMA representation and its equivalent echelon form. However, in the present case it is fairly easy to relate the representations (12.1.20) and (12.1.21)/(12.1.22). Premultiplying (12.1.20) by the operator

$$\begin{bmatrix} 1 & 0 & 0 \\ 0 & 1 & \theta_5 L \\ 0 & 0 & 1 \end{bmatrix} \quad (12.1.23)$$

results in a VAR operator

$$\begin{bmatrix} 1-\theta_1 L & -\theta_2 L & 0 \\ 0 & 1-\theta_3 L - \theta_4 L^2 & 0 \\ 0 & 0 & 1 \end{bmatrix}$$

and the MA operator changes accordingly. Notice that the operator (12.1.23) has constant determinant and, of course, the resulting VARMA model is equivalent to (12.1.20). The relation between its coefficients and those of the echelon representation (12.1.21)/(12.1.22) is obvious:

$$\alpha_{11,1} = \theta_1, \quad \alpha_{12,1} = \theta_2, \quad \theta_{13,1} = 0,$$
$$\alpha_{21,2} = 0, \quad \alpha_{22,1} = \theta_3, \quad \alpha_{22,2} = \theta_4, \quad \alpha_{23,2} = 0,$$
$$\alpha_{31,1} = \alpha_{32,0} = \alpha_{32,1} = \alpha_{33,1} = 0,$$

and the relation between (12.1.22) and the coefficients of (12.1.20) is also apparent. Of course, if the zero coefficients are known, then this knowledge may be used to reduce the number of free coefficients in the echelon form.

In this example, the unrestricted final equations form has 3 AR coefficients and 36 MA coefficients. Thus, the unrestricted form contains 39 parameters, apart from white noise covariance coefficients. In contrast, the unrestricted echelon form (12.1.21)/(12.1.22) has only 23 free parameters and is therefore preferable in terms of parameter parsimony. Note that, in practice, the true coefficient values are unknown and we pick an identified structure, for example, a final equations form or an echelon form. At that stage, further parameter restrictions may not be available. Hence, if (12.1.20) is the actual data generation process we may pick a VARMA(3, 4) model with scalar AR operator if we decide to go with a final equations representation and we may choose the model (12.1.21)/(12.1.22) if we decide to use an echelon form representation. Obviously, the latter choice results in a more parsimonious parameterization. As mentioned earlier, for estimation purposes the more parsimonious representation is advantageous.

Although $A_0 \neq I$ in the previous example, it should be understood that in many echelon representations $A_0 = M_0 = I_K$. In particular, if the row degrees $p_1 = \cdots = p_K = p$, all $p_{ki} = p, i, k = 1, \ldots, K$, and the echelon form is easily seen to be a standard VARMA(p,p) model with $A_0 = M_0 = I_K$. We are now ready to turn to the actual estimation of the parameters of an identified VARMA model and we shall discuss its Gaussian likelihood function next.

12.2 The Gaussian Likelihood Function

For maximum likelihood (ML) estimation the likelihood function is needed. We will now derive useful approximations to the likelihood function of a Gaussian VARMA(p,q) process. Special case MA processes will be considered first.

12.2.1 The Likelihood Function of an MA(1) Process

Because a zero mean MA(1) process is the simplest member of the finite order MA family, we use that as a starting point. Hence, we assume to have a sample $y_1, \ldots, y_T$ which is generated by the Gaussian, K-dimensional, invertible MA(1) process

$$y_t = u_t + M_1 u_{t-1}, \tag{12.2.1}$$

where u_t is a Gaussian white noise process with covariance matrix Σ_u. Thus,

$$\mathbf{y} := \begin{bmatrix} y_1 \\ \vdots \\ y_T \end{bmatrix} = \overline{\mathfrak{M}}_1 \begin{bmatrix} u_0 \\ u_1 \\ \vdots \\ u_T \end{bmatrix},$$

where

$$\overline{\mathfrak{M}}_1 := \begin{bmatrix} M_1 & I_K & 0 & \cdots & 0 & 0 \\ 0 & M_1 & I_K & & 0 & 0 \\ \vdots & & \ddots & \ddots & & \vdots \\ \vdots & & & \ddots & \ddots & \vdots \\ 0 & 0 & 0 & \cdots & M_1 & I_K \end{bmatrix} \tag{12.2.2}$$

is a $(KT \times K(T+1))$ matrix. Using that u_t is Gaussian white noise and, thus,

$$\begin{bmatrix} u_0 \\ u_1 \\ \vdots \\ u_T \end{bmatrix} \sim \mathcal{N}(0, I_{T+1} \otimes \Sigma_u),$$

if follows that

$$\mathbf{y} \sim \mathcal{N}(0, \overline{\mathfrak{M}}_1 (I_{T+1} \otimes \Sigma_u) \overline{\mathfrak{M}}_1')$$

and the likelihood function is seen to be

$$l(M_1, \Sigma_u | \mathbf{y})$$
$$\propto |\overline{\mathfrak{M}}_1 (I_{T+1} \otimes \Sigma_u) \overline{\mathfrak{M}}_1'|^{-1/2} \exp\{-\tfrac{1}{2} \mathbf{y}' [\overline{\mathfrak{M}}_1 (I_{T+1} \otimes \Sigma_u) \overline{\mathfrak{M}}_1']^{-1} \mathbf{y}\}, \tag{12.2.3}$$

where $\propto$ stands for "is proportional to". In other words, we have dropped a multiplicative constant from the likelihood function which does not change the maximizing values of M_1 and Σ_u.

It is inconvenient that this function involves the determinant and the inverse of a $(KT \times KT)$ matrix. A simpler form is obtained if u_0 is set to zero, that is, the MA(1) process is assumed to be started up with a nonrandom fixed vector $u_0 = 0$. In that case,

$$\mathbf{y} = \mathfrak{M}_1 \mathbf{u},$$

where

$$\mathfrak{M}_1 := \begin{bmatrix} I_K & 0 & \cdots & 0 & 0 \\ M_1 & I_K & & 0 & 0 \\ & \ddots & \ddots & & \vdots \\ & & \ddots & \ddots & \vdots \\ 0 & 0 & \cdots & M_1 & I_K \end{bmatrix}_{(KT \times KT)} \quad \text{and} \quad \mathbf{u} := \begin{bmatrix} u_1 \\ \vdots \\ u_T \end{bmatrix}_{(KT \times 1)}. \quad (12.2.4)$$

The likelihood function is then proportional to

$$\begin{aligned} l_0(M_1, \Sigma_u | \mathbf{y}) &= |\mathfrak{M}_1(I_T \otimes \Sigma_u)\mathfrak{M}_1'|^{-1/2} \exp\{-\tfrac{1}{2}\mathbf{y}'[\mathfrak{M}_1(I_T \otimes \Sigma_u)\mathfrak{M}_1']^{-1}\mathbf{y}\} \\ &= |\Sigma_u|^{-T/2} \exp\{-\tfrac{1}{2}\mathbf{y}'\mathfrak{M}_1'^{-1}(I_T \otimes \Sigma_u^{-1})\mathfrak{M}_1^{-1}\mathbf{y}\} \\ &= |\Sigma_u|^{-T/2} \exp\left\{-\frac{1}{2}\sum_{t=1}^{T} u_t' \Sigma_u^{-1} u_t\right\}, \end{aligned} \quad (12.2.5)$$

where it has been used that $|\mathfrak{M}_1| = 1$ and

$$\begin{aligned} \mathfrak{M}_1^{-1} &= \begin{bmatrix} I_K & 0 & \cdots & 0 & 0 \\ -M_1 & I_K & & 0 & 0 \\ (-M_1)^2 & -M_1 & \ddots & 0 & 0 \\ \vdots & \vdots & \ddots & \ddots & \vdots \\ (-M_1)^{T-1} & (-M_1)^{T-2} & \cdots & -M_1 & I_K \end{bmatrix} \\ &= \begin{bmatrix} I_K & 0 & \cdots & 0 \\ -\Pi_1 & I_K & & 0 \\ \vdots & & \ddots & \vdots \\ -\Pi_{T-1} & -\Pi_{T-2} & \cdots & I_K \end{bmatrix}, \end{aligned}$$

where the $\Pi_i = -(-M_1)^i$ are the coefficients of the pure VAR representation of the process. By successive substitution, the MA(1) process in (12.2.1) can be rewritten as

$$y_t + \sum_{i=1}^{t-1}(-M_1)^i y_{t-i} + (-M_1)^t u_0 = u_t. \quad (12.2.6)$$

Thus, if $u_0 = 0$,

$$u_t = y_t + \sum_{i=1}^{t-1}(-M_1)^i y_{t-i},$$

from which the last expression in (12.2.5) is obtained.

The equation (12.2.6) also shows that, for large t, the assumption regarding u_0 becomes inconsequential because, for an invertible process, M_1^t approaches zero as $t \to \infty$. The impact of u_0 disappears more rapidly for processes for which M_1^t goes to zero more rapidly as t gets large. In other words, if all eigenvalues of M_1 are close to zero or, equivalently, all roots of $\det(I_K + M_1 z)$ are far outside the unit circle, then the impact of u_0 is lower than for processes with roots close to the unit circle. In summary, the likelihood approximation in (12.2.5) will improve as the sample size gets large and will become exact as $T \to \infty$. In small samples, it is better for processes with roots of $\det(I_K + M_1 z)$ far away from the unit circle than for those with roots close to the noninvertibility region. Because we will be concerned predominantly with large sample properties in the following, we will often work with likelihood approximations such as l_0 in (12.2.5).

12.2.2 The MA(q) Case

A similar reasoning as for MA(1) processes can also be employed for higher order MA processes. Suppose the generation process of y_t has a zero mean MA(q) representation

$$y_t = u_t + M_1 u_{t-1} + \cdots + M_q u_{t-q}. \qquad (12.2.7)$$

Then

$$\mathbf{y} = \overline{\mathfrak{M}}_q \begin{bmatrix} u_{-q+1} \\ \vdots \\ u_0 \\ u_1 \\ \vdots \\ u_T \end{bmatrix},$$

where

$$\overline{\mathfrak{M}}_q := \begin{bmatrix} M_q & M_{q-1} & \cdots & M_1 & I_K & 0 & \cdots & \cdots & 0 \\ 0 & M_q & \cdots & M_2 & M_1 & I_K & & & 0 \\ \vdots & & \ddots & \ddots & \ddots & \ddots & \ddots & & \vdots \\ \vdots & & & \ddots & \ddots & \ddots & \ddots & \ddots & \vdots \\ 0 & 0 & \cdots & & M_q & \cdots & M_2 & M_1 & I_K \end{bmatrix} \qquad (12.2.8)$$

is a $(KT \times K(T+q))$ matrix and the exact likelihood for a sample of size T is seen to be

$$l(M_1, \ldots, M_q, \Sigma_u|\mathbf{y}) \propto |\overline{\mathfrak{M}}_q(I_{T+q} \otimes \Sigma_u)\overline{\mathfrak{M}}_q'|^{-1/2}$$
$$\times \exp\{-\tfrac{1}{2}\mathbf{y}'[\overline{\mathfrak{M}}_q(I_{T+q} \otimes \Sigma_u)\overline{\mathfrak{M}}_q']^{-1}\mathbf{y}\}. \quad (12.2.9)$$

Again a convenient approximation to the likelihood function is obtained by setting $u_{-q+1} = \cdots = u_0 = 0$. In that case, the likelihood is, apart from a multiplicative constant,

$$l_0(M_1, \ldots, M_q, \Sigma_u|\mathbf{y}) = |\Sigma_u|^{-T/2} \exp\{-\tfrac{1}{2}\mathbf{y}'[\mathfrak{M}_q'^{-1}(I_T \otimes \Sigma_u^{-1})\mathfrak{M}_q^{-1}]\mathbf{y}\}, \quad (12.2.10)$$

where

$$\mathfrak{M}_q := \begin{bmatrix} I_K & 0 & \cdots & \cdots & \cdots & \cdots & 0 & 0 \\ M_1 & I_K & & & & & 0 & 0 \\ M_2 & M_1 & \ddots & & & & 0 & 0 \\ \vdots & \vdots & \ddots & \ddots & & & & \vdots \\ M_q & M_{q-1} & & \ddots & \ddots & & & \vdots \\ 0 & M_q & & & \ddots & \ddots & & \\ \vdots & & \ddots & & & \ddots & \ddots & \\ 0 & 0 & \cdots & M_q & \cdots & \cdots & M_1 & I_K \end{bmatrix} \quad (12.2.11)$$

and, hence,

$$\mathfrak{M}_q^{-1} = \begin{bmatrix} I_K & 0 & \cdots & 0 \\ -\Pi_1 & I_K & & 0 \\ \vdots & \vdots & \ddots & \vdots \\ -\Pi_{T-1} & -\Pi_{T-2} & \cdots & I_K \end{bmatrix}.$$

Here the Π_i are the coefficient matrices of the pure VAR representation of the process y_t. Thus, the Π_i can be computed recursively as in Section 11.2 of Chapter 11.

An alternative expression for the approximate likelihood is easily seen to be

$$l_0(M_1, \ldots, M_q, \Sigma_u|\mathbf{y}) = |\Sigma_u|^{-T/2} \exp\left\{-\frac{1}{2}\sum_{t=1}^T u_t'\Sigma_u^{-1}u_t\right\}, \quad (12.2.12)$$

where

$$u_t = y_t - \sum_{i=1}^{t-1} \Pi_i y_{t-i}.$$

Again, the likelihood approximation will be quite precise if T is reasonably large and the roots of $\det(I_K + M_1 z + \cdots + M_q z^q)$ are not close to the unit circle.

Although we will work with likelihood approximations in the following, it is perhaps worth noting that an expression for the exact likelihood of an MA(q) process can be derived that is more manageable than the one in (12.2.9) (see, e.g., Hillmer & Tiao (1979), Kohn (1981)).

12.2.3 The VARMA(1, 1) Case

Before we tackle general mixed VARMA models, we shall consider the simplest candidate, namely a Gaussian zero mean, stationary, stable, and invertible VARMA(1, 1) process,

$$y_t = A_1 y_{t-1} + u_t + M_1 u_{t-1}. \tag{12.2.13}$$

Assuming that we have a sample $y_1, \ldots, y_T$, generated by this process and defining

$$\mathfrak{A}_p := \begin{bmatrix} I_K & 0 & \cdots & & & 0 & 0 \\ -A_1 & I_K & & & & 0 & 0 \\ -A_2 & -A_1 & \ddots & & & 0 & 0 \\ \vdots & \vdots & & \ddots & & \vdots & \vdots \\ -A_p & -A_{p-1} & & & & 0 & 0 \\ 0 & -A_p & & & \ddots & 0 & 0 \\ \vdots & & & & & \vdots & \vdots \\ 0 & 0 & & & \ddots & I_K & 0 \\ 0 & 0 & \cdots & -A_p & \cdots & -A_1 & I_K \end{bmatrix}, \tag{12.2.14}$$

we get

$$\mathfrak{A}_1 \begin{bmatrix} y_1 \\ \vdots \\ y_T \end{bmatrix} + \begin{bmatrix} -A_1 y_0 \\ 0 \\ \vdots \\ 0 \end{bmatrix} = \overline{\mathfrak{M}}_1 \begin{bmatrix} u_0 \\ u_1 \\ \vdots \\ u_T \end{bmatrix}.$$

Hence, for given, fixed presample values y_0,

$$\mathbf{y} = \begin{bmatrix} y_1 \\ \vdots \\ y_T \end{bmatrix} \sim \mathcal{N}(\mathfrak{A}_1^{-1} y_0, \mathfrak{A}_1^{-1} \overline{\mathfrak{M}}_1 (I_{T+1} \otimes \Sigma_u) \overline{\mathfrak{M}}_1' \mathfrak{A}_1'^{-1}), \tag{12.2.15}$$

where

$$\mathbf{y}_0 := \begin{bmatrix} A_1 y_0 \\ 0 \\ \vdots \\ 0 \end{bmatrix}.$$

The corresponding likelihood function, conditional on y_0, is

$$\begin{aligned} l(A_1, M_1, \Sigma_u | \mathbf{y}, y_0) & \\ \propto\ & |\mathfrak{A}_1^{-1}\overline{\mathfrak{M}}_1(I_{T+1} \otimes \Sigma_u)\overline{\mathfrak{M}}_1'\mathfrak{A}_1'^{-1}|^{-1/2} \\ & \times \exp\{-\tfrac{1}{2}(\mathbf{y} - \mathfrak{A}_1^{-1}\mathbf{y}_0)'\mathfrak{A}_1'[\overline{\mathfrak{M}}_1(I_{T+1} \otimes \Sigma_u)\overline{\mathfrak{M}}_1']^{-1}\mathfrak{A}_1(\mathbf{y} - \mathfrak{A}_1^{-1}\mathbf{y}_0)\} \\ =\ & |\overline{\mathfrak{M}}_1(I_{T+1} \otimes \Sigma_u)\overline{\mathfrak{M}}_1'|^{-1/2} \\ & \times \exp\{-\tfrac{1}{2}(\mathfrak{A}_1\mathbf{y} - \mathbf{y}_0)'[\overline{\mathfrak{M}}_1(I_{T+1} \otimes \Sigma_u)\overline{\mathfrak{M}}_1']^{-1}(\mathfrak{A}_1\mathbf{y} - \mathbf{y}_0)\}, \quad (12.2.16) \end{aligned}$$

where $|\mathfrak{A}_1| = 1$ has been used.

With the same arguments as in the pure MA case, a simple approximation is obtained by setting $u_0 = y_0 = 0$. Then we get

$$\begin{aligned} l_0(A_1, M_1, \Sigma_u) & = |\Sigma_u|^{-T/2} \exp\{-\tfrac{1}{2}(\mathfrak{M}_1^{-1}\mathfrak{A}_1\mathbf{y})'(I_T \otimes \Sigma_u^{-1})\mathfrak{M}_1^{-1}\mathfrak{A}_1\mathbf{y}\} \\ & = |\Sigma_u|^{-T/2} \exp\left\{-\frac{1}{2}\sum_{t=1}^T u_t' \Sigma_u^{-1} u_t\right\}, \quad (12.2.17) \end{aligned}$$

where

$$u_t = y_t - \sum_{i=1}^{t-1} \Pi_i y_{t-i} \quad (12.2.18)$$

and the Π_i are the coefficient matrices of the pure VAR representation, that is, for the present case $\Pi_i = (-1)^{i-1}(M_1^i + M_1^{i-1}A_1)$, $i = 1, 2, \ldots$ (see Section 11.3.1). Note that in writing the likelihood approximation l_0 we have dropped the conditions $\mathbf{y}$ and y_0 for notational simplicity.

The effect of starting up the process with $y_0 = u_0 = 0$ is quite easily seen in (12.2.18), namely, for observation y_t, the infinite order pure VAR representation is truncated at lag $t - 1$. Such a truncation has little effect if the sample size is large and the roots of the MA operator are not close to the unit circle.

12.2.4 The General VARMA(p, q) Case

Now suppose a sample $y_1, \ldots, y_T$ is generated by the Gaussian K-dimensional, stable, invertible VARMA(p, q) process

$$\begin{aligned} A_0(y_t - \mu) & = A_1(y_{t-1} - \mu) + \cdots + A_p(y_{t-p} - \mu) \\ & \quad + A_0 u_t + M_1 u_{t-1} + \cdots + M_q u_{t-q} \end{aligned} \quad (12.2.19)$$

with mean vector μ and nonsingular white noise covariance matrix Σ_u. Notice that A_0 appears as the coefficient matrix of y_t and of u_t as in the echelon form. Thus, the echelon form is covered by our treatment of the general VARMA(p,q) case. We have chosen the mean-adjusted form of the process because this form has certain advantages in ML estimation, as we will see later.

Usually some elements of the coefficient matrices will be zero or obey some other type of restrictions. Therefore, to be realistic, we define

$$\alpha_0 := \text{vec}(A_0) \text{ and } \boldsymbol{\beta} := \text{vec}[A_1, \ldots, A_p, M_1, \ldots, M_q] \qquad (12.2.20)$$

and assume that these coefficients are linearly related to an $(N \times 1)$ parameter vector $\boldsymbol{\gamma}$, that is,

$$\begin{bmatrix} \alpha_0 \\ \boldsymbol{\beta} \end{bmatrix} = R\boldsymbol{\gamma} + r \qquad (12.2.21)$$

for a suitable, known $(K^2(p+q+1) \times N)$ matrix R and a known $K^2(p+q+1)$-vector r. For example, for a bivariate ARMA$_E(1,0)$ process with Kronecker indices $p_1 = 1$ and $p_2 = 0$,

$$\begin{bmatrix} 1 - \alpha_{11,1}L & 0 \\ -\alpha_{21,0} & 1 \end{bmatrix}(y_t - \mu) = \begin{bmatrix} 1 + m_{11,1}L & m_{12,1}L \\ -\alpha_{21,0} & 1 \end{bmatrix} u_t$$

or

$$\begin{bmatrix} 1 & 0 \\ -\alpha_{21,0} & 1 \end{bmatrix}(y_t - \mu) = \begin{bmatrix} \alpha_{11,1} & 0 \\ 0 & 0 \end{bmatrix}(y_{t-1} - \mu) + \begin{bmatrix} 1 & 0 \\ -\alpha_{21,0} & 1 \end{bmatrix} u_t$$
$$+ \begin{bmatrix} m_{11,1} & m_{12,1} \\ 0 & 0 \end{bmatrix} u_{t-1},$$

we have

$$\alpha_0 = \begin{bmatrix} 1 \\ -\alpha_{21,0} \\ 0 \\ 1 \end{bmatrix}, \quad \boldsymbol{\beta} = \begin{bmatrix} \alpha_{11,1} \\ 0 \\ 0 \\ 0 \\ m_{11,1} \\ 0 \\ m_{12,1} \\ 0 \end{bmatrix}, \quad R = \begin{bmatrix} 0 & 0 & 0 & 0 \\ -1 & 0 & 0 & 0 \\ 0 & 0 & 0 & 0 \\ 0 & 0 & 0 & 0 \\ 0 & 1 & 0 & 0 \\ 0 & 0 & 0 & 0 \\ 0 & 0 & 0 & 0 \\ 0 & 0 & 0 & 0 \\ 0 & 0 & 1 & 0 \\ 0 & 0 & 0 & 0 \\ 0 & 0 & 0 & 1 \\ 0 & 0 & 0 & 0 \end{bmatrix},$$

$$\boldsymbol{\gamma} = \begin{bmatrix} \alpha_{21,0} \\ \alpha_{11,1} \\ m_{11,1} \\ m_{12,1} \end{bmatrix}, \quad \text{and} \quad r = \begin{bmatrix} 1 \\ 0 \\ 0 \\ 1 \\ 0 \\ 0 \\ \vdots \\ 0 \end{bmatrix}.$$

Similarly, for the final equations form

$$(1 - \alpha_1 L)(y_t - \mu) = \begin{bmatrix} 1 + m_{11}L & m_{12}L \\ m_{21}L & 1 + m_{22}L \end{bmatrix} u_t$$

or

$$\begin{bmatrix} 1 & 0 \\ 0 & 1 \end{bmatrix}(y_t - \mu) = \begin{bmatrix} \alpha_1 & 0 \\ 0 & \alpha_1 \end{bmatrix}(y_{t-1} - \mu) + \begin{bmatrix} 1 & 0 \\ 0 & 1 \end{bmatrix} u_t + \begin{bmatrix} m_{11} & m_{12} \\ m_{21} & m_{22} \end{bmatrix} u_{t-1},$$

we get

$$\boldsymbol{\alpha}_0 = \begin{bmatrix} 1 \\ 0 \\ 0 \\ 1 \end{bmatrix}, \quad \boldsymbol{\beta} = \begin{bmatrix} \alpha_1 \\ 0 \\ 0 \\ \alpha_1 \\ m_{11} \\ m_{21} \\ m_{12} \\ m_{22} \end{bmatrix}, \quad R = \begin{bmatrix} 0 & 0 & 0 & 0 & 0 \\ 0 & 0 & 0 & 0 & 0 \\ 0 & 0 & 0 & 0 & 0 \\ 0 & 0 & 0 & 0 & 0 \\ 1 & 0 & 0 & 0 & 0 \\ 0 & 0 & 0 & 0 & 0 \\ 0 & 0 & 0 & 0 & 0 \\ 1 & 0 & 0 & 0 & 0 \\ 0 & 1 & 0 & 0 & 0 \\ 0 & 0 & 1 & 0 & 0 \\ 0 & 0 & 0 & 1 & 0 \\ 0 & 0 & 0 & 0 & 1 \end{bmatrix},$$

$$\boldsymbol{\gamma} = \begin{bmatrix} \alpha_1 \\ m_{11} \\ m_{21} \\ m_{12} \\ m_{22} \end{bmatrix}, \quad \text{and} \quad r = \begin{bmatrix} 1 \\ 0 \\ 0 \\ 1 \\ 0 \\ \vdots \\ 0 \end{bmatrix}.$$

The likelihood function is a function of $\mu, \boldsymbol{\gamma}$, and Σ_u. Its exact form, given fixed initial values $y_{-p+1}, \ldots, y_0$, can be derived analogously to the previously considered special cases (see Problem 12.4 and Hillmer & Tiao (1979)). Here we will just give the likelihood approximation obtained by assuming

$$y_{-p+1} - \mu = \cdots = y_0 - \mu = u_{-q+1} = \cdots = u_0 = 0.$$

Apart from a multiplicative constant, we get

$$l_0(\mu, \gamma, \Sigma_u) = |\Sigma_u|^{-T/2} \exp\left\{-\frac{1}{2}\sum_{t=1}^{T} u_t(\mu, \gamma)' \Sigma_u^{-1} u_t(\mu, \gamma)\right\}, \qquad (12.2.22)$$

where

$$u_t(\mu, \gamma) = (y_t - \mu) - \sum_{i=1}^{t-1} \Pi_i(\gamma)(y_{t-i} - \mu), \qquad (12.2.23)$$

with the $\Pi_i(\gamma)$'s being again the coefficient matrices of the pure VAR representation of y_t. We have indicated that these matrices are determined by the parameter vector γ. Formally the likelihood approximation has the same appearance as in the special cases. Of course, the u_t's are now potentially more complicated functions of the parameters.

It is perhaps worth noting that the uniqueness or identification problem discussed in Section 12.1 is reflected in the likelihood function. If the model is parameterized in a unique way, for instance, in final equations form or echelon form, the likelihood function has a locally unique maximum. This property is of obvious importance to guarantee unique ML estimators. Note, however, that the likelihood function in general has more than one local maximum. A more detailed discussion of the properties of the likelihood function can be found in Deistler & Pötscher (1984).

The next section focuses on the maximization of the approximate likelihood function (12.2.22) or, equivalently, the maximization of its logarithm,

$$\ln l_0(\mu, \gamma, \Sigma_u) = -\frac{T}{2}\ln|\Sigma_u| - \frac{1}{2}\sum_{t=1}^{T} u_t(\mu, \gamma)' \Sigma_u^{-1} u_t(\mu, \gamma). \qquad (12.2.24)$$

12.3 Computation of the ML Estimates

In the pure finite order VAR case considered in Chapters 3 and 5, we have obtained the ML estimates by solving the normal equations. In the presently considered VARMA(p, q) case, we may use the same principle. In other words, we determine the first order partial derivatives of the log-likelihood function or rather its approximation given in (12.2.24) and equate them to zero. We will obtain the normal equations in Section 12.3.1. It turns out that they are nonlinear in the parameters and we discuss algorithms for solving the ML optimization problem in Section 12.3.2. The optimization procedures are iterative algorithms that require starting-up values or preliminary estimates for the parameters. A possible choice of initial estimates is proposed in Section 12.3.4. One of the optimization algorithms involves the information matrix which is given in Section 12.3.3. An example is discussed in Section 12.3.5.

12.3.1 The Normal Equations

In order to set up the normal equations corresponding to the approximate log-likelihood given in (12.2.24), we derive the first order partial derivatives with respect to all the parameters μ, γ, and Σ_u.

$$\frac{\partial \ln l_0}{\partial \mu'} = -\sum_{t=1}^{T} u_t' \Sigma_u^{-1} \frac{\partial u_t}{\partial \mu'} = \sum_{t=1}^{T} u_t' \Sigma_u^{-1} \left[I_K - \sum_{i=1}^{t-1} \Pi_i(\gamma) \right], \quad (12.3.1)$$

$$\frac{\partial \ln l_0}{\partial \gamma'} = -\sum_{t=1}^{T} u_t' \Sigma_u^{-1} \frac{\partial u_t}{\partial \gamma'}. \quad (12.3.2)$$

A recursive formula for computing the $\partial u_t / \partial \gamma'$ is given in the following lemma.

Lemma 12.1

Suppose $\mu = 0$ and let

$$u_t = y_t - A_0^{-1}[A_1 y_{t-1} + \cdots + A_p y_{t-p} + M_1 u_{t-1} + \cdots + M_q u_{t-q}], \quad (12.3.3)$$

$$\alpha_0 := \text{vec}(A_0),$$

$$\beta := \text{vec}[A_1, \ldots, A_p, M_1, \ldots, M_q],$$

and suppose

$$\begin{bmatrix} \alpha_0 \\ \beta \end{bmatrix} = R\gamma + r, \quad (12.3.4)$$

where R is a known $(K^2(p+q+1) \times N)$ matrix, r is a known $K^2(p+q+1)$-dimensional vector, and γ is an $(N \times 1)$ vector of unknown parameters. Then, defining $\partial u_0 / \partial \gamma' = \partial u_{-1} / \partial \gamma' = \cdots = \partial u_{-q+1} / \partial \gamma' = 0$ and $y_0 = \cdots = y_{-p+1} = u_0 = \cdots = u_{-q+1} = 0$,

$$\begin{aligned}
\frac{\partial u_t}{\partial \gamma'} &= \{(A_0^{-1}[A_1 y_{t-1} + \cdots + A_p y_{t-p} \\
&\quad + M_1 u_{t-1} + \cdots + M_q u_{t-q}])' \otimes A_0^{-1}\}[I_{K^2} : 0 : \cdots : 0]R \\
&\quad - [(y_{t-1}', \ldots, y_{t-p}', u_{t-1}', \ldots, u_{t-q}') \otimes A_0^{-1}][0 : I_{K^2(p+q)}]R \\
&\quad - A_0^{-1} \left[M_1 \frac{\partial u_{t-1}}{\partial \gamma'} + \cdots + M_q \frac{\partial u_{t-q}}{\partial \gamma'} \right], \quad (12.3.5)
\end{aligned}$$

for $t = 1, \ldots, T$. ∎

Replacing y_t with $y_t - \mu$ in this lemma, the expression in (12.3.5) can be used for recursively computing the $\partial u_t / \partial \gamma'$ required in (12.3.2).

Proof:

$$\frac{\partial u_t}{\partial \gamma'} = -[(A_1 y_{t-1} + \cdots + A_p y_{t-p} + M_1 u_{t-1} + \cdots + M_q u_{t-q})' \otimes I_K]$$

$$\times \frac{\partial \operatorname{vec}(A_0^{-1})}{\partial \gamma'}$$

$$- [(y'_{t-1}, \ldots, y'_{t-p}, u'_{t-1}, \ldots, u'_{t-q}) \otimes A_0^{-1}]$$

$$\times \frac{\partial \operatorname{vec}[A_1, \ldots, A_p, M_1, \ldots, M_q]}{\partial \gamma'}$$

$$- A_0^{-1}[A_1, \ldots, A_p, M_1, \ldots, M_q] \partial \begin{bmatrix} \begin{bmatrix} y_{t-1} \\ \vdots \\ y_{t-p} \\ u_{t-1} \\ \vdots \\ u_{t-q} \end{bmatrix} \end{bmatrix} / \partial \gamma'. \qquad (12.3.6)$$

The lemma follows by noting that

$$\frac{\partial \operatorname{vec}(A_0^{-1})}{\partial \gamma'} = \frac{\partial \operatorname{vec}(A_0^{-1})}{\partial \alpha_0'} \frac{\partial \alpha_0}{\partial \gamma'} = -[(A_0^{-1})' \otimes A_0^{-1}][I_{K^2} : 0 : \cdots : 0]R \qquad (12.3.7)$$

(see Rule (9) of Appendix A.13). ∎

The partial derivatives of the approximate log-likelihood with respect to the elements of Σ_u are

$$\frac{\partial \ln l_0}{\partial \Sigma_u} = -\frac{T}{2}\Sigma_u^{-1} + \frac{1}{2}\Sigma_u^{-1}\left(\sum_{t=1}^{T} u_t u'_t\right)\Sigma_u^{-1} \qquad (12.3.8)$$

(see Problem 12.5). Setting this expression to zero and solving for Σ_u gives

$$\widetilde{\Sigma}_u(\mu, \gamma) = \frac{1}{T}\sum_{t=1}^{T} u_t(\mu, \gamma)u_t(\mu, \gamma)'. \qquad (12.3.9)$$

Substituting for Σ_u in (12.3.1) and (12.3.2) and setting to zero results in a generally nonlinear set of normal equations which may be solved by numerical methods. Before we discuss a possible algorithm, it may be worth pointing out that by substituting $\widetilde{\Sigma}_u(\mu, \gamma)$ for Σ_u in $\ln l_0$, we get

$$\begin{aligned}
\ln l_0(\mu, \gamma) &= -\frac{T}{2}\ln|\widetilde{\Sigma}_u(\mu, \gamma)| - \frac{1}{2}\operatorname{tr}\left(\widetilde{\Sigma}_u(\mu, \gamma)^{-1}\sum_{t=1}^{T} u_t(\mu, \gamma)u_t(\mu, \gamma)'\right) \\
&= -\frac{T}{2}\ln|\widetilde{\Sigma}_u(\mu, \gamma)| - \frac{TK}{2}. \qquad (12.3.10)
\end{aligned}$$

Thus, instead of maximizing $\ln l_0$ we may equivalently minimize

$$\ln|\widetilde{\Sigma}_u(\mu, \gamma)| \quad \text{or} \quad |\widetilde{\Sigma}_u(\mu, \gamma)|. \qquad (12.3.11)$$

12.3.2 Optimization Algorithms

The problem of optimizing (minimizing or maximizing) a function arises not only in ML estimation but also in various other contexts. Therefore, general algorithms have been developed. Following Judge et al. (1985, Section B.2), we will give a brief introduction to so-called gradient algorithms and then address the specific problem at hand. With the objective in mind that we want to find the coefficient values that minimize $-\ln l_0$ or $\ln|\widetilde{\Sigma}_u(\mu, \gamma)|$, we assume that the problem is to minimize a twice continuously differentiable, scalar valued function $h(\gamma)$, where γ is some $(N \times 1)$ vector.

Given a vector γ_i in the parameter space, we are looking for a direction (vector) $\mathbf{d}$ in which the objective function declines. Then we can perform a step of length s, say, in that direction which will take us downhill. In other words, we seek an appropriate *step direction* $\mathbf{d}$ and a step length s such that

$$h(\gamma_i + s\mathbf{d}) < h(\gamma_i). \tag{12.3.12}$$

If $\mathbf{d}$ is a downhill direction, a small step in that direction will always decrease the objective function. Thus, we are seeking a $\mathbf{d}$ such that $h(\gamma_i + s\mathbf{d})$ is a decreasing function of s, for s sufficiently close to zero. In other words, $\mathbf{d}$ must be such that

$$0 > \left.\frac{dh(\gamma_i + s\mathbf{d})}{ds}\right|_{s=0} = \left[\left.\frac{\partial h(\gamma)}{\partial \gamma'}\right|_{\gamma_i}\right]\left[\left.\frac{\partial(\gamma_i + s\mathbf{d})}{\partial s}\right|_{s=0}\right] = \left[\left.\frac{\partial h(\gamma)}{\partial \gamma'}\right|_{\gamma_i}\right]\mathbf{d}.$$

Using the abbreviation

$$\mathbf{h}_i := \left.\frac{\partial h(\gamma)}{\partial \gamma}\right|_{\gamma_i}$$

for the gradient of $h(\gamma)$ at γ_i, a possible choice of $\mathbf{d}$ is

$$\mathbf{d} = -D_i \mathbf{h}_i,$$

where D_i is any positive definite matrix. With this choice of $\mathbf{d}$,

$$\mathbf{h}'_i \mathbf{d} = -\mathbf{h}'_i D_i \mathbf{h}_i < 0$$

if $\mathbf{h}_i \neq 0$. Because the gradient is zero at a local minimum of the function, we hope to have reached the minimum once $\mathbf{h}_i = 0$ and, hence, $\mathbf{d} = 0$. The general form of an iteration of a *gradient algorithm* is therefore

$$\gamma_{i+1} = \gamma_i - s_i D_i \mathbf{h}_i, \tag{12.3.13}$$

where s_i denotes the step length in the i-th iteration and D_i is a positive definite *direction matrix*. The name "gradient algorithm" stems from the fact that the gradient $\mathbf{h}_i$ is involved in the choice of the step direction. Many such algorithms have been proposed in the literature (see, for example, Judge et al. (1985, Section B.2)). They differ in their choice of the direction matrix D_i and the step length s_i.

To motivate the choice of the D_i matrix that will be considered in the ML algorithm presented below, we expand the objective function $h(\boldsymbol{\gamma})$ in a Taylor series about $\boldsymbol{\gamma}_i$ (see Appendix A.13, Proposition A.3),

$$h(\boldsymbol{\gamma}) \approx h(\boldsymbol{\gamma}_i) + \mathbf{h}'_i(\boldsymbol{\gamma} - \boldsymbol{\gamma}_i) + \tfrac{1}{2}(\boldsymbol{\gamma} - \boldsymbol{\gamma}_i)' H_i (\boldsymbol{\gamma} - \boldsymbol{\gamma}_i), \tag{12.3.14}$$

where

$$H_i := \left. \frac{\partial^2 h}{\partial \boldsymbol{\gamma} \partial \boldsymbol{\gamma}'} \right|_{\boldsymbol{\gamma}_i}$$

is the Hessian matrix of second order partial derivatives of $h(\boldsymbol{\gamma})$, evaluated at $\boldsymbol{\gamma}_i$. If $h(\boldsymbol{\gamma})$ were a quadratic function, the right-hand side of (12.3.14) were exactly equal to $h(\boldsymbol{\gamma})$ and the first order conditions for a minimum would result by taking first order partial derivatives of the right-hand side and setting to zero:

$$\mathbf{h}'_i + H_i(\boldsymbol{\gamma} - \boldsymbol{\gamma}_i)' = 0$$

or

$$\boldsymbol{\gamma} = \boldsymbol{\gamma}_i - H_i^{-1} \mathbf{h}_i.$$

Thus, if $h(\boldsymbol{\gamma})$ were a quadratic function, starting from any vector $\boldsymbol{\gamma}_i$, we would reach the minimum in one step of length $s_i = 1$ by choosing the inverse Hessian as the direction matrix. In general, if $h(\boldsymbol{\gamma})$ is not a quadratic function, then the choice $D_i = H_i^{-1}$ is still reasonable once we are close to the minimum. Recall that a positive definite Hessian is the second order condition for a local minimum. Therefore, the inverse Hessian qualifies as a direction matrix. A gradient algorithm with the inverse Hessian as the direction matrix is called a *Newton* or *Newton-Raphson algorithm*.

From the previous subsection, we know that the first order partial derivatives of our objective function $-\ln l_0$ are quite complicated and, thus, finding the Hessian matrix of second order partial derivatives is even more complicated. Therefore we approximate the Hessian by an estimate of the information matrix,

$$\mathcal{I}(\boldsymbol{\gamma}) := E\left[\frac{\partial^2(-\ln l_0)}{\partial \boldsymbol{\gamma} \partial \boldsymbol{\gamma}'}\right], \tag{12.3.15}$$

which is the expected value of the Hessian matrix. The estimate of $\mathcal{I}(\boldsymbol{\gamma})$ will be denoted by $\widehat{\mathcal{I}}(\boldsymbol{\gamma})$. A computable expression will be given in the next subsection. Because the true parameter vector $\boldsymbol{\gamma}$ is unknown, $\widehat{\mathcal{I}}(\boldsymbol{\gamma}_i)$ is used as an estimate of $\mathcal{I}(\boldsymbol{\gamma})$ in the i-th iteration step. Hence, for given mean vector μ

and white noise covariance matrix Σ_u, we get a minimization algorithm with i-th iteration step

$$\gamma_{i+1} = \gamma_i - s_i \widehat{\mathcal{I}}(\gamma_i)^{-1} \left[\frac{\partial(-\ln l_0)}{\partial \gamma} \bigg|_{\gamma_i} \right]. \tag{12.3.16}$$

This algorithm is called the *scoring algorithm*.

As it stands, we still need some more information before we can execute this algorithm. First, we need a starting-up vector γ_1 for the first iteration. This vector should be close to the minimizing vector to ensure that $\widehat{\mathcal{I}}(\gamma_1)$ is positive definite and we make good progress towards the minimum even in the first iteration. We will consider one possible choice in Section 12.3.4.

Second, we have to choose the step length s_i. There are various possible alternatives (see, e.g., Judge et al. (1985, Section B.2)). Because we are just interested in the main principles of the algorithm, we will ignore the problem here and choose $s_i = 1$.

Third, the algorithm provides an ML estimate of γ, conditional on some given Σ_u matrix and mean vector μ, because both the information matrix and the gradient vector involve these quantities. They are usually also unknown. As in the pure finite order VAR case, it can be shown that the sample mean

$$\overline{y} = \frac{1}{T} \sum_{t=1}^{T} y_t$$

is an estimator for μ which has the same asymptotic properties as the ML estimator. Therefore, ML estimation of γ and Σ_u is often done conditionally on $\mu = \overline{y}$. In other words, the sample mean is subtracted from the data before the VARMA coefficients are estimated.

There are different ways to handle the unknown Σ_u matrix. From (12.3.9), we know that

$$\widetilde{\Sigma}_u(\mu, \gamma) = \frac{1}{T} \sum_{t=1}^{T} u_t(\mu, \gamma) u_t(\mu, \gamma)'.$$

Therefore, one possibility is to use $\Sigma_i := \widetilde{\Sigma}_u(\overline{y}, \gamma_i)$ in the i-th iteration. Equivalently, the minimization algorithm can be applied to $\ln |\widetilde{\Sigma}_u(\overline{y}, \gamma)|$.

A number of computer program packages contain exact or approximate ML algorithms which may be used in practice. The foregoing algorithm is just meant to demonstrate some basic principles. Modifications in actual applications may result in improved convergence properties. Slow convergence or no convergence at all may be the consequence of working with VARMA orders or Kronecker indices which are larger than the true ones and, hence, with an overparameterized model.

12.3.3 The Information Matrix

In the scoring algorithm described previously, an estimate of the information matrix is needed. To see how that can be obtained, we consider the second order partial derivatives of $-\ln l_0$,

$$\frac{\partial^2(-\ln l_0)}{\partial \gamma \partial \gamma'} = \partial\left[\sum_{t=1}^{T} \frac{\partial u_t'}{\partial \gamma} \Sigma_u^{-1} u_t\right] \bigg/ \partial \gamma' \quad \text{(see (12.3.2))}$$

$$= \sum_{t=1}^{T} \frac{\partial u_t'}{\partial \gamma} \Sigma_u^{-1} \frac{\partial u_t}{\partial \gamma'} + (u_t' \Sigma_u^{-1} \otimes I) \frac{\partial \operatorname{vec}[\partial u_t'/\partial \gamma]}{\partial \gamma'}.$$

Taking the expectation of this expression, the last term vanishes because $E(u_t) = 0$ and $u_t' \Sigma_u^{-1} \otimes I$ is independent of

$$\frac{\partial \operatorname{vec}[\partial u_t'/\partial \gamma]}{\partial \gamma'}$$

as this term does not contain current y_t or u_t variables (see Lemma 12.1). Hence,

$$E\left[\frac{\partial^2(-\ln l_0)}{\partial \gamma \partial \gamma'}\right] = \sum_{t=1}^{T} E\left[\frac{\partial u_t'}{\partial \gamma} \Sigma_u^{-1} \frac{\partial u_t}{\partial \gamma'}\right].$$

Estimating the expected value in the usual way by the sample average gives an estimator

$$\frac{1}{T} \sum_{t=1}^{T} \frac{\partial u_t'}{\partial \gamma} \Sigma_u^{-1} \frac{\partial u_t}{\partial \gamma'}$$

for

$$E\left[\frac{\partial u_t'}{\partial \gamma} \Sigma_u^{-1} \frac{\partial u_t}{\partial \gamma'}\right].$$

These considerations suggest the estimator

$$\widehat{\mathcal{I}}(\gamma) = \sum_{t=1}^{T} \frac{\partial u_t(\bar{y}, \gamma)'}{\partial \gamma} \Sigma_u^{-1} \frac{\partial u_t(\bar{y}, \gamma)}{\partial \gamma'} \tag{12.3.17}$$

for the information matrix $\mathcal{I}(\gamma)$. In the i-th iteration of the scoring algorithm, we evaluate this estimator for $\gamma = \gamma_i$. The quantities $\partial u_t/\partial \gamma'$ may be obtained recursively as in Lemma 12.1 to make this estimator operational.

If γ is the true parameter value, the asymptotic information matrix equals plim $\widehat{\mathcal{I}}(\gamma)/T$. Thus, if we have a consistent estimator $\tilde{\gamma}$ of γ, $\widehat{\mathcal{I}}(\tilde{\gamma})/T$ is a consistent estimator of the asymptotic information matrix, that is,

$$\mathcal{I}_a(\gamma) = \operatorname{plim} \widehat{\mathcal{I}}(\tilde{\gamma})/T. \tag{12.3.18}$$

In Section 12.4, we will see that the inverse of this matrix, if it exists, is the asymptotic covariance matrix of the ML estimator for γ. If a nonidentified structure is used, this problem is reflected in the asymptotic information matrix being singular. Hence, it is important at this stage to have an identified version of a VARMA model.

12.3.4 Preliminary Estimation

The coefficients of a VARMA(p, q) model in standard form,

$$y_t = A_1 y_{t-1} + \cdots + A_p y_{t-p} + u_t + M_1 u_{t-1} + \cdots + M_q u_{t-q},$$

could be estimated by multivariate LS, if the lagged u_t were given. We assume that the sample mean $\bar{y}$ has been subtracted previously. It is therefore neglected here. In deriving preliminary estimators for the other parameters, the idea is to fit a long pure autoregression first and then use estimated residuals in place of the true residuals. Hence, we fit a VAR(n) model

$$y_t = \sum_{i=1}^{n} \Pi_i(n) y_{t-i} + u_t(n),$$

where n is larger than p and q. From that estimation, we compute estimated residuals

$$\widehat{u}_t(n) := y_t - \sum_{i=1}^{n} \widehat{\Pi}_i(n) y_{t-i}, \qquad (12.3.19)$$

where $\widehat{\Pi}_i(n)$ are the multivariate LS estimators. Then we set up a multivariate regression model

$$Y = [A : M] X_n + U^0, \qquad (12.3.20)$$

where $Y := [y_1, \ldots, y_T]$, $A := [A_1, \ldots, A_p]$, $M := [M_1, \ldots, M_q]$,

$$X_n := [Y_{0,n}, \ldots, Y_{T-1,n}] \quad \text{with} \quad Y_{t,n} := \begin{bmatrix} y_t \\ \vdots \\ y_{t-p+1} \\ \widehat{u}_t(n) \\ \vdots \\ \widehat{u}_{t-q+1}(n) \end{bmatrix} \quad (K(p+q) \times 1)$$

and U^0 is a $(K \times T)$ matrix of residuals. Usually restrictions will be imposed on the parameters A and M of the model, for instance, if the model is given in final equations form. Additional restrictions may also be available. Suppose the restrictions are such that there exists a matrix R and a vector γ satisfying

$$\text{vec}[A:M] = R\gamma. \tag{12.3.21}$$

Applying the vec operator to (12.3.20) and substituting $R\gamma$ for $\text{vec}[A:M]$ gives

$$\text{vec}(Y) = (X'_n \otimes I_K)R\gamma + \text{vec}(U^0) \tag{12.3.22}$$

and the LS estimator of γ is known to be

$$\widehat{\gamma}(n) = [R'(X_n X'_n \otimes I_K)R]^{-1} R'(X_n \otimes I_K)\text{vec}(Y) \tag{12.3.23}$$

(see Chapter 5, Section 5.2). This estimator may be used as an initial vector γ_1 in the ML algorithm described in the previous subsections.

Using this estimator, a new set of residuals may be obtained as

$$\text{vec}(\widehat{U}^0) = \text{vec}(Y) - (X'_n \otimes I_K)R\widehat{\gamma}(n)$$

which may be used to obtain a white noise covariance estimator

$$\widetilde{\Sigma}_u(n) = \widehat{U}^0 \widehat{U}^{0\prime}/T. \tag{12.3.24}$$

This estimator may be used in place of Σ_u in the initial round of the iterative optimization algorithm described earlier.

Alternatively, instead of the LS estimator (12.3.23), we may use an EGLS estimator,

$$\widehat{\widehat{\gamma}}(n) = [R'(X_n X'_n \otimes \widetilde{\Sigma}_u)R]^{-1} R'(X_n \otimes \widetilde{\Sigma}_u)\text{vec}(Y),$$

with $\widetilde{\Sigma}_u(n)$ in place of $\widetilde{\Sigma}_u$ or a white noise covariance matrix estimator based on the residuals $\widehat{u}_t(n)$.

The echelon form of a VARMA(p,q) process may be of the more general type

$$A_0 y_t = A_1 y_{t-1} + \cdots + A_p y_{t-p} + A_0 u_t + M_1 u_{t-1} + \cdots + M_q u_{t-q}, \tag{12.3.25}$$

where A_0 is a lower triangular matrix with unit diagonal. To handle this case, we proceed in a similar manner as in the standard case and substitute the residuals $\widehat{u}_t(n)$ for the lagged u_t and for current residuals from other equations. In other words, in the k-th equation we substitute estimation residuals for $u_{it}, i < k$. Because A_0 is the coefficient matrix for both y_t and u_t, we define

$$X_n^c := [Y_{0,n}^c, \ldots, Y_{T-1,n}^c], \quad \text{where} \quad Y_{t,n}^c := \begin{bmatrix} y_{t+1} - \widehat{u}_{t+1}(n) \\ Y_{t,n} \end{bmatrix}$$

and we pick a restriction matrix R_c and a vector γ_c such that

$$R_c \gamma_c = \text{vec}[I_K - A_0, A, M].$$

Hence,

$$\text{vec}(Y) = (X_n^{c\prime} \otimes I_K) R_c \gamma_c + \text{vec}(U^0)$$

and the LS estimator of γ_c becomes

$$\widehat{\gamma}_c(n) = [R_c'(X_n^c X_n^{c\prime} \otimes I_K) R_c]^{-1} R_c'(X_n^c \otimes I_K) \text{vec}(Y).$$

The starting-up estimator of Σ_u is then obtained from the residuals of this regression. It is possible that the VARMA process corresponding to these coefficients is unstable or noninvertible. Especially in the latter case, modifications are desirable (see Hannan & Kavalieris (1984), Hannan & Deistler (1988)).

To see more clearly what is being done in this preliminary estimation procedure, let us look at an example. Suppose the bivariate VARMA(1, 1) echelon form model from (12.1.19) with Kronecker indices $(p_1, p_2) = (1, 0)$ is to be estimated:

$$\begin{aligned} y_{1,t} &= \alpha_{11,1} y_{1,t-1} + u_{1,t} + m_{11,1} u_{1,t-1} + m_{12,1} u_{2,t-1}, \\ y_{2,t} &= \alpha_{21,0} y_{1,t} - \alpha_{21,0} u_{1,t} + u_{2,t} = \alpha_{21,0}(y_{1,t} - u_{1,t}) + u_{2,t}. \end{aligned} \quad (12.3.26)$$

We assume that the sample mean has been removed previously. The parameters in the first equation are estimated by applying LS to

$$\begin{bmatrix} y_{1,1} \\ \vdots \\ y_{1,T} \end{bmatrix} = \begin{bmatrix} y_{1,0} & \widehat{u}_{1,0}(n) & \widehat{u}_{2,0}(n) \\ \vdots & \vdots & \vdots \\ y_{1,T-1} & \widehat{u}_{1,T-1}(n) & \widehat{u}_{2,T-1}(n) \end{bmatrix} \begin{bmatrix} \alpha_{11,1} \\ m_{11,1} \\ m_{12,1} \end{bmatrix} + \begin{bmatrix} u_{1,1} \\ \vdots \\ u_{1,T} \end{bmatrix},$$

or, using obvious notation, to

$$y_{(1)} = X_{(1)} \gamma_1 + u_{(1)}.$$

Here the $\widehat{u}_{i,t}(n)$ are the residuals from the estimated long VAR model of order n. The LS estimator of γ_1 is $\widehat{\gamma}_1 = (X_{(1)}' X_{(1)})^{-1} X_{(1)}' y_{(1)}$.

Similarly, $\alpha_{21,0}$ is estimated by applying LS to

$$\begin{bmatrix} y_{2,1} \\ \vdots \\ y_{2,T} \end{bmatrix} = \begin{bmatrix} y_{1,1} - \widehat{u}_{1,1}(n) \\ \vdots \\ y_{1,T} - \widehat{u}_{1,T}(n) \end{bmatrix} \alpha_{21,0} + \begin{bmatrix} u_{2,1} \\ \vdots \\ u_{2,T} \end{bmatrix}.$$

In this case, it would be possible to use the residuals of the first regression instead of the $\widehat{u}_{1,t}(n)$ which are the residuals from the long VAR. However, we have chosen to use the latter in the preliminary estimation procedure.

In the foregoing, we have so far ignored the problem of choosing presample values for the estimation. Two alternative choices are reasonable. Either all presample values are replaced by zero or some y_t values at the beginning of the sample are set aside as presample values and the presample values for the residuals are replaced by zero.

The initial estimators obtained in the foregoing procedure can be shown to be consistent under general conditions if n goes to infinity with the sample size (see Hannan & Kavalieris (1984), Hannan & Deistler (1988), Poskitt (1992)). We will discuss the situation, where VAR processes of increasing order are fitted to a potentially infinite order process, in Chapter 15 and therefore we do not give details here.

12.3.5 An Illustration

We illustrate the estimation procedure using the income (y_1) and consumption (y_2) data from File E1. As in previous chapters, we use first differences of logarithms of the data from 1960 to 1978. In this case, we subtract the sample mean at an initial stage and denote the mean-adjusted income and consumption variables by y_{1t} and y_{2t}, respectively. We assume a VARMA(2, 2) model in echelon form with Kronecker indices $\mathbf{p} = (p_1, p_2) = (0, 2)$ [ARMA$_E(0, 2)$],

$$\begin{bmatrix} y_{1,t} \\ y_{2,t} \end{bmatrix} = \begin{bmatrix} 0 & 0 \\ 0 & \alpha_{22,1} \end{bmatrix} \begin{bmatrix} y_{1,t-1} \\ y_{2,t-1} \end{bmatrix} + \begin{bmatrix} 0 & 0 \\ 0 & \alpha_{22,2} \end{bmatrix} \begin{bmatrix} y_{1,t-2} \\ y_{2,t-2} \end{bmatrix} + \begin{bmatrix} u_{1,t} \\ u_{2,t} \end{bmatrix}$$
$$+ \begin{bmatrix} 0 & 0 \\ m_{21,1} & m_{22,1} \end{bmatrix} \begin{bmatrix} u_{1,t-1} \\ u_{2,t-1} \end{bmatrix} + \begin{bmatrix} 0 & 0 \\ m_{21,2} & m_{22,2} \end{bmatrix} \begin{bmatrix} u_{1,t-2} \\ u_{2,t-2} \end{bmatrix}.$$
(12.3.27)

In the next chapter, it will become apparent why this model is chosen. It implies that the first variable (income) is white noise ($y_{1t} = u_{1t}$). Given the subset VAR models of Chapter 5 (Table 5.1), this specification does not appear to be totally unreasonable. The second equation in (12.3.27) describes consumption as a function of lagged consumption, lagged income ($u_{1,t-i} = y_{1,t-i}$), and a moving average term involving lagged residuals $u_{2,t}$.

Eventually we use a sample from 1960.2 ($t = 1$) to 1978.4 ($t = 75$), that is, $T = 75$. In the preliminary estimation of the model (12.3.27), we estimate a VAR(8) model first, using 8 presample values. Then, using two more presample values, we run a regression of y_{2t} on its own lags and lagged $\widehat{u}_{it}(8)$. More precisely, the regression model is

$$\begin{bmatrix} y_{2,11} \\ \vdots \\ y_{2,T} \end{bmatrix}$$
$$= \begin{bmatrix} y_{2,10} & y_{2,9} & \widehat{u}_{1,10}(8) & \widehat{u}_{2,10}(8) & \widehat{u}_{1,9}(8) & \widehat{u}_{2,9}(8) \\ \vdots & \vdots & \vdots & \vdots & \vdots & \vdots \\ y_{2,T-1} & y_{2,T-2} & \widehat{u}_{1,T-1}(8) & \widehat{u}_{2,T-1}(8) & \widehat{u}_{1,T-2}(8) & \widehat{u}_{2,T-2}(8) \end{bmatrix} \gamma$$
$$+ \begin{bmatrix} u_{2,11} \\ \vdots \\ u_{2,T} \end{bmatrix},$$

where $\gamma := (\alpha_{22,1}, \alpha_{22,2}, m_{21,1}, m_{22,1}, m_{21,2}, m_{22,2})'$. In this particular case, we could have substituted y_{1t} for $\widehat{u}_{1t}(8)$ because the model implies $y_{1t} = u_{1t}$.

We have not done so, however, but we have used the residuals from the long autoregression. The resulting preliminary parameter estimates

$$\widetilde{\gamma}_1 = (\widetilde{\alpha}_{22,1}(1), \ldots, \widetilde{m}_{22,2}(1))'$$

are given in Table 12.1.

Table 12.1. Iterative estimates of the income/consumption system

	$\widetilde{\gamma}_i$						
i	$\widetilde{\alpha}_{22,1}$	$\widetilde{\alpha}_{22,2}$	$\widetilde{m}_{21,1}$	$\widetilde{m}_{22,1}$	$\widetilde{m}_{21,2}$	$\widetilde{m}_{22,2}$	$|\widetilde{\Sigma}_u(\widetilde{\gamma}_i)| \times 10^8$
1	0.020	0.395	0.296	-0.367	0.181	-0.224	0.872564
2	-0.178	0.492	0.331	-0.527	0.175	-0.015	0.942791
3	0.072	0.117	0.305	-0.589	0.191	0.065	0.779788
4	0.202	0.078	0.311	-0.731	0.146	0.147	0.776107
5	0.219	0.063	0.312	-0.744	0.142	0.158	0.775959
6	0.224	0.062	0.313	-0.748	0.140	0.159	0.775952
$\vdots$							
10	0.225	0.061	0.313	-0.750	0.140	0.160	0.775951

We use these estimates to start the scoring algorithm. For our particular example, the i-th iteration proceeds as follows:

(1) Compute residuals

$$\widetilde{u}_t(i) = y_t - \widetilde{A}_1(i)y_{t-1} - \widetilde{A}_2(i)y_{t-2} - \widetilde{M}_1(i)\widetilde{u}_{t-1}(i) - \widetilde{M}_2(i)\widetilde{u}_{t-2}(i)$$

recursively, for $t = 1, 2, \ldots, T$, with $\widetilde{u}_{-1}(i) = \widetilde{u}_0(i) = y_{-1} = y_0 = 0$ and

$$\widetilde{A}_1(i) = \begin{bmatrix} 0 & 0 \\ 0 & \widetilde{\alpha}_{22,1}(i) \end{bmatrix}, \quad \widetilde{A}_2(i) = \begin{bmatrix} 0 & 0 \\ 0 & \widetilde{\alpha}_{22,2}(i) \end{bmatrix},$$

$$\widetilde{M}_1(i) = \begin{bmatrix} 0 & 0 \\ \widetilde{m}_{21,1}(i) & \widetilde{m}_{22,1}(i) \end{bmatrix}, \quad \widetilde{M}_2(i) = \begin{bmatrix} 0 & 0 \\ \widetilde{m}_{21,2}(i) & \widetilde{m}_{22,2}(i) \end{bmatrix}.$$

(2) Compute the partial derivatives $\widetilde{\partial} u_t / \partial \gamma$ recursively as

$$\frac{\widetilde{\partial} u_t}{\partial \gamma'}(i) = -\begin{bmatrix} 0 & 0 & 0 & 0 & 0 & 0 \\ y_{2,t-1} & y_{2,t-2} & \widetilde{u}_{2,t-1}(i) & \widetilde{u}_{2,t-1}(i) & \widetilde{u}_{1,t-2}(i) & \widetilde{u}_{2,t-2}(i) \end{bmatrix}$$
$$- \widetilde{M}_1(i)\frac{\widetilde{\partial} u_{t-1}}{\partial \gamma'}(i) - \widetilde{M}_2(i)\frac{\widetilde{\partial} u_{t-2}}{\partial \gamma'}(i)$$

for $t = 1, 2, \ldots, T$, with

$$\frac{\widetilde{\partial} u_{-1}}{\partial \gamma'}(i) = \frac{\widetilde{\partial} u_0}{\partial \gamma'}(i) = 0.$$

(3) Compute
$$\widetilde{\Sigma}_u(\widetilde{\gamma}_i) = \frac{1}{T}\sum_{t=1}^{T}\widetilde{u}_t(i)\widetilde{u}_t(i)',$$

$$\widehat{\mathcal{I}}(\widetilde{\gamma}_i) = \sum_{t=1}^{T}\frac{\partial \widetilde{u}_t'}{\partial \gamma}(i)\widetilde{\Sigma}_u(\widetilde{\gamma}_i)^{-1}\frac{\partial \widetilde{u}_t}{\partial \gamma'}(i),$$

and
$$\left.\frac{\partial(-\ln l_0)}{\partial \gamma'}\right|_{\widetilde{\gamma}_i} = \sum_{t=1}^{T}\widetilde{u}_t(i)'\widetilde{\Sigma}_u(\widetilde{\gamma}_i)^{-1}\frac{\partial \widetilde{u}_t}{\partial \gamma'}(i).$$

(4) Perform the iteration step
$$\widetilde{\gamma}_{i+1} = \widetilde{\gamma}_i - \widehat{\mathcal{I}}(\widetilde{\gamma}_i)^{-1}\left[\left.\frac{\partial(-\ln l_0)}{\partial \gamma}\right|_{\widetilde{\gamma}_i}\right].$$

Some estimates obtained in these iterations are also given in Table 12.1 together with $|\widetilde{\Sigma}_u(\widetilde{\gamma}_i)|$. After a few iterations the latter quantity approximately reaches its minimum and, thus, $-\ln l_0$ obtains its minimum. After the tenth iteration there is not much change in the $\widetilde{\gamma}_i$ and $|\widetilde{\Sigma}_u(\widetilde{\gamma}_i)|$ in further steps. We work with $\widetilde{\gamma}_{10}$ in the following.

The determinantal polynomial of the MA operator for $i=10$ is
$$|I_2 + \widetilde{M}_1(10)z + \widetilde{M}_2(10)z^2| = 1 + \widetilde{m}_{22,1}(10)z + \widetilde{m}_{22,2}(10)z^2$$
$$= 1 - .750z + .160z^2$$

which has roots that are clearly outside the unit circle. Thus, the estimated MA operator is invertible. Also, the determinant of the estimated VAR polynomial,
$$|I_2 - \widetilde{A}_1(10)z - \widetilde{A}_2(10)z^2| = 1 - \widetilde{\alpha}_{22,1}(10)z - \widetilde{\alpha}_{22,2}(10)z^2$$
$$= 1 - .225z - .061z^2,$$

is easily seen to have its roots outside the unit circle. Hence, the estimated VARMA process is stable and invertible.

Generally, computing the ML estimates is not always easy. Therefore, other estimation methods were also proposed in the literature (e.g., Koreisha & Pukkila (1987), van Overschee & DeMoor (1994), Kapetanios (2003)).

12.4 Asymptotic Properties of the ML Estimators

12.4.1 Theoretical Results

In this section, the asymptotic properties of the ML estimators are given. We will not prove the main result but refer the reader to Hannan (1979), Dunsmuir & Hannan (1976), Hannan & Deistler (1988), and Kohn (1979) for further discussions and proofs.

Proposition 12.1 (*Asymptotic Properties of ML Estimators*)
Let y_t be a K-dimensional, stationary Gaussian process with stable and invertible VARMA(p, q) representation

$$A_0(y_t-\mu) = A_1(y_{t-1}-\mu)+\cdots+A_p(y_{t-p}-\mu)+A_0 u_t+M_1 u_{t-1}+\cdots+M_q u_{t-q}, \tag{12.4.1}$$

where u_t is Gaussian white noise with nonsingular covariance matrix Σ_u. Suppose the VAR and MA operators are left-coprime and either in final equations form or in echelon form with possibly linear restrictions on the coefficients so that the coefficient matrices $A_0, A_1, \ldots, A_p, M_1, \ldots, M_q$ depend on a set of unrestricted parameters γ as in (12.2.21). Let $\widetilde{\mu}, \widetilde{\gamma}$, and $\widetilde{\Sigma}_u$ be the ML estimators of μ, γ, and Σ_u, respectively, and denote vech(Σ_u) and vech$(\widetilde{\Sigma}_u)$ by σ and $\widetilde{\sigma}$, respectively. Then all three ML estimators are consistent and asymptotically normally distributed,

$$\sqrt{T}\begin{bmatrix} \widetilde{\mu}-\mu \\ \widetilde{\gamma}-\gamma \\ \widetilde{\sigma}-\sigma \end{bmatrix} \xrightarrow{d} \mathcal{N}\left(0, \begin{bmatrix} \Sigma_{\widetilde{\mu}} & 0 & 0 \\ 0 & \Sigma_{\widetilde{\gamma}} & 0 \\ 0 & 0 & \Sigma_{\widetilde{\sigma}} \end{bmatrix}\right), \tag{12.4.2}$$

where

$$\Sigma_{\widetilde{\mu}} = A(1)^{-1}M(1)\Sigma_u M(1)'A(1)'^{-1},$$

$$\Sigma_{\widetilde{\gamma}} = \mathcal{I}_a(\gamma)^{-1} = \text{plim}\left[\frac{1}{T}\sum_{t=1}^{T}\frac{\partial u_t'}{\partial \gamma}\Sigma_u^{-1}\frac{\partial u_t}{\partial \gamma'}\right]^{-1}$$

with $\partial u_t/\partial \gamma'$ as given in Lemma 12.1, and

$$\Sigma_{\widetilde{\sigma}} = 2\mathbf{D}_K^+(\Sigma_u \otimes \Sigma_u)\mathbf{D}_K^{+\prime}$$

with $\mathbf{D}_K^+ = (\mathbf{D}_K'\mathbf{D}_K)^{-1}\mathbf{D}_K'$ and $\mathbf{D}_K$ is the $(K^2 \times \frac{1}{2}K(K+1))$ duplication matrix. The covariance matrix in (12.4.2) is consistently estimated by replacing the unknown quantities by their ML estimators. ■

Some remarks on this proposition may be worthwhile.

Remark 1 The results of the proposition do not change if the ML estimator $\widetilde{\mu}$ is replaced by the sample mean $\bar{y}$ and $\widetilde{\gamma}$ and $\widetilde{\sigma}$ are ML estimators conditional on $\bar{y}$, that is, $\widetilde{\gamma}$ and $\widetilde{\sigma}$ are obtained by replacing μ by $\bar{y}$ in the ML algorithm. One consequence of this result is that asymptotically the sample mean is a fully efficient estimator of μ. ■

Remark 2 The proposition is formulated for final equations or echelon form VARMA models. Its statement remains true for other uniquely identified structures. ■

Remark 3 Because the covariance matrix of the asymptotic distribution in (12.4.2) is block-diagonal, the estimators of μ, γ, and Σ_u are asymptotically independent. ■

Remark 4 Much of the proposition remains valid even if y_t is not normally distributed. In that case the estimators obtained by maximizing the Gaussian likelihood function are quasi ML estimators. If u_t is independent standard white noise (see Chapter 3, Definition 3.1), $\widetilde{\gamma}$ and $\bar{y}$ maintain their asymptotic properties. The covariance matrix of $\widetilde{\sigma}$ may be different from the one given in Proposition 12.1. ■

Remark 5 The results of the proposition remain valid under general conditions if instead of the ML estimator $\widetilde{\gamma}$ an estimator is used which is obtained from one iteration of the scoring algorithm outlined in Section 12.3.2, starting from the preliminary estimator of Section 12.3.4. Thus, one possible approach to estimating the parameters of a VARMA model is to compute the sample mean $\bar{y}$ first and use that as an estimator of μ. Then the preliminary estimator for γ may be computed as described in Section 12.3.4 and that estimator is used as the initial vector in the optimization algorithm of Section 12.3.2. Then just one step of the form (12.3.16) is performed with $s_i = s_1 = 1$. The resulting estimators $\widetilde{\gamma}_2$ and $\widetilde{\Sigma}_u(\bar{y}, \widetilde{\gamma}_2)$ may then be used instead of $\widetilde{\gamma}$ and $\widetilde{\Sigma}_u$ in Proposition 12.1. Under general conditions, they have the same *asymptotic* distributions as the actual ML estimators. Of course, this possibility is a computationally attractive way to estimate the coefficients of a VARMA model. In general, the small sample properties of the resulting estimators are not the same as those of the ML estimators, however. ■

Remark 6 Because often the final equations form involves more parameters than the echelon form, unrestricted estimation of the former may result in inefficient estimators. Intuitively, if we start from the echelon form and determine the corresponding final equations form, the coefficients of the latter are seen to satisfy restrictions that could be imposed to obtain more efficient estimators. ■

In the following sections, we will occasionally be interested in the asymptotic distribution of the coefficients of the standard representation of the process,

$$(y_t - \mu) = A_1(y_{t-1} - \mu) + \cdots + A_p(y_{t-p} - \mu) + u_t + M_1 u_{t-1} + \cdots + M_q u_{t-q}. \tag{12.4.3}$$

The coefficients are functions of γ and their asymptotic distributions follow in the usual way. Let

$$\boldsymbol{\alpha} := \text{vec}[A_1, \ldots, A_p] \quad \text{and} \quad \mathbf{m} := \text{vec}[M_1, \ldots, M_q],$$

then

$$\begin{bmatrix} \boldsymbol{\alpha} \\ \mathbf{m} \end{bmatrix} = \begin{bmatrix} \boldsymbol{\alpha}(\gamma) \\ \mathbf{m}(\gamma) \end{bmatrix}.$$

The ML estimators are

$$\begin{bmatrix} \tilde{\boldsymbol{\alpha}} \\ \tilde{\mathbf{m}} \end{bmatrix} = \begin{bmatrix} \boldsymbol{\alpha}(\tilde{\boldsymbol{\gamma}}) \\ \mathbf{m}(\tilde{\boldsymbol{\gamma}}) \end{bmatrix}.$$

They are consistent and asymptotically normal,

$$\sqrt{T}\left(\begin{bmatrix} \tilde{\boldsymbol{\alpha}} \\ \tilde{\mathbf{m}} \end{bmatrix} - \begin{bmatrix} \boldsymbol{\alpha} \\ \mathbf{m} \end{bmatrix}\right) \xrightarrow{d} \mathcal{N}\left(0, \Sigma_{[\tilde{\boldsymbol{\alpha}}]} = \begin{bmatrix} \frac{\partial \boldsymbol{\alpha}}{\partial \boldsymbol{\gamma}'} \\ \frac{\partial \mathbf{m}}{\partial \boldsymbol{\gamma}'} \end{bmatrix} \Sigma_{\tilde{\boldsymbol{\gamma}}} \begin{bmatrix} \frac{\partial \boldsymbol{\alpha}'}{\partial \boldsymbol{\gamma}}, \frac{\partial \mathbf{m}'}{\partial \boldsymbol{\gamma}} \end{bmatrix}\right). \tag{12.4.4}$$

If $A_0 = I_K$, $\boldsymbol{\alpha}$ and $\mathbf{m}$ will often be linearly related to $\boldsymbol{\gamma}$ and we get the following corollary of Proposition 12.1.

Corollary 12.1.1

Under the conditions of Proposition 12.1, if

$$\begin{bmatrix} \boldsymbol{\alpha} \\ \mathbf{m} \end{bmatrix} = R\boldsymbol{\gamma} + r,$$

$$\sqrt{T}\left(\begin{bmatrix} \tilde{\boldsymbol{\alpha}} \\ \tilde{\mathbf{m}} \end{bmatrix} - \begin{bmatrix} \boldsymbol{\alpha} \\ \mathbf{m} \end{bmatrix}\right) \xrightarrow{d} \mathcal{N}(0, R\Sigma_{\tilde{\boldsymbol{\gamma}}}R')$$

and $\tilde{\boldsymbol{\alpha}}$ and $\tilde{\mathbf{m}}$ are asymptotically independent of $\bar{y}$, $\tilde{\mu}$, and $\tilde{\sigma}$. ∎

The remarks following the proposition also apply for the corollary. For illustrative purposes, consider the bivariate $\text{ARMA}_E(0,1)$ model,

$$\begin{bmatrix} 1 & 0 \\ 0 & 1 - \alpha_{22,1}L \end{bmatrix} y_t = \begin{bmatrix} 1 & 0 \\ m_{21,1}L & 1 + m_{22,1}L \end{bmatrix} u_t \tag{12.4.5}$$

or

$$y_t = \begin{bmatrix} 0 & 0 \\ 0 & \alpha_{22,1} \end{bmatrix} y_{t-1} + u_t + \begin{bmatrix} 0 & 0 \\ m_{21,1} & m_{22,1} \end{bmatrix} u_{t-1}.$$

In this case,

$$\boldsymbol{\alpha} = \begin{bmatrix} 0 \\ 0 \\ 0 \\ \alpha_{22,1} \end{bmatrix}, \quad \mathbf{m} = \begin{bmatrix} 0 \\ m_{21,1} \\ 0 \\ m_{22,1} \end{bmatrix}, \quad \boldsymbol{\gamma} = \begin{bmatrix} \alpha_{22,1} \\ m_{21,1} \\ m_{22,1} \end{bmatrix},$$

$$R = \begin{bmatrix} 0 & 0 & 0 \\ 0 & 0 & 0 \\ 0 & 0 & 0 \\ 1 & 0 & 0 \\ 0 & 0 & 0 \\ 0 & 1 & 0 \\ 0 & 0 & 0 \\ 0 & 0 & 1 \end{bmatrix}, \quad \text{and} \quad r = \begin{bmatrix} 0 \\ 0 \\ 0 \\ 0 \\ 0 \\ 0 \\ 0 \\ 0 \end{bmatrix}.$$

12.4 Asymptotic Properties of the ML Estimators

If the VARMA model is not in standard form originally, we premultiply by A_0^{-1} to get

$$\begin{aligned} y_t - \mu &= A_0^{-1}A_1(y_{t-1} - \mu) + \cdots + A_0^{-1}A_p(y_{t-p} - \mu) \\ &\quad + u_t + A_0^{-1}M_1 u_{t-1} + \cdots + A_0^{-1}M_q u_{t-q}. \end{aligned} \quad (12.4.6)$$

In this case, it is more reasonable to assume that

$$\boldsymbol{\beta}_0 := \text{vec}[A_0, A_1, \ldots, A_p, M_1, \ldots, M_q] \quad (12.4.7)$$

is linearly related to $\boldsymbol{\gamma}$, say,

$$\boldsymbol{\beta}_0 = R\boldsymbol{\gamma} + r. \quad (12.4.8)$$

Then it follows for

$$\boldsymbol{\alpha} := \text{vec}[A_0^{-1}A_1, \ldots, A_0^{-1}A_p] = \text{vec}(A_0^{-1}[A_1, \ldots, A_p]) \quad (12.4.9)$$

and

$$\mathbf{m} := \text{vec}(A_0^{-1}[M_1, \ldots, M_q]), \quad (12.4.10)$$

that

$$\begin{bmatrix} \dfrac{\partial \boldsymbol{\alpha}}{\partial \boldsymbol{\gamma}'} \\ \dfrac{\partial \mathbf{m}}{\partial \boldsymbol{\gamma}'} \end{bmatrix} = \begin{bmatrix} \dfrac{\partial \boldsymbol{\alpha}}{\partial \boldsymbol{\beta}_0'} \\ \dfrac{\partial \mathbf{m}}{\partial \boldsymbol{\beta}_0'} \end{bmatrix} \dfrac{\partial \boldsymbol{\beta}_0}{\partial \boldsymbol{\gamma}'} = \begin{bmatrix} \dfrac{\partial \boldsymbol{\alpha}}{\partial \boldsymbol{\beta}_0'} \\ \dfrac{\partial \mathbf{m}}{\partial \boldsymbol{\beta}_0'} \end{bmatrix} R.$$

Hence, we need to evaluate $\partial\boldsymbol{\alpha}/\partial\boldsymbol{\beta}_0'$ and $\partial\mathbf{m}/\partial\boldsymbol{\beta}_0'$ to obtain the asymptotic covariance matrix of the standard form coefficients.

$$\begin{aligned} \dfrac{\partial \boldsymbol{\alpha}}{\partial \boldsymbol{\beta}_0'} &= (I_{Kp} \otimes A_0^{-1}) \dfrac{\partial \text{vec}[A_1, \ldots, A_p]}{\partial \boldsymbol{\beta}_0'} + \left(\begin{bmatrix} A_1' \\ \vdots \\ A_p' \end{bmatrix} \otimes I_K \right) \dfrac{\partial \text{vec}(A_0^{-1})}{\partial \boldsymbol{\beta}_0'} \\ &= (I_{Kp} \otimes A_0^{-1})[0 : I_{K^2 p} : 0] \\ &\quad - \left(\begin{bmatrix} A_1' \\ \vdots \\ A_p' \end{bmatrix} \otimes I_K \right) ((A_0^{-1})' \otimes A_0^{-1}) \dfrac{\partial \text{vec}(A_0)}{\partial \boldsymbol{\beta}_0'} \end{aligned}$$

(see Rule 9 of Appendix A.13)

$$\begin{aligned} &= (I_{Kp} \otimes A_0^{-1})[0 : I_{K^2 p} : 0] \\ &\quad - \left(\begin{bmatrix} (A_0^{-1}A_1)' \\ \vdots \\ (A_0^{-1}A_p)' \end{bmatrix} \otimes A_0^{-1} \right) [I_{K^2} : 0]. \end{aligned} \quad (12.4.11)$$

A similar expression is obtained for $\partial\mathbf{m}/\partial\boldsymbol{\beta}_0'$. This result is summarized in the next corollary.

Corollary 12.1.2

Under the conditions of Proposition 12.1, if $\boldsymbol{\beta}_0$ is as defined in (12.4.7) and satisfies the restrictions in (12.4.8) and $\boldsymbol{\alpha}$ and $\mathbf{m}$ are the coefficients of the standard form VARMA representation defined in (12.4.9) and (12.4.10), respectively, with ML estimators $\tilde{\boldsymbol{\alpha}}$ and $\tilde{\mathbf{m}}$, then

$$\sqrt{T}\left(\begin{bmatrix} \tilde{\boldsymbol{\alpha}} \\ \tilde{\mathbf{m}} \end{bmatrix} - \begin{bmatrix} \boldsymbol{\alpha} \\ \mathbf{m} \end{bmatrix}\right) \xrightarrow{d} \mathcal{N}\left(0, \Sigma_{\begin{bmatrix} \tilde{\boldsymbol{\alpha}} \\ \tilde{\mathbf{m}} \end{bmatrix}} = \begin{bmatrix} H_{\boldsymbol{\alpha}} \\ H_{\mathbf{m}} \end{bmatrix} R \Sigma_{\tilde{\gamma}} R' [H'_{\boldsymbol{\alpha}} : H'_{\mathbf{m}}]\right),$$

where

$$H_{\boldsymbol{\alpha}} := \frac{\partial \boldsymbol{\alpha}}{\partial \boldsymbol{\beta}'_0} \quad (K^2 p \times K^2(p+q+1))$$

$$= (I_{Kp} \otimes A_0^{-1})[\underbrace{0}_{(K^2 p \times K^2)} : I_{K^2 p} : \underbrace{0}_{(K^2 p \times K^2 q)}]$$

$$- \left(\begin{bmatrix} (A_0^{-1} A_1)' \\ \vdots \\ (A_0^{-1} A_p)' \end{bmatrix} \otimes A_0^{-1}\right) [I_{K^2} : 0]$$

and

$$H_{\mathbf{m}} := \frac{\partial \mathbf{m}}{\partial \boldsymbol{\beta}'_0} \quad (K^2 q \times K^2(p+q+1))$$

$$= (I_{Kq} \otimes A_0^{-1})[0 : I_{K^2 q}] - \left(\begin{bmatrix} (A_0^{-1} M_1)' \\ \vdots \\ (A_0^{-1} M_q)' \end{bmatrix} \otimes A_0^{-1}\right) [I_{K^2} : 0]. \quad \blacksquare$$

Again an example may be worthwhile. Consider the following bivariate $ARMA_E(2,1)$ process with some zero restrictions placed on the coefficients (see also Problem 12.3):

$$\begin{bmatrix} 1 - \alpha_{11,1} L - \alpha_{11,2} L^2 & 0 \\ -\alpha_{21,0} - \alpha_{21,1} L & 1 - \alpha_{22,1} L \end{bmatrix} y_t = \begin{bmatrix} 1 & 0 \\ -\alpha_{21,0} & 1 + m_{22,1} L \end{bmatrix} u_t \quad (12.4.12)$$

or

$$\begin{bmatrix} 1 & 0 \\ -\alpha_{21,0} & 1 \end{bmatrix} y_t = \begin{bmatrix} \alpha_{11,1} & 0 \\ \alpha_{21,1} & \alpha_{22,1} \end{bmatrix} y_{t-1} + \begin{bmatrix} \alpha_{11,2} & 0 \\ 0 & 0 \end{bmatrix} y_{t-2}$$

$$+ \begin{bmatrix} 1 & 0 \\ -\alpha_{21,0} & 1 \end{bmatrix} u_t + \begin{bmatrix} 0 & 0 \\ 0 & m_{22,1} \end{bmatrix} u_{t-1}.$$

Hence,

$$\boldsymbol{\beta}_0 = \begin{bmatrix} 1 \\ -\alpha_{21,0} \\ 0 \\ 1 \\ \alpha_{11,1} \\ \alpha_{21,1} \\ 0 \\ \alpha_{22,1} \\ \alpha_{11,2} \\ 0 \\ 0 \\ 0 \\ 0 \\ 0 \\ 0 \\ m_{22,1} \end{bmatrix}, \quad R = \begin{bmatrix} 0 & 0 & 0 & 0 & 0 & 0 \\ -1 & 0 & 0 & 0 & 0 & 0 \\ 0 & 0 & 0 & 0 & 0 & 0 \\ 0 & 0 & 0 & 0 & 0 & 0 \\ 0 & 1 & 0 & 0 & 0 & 0 \\ 0 & 0 & 1 & 0 & 0 & 0 \\ 0 & 0 & 0 & 0 & 0 & 0 \\ 0 & 0 & 0 & 1 & 0 & 0 \\ 0 & 0 & 0 & 0 & 1 & 0 \\ 0 & 0 & 0 & 0 & 0 & 0 \\ 0 & 0 & 0 & 0 & 0 & 0 \\ 0 & 0 & 0 & 0 & 0 & 0 \\ 0 & 0 & 0 & 0 & 0 & 0 \\ 0 & 0 & 0 & 0 & 0 & 0 \\ 0 & 0 & 0 & 0 & 0 & 0 \\ 0 & 0 & 0 & 0 & 0 & 1 \end{bmatrix}, \quad r = \begin{bmatrix} 1 \\ 0 \\ 0 \\ 1 \\ 0 \\ 0 \\ 0 \\ 0 \\ 0 \\ 0 \\ 0 \\ 0 \\ 0 \\ 0 \\ 0 \\ 0 \end{bmatrix},$$

$$\boldsymbol{\gamma} = \begin{bmatrix} \alpha_{21,0} \\ \alpha_{11,1} \\ \alpha_{21,1} \\ \alpha_{22,1} \\ \alpha_{11,2} \\ m_{22,1} \end{bmatrix}.$$

Furthermore,

$$A_0^{-1} = \begin{bmatrix} 1 & 0 \\ -\alpha_{21,0} & 1 \end{bmatrix}^{-1} = \begin{bmatrix} 1 & 0 \\ \alpha_{21,0} & 1 \end{bmatrix}.$$

Thus,

$$\boldsymbol{\alpha} = \mathrm{vec}[A_0^{-1} A_1, A_0^{-1} A_2]$$

$$= \mathrm{vec} \begin{bmatrix} \alpha_{11,1} & 0 & \alpha_{11,2} & 0 \\ \alpha_{11,1}\alpha_{21,0} + \alpha_{21,1} & \alpha_{22,1} & \alpha_{11,2}\alpha_{21,0} & 0 \end{bmatrix}$$

$$= \begin{bmatrix} \alpha_{11,1} \\ \alpha_{11,1}\alpha_{21,0} + \alpha_{21,1} \\ 0 \\ \alpha_{22,1} \\ \alpha_{11,2} \\ \alpha_{11,2}\alpha_{21,0} \\ 0 \\ 0 \end{bmatrix}$$

and

$$\mathbf{m} = \text{vec}[A_0^{-1} M_1] = \text{vec}\begin{bmatrix} 0 & 0 \\ 0 & m_{22,1} \end{bmatrix} = \begin{bmatrix} 0 \\ 0 \\ 0 \\ m_{22,1} \end{bmatrix}.$$

Consequently,

$$\frac{\partial \boldsymbol{\alpha}}{\partial \boldsymbol{\gamma}'} = H_\alpha R = \begin{bmatrix} 0 & 1 & 0 & 0 & 0 & 0 \\ \alpha_{11,1} & \alpha_{21,0} & 1 & 0 & 0 & 0 \\ 0 & 0 & 0 & 0 & 0 & 0 \\ 0 & 0 & 0 & 1 & 0 & 0 \\ 0 & 0 & 0 & 0 & 1 & 0 \\ \alpha_{11,2} & 0 & 0 & 0 & \alpha_{21,0} & 0 \\ 0 & 0 & 0 & 0 & 0 & 0 \\ 0 & 0 & 0 & 0 & 0 & 0 \end{bmatrix} \tag{12.4.13}$$

and

$$\frac{\partial \mathbf{m}}{\partial \boldsymbol{\gamma}'} = H_\mathbf{m} R = \begin{bmatrix} 0 & 0 & 0 & 0 & 0 & 0 \\ 0 & 0 & 0 & 0 & 0 & 0 \\ 0 & 0 & 0 & 0 & 0 & 0 \\ 0 & 0 & 0 & 0 & 0 & 1 \end{bmatrix} \tag{12.4.14}$$

(see also Problem 12.7).

12.4.2 A Real Data Example

In their general form, the results may look more complicated than they usually are. Therefore, considering our income/consumption example from Section 12.3.5 again may be helpful. For the VARMA$(2,2)$ model with Kronecker indices $(0,2)$ given in (12.3.27), the parameters are

$$\boldsymbol{\gamma} = (\alpha_{22,1}, \alpha_{22,2}, m_{21,1}, m_{22,1}, m_{21,2}, m_{22,2})'.$$

The ML estimates are given in Table 12.1. Using $\widetilde{\boldsymbol{\gamma}} = \widetilde{\boldsymbol{\gamma}}_{10}$ from that table, an estimate of $\mathcal{I}(\boldsymbol{\gamma})$ is obtained from the iterations described in Section 12.3.5, that is, we use $\widehat{\mathcal{I}}(\widetilde{\boldsymbol{\gamma}}_{10}) = \widehat{\mathcal{I}}(\widetilde{\boldsymbol{\gamma}})$. The square roots of the diagonal elements of $\widehat{\mathcal{I}}(\widetilde{\boldsymbol{\gamma}})^{-1}$ are estimates of the standard errors of the elements of $\widetilde{\boldsymbol{\gamma}}$. Giving the estimated standard errors in parentheses, we get

$$\widetilde{\boldsymbol{\gamma}} = \begin{bmatrix} .225 & (.252) \\ .061 & (.166) \\ .313 & (.090) \\ -.750 & (.274) \\ .140 & (.141) \\ .160 & (.233) \end{bmatrix}. \tag{12.4.15}$$

As mentioned in Remark 5 of Section 12.4.1, an alternative, *asymptotically equivalent* estimator is obtained by iterating just once. In the present example that leads to estimates

$$\widetilde{\gamma}_2 = \begin{bmatrix} -.178 & (.165) \\ .492 & (.133) \\ .331 & (.099) \\ -.527 & (.172) \\ .175 & (.127) \\ -.015 & (.152) \end{bmatrix}. \tag{12.4.16}$$

These estimates are somewhat different from those in (12.4.15). However, given the sampling variability reflected in the estimated standard errors, the differences in most of the parameter estimates are not substantial.

Under a two-standard error criterion, only two of the coefficients in (12.4.15) are significantly different from zero. As a consequence, one may wish to restrict some of the coefficients to zero and thereby further reduce the parameter space. We will not do so at this stage but consider the estimates of $\boldsymbol{\alpha}$ and $\mathbf{m}$ implied by $\widetilde{\gamma}$ given in (12.4.15) (see, however, Problem 12.10):

$$\widetilde{\boldsymbol{\alpha}} = \text{vec}[\widetilde{A}_1, \widetilde{A}_2] = \begin{bmatrix} 0 \\ 0 \\ 0 \\ .225(.252) \\ 0 \\ 0 \\ 0 \\ .061(.166) \end{bmatrix},$$

$$\widetilde{\mathbf{m}} = \text{vec}[\widetilde{M}_1, \widetilde{M}_2] = \begin{bmatrix} 0 \\ .313(.090) \\ 0 \\ -.750(.274) \\ 0 \\ .140(.141) \\ 0 \\ .160(.233) \end{bmatrix}. \tag{12.4.17}$$

The standard errors are, of course, not affected by adding a few zero elements. A more elaborate but still simple computation becomes necessary to obtain the standard errors if $A_0 \neq I_K$ (see Corollary 12.1.2).

12.5 Forecasting Estimated VARMA Processes

With respect to forecasting with estimated processes, in principle, the same arguments apply for VARMA models that have been put forward for pure

VAR models. Suppose that the generation process of a multiple time series of interest admits a VARMA(p, q) representation,

$$y_t - \mu = A_1(y_{t-1} - \mu) + \cdots + A_p(y_{t-p} - \mu) + u_t + M_1 u_{t-1} + \cdots + M_q u_{t-q},$$
(12.5.1)

and denote by $\widehat{y}_t(h)$ the h-step ahead forecast (with nonzero mean) at origin t given in Section 11.5, based on estimated rather than known coefficients. For instance, using the pure VAR representation of the process,

$$\widehat{y}_t(h) = \widehat{\mu} + \sum_{i=1}^{h-1} \widehat{\Pi}_i(\widehat{y}_t(h-i) - \widehat{\mu}) + \sum_{i=h}^{\infty} \widehat{\Pi}_i(y_{t+h-i} - \widehat{\mu}).$$
(12.5.2)

For practical purposes, one would, of course, truncate the infinite sum. For the moment we will, however, consider the infinite sum. For this predictor, the forecast error is

$$y_{t+h} - \widehat{y}_t(h) = [y_{t+h} - y_t(h)] + [y_t(h) - \widehat{y}_t(h)],$$

where $y_t(h)$ is the optimal forecast based on known coefficients and the two terms on the right-hand side are uncorrelated as the first one can be written in terms of u_s with $s > t$ and the second one contains y_s with $s \leq t$, if the parameter estimators are based on y_s with $s \leq t$ only. Thus, the forecast MSE becomes

$$\begin{aligned} \Sigma_{\widehat{y}}(h) &= \text{MSE}[y_t(h)] + \text{MSE}[y_t(h) - \widehat{y}_t(h)] \\ &= \Sigma_y(h) + E[y_t(h) - \widehat{y}_t(h)][y_t(h) - \widehat{y}_t(h)]'. \end{aligned}$$
(12.5.3)

Formally, this is the same expression that was obtained for finite order VAR processes and, using the same arguments as in that case, we approximate the $\text{MSE}[y_t(h) - \widehat{y}_t(h)]$ by $\Omega(h)/T$, where

$$\Omega(h) = E\left[\frac{\partial y_t(h)}{\partial \eta'} \Sigma_{\widetilde{\eta}} \frac{\partial y_t(h)'}{\partial \eta}\right],$$
(12.5.4)

η is the vector of estimated coefficients, and $\Sigma_{\widetilde{\eta}}$ is its asymptotic covariance matrix. If ML estimation is used and

$$\eta = \begin{bmatrix} \mu \\ \alpha \\ \mathbf{m} \end{bmatrix},$$

where $\boldsymbol{\alpha} = \text{vec}[A_1, \ldots, A_p]$ and $\mathbf{m} = \text{vec}[M_1, \ldots, M_q]$, we have from Proposition 12.1 and Corollaries 12.1.1 and 12.1.2,

$$\Sigma_{\widetilde{\eta}} = \begin{bmatrix} \Sigma_{\widetilde{\mu}} & 0 \\ 0 & \Sigma_{\left[\widetilde{\overset{\alpha}{\mathbf{m}}}\right]} \end{bmatrix}.$$

Thus,

$$\frac{\partial y_t(h)}{\partial \boldsymbol{\eta}'} \Sigma_{\tilde{\boldsymbol{\eta}}} \frac{\partial y_t(h)'}{\partial \boldsymbol{\eta}} = \frac{\partial y_t(h)}{\partial \mu'} \Sigma_{\tilde{\mu}} \frac{\partial y_t(h)'}{\partial \mu} + \frac{\partial y_t(h)}{\partial [\boldsymbol{\alpha}', \mathbf{m}']} \Sigma_{[\tilde{\boldsymbol{\alpha}}]} \frac{\partial y_t(h)'}{\partial \begin{bmatrix} \boldsymbol{\alpha} \\ \mathbf{m} \end{bmatrix}}.$$

Hence, in order to get an expression for $\Omega(h)$ we need the partial derivatives of $y_t(h)$ with respect to $\mu, \boldsymbol{\alpha}$, and $\mathbf{m}$. They are given in the next lemma.

Lemma 12.2

If y_t is a process with stable and invertible VARMA(p,q) representation (12.5.1) and pure VAR representation

$$y_t = \mu + \sum_{i=1}^{\infty} \Pi_i (y_{t-i} - \mu) + u_t,$$

we have

$$\frac{\partial y_t(h)}{\partial \mu'} = \begin{cases} \left(I_K - \sum_{i=1}^{\infty} \Pi_i \right), & \text{for } h = 1, \\ \left(I_K - \sum_{i=1}^{\infty} \Pi_i \right) + \sum_{i=1}^{h-1} \Pi_i \frac{\partial y_t(h-i)}{\partial \mu'}, & h = 2, 3, \ldots, \end{cases}$$

$$\frac{\partial y_t(h)}{\partial [\boldsymbol{\alpha}', \mathbf{m}']} = \sum_{i=1}^{h-1} [(y_t(h-i) - \mu)' \otimes I_K] \frac{\partial \text{vec}(\Pi_i)}{\partial [\boldsymbol{\alpha}', \mathbf{m}']} + \sum_{i=1}^{h-1} \Pi_i \frac{\partial y_t(h-i)}{\partial [\boldsymbol{\alpha}', \mathbf{m}']}$$

$$+ \sum_{i=h}^{\infty} [(y_{t+h-i} - \mu)' \otimes I_K] \frac{\partial \text{vec}(\Pi_i)}{\partial [\boldsymbol{\alpha}', \mathbf{m}']}, \quad \text{for } h = 1, 2, \ldots,$$

with

$$\frac{\partial \text{vec}(\Pi_i)}{\partial [\boldsymbol{\alpha}', \mathbf{m}']} = -\sum_{j=0}^{i-1} [H'(\mathbf{M}')^{i-1-j} \otimes J\mathbf{M}^j] \begin{bmatrix} 0 & I_{Kq} \otimes J' \\ I_{Kp} \otimes J' & 0 \end{bmatrix},$$

where H, $\mathbf{M}$, and J are as defined in Chapter 11, Section 11.3.2, (11.3.13). In other words, H, $\mathbf{M}$, and J are defined so that $-\Pi_i = J\mathbf{M}^i H$. ∎

The proof of this lemma is left as an exercise (see Problem 12.8). The formulas given in this lemma can be used for recursively computing the partial derivatives of $y_t(h)$ with respect to the VARMA coefficients for $h = 1, 2, \ldots$.

An estimator of $\Omega(h)$ is obtained by replacing all unknown quantities by their respective estimators and truncating the infinite sum or, equivalently, replacing $y_t - \mu$ by zero for $t \leq 0$. Denoting the resulting estimated partial derivatives by

$$\frac{\widehat{\partial y_t(h)}}{\partial \mu'} \quad \text{and} \quad \frac{\widehat{\partial y_t(h)}}{\partial [\boldsymbol{\alpha}', \mathbf{m}']},$$

an estimator for $\Omega(h)$ is

$$\widehat{\Omega}(h) = \frac{1}{T} \sum_{t=1}^{T} \left[\frac{\widehat{\partial y_t(h)}}{\partial \mu'} \widehat{\Sigma}_{\widetilde{\mu}} \frac{\widehat{\partial y_t(h)}'}{\partial \mu} + \frac{\widehat{\partial y_t(h)}}{\partial [\alpha', \mathbf{m}']} \widehat{\Sigma}_{[\frac{\widetilde{\alpha}}{\mathbf{m}}]} \frac{\widehat{\partial y_t(h)}'}{\partial \begin{bmatrix} \alpha \\ \mathbf{m} \end{bmatrix}} \right], \qquad (12.5.5)$$

where $\widehat{\Sigma}_{\widetilde{\mu}}$ and

$$\widehat{\Sigma}_{[\frac{\widetilde{\alpha}}{\mathbf{m}}]}$$

are estimators of $\Sigma_{\widetilde{\mu}}$ and

$$\Sigma_{[\frac{\widetilde{\alpha}}{\mathbf{m}}]},$$

respectively (see Corollaries 12.1.1 and 12.1.2 for the latter matrix). An estimator of the forecast MSE matrix (12.5.3) is then

$$\widehat{\Sigma}_{\widehat{y}}(h) = \widehat{\Sigma}_y(h) + \frac{1}{T}\widehat{\Omega}(h), \qquad (12.5.6)$$

where the estimator $\widehat{\Sigma}_y(h)$ is again obtained by replacing unknown quantities by their respective estimators.

With these results in hand, forecast intervals can be set up, under Gaussian assumptions, just as in the finite order VAR case discussed in Chapters 2 and 3.

12.6 Estimated Impulse Responses

As mentioned in Section 11.7.2, the impulse responses of a VARMA(p,q) process are the coefficients of pure MA representations. For instance, if the process is in standard form, the forecast error impulse responses are

$$\Phi_i = J\mathbf{A}^i H \qquad (12.6.1)$$

with J, $\mathbf{A}$, and H as defined in Section 11.3.2 (see (11.3.10)). Other quantities of interest may be the elements of $\Theta_i = \Phi_i P$, where P is a lower triangular Choleski decomposition of Σ_u, the white noise covariance matrix. Also forecast error variance components and accumulated impulse responses may be of interest. All these quantities are estimated in the usual way from the estimated coefficients of the process. For example, $\widehat{\Phi}_i = J\widehat{\mathbf{A}}^i H$, where $\widehat{\mathbf{A}}$ is obtained from $\mathbf{A}$ by replacing the A_i and M_j by estimators $\widehat{A}_i$ and $\widehat{M}_j$. The asymptotic distributions of the estimated quantities follow immediately from Proposition 3.6, which is formulated for the finite order VAR case. The only modifications that we have to make to accommodate the VARMA(p,q) case are to replace α by

$$\boldsymbol{\beta} := \text{vec}[A_1, \ldots, A_p, M_1, \ldots, M_q] = \begin{bmatrix} \boldsymbol{\alpha} \\ \mathbf{m} \end{bmatrix},$$

replace $\Sigma_{\tilde{\alpha}}$ by $\Sigma_{\tilde{\beta}}$ and specify

$$\begin{aligned} G_i &= \frac{\partial \text{vec}(\Phi_i)}{\partial \boldsymbol{\beta}'} = (H' \otimes J) \frac{\partial \text{vec}(\mathbf{A}^i)}{\partial \boldsymbol{\beta}'} \\ &= (H' \otimes J) \left[\sum_{m=0}^{i-1} (\mathbf{A}')^{i-1-m} \otimes \mathbf{A}^m \right] \frac{\partial \text{vec}(\mathbf{A})}{\partial \boldsymbol{\beta}'} \\ &= \sum_{m=0}^{i-1} H'(\mathbf{A}')^{i-1-m} \otimes J\mathbf{A}^m J'. \end{aligned} \qquad (12.6.2)$$

With these modifications of Proposition 3.6, the asymptotic distributions of all the quantities of interest are available. Of course, all the caveats of Proposition 3.6 apply here too. In principle, structural impulse responses, as discussed in Chapter 9, may be of interest as well. They are typically not based on VARMA models, however.

12.7 Exercises

Problem 12.1
Are the operators

$$\begin{bmatrix} 1 - 0.5L & 0.3L \\ 0 & 1 \end{bmatrix} \text{ and } \begin{bmatrix} 1 - 0.2L & 1.3L - 0.44L^2 \\ 0.5L & 1 + 0.2L \end{bmatrix}$$

left-coprime? (Hint: Show that the first operator is a common factor.)

Problem 12.2
Write the bivariate process

$$\begin{bmatrix} 1 - \beta_1 L & 0 \\ \beta_2 L^2 & 1 - \beta_3 L \end{bmatrix} y_t = \begin{bmatrix} 1 - \beta_1 L & 0 \\ \beta_4 L & 1 \end{bmatrix} u_t$$

in final equations form and in echelon form.

Problem 12.3
Show that (12.4.12) is an echelon form representation.

Problem 12.4
Derive the likelihood function for a general Gaussian VARMA(p,q) model given fixed but not necessarily zero initial vectors $y_{-p+1}, \ldots, y_0$. Do not assume that $u_{-q+1} = \cdots = u_0 = 0$!

Problem 12.5
Identify the rules from Appendix A that are used in deriving the partial derivatives in (12.3.8).

Problem 12.6
Suppose that $\ln|\widetilde{\Sigma}_u(\mu,\gamma)|$ given in (12.3.11) is to be minimized with respect to γ. Show that the resulting normal equations are

$$\frac{\partial \ln|\widetilde{\Sigma}_u(\mu,\gamma)|}{\partial \gamma'} = \frac{2}{T}\sum_{t=1}^{T} u_t'\Sigma_u^{-1}\frac{\partial u_t}{\partial \gamma'}.$$

Thus, the normal equations are equivalent to those obtained from the log-likelihood function.

Problem 12.7
Consider the bivariate VARMA(2, 1) process given in (12.4.12) and set up the matrices H_α and H_m according to their general form given in Corollary 12.1.2. Show that $H_\alpha R$ and $H_m R$ are identical to the matrices specified in (12.4.13) and (12.4.14), respectively.

Problem 12.8
Prove Lemma 12.2. (Hint: Use Rule (8) of Appendix A.13.)

Problem 12.9
Derive the asymptotic covariance matrices of the impulse responses and forecast error variance components obtained from an estimated VARMA process. (Hint: Use the suggestion given in Section 12.6.)

Problem 12.10
Consider the income/consumption example of Section 12.3.5 and determine preliminary and full ML estimates for the parameters of the model

$$\begin{bmatrix} y_{1,t} \\ y_{2,t} \end{bmatrix} = \begin{bmatrix} 0 & 0 \\ 0 & \alpha_{22,2} \end{bmatrix}\begin{bmatrix} y_{1,t-2} \\ y_{2,t-2} \end{bmatrix}$$
$$+ \begin{bmatrix} u_{1,t} \\ u_{2,t} \end{bmatrix} + \begin{bmatrix} 0 & 0 \\ m_{21,1} & m_{22,1} \end{bmatrix}\begin{bmatrix} u_{1,t-1} \\ u_{2,t-1} \end{bmatrix}$$
$$+ \begin{bmatrix} 0 & 0 \\ m_{21,2} & 0 \end{bmatrix}\begin{bmatrix} u_{1,t-2} \\ u_{2,t-2} \end{bmatrix}.$$

13

Specification and Checking the Adequacy of VARMA Models

13.1 Introduction

A great number of strategies has been suggested for specifying VARMA models. There is not a single one that has become a standard like the Box & Jenkins (1976) approach in the univariate case. None of the multivariate procedures is in widespread use for modelling moderate or high-dimensional economic time series. Some are mainly based on a subjective assessment of certain characteristics of a process such as the autocorrelations and partial autocorrelations. A decision on specific orders and constraints on the coefficient matrices is then based on these quantities. Other methods rely on a mixture of statistical testing, use of model selection criteria and personal judgement of the analyst. Again other procedures are based predominantly on statistical model selection criteria and, in principle, they could be performed automatically by a computer. Automatic procedures have the advantage that their statistical properties can possibly be derived rigorously. In actual applications, some kind of mixture of different approaches is often used. In other words, the expertise and prior knowledge of an analyst will usually not be abolished in favor of purely statistical procedures. Models suggested by different types of criteria and procedures will be judged and evaluated by an expert before one or more candidates are put to a specific use such as forecasting. The large amount of information in a number of moderately long time series makes it usually necessary to condense the information considerably before essential features of a system become visible.

In the following, we will outline procedures for specifying the final equations form and the echelon form of a VARMA process. We do not claim that these procedures are superior to other approaches. They are just meant to illustrate what is involved in the specification of VARMA models. The specification strategies for both forms could be turned into automatic algorithms. On the other hand, they also leave room for human intervention if desired. In Section 13.4, some references for other specification strategies are given and model checking is discussed briefly in Section 13.5. A critique of VARMA modelling is given in Section 13.6.

13.2 Specification of the Final Equations Form

13.2.1 A Specification Procedure

Historically, procedures for specifying final equations VARMA representations were among the earlier strategies for modelling systems of economic time series (see, for example, Zellner & Palm (1974), Wallis (1977)). The objective is to find the orders p and q of the representation

$$\alpha(L)y_t = M(L)u_t, \qquad (13.2.1)$$

where

$$\alpha(L) := 1 - \alpha_1 L - \cdots - \alpha_p L^p$$

is a (1×1) scalar operator,

$$M(L) := I_K + M_1 L + \cdots + M_q L^q$$

is a $(K \times K)$ matrix operator and it is assumed that the process mean has been removed in a previous step of the analysis.

If a K-dimensional system $y_t = (y_{1t}, \ldots, y_{Kt})'$ has a VARMA representation of the form (13.2.1), then it follows that each component series has a univariate ARMA representation

$$\alpha(L)y_{kt} = \overline{m}_k(L)v_{kt}, \quad k = 1, \ldots, K,$$

where $\overline{m}_k(L)$ is an operator of degree at most q because the k-th row of $M(L)u_t$ is

$$m_{k1}(L)u_{1t} + \cdots + m_{kK}(L)u_{Kt}.$$

In other words, it is a sum of MA(q) processes which is known to have an MA representation of degree at most q (see Proposition 11.1). Thus, each component series of y_t has the same AR operator and an MA operator of degree at most q. In general, at least one of the component series will have MA degree q because a reduction of the MA order of all component series requires a very special set of parameters which is not regarded as likely in practice. This fact is used in specifying the final form VARMA representation by first determining univariate component models and then putting them together in a joint model. Specifically, the following specification strategy is used.

STAGE I: Specify univariate models

$$\alpha_k(L)y_{kt} = m_k(L)v_{kt}$$

for the components of y_t. Here

$$\alpha_k(L) := 1 - \alpha_{k1}L - \cdots - \alpha_{kp_k}L^{p_k}$$

is of order p_k,

$$m_k(L) := 1 + m_{k1}L + \cdots + m_{kq_k}L^{q_k}$$

is of order q_k, and v_{kt} is a univariate white noise process. ∎

The Box & Jenkins (1976) strategy for specifying univariate ARMA models may be used at this stage. Alternatively, some automatic procedure or criterion such as the one proposed by Hannan & Rissanen (1982) or Poskitt (1987) may be applied.

STAGE II: Determine a common AR operator $\alpha(L)$ for all component processes, specify the corresponding MA orders and choose the degree q of the joint MA operator as the maximum of the individual MA degrees obtained in this way. ∎

At this stage, a common AR operator may, for example, be obtained as the product of the individual operators, that is,

$$\alpha(L) = \alpha_1(L) \cdots \alpha_K(L).$$

In this case, the k-th component process is multiplied by

$$\prod_{i=1, i \neq k}^{K} \alpha_i(L)$$

and $\alpha(L)$ has degree $p = \sum_{i=1}^{K} p_i$, while the MA operator

$$\overline{m}_k(L) = m_k(L) \prod_{i=1, i \neq k}^{K} \alpha_i(L)$$

has degree

$$q_k + \sum_{i=1, i \neq k}^{K} p_i.$$

The joint MA operator of the VARMA representation (13.2.1) is then assumed to have degree

$$\max_k \left(q_k + \sum_{i=1, i \neq k}^{K} p_i \right). \tag{13.2.2}$$

Of course, the $\alpha_k(L)$, $k = 1, \ldots, K$, may have common factors. In that case, a joint AR operator $\alpha(L)$ with degree much lower than $\sum_{i=1}^{K} p_i$ may be possible. Correspondingly, the joint MA operator may have degree lower than (13.2.2). Suppose, for instance, that $K = 3$ and

$$\alpha_1(L) = 1 - \alpha_{11}L, \qquad m_1(L) = 1 + m_{11}L,$$
$$\alpha_2(L) = 1 - \alpha_{21}L - \alpha_{22}L^2, \qquad m_2(L) = 1 + m_{21}L,$$
$$\alpha_3(L) = 1 - \alpha_{31}L, \qquad m_3(L) = 1.$$

Now a joint AR operator is

$$\alpha(L) = \alpha_1(L)\alpha_2(L)\alpha_3(L),$$

which has degree 4. However, if $\alpha_2(L)$ can be factored as

$$\alpha_2(L) = (1 - \alpha_{11}L)(1 - \alpha_{31}L) = \alpha_1(L)\alpha_3(L),$$

then a common AR operator $\alpha(L) = \alpha_2(L)$ may be chosen and we get univariate models

$$\alpha(L)y_{1t} = \alpha_3(L)m_1(L)v_{1t} \quad [\text{ARMA}(2,2)],$$
$$\alpha(L)y_{2t} = m_2(L)v_{2t} \quad [\text{ARMA}(2,1)], \qquad (13.2.3)$$
$$\alpha(L)y_{3t} = \alpha_1(L)m_3(L)v_{3t} \quad [\text{ARMA}(2,1)].$$

The maximum of the individual MA degrees is chosen as the joint MA degree, that is, $q = 2$ and, of course,

$$p = \text{degree}(\alpha(L)) = 2.$$

A problem that should be noticed from this discussion and example is that the degrees p and q determined in this way may be quite large. It is conceivable that $p = \sum_{i=1}^{K} p_i$ is the smallest possible AR order for the final equations form representation and the corresponding MA degree may be quite substantial too. This, clearly, can be a disadvantage as unduely many parameters can cause trouble in a final estimation algorithm and may lead to imprecise forecasts and impulse responses.

Often it may be possible to impose restrictions on the AR and MA operators in (13.2.1). This modification may either be done in a third stage of the procedure or it may be incorporated in Stages I and/or II, depending on the type of information available. Restrictions may be obtained with the help of statistical tools such as testing the significance of single coefficients or groups of parameters. Alternatively, restrictions may be implied by subject matter theory. Zellner & Palm (1974) give a detailed example where both types of restrictions are used.

Perhaps because of the potentially great number of parameters, final form modelling has not become very popular. It can only be recommended if it results in a reasonably parsimonious parameterization.

13.2.2 An Example

For illustrative purposes, we consider a bivariate system consisting of first differences of logarithms of income (y_1) and consumption (y_2). We use again the data from File E1 up to the fourth quarter of 1978. If the 3-dimensional system involving investment in addition really were generated by a VAR(2) process, as assumed in Chapters 3 and 4, it is quite possible that the subprocess consisting of income and consumption only has a mixed VARMA generation process with nontrivial MA part (see Section 11.6.1). Moreover, the marginal univariate processes for y_1 and y_2 may be of a mixed ARMA type. However, we found that the subset AR(3) models (with standard errors in parentheses)

$$(1 - \underset{(.113)}{.245} L^3) y_{1t} = \underset{(.003)}{.015} + v_{1t} \tag{13.2.4}$$

and

$$(1 - \underset{(.111)}{.309} L^2 - \underset{(.111)}{.187} L^3) y_{2t} = \underset{(.004)}{.010} + v_{2t} \tag{13.2.5}$$

fit the data quite well. For illustrative purposes, we will therefore proceed from these models. The reader may try to find better models and repeat the analysis with them.

Generally, a (1×1) scalar operator

$$\gamma(L) = 1 - \gamma_1 L - \cdots - \gamma_p L^p$$

of degree p can be factored in p components,

$$\gamma(L) = (1 - \lambda_1 L) \cdots (1 - \lambda_p L),$$

where $\lambda_1, \ldots, \lambda_p$ are the reciprocals of the roots of $\gamma(z)$. Thus, the two AR operators from (13.2.4) and (13.2.5) can be factored as

$$\begin{aligned} \alpha_1(L) &= 1 - .245 L^3 \\ &= (1 - .626L)(1 + (.313 + .542i)L)(1 + (.313 - .542i)L) \end{aligned} \tag{13.2.6}$$

and

$$\begin{aligned} \alpha_2(L) &= 1 - .309 L^2 - .187 L^3 \\ &= (1 - .747L)(1 + (.374 + .332i)L)(1 + (.374 - .332i)L), \end{aligned} \tag{13.2.7}$$

where i denotes the imaginary part of the complex numbers. None of the factors in (13.2.6) is very close to any of the factors in (13.2.7). Thus, models with common AR operator may be of the form

$$\alpha_1(L)\alpha_2(L)y_{1t} = \alpha_2(L)v_{1t}$$

and

$$\alpha_1(L)\alpha_2(L)y_{2t} = \alpha_1(L)v_{2t}.$$

With the arguments of the previous subsection, the resulting bivariate final equations model is a VARMA(6, 3) process,

$$\alpha_1(L)\alpha_2(L)\begin{bmatrix} y_{1t} \\ y_{2t} \end{bmatrix} = (I_3 + M_1 L + M_2 L^2 + M_3 L^3)\begin{bmatrix} u_{1t} \\ u_{2t} \end{bmatrix}. \qquad (13.2.8)$$

Obviously, this model involves very many parameters and is therefore unattractive. In fact, such a heavily parameterized model may cause numerical problems when full maximum likelihood estimation is attempted. It is possible, if not likely, that some parameters turn out to be insignificant and could be set to zero. However, the significance of parameters is commonly judged on the basis of their standard errors or t-ratios. These quantities become available in the ML estimation round which, as we have argued, may be problematic in the present case.

Given the estimation uncertainty, one may argue that the real factors in the operators $\alpha_1(L)$ and $\alpha_2(L)$ may be identical. Proceeding under that assumption results in a VARMA(5, 2) final equations form. Such a model is more parsimonious and has therefore more appeal than (13.2.8). Still it involves a considerable number of parameters. This example illustrates why final equations modelling, although relatively simple, does not enjoy much popularity. For higher-dimensional models, the problem of heavy parameterization becomes even more severe because the number of parameters is likely to increase rapidly with the dimension of the system. We will now present procedures for specifying echelon forms.

13.3 Specification of Echelon Forms

In specifying an echelon VARMA representation, the objective is to find the Kronecker indices and possibly impose some further restrictions on the parameters. For a K-dimensional process, there are K Kronecker indices. Different strategies have been proposed for their specification. We will discuss some of them in the following. Once the Kronecker indices are determined, further restrictions may be imposed, for instance, on the basis of significance tests for individual coefficients or groups of parameters.

In the first subsection below, we will discuss a procedure for specifying the Kronecker indices which is usually not feasible in practice. It is nevertheless useful to study that procedure because the feasible strategies considered in Subsections 13.3.2–13.3.4 may be regarded as approximations or short-cuts of that procedure with similar asymptotic properties. In Subsection 13.3.2, we present a procedure which is easy to carry out for systems with small Kronecker indices and low dimensions. It is quite costly for higher-dimensional systems, though. For such systems a specification strategy inspired by Hannan & Kavalieris (1984) or a procedure due to Poskitt (1992) may be more

13.3 Specification of Echelon Forms

appealing. These approaches are considered in Subsections 13.3.3 and 13.3.4, respectively. The material discussed in this section is covered in more depth and more rigorously in Hannan & Deistler (1988, Chapters 5, 6, 7) and Poskitt (1992).

13.3.1 A Procedure for Small Systems

If it is known that the generation process of a given K-dimensional multiple time series admits an echelon VARMA representation with Kronecker indices $p_k \leq p_{\max}$, $k = 1, \ldots, K$, where $p_{\max}$ is a prespecified number, then, in theory, it is possible to evaluate the maximum log-likelihood for all sets of Kronecker indices $\mathbf{p} = (p_1, \ldots, p_K)$ with $p_k \leq p_{\max}$ and choose the set $\widehat{\mathbf{p}}$ that optimizes a specific criterion. This approach is completely analogous to the specification of the VAR order in the finite order VAR case considered in Section 4.3. In that section, we have discussed the possibility to consistently estimate the VAR order with such an approach. It turns out that a similar result can be obtained for the present more general VARMA case.

Before we give further details, it may be worth emphasizing, however, that in the VARMA case, such a specification strategy is generally not feasible in practice because the maximization of the log-likelihood is usually quite costly and, for systems with moderate or high dimensions, an enormous number of likelihood maximizations would be required. For instance, for a five-dimensional system, evaluating the maximum of the log-likelihood for all vectors of Kronecker indices $\mathbf{p} = (p_1, \ldots, p_5)$ with $p_k \leq 8$ requires $9^5 = 59{,}049$ likelihood maximizations. Despite this practical problem, we discuss the theoretical properties of this procedure to provide a basis for the following subsections.

Let us denote by $\widetilde{\Sigma}(\mathbf{p})$ the ML estimator of the white noise covariance matrix Σ_u obtained for a set of Kronecker indices $\mathbf{p}$. Furthermore, let

$$\mathrm{Cr}(\mathbf{p}) := \ln|\widetilde{\Sigma}_u(\mathbf{p})| + c_T d(\mathbf{p})/T \qquad (13.3.1)$$

be a criterion to be minimized over all sets of Kronecker indices $\mathbf{p} = (p_1, \ldots, p_K)$, $p_k \leq p_{\max}$. Here $d(\mathbf{p})$ is the number of freely varying parameters in the $\mathrm{ARMA}_E(\mathbf{p})$ form. For example, for a bivariate system with Kronecker indices $\mathbf{p} = (p_1, p_2) = (1, 0)$, the $\mathrm{ARMA}_E(1, 0)$ form is

$$\begin{bmatrix} 1 & 0 \\ -\alpha_{21,0} & 1 \end{bmatrix} \begin{bmatrix} y_{1,t} \\ y_{2,t} \end{bmatrix} = \begin{bmatrix} \alpha_{11,1} & 0 \\ 0 & 0 \end{bmatrix} \begin{bmatrix} y_{1,t-1} \\ y_{2,t-1} \end{bmatrix} + \begin{bmatrix} 1 & 0 \\ -\alpha_{21,0} & 1 \end{bmatrix} \begin{bmatrix} u_{1,t} \\ u_{2,t} \end{bmatrix}$$
$$+ \begin{bmatrix} m_{11,1} & m_{12,1} \\ 0 & 0 \end{bmatrix} \begin{bmatrix} u_{1,t-1} \\ u_{2,t-1} \end{bmatrix}.$$

Thus, $d(1,0) = 4$. In (13.3.1), c_T is a sequence indexed by the sample size T.

In general, if models are included in the search procedure for which all Kronecker indices exceed the true ones, the estimation of unidentified models is required for which cancellation of the VAR and MA operators is possible.

This over-specification is not necessarily a problem for evaluating the criterion in (13.3.1) because we only need the maximum log-likelihood or rather $\ln|\widetilde{\Sigma}_u(\mathbf{p})|$ in that criterion. That quantity can be determined even if the corresponding VARMA coefficients are meaningless. The coefficients cannot and should not be interpreted, however.

Note that the criterion (13.3.1) is very similar to that considered in Proposition 4.2 of Chapter 4. In that proposition, the consistency or inconsistency of a criterion is seen to depend on the choice of the sequence c_T. Hannan (1981) and Hannan & Deistler (1988, Chapter 5, Section 5) showed that a criterion such as the one in (13.3.1) provides a consistent estimator of the true set of Kronecker indices if c_T is a nondecreasing function of T satisfying

$$c_T \to \infty \quad \text{and} \quad c_T/T \to 0 \quad \text{as} \quad T \to \infty, \tag{13.3.2}$$

and the true data generation process satisfies some weak conditions. If, in addition,

$$c_T/2\ln\ln T > 1 \tag{13.3.3}$$

eventually as $T \to \infty$, the procedure provides a strongly consistent estimator of the true Kronecker indices. The conditions for the VARMA process are, for instance, satisfied if the white noise process u_t is identically distributed standard white noise (see Definition 3.1) and the true data generation process admits a stable and invertible ARMA_E representation with Kronecker indices not greater than $p_{\max}$. This result extends Proposition 4.2 to the VARMA case.

Implications of this result are that the Schwarz criterion with $c_T = \ln T$,

$$\text{SC}(\mathbf{p}) := \ln|\widetilde{\Sigma}_u(\mathbf{p})| + d(\mathbf{p})\ln T/T, \tag{13.3.4}$$

is strongly consistent and that the Hannan-Quinn criterion, using the borderline penalty term $c_T = 2\ln\ln T$,

$$\text{HQ}(\mathbf{p}) := \ln|\widetilde{\Sigma}_u(\mathbf{p})| + 2d(\mathbf{p})\ln\ln T/T, \tag{13.3.5}$$

is consistent. Hannan & Deistler (1988) also showed that

$$\text{AIC}(\mathbf{p}) := \ln|\widetilde{\Sigma}_u(\mathbf{p})| + 2d(\mathbf{p})/T \tag{13.3.6}$$

with $c_T = 2$ is not a consistent criterion. Again these results are similar to those for the finite order VAR case.

As in that case, it is worth emphasizing that these results do not necessarily imply the inferiority of AIC or HQ. In small samples, these criteria may be preferable. They may, in fact, provide superior models for a specific analysis of interest. Also, in practice, the actual data generation mechanism will usually not really admit a VARMA representation. Recall that the best we can hope for is that our model is a good approximation to the true data generation process. In that case, the relevance of the consistency property is of course doubtful.

Again, the specification strategy presented in the foregoing is not likely to have much practical appeal as it is computationally too burdensome. In the next subsections, more practical modifications are discussed.

13.3.2 A Full Search Procedure Based on Linear Least Squares Computations

The Procedure

A major obstacle for using the procedure described in the previous subsection is the requirement to maximize the log-likelihood various times. This maximization is costly because for mixed VARMA models the log-likelihood function is nonlinear in the parameters and iterative optimization algorithms have to be employed. Because we just need an estimator of Σ_u for the evaluation of model selection criteria such as (13.3.1), an obvious modification of the procedure would use an estimator that avoids the nonlinear optimization problem. Such an estimator may be obtained from the preliminary estimation procedure described in Chapter 12, Section 12.3.4. Therefore, a specification of the Kronecker indices may proceed in the following stages.

STAGE I: Fit a long VAR process of order n, say, to the data and obtain the estimated residual vectors $\widehat{u}_t(n)$, $t = 1, \ldots, T$. ∎

The choice of n could be based on an order selection criterion such as AIC. In any case, n has to be greater than the largest Kronecker index $p_{\max}$ to be considered in the next stage of the procedure.

STAGE II: Using the residuals $\widehat{u}_t(n)$ from Stage I, compute the preliminary estimator of Section 12.3.4 for all sets of Kronecker indices **p** with $p_k \leq p_{\max}$, where the latter number is a prespecified upper bound for the Kronecker indices. Determine all corresponding estimators $\widetilde{\Sigma}_u(\mathbf{p})$ based on the residuals of the preliminary estimations (see (12.3.24)). (Here we suppress the order n from the first stage for notational convenience because the same n is used for all $\widetilde{\Sigma}_u(\mathbf{p})$ at this stage.) Choose the estimator $\widehat{\mathbf{p}}$ which minimizes a prespecified criterion of the form (13.3.1). ∎

The choice of the criterion $\text{Cr}(\mathbf{p})$ is left to the researcher. SC, HQ, and AIC from (13.3.4)–(13.3.6) are possible candidates. Stage II could be iterated by using the residuals from a previous run through Stage II instead of the residuals from Stage I. Once an estimate $\widehat{\mathbf{p}}$ of the Kronecker indices is determined, the ML estimates conditional on $\widehat{\mathbf{p}}$ may be computed in a final stage.

STAGE III: Estimate the echelon form VARMA model with Kronecker indices $\widehat{\mathbf{p}}$ by maximizing the Gaussian log-likelihood function or by just one step of the scoring algorithm (see Section 12.3.2). ∎

Hannan & Deistler (1988, Chapter 6) discussed conditions under which this procedure provides a consistent estimator of the Kronecker indices and VARMA parameter estimators that have the same asymptotic properties as the estimators obtained for given, known, true Kronecker indices (see Proposition 12.1). In addition to our usual assumptions for the VARMA process such as stability and invertibility, the required assumptions relate to the criteria for choosing the VAR order in Stage I and the Kronecker indices in Stage II. These conditions are asymptotic conditions that leave some room for the actual choice in small samples. Any criterion from (13.3.4)–(13.3.6) may be a reasonable choice in practice.

The procedure still involves extensive computations, unless the dimension K of the underlying multiple time series and $p_{\max}$ are small. For example, for a five-dimensional system with $p_{\max} = 8$ we still have to perform $9^5 = 59{,}049$ estimations in order to compare all feasible models. Although these estimations involve linear least squares computations only, the computational costs may be substantial. Therefore, we outline two less costly procedures in the following subsections. For small systems, the present procedure is a reasonable choice. We give an example next.

An Example

We consider again the income/consumption example from Section 13.2.2. In the first stage of our procedure, we fit a VAR(8) model ($n = 8$) and we use the residuals, $\widehat{u}_t(8)$, at the next stage. The choice of $n = 8$ is to some extent arbitrary. We have chosen a fairly high order to gain flexibility for the Kronecker indices considered at Stage II. Recall that n must exceed all Kronecker indices to be considered subsequently. Using the procedure described as Stage II, we have estimated models with Kronecker indices $p_k \leq p_{\max} = 4$ and we have determined the corresponding values of the criteria AIC and HQ. They are given in Tables 13.1 and 13.2, respectively. Both criteria reach their minimum for $\mathbf{p} = (p_1, p_2) = (0, 2)$. The $\text{ARMA}_E(0, 2)$ form is precisely the model estimated in Chapter 12, Section 12.3.5. Replacing the parameters by their ML estimates with estimated standard errors in parentheses, we have

$$\begin{bmatrix} y_{1,t} \\ y_{2,t} \end{bmatrix} = \begin{bmatrix} 0 & 0 \\ 0 & .225 \\ & (.252) \end{bmatrix} \begin{bmatrix} y_{1,t-1} \\ y_{2,t-1} \end{bmatrix} + \begin{bmatrix} 0 & 0 \\ 0 & .061 \\ & (.166) \end{bmatrix} \begin{bmatrix} y_{1,t-2} \\ y_{2,t-2} \end{bmatrix}$$

$$+ \begin{bmatrix} \widehat{u}_{1,t} \\ \widehat{u}_{2,t} \end{bmatrix} + \begin{bmatrix} 0 & 0 \\ .313 & -.750 \\ (.090) & (.274) \end{bmatrix} \begin{bmatrix} \widehat{u}_{1,t-1} \\ \widehat{u}_{2,t-1} \end{bmatrix}$$

$$+ \begin{bmatrix} 0 & 0 \\ .140 & .160 \\ (.141) & (.233) \end{bmatrix} \begin{bmatrix} \widehat{u}_{1,t-2} \\ \widehat{u}_{2,t-2} \end{bmatrix}.$$

Obviously, some of the parameters are quite small compared to their estimated standard errors. In such a situation, one may want to impose further zero constraints on the parameters. Because $\widehat{\alpha}_{22,1}, \widehat{\alpha}_{22,2}$, and $\widehat{m}_{22,2}$ have the smallest t-ratios in absolute terms, we restrict these estimates to zero and reestimate the model. The resulting system obtained by ML estimation is

$$\begin{bmatrix} y_{1,t} \\ y_{2,t} \end{bmatrix} = \begin{bmatrix} \widehat{u}_{1,t} \\ \widehat{u}_{2,t} \end{bmatrix} + \begin{bmatrix} 0 & 0 \\ \underset{(.088)}{.308} & \underset{(.104)}{-.475} \end{bmatrix} \begin{bmatrix} \widehat{u}_{1,t-1} \\ \widehat{u}_{2,t-1} \end{bmatrix}$$
$$+ \begin{bmatrix} 0 & 0 \\ \underset{(.076)}{.302} & 0 \end{bmatrix} \begin{bmatrix} \widehat{u}_{1,t-2} \\ \widehat{u}_{2,t-2} \end{bmatrix}.$$

Now all parameters are significant under a two-standard error criterion.

Table 13.1. AIC values of ARMA$_E(p_1, p_2)$ models for the income/consumption data

p_2	0	1	2	3	4
			p_1		
0	−16.83	−18.41	−18.30	−18.25	−18.15
1	−18.50	−18.42	−18.30	−18.23	−18.13
2	−18.64*	−18.55	−18.42	−18.29	−18.19
3	−18.57	−18.50	−18.37	−18.27	−18.19
4	−18.47	−18.38	−18.27	−18.20	−18.05

*Minimum

Table 13.2. HQ values of ARMA$_E(p_1, p_2)$ models for the income/consumption data

p_2	0	1	2	3	4
			p_1		
0	−16.83	−18.35	−18.21	−18.12	−17.98
1	−18.46	−18.31	−18.14	−18.03	−17.89
2	−18.56*	−18.41	−18.21	−18.03	−17.88
3	−18.45	−18.32	−18.12	−17.95	−17.82
4	−18.31	−18.16	−17.98	−17.84	−17.63

*Minimum

13.3.3 Hannan-Kavalieris Procedure

A full search procedure for the optimal Kronecker indices, as in Stage II of the previous subsection, involves a substantial amount of computation work if the dimension K of the time series considered is large or if the upper bound $p_{\max}$ for the Kronecker indices is high. For instance, if monthly data are considered and lags of at least one year are deemed necessary, $p_{\max} \geq 12$ is required. Even if the system involves just three variables ($K = 3$) the number of models to be compared is vast, namely $13^3 = 2197$. Therefore, shortcuts for Stage II of the previous subsection were proposed. The first one we present here is inspired by discussions of Hannan & Kavalieris (1984). Therefore, we call it the Hannan-Kavalieris procedure although these authors proposed a more sophisticated approach. In particular, they discussed a number of computational simplifications (see also Hannan & Deistler (1988, Chapter 6)). The following modification of Stage II may be worth trying.

STAGE II (HK): Based on the covariance estimators obtained from the preliminary estimation procedure of Section 12.3.4, find the Kronecker indices, say $\mathbf{p}^{(1)} = p^{(1)}(1, \ldots, 1)$, that minimize a prespecified criterion of the type $\text{Cr}(\mathbf{p})$ in (13.3.1) over $\mathbf{p} = p(1, \ldots, 1)$, $p = 0, \ldots, p_{\max}$, that is, all Kronecker indices are identical in this first step. Then the last index p_K is varied between 0 and $p^{(1)}$ while all other indices are fixed at $p^{(1)}$. We denote the optimal value of p_K by $\widehat{p}_K$, that is, $\widehat{p}_K$ minimizes the prespecified criterion. Then we proceed in the same way with p_{K-1} and so on. More generally, $\widehat{p}_k$ is chosen such that

$$\text{Cr}(p^{(1)}, \ldots, p^{(1)}, \widehat{p}_k, \ldots, \widehat{p}_K)$$
$$= \min\{\text{Cr}(p^{(1)}, \ldots, p^{(1)}, p, \widehat{p}_{k+1}, \ldots, \widehat{p}_K) | p = 0, \ldots, p^{(1)}\}.$$

∎

This modification reduces the computational burden considerably. Just to give an example, for $K = 5$ and $p_{\max} = 8$, at most $9 + 5 \cdot 9 = 54$ models have to be estimated. If $p^{(1)}$ is small, then the number may be substantially lower. For comparison purposes, we repeat that the number of estimations in a full search procedure would be $9^5 = 59,049$.

To illustrate the procedure, consider the following panel of criterion values for Kronecker indices (p_1, p_2):

		p_1		
p_2	0	1	2	3
0	3.48	3.28	3.26	3.27
1	3.25	3.23	3.14	3.20
2	3.23	3.21	3.15	3.19
3	3.24	3.20	3.21	3.18

The minimum value on the main diagonal is obtained for $p^{(1)} = 2$ with $\text{Cr}(p^{(1)}, p^{(1)}) = 3.15$. Going upward from $(p_1, p_2) = (2, 2)$, the minimum is seen to be $\text{Cr}(2, 1) = 3.14$. Turning left from $(p_1, p_2) = (2, 1)$, a further reduction of the criterion value is not obtained. Therefore, the estimate for the Kronecker indices is $\widehat{\mathbf{p}} = (2, 1)$.

In this case, we have actually found the overall minimum of all criterion values in the panel, that is,

$$\mathrm{Cr}(2,1) = \min\{\mathrm{Cr}(p_1,p_2)|p_i = 0,1,2,3\}.$$

In general, the Hannan-Kavalieris procedure will not lead to the overall minimum. For instance, for our bivariate income/consumption example from Section 13.3.2, we can find the HK estimates $\widehat{\mathbf{p}}$ from Tables 13.1 and 13.2. For example, on the main diagonal of the panel in Table 13.2, the HQ criterion assumes its minimum for $(p_1, p_2) = (1, 1)$ and $\widehat{\mathbf{p}}(\mathrm{HQ}) = (1, 0)$. Clearly, this result differs from the estimate $\widehat{\mathbf{p}} = (0, 2)$ that was obtained in the full search procedure.

Under suitable conditions for the model selection criteria, the HK procedure is consistent. Hannan & Deistler (1988) also discussed the consequences of the true data generation process being not in the class of considered processes.

13.3.4 Poskitt's Procedure

Another short-cut version of Stage II of the model specification procedure was suggested by Poskitt (1992). It capitalizes on the important property of echelon forms that the restrictions for the k-th equation implied by a set of Kronecker indices $\mathbf{p}$ are determined by the indices $p_i \leq p_k$. They do not depend on the specific values of the p_j which are greater than p_k. The proposed modification of Stage II is based on separate LS estimations of each of the K equations of the system. The estimation is similar to the preliminary estimation method outlined in Section 12.3.4, that is, it uses the residuals of a long autoregression from Stage I instead of the true u_t's. A model selection criterion of the form

$$\mathrm{Cr}_k(\mathbf{p}) := \ln \widetilde{\sigma}_k^2(\mathbf{p}) + c_T d_k(\mathbf{p})/T \qquad (13.3.7)$$

is then evaluated for each of the K equations separately. Here $\widetilde{\sigma}_k^2(\mathbf{p})$ is the residual variance estimate such that $T\widetilde{\sigma}_k^2(\mathbf{p})$ is the residual sum of squares of the k-th equation in a system with Kronecker indices $\mathbf{p}$, $d_k(\mathbf{p})$ is the number of freely varying parameters in the k-th equation, and c_T is a number that depends on the sample size T. Of course, (13.3.7) is the single equation analogue of the systems criterion (13.3.1). Stage II of Poskitt's procedure proceeds then as follows:

STAGE II (P): Determine the required values $\mathrm{Cr}_k(\mathbf{p})$ and choose the estimates $\widehat{p}_k$ of the Kronecker indices according to the following rule:

If $\mathrm{Cr}_k(0,\ldots,0) \geq \mathrm{Cr}_k(1,\ldots,1)$ for all $k = 1,\ldots,K$, compute $\mathrm{Cr}_k(2,\ldots,2)$, $k = 1,\ldots,K$, and compare to $\mathrm{Cr}_k(1,\ldots,1)$. If the $\mathrm{Cr}_k(2,\ldots,2)$ are all not greater than the corresponding $\mathrm{Cr}_k(1,\ldots,1)$, proceed to $\mathrm{Cr}_k(3,\ldots,3)$ and so on.

If at some stage

$$\mathrm{Cr}_k(j-1,\ldots,j-1) \geq \mathrm{Cr}_k(j,\ldots,j)$$

does *not* hold *for all* k, choose $\widehat{p}_k = j-1$ for all k with

$$\mathrm{Cr}_k(j-1,\ldots,j-1) < \mathrm{Cr}_k(j,\ldots,j).$$

The $\widehat{p}_k$ obtained in this way are fixed in all the following steps. We continue by increasing the remaining indices and comparing the criteria for those equations for which the Kronecker indices are not yet fixed. Here it is important that the restrictions for the k-th equation do not depend on the Kronecker indices $p_i > p_k$ which are chosen in subsequent steps. ∎

To make the procedure a bit more transparent, it may be helpful to consider an example. Suppose that interest centers on a three-dimensional system, that is, $K = 3$. First $\mathrm{Cr}_k(0,0,0)$ and $\mathrm{Cr}_k(1,1,1)$ are computed for $k = 1,2,3$. Suppose

$$\mathrm{Cr}_k(0,0,0) \geq \mathrm{Cr}_k(1,1,1), \quad \text{for } k = 1,2,3.$$

Then we evaluate $\mathrm{Cr}_k(2,2,2)$, $k = 1,2,3$. Suppose

$$\mathrm{Cr}_1(1,1,1) < \mathrm{Cr}_1(2,2,2)$$

and

$$\mathrm{Cr}_k(1,1,1) \geq \mathrm{Cr}_k(2,2,2), \quad \text{for } k = 2,3.$$

Then $\widehat{p}_1 = 1$ is fixed and $\mathrm{Cr}_k(1,2,2)$ is compared to $\mathrm{Cr}_k(1,3,3)$ for $k = 2,3$. Suppose

$$\mathrm{Cr}_2(1,2,2) \geq \mathrm{Cr}_2(1,3,3) \quad \text{and} \quad \mathrm{Cr}_3(1,2,2) < \mathrm{Cr}_3(1,3,3).$$

Then we fix $\widehat{p}_3 = 2$ and compare $\mathrm{Cr}_2(1,3,2)$ to $\mathrm{Cr}_2(1,4,2)$ and so on until p_2 can also be fixed because no further reduction of the criterion $\mathrm{Cr}_2(\cdot)$ is obtained in one step. It is important to note that for each index only the first local minimum of the corresponding criterion is searched for. We are not seeking a global minimum over all $\mathbf{p}$ with p_k less than some prespecified upper bound. For moderate or large systems, the present procedure has the advantage of involving a very reasonable amount of computation work only.

Poskitt (1992) derived the properties of the Kronecker indices and the VARMA coefficients estimated by this procedure. He gave conditions under which the Kronecker indices are estimated consistently and the final VARMA parameter estimators have the asymptotic distribution given in Proposition 12.1. Assuming that the true data generation process can indeed be described by a stable, invertible ARMA_E representation with a finite set of Kronecker indices, the conditions imposed by Poskitt relate to the distribution of the white noise process u_t, to the choice of n, and to the criteria $\text{Cr}_k(\mathbf{p})$.

With respect to the process or white noise distribution, the assumptions are satisfied, for example, if u_t is Gaussian. In fact, for most results it suffices that u_t is standard white noise. An exception is the asymptotic distribution of the white noise covariance estimator $\widetilde{\Sigma}_u$. It may change if u_t has a nonnormal distribution.

The VAR order n at Stage I is assumed to go to infinity with the sample size at a certain rate. In practice, the order selection criteria AIC or HQ may be used at Stage I. It must be guaranteed, however, that n is greater than the Kronecker indices considered at Stage II(P).

Poskitt (1992) also discussed a modification of his algorithm that appears to have some practical advantages. We do not go into that procedure here but recommend that the interested reader examines the relevant literature. The message from the present discussion should be that consistent and feasible strategies for estimating the Kronecker indices exist. Poskitt also discussed the case where the true data generation process is not in the class of VARMA processes considered in the specification procedure. He derived some asymptotic results for this case as well.

In summary, a full search procedure is feasible for low-dimensional systems if the maximum for the Kronecker indices is small or moderate. For high-dimensional systems and/or large upper bounds of the Kronecker indices, the Hannan-Kavalieris procedure or Poskitt's specification strategy are preferable from a computational point of view. The relative performance in small samples is so far unknown in general. It is left to the individual researcher to decide on a specific specification procedure with his or her available resources and perhaps the objective of the analysis in mind. Of course, it is legitimate to try different strategies and criteria and compare the resulting models and the implications for the subsequent analysis.

13.4 Remarks on Other Specification Strategies for VARMA Models

A number of other specification strategies for VARMA processes were proposed and investigated in the literature based on representations other than the final equations and echelon forms. Examples are Quenouille (1957), Tiao & Box (1981), Jenkins & Alavi (1981), Aoki (1987), Cooper & Wood (1982), Granger & Newbold (1986), Akaike (1976), Tiao & Tsay (1989), Tsay (1989a, b), to list just a few. Some of these strategies are based on subjective criteria. As mentioned earlier, none of these procedures seems to be in common use for analyzing economic time series and none of them has become *the* standard procedure. So far, few VARMA analyses of higher-dimensional time series are reported in the literature. Given this state of affairs, it is difficult to give well-founded recommendations as to which strategy to use in any particular situation. Those familiar with the Box-Jenkins approach for univariate time series modelling will be aware of the problems that can arise even in the univariate case if the investigator has to decide on a model on the basis of statistics such as the autocorrelations and partial autocorrelations. Therefore, it is an obvious advantage to have an automatic or semiautomatic procedure if one feels uncertain about the interpretation of statistical quantities related to specific characteristics of a process and if little or no prior information is available. On the other hand, if firmly based prior information about the data generation process is available, then it may be advantageous to use that at an early stage and depart from automatic procedures.

13.5 Model Checking

Prominent candidates in the model checking tool-kit are tests of statistical hypotheses. All three testing principles, LR (likelihood ratio), LM (Lagrange multiplier), and Wald tests (see Appendix C.7) can be applied in principle in the VARMA context. Because estimation requires iterative procedures, it is often desirable to estimate just one model. Hence, LR tests which require estimation under both the null and alternative hypotheses are often unattractive. In finite order VAR modelling, the unrestricted version is usually relatively easy to estimate and therefore it makes sense to use Wald tests in the pure VAR case because these tests are based on the unconstrained estimator. In contrast, the restricted estimator is often easier to obtain in the VARMA context when models with nontrivial MA part are considered. In this situation, LM tests have an obvious advantage because the LM statistic involves the restricted estimator only. Of course, the restricted estimator is especially easy to determine if the constrained model is a pure, finite order VAR process. We will briefly discuss LM tests in the following. For further discussions and proofs the reader is referred to Kohn (1979), Hosking (1981b), and Poskitt & Tremayne (1982).

13.5.1 LM Tests

Suppose we wish to test

$$H_0 : \varphi(\boldsymbol{\beta}) = 0 \quad \text{against} \quad H_1 : \varphi(\boldsymbol{\beta}) \neq 0, \tag{13.5.1}$$

where $\boldsymbol{\beta}$ is an M-dimensional parameter vector and $\varphi(\cdot)$ is a twice continuously differentiable function with values in the N-dimensional Euclidean space. In other words, $\varphi(\boldsymbol{\beta})$ is an $(N \times 1)$ vector and we assume that the matrix $\partial \varphi / \partial \boldsymbol{\beta}'$ of first order partial derivatives has rank N at the true parameter vector. In this setup, we consider the case where the restrictions relate to the VARMA coefficients only. Moreover, we assume that the conditions of Proposition 12.1 are satisfied.

13.5 Model Checking

For instance, in the bivariate zero mean VARMA(1,1) model with Kronecker indices $(p_1, p_2) = (1, 1)$,

$$\begin{bmatrix} y_{1,t} \\ y_{2,t} \end{bmatrix} = \begin{bmatrix} \alpha_{11,1} & \alpha_{12,1} \\ \alpha_{21,1} & \alpha_{22,1} \end{bmatrix} \begin{bmatrix} y_{1,t-1} \\ y_{2,t-1} \end{bmatrix}$$
$$+ \begin{bmatrix} u_{1,t} \\ u_{2,t} \end{bmatrix} + \begin{bmatrix} m_{11,1} & m_{12,1} \\ m_{21,1} & m_{22,1} \end{bmatrix} \begin{bmatrix} u_{1,t-1} \\ u_{2,t-1} \end{bmatrix}, \quad (13.5.2)$$

with $\boldsymbol{\beta}' = (\alpha_{11,1}, \alpha_{21,1}, \alpha_{12,1}, \alpha_{22,1}, m_{11,1}, m_{21,1}, m_{12,1}, m_{22,1})$, one may wish to test that the MA degree is zero, that is,

$$\varphi(\boldsymbol{\beta}) = \begin{bmatrix} m_{11,1} \\ m_{21,1} \\ m_{12,1} \\ m_{22,1} \end{bmatrix} = \begin{bmatrix} 0 \\ 0 \\ 0 \\ 0 \end{bmatrix}.$$

The corresponding matrix of partial derivatives is

$$\frac{\partial \varphi}{\partial \boldsymbol{\beta}'} = [0 : I_4]$$

which obviously has rank $N = 4$.

As another example, suppose we wish to test for Granger-causality from y_{2t} to y_{1t} in the model (13.5.2). In that case,

$$\varphi(\boldsymbol{\beta}) = \begin{bmatrix} \alpha_{12,1} + m_{12,1} \\ \alpha_{22,1} m_{12,1} - \alpha_{12,1} m_{22,1} \end{bmatrix} = \begin{bmatrix} 0 \\ 0 \end{bmatrix} \quad (13.5.3)$$

(see Remark 1 of Section 11.7.1). The corresponding matrix of partial derivatives is

$$\frac{\partial \varphi}{\partial \boldsymbol{\beta}'} = \begin{bmatrix} 0 & 0 & 1 & 0 & 0 & 0 & 1 & 0 \\ 0 & 0 & -m_{22,1} & m_{12,1} & 0 & 0 & \alpha_{22,1} & -\alpha_{12,1} \end{bmatrix}.$$

This matrix may have rank 1 under special conditions. In particular, this occurs if $\alpha_{12,1} = m_{12,1} = 0$ and $\alpha_{22,1} = -m_{22,1}$.

The LM statistic for testing (13.5.1) is

$$\lambda_{LM} := s(\widetilde{\boldsymbol{\beta}}_r)' \widetilde{\mathcal{I}}_a(\widetilde{\boldsymbol{\beta}}_r, \widetilde{\Sigma}_u^r)^{-1} s(\widetilde{\boldsymbol{\beta}}_r)/T, \quad (13.5.4)$$

where

$$s(\widetilde{\boldsymbol{\beta}}_r) = \frac{\partial \ln l_0}{\partial \boldsymbol{\beta}} \bigg|_{\widetilde{\boldsymbol{\beta}}_r} = \sum_{t=1}^{T} \left[\frac{\partial u_t(\overline{y}, \boldsymbol{\beta})'}{\partial \boldsymbol{\beta}} \bigg|_{\widetilde{\boldsymbol{\beta}}_r} \right] (\widetilde{\Sigma}_u^r)^{-1} \widetilde{u}_t(\overline{y}, \widetilde{\boldsymbol{\beta}}_r) \quad (13.5.5)$$

is the score vector evaluated at the restricted estimator $\widetilde{\boldsymbol{\beta}}_r$ and

$$\widetilde{\mathcal{I}}_a(\widetilde{\boldsymbol{\beta}}_r, \widetilde{\Sigma}_u^r) = \frac{1}{T} \sum_{t=1}^{T} \left[\frac{\partial u_t(\overline{y}, \boldsymbol{\beta})'}{\partial \boldsymbol{\beta}} \bigg|_{\widetilde{\boldsymbol{\beta}}_r} \right] (\widetilde{\Sigma}_u^r)^{-1} \left[\frac{\partial u_t(\overline{y}, \boldsymbol{\beta})}{\partial \boldsymbol{\beta}'} \bigg|_{\widetilde{\boldsymbol{\beta}}_r} \right] \quad (13.5.6)$$

is an estimator of the asymptotic information matrix based on the restricted estimator $\widetilde{\boldsymbol{\beta}}_r$. Here

$$\widetilde{\Sigma}_u^r = \frac{1}{T}\sum_{t=1}^{T} \widetilde{u}_t(\overline{y},\widetilde{\boldsymbol{\beta}}_r)\widetilde{u}_t(\overline{y},\widetilde{\boldsymbol{\beta}}_r)'.$$

Note that in contrast to Appendix C, Section C.7, an estimator of the *asymptotic* information matrix rather than the information matrix is used in (13.5.4). Therefore T appears in the denominator. If H_0 is true, the statistic λ_{LM} has an asymptotic $\chi^2(N)$-distribution under general conditions.

The LM test is especially suitable for model checking because testing larger VAR or MA orders against a maintained model is particularly easy. A new estimation is not required as long as the null hypothesis does not change. For instance, if we wish to test a given VARMA(p,q) specification against a VARMA$(p+s,q)$ or a VARMA$(p,q+s)$ model, we just need an estimator of the coefficients of the VARMA(p,q) process. Note, however, that a VARMA(p,q) cannot be tested against a VARMA$(p+s,q+s)$, that is, we cannot increase both the VAR and MA orders simultaneously because the VARMA$(p+s,q+s)$ model will not be identified (cancellation is possible!) if the null hypothesis is true. In that case, the LM statistic will not have its usual asymptotic χ^2-distribution.

13.5.2 Residual Autocorrelations and Portmanteau Tests

Alternative tools for model checking are the residual autocorrelations and portmanteau tests. The asymptotic distributions of the residual autocorrelations of estimated VARMA models were discussed by Hosking (1980), Li & McLeod (1981), and Poskitt & Tremayne (1982), among others. We do not give the details here but just mention that the resulting standard errors of autocorrelations at large lags obtained from asymptotic considerations are approximately $1/\sqrt{T}$, while they may be much smaller for low lags, just as for pure finite order VAR processes.

The modified portmanteau statistic is

$$\bar{Q}_h := T^2 \sum_{i=1}^{h}(T-i)^{-1}\,\mathrm{tr}(\widehat{C}_i'\widehat{C}_0^{-1}\widehat{C}_i\widehat{C}_0^{-1}), \qquad (13.5.7)$$

where

$$\widehat{C}_i := \frac{1}{T}\sum_{t=i+1}^{T} \widetilde{u}_t(\overline{y},\widetilde{\boldsymbol{\beta}})\widetilde{u}_{t-i}(\overline{y},\widetilde{\boldsymbol{\beta}})'$$

and the $\widetilde{u}_t(\overline{y},\widetilde{\boldsymbol{\beta}})$'s are the residuals of an estimated VARMA model, as before. Under general conditions, $\bar{Q}_h$ has an *approximate* asymptotic χ^2-distribution. The degrees of freedom are obtained by subtracting the number of freely estimated VARMA coefficients from K^2h.

13.5.3 Prediction Tests for Structural Change

In the pure VAR case, we have considered prediction tests for structural change as model checking devices. If the data generation process is Gaussian, the two tests introduced in Chapter 4, Section 4.6.2, may be applied in the VARMA case as well with minor modifications.

The statistics based on h-step ahead forecasts only are of the form

$$\overline{\tau}_h := \widehat{e}_T(h)' \widehat{\Sigma}_{\widehat{y}}(h)^{-1} \widehat{e}_T(h), \qquad (13.5.8)$$

where $\widehat{e}_T(h) = y_{T+h} - \widehat{y}_T(h)$ is the error vector of an h-step forecast based on an estimated VARMA(p,q) process and $\widehat{\Sigma}_{\widehat{y}}(h)$ is an estimator of the corresponding MSE matrix (see Section 12.5). The statistic may be applied in conjunction with an $F(K, T - K(p+q) - 1)$-distribution. The denominator degrees of freedom may be used even if constraints are imposed on the VARMA coefficients because the F-distribution is just chosen as a small sample approximation to a $\chi^2(K)/K$ distribution. Its justification comes from the fact that $F(K, T - s)$ converges to $\chi^2(K)/K$ for any fixed constant s, as T approaches infinity. Thus, any constant that is subtracted from T in the denominator degrees of freedom of the F-distribution, is justified on the same asymptotic grounds. It is not clear which choice is best from a small sample point of view.

The other statistic considered in Section 4.6.2 is based on 1- to h-step forecasts and, for the present case, it may be modified as

$$\overline{\lambda}_h := T \sum_{i=1}^{h} \widehat{u}'_{T+i} \widetilde{\Sigma}_u^{-1} \widehat{u}_{T+i} / [(T + K(p+q) + 1)Kh] \qquad (13.5.9)$$

and its approximate distribution for a structurally stable Gaussian process is $F(Kh, T - K(p+q) - 1)$. Here $\widehat{u}_{T+i} = y_{T+i} - \widehat{y}_{T+i-1}(1)$ and $\widetilde{\Sigma}_u$ is the ML estimator of Σ_u. Note that the LS estimator of Σ_u was used in Section 4.6.2 instead. Again, there is not much theoretical justification for the choice of the denominator in (13.5.9) and for the denominator degrees of freedom in the approximating F-distribution. More detailed investigations of the small sample distribution of $\overline{\lambda}_h$ are required before firmly based recommendations regarding modifications of the statistic are possible. Here we have just used the direct analogue of the finite order pure VAR case.

It is also possible to fit a finite order VAR process to data generated by a mixed VARMA process and base the prediction tests on forecasts from that model. In Chapter 15, it will be shown that such an approach is theoretically sound under general conditions.

13.6 Critique of VARMA Model Fitting

In this and the previous two chapters, much of the analysis is based on the assumption that the true data generation mechanism is from the VARMA(p,q)

class. In practice, any such model is just an approximation to the actual data generation process. Therefore, the model selection task is not really the problem of finding the true structure but of finding a good or useful approximation to the real life mechanism. Despite this fact, it is sometimes helpful to assume a specific true process or process class to be able to derive, under ideal conditions, the statistical properties of the procedures used. One then hopes that the actual properties of a procedure in a particular practical situation are at least similar to those obtained under ideal conditions.

Against this background, one may wonder whether it is sufficient or even preferable to approximate the generation process of a given multiple time series by a finite order VAR(p) process rather than go through the painstaking specification and estimation of a mixed VARMA model. Clearly, the estimation of VARMA models is in general more complicated than that of finite order VAR models. Moreover, the specification of VAR models by statistical methods is much simpler than that of VARMA models. Are there still situations where it is reasonable to consider the more complicated VARMA models? The answer to this question is in the affirmative. For instance, if subject matter theory suggests a VARMA model with nontrivial MA part, it is often necessary to work with such a specification to answer the questions of interest or derive the relevant results. Also, in some cases, a VARMA approximation may be more parsimonious in terms of the number of parameters involved than an appropriate finite order VAR approximation. In such cases, the VARMA approximation may, for instance, result in more efficient forecasts that justify the costly specification and estimation procedures. The future attractiveness of VARMA models will depend on the easy availability of efficient and robust estimation and specification procedures that reduce the costs to an acceptable level.

In Chapter 15, we will follow another road and explicitly assume that just an approximating and not a true VAR(p) model is fitted. Assumptions will be provided that allow the derivation of statistical properties in that case. So far, we have considered stable, stationary VARMA processes. In the next chapter, extensions for integrated and cointegrated variables will be considered.

13.7 Exercises

Problem 13.1
At the first stage of a final equations form specification procedure, the following two univariate models were obtained:

$$(1 + 0.3L - 0.4L^2)y_{1t} = (1 + 0.6L)v_{1t},$$
$$(1 - 0.5L)y_{2t} = (1 + 0.6L)v_{2t}.$$

Which orders do you choose for the bivariate final equations VARMA representation of $(y_{1t}, y_{2t})'$?

13.7 Exercises

Problem 13.2

At Stage II of a specification procedure for an echelon form of a bivariate system, the following values of the HQ criterion are obtained:

			p_1		
p_2	0	1	2	3	4
0	2.1	1.9	1.5	1.5	1.6
1	1.8	1.7	1.4	1.2	1.3
2	1.7	1.4	1.3	1.4	1.4
3	1.7	1.4	1.3	1.4	1.5
4	1.8	1.7	1.6	1.5	1.5

Choose an estimate $(\hat{p}_1, \hat{p}_2)'$ by the Hannan-Kavalieris procedure. Interpret the estimate in the light of a full search procedure.

Problem 13.3

At the second stage of the Poskitt procedure for a bivariate model, the specification criteria $\text{Cr}_1(p_1, p_2)$ and $\text{Cr}_2(p_1, p_2)$ assume the following values:

$\text{Cr}_1(p_1, p_2)$						$\text{Cr}_2(p_1, p_2)$				
		p_1						p_1		
p_2	0	1	2	3	p_2	0	1	2	3
0	3.5	2.5	1.7	1.8	0	4.2	3.2	3.2	3.2
1	3.5	1.5	1.8	1.7	1	3.5	1.8	1.9	1.9
2	3.5	1.5	1.8	1.9	2	3.1	1.9	1.7	1.6
3	3.5	1.5	1.8	1.4	3	3.4	2.1	1.8	1.9

Use the Poskitt strategy to find an estimate $(\hat{p}_1, \hat{p}_2)$ of the Kronecker indices.

The following problems require the use of a computer. They are based on the *first differences* of the U.S. investment data given in File E2.

Problem 13.4

Determine a final equations form VARMA model for the U.S. investment data for the years 1947–1968 using the specification strategy described in Section 13.2.1.

Problem 13.5

Determine an ARMA_E model for the U.S. investment data using the specification strategy described in Section 13.3.2 with $n = 6$ and based on the HQ criterion. Compare the model to the final equations form model from Problem 13.4.

Problem 13.6

Compute forecasts for the investment series for the years 1969 and 1970 based on (i) the final equations form VARMA model, (ii) the ARMA_E model, and (iii) a bivariate VAR(1) model. Compare the forecasts to the true values and interpret.

Problem 13.7
Compute Φ_i and Θ_i impulse responses from the two models obtained in Problems 13.4 and 13.5, compare and interpret them.

Problem 13.8
Specify a univariate ARMA model for the sum $z_t = y_{1t} + y_{2t}$ of the two investment series for the years 1947–1968. Is the univariate ARMA model compatible with the bivariate echelon form model specified in Problem 13.5? (Hint: Use the results of Section 11.6.1.)

Problem 13.9
Evaluate forecasts for the z_t series of the previous problem for the years 1969–1970 and compare them to forecasts obtained by aggregating the bivariate forecasts from the ARMA_E model of Problem 13.5.

14
Cointegrated VARMA Processes

14.1 Introduction

So far, we have concentrated on stationary VARMA processes for $I(0)$ variables. In this chapter, the variables are allowed to be $I(1)$ and may be cointegrated. As we have seen in Chapter 12, one of the problems in dealing with VARMA models is the nonuniqueness of their parameterization. For inference purposes, it is necessary to focus on a unique representation of a DGP. For stationary VARMA processes, we have considered the echelon form to tackle the identification problem. In the next section, this representation of a VARMA process will be combined with the error correction (EC) form. Thereby it is again possible to separate the long-run cointegration relations from the short-term dynamics. The resulting representation turns out to be a convenient framework for modelling cointegrated variables.

The representation of a VARMA process considered in this chapter is characterized by the cointegrating rank and the Kronecker indices. When these quantities are given, the model can be estimated. Estimation procedures and their asymptotic properties are considered in Section 14.3. A procedure for specifying the Kronecker indices and the cointegrating rank from a given multiple time series will be discussed in Section 14.4. The forecasting aspects of our models will be addressed briefly in Section 14.5 and an example is given in Section 14.6.

In this chapter, an introductory treatment of cointegrated VARMA models is given. The chapter draws on material from Lütkepohl & Claessen (1997) who introduced the error correction echelon form of a VARMA process, Poskitt & Lütkepohl (1995) and Poskitt (2003) who presented estimation and specification procedures for such models, as well as Bartel & Lütkepohl (1998) who explored the small sample properties of some of the procedures. Further references to more advanced treatments of specific issues will be given throughout the chapter.

14.2 The VARMA Framework for $I(1)$ Variables

14.2.1 Levels VARMA Models

In this chapter, it is assumed that some or all of the variables of interest are $I(1)$ variables, whereas the remaining ones are again $I(0)$. Moreover, the variables may be cointegrated. Thus, we consider the situation that was discussed extensively in Part II. In contrast to the framework of that part, we now assume that the DGP of $y_t = (y_{1t}, \ldots, y_{Kt})'$ is from the VARMA class,

$$A_0 y_t = A_1 y_{t-1} + \cdots + A_p y_{t-p} + M_0 u_t + M_1 u_{t-1} + \cdots + M_p u_{t-p}, \ t = 1, 2, \ldots, \tag{14.2.1}$$

or

$$A(L)y_t = M(L)u_t, \quad t = 1, 2, \ldots, \tag{14.2.2}$$

where $u_t = y_t = 0$ for $t \leq 0$ is assumed for convenience and, as usual, u_t is a white noise process with zero mean and nonsingular, time invariant covariance matrix $E(u_t u_t') = \Sigma_u$. Moreover, in (14.2.2) the VAR operator is

$$A(L) := A_0 - A_1 L - \cdots - A_p L^p$$

and the MA operator is

$$M(L) := M_0 + M_1 L + \cdots + M_p L^p.$$

The zero order matrices A_0 and M_0 are assumed to be nonsingular and some of the coefficient matrices may be zero so that the AR or MA order may actually be less than p. The matrix polynomials are assumed to satisfy

$$\det A(z) \neq 0, \ |z| \leq 1, z \neq 1, \quad \text{and} \quad \det M(z) \neq 0, \ |z| \leq 1. \tag{14.2.3}$$

The second part of this condition is the usual invertibility condition for the MA operator. As in the pure VAR case, we allow the VAR operator $A(z)$ to have roots for $z = 1$ to account for integrated and cointegrated components of y_t. As mentioned previously, all component series are at most $I(1)$, that is, Δy_t is stationary or at least asymptotically stationary.

Notice that there are no deterministic terms in our model. For the introductory treatment of the present chapter, this setup is convenient. Of course, in applied work, deterministic terms will usually be required. Although adding such terms is formally straightforward, it is known from the discussion in Chapter 6, Section 6.4, and Chapter 7, Section 7.2.4, that the implications of such terms in models with integrated variables are more complicated than in the stationary case, in particular with respect to statistical inference.

In this context, it may also be worth emphasizing that the zero initial value assumption ($u_t = y_t = 0$ for $t \leq 0$) is not altogether innocent. Allowing

14.2 The VARMA Framework for $I(1)$ Variables

for more general initial values will result in additional complications which we intend to avoid here. Further comments on these issues will be provided later.

Under our assumptions of zero initial values, the process has the pure VAR representation

$$y_t = \sum_{i=1}^{t-1} \Pi_i y_{t-i} + u_t, \tag{14.2.4}$$

where

$$\Pi(z) = \sum_{i=1}^{\infty} \Pi_i z^i = M(z)^{-1} A(z),$$

as in Section 11.3. Notice that the inverse of $M(z)$ exists under our invertibility assumption (14.2.3). The process also has a pure MA representation

$$y_t = \sum_{i=0}^{t-1} \Phi_i u_{t-i},$$

where

$$\Phi(z) = \sum_{i=1}^{\infty} \Phi_i z^i = A(z)^{-1} M(z).$$

Here the inverse of $A(z)$ is defined only in a small neighborhood of zero and, in particular,

$$\sum_{i=1}^{n} \Phi_i$$

may diverge for $n \to \infty$. Our MA representation is still valid due to the zero initial value assumption. The VAR and MA representations of the process show that the uniqueness of the VARMA representation can be discussed in the same way as in Chapter 12. We have to find restrictions for $A(L)$ and $M(L)$ such that a unique relation between $[A(L) : M(L)]$ and $M(L)^{-1} A(L)$ is obtained. From Chapter 12, we know already that the echelon form restrictions can be used for that purpose. In the present situation, a slight modification turns out to be useful. We will present it in the next subsection.

If zero initial values are not assumed, the initial values may also help in identifying the model. A discussion of how initial values can contribute to uniquely identifying a VARMA process with $I(1)$ variables is provided by Poskitt (2004). The problem is, however, that the initial values of the u_t will usually be unknown in practice and may not be available for identification.

14.2.2 The Reverse Echelon Form

In order to obtain a unique representation, we use similar restrictions as in Definition 12.2. We will, however, reverse the roles of $A(L)$ and $M(L)$ in this case, as proposed by Lütkepohl & Claessen (1997). In other words, we now impose the restrictions placed on the VAR operator in Definition 12.2 on $M(L)$ and similarly, the restrictions for $M(L)$ in that definition will now be imposed on the VAR operator. This modification will turn out to be convenient in combining the restrictions with the error correction form. We denote the kl-th elements of $A(z)$ and $M(z)$ by $\alpha_{kl}(z)$ and $m_{kl}(z)$, respectively, and impose the constraints specified in the following definition.

Definition 14.1 (*Reverse Echelon Form*)
The VARMA representation (14.2.1) is in *reverse echelon form* if $A(L)$ and $M(L)$ satisfy the following restrictions: The operator $[A(z) : M(z)]$ is left-coprime,

$$m_{kk}(L) = 1 + \sum_{i=1}^{p_k} m_{kk,i} L^i, \quad \text{for } k = 1, \ldots, K, \tag{14.2.5}$$

$$m_{kl}(L) = \sum_{i=p_k-p_{kl}+1}^{p_k} m_{kl,i} L^i, \quad \text{for } k \neq l, \tag{14.2.6}$$

and

$$\alpha_{kl}(L) = \alpha_{kl,0} - \sum_{i=1}^{p_k} \alpha_{kl,i} L^i, \quad \text{with } \alpha_{kl,0} = m_{kl,0} \quad \text{for } k, l = 1, \ldots, K. \tag{14.2.7}$$

Here

$$p_{kl} = \begin{cases} \min(p_k + 1, p_l) & \text{for } k \geq l, \\ \min(p_k, p_l) & \text{for } k < l, \end{cases} \quad k, l = 1, \ldots, K.$$

The row degrees p_k in this representation are again called *Kronecker indices*. In (14.2.1), $p = \max(p_1, \ldots, p_K)$, that is, p is the maximum row degree or Kronecker index. ARMA$_{RE}(p_1, \ldots, p_K)$ denotes a reverse echelon form with Kronecker indices $p_1, \ldots, p_K$. ∎

It was argued by Poskitt (2004) that the initial conditions may contribute to a unique representation of an integrated VARMA process in such a way that $[A(z) : M(z)]$ does not have to be left-coprime. In that case, the mapping from the set of operators $[A(z) : M(z)]$ to the set of admissible transfer functions $\Phi(z)$ may not be one-to-one, whereas in the present formulation which requires left-coprimeness of $[A(z) : M(z)]$, a one-to-one mapping may be obtained using the reverse echelon form restrictions.

14.2 The VARMA Framework for $I(1)$ Variables

To see the difference to the ARMA$_E$ form discussed in Section 12.1, consider a three-dimensional process with Kronecker indices $(p_1, p_2, p_3) = (1, 2, 1)$ as in (12.1.21)/(12.1.22). In this case,

$$[p_{kl}] = \begin{bmatrix} 1 & 1 & 1 \\ 1 & 2 & 1 \\ 1 & 2 & 1 \end{bmatrix}.$$

Hence, an ARMA$_{RE}(1,2,1)$ has the following form:

$$\begin{bmatrix} 1 & 0 & 0 \\ 0 & 1 & 0 \\ 0 & \alpha_{32,0} & 1 \end{bmatrix} y_t$$

$$= \begin{bmatrix} \alpha_{11,1} & \alpha_{12,1} & \alpha_{13,1} \\ \alpha_{21,1} & \alpha_{22,1} & \alpha_{23,1} \\ \alpha_{31,1} & \alpha_{32,1} & \alpha_{33,1} \end{bmatrix} y_{t-1} + \begin{bmatrix} 0 & 0 & 0 \\ \alpha_{21,2} & \alpha_{22,2} & \alpha_{23,2} \\ 0 & 0 & 0 \end{bmatrix} y_{t-2}$$

$$+ \begin{bmatrix} 1 & 0 & 0 \\ 0 & 1 & 0 \\ 0 & \alpha_{32,0} & 1 \end{bmatrix} u_t + \begin{bmatrix} m_{11,1} & m_{12,1} & m_{13,1} \\ 0 & m_{22,1} & 0 \\ m_{31,1} & m_{32,1} & m_{33,1} \end{bmatrix} u_{t-1} \quad (14.2.8)$$

$$+ \begin{bmatrix} 0 & 0 & 0 \\ m_{21,2} & m_{22,2} & m_{23,2} \\ 0 & 0 & 0 \end{bmatrix} u_{t-2}.$$

Clearly, in this representation the autoregressive operator is unrestricted except for the constraints imposed by the maximum row degrees or Kronecker indices and the zero order matrix ($A_0 = M_0$), whereas zero restrictions are placed on the moving average coefficient matrices attached to low lags of the u_t. For example, in (14.2.8), there are two zero restrictions on M_1. A comparison with the representation in (12.1.21)/(12.1.22) shows that the restrictions imposed on A_1 in (12.1.21) correspond to those imposed on M_1 in (14.2.8).

14.2.3 The Error Correction Echelon Form

The EC form may be obtained from (14.2.1) by subtracting $A_0 y_{t-1}$ on both sides and rearranging terms, as for the VECM form of a VAR model in Section 6.3:

$$\begin{aligned} A_0 \Delta y_t &= \Pi y_{t-1} + \Gamma_1 \Delta y_{t-1} + \cdots + \Gamma_{p-1} \Delta y_{t-p+1} \\ &\quad + M_0 u_t + M_1 u_{t-1} + \cdots + M_p u_{t-p} \end{aligned} \quad (14.2.9)$$

where

$$\Pi = -(A_0 - A_1 - \cdots - A_p)$$

and

$$\Gamma_i = -(A_{i+1} + \cdots + A_p), \quad i = 1, \ldots, p-1.$$

Again, Πy_{t-1} is the error correction term and $r = \text{rk}(\Pi)$ is the cointegrating rank of the system.

If the operators $A(L)$ and $M(L)$ satisfy the reverse echelon from restrictions, it is easily seen that the Γ_i satisfy similar identifying constraints as the A_i. More precisely, Γ_i obeys the same zero restrictions as A_{i+1} for $i = 1, \ldots, p-1$, because a zero restriction on an element $\alpha_{kl,i}$ of A_i implies that the corresponding elements $\alpha_{kl,j}$ of A_j are also zero for $j > i$. For the same reason, the zero restrictions on Π are the same as those on $A_0 - A_1$. This means in particular that there are no echelon form zero restrictions on Π if all Kronecker indices $p_k \geq 1$, $k = 1, \ldots, K$, because in that case the reverse echelon form does not impose zero restrictions on A_1. On the other hand, if some Kronecker indices are zero, this fact has implications for the integration and cointegration structure of the variables. A specific analysis of the relations between the variables is called for in that case. Denoting by ϱ the number of Kronecker indices which are zero, it is not difficult to see that

$$\text{rk}(\Pi) \geq \varrho \tag{14.2.10}$$

(see Problem 14.1). This result has to be taken into account in the procedure for specifying the cointegrating rank of a VARMA system, as discussed in Section 14.4.

An EC model which satisfies the reverse echelon from restrictions will be called an EC-ARMA$_{RE}$ form in the following. As an example, consider again the system (14.2.8). Its EC-ARMA$_{RE}$ form is

$$\begin{bmatrix} 1 & 0 & 0 \\ 0 & 1 & 0 \\ 0 & \alpha_{32,0} & 1 \end{bmatrix} \Delta y_t$$

$$= \begin{bmatrix} \pi_{11} & \pi_{12} & \pi_{13} \\ \pi_{21} & \pi_{22} & \pi_{23} \\ \pi_{31} & \pi_{32} & \pi_{33} \end{bmatrix} y_{t-1} + \begin{bmatrix} 0 & 0 & 0 \\ \gamma_{21,1} & \gamma_{22,1} & \gamma_{23,1} \\ 0 & 0 & 0 \end{bmatrix} \Delta y_{t-1}$$

$$+ \begin{bmatrix} 1 & 0 & 0 \\ 0 & 1 & 0 \\ 0 & \alpha_{32,0} & 1 \end{bmatrix} u_t + \begin{bmatrix} m_{11,1} & m_{12,1} & m_{13,1} \\ 0 & m_{22,1} & 0 \\ m_{31,1} & m_{32,1} & m_{33,1} \end{bmatrix} u_{t-1}$$

$$+ \begin{bmatrix} 0 & 0 & 0 \\ m_{21,2} & m_{22,2} & m_{23,2} \\ 0 & 0 & 0 \end{bmatrix} u_{t-2}.$$

As a further example, consider the three-dimensional ARMA$_{RE}(0,0,1)$ model

$$y_t = \begin{bmatrix} 0 & 0 & 0 \\ 0 & 0 & 0 \\ \alpha_{31,1} & \alpha_{32,1} & \alpha_{33,1} \end{bmatrix} y_{t-1}$$

$$+ u_t + \begin{bmatrix} 0 & 0 & 0 \\ 0 & 0 & 0 \\ m_{31,1} & m_{32,1} & m_{33,1} \end{bmatrix} u_{t-1}. \tag{14.2.11}$$

Its EC-ARMA$_{RE}$ form is

$$\Delta y_t = \begin{bmatrix} -1 & 0 & 0 \\ 0 & -1 & 0 \\ \pi_{31} & \pi_{32} & \pi_{33} \end{bmatrix} y_{t-1} + u_t + \begin{bmatrix} 0 & 0 & 0 \\ 0 & 0 & 0 \\ m_{31,1} & m_{32,1} & m_{33,1} \end{bmatrix} u_{t-1}.$$

Obviously, the rank of

$$\Pi = \begin{bmatrix} -1 & 0 & 0 \\ 0 & -1 & 0 \\ \pi_{31} & \pi_{32} & \pi_{33} \end{bmatrix}$$

is at least 2 and, thus, the cointegrating rank in this case is also at least 2.

Specifying an EC-ARMA$_{RE}$ model requires that the cointegrating rank r is determined, the Kronecker indices $p_1, \ldots, p_K$ are obtained and possibly further over identifying zero restrictions are placed on the coefficient matrices Γ_i and M_j. Before we consider strategies for these tasks, we discuss the estimation of EC-ARMA$_{RE}$ models for given cointegrating rank and Kronecker indices in the next section.

14.3 Estimation

14.3.1 Estimation of ARMA$_{RE}$ Models

For given Kronecker indices, an ARMA$_{RE}$ model can be estimated even if the cointegrating rank is unknown. Under Gaussian assumptions, ML estimation can be used. The estimators may be determined by maximizing a log-likelihood function as in (12.2.24),

$$\ln l_0(\gamma, \Sigma_u) = -\frac{T}{2} \ln |\Sigma_u| - \frac{1}{2} \sum_{t=1}^{T} u_t(\gamma)' \Sigma_u^{-1} u_t(\gamma), \tag{14.3.1}$$

where an additive constant is dropped and zero initial conditions are assumed so that

$$u_t(\gamma) = y_t - \sum_{i=1}^{t-1} \Pi_i(\gamma) y_{t-i}.$$

If the initial values are nonzero, $\ln l_0$ is just an approximate log-likelihood. Here γ contains all unrestricted autoregressive and moving average parameters, as in Section 12.2, and maximization may proceed by an iterative procedure, as in Section 12.3. Starting values are required for such an algorithm.

The preliminary estimator presented in Chapter 12, Section 12.3.4, can be used for that purpose (e.g., Poskitt (2003)).

As in the case of a cointegrated VAR model, the ML estimators have asymptotic properties which are in some respects different from those obtained in the stationary case. Roughly speaking, they are the same that would be obtained if the true cointegration matrix were known. Thus, if $0 < r < K$, generally the ML estimator $\widetilde{\gamma}$ is consistent and

$$\sqrt{T}(\widetilde{\gamma} - \gamma) \xrightarrow{d} \mathcal{N}(0, \Sigma_{\widetilde{\gamma}}),$$

where the covariance matrix $\Sigma_{\widetilde{\gamma}}$ is singular. These results follow from Yap & Reinsel (1995) and also hold under suitable alternative conditions if y_t is not Gaussian.

If the cointegrating rank is known, it is often desirable to estimate the EC-ARMA$_{RE}$ form of the process because it also provides estimates of the cointegration relations which may well be of major interest. Therefore, estimation of these models will be considered next.

14.3.2 Estimation of EC-ARMA$_{RE}$ Models

If identifying restrictions are imposed on the cointegration matrix, then estimation of the EC-ARMA$_{RE}$ form can also be done by Gaussian ML based on a log-likelihood function similar to (14.3.1), where γ now contains the free parameters of the EC-ARMA$_{RE}$ form. An alternative approach would be to estimate the cointegration matrix β first by reduced rank regression or an EGLS procedure based on a long VAR(n) model, as in Section 7.2. The properties of this estimator will be discussed further in Chapter 15, where fitting approximate VAR models is discussed. For the present purposes, it is sufficient to note that this estimator, say $\widehat{\beta}$, may be used in an ML procedure which estimates the other parameters by maximizing the log-likelihood function conditionally on $\widehat{\beta}$. In other words, the cointegration parameters are fixed at the first stage estimator $\widehat{\beta}$ of the cointegration matrix β and then the log-likelihood is maximized with respect to the other parameters. The resulting estimators have the same asymptotic properties as the full ML estimators (see Yap & Reinsel (1995)).

Starting values for the other parameters that may be used as initial values for an iterative procedure to maximize the log-likelihood function, may be determined in an analogous way as in Section 12.3.4. The short-run and loading parameter estimators have an asymptotic normal distribution which is the same as if the cointegration matrix β were known. This result, of course, is analogous to the pure VAR case considered in Chapter 7 (see also Phillips (1991, Remark (n)) and Yap & Reinsel (1995)).

The previous discussion assumes given Kronecker indices and possibly a known cointegrating rank. Statistical procedures for specifying these quantities will be discussed next.

14.4 Specification of EC-ARMA$_{RE}$ Models

14.4.1 Specification of Kronecker Indices

For stationary processes, proposals for specifying the Kronecker indices of an ARMA$_E$ model were discussed in Section 13.3. The strategies for specifying the Kronecker indices of cointegrated ARMA$_{RE}$ forms presented in this section were proposed by Poskitt & Lütkepohl (1995) and Poskitt (2003). In the latter article, it is also argued that they result in consistent estimators of the Kronecker indices under suitable conditions. In a simulation study, Bartel & Lütkepohl (1998) found that they worked reasonably well in small samples, at least for the processes explored in their Monte Carlo study.

The specification procedures may be partitioned in two stages. The first stage is the same as for the procedures for stationary processes discussed in Section 13.3.2 and consists of fitting a long autoregression by least squares in order to obtain estimates of the unobservable innovations u_t, $t = 1, \ldots, T$.

STAGE I: Use multivariate LS estimation to fit a long VAR(n) process to the data to obtain residuals $\widehat{u}_t(n)$. ∎

These residuals are then substituted for the unknown lagged u_t's in the individual equations of an ARMA$_{RE}$ form which may then be estimated by linear LS procedures. Based on the equations estimated in this way, a choice of the Kronecker indices is made using model selection criteria. Poskitt & Lütkepohl (1995), Guo, Huang & Hannan (1990), and Huang & Guo (1990) showed that the estimated residuals $\widehat{u}_t(n)$ are "good" estimates of the true residuals if n approaches infinity at a suitable rate, as T goes to infinity (see Lemma 3.1 of Poskitt & Lütkepohl (1995) for details).

The methods presented in the following differ in the way they choose the Kronecker indices in the next step. An obvious idea may be to search over all models associated with Kronecker indices

$$\{(p_1, \ldots, p_K) | 0 \leq p_k \leq p_{\max}, k = 1, \ldots, K\}$$

for some prespecified upper bound $p_{\max}$ and choose the set of Kronecker indices which optimizes some model selection criterion, as in Section 13.3.2 for the stationary case. The two procedures presented in the following are more efficient computationally and they are similar to Poskitt's procedure presented in Section 13.3.4. The first variant uses linear regressions to estimate the individual equations separately for different lag lengths. A choice of the optimal lag length is then based on some prespecified criterion similar to those considered for the stationary case. The following formal description of the procedure is taken from Poskitt & Lütkepohl (1995).

STAGE II: Proceed in the following steps.

(ia) For $m = 0$, set $T\widetilde{\sigma}_k^2(m)$ equal to the residual sum of squares from the regression of y_{kt} on a constant and $(y_{jt} - \widehat{u}_{jt}(n))$, $j = 1, \ldots, K$, $j \neq k$. For $m = 1, \ldots, p_{\max} \leq n$, regress y_{kt} on a constant, $(y_{jt} - \widehat{u}_{jt}(n))$, $j = 1, \ldots, K$, $j \neq k$, and y_{t-s} and $\widehat{u}_{t-s}(n)$, $s = 1, \ldots, m$, and determine the residual sums of squares, $T\widetilde{\sigma}_k^2(m)$, for $k = 1, \ldots, K$.

(ib) For $k = 1, \ldots, K$, compute a selection criterion of the form

$$\mathrm{Cr}_k(m) = \ln \widetilde{\sigma}_k^2(m) + c_T m/T, \quad m = 0, 1, \ldots, p_{\max},$$

where c_T is a function of T which will be specified later.

(ii) Set the estimate of the k-th Kronecker index equal to

$$\widehat{p}_k = \underset{0 \leq m \leq p_{\max}}{\arg\min} \; \mathrm{Cr}_k(m), \quad k = 1, \ldots, K.$$

∎

In the regressions in Step (ia), restrictions from the echelon structure are not explicitly taken into account, because for each value of m, the algorithm implicitly assumes that the current index under consideration is the smallest and, thus, no restrictions are imported from other equations. Still, the k-th equation will be misspecified whenever m is less than the true Kronecker index because in that case, lagged values required for a correct specification are omitted. On the other hand, if m is greater than the true Kronecker index, the k-th equation will be correctly specified but may include redundant parameters and variables. Therefore, it is intuitively plausible that for an appropriate choice of c_T, the criterion function $\mathrm{Cr}_k(m)$ will be minimized asymptotically when m is equal to the true Kronecker index. For practical purposes, possible choices of c_T are $c_T = n \ln T$ or $c_T = n^2$.

At Stage II, values for n, $p_{\max}$, and c_T have to be chosen. The theoretical consistency results stated in Poskitt (2003) are quite general and provide an asymptotic justification for many different values of these quantities. The following choices may be considered in practice:

- Choose n by AIC or use $n = \max\{(\ln T)^a, \widehat{p}(\mathrm{AIC})\}$, where $a > 1$.
- Choose $p_{\max} = \frac{1}{2}n$.
- Choose $c_T = n \ln T$ or $c_T = n^2$.

Poskitt & Lütkepohl (1995) also proposed a modification of Stage II which permits to take into account coefficient restrictions derived from those equations in the system which have smaller Kronecker indices. In that modification, after running through Stage II for the first time, we fix the smallest Kronecker index and repeat Stage II, but search only those equations which are found to have indices larger than the smallest. In this second application of Stage II, the restrictions implied by the smallest Kronecker index found in the first round are taken into account when the second smallest index is determined. We proceed in this way by fixing the smallest Kronecker index found in each successive round until all the Kronecker indices have been specified. In this procedure, the variables are ordered in such a way that the Kronecker indices of the final system are ordered from largest to smallest. That is, the variable whose equation is associated with the smallest Kronecker index is placed last in the list of variables. The one with the second smallest Kronecker index is assigned the next to the last place and so on. For details, see Poskitt & Lütkepohl (1995) and Poskitt (2003).

It should be understood that the Kronecker indices found in such a procedure for a given time series of finite length can only be expected to be a reasonable starting point for a more refined analysis of the system under consideration. Based on the specified Kronecker indices, a more efficient procedure for estimating the parameters may be applied and the model may be modified subsequently.

So far, we have not discussed the choice of the cointegrating rank. In practice, of course, this quantity is unlikely to be known. Comments on the estimation of r will be given in the following subsection.

14.4.2 Specification of the Cointegrating Rank

Saikkonen & Luukkonen (1997) and Lütkepohl & Saikkonen (1999b) showed that Johansen's LR tests for the cointegrating rank (see Section 8.2) maintain their asymptotic properties even if a finite order VAR process is fitted although the true underlying process has an infinite order VAR structure. Consequently, these tests may be applied at Stage I of the present specification procedure. The cointegrating rank is then determined independently of the Kronecker indices. Alternatively, Yap & Reinsel (1995) extended the likelihood ratio principle to VARMA processes and developed cointegration rank tests under the assumption that identified versions of $A(z)$ and $M(z)$ are used. Thus, these tests may be applied once the Kronecker indices have been specified. Whatever approach is adopted, for our purposes the following modification is noteworthy.

If a Kronecker index $p_k = 0$, the variable y_{kt} inherits all of its dynamics from other variables in the system and it is known from (14.2.10) that the cointegrating rank $r \geq \varrho$, the number of zero Kronecker indices. Hence, the testing procedure proceeds by considering only null hypotheses where r is greater than or equal to ϱ. In other words, the following sequence of null hypotheses is tested: $H_0 : r = \varrho$, $H_0 : r = \varrho + 1$, ..., $H_0 : r = K - 1$. The estimator of r is chosen such that it is the smallest value for which H_0 cannot be rejected.

Once a model has been estimated, some checks for model adequacy are in order and possible further model reductions or modifications may be called for. For instance, insignificant parameter estimates may be restricted to zero. Here it is convenient that the t-ratios of the short-run parameters have their usual asymptotic standard normal distributions under the null hypothesis, due to the asymptotic normal distribution of the ML estimators. Thus, they can be used for significance tests in the usual way and may help to place over identifying restrictions on the parameters. Moreover, a detailed analysis of the residual properties should be performed to reveal possible model deficiencies. The checks for model adequacy described in Chapters 4 and 8 can be used here as well with appropriate modifications.

14.5 Forecasting Cointegrated VARMA Processes

Forecasting cointegrated VARMA processes proceeds completely analogously to forecasting stationary VARMA processes. The same formulas can be used. Like for pure VAR models, the properties of the forecasts will be different, however. In particular, the forecast error covariance matrices will be unbounded for increasing forecast horizon. Hence, also forecast intervals will be unbounded in length. In this respect, the properties of the forecasts are analogous to those of cointegrated pure VAR processes. The reader is referred to Section 6.5 for details.

14.6 An Example

For illustrative purposes, we use an example from Lütkepohl & Claessen (1997), based on the U.S. macroeconomic data which were also considered in Section 7.4.3. The data are available in File E3. It consists of 136 quarterly observations for the years 1954.1 to 1983.4 of the real money stock M1 (y_{1t}), GNP in billions of 1982 dollars (y_{2t}), the discount interest rate on new issues of 91-day treasury bills (y_{3t}), and the yield on long term (20 years) treasury bonds (y_{4t}). Logarithms of seasonally adjusted GNP and M1 data are used. Thus, y_t is a four-dimensional vector. Notice that we do not use the full sample period covered in File E3 but truncate the data for the last four years. The reason is that in the exercises readers are asked to perform a forecast comparison based on the model presented in the following. The data for the years 1984–1987 are set aside for this comparison.

Following the procedure outlined in Section 14.4.2, the cointegrating rank may be determined with LR type tests applied to a long VAR model. After running through an extensive specification procedure, Lütkepohl & Claessen (1997) finally specified an EC-ARMA$_{RE}(2,1,1,1)$ model with cointegrating rank 1 for the data generation process of this system and obtained the following estimated model

$$\begin{bmatrix} 1 & 0 & 0 & 0 \\ -.509 & 1 & 0 & 0 \\ {\scriptstyle (.117)} & & & \\ -.099 & 0 & 1 & 0 \\ {\scriptstyle (.105)} & & & \\ .084 & 0 & 0 & 1 \\ {\scriptstyle (.043)} & & & \end{bmatrix} \Delta y_t$$

$$= \begin{bmatrix} .091 \\ (.035) \\ .216 \\ (.060) \\ .190 \\ (.056) \\ .055 \\ (.022) \end{bmatrix} + \begin{bmatrix} -.039 \\ (.015) \\ -.090 \\ (.026) \\ -.082 \\ (.024) \\ -.023 \\ (.010) \end{bmatrix} [1, -.343, -16.72, 19.35] y_{t-1}$$

$$+ \begin{bmatrix} .810 & .074 & -.682 & -.507 \\ (.084) & (.069) & (.101) & (.192) \\ 0 & 0 & 0 & 0 \\ 0 & 0 & 0 & 0 \\ 0 & 0 & 0 & 0 \end{bmatrix} \Delta y_{t-1}$$

$$+ \begin{bmatrix} 1 & 0 & 0 & 0 \\ -.509 & 1 & 0 & 0 \\ (.117) & & & \\ -.099 & 0 & 1 & 0 \\ (.105) & & & \\ .084 & 0 & 0 & 1 \\ (.043) & & & \end{bmatrix} u_t$$

$$+ \begin{bmatrix} -.478 & 0 & 0 & 0 \\ (.109) & & & \\ -.101 & .006 & .339 & .898 \\ (.113) & (.091) & (.144) & (.258) \\ .160 & .123 & .377 & .154 \\ (.084) & (.070) & (.106) & (.202) \\ .037 & .043 & .093 & -.070 \\ (.045) & (.036) & (.057) & (.103) \end{bmatrix} u_{t-1}$$

$$+ \begin{bmatrix} .082 & -.022 & .205 & .646 \\ (.091) & (.073) & (.123) & (.220) \\ 0 & 0 & 0 & 0 \\ 0 & 0 & 0 & 0 \\ 0 & 0 & 0 & 0 \end{bmatrix} u_{t-2}. \qquad (14.6.1)$$

Estimated standard errors are given in parentheses. The cointegration vector $\widehat{\beta}' = [1, -.343, -16.72, 19.35]$ was obtained by estimating a VECM with one lagged difference of y_t and with cointegrating rank 1, using the ML procedure presented in Section 7.2.3 and normalizing the first element of $\widehat{\beta}$ to be 1.

Some of the parameter values in (14.6.1) are quite small compared to their estimated standard errors. In particular, some of them are not significant under a two-standard error criterion. Therefore, zero restrictions were placed on the coefficients and the following estimated model was obtained:

$$\begin{bmatrix} 1 & 0 & 0 & 0 \\ -.476 & 1 & 0 & 0 \\ (.107) & & & \\ 0 & 0 & 1 & 0 \\ .145 & 0 & 0 & 1 \\ (.032) & & & \end{bmatrix} \Delta y_t$$

$$= \begin{bmatrix} .094 \\ (.035) \\ .219 \\ (.056) \\ .207 \\ (.057) \\ .069 \\ (.020) \end{bmatrix} + \begin{bmatrix} -.042 \\ (.016) \\ -.096 \\ (.026) \\ -.094 \\ (.026) \\ -.031 \\ (.009) \end{bmatrix} [1, -.343, -16.72, 19.35] y_{t-1}$$

$$+ \begin{bmatrix} .772 & .087 & .788 & .198 \\ (.063) & (.067) & (.100) & (.195) \\ 0 & 0 & 0 & 0 \\ 0 & 0 & 0 & 0 \\ 0 & 0 & 0 & 0 \end{bmatrix} \Delta y_{t-1}$$

$$+ \begin{bmatrix} 1 & 0 & 0 & 0 \\ -.476 & 1 & 0 & 0 \\ (.107) & & & \\ 0 & 0 & 1 & 0 \\ .145 & 0 & 0 & 1 \\ (.032) & & & \end{bmatrix} u_t$$

$$+ \begin{bmatrix} -.640 & 0 & 0 & 0 \\ (.082) & & & \\ 0 & 0 & 0 & 1.105 \\ & & & (.297) \\ .331 & 0 & .339 & 0 \\ (.104) & & (.083) & \\ .162 & 0 & .110 & -.323 \\ (.042) & & (.054) & (.095) \end{bmatrix} u_{t-1}$$

$$+ \begin{bmatrix} .107 & 0 & .233 & .984 \\ (.081) & & (.114) & (.196) \\ 0 & 0 & 0 & 0 \\ 0 & 0 & 0 & 0 \\ 0 & 0 & 0 & 0 \end{bmatrix} u_{t-2}. \quad (14.6.2)$$

This example is just meant to illustrate that the procedures presented in this chapter are indeed feasible in practice. The reader is encouraged to perform a forecast comparison of the model presented here with pure VAR models and VECMs for the data (see Problem 14.6).

14.7 Exercises

14.7.1 Algebraic Exercises

Problem 14.1
Show that in the model (14.2.9), rk(Π) $\geq \varrho$, where ϱ is the number of Kronecker indices which are zero. (Hint: Consider the matrix $A_0 - A_1$).

Problem 14.2
Write down an ARMA$_{RE}$(2, 1, 2) model explicitly in matrix form and also write down the corresponding EC-ARMA$_{RE}$ form.

Problem 14.3
Consider the following EC-ARMA$_{RE}$ model:

$$\begin{bmatrix} 1 & 0 \\ \alpha_{21,0} & 1 \end{bmatrix} \Delta y_t = \begin{bmatrix} \pi_{11} & \pi_{12} \\ \pi_{21} & \pi_{22} \end{bmatrix} y_{t-1} + \begin{bmatrix} \gamma_{11,1} & \gamma_{12,1} \\ 0 & 0 \end{bmatrix} \Delta y_{t-1}$$

$$+ \begin{bmatrix} 1 & 0 \\ \alpha_{21,0} & 1 \end{bmatrix} u_t + \begin{bmatrix} m_{11,1} & 0 \\ m_{21,1} & m_{22,1} \end{bmatrix} u_{t-1}.$$

(a) Write the model in ARMA$_{RE}$ form.
(b) Specify the Kronecker indices.
(c) How many over-identifying restrictions are present in this model?
(d) Write the model in pure VAR form.

14.7.2 Numerical Exercises

The following problems are based on the U.S. data given in File E3, as described in Section 14.6. The variables are defined in the same way as in that section. Thus, a system of dimension four is considered.

Problem 14.4
Fit a pure VAR model to the four-dimensional data set without considering integration and cointegration properties of the variables. Use only the data for 1954.1–1983.4 for modelling and estimation. Compute forecasts from the model for the period 1984.1–1987.4.

Problem 14.5
Use the following steps in constructing VECMs for the period 1954.1–1983.4 and computing forecasts of the four variables for the period 1984.1–1987.4.

(a) Determine the cointegrating rank of the system.
(b) Estimate the cointegration relation(s) with the reduced rank ML and the EGLS methods discussed in Chapter 7.
(c) Construct subset VECMs based on the estimated cointegration relations from the previous step.
(d) Confirm that the models obtained in the previous steps are adequate representations of the data generation process.
(e) Compute forecasts from your model for the period 1984.1–1987.4.

Problem 14.6
Compare the forecasts obtained in Problems 14.4 and 14.5 with those from the EC-ARMA$_{RE}$ model (14.6.2) on the basis of the MSEs. Discuss the results. (Hint: See Lütkepohl & Claessen (1997).)

15
Fitting Finite Order VAR Models to Infinite Order Processes

15.1 Background

In the previous chapters, we have derived properties of models, estimators, forecasts, and test statistics under the assumption of a true model. We have also argued that such an assumption is virtually never fulfilled in practice. In other words, in practice, all we can hope for is a model that provides a useful approximation to the actual data generation process of a given multiple time series. In this chapter, we will, to some extent, take into account this state of affairs and assume that an approximating rather than a true model is fitted. Specifically, we assume that the true data generation process is an infinite order VAR process and, for a given sample size T, a finite order VAR(p) is fitted to the data.

In practice, it is likely that a higher order VAR model is considered if the sample size or time series length is larger. In other words, the order p increases with the sample size T. If an order selection criterion is used in choosing the VAR order, the maximum order to be considered is likely to depend on T. This again implies that the actual order chosen depends on the sample size because it will depend on the maximum order. In summary, the actual order selected may be regarded as a function of the sample size T. In order to derive statistical properties of estimators and forecasts, we will make this assumption in the following. More precisely, we will assume that the VAR order goes to infinity with the sample size. Under that assumption, an asymptotic theory has been developed that will be discussed in this chapter.

In Section 15.2, the assumptions for the underlying true process and for the order of the process fitted to the data are specified in detail and asymptotic estimation results are provided for stable processes. In Section 15.3, the consequences for forecasting are discussed and impulse response analysis is considered in Section 15.4. Our standard investment/income/consumption example is used to contrast the present approach to that considered in Chapter 3, where a true finite order process is assumed. Finally, in Section 15.5, extensions to cointegrated processes are discussed.

15.2 Multivariate Least Squares Estimation

Suppose the generation process of a given multiple time series is a stationary, stable, K-dimensional, infinite order VAR process,

$$y_t = \sum_{i=1}^{\infty} \Pi_i y_{t-i} + u_t, \qquad (15.2.1)$$

with absolutely summable Π_i, that is,

$$\sum_{i=1}^{\infty} \|\Pi_i\| < \infty \qquad (15.2.2)$$

(see Appendix C.3) and canonical MA representation

$$y_t = \sum_{i=0}^{\infty} \Phi_i u_{t-i}, \quad \Phi_0 = I_K, \qquad (15.2.3)$$

satisfying

$$\det\left(\sum_{i=0}^{\infty} \Phi_i z^i\right) \neq 0 \quad \text{for } |z| \leq 1 \quad \text{and} \quad \sum_{i=1}^{\infty} i^{1/2} \|\Phi_i\| < \infty. \qquad (15.2.4)$$

The zero mean assumption implied by these conditions is not essential and is imposed for convenience only. Stable, invertible VARMA processes satisfy the foregoing conditions. The assumptions allow for more general processes, however. Of course, the generation process may also be a stable, finite order VAR(p) in which case $\Pi_i = 0$ for $i > p$.

We have argued in the previous section that in practice the true structure will usually be unknown and the investigator may consider fitting a finite order VAR process with the VAR order depending on the length T of the available time series. For this situation, Lewis & Reinsel (1985) have shown consistency and asymptotic normality of the multivariate LS estimators. For univariate processes, similar results were discussed earlier by Berk (1974) and Bhansali (1978).

To state these results formally, we use the following notation:

$$\Pi(n) := [\Pi_1, \ldots, \Pi_n],$$

$$\boldsymbol{\pi}(n) := \text{vec } \Pi(n).$$

Fitting a VAR(n) process, the i-th estimated coefficient matrix is denoted by $\widehat{\Pi}_i(n)$,

$$\widehat{\Pi}(n) := [\widehat{\Pi}_1(n), \ldots, \widehat{\Pi}_n(n)],$$

and

$$\widehat{\boldsymbol{\pi}}(n) := \text{vec } \widehat{\Pi}(n).$$

Now we can state a result of Lewis & Reinsel (1985).

15.2 Multivariate Least Squares Estimation

Proposition 15.1 (*Properties of the LS Estimator of an Approximating VAR Model*)
Let the multiple time series $y_1, \ldots, y_T$ be generated by a potentially infinite order VAR process satisfying (15.2.1)–(15.2.4) with standard white noise u_t. Suppose finite order VAR(n_T) processes are fitted by multivariate LS and assume that the order n_T depends upon the sample size T such that

$$n_T \to \infty, \quad n_T^3/T \to 0, \quad \text{and} \quad \sqrt{T} \sum_{i=n_T+1}^{\infty} \|\Pi_i\| \to 0 \quad \text{as} \quad T \to \infty. \tag{15.2.5}$$

Furthermore, let c_1, c_2 be positive constants and $\mathbf{f}(n)$ a sequence of $(K^2 n \times 1)$ vectors such that

$$0 < c_1 \leq \mathbf{f}(n)'\mathbf{f}(n) \leq c_2 < \infty \quad \text{for } n = 1, 2, \ldots.$$

Then

$$\frac{\sqrt{T - n_T}\, \mathbf{f}(n_T)'[\widehat{\boldsymbol{\pi}}(n_T) - \boldsymbol{\pi}(n_T)]}{[\mathbf{f}(n_T)'(\Gamma_{n_T}^{-1} \otimes \Sigma_u)\mathbf{f}(n_T)]^{1/2}} \xrightarrow{d} \mathcal{N}(0, 1), \tag{15.2.6}$$

where

$$\Gamma_n := E\left(\begin{bmatrix} y_t \\ \vdots \\ y_{t-n+1} \end{bmatrix} [y_t', \ldots, y_{t-n+1}']\right). \tag{15.2.7}$$

∎

Remark 1 The assumption (15.2.5) means that, although the VAR order has to go to infinity with the sample size, it has to do so at a much slower rate because $n_T^3/T \to 0$. The requirement

$$\sqrt{T} \sum_{i=n_T+1}^{\infty} \|\Pi_i\| \to 0 \tag{15.2.8}$$

is always satisfied if y_t is actually a finite order VAR process and $n_T \to \infty$. For infinite order VAR processes, this condition implies a lower bound for the rate at which n_T goes to infinity. To see this, consider the *univariate* MA(1) process

$$y_t = u_t - m u_{t-1},$$

where $0 < |m| < 1$ to ensure invertibility. Its AR representation is

$$y_t = -\sum_{i=1}^{\infty} m^i y_{t-i} + u_t$$

and condition (15.2.8) becomes

$$\sqrt{T} \sum_{i=n+1}^{\infty} |m^i| = \sqrt{T} |m|^{n+1} \sum_{i=0}^{\infty} |m|^i$$

$$= \sqrt{T} \frac{|m|^{n+1}}{1-|m|} \xrightarrow[T \to \infty]{} 0. \qquad (15.2.9)$$

Here the subscript T has been dropped from n_T for notational simplicity. In this example, $n_T = T^{1/\epsilon}$ with $\epsilon > 3$ is a possible choice for the sequence n_T that satisfies both (15.2.9) and $n_T^3/T \to 0$. On the other hand, $n_T = \ln \ln T$ is not a permissible choice because in this case

$$\sqrt{T}|m|^{n_T+1}$$

does not approach zero as $T \to \infty$. This result is easily established by considering the logarithm of (15.2.9),

$$\tfrac{1}{2} \ln T + (n_T + 1) \ln |m| - \ln(1 - |m|),$$

which goes to infinity for $n_T = \ln \ln T$.

In summary, (15.2.8) is a lower bound and $n_T^3/T \to 0$ establishes an upper bound for the rate at which n_T has to go to infinity with the sample size T. ∎

Remark 2 Proposition 15.1 implies that for fixed m,

$$\sqrt{T - n_T} \operatorname{vec}([\widehat{\Pi}_1(n_T), \ldots, \widehat{\Pi}_m(n_T)] - [\Pi_1, \ldots, \Pi_m])$$

has an asymptotic multivariate normal distribution with mean zero and covariance matrix $V \otimes \Sigma_u$, where V is obtained as follows: Let V_n be the upper left-hand $(Km \times Km)$ block of the inverse of Γ_n, for $n \geq m$. Then $V = \lim_{n \to \infty} V_n$. Loosely speaking, V is the upper left-hand $(Km \times Km)$ block of the inverse of the infinite order matrix

$$E\left(\begin{bmatrix} y_t \\ y_{t-1} \\ \vdots \end{bmatrix} [y_t', y_{t-1}', \ldots]\right).$$

Thus, the result can be used for inference on a finite number of parameters. It is also possible, however, to use the result from Proposition 15.1 to construct tests for hypotheses involving an infinite number of restrictions. Such hypotheses can arise in studying Granger-causality in infinite order VAR processes. This case was considered explicitly by Lütkepohl & Poskitt (1996). ∎

Remark 3 If the data generation process has nonzero mean originally, the sample mean $\bar{y}$ may be subtracted initially from the data. It is asymptotically independent of the $\widehat{\Pi}_i(n_T)$ and has an asymptotic normal distribution,

15.2 Multivariate Least Squares Estimation

$$\sqrt{T}(\bar{y} - \mu) \xrightarrow{d} \mathcal{N}(0, \Sigma_{\bar{y}}),$$

where

$$\Sigma_{\bar{y}} = \left(\sum_{i=0}^{\infty} \Phi_i\right) \Sigma_u \left(\sum_{i=0}^{\infty} \Phi_i\right)'.$$

∎

A corresponding result from Lütkepohl & Poskitt (1991) for the white noise covariance matrix is stated next.

Proposition 15.2 (*Asymptotic Properties of the White Noise Covariance Matrix Estimator*)
Let

$$\widehat{u}_t(n) := y_t - \sum_{i=1}^{n} \widehat{\Pi}_i(n) y_{t-i}, \quad t = 1, \ldots, T,$$

be the multivariate LS residuals from a VAR(n) model fitted to a multiple time series of length T, let

$$\widetilde{\Sigma}_u(n) := \frac{1}{T} \sum_{t=1}^{T} \widehat{u}_t(n) \widehat{u}_t(n)'$$

be the corresponding estimator of the white noise covariance matrix and let $U := [u_1, \ldots, u_T]$ so that

$$\frac{1}{T} UU' = \frac{1}{T} \sum_{t=1}^{T} u_t u_t'$$

is an estimator of Σ_u based on the true white noise process u_t. Then, under the conditions of Proposition 15.1,

$$\text{plim } \sqrt{T}(\widetilde{\Sigma}_u(n_T) - T^{-1}UU') = 0.$$

∎

We know from Chapter 3, Propositions 3.2 and 3.4, that, for a Gaussian process, $T^{-1}UU'$ has an asymptotic normal distribution,

$$\sqrt{T} \, \text{vech}(T^{-1}UU' - \Sigma_u) \xrightarrow{d} \mathcal{N}(0, 2\mathbf{D}_K^+(\Sigma_u \otimes \Sigma_u)\mathbf{D}_K^{+\prime}), \qquad (15.2.10)$$

where, as usual, $\mathbf{D}_K^+ = (\mathbf{D}_K'\mathbf{D}_K)^{-1}\mathbf{D}_K'$ is the Moore-Penrose inverse of the $(K^2 \times \frac{1}{2}K(K+1))$ duplication matrix $\mathbf{D}_K$. Using Proposition C.2(2) of Appendix C.1, Proposition 15.2 implies that

$$\sqrt{T}\,\text{vech}(\widetilde{\Sigma}_u(n_T) - \Sigma_u)$$

has precisely the same asymptotic distribution as the one in (15.2.10). Obviously, this distribution does not depend on the VAR structure of y_t or the VAR coefficients. In addition, the estimator $\widetilde{\Sigma}_u(n_T)$ is asymptotically independent of $\widehat{\pi}(n_T)$. In the following, the consequences of these results for prediction and impulse response analysis will be discussed.

15.3 Forecasting

15.3.1 Theoretical Results

Suppose the VAR(n_T) model estimated in the previous section is used for forecasting. In that case, the usual h-step forecast at origin T, $\widetilde{y}_T(h)$, can be computed recursively for $h = 1, 2, \ldots$, using

$$\widetilde{y}_T(h) = \sum_{i=1}^{n_T} \widehat{\Pi}_i(n_T)\widetilde{y}_T(h-i), \tag{15.3.1}$$

where $\widetilde{y}_T(j) := y_{T+j}$ for $j \leq 0$ (see Section 3.5). We use the notation

$$\widetilde{\mathbf{y}}_T(h) := \begin{bmatrix} \widetilde{y}_T(1) \\ \vdots \\ \widetilde{y}_T(h) \end{bmatrix}, \quad \mathbf{y}_T(h) := \begin{bmatrix} y_T(1) \\ \vdots \\ y_T(h) \end{bmatrix}, \quad \mathbf{y}_{T,h} := \begin{bmatrix} y_{T+1} \\ \vdots \\ y_{T+h} \end{bmatrix}$$

and

$$\mathbf{\Sigma_y}(h) := E\left\{[\mathbf{y}_{T,h} - \mathbf{y}_T(h)][\mathbf{y}_{T,h} - \mathbf{y}_T(h)]'\right\},$$

where $y_T(j)$, $j = 1, \ldots, h$, is the optimal j-step forecast at origin T based on the infinite past, that is,

$$y_T(j) = \sum_{i=1}^{\infty} \Pi_i y_T(j-i)$$

with $y_T(i) := y_{T+i}$ for $i \leq 0$ (see Section 11.5). The following result is also essentially due to Lewis & Reinsel (1985) (see also Lütkepohl (1987, Section 3.3, Proposition 3.2)).

Proposition 15.3 (*Asymptotic Distributions of Estimated Forecasts*)
Under the conditions of Proposition 15.1, if y_t is a Gaussian process and if independent processes with identical stochastic structures are used for estimation and forecasting, respectively, then

$$\sqrt{\frac{T}{n_T}}[\widetilde{\mathbf{y}}_T(h) - \mathbf{y}_T(h)] \xrightarrow{d} \mathcal{N}(0, K\mathbf{\Sigma_y}(h))$$

for $h = 1, 2, \ldots$. ∎

Remark 1 The proposition implies that for large samples the forecast vector $\widetilde{\mathbf{y}}_T(h)$ has approximate MSE matrix

$$\boldsymbol{\Sigma}_{\widetilde{\mathbf{y}}}(h) = \left(1 + \frac{Kn_T}{T}\right)\boldsymbol{\Sigma}_{\mathbf{y}}(h). \tag{15.3.2}$$

This result can be seen by noting that

$$E\{[\mathbf{y}_{T,h} - \widetilde{\mathbf{y}}_T(h)][\mathbf{y}_{T,h} - \widetilde{\mathbf{y}}_T(h)]'\}$$
$$= E\{[\mathbf{y}_{T,h} - \mathbf{y}_T(h)][\mathbf{y}_{T,h} - \mathbf{y}_T(h)]'\}$$
$$+ E\{[\mathbf{y}_T(h) - \widetilde{\mathbf{y}}_T(h)][\mathbf{y}_T(h) - \widetilde{\mathbf{y}}_T(h)]'\}$$

and approximating the last term via the asymptotic result of Proposition 15.3. ∎

Remark 2 An approximation for the MSE matrix of an h-step forecast $\widetilde{y}_T(h)$ follows directly from (15.3.2),

$$\Sigma_{\widetilde{y}}(h) = \left(1 + \frac{Kn_T}{T}\right)\Sigma_y(h), \quad h = 1, 2, \ldots. \tag{15.3.3}$$

In Section 3.5.1, we have obtained an approximate MSE matrix

$$\Sigma_{\widehat{y}}(h) = \Sigma_y(h) + \frac{1}{T}\Omega(h) \tag{15.3.4}$$

for an h-step forecast based on an estimated VAR process with known finite order. If in Chapter 3 the process mean is known to be zero and is not estimated, it can be shown that $\Omega(h)$ approaches zero as $h \to \infty$. In other words, the MSE part due to estimation variability goes to zero as the forecast horizon increases. The same does not hold in the present case. In fact, the $\Sigma_{\widetilde{y}}(h)$'s are monotonically nondecreasing for growing h, that is,

$$\Sigma_{\widetilde{y}}(h) \geq \Sigma_{\widetilde{y}}(i), \quad \text{for } h \geq i.$$

The explanation for this result is that, under the present assumptions, increasingly many parameters are estimated with growing sample size. For a zero mean VAR process with known finite order, the optimal forecast approaches the process mean of zero when the forecast horizon gets large and, thus, the estimated VAR parameters do not contribute to the forecast uncertainty for long-run forecasts. The same is not true under the present conditions, where the VAR order goes to infinity. ∎

Remark 3 We have also seen in Section 3.5.2 that $\Omega(1) = (Kp+1)\Sigma_u$ for a K-dimensional VAR(p) process with estimated intercept term. It is easy to see that, if the process mean is known to be zero and the mean term is not estimated, $\Omega(1) = Kp\Sigma_u$. Hence, in that case,

$$\Sigma_{\widehat{y}}(1) = \Sigma_y(1) + \frac{Kp}{T}\Sigma_u = \Sigma_u + \frac{Kp}{T}\Sigma_u = \Sigma_{\widetilde{y}}(1),$$

if $n_T = p$. In other words, for 1-step ahead forecasts, the two MSE approximations are identical if the same VAR orders are used in both approaches. It is easy to see that the same does not hold in general for predictions more than 1 step ahead (see Problem 15.2). ∎

Remark 4 Because forecasts can be obtained from finite order approximations to infinite order VAR processes, we may also base the prediction tests for structural change considered in Sections 4.6.2 and 13.5.3 on such approximations. Of course, in that case the MSE approximation implied by Proposition 15.3 should be used in setting up the test statistics. For instance, a test statistic based on h-step forecasts would be

$$\widetilde{\tau}_h = (y_{T+h} - \widetilde{y}_T(h))' \widetilde{\Sigma}_{\widetilde{y}}(h)^{-1} (y_{T+h} - \widetilde{y}_T(h)),$$

where $\widetilde{\Sigma}_{\widetilde{y}}(h)$ is an estimator of $\Sigma_{\widetilde{y}}(h)$. ∎

Remark 5 If y_t is a process with nonzero mean vector μ, then the sample mean may be subtracted from the original data and the previous analysis may be performed with the mean-adjusted data. If the sample mean is added to the forecasts, an extra term should be added to the approximate MSE matrix. A term similar to that resulting from an estimated mean term in a finite order VAR setting with known order may be added (see Problem 3.9, Chapter 3). ∎

15.3.2 An Example

To illustrate the effects of approximating a potentially infinite order VAR process by a finite order model, we use again the West German investment/income/consumption data from File E1. The variables y_1, y_2, and y_3 are defined as in Chapter 3, Section 3.2.3, and we use the same sample period 1960–1978 and a VAR order $n_T = 2$. That is, we assume that the VAR order depends on the sample size in such a way that $n_T = 2$ for $T = 73$. Note that the condition (15.2.5) for the VAR order is an asymptotic condition that leaves open the actual choice in finite samples. Therefore, we choose the VAR order that was suggested by the AIC criterion in Chapter 4 and, thus, we use the same VAR order as in Chapter 3. As a consequence, the point forecasts obtained under our present assumptions are the same one gets from a mean-adjusted model under the conditions of Chapter 3. The interval forecasts obtained under the different sets of assumptions are different for $h > 1$, however, because the approximate MSE matrices are different. We have estimated $\Sigma_{\widetilde{y}}(h)$ by

$$\widetilde{\Sigma}_{\widetilde{y}}(h) = \left(1 + \frac{3n_T}{T}\right) \sum_{i=0}^{h-1} \widehat{\Phi}_i \widehat{\Sigma}_u \widehat{\Phi}_i' + \frac{1}{T} \widehat{G}_y(h), \tag{15.3.5}$$

where the $\widehat{\Phi}_i$'s and $\widehat{\Sigma}_u$ are obtained from the VAR(2) estimates, as in Section 3.5.3, and $\widehat{G}_y(h)/T$ is a term that takes account of the fact that the mean term is estimated in addition to the VAR coefficients. It is the same term that is used if a VAR(2) process with true order $p = 2$ is assumed and the model is estimated in mean-adjusted form (see Problem 3.9).

Table 15.1. Interval forecasts from a VAR(2) model for the investment/income/consumption example series based on different asymptotic theories

variable	forecast horizon	point forecast	95% interval forecasts	
			based on known order assumption	based on infinite order assumption
investment	1	−.010	[−.105, .085]	[−.105, .085]
	2	.012	[−.087, .110]	[−.088, .112]
	3	.022	[−.075, .119]	[−.078, .122]
	4	.013	[−.084, .111]	[−.088, .114]
income	1	.020	[−.004, .044]	[−.004, .044]
	2	.020	[−.004, .045]	[−.005, .045]
	3	.017	[−.007, .042]	[−.008, .042]
	4	.021	[−.004, .045]	[−.005, .047]
consumption	1	.022	[.002, .041]	[.002, .041]
	2	.015	[−.005, .035]	[−.005, .035]
	3	.020	[−.002, .042]	[−.002, .042]
	4	.019	[−.003, .041]	[−.003, .041]

We have used the approximate forecast MSEs from (15.3.5) to set up forecast intervals under Gaussian assumptions and give them in Table 15.1. For comparison purposes we also give forecast intervals obtained from a VAR(2) process in mean-adjusted form based on the asymptotic theory of Chapter 3, assuming that the true order is $p = 2$. As we know from Remark 3 in Section 15.3.1, the 1-step forecast MSEs are the same under the two competing assumptions. For larger forecast horizons, most of the intervals based on the infinite order assumption become slightly wider than those based on the known finite order assumption, as expected on the basis of Remark 2 in Section 15.3.2. For our sample size, the differences are quite small, though.

Which of the two sets of forecast intervals should we use in practice? This question is difficult to answer. Assuming a known finite VAR order is, of course, more restrictive and less realistic than the assumption of an unknown and possibly infinite order. The additional uncertainty introduced by the latter assumption is reflected in the wider forecast intervals. It may be worth noting, however, that such a result is not necessarily obtained in all practical situations. In other words, there may be time series and generation processes for which the infinite order assumption actually leads to smaller forecast intervals than the assumption of a known finite VAR order (see Problem 15.2).

Under both sets of assumptions, the MSE approximations are derived from asymptotic theory and little is known about the small sample quality of these approximations. Both approaches are based on a set of assumptions that may not hold in practice. Notably the stationarity and normality assumptions may be doubtful in many practical situations. Given all these reservations, there is still one argument in favor of the present approach, assuming a potentially infinite VAR order. For $h > 1$, the MSE approximation in (15.3.3) is generally simpler to compute than the one obtained in Chapter 3.

15.4 Impulse Response Analysis and Forecast Error Variance Decompositions

15.4.1 Asymptotic Theory

For a researcher who does not know the true structure of the data generating process, it is possible to base an impulse response analysis or forecast error variance decomposition on an approximating finite order VAR process. Given the results of Section 15.2, we can now study the consequences of such an approach. As in Sections 2.3.2 and 2.3.3, the quantities of interest here are the forecast error impulse responses,

$$\Phi_i = \sum_{j=1}^{i} \Phi_{i-j} \Pi_j, \quad i = 1, 2, \ldots, \quad \Phi_0 = I_K,$$

the accumulated forecast error impulse responses,

$$\Psi_m = \sum_{i=0}^{m} \Phi_i, \quad m = 0, 1, \ldots,$$

the responses to orthogonalized impulses,

$$\Theta_i = \Phi_i P, \quad i = 0, 1, \ldots,$$

where P is the lower triangular matrix obtained by a Choleski decomposition of Σ_u, the accumulated orthogonalized impulse responses,

$$\Xi_m = \sum_{i=0}^{m} \Theta_i, \quad m = 0, 1, \ldots,$$

and the forecast error variance components,

$$\omega_{jk,h} = \sum_{i=0}^{h-1} (e_j' \Theta_i e_k)^2 / \mathrm{MSE}_j(h), \quad h = 1, 2, \ldots,$$

where e_k is the k-th column of I_K and

15.4 Impulse Response Analysis and Forecast Error Variance Decompositions 541

$$\text{MSE}_j(h) = \sum_{i=0}^{h-1} e_j' \Phi_i \Sigma_u \Phi_i' e_j$$

is the j-th diagonal element of the MSE matrix, $\Sigma_y(h)$, of an h-step forecast.

Estimators of these quantities are obtained from the $\widehat{\Pi}_i(n_T)$ and $\widetilde{\Sigma}_u(n_T)$ in the obvious way. For instance, estimators for the Φ_i's are obtained recursively as

$$\widetilde{\Phi}_i(n_T) = \sum_{j=1}^{i} \widetilde{\Phi}_{i-j}(n_T) \widehat{\Pi}_j(n_T), \quad i = 1, 2, \ldots,$$

with $\widetilde{\Phi}_0(n_T) = I_K$, and

$$\widetilde{\Theta}_i(n_T) = \widetilde{\Phi}_i(n_T) \widetilde{P}(n_T), \quad i = 0, 1, \ldots,$$

are estimators of the Θ_i's. Here $\widetilde{P}(n_T)$ is the unique lower triangular matrix with positive main diagonal for which

$$\widetilde{P}(n_T) \widetilde{P}(n_T)' = \widetilde{\Sigma}_u(n_T).$$

The asymptotic distributions of all the estimators are given in the next proposition. Proofs, based on Propositions 15.1 and 15.2, are given by Lütkepohl (1988a) and Lütkepohl & Poskitt (1991).

Proposition 15.4 (*Asymptotic Distributions of Impulse Responses*)
Under the conditions of Proposition 15.2, the impulse responses and forecast error variance components have the following asymptotic normal distributions:

$$\sqrt{T} \, \text{vec}(\widetilde{\Phi}_i(n_T) - \Phi_i) \xrightarrow{d} \mathcal{N}\left(0, \Sigma_u^{-1} \otimes \sum_{j=0}^{i-1} \Phi_j \Sigma_u \Phi_j'\right), \quad i = 1, 2, \ldots; \tag{15.4.1}$$

$$\sqrt{T} \, \text{vec}(\widetilde{\Psi}_m(n_T) - \Psi_m) \xrightarrow{d} \mathcal{N}\left(0, \Sigma_u^{-1} \otimes \sum_{k=1}^{m} \sum_{l=1}^{m} \sum_{j=0}^{l-1} \Phi_j \Sigma_u \Phi_{k-l+j}'\right), \tag{15.4.2}$$

$m = 1, 2, \ldots$, with $\Phi_j := 0$ for $j < 0$;

$$\sqrt{T} \, \text{vec}(\widetilde{\Theta}_i(n_T) - \Theta_i) \xrightarrow{d} \mathcal{N}(0, \Omega_\theta(i)), \quad i = 0, 1, \ldots, \tag{15.4.3}$$

where

$$\Omega_\theta(i) = \left(I_K \otimes \sum_{j=0}^{i-1} \Phi_j \Sigma_u \Phi_j'\right) + (I_K \otimes \Phi_i) H \Sigma_{\widetilde{\sigma}} H'(I_K \otimes \Phi_i'),$$

$H = \mathbf{L}_K'[\mathbf{L}_K(I_{K^2} + \mathbf{K}_{KK})(P \otimes I_K)\mathbf{L}_K']^{-1}$,

$\mathbf{L}_K$ is the $(\frac{1}{2}K(K+1) \times K^2)$ elimination matrix,

$\mathbf{K}_{KK}$ is the $(K^2 \times K^2)$ commutation matrix,

and $\Sigma_{\widetilde{\sigma}}$ is the asymptotic covariance matrix of $\sqrt{T}$ vech$(T^{-1} \sum_{t=1}^T u_t u_t' - \Sigma_u)$;

$$\sqrt{T} \text{ vec}(\widetilde{\Xi}_m(n_T) - \Xi_m) \xrightarrow{d} \mathcal{N}(0, \Omega_\xi(m)), \quad m = 1, 2, \ldots, \tag{15.4.4}$$

where

$$\Omega_\xi(m) = \sum_{k=0}^m \sum_{l=0}^m \left[I_K \otimes \sum_{j=0}^{l-1} \Phi_j \Sigma_u \Phi_{k-l+j}' + (I_K \otimes \Phi_l) H \Sigma_{\widetilde{\sigma}} H'(I_K \otimes \Phi_k')\right]$$

with $\Phi_j := 0$ for $j < 0$;

$$\sqrt{T}(\widetilde{\omega}_{jk,h}(n_T) - \omega_{jk,h}) \xrightarrow{d} \mathcal{N}(0, \sigma_{jk,h}^2), \quad h = 1, 2, \ldots, \quad j, k = 1, \ldots, K, \tag{15.4.5}$$

where

$$\sigma_{jk,h}^2 = \sum_{l=0}^{h-1} \sum_{m=0}^{h-1} g_{jk,h}(l) \left[I_K \otimes \sum_{i=0}^{m-1} \Phi_i \Sigma_u \Phi_{l-m+i}' \right.$$
$$\left. + (I_K \otimes \Phi_m) H \Sigma_{\widetilde{\sigma}} H'(I_K \otimes \Phi_l')\right] g_{jk,h}(m)'$$

with

$$g_{jk,h}(m) = 2\Big[(e_k' \otimes e_j')(e_j' \Theta_m e_k) \text{MSE}_j(h)$$
$$- (e_j' \Theta_m \otimes e_j') \sum_{i=0}^{h-1}(e_j' \Theta_i e_k)^2 \Big] \Big/ \text{MSE}_j(h)^2.$$

■

Remark 1 In the proposition, it is ignored that $\sigma_{jk,h}^2$ may be zero, in which case the asymptotic normal distribution is degenerate. In particular, $\sigma_{jk,h}^2 = 0$ if $\omega_{jk,h} = 0$. This result is easily seen by noting that $\omega_{jk,h}$ is zero if and only if $\theta_{jk,0} = \cdots = \theta_{jk,h-1} = 0$, where $\theta_{jk,m}$ is the jk-th element of Θ_m. Thus, the asymptotic distribution in (15.4.5) is not immediately useful for testing

$$H_0 : \omega_{jk,h} = 0 \quad \text{against} \quad H_1 : \omega_{jk,h} \neq 0, \tag{15.4.6}$$

15.4 Impulse Response Analysis and Forecast Error Variance Decompositions

which is a set of hypotheses of particular interest in practice. The significance of $\omega_{jk,h}$ may be checked, however, by testing $\theta_{jk,0} = \cdots = \theta_{jk,h-1} = 0$. Using a minor generalization of (15.4.3), this hypothesis can be tested (see Lütkepohl & Poskitt (1991)). ∎

Remark 2 In sharp contrast to the case where the VAR order is assumed to be known and finite (see Proposition 3.6), the asymptotic variances of all impulse responses are nonzero in the present case. Another difference between the finite and infinite order VAR cases is that in the former the asymptotic standard errors of the Φ_i and Θ_i go to zero as i increases, while the covariance matrix in (15.4.1) is a nondecreasing function of i and the covariance matrix in (15.4.3) is bounded away from zero, for $i > 0$. ∎

Remark 3 For $i = 1$, the asymptotic covariance matrix of $\widetilde{\Phi}_1(n_T)$ in (15.4.1) is $\Sigma_u^{-1} \otimes \Sigma_u$. It can be shown that the same asymptotic covariance matrix is obtained for $\widehat{\Phi}_1$ from Proposition 3.6, if a VAR(n) process is fitted with n greater than the true order p (see Lütkepohl (1988a)). A similar result is obtained for $\widetilde{\Theta}_i(n_T)$ and $\widehat{\Theta}_i$ for $i = 0, 1$ (see Problem 15.4). ∎

Remark 4 The results in Proposition 15.4 can also be used to construct tests for zero impulse responses. This case was considered by Lütkepohl (1996b). ∎

Remark 5 Although forecast error and orthogonalized impulse responses are considered only in Proposition 15.4, similar results can also be obtained for the structural impulse responses discussed in Chapter 9. ∎

15.4.2 An Example

To illustrate the consequences of the finite and infinite VAR order assumptions, we use again the VAR(2) model for the investment/income/consumption data. Of course, the same estimated impulse responses are obtained as in Section 3.7. (The intercept form of the model is used now.) The standard errors are different, however. In Figures 15.1 and 15.2, consumption responses to income impulses are depicted and the two-standard error bounds obtained from both sets of assumptions are shown. In both figures, the two-standard error bounds based on Proposition 3.6 decline almost to zero for longer lags while the two-standard error bounds from Proposition 15.4 are seen to grow with the time lag. This behavior reflects the additional estimation uncertainty that results from assuming that the VAR order goes to infinity with the sample size. Thereby more and more parameters are estimated as the sample size gets large.

In Table 15.2, forecast error variance decompositions of the system are shown. Again most standard errors based on the infinite VAR assumption are slightly larger than those from Chapter 3, which are also given in the table. Although it is tempting to use the estimated standard errors in checking the

544 15 Fitting Finite Order VAR Models to Infinite Order Processes

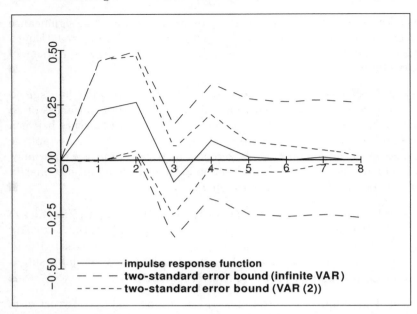

Fig. 15.1. Estimated responses of consumption to a forecast error impulse in income with two-standard error bounds based on finite and infinite VAR order assumptions.

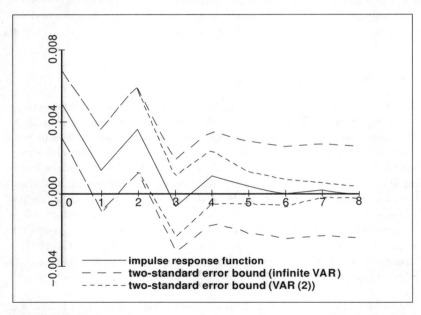

Fig. 15.2. Estimated responses of consumption to an orthogonalized impulse in income with two-standard error bounds based on finite and infinite VAR order assumptions.

Table 15.2. Forecast error variance decompositions of the investment/income/consumption system with standard errors from two different asymptotic theories

forecast error in	forecast horizon h	proportions of forecast error variance, h periods ahead, accounted for by innovations in[a]		
		investment $\omega_{j1,h}$	income $\omega_{j2,h}$	consumption $\omega_{j3,h}$
investment ($j=1$)	1	1.000(.000)[.000]	.000(.000)[.000]	.000(.000)[.000]
	2	.960(.042)[.044]	.018(.030)[.030]	.023(.031)[.033]
	3	.946(.042)[.045]	.028(.033)[.033]	.026(.029)[.032]
	4	.941(.045)[.048]	.029(.031)[.032]	.030(.032)[.036]
	8	.938(.048)[.050]	.031(.032)[.034]	.032(.035)[.039]
income ($j=2$)	1	.018(.031)[.031]	.983(.031)[.031]	.000(.000)[.000]
	2	.060(.054)[.053]	.908(.063)[.064]	.032(.037)[.040]
	3	.070(.057)[.058]	.896(.066)[.068]	.035(.039)[.041]
	4	.068(.056)[.057]	.892(.067)[.069]	.039(.041)[.045]
	8	.069(.057)[.058]	.891(.068)[.070]	.040(.041)[.045]
consumption ($j=3$)	1	.080(.061)[.061]	.273(.086)[.086]	.647(.090)[.090]
	2	.077(.059)[.059]	.274(.082)[.082]	.649(.088)[.088]
	3	.130(.080)[.080]	.334(.089)[.091]	.537(.091)[.092]
	4	.129(.079)[.079]	.335(.088)[.090]	.536(.089)[.091]
	8	.129(.080)[.081]	.340(.089)[.092]	.532(.091)[.093]

[a] Estimated standard error based on a finite known VAR order assumption in parentheses and estimated standard error based on an infinite VAR order assumption in brackets.

significance of individual forecast error variance components, we know from Remark 1 of Section 15.4.1 that they are not useful for that purpose because the asymptotic standard errors from Proposition 15.4 corresponding to zero forecast error variance components are zero.

15.5 Cointegrated Infinite Order VARs

In Chapter 6, it was discussed that assuming a fixed finite starting date is advantageous if integrated variables are considered. Therefore, some modifications will be necessary in defining infinite order processes for integrated variables. Details of the model setup will be given in Section 15.5.1. The properties of estimators of the parameters of such models are considered in Section 15.5.2, and testing for the cointegrating rank will be discussed in Section 15.5.3.

15.5.1 The Model Setup

The general framework presented in the following is that of Saikkonen (1992) and Saikkonen & Lütkepohl (1996). Given a K-dimensional system of time series variables y_t with cointegrating rank r, we assume that the variables are arranged such that for $t = 1, 2, \ldots$,

$$y_t^{(1)} = -\beta'_{(K-r)} y_t^{(2)} + z_t^{(1)},$$
$$\Delta y_t^{(2)} = z_t^{(2)}, \tag{15.5.1}$$

where $y_t^{(1)}$ and $y_t^{(2)}$ are $(r \times 1)$ and $((K-r) \times 1)$, respectively, such that

$$y_t = \begin{bmatrix} y_t^{(1)} \\ y_t^{(2)} \end{bmatrix},$$

as in the triangular representation discussed in Section 6.3 (see (6.3.10)). Hence, $\beta_{(K-r)}$ is $((K-r) \times r)$ such that

$$\begin{bmatrix} I_r \\ \beta_{(K-r)} \end{bmatrix}$$

is the cointegration matrix and

$$z_t = \begin{bmatrix} z_t^{(1)} \\ z_t^{(2)} \end{bmatrix}$$

is a strictly stationary process with $E(z_t) = 0$ and positive definite covariance matrix $\Sigma_z = E(z_t z_t')$. As a further technical condition which is needed in some proofs, we also assume that z_t has a continuous spectral density matrix which is positive definite at zero frequency. For a discussion of spectral density matrices of vector processes see, e.g., Fuller (1976, Section 4.4). The initial vector y_0 is assumed to be such that the process Δy_t is stationary.

In matrix form, the process y_t may be written as

$$\begin{bmatrix} I_r & \beta'_{(K-r)} \\ 0 & I_{K-r} \end{bmatrix} y_t = \begin{bmatrix} 0 & 0 \\ 0 & I_{K-r} \end{bmatrix} y_{t-1} + z_t. \tag{15.5.2}$$

Multiplying by the inverse of the left-hand matrix,

$$\begin{bmatrix} I_r & -\beta'_{(K-r)} \\ 0 & I_{K-r} \end{bmatrix},$$

subtracting y_{t-1} on both sides of the equation and rearranging terms gives

$$\Delta y_t = -\begin{bmatrix} I_r & \beta'_{(K-r)} \\ 0 & 0 \end{bmatrix} y_{t-1} + v_t = -\begin{bmatrix} \beta' \\ 0 \end{bmatrix} y_{t-1} + v_t, \tag{15.5.3}$$

where $\beta' = [I_r : \beta'_{(K-r)}]$ and

$$v_t = \begin{bmatrix} I_r & -\beta'_{(K-r)} \\ 0 & I_{K-r} \end{bmatrix} z_t$$

is a stationary process. It is assumed to have an infinite order VAR representation,

$$v_t = \sum_{j=1}^{\infty} G_j v_{t-j} + u_t, \quad t \in \mathbb{Z}, \tag{15.5.4}$$

where u_t is again standard white noise. Notice that, due to the stationarity of v_t, there is no problem in defining it for all integers t. Moreover, because the process v_t is stationary, it also has an MA representation for which we could make similar assumptions as in (15.2.4). We do not need that representation here, however, and therefore we formulate the required assumptions directly for the VAR coefficients. In particular, the G_j's are assumed to satisfy

$$\det \left(I_K - \sum_{j=1}^{\infty} G_j z^j \right) \neq 0 \text{ for } |z| \leq 1 \quad \text{and} \quad \sum_{j=1}^{\infty} j \|G_j\| < \infty. \tag{15.5.5}$$

This condition imposes weak restrictions on the autocorrelation structure of the process v_t and is, for example, satisfied for VARMA processes. From the previous assumptions, it follows that if the infinite order VAR is approximated by a finite order process, the approximation error gets sufficiently small for our purposes, if the order of the approximating process is chosen as in (15.2.5) with $\sqrt{T} \sum_{i=n_T+1}^{\infty} \|G_j\| \to 0$ as $T \to \infty$.

Defining

$$G_j^* := -(G_{j+1} + \cdots + G_n), \quad j = 0, 1, \ldots, n-1,$$

and

$$G_{n-1}^*(L) := \sum_{j=0}^{n-1} G_j^* L^j,$$

it follows that

$$G_n(L) := I_K - \sum_{j=1}^{n} G_j L^j = G_n(1) - G_{n-1}^*(L)(1-L) \tag{15.5.6}$$

(see Problem 15.6). Multiplying (15.5.3) by $G_n(L)$ and rearranging terms gives the VECM representation

$$\Delta y_t = \alpha \beta' y_{t-1} + \sum_{j=1}^{n} \Gamma_j \Delta y_{t-j} + e_t, \quad t = n+1, n+2, \ldots, \tag{15.5.7}$$

where

$$\alpha := -\left(I_K - \sum_{j=1}^{n} G_j\right) \begin{bmatrix} I_r \\ 0 \end{bmatrix},$$

$$e_t := u_t + \sum_{j=n+1}^{\infty} G_j v_{t-j}$$

and

$$\sum_{j=1}^{n} \Gamma_j L^j = \sum_{j=1}^{n} G_j L^j + G_{n-1}^*(L) \begin{bmatrix} \beta' \\ 0 \end{bmatrix} L$$

$$= \sum_{j=1}^{n} \left(G_j + G_{j-1}^* \begin{bmatrix} \beta' \\ 0 \end{bmatrix}\right) L^j.$$

Hence,

$$\Gamma_j = G_j - (G_j + \cdots + G_n) \begin{bmatrix} \beta' \\ 0 \end{bmatrix}, \quad j = 1, \ldots, n$$

(see Problem 15.7). Although this fact is not specifically indicated, the coefficient matrices α and Γ_j, $j = 1, \ldots, n$, depend on n. In particular, $\Gamma_n = [0 : \Gamma_{n2}]$, where Γ_{n2} is $(K \times (K - r))$. It can be shown that assumption (15.5.5) implies that the Γ_i's are absolutely summable, that is, $\lim_{n\to\infty} \sum_{j=1}^{n} \|\Gamma_j\|$ exists, and the process y_t is well-defined (Phillips & Solo (1992, 2.1 Lemma)).

Rearranging terms, the VECM (15.5.7) can also be rewritten in levels VAR form as

$$y_t = \sum_{j=1}^{n+1} \Pi_j y_{t-j} + e_t, \quad t = n+1, n+2, \ldots, \tag{15.5.8}$$

where

$$\Pi_1 = I_K + \alpha\beta' + \Gamma_1 = I_K + G_1 - \begin{bmatrix} \beta' \\ 0 \end{bmatrix},$$

$$\Pi_j = \Gamma_j - \Gamma_{j-1} = G_j - [0 : G_{j-1,1}\beta'_{(K-r)} - G_{j-1,2}], \quad j = 2, \ldots, n,$$

$$\Pi_{n+1} = -\Gamma_n.$$

Here $G_{j-1,1}$ and $G_{j-1,2}$ are submatrices of G_{j-1} consisting of the first r and last $K - r$ columns, respectively. Thus, although the Γ_j depend on n, the same is not true for the Π_j, except for Π_{n+1}. In the following subsection, the asymptotic properties of the LS estimators of the VECM and the levels VAR representations will be considered.

15.5.2 Estimation

Suppose a levels VAR($n_T + 1$) model of the form (15.5.8) is estimated by multivariate LS based on a sample of size T. Notice that the order of the model now depends explicitly on the sample size T, as in the case where stationary processes were approximated by finite order VARs, discussed earlier in this chapter. We assume again that the VAR order goes to infinity with the sample size although at a smaller rate than T. The following proposition, which is similar to Theorem 2 of Saikkonen & Lütkepohl (1996), gives the details. In stating the proposition, the LS estimators are denoted by $\widehat{\Pi}_j$, $\Pi(n) := [\Pi_1, \ldots, \Pi_n]$, and $\widehat{\Pi}(n) := [\widehat{\Pi}_1, \ldots, \widehat{\Pi}_n]$, as before. Now we can present the result.

Proposition 15.5 (*Asymptotic Distribution of the LS Estimator of the VAR Coefficients*)
Suppose that finite order VAR(n_T+1) processes are fitted by multivariate LS to a multiple time series generated by the process specified in Section 15.5.1 and assume that the order n_T depends on the sample size T such that

$$n_T \to \infty, \quad n_T^3/T \to 0, \quad \text{and} \quad \sqrt{T} \sum_{i=n_T+1}^{\infty} \|G_i\| \to 0 \quad \text{as} \quad T \to \infty. \tag{15.5.9}$$

Furthermore, let c_1, c_2 be fixed constants and $\mathbf{f}(n)$ a sequence of nonzero $((Kr + K^2 n) \times 1)$ vectors such that

$$0 < c_1 \leq \mathbf{f}(n)'\mathbf{f}(n) \leq c_2 < \infty \quad \text{for } n = 1, 2, \ldots.$$

Then

$$\frac{\sqrt{T - n_T}\, \mathbf{f}(n_T)'[\widehat{\boldsymbol{\pi}}(n_T) - \boldsymbol{\pi}(n_T)]}{[\mathbf{f}(n_T)'(H'_{n_T}\Gamma^{-1}_{n_T,VECM}H_{n_T} \otimes \Sigma_u)\mathbf{f}(n_T)]^{1/2}} \xrightarrow{d} \mathcal{N}(0,1), \tag{15.5.10}$$

where $\boldsymbol{\pi}(n) := \text{vec}\,\Pi(n)$ and $\widehat{\boldsymbol{\pi}}(n) := \text{vec}\,\widehat{\Pi}(n)$, as in Section 15.2, H_{n_T} is a $((r + Kn_T) \times Kn_T)$ matrix defined such that

$$[\Pi_1 : \cdots : \Pi_{n_T}] = [\boldsymbol{\alpha} : \boldsymbol{\Gamma}_1 : \cdots : \boldsymbol{\Gamma}_{n_T}]H_{n_T} + [I_K : 0 : \cdots : 0]$$

and

$$\Gamma_{n_T,VECM} := E\left(\begin{bmatrix} u^{(1)}_{t-1} \\ \Delta y_{t-1} \\ \vdots \\ \Delta y_{t-n_T} \end{bmatrix} [u^{(1)'}_{t-1}, \Delta y'_{t-1}, \ldots, \Delta y'_{t-n_T}]\right). \tag{15.5.11}$$

Here $u^{(1)}_{t-1}$ denotes the vector of the first r components of u_{t-1}. ∎

This proposition can be proven analogously to Theorem 2 in Saikkonen & Lütkepohl (1996). Clearly, the proposition is similar to Proposition 15.1. Note, however, that in the present proposition, only the first n_T coefficient matrices are considered, although a VAR(n_T+1) process is fitted to the data. Dropping the last lag in deriving the asymptotic distribution of the estimators ensures that standard asymptotic properties are obtained. This devise was also used in Section 7.6.3 in deriving a Wald test for Granger-causality in a finite order cointegrated VAR context.

Consider now the VECM with n_T lagged differences,

$$\Delta y_t = \mathbf{\Pi} y_{t-1} + \sum_{j=1}^{n_T} \mathbf{\Gamma}_j \Delta y_{t-j} + e_t. \tag{15.5.12}$$

Suppose that the model is also estimated by multivariate LS based on a sample of size T. The estimators are denoted by $\widehat{\mathbf{\Pi}}$ and $\widehat{\mathbf{\Gamma}}_j$ and the residuals are signified as $\widehat{u}_t(n_T)$. Using this notation,

$$\widetilde{\Sigma}_u = \frac{1}{T - n_T - 1} \sum_{t=n_T+2}^{T} \widehat{u}_t(n_T) \widehat{u}_t(n_T)' \tag{15.5.13}$$

is an estimator of the white noise covariance matrix Σ_u. The loading matrix $\boldsymbol{\alpha}$ may be estimated as $\widehat{\boldsymbol{\alpha}} = \widehat{\mathbf{\Pi}}_1$, where the latter matrix consists of the first r columns of $\widehat{\mathbf{\Pi}}$, as in the EGLS procedure presented in Section 7.2.2. As in that procedure, the matrix $\boldsymbol{\beta}'_{(K-r)}$ may be estimated as

$$\widehat{\boldsymbol{\beta}}'_{(K-r)} = (\widehat{\boldsymbol{\alpha}}' \widetilde{\Sigma}_u^{-1} \widehat{\boldsymbol{\alpha}})^{-1} \widehat{\boldsymbol{\alpha}}' \widetilde{\Sigma}_u^{-1} \widehat{\mathbf{\Pi}}_2, \tag{15.5.14}$$

where $\widehat{\mathbf{\Pi}}_2$ consists of the last $K - r$ columns of $\widehat{\mathbf{\Pi}}$ (see Remark 4 of Section 7.2.2). The next proposition summarizes the asymptotic properties of the estimators. Proofs can be found in Saikkonen (1992) and Saikkonen & Lütkepohl (1996).

Proposition 15.6 (*Asymptotic Distribution of VECM Estimators*)
Under the conditions of Proposition 15.5,

$$T(\widehat{\boldsymbol{\beta}}'_{(K-r)} - \boldsymbol{\beta}'_{(K-r)}) \xrightarrow{d} \left(\int_0^1 \mathbf{W}^{\#}_{K-r} d\mathbf{W}^{\#'}_r \right)' \left(\int_0^1 \mathbf{W}^{\#}_{K-r} \mathbf{W}^{\#'}_{K-r} ds \right)^{-1}, \tag{15.5.15}$$

where $\mathbf{W}^{\#}_{K-r}$ and $\mathbf{W}^{\#}_r$ are independent $(K-r)$- and r-dimensional Wiener processes, respectively, as in Proposition 7.2. Furthermore,

$$\frac{\sqrt{T - n_T} \, \mathbf{f}(n_T)'[\widehat{\gamma}(n_T) - \gamma(n_T)]}{[\mathbf{f}(n_T)'(\Gamma_{n_T,VECM}^{-1} \otimes \Sigma_u) \mathbf{f}(n_T)]^{1/2}} \xrightarrow{d} \mathcal{N}(0, 1), \tag{15.5.16}$$

where $\gamma(n) := \text{vec}[\boldsymbol{\alpha} : \mathbf{\Gamma}_1 : \cdots : \mathbf{\Gamma}_n]$ and $\widehat{\gamma}(n) := \text{vec}[\widehat{\boldsymbol{\alpha}} : \widehat{\mathbf{\Gamma}}_1 : \cdots : \widehat{\mathbf{\Gamma}}_n]$. ∎

Notice that the asymptotic distribution of the cointegration parameters in (15.5.15) is the same as that of the corresponding ML estimator for Gaussian finite order VAR processes, as discussed in Section 7.2 (see in particular Proposition 7.2 and Remark 3 for Proposition 7.4). The loading and short-run parameters have asymptotic properties similar to those of finite order processes as well. Their asymptotic properties are the same that would be obtained if the true β matrix were known and used in the estimation procedure.

Moreover, Saikkonen & Lütkepohl (1996) showed that the white noise covariance matrix estimator $\widetilde{\Sigma}_u$ also has similar asymptotic properties as in the finite order case. Furthermore, they stated slightly more general versions of Propositions 15.5 and 15.6 and discussed how the results can be used in testing hypotheses of parameter restrictions. In particular, they considered the case of testing for Granger-causality. They also discussed adding an intercept term to the model. Saikkonen & Lütkepohl (2000a) presented extensions which can be used in deriving, for example, asymptotic properties of impulse responses in the present framework.

In practice, the cointegrating rank is usually unknown and has to be determined from the given multiple time series. How to do so in the present framework of an infinite order process is discussed next.

15.5.3 Testing for the Cointegrating Rank

In Section 8.2, we have discussed testing the cointegrating rank of a finite order VAR process by considering pairs of hypotheses of the form

$$H_0 : \text{rk}(\Pi) = r_0 \quad \text{against} \quad H_1 : r_0 < \text{rk}(\Pi) \leq r_1. \tag{15.5.17}$$

In particular, the cases $r_1 = r_0 + 1$ and $r_1 = K$ were discussed and suitable likelihood ratio tests were introduced. Suppose now that the test statistics are computed in precisely the same way as in Section 8.2.1, based on the VECM (15.5.12) with n_T lagged differences of y_t. In other words, we compute the statistic as if (15.5.12) were a Gaussian process with lag order n_T. To emphasize the dependence on the lag order, we denote the test statistic corresponding to the pair of hypotheses in (15.5.17) by $\lambda_{LR}^{(n_T)}(r_0, r_1)$. Lütkepohl & Saikkonen (1999b) proved the following result.

Proposition 15.7 (*Asymptotic Distributions of Tests for the Cointegrating Rank*)
Suppose y_t is generated by an infinite order process as described in Section 15.5.1. Moreover, suppose that

$$n_T \to \infty \quad \text{and} \quad n_T^3/T \to 0 \quad \text{as} \quad T \to \infty. \tag{15.5.18}$$

Then $\lambda_{LR}^{(n_T)}(r_0, r_0 + 1)$ and $\lambda_{LR}^{(n_T)}(r_0, K)$ have the same limiting null distributions as for a Gaussian finite order process given in Proposition 8.2. ∎

Notice that in this proposition we just have an upper bound for the rate at which the lag order n_T has to go to infinity. No lower bound for the rate of divergence is needed. In fact, Lütkepohl & Saikkonen (1999b) considered also processes with nonzero mean term and, in addition, they treated the case where the lag order is chosen by some model selection procedure instead of a deterministic rule derived from (15.5.18). In summary, these results show that as far as asymptotic theory is concerned, the cointegrating rank of an $I(1)$ process may be chosen on the basis of a finite order approximation rather than a correctly specified model. This result is not only important because, in practice, models are usually just approximations to the true DGP and, hence, allowing explicitly for some approximation error is more realistic, it is also important because we have proposed this approach for choosing the cointegrating rank of a VARMA process in Chapter 14, Section 14.4.2.

15.6 Exercises

Problem 15.1
For the invertible MA(1) process $y_t = u_t + M u_{t-1}$ and $n = 1, 2$, determine the matrix Γ_n defined in (15.2.7).

Problem 15.2
Suppose the true data generation mechanism is a univariate AR(1) process, $y_t = \alpha y_{t-1} + u_t$. Assume that a univariate AR(1) is indeed fitted to the data and compare the resulting approximate forecast MSEs $\Sigma_{\hat{y}}(h)$ (given in Section 3.5) and $\Sigma_{\tilde{y}}(h)$ (given in Section 15.3.1) for $h = 1, 2, \ldots$. (Hint: See Lütkepohl (1987, pp. 76, 77).)

Problem 15.3
Suppose the true data generation process is an invertible MA(1), as in Problem 15.1. Write down explicit expressions for the asymptotic covariance matrices of $\widetilde{\Phi}_i(n_T)$, $\widetilde{\Theta}_i(n_T)$, $i = 1, 2$, and of $\widetilde{\Psi}_m(n_T)$, $\widetilde{\Xi}_m(n_T)$, $m = 1, 2$.

Problem 15.4
Let $\widetilde{\Theta}_i(n_T)$ and $\widehat{\Theta}_i$ be estimators of the orthogonalized impulse responses Θ_i obtained under the conditions of Propositions 15.4 and 3.6, respectively. If the true data generation mechanism is a finite order VAR(p) process and the actual process fitted to the data has order $n_T > p$, show that the asymptotic covariance matrices in (15.4.3) and (3.7.8) are identical for $i = 0, 1$.

Problem 15.5
Consider the investment/income/consumption system of Section 15.3.2 and fit a VAR(4) process to the data.

(a) Determine 95% interval forecasts for all three variables and forecast horizons $h = 1, 2, 3$ under the assumption of a known true VAR order of $p = 4$ and under the assumption of an infinite order true generation process.

(b) Determine Φ_i and Θ_i impulse responses and their asymptotic standard errors for $i = 1, 2, 3, 4$ under both the assumption of a finite and an infinite true VAR order. Compare the estimated standard errors obtained under the two alternative scenarios for all variables.

Problem 15.6
Show that the relation in (15.5.6) holds.

Problem 15.7
Derive the model representation (15.5.7). (Hint: See Saikkonen (1992, Section 2).)

Part V

Time Series Topics

16
Multivariate ARCH and GARCH Models

16.1 Background

In the previous chapters, we have discussed modelling the conditional mean of the data generation process of a multiple time series, conditional on the past at each particular time point. In that context, the variance or covariance matrix of the conditional distribution was assumed to be time invariant. In fact, in much of the discussion, the residuals or forecast errors were assumed to be independent white noise. Such a simplification is useful and justified in many applications.

There are also situations, however, when such an assumption is problematic, for instance, when financial time series are being analyzed. To see this, consider the monthly returns of the DAX (German stock index) for the period 1965–1995 depicted in Figure 16.1. The autocorrelations are all within the $\pm 2/\sqrt{T}$ band and, hence, in accordance with the results discussed in Chapter 4, Section 4.4, one may conclude that the returns are not autocorrelated. If they were not only uncorrelated but also independent, then their squares were independent too. That this is not the case is clearly seen in the third panel of Figure 16.1, where also the autocorrelations of the squared returns are given. Consequently, in this case, assuming independent observations or, equivalently, independent residuals in the AR(0) model $y_t = \nu + u_t$ is clearly problematic. Because we have used the independence assumption in deriving the $\pm 2/\sqrt{T}$ confidence bounds in Chapter 4, Section 4.4, the conclusion of uncorrelated returns may also be questioned in this case.

The correlations in the squares of the DAX returns shown in Figure 16.1 indicate that there is conditional heteroskedasticity. With a little imagination, it can also be seen in the figure that the volatility in the DAX returns changes over time. It is lower in the first half of the sample period than in the second half. Similar characteristics in many time series, in particular in financial market series, have motivated the development of specific models for conditionally heteroskedastic data.

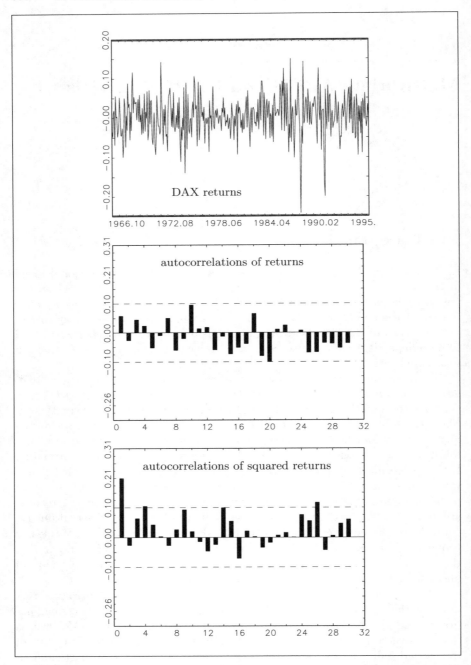

Fig. 16.1. Monthly DAX returns for the years 1965–1995 with autocorrelations and autocorrelations of squared returns.

It may be tempting to argue that the conditional mean is the optimal forecast and, hence, changes in volatility are of less importance from a forecasting point of view. This position ignores, however, that the forecast error variances, that is, the variances of the conditional distributions are needed for setting up forecast intervals. Taking into account conditional heteroskedasticity is therefore important also when forecasts of the variables under investigation are desired. Moreover, for example in financial analysis, forecasts of the future volatility of a series under consideration are often of interest to assess the risk associated with certain assets. In that case, variance forecasts are of direct interest, of course. Furthermore, the volatility in a market and, hence, the risk associated with investments in a particular market may have a direct effect on the expectations of the market participants. Hence, there may be a feedback from the second to the first moments. Therefore, the emphasis on a more detailed modelling of the volatility of time series was a natural development which was boosted by Engle's (1982) invention of ARCH (*autoregressive conditional heteroskedasticity*) models. By now the acronym ARCH stands for a wide range of models for changing conditional volatility. Moreover, there is also some literature on multivariate extensions which are the central topic of this chapter.

Because many series have a close relationship, it is obvious to conjecture that an increase, say, in the volatility of one series may have an impact on the volatility of another series as well. For example, this may occur in exchange rates of different currencies, in interest rates for bonds of different times to maturity, or in returns on stocks in a specific segment of the market. Therefore, multivariate models for conditional heteroskedasticity are of interest.

In the following, a brief review of some facts on univariate ARCH and generalized ARCH (GARCH) models is given and then multivariate extensions will be discussed. Part of this chapter reports results from an article by Engle & Kroner (1995). There are also a number of review articles which cover multivariate ARCH and GARCH models among other things. Examples are Bollerslev, Engle & Nelson (1994), Bera & Higgins (1993), Bauwens, Laurent & Rombouts (2004), Bollerslev, Chou & Kroner (1992), and Pagan (1996). The latter two articles also survey some of the applied literature.

16.2 Univariate GARCH Models

16.2.1 Definitions

Consider the univariate serially uncorrelated, zero mean process u_t. For instance, u_t may represent the residuals of an autoregressive process. The u_t are said to follow an *autoregressive conditionally heteroskedastic process of order q* (ARCH(q)) if the conditional distribution of u_t, given its past $\Omega_{t-1} := \{u_{t-1}, u_{t-2}, \dots\}$, has zero mean and the conditional variance is

$$\sigma^2_{t|t-1} := \text{Var}(u_t|\Omega_{t-1}) = E(u_t^2|\Omega_{t-1}) = \gamma_0 + \gamma_1 u_{t-1}^2 + \cdots + \gamma_q u_{t-q}^2, \quad (16.2.1)$$

that is, $u_t|\Omega_{t-1} \sim (0, \sigma^2_{t|t-1})$. Another, sometimes quite useful way to define an ARCH process is to specify

$$u_t = \sigma_{t|t-1}\varepsilon_t, \quad \varepsilon_t \sim \text{i.i.d.}(0,1). \tag{16.2.2}$$

Here the i.i.d. assumption for ε_t is slightly more restrictive than the previous definition which makes statements about the first two moments of the conditional distribution only. In the following, the definition (16.2.2) will be used. The u_t's, generated in this way, will be serially uncorrelated with mean zero.

Originally, Engle (1982), in his seminal paper on ARCH models, assumed the conditional distribution to be normal so that

$$\varepsilon_t \sim \text{i.i.d.}\,\mathcal{N}(0,1) \quad \text{and} \quad u_t|\Omega_{t-1} \sim \mathcal{N}(0, \sigma^2_{t|t-1}). \tag{16.2.3}$$

Although different distributions were considered later as well, even with this special distributional assumption the model is capable of generating series with characteristics similar to those of many observed time series. In particular, it is capable to generate series with volatility clustering and outliers similar to the DAX series in Figure 16.1. Even if the conditional distribution underlying an ARCH(q) model is normal, the unconditional distribution will generally be nonnormal. In particular, it is *leptokurtic*, that is, it has more mass around zero and in the tails than the normal distribution and, hence, it can produce occasional outliers.

It turns out, however, that, for many series, ARCH processes with fairly large orders are necessary to capture the dynamics in the conditional variances. Therefore, Bollerslev (1986) and Taylor (1986) proposed to gain greater parsimony by extending the model in a similar manner as the AR model when moving to mixed ARMA models. They suggested the *generalized ARCH* (GARCH) model with conditional variances given by

$$\sigma^2_{t|t-1} = \gamma_0 + \gamma_1 u^2_{t-1} + \cdots + \gamma_q u^2_{t-q} + \beta_1 \sigma^2_{t-1|t-2} + \cdots + \beta_m \sigma^2_{t-m|t-m-1}. \tag{16.2.4}$$

These models are briefly denoted by GARCH(q, m). They generate processes with existing *un*conditional variance if and only if the coefficient sum

$$\gamma_1 + \cdots + \gamma_q + \beta_1 + \cdots + \beta_m < 1. \tag{16.2.5}$$

If this condition is satisfied, u_t has a constant *un*conditional variance given by

$$\sigma^2_u = \frac{\gamma_0}{1 - \gamma_1 - \cdots - \gamma_q - \beta_1 - \cdots - \beta_m}. \tag{16.2.6}$$

The similarity of GARCH models and ARMA models for the conditional mean can be seen by defining $v_t := u_t^2 - \sigma^2_{t|t-1}$, substituting $u_t^2 - v_t$ for $\sigma^2_{t|t-1}$ in (16.2.4) and rearranging terms. Thereby we get

$$u_t^2 = \gamma_0 + (\beta_1 + \gamma_1)u_{t-1}^2 + \cdots + (\beta_q + \gamma_q)u_{t-q}^2$$
$$+ v_t - \beta_1 v_{t-1} - \cdots - \beta_m v_{t-m} \qquad (16.2.7)$$

which is formally an ARMA(q, m) model for u_t^2. Here it is assumed without loss of generality that $q \geq m$ and $\beta_j := 0$ for $j > m$.

16.2.2 Forecasting

Although the conditional expectation of the process u_t given Ω_{t-h} is zero and, hence, the optimal h-step forecasts are all zero for $h = 1, 2, \ldots$, there is an important difference to the situation where the u_t are an independent white noise process. If the u_t are Gaussian $\mathcal{N}(0, \sigma_u^2)$, a 1-step ahead $(1-\alpha)100\%$ forecast interval has the form

$$u_t(1) \pm c_{1-\alpha/2}\sigma_u,$$

where $u_t(1)$ denotes the forecast, as usual, and $c_{1-\alpha/2}$ is the relevant $1-\alpha/2$ percentage point of the normal distribution (see Section 2.2.3). Thus, the forecast intervals are of constant width, regardless of the forecast origin t. In contrast, if u_t is a GARCH(q, m) process, the correct 1-step ahead $(1-\alpha)100\%$ forecast interval is

$$u_t(1) \pm c_{1-\alpha/2}\sigma_{t+1|t}, \qquad (16.2.8)$$

where the length depends on the history of the process because the conditional standard deviation, $\sigma_{t+1|t}$, varies over time.

To illustrate this phenomenon, suppose the mean-adjusted DAX returns were generated by a GARCH(1, 1) model with conditionally normal components and conditional variances

$$\sigma_{t|t-1}^2 = 0.0003 + 0.120u_{t-1}^2 + 0.771\sigma_{t-1|t-2}^2.$$

This model was actually fitted to the monthly DAX returns by Lütkepohl (1997) for the period 1960–1991. The 1-step ahead 95% forecast intervals are shown in Figure 16.2. The unconditional variance is in this case

$$\sigma_u^2 = \frac{\gamma_0}{1 - \gamma_1 - \beta_1} = \frac{0.0003}{1 - 0.120 - 0.771} = 2.75 \times 10^{-3}.$$

Assuming mistakenly that the data is i.i.d. normal and using the foregoing white noise variance, results in the constant forecast intervals also shown in Figure 16.2. It is important to note the implications of these results. The constant intervals completely ignore the variations in volatility, whereas the GARCH intervals clearly reflect the greater forecast uncertainty in times of high volatility and are narrower in times where the stock market is less volatile.

As mentioned earlier, if the residuals follow a normal GARCH process, the unconditional distribution of the observations will generally be nonnormal.

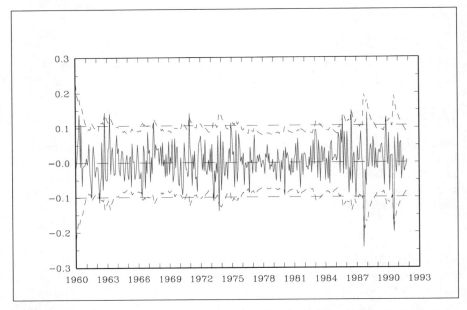

Fig. 16.2. 95% 1-step ahead forecast intervals for the DAX returns obtained under GARCH (- - -) and constant (- -) variance assumptions.

Hence, the constant forecast intervals which have been computed under normality assumptions may not have the desired 95% probability content because of the false distributional assumption. The nonnormal unconditional distribution of GARCH processes also complicates multi-step interval forecasting. Formulas and properties of multi-step forecasts were discussed by Baillie & Bollerslev (1992). Without going into details, it may be worth noting that for a stationary process, when the forecast horizon increases, the optimal forecast will always approach the process mean with the unconditional variance being the forecast error variance and the forecast error distribution approaching the unconditional process distribution, which will generally be nonnormal if the conditional distribution is normal.

We will now discuss how to extend these concepts to the case of vector processes. In that context, we will also address the issue of estimating the parameters of a GARCH model.

16.3 Multivariate GARCH Models

Multivariate extensions of ARCH and GARCH models may be defined in principle similarly to VAR and VARMA models. Early articles on multivariate ARCH and GARCH models are Engle, Granger & Kraft (1986), Diebold & Nerlove (1989), Bollerslev, Engle & Wooldridge (1988). There are a number of

16.3.1 Multivariate ARCH

Suppose that $u_t = (u_{1t}, \ldots, u_{Kt})'$ is a K-dimensional zero mean, serially uncorrelated process which may be the residual process of some dynamic model and which can be represented as

$$u_t = \Sigma_{t|t-1}^{1/2} \varepsilon_t, \qquad (16.3.1)$$

where ε_t is K-dimensional i.i.d. white noise, $\varepsilon_t \sim$ i.i.d. $(0, I_K)$, and $\Sigma_{t|t-1}$ is the conditional covariance matrix of u_t, given $u_{t-1}, u_{t-2}, \ldots$. As usual, $\Sigma_{t|t-1}^{1/2}$ is the symmetric positive definite square root of $\Sigma_{t|t-1}$ (see Appendix A.9.2 for details on the square root of a positive definite matrix). Obviously, the u_t's have a conditional distribution, given $\Omega_{t-1} := \{u_{t-1}, u_{t-2}, \ldots\}$, of the form

$$u_t | \Omega_{t-1} \sim (0, \Sigma_{t|t-1}). \qquad (16.3.2)$$

They represent a multivariate ARCH(q) process if

$$\text{vech}(\Sigma_{t|t-1}) = \gamma_0 + \Gamma_1 \text{vech}(u_{t-1} u_{t-1}') + \cdots + \Gamma_q \text{vech}(u_{t-q} u_{t-q}'), \qquad (16.3.3)$$

where vech again denotes the half-vectorization operator which stacks the columns of a square matrix from the diagonal downwards in a vector, γ_0 is a $\frac{1}{2}K(K+1)$-dimensional vector of constants and the Γ_j's are $(\frac{1}{2}K(K+1) \times \frac{1}{2}K(K+1))$ coefficient matrices. Different conditional distributions have been assumed and analyzed. For example, a multivariate normal conditional distribution may be considered, i.e., $\varepsilon_t \sim \mathcal{N}(0, I_K)$, so that $u_t | \Omega_{t-1} \sim \mathcal{N}(0, \Sigma_{t|t-1})$. Although this distribution is perhaps not the most suitable one for many financial time series, it will play a role when parameter estimation is discussed in Section 16.4. Conditional distributions of processes representing financial time series are often better represented by more heavy-tailed distributions such as t-distributions with a small degrees of freedom parameter.

As an example, consider a bivariate ($K = 2$) ARCH(1) process,

$$\text{vech}\begin{bmatrix} \sigma_{11,t|t-1} & \sigma_{12,t|t-1} \\ \sigma_{12,t|t-1} & \sigma_{22,t|t-1} \end{bmatrix} = \begin{bmatrix} \sigma_{11,t|t-1} \\ \sigma_{12,t|t-1} \\ \sigma_{22,t|t-1} \end{bmatrix}$$

$$= \begin{bmatrix} \gamma_{10} \\ \gamma_{20} \\ \gamma_{30} \end{bmatrix} + \begin{bmatrix} \gamma_{11} & \gamma_{12} & \gamma_{13} \\ \gamma_{21} & \gamma_{22} & \gamma_{23} \\ \gamma_{31} & \gamma_{32} & \gamma_{33} \end{bmatrix} \begin{bmatrix} u_{1,t-1}^2 \\ u_{1,t-1} u_{2,t-1} \\ u_{2,t-1}^2 \end{bmatrix}.$$

Obviously, even this simple model for a bivariate series has a fair number of parameters which makes it difficult to handle. In particular, the implications of a general model of this type for the relationships between the variables and their higher order moment properties are not obvious. Therefore, more restricted models have been proposed. For instance, Bollerslev et al. (1988) considered diagonal ARCH processes where the Γ_j matrices are all diagonal. In the first order case, the model has the form

$$\begin{bmatrix} \sigma_{11,t|t-1} \\ \sigma_{12,t|t-1} \\ \sigma_{22,t|t-1} \end{bmatrix} = \begin{bmatrix} \gamma_{10} \\ \gamma_{20} \\ \gamma_{30} \end{bmatrix} + \begin{bmatrix} \gamma_{11} & 0 & 0 \\ 0 & \gamma_{22} & 0 \\ 0 & 0 & \gamma_{33} \end{bmatrix} \begin{bmatrix} u_{1,t-1}^2 \\ u_{1,t-1}u_{2,t-1} \\ u_{2,t-1}^2 \end{bmatrix}.$$

Even simple processes of this type can generate rich volatility dynamics. Still, despite their simpler structure, processes of this type involve nontrivial technical problems. One of them is that the parameters have to be such that the conditional covariance matrices $\Sigma_{t|t-1}$ are all positive definite. To guarantee this property, Baba, Engle, Kraft & Kroner (1990) and Engle & Kroner (1995) investigated the following variant of a multivariate ARCH model,

$$\Sigma_{t|t-1} = \Gamma_0^* + \Gamma_1^{*\prime} u_{t-1} u_{t-1}' \Gamma_1^* + \cdots + \Gamma_q^{*\prime} u_{t-q} u_{t-q}' \Gamma_q^*, \qquad (16.3.4)$$

where the Γ_j^*'s are each $(K \times K)$ matrices. This particular multivariate model has been christened *BEKK model*. Here the $\Sigma_{t|t-1}$ are positive definite if Γ_0^* has this property which may be enforced by writing it in a product form, $\Gamma_0^* = C_0^{*\prime} C_0^*$ with triangular C_0^* matrix. Another advantage of this model is that it is relatively parsimonious. For instance, for a bivariate process with $K = 2$ and $q = 1$, there are only 7 parameters, whereas the full model has 12 coefficients. Moreover, in contrast to the diagonal model, it can produce quite rich interactions between the conditional second order moments.

16.3.2 MGARCH

In principle, multivariate ARCH models may be generalized in the same way as in the univariate case. In the multivariate GARCH (MGARCH) model for u_t, the conditional covariance matrices have the form

$$\text{vech}(\Sigma_{t|t-1}) = \gamma_0 + \sum_{j=1}^{q} \Gamma_j \text{vech}(u_{t-j} u_{t-j}') + \sum_{j=1}^{m} G_j \text{vech}(\Sigma_{t-j|t-j-1}),$$

(16.3.5)

where the G_j's are also fixed ($\frac{1}{2}K(K+1) \times \frac{1}{2}K(K+1)$) coefficient matrices. For example, for a bivariate GARCH(1,1) model,

$$\text{vech} \begin{bmatrix} \sigma_{11,t|t-1} & \sigma_{12,t|t-1} \\ \sigma_{12,t|t-1} & \sigma_{22,t|t-1} \end{bmatrix} = \begin{bmatrix} \sigma_{11,t|t-1} \\ \sigma_{12,t|t-1} \\ \sigma_{22,t|t-1} \end{bmatrix}$$

$$= \begin{bmatrix} \gamma_{10} \\ \gamma_{20} \\ \gamma_{30} \end{bmatrix} + \begin{bmatrix} \gamma_{11} & \gamma_{12} & \gamma_{13} \\ \gamma_{21} & \gamma_{22} & \gamma_{23} \\ \gamma_{31} & \gamma_{32} & \gamma_{33} \end{bmatrix} \begin{bmatrix} u_{1,t-1}^2 \\ u_{1,t-1}u_{2,t-1} \\ u_{2,t-1}^2 \end{bmatrix}$$

$$+ \begin{bmatrix} g_{11} & g_{12} & g_{13} \\ g_{21} & g_{22} & g_{23} \\ g_{31} & g_{32} & g_{33} \end{bmatrix} \begin{bmatrix} \sigma_{11,t-1|t-2} \\ \sigma_{12,t-1|t-2} \\ \sigma_{22,t-1|t-2} \end{bmatrix}.$$

A VARMA representation of an MGARCH process may be obtained analogously to the univariate case (see (16.2.7)) by defining $\mathbf{x}_t := \text{vech}(u_t u_t')$ and $v_t := \mathbf{x}_t - \text{vech}(\Sigma_{t|t-1})$. Using these specifications and substituting $\mathbf{x}_t - v_t$ for $\text{vech}(\Sigma_{t|t-1})$, (16.3.5) can be rewritten as

$$\mathbf{x}_t = \gamma_0 + \sum_{j=1}^{\max(q,m)} (\Gamma_j + G_j)\mathbf{x}_{t-j} + v_t - \sum_{j=1}^{m} G_j v_{t-j},$$

where $\Gamma_j = 0$ for $j > q$ and $G_j = 0$ for $j > m$. This representation is occasionally useful in deriving properties of MGARCH processes (e.g., Section 16.6.1).

Engle & Kroner (1995) showed that the MGARCH process u_t with conditional covariances as given in (16.3.5) is stationary if and only if all eigenvalues of the matrix

$$\sum_{j=1}^{q} \Gamma_j + \sum_{j=1}^{m} G_j \qquad (16.3.6)$$

have modulus less than one.

The parameter space of an MGARCH model has a large dimension in general and needs to be restricted to guarantee uniqueness of the representation and to obtain suitable properties of the conditional covariances. To reduce the parameter space, Bollerslev et al. (1988) discussed diagonal MGARCH models, where the Γ_j's and G_i's in (16.3.5) are diagonal matrices. Alternatively, a BEKK GARCH model of the following form may be useful:

$$\Sigma_{t|t-1} = C_0^{*\prime} C_0^* + \sum_{n=1}^{N} \sum_{j=1}^{q} \Gamma_{jn}^{*\prime} u_{t-j} u_{t-j}' \Gamma_{jn}^* + \sum_{n=1}^{N} \sum_{j=1}^{m} G_{jn}^{*\prime} \Sigma_{t-j|t-j-1} G_{jn}^*,$$

(16.3.7)

where again C_0^* is a triangular $(K \times K)$ matrix and the coefficient matrices Γ_{jn}^*, G_{jn}^* are also $(K \times K)$. Given the similarity of MGARCH and VARMA models, it is clear from Chapter 12, Section 12.1, that restrictions have to be imposed on the coefficient matrices to ensure uniqueness of the parameterization. Engle & Kroner (1995) gave the following properties of BEKK GARCH models which also address the uniqueness problem:

(1) From the stationarity condition (16.3.6), the BEKK model is seen to be stationary if all eigenvalues of the matrix

$$\sum_{n=1}^{N}\sum_{j=1}^{q} \Gamma_{jn}^{*\prime} \otimes \Gamma_{jn}^{*\prime} + \sum_{n=1}^{N}\sum_{j=1}^{m} G_{jn}^{*\prime} \otimes G_{jn}^{*\prime} \tag{16.3.8}$$

have modulus less than one.

(2) The BEKK model nests all positive definite diagonal GARCH models, that is, every diagonal GARCH model with positive definite conditional covariance matrices has a BEKK representation.

(3) The BEKK model (16.3.7) generates positive definite covariance matrices $\Sigma_{t|t-1}$ if $\Sigma_{0|-1}, \Sigma_{-1|-2}, \ldots, \Sigma_{-m+1|-m}$ are positive definite and if at least one of the matrices C_0^*, G_{jn}^*, $j = 1, \ldots, m$, $n = 1, \ldots, N$, is nonsingular (see Engle & Kroner (1995, Proposition 2.5)).

(4) In the class of BEKK GARCH(1,1) models with $N = 1$, the representation

$$\Sigma_{t|t-1} = C_0^{*\prime}C_0^* + \Gamma_{11}^{*\prime} u_{t-1} u_{t-1}^\prime \Gamma_{11}^* + G_{11}^{*\prime} \Sigma_{t-1|t-2} G_{11}^*$$

is unique if all diagonal elements of C_0^* are positive and $\gamma_{11,1}^*, g_{11,1}^* > 0$. Here $\gamma_{11,1}^*$ and $g_{11,1}^*$ represent the upper left-hand elements of Γ_{11}^* and G_{11}^*, respectively.

(5) For a more general BEKK GARCH(1,1) model with

$$\Sigma_{t|t-1} = C_0^{*\prime}C_0^* + \sum_{n=1}^{N} \Gamma_{1n}^{*\prime} u_{t-1} u_{t-1}^\prime \Gamma_{1n}^* + \sum_{n=1}^{N} G_{1n}^{*\prime} \Sigma_{t-1|t-2} G_{1n}^*,$$

uniqueness is achieved by the following restrictions:

(a) All diagonal elements of C_0^* are positive.
(b) $\Gamma_{1n}^* = G_{1n}^* = 0$ for $n > K^2$.
(c) In the matrices $\Gamma_{1n_j}^*$ with $n_j = K(j-1)+1, \ldots, Kj$, and $j = 1, \ldots, K$, the first $j-1$ columns and the first $n_j - K(j-1) - 1$ rows are zero. Moreover, the lower right hand element of $\Gamma_{1n_j}^*$, $\gamma_{KK,n_j}^* > 0$.
(d) Restrictions analogous to those for the Γ_{1n}^* also hold for the G_{1n}^*.

(Engle & Kroner (1995, Proposition 2.3)).

For illustrative purposes, suppose $K = 3$ so that $N = K^2 = 9$ and $n_1 = 1, 2, 3$; $n_2 = 4, 5, 6$; $n_3 = 7, 8, 9$. Hence, a unique representation is obtained if the zero restrictions shown in the following matrices are imposed:

$$\Gamma_{11}^* = \begin{bmatrix} \gamma_{11,1}^* & \gamma_{12,1}^* & \gamma_{13,1}^* \\ \gamma_{21,1}^* & \gamma_{22,1}^* & \gamma_{23,1}^* \\ \gamma_{31,1}^* & \gamma_{32,1}^* & \gamma_{33,1}^* \end{bmatrix}, \quad \Gamma_{12}^* = \begin{bmatrix} 0 & 0 & 0 \\ \gamma_{21,2}^* & \gamma_{22,2}^* & \gamma_{23,2}^* \\ \gamma_{31,2}^* & \gamma_{32,2}^* & \gamma_{33,2}^* \end{bmatrix},$$

$$\Gamma_{13}^* = \begin{bmatrix} 0 & 0 & 0 \\ 0 & 0 & 0 \\ \gamma_{31,3}^* & \gamma_{32,3}^* & \gamma_{33,3}^* \end{bmatrix}, \quad \Gamma_{14}^* = \begin{bmatrix} 0 & \gamma_{12,4}^* & \gamma_{13,4}^* \\ 0 & \gamma_{22,4}^* & \gamma_{23,4}^* \\ 0 & \gamma_{32,4}^* & \gamma_{33,4}^* \end{bmatrix},$$

$$\Gamma_{15}^* = \begin{bmatrix} 0 & 0 & 0 \\ 0 & \gamma_{22,5}^* & \gamma_{23,5}^* \\ 0 & \gamma_{32,5}^* & \gamma_{33,5}^* \end{bmatrix}, \quad \Gamma_{16}^* = \begin{bmatrix} 0 & 0 & 0 \\ 0 & 0 & 0 \\ 0 & \gamma_{32,6}^* & \gamma_{33,6}^* \end{bmatrix},$$

$$\Gamma_{17}^* = \begin{bmatrix} 0 & 0 & \gamma_{13,7}^* \\ 0 & 0 & \gamma_{23,7}^* \\ 0 & 0 & \gamma_{33,7}^* \end{bmatrix}, \quad \Gamma_{18}^* = \begin{bmatrix} 0 & 0 & 0 \\ 0 & 0 & \gamma_{23,8}^* \\ 0 & 0 & \gamma_{33,8}^* \end{bmatrix},$$

$$\Gamma_{19}^* = \begin{bmatrix} 0 & 0 & 0 \\ 0 & 0 & 0 \\ 0 & 0 & \gamma_{33,9}^* \end{bmatrix}.$$

The same zero restrictions are also imposed on the G_{1n}^*. Of course, in a specific case, there may be further zero restrictions on the coefficient matrices. In particular, N may be less than K^2. An example of this type is given in Section 16.4.2.

Given the correspondence between GARCH and VARMA models, it should be clear from the discussion of uniqueness of VARMA representations in Chapter 12 that a unique parameterization of a multivariate GARCH representation is not a trivial matter. Whether the constraints given here are the most operational ones in practice remains to be seen. If a unique representation is set up, estimation becomes possible. This issue will be discussed in Section 16.4.

16.3.3 Other Multivariate ARCH and GARCH Models

Although the BEKK model with low orders may be a relatively parsimonious representation of the conditional covariance structure of a process, the number of parameters still grows quickly with the dimension of the underlying system. Therefore, in practice, it is only feasible if systems with just a few variables are under consideration and further simplifications were proposed to alleviate modelling of higher dimensional processes. Some of them can be viewed as special BEKK models. For example, Lin (1992) specified a *factor GARCH model*, where the Γ_{1n}^*'s and G_{1n}^*'s in a BEKK GARCH(1,1) model are of the form

$$\Gamma_{1n}^* = \gamma_n \eta_n \xi_n' \quad \text{and} \quad G_{1n}^* = g_n \eta_n \xi_n', \quad n = 1, \ldots, N. \tag{16.3.9}$$

Here γ_n and g_n are scalars and η_n and ξ_n are $(K \times 1)$ vectors satisfying $\xi_n' \xi_n = 1$, $\eta_n' \xi_n = 1$ for $n = 1, \ldots, N$ and $\eta_n' \xi_k = 0$ for $n \neq k$. Thus, the Γ_{1n}^*'s and G_{1n}^*'s have all rank 1.

In some proposals, the conditional covariance matrix has the form

$$\Sigma_{t|t-1} = Q H_{t|t-1} Q', \tag{16.3.10}$$

where Q is $(K \times K)$ and does not depend on t, whereas $H_{t|t-1}$ is a positive definite $(K \times K)$ matrix which may depend on t. For example, Vrontos, Dellaportas & Politis (2003) proposed to use a triangular matrix Q and specified

$$H_{t|t-1} = \mathrm{diag}(\sigma^2_{1t|t-1}, \ldots, \sigma^2_{Kt|t-1}) \qquad (16.3.11)$$

to be a diagonal matrix with univariate GARCH conditional variances $\sigma^2_{kt|t-1}$ on the diagonal. A closely related model, the so-called *generalized orthogonal GARCH model*, was proposed by van der Weide (2002).

Clearly, restricting the second moment dynamics to a transformation of univariate GARCH models as in (16.3.11) is restrictive and, in particular, it limits the covariance dynamics in a potentially undesired way. Therefore, the alternative specification

$$\Sigma_{t|t-1} = D_t R_t D_t \qquad (16.3.12)$$

was proposed, where restrictions of different forms are specified for the $(K \times K)$ matrices D_t and R_t. For example, if $R_t = R$ is a time invariant correlation matrix and $D_t = \mathrm{diag}(\sigma_{1t|t-1}, \ldots, \sigma_{Kt|t-1})$ is a diagonal matrix with time varying conditional standard deviations on the diagonal, Bollerslev's (1990) *constant conditional correlation (CCC) MGARCH model* is obtained. Clearly, in this model, the time invariant R is the correlation matrix corresponding to the covariance matrix $\Sigma_{t|t-1}$ for all t. Engle (2002) extended the model by allowing for richer dynamics and proposed the so-called *dynamic conditional correlation (DCC) model*. A related model was also proposed by Tse & Tsui (2002).

In financial markets, it has been observed frequently that positive and negative shocks or news have quite different effects (Black (1976)). This so-called *leverage effect* can be introduced in different ways in MGARCH models. For example, Hafner & Herwartz (1998b) and Herwartz & Lütkepohl (2000) generalized a univariate proposal by Glosten, Jagannathan & Runkle (1993) and replaced

$$\Gamma^{*\prime}_{11} u_{t-1} u'_{t-1} \Gamma^*_{11} \quad \text{by} \quad \Gamma^{*\prime}_{11} u_{t-1} u'_{t-1} \Gamma^*_{11} + \Gamma^{*\prime}_{-} u_{t-1} u'_{t-1} \Gamma^*_{-} \mathbb{I} \left(\sum_{k=1}^{K} u_{kt} < 0 \right) \qquad (16.3.13)$$

in a BEKK model with $N = 1$. Here $\mathbb{I}(\cdot)$ denotes an indicator function which takes the value 1 if the argument is valid and 0 otherwise and Γ^*_{-} is an additional $(K \times K)$ coefficient matrix. Another approach to allow for asymmetry is to use the so-called *exponential GARCH* (EGARCH) model proposed by Nelson (1991). A multivariate version was considered by Braun, Nelson & Sunier (1995).

A range of other models was also proposed and the literature on MGARCH models has grown rapidly over the last years. A recent survey was provided by Bauwens et al. (2004), where more information on the aforementioned models, further proposals and references can be found.

16.4 Estimation

16.4.1 Theory

Using Bayes' theorem, the joint density function of $u_1, \ldots, u_T$ is $f(u_1, \ldots, u_T)$ $= f(u_1)f(u_2|u_1) \cdots f(u_T|u_{T-1}, \ldots, u_1)$. Thus, if in (16.3.1) $\varepsilon_t \sim$ i.i.d. $\mathcal{N}(0, I_K)$ so that the conditional distribution of u_t given Ω_{t-1} is Gaussian and if the u_t are observed quantities, the log-likelihood function of the general GARCH model described by (16.3.5), for a sample $u_1, \ldots, u_T$, is given by

$$\ln l(\delta) = \sum_{t=1}^{T} \ln l_t(\delta), \qquad (16.4.1)$$

where $\delta = \text{vec}(\gamma_0, \Gamma_1, \ldots, \Gamma_q, G_1, \ldots, G_m)$ is the vector of unknown parameters and

$$\ln l_t(\delta) = -\frac{K}{2}\ln 2\pi - \frac{1}{2}\ln|\Sigma_{t|t-1}| - \frac{1}{2}u_t'\Sigma_{t|t-1}^{-1}u_t, \quad t = 1, \ldots, T, \quad (16.4.2)$$

where the required initial values for specifying $\Sigma_{t|t-1}$ are assumed to be available. Similarly, the log-likelihood may be set up for special cases such as diagonal or BEKK models.

The likelihood function may be maximized with respect to the parameters δ by using numerical methods. A closed form solution does not exist because of the nonlinearity of the function. For uniqueness of the maximum and, hence, the existence of a unique ML estimator, it is important that an identified, unique parameterization is used, e.g., the BEKK form of the model with the restrictions discussed in Section 16.3.2. Of course, if the log-likelihood function (16.4.1)/(16.4.2) is used although the true distribution of the ε_t is nonnormal, the resulting estimators will just be quasi ML estimators. Comte & Lieberman (2003) showed that quasi ML estimators have the following properties.

Proposition 16.1 (*Properties of Quasi ML Estimators of GARCH Models*) Let u_t be a BEKK GARCH process satisfying the following conditions:

(a) The parameter space is compact and identification restrictions are imposed.
(b) The eigenvalues of the matrix (16.3.6) have modulus less than one.
(c) $\varepsilon_t = (\varepsilon_{1t}, \ldots, \varepsilon_{Kt})' \sim$ i.i.d. $(0, I_K)$ with $\varepsilon_{it}, \varepsilon_{jt}$ independent for $i \neq j$ $(i, j = 1, \ldots, K)$ and such that u_t admits moments of at least order 8. Moreover, the ε_t are continuous random variables with a density which is positive in a neighborhood of the origin.
(d) The initial values $u_t, t \leq 0$, are such that the process u_t is strictly stationary.

Then the quasi ML estimator $\widetilde{\delta}$ of δ obtained by maximizing the Gaussian likelihood function exists and is strongly consistent,

$\widetilde{\delta} \stackrel{a.s.}{\to} \delta.$

Moreover, $\widetilde{\delta}$ has an asymptotic normal distribution,

$$\sqrt{T}(\widetilde{\delta} - \delta) \stackrel{d}{\to} \mathcal{N}(0, C_1^{-1} C_0 C_1^{-1}), \tag{16.4.3}$$

where

$$C_1 = -E\left(\frac{\partial^2 \ln l_t(\delta)}{\partial \delta \partial \delta'}\right) \quad \text{and} \quad C_0 = E\left(\frac{\partial \ln l_t(\delta)}{\partial \delta} \frac{\partial \ln l_t(\delta)}{\partial \delta'}\right). \tag{16.4.4}$$

■

A number of comments are worth making regarding this proposition.

Remark 1 It can be shown that $C_0 = C_1$ if ε_t is normally distributed. Hence, in this case, the asymptotic distribution in (16.4.3) becomes $\mathcal{N}(0, C_1^{-1})$, that is, the covariance matrix is the inverse asymptotic information matrix. ■

Remark 2 The condition of a compact parameter space is typical for nonlinear estimation problems. Although not totally satisfactory, it is not regarded as very problematic because the compact subset of the Euclidean space to which it refers may be so large that the condition is not really restrictive. The assumption regarding the initial values is also not restrictive if the stationarity of the process is accepted. It can be replaced by the assumption that the initial values are fixed, nonstochastic values. ■

Remark 3 In contrast, the assumptions regarding the ε_t are not fully satisfactory. In particular, the requirement that moments of order 8 have to exist for u_t is undesirable for financial time series where the existence of higher order moments is regarded as problematic. On the other hand, the theorem improves on previously available results which shows how difficult it is to derive asymptotic properties of the estimators of MGARCH processes. A number of other authors have derived more specialized results, notably for univariate processes (see the review articles mentioned at the end of Section 16.1). For multivariate GARCH processes, consistency of the quasi ML estimators was shown by Jeantheau (1998) under the main assumption of a strictly stationary and ergodic process. Ling & McAleer (2003) derived asymptotic normality of quasi ML estimators under less restrictive moment assumptions for a VARMA process with CCC GARCH residuals. ■

Typically, the u_t are residuals of some dynamic model. Suppose they are the errors of a VAR(p) process, possibly with integrated or cointegrated variables. Thus, we have a model of the form

$$y_t = \nu + A_1 y_{t-1} + \cdots + A_p y_{t-p} + u_t.$$

In this case, the VAR parameters have to be estimated in addition to the coefficients associated with the u_t process. Setting up the corresponding Gaussian likelihood or quasi likelihood function is not difficult. However, the optimization may be a formidable task. Assuming that the numerical problems can be solved, there is some hope that the asymptotics can also be resolved because, under quite general conditions, the asymptotic information matrix of the VAR parameters and the GARCH parameters is block diagonal so that the estimators of the VAR coefficients are asymptotically independent of the GARCH parameter estimators. This result also suggests a two step estimation procedure in which the VAR coefficients $\nu, A_1, \ldots, A_p$ are estimated by LS or, if restrictions are imposed on the parameters, by EGLS and then a GARCH model is fitted to the residuals of the first stage estimation.

Given that normality of the conditional distribution of the u_t is often difficult to justify, in particular, in financial applications, it may also be worth pointing out that ML estimation with other distributions has been studied. The survey by Bauwens et al. (2004) provides further information and references on these issues as well as computational aspects of ML and quasi ML estimation.

16.4.2 An Example

Two series of daily stock returns (first differences of ln prices) will be used to illustrate the previous theoretical considerations. In particular, returns of VW (Volkswagen) common stock (y_{1t}) and preference stock (y_{2t}) for the period January 1987–December 1992 (1579 observations) are used.[1] The two series are plotted in Figure 16.3. The corresponding logarithms of the price series are both strongly related to the performance of the VW company and, hence, they are likely to be related to each other. Therefore, it makes sense to analyze the stocks as a bivariate series. The ln price series were previously analyzed by Herwartz & Lütkepohl (2000). In contrast to these authors, we consider the bivariate series $y_t = (y_{1t}, y_{2t})'$ of returns. Although the ln prices may be cointegrated, a preliminary analysis has shown that there is weak evidence of cointegration at best. Therefore, it seems justified to focus on the returns in the following.

The two series of stock returns display some changes in their volatility and there are also some unusually large (in absolute value) observations. Such values are often classified as outliers. Thus, based on the graphs in Figure 16.3, one may not expect the series to be generated by a Gaussian process and ARCH or GARCH models may be used to capture the volatility dynamics.

Fitting VAR(p) models of increasing order to the bivariate series y_t, it turns out that AIC and HQ recommend an order of $p = 3$ while SC suggests $p = 0$. Therefore, the residuals of the following estimated VAR(3) model (with t-values in parentheses) will be used in the following bivariate GARCH analysis:

---

[1] The price series are from Deutsche Finanzdatenbank Karlsruhe.

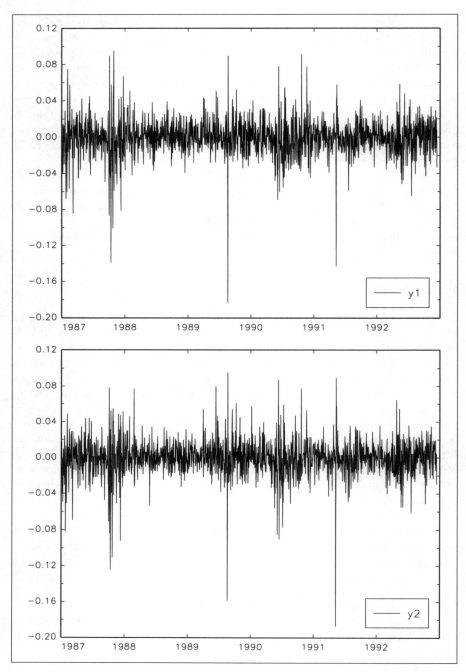

Fig. 16.3. Daily returns of VW common (y_1) and preference stock (y_2).

16.4 Estimation   573

$$y_t = \begin{bmatrix} -0.24 \times 10^{-3} \\ (-0.5) \\ -0.42 \times 10^{-3} \\ (-0.8) \end{bmatrix} + \begin{bmatrix} -0.00 & 0.02 \\ (-0.0) & (0.3) \\ 0.12 & -0.13 \\ (2.0) & (-2.3) \end{bmatrix} y_{t-1}$$

$$+ \begin{bmatrix} -0.18 & 0.16 \\ (-3.2) & (2.8) \\ -0.08 & 0.03 \\ (-1.3) & (0.6) \end{bmatrix} y_{t-2} + \begin{bmatrix} -0.08 & 0.11 \\ (-1.3) & (1.9) \\ -0.01 & 0.01 \\ (-0.2) & (0.1) \end{bmatrix} y_{t-3} + \widehat{u}_t. \quad (16.4.5)$$

The residual series are plotted in Figure 16.4. They still show volatility clusters and outliers. Hence, there may be conditional heteroskedasticity in the residuals of model (16.4.5). In that case, it may not be a good strategy to choose the VAR order first by one of our standard model selection criteria, as we have done it here. Alternatively, it may be preferable to derive criteria that allow a simultaneous determination of the joint model for the conditional first and second moments (see Brooks & Burke (2003)). We will nevertheless use the residuals from the model (16.4.5) in the subsequent analysis for illustrative purposes.

Based on the residuals of the model (16.4.5), the following BEKK GARCH-$(1,1)$ model was estimated (with t-values in parentheses):

$$\Sigma_{t|t-1} = \begin{bmatrix} 0.004 & 0.005 \\ (2.6) & (3.2) \\ 0 & 0.003 \\ & (5.5) \end{bmatrix} \begin{bmatrix} 0.004 & 0 \\ (2.6) & \\ 0.005 & 0.003 \\ (3.2) & (5.5) \end{bmatrix}$$

$$+ \begin{bmatrix} 0.254 & -0.004 \\ (1.9) & (-0.0) \\ 0.040 & 0.332 \\ (0.1) & (1.1) \end{bmatrix} \widehat{u}_{t-1} \widehat{u}'_{t-1} \begin{bmatrix} 0.254 & 0.040 \\ (1.9) & (0.1) \\ -0.004 & 0.332 \\ (-0.0) & (1.1) \end{bmatrix}$$

$$+ \begin{bmatrix} 0.941 & 0.023 \\ (7.8) & (0.2) \\ -0.019 & 0.864 \\ (-0.2) & (17.6) \end{bmatrix} \Sigma_{t-1|t-2} \begin{bmatrix} 0.941 & -0.019 \\ (7.8) & (-0.2) \\ 0.023 & 0.864 \\ (0.2) & (17.6) \end{bmatrix}.$$

(16.4.6)

Note that the uniqueness conditions mentioned in Section 16.3.2 are satisfied here. It is not clear, however, that in the present situation the t-ratios have standard normal limiting distributions because the assumptions of Proposition 16.1 are violated. In particular, we are working with residuals from a previously fitted model rather than with original observations. Still the sizes of the t-ratios underneath the coefficient estimates indicate some interaction in the conditional second moments.

It would be helpful to have tools for checking the model quality and for analyzing the relationships summarized in the model. For model checking, the estimated $\varepsilon_t = \Sigma_{t|t-1}^{-1/2} u_t$ can be used. Standardized estimated ε_t (divided

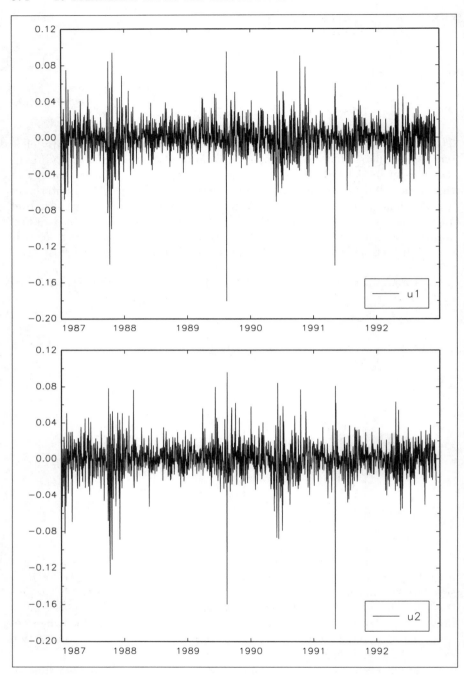

Fig. 16.4. Residual series of model (16.4.5).

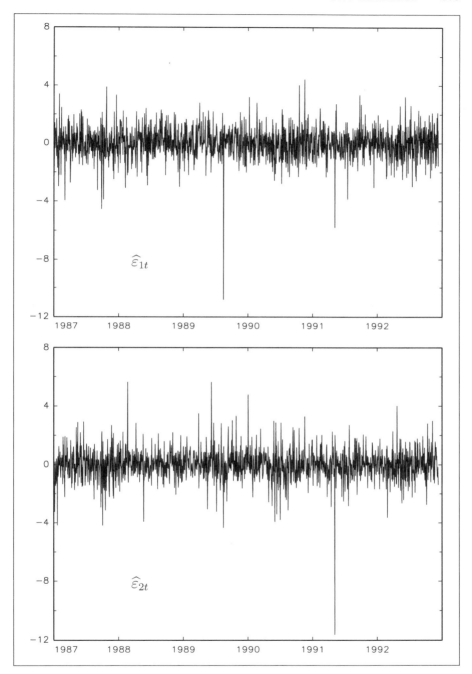

Fig. 16.5. Standardized residuals of model (16.4.6) ($\widehat{\varepsilon}_{1t}$ upper panel, $\widehat{\varepsilon}_{2t}$ lower panel).

by their estimated standard deviations) are plotted in Figure 16.5. Volatility clusters are not quite so obvious anymore as in Figure 16.4. On the other hand, outliers are still present which sheds doubt on the normality of the ε_t. Some tests for model adequacy will be discussed in the next section.

16.5 Checking MGARCH Models

16.5.1 ARCH-LM and ARCH-Portmanteau Tests

Before an MGARCH model is fitted to the residuals of a VAR or VECM, one may want to check if ARCH effects are present in the residuals. An LM test is a standard tool for this purpose (e.g., Doornik & Hendry (1997)). The idea is to consider the auxiliary model

$$\text{vech}(u_t u_t') = \beta_0 + B_1 \text{vech}(u_{t-1} u_{t-1}') + \cdots + B_q \text{vech}(u_{t-q} u_{t-q}') + \text{error}_t, \tag{16.5.1}$$

where β_0 is $\frac{1}{2}K(K+1)$-dimensional and the B_j's are $(\frac{1}{2}K(K+1) \times \frac{1}{2}K(K+1))$ coefficient matrices $(j = 1, \ldots, q)$. If all the B_j matrices are zero, there is no ARCH in the residuals. Therefore, the pair of hypotheses

$$H_0 : B_1 = \cdots = B_q = 0 \quad \text{versus} \quad H_1 : B_1 \neq 0 \text{ or } \cdots \text{ or } B_q \neq 0, \tag{16.5.2}$$

is checked. It turns out that the corresponding LM statistic can be determined by replacing all unknown u_t's in (16.5.1) by estimated residuals from a VAR or VECM, say, and estimating the parameters in the resulting auxiliary model by LS. Denoting the resulting residual covariance matrix estimator based on (16.5.1) by $\widehat{\Sigma}_{\text{vech}}$ and the corresponding matrix obtained for $q = 0$ by $\widehat{\Sigma}_0$, the relevant LM statistic can be shown to be of the form

$$LM_{MARCH}(q) = \frac{1}{2}TK(K+1) - T\text{tr}(\widehat{\Sigma}_{\text{vech}} \widehat{\Sigma}_0^{-1}). \tag{16.5.3}$$

Under the null hypothesis, the statistic has an asymptotic $\chi^2(qK^2(K+1)^2/4)$-distribution, if u_t satisfies standard conditions (see Doornik & Hendry (1997, Sec. 10.9.2.4)).

In (16.5.1), each of the B_j matrices is of dimension $(\frac{1}{2}K(K+1) \times \frac{1}{2}K(K+1))$ and, hence, the auxiliary model involves a large number of parameters even if the order q and the dimension of the process K are only moderate. Therefore the test is not suitable to check for large q, unless the sample size is very large too. It is possible, however, to apply the test to each of the K residual series individually.

From the VARMA representation of an MGARCH process in Section 16.3.2, it can be seen that there is no ARCH in the process u_t, if the process $\mathbf{x}_t := \text{vech}(u_t u_t')$ has no serial correlation. This observation suggests that we may apply a portmanteau test to $\mathbf{x}_t$ to check for ARCH in u_t. Thus, one may use

$$Q_h^{ARCH} := T \sum_{i=1}^{h} \operatorname{tr}(C_i' C_0^{-1} C_i C_0^{-1}) \tag{16.5.4}$$

or the associated modified version

$$\bar{Q}_h^{ARCH} := T^2 \sum_{i=1}^{h} (T-i)^{-1} \operatorname{tr}(C_i' C_0^{-1} C_i C_0^{-1}) \tag{16.5.5}$$

where now $C_i = T^{-1} \sum_{t=i+1}^{T} (\mathbf{x}_t - \bar{\mathbf{x}})(\mathbf{x}_{t-i} - \bar{\mathbf{x}})'$ $(i = 0, 1, \ldots, h)$.

The asymptotic χ^2-distributions of these tests follow from the results in Section 4.4.3, if $\mathbf{x}_t$ is indeed white noise. In practice, it will usually be replaced by a quantity based on estimation residuals $\hat{u}_t$. A rigorous treatment of the properties of the statistics in that case seems to be still missing. In principle, the ARCH-portmanteau test can also be applied to the individual residual series.

16.5.2 LM and Portmanteau Tests for Remaining ARCH

In practice, it is also useful to check for remaining ARCH in the residuals of a fitted ARCH or GARCH model. Such tests are of particular importance in the present context because low order multivariate models are typically fitted as a first attempt to account for conditionally heteroskedastic residuals. Higher order models often have an excessive number of parameters and the estimates are difficult to compute numerically. Therefore, it makes sense to start with low order models and increase the order only if the low order model cannot capture the second order moment dynamics in the data properly. Hence, tests for remaining ARCH in the residuals of an MGARCH model are needed.

Both the ARCH-LM and the ARCH-portmanteau tests have been used for this purpose. In that case, the u_t's in the $\mathbf{x}_t$ vectors are replaced by the estimated ε_t from (16.3.1). In other words, $\tilde{\varepsilon}_t := \tilde{\Sigma}_{t|t-1}^{-1/2} \tilde{u}_t$ is used instead of u_t. Here ML estimators are signified by a tilde. Whereas the LM tests maintain their validity under general conditions in the present situation (Engle & Kroner (1995)), the same is not true for the portmanteau tests. They have still found widespread use in applied work (see Tse & Tsui (1999) for references and further discussion).

Again, it may be useful to apply the tests not only to the multivariate residual vectors but also to the univariate components separately. There are also other tests for remaining ARCH which have a sounder theoretical basis than the portmanteau tests (see Bauwens et al. (2004) for a review and Lundbergh & Teräsvirta (2002) for a discussion of the univariate case).

16.5.3 Other Diagnostic Tests

Other diagnostic tools for checking the validity of fitted MGARCH models are also available. In fact, some of the residual diagnostics for VAR models are also applicable here. Instead of the u_t's, the ε_t's should now be used as the basic residuals. For example, they can be used to perform tests for nonnormality. Although the necessary extensions are in many cases possible, some care is needed in the present context. It can by no means be taken for granted that all the procedures work properly. The case of normality tests and related caveats when they are applied to GARCH residuals was discussed by Fiorentini, Sentana & Calzolari (2004).

16.5.4 An Example

As an example, we consider again the VW stock returns. In Section 16.4.2, we have fitted an MGARCH(1, 1) model because the residuals of the model (16.4.5) appeared to have volatility clusters. Now we can use ARCH-LM tests and formally test for conditional heteroskedasticity of the u_t's. Some results are presented in Table 16.1. Both bivariate and univariate tests applied to the individual residual series clearly reject the no-ARCH null hypothesis. Thus, there is strong evidence in favor of conditionally heteroskedastic residuals. We do not present results of the ARCH-portmanteau test because its validity is not clear.

Table 16.1. ARCH-LM tests for $\widehat{u}_t$ residuals from (16.4.5)

test	bivariate		$\widehat{u}_{1t}$		$\widehat{u}_{2t}$	
	$LM(1)$	$LM(4)$	$LM(1)$	$LM(4)$	$LM(1)$	$LM(4)$
test value	147.4	245.5	54.9	62.1	56.1	70.8
asymptotic distribution	$\chi^2(9)$	$\chi^2(36)$	$\chi^2(1)$	$\chi^2(4)$	$\chi^2(1)$	$\chi^2(4)$
p-value	0.00	0.00	0.00	0.00	0.00	0.00

Of course, the fact that there may be ARCH in the residuals does not necessarily mean that an MGARCH(1, 1) process is a suitable model. Therefore, we also applied tests for remaining ARCH to the residuals $\widetilde{\varepsilon}_t = \widetilde{\Sigma}_{t|t-1}^{-1/2}\widetilde{u}_t$ based on (16.4.6). The test results are given in Table 16.2. Now none of the ARCH-LM tests rejects the null hypothesis at conventional significance levels. On the other hand, applying nonnormality tests confirms what could have been conjectured by looking at the residuals in Figure 16.5, namely that, due to the outliers, normality of the conditional distribution is not likely to be a reasonable assumption. Thus, it may be worth trying some other distribution for the ε_t or some other model than the standard BEKK GARCH(1, 1) we have presented in Section 16.4.2.

Table 16.2. ARCH-LM and nonnormality ($\widehat{\lambda}_{sk}$) tests for $\widehat{\varepsilon}_t$ residuals associated with (16.4.6)

	bivariate			$\widehat{\varepsilon}_{1t}$			$\widehat{\varepsilon}_{2t}$		
test	$LM(1)$	$LM(4)$	$\widehat{\lambda}_{sk}$	$LM(1)$	$LM(4)$	$\widehat{\lambda}_{sk}$	$LM(1)$	$LM(4)$	$\widehat{\lambda}_{sk}$
test value	10.43	33.48	12770	0.067	0.364	4440	0.099	0.385	10314
asymp. distr.	$\chi^2(9)$	$\chi^2(36)$	$\chi^2(4)$	$\chi^2(1)$	$\chi^2(4)$	$\chi^2(2)$	$\chi^2(1)$	$\chi^2(4)$	$\chi^2(2)$
p-value	0.32	0.59	0.00	0.80	0.99	0.00	0.75	0.98	0.00

16.6 Interpreting GARCH Models

16.6.1 Causality in Variance

As we have seen in Chapter 2, Section 2.3.1, Granger's definition of causality is based on forecasts. We have also seen in Chapter 2 that, under suitable conditions, optimal forecasts are obtained as conditional expectations. Therefore, Granger-causality may be defined in terms of conditional expectations. In other words, we may define a time series variable x_t to be causal for z_t, if

$$E(z_{t+1}|z_t, z_{t-1}, \ldots) \neq E(z_{t+1}|z_t, z_{t-1}, \ldots, x_t, x_{t-1}, \ldots). \qquad (16.6.1)$$

This definition suggests a direct extension to higher order conditional moments. We define x_t to be *causal* for z_t *in r-th moment* if

$$E(z_{t+1}^r|z_t, z_{t-1}, \ldots) \neq E(z_{t+1}^r|z_t, z_{t-1}, \ldots, x_t, x_{t-1}, \ldots). \qquad (16.6.2)$$

Thus, (16.6.1) defines causality in mean and considering the central second moments in (16.6.2) gives a definition of causality in variance which is analogous to the previous definition of Granger-causality. The interpretation is also analogous to that of Granger-causality in mean. In other words, if x_t is causal-in-variance for z_t, the conditional volatility of z_t can be predicted more precisely by taking into account present and past information in x_t than without taking this information into account.

If the conditional covariance structure can be described by multivariate ARCH or MGARCH models, the restrictions implied by these definitions are also similar to those for Granger-causality in VAR and VARMA models (see Comte & Lieberman (2000)). In other words, they can be described in terms of zero restrictions on the ARCH or MGARCH parameters. Depending on the specific parameterization of the MGARCH model, the restrictions can be nonlinear in the present situation, however. Tests for causality in variance were proposed and investigated by Cheung & Ng (1996), Hong (2001), and Pantelidis & Pittis (2004).

It is also possible to generalize the causality definition and specify, for example, conditions for both the conditional first and second order moments (e.g., Granger, Robins & Engle (1986)). More generally, one may consider the full conditional distributions rather than just specific moments. In other words, one may define x_t to be causal for z_t if

$$F_{z_{t+1}|z_t,z_{t-1},\ldots}(\cdot) \neq F_{z_{t+1}|z_t,z_{t-1},\ldots,x_t,x_{t-1},\ldots}(\cdot), \tag{16.6.3}$$

where $F_{z|x}(\cdot)$ denotes the conditional distribution function of z given x. Generalizing these concepts to the case where x_t and z_t are vectors is theoretically straightforward, as in the case of Granger-causality in mean.

16.6.2 Conditional Moment Profiles and Generalized Impulse Responses

Impulse responses were used among other tools for analyzing the relations between the variables of linear models such as VARs and VECMs. In linear models, they have the advantage of being time invariant and their shape is invariant to the size and direction of the impulses. These features enable the analyst to represent the reactions of the variables to impulses hitting the system in a small set of graphs. GARCH models are nonlinear models, however. In such models, the situation is quite different. In general, in a nonlinear model, the marginal effect of an impulse will depend on the state of the system when the impulse arrives. Thus, it depends on the history of the variables and it may be different in each time point during the sample. Moreover, the shape of the impulse responses will generally depend on the size and direction of the impulse. For example, quite different reactions may be obtained from positive and negative impulses. In a linear model, a negative impulse of one unit induces the same responses of the variables with opposite sign as a positive impulse of one unit. In contrast, in a nonlinear model, a positive impulse may, e.g., induce almost no reaction of the variables whereas a corresponding negative impulse hitting the system at the same state may lead to a strong reaction. These features are quite plausible in some systems. For example, if the impulses represent news arriving in a financial market, positive news may have a quite different effect than negative news. Hence, nonlinear models clearly have their attractive features for describing economic systems or phenomena.

Still, the greater flexibility of nonlinear models makes them more difficult to interpret properly. In fact, it is not obvious how to define impulse responses of nonlinear models in a meaningful manner. Gallant, Rossi & Tauchen (1993) proposed so-called *conditional moment profiles* which may give useful information on important features and implications of nonlinear multiple time series models. In the spirit of their definition, we consider quantities of the general form

$$E[g(y_{t+h})|y_t + \xi, \Omega_{t-1}] - E[g(y_{t+h})|y_t, \Omega_{t-1}], \quad h = 1, 2, \ldots, \tag{16.6.4}$$

where $g(\cdot)$ denotes some function of interest, ξ represents the impulses hitting the system at time t, and $\Omega_{t-1} := (y_{t-1}, y_{t-2}, \ldots)$ denotes the history of the variables at time t. In other words, the conditional expectation of some quantity of interest, given the history of y_t in period t, is compared to the conditional expectation that is obtained if a shock ξ occurs at time t. For example, defining

$$g(y_{t+h}) = [y_{t+h} - E(y_{t+h}|\Omega_{t+h-1})][y_{t+h} - E(y_{t+h}|\Omega_{t+h-1})]' \qquad (16.6.5)$$

results in conditional volatility profiles, which may be compared to a baseline profile obtained for a specific history of the process and a zero impulse. Clearly, in general the conditional moment profiles depend on the history Ω_{t-1} as well as the impulse ξ. Similar quantities were also considered by Koop, Pesaran & Potter (1996) who called them *generalized impulse responses*.

If models with ARCH or MGARCH errors are used to describe the volatility dynamics of a financial market, the volatility resulting from the arrival of news may be of interest (see Engle & Ng (1993)). In this case, using the function (16.6.5) and comparing conditional covariance matrices

$$\Sigma_{t+h|t} = E\{[y_{t+h} - E(y_{t+h}|\Omega_{t+h-1})][y_{t+h} - E(y_{t+h}|\Omega_{t+h-1})]'|y_t, \Omega_{t-1}\},$$

based on the actual history at time t, to

$$\Sigma^{\xi}_{t+h|t} = E\{[y_{t+h} - E(y_{t+h}|\Omega_{t+h-1})][y_{t+h} - E(y_{t+h}|\Omega_{t+h-1})]'|y_t+\xi, \Omega_{t-1}\},$$

for $h = 1, 2, \ldots$, may give an impression of the reactions of the market under consideration. For instance, for the BEKK GARCH(1, 1) model, we get

$$\Sigma_{t+1|t} = C_0^{*'}C_0^* + \Gamma_{11}^{*'}E(u_t u_t'|y_t, \Omega_{t-1})\Gamma_{11}^* + G_{11}^{*'}\Sigma_{t|t-1}G_{11}^*. \qquad (16.6.6)$$

The quantities in (16.6.6) are usually computed using the estimates of the conditional mean equation and the relevant GARCH volatility model. The matrix $E(u_t u_t'|y_t, \Omega_{t-1}) = u_t u_t'$ is replaced by $\hat{u}_t \hat{u}_t'$, where the $\hat{u}_t$ are typically residuals from estimating the conditional mean model. If the corresponding quantities related to an impulse ξ are considered, the impulse is simply added to the $\hat{u}_t$. Because $\Sigma_{t+h|t}, h = 2, 3, \ldots$, is a convenient estimator for $E(u_{t+h}u_{t+h}'|y_t, \Omega_{t-1})$, recursive forecasts of future volatility, conditional on information which is available at time t, are computed as:

$$\Sigma_{t+h|t} = C_0^{*'}C_0^* + \Gamma_{11}^{*'}\Sigma_{t+h-1|t}\Gamma_{11}^* + G_{11}^{*'}\Sigma_{t+h-1|t}G_{11}^*, \quad h = 2, 3, \ldots. \qquad (16.6.7)$$

From the conditional covariance matrices, conditional moment profiles are obtained as differences

$$\begin{bmatrix} \phi_{11t,h}(\xi) & \cdots & \phi_{1Kt,h}(\xi) \\ \vdots & \ddots & \vdots \\ \phi_{1Kt,h}(\xi) & \cdots & \phi_{KKt,h}(\xi) \end{bmatrix} = \Sigma^{\xi}_{t+h|t} - \Sigma_{t+h|t}. \qquad (16.6.8)$$

Although these quantities may be interesting to look at, they depend on t, h, and ξ. Hence, there is a separate impulse response function for each given t and ξ. In empirical work, it will therefore be necessary to summarize the wealth of information in the conditional moment profiles in a meaningful way. In a study of two stock price series, Herwartz & Lütkepohl (2000), for example, considered the following summary statistics:

- Averages over all histories for different impulse vectors ξ, $\bar{\phi}_{ij.,h}(\xi) = T^{-1} \sum_{t=1}^{T} \phi_{ijt,h}(\xi)$.
- Averages over a large range of different impulse vectors ξ_r, $\bar{\phi}_{ijt,h}(\cdot) = R^{-1} \sum_{r=1}^{R} \phi_{ijt,h}(\xi_r)$, for given values of t and h. Here R is the number of impulses considered. The impulse ξ_r may, for instance, be obtained from the estimated model residuals.

Although these summary statistics condense the information in the conditional moment profiles considerably, they are still a rich source of information which can be presented in graphs or further condensed by fitting nonparametric density functions or using other summary statistics (see Herwartz & Lütkepohl (2000)).

Of course, in practice, an additional obstacle is that the actual data generation process is unknown and estimated models are available at best. In that case, the conditional moment profiles or generalized impulse responses will be computed from estimated quantities only. They are therefore also estimates and it would be useful to have measures for their sampling variability. It is not clear how this additional information is computed and presented in the best way in practice. In any case, if only the estimated quantities are available and presented, it is useful to keep in mind these further limitations when the results are interpreted.

It is naturally of interest to better understand what the various models for conditional volatility can tell us about the relations between variables and, hence, about what is actually going on in a particular market or segment of the economy. Therefore it is not surprising that the interpretation of MGARCH models is a field of active research. Some important recent contributions in addition to those noted earlier are Engle & Ng (1993), Lin (1997), and Hafner & Herwartz (1998a).

16.7 Problems and Extensions

There are a number of problems associated with ARCH and GARCH modelling. Some of them have been mentioned in earlier sections of this chapter but may be worth emphasizing again. In addition, there are some problems which we have not addressed so far.

First, due to the highly nonlinear form of the log-likelihood function and the potentially large number of parameters in a multivariate GARCH model which have to satisfy a number of restrictions, computing ML estimates is a difficult task. Therefore it is highly desirable to develop fast and robust optimization algorithms which work well under these particular conditions. A review and comparison of some of the available software was provided by Brooks, Burke & Persand (2003).

Secondly, a sound analysis of conditional heteroskedasticity in a multivariate time series context requires that at least the asymptotic properties of the estimators are known. As we have seen in Section 16.4.1, some asymptotic theory is available for quasi ML estimators of specific MGARCH models. Unfortunately, the required conditions are not satisfactory in all situations. Hence, developing asymptotic theory under more general conditions is desirable.

Third, a toolkit for model specification and model checking is available, as we have seen in Section 16.5. There are some open questions regarding the statistical properties of these tools, however. Moreover, given the wealth of possible model specifications, some more refined tools are desirable that help the analyst to find the best specification for a particular data set and analysis objective and for discriminating between alternative models.

Fourth, although a range of proposals have been made on how to interpret multivariate GARCH models, the available tools leave room for improvements. The nonlinearity of these models makes it more difficult to extract the essential features than in linear models for the conditional mean.

Finally, there are many features in financial and other economic data which are not described well by the GARCH models considered in this chapter. Therefore, a range of other models have been proposed that can capture specific aspects of the distributional properties of financial series in a more satisfactory way. For example, exogenous variables may be included in a multivariate GARCH model (see Engle & Kroner (1995)). Also, as mentioned in Section 16.1, the volatility in a series may have an impact on the conditional mean. To account for this possibility, it may be useful to allow conditional variances to enter the conditional mean function (Engle, Lilien & Robins (1987)). These so-called *ARCH-in-mean (ARCH-M) models* may also be generalized to the multivariate case.

Stochastic volatility models represent another approach to modelling time-varying volatility. In this approach, the conditional covariance matrix depends on an unobserved latent process and not on past observations as in the ARCH model. For instance, in the univariate case, letting $\varepsilon_t \sim$ i.i.d. $\mathcal{N}(0,1)$ and specifying $u_t := \sigma_t \varepsilon_t$, the logarithm of the conditional standard deviation is assumed to be generated as

$$\ln \sigma_t = \varphi \ln \sigma_{t-1} + \eta \kappa_t,$$

where $\kappa_t \sim$ i.i.d. $\mathcal{N}(0,1)$ and φ and η are constant parameters. A survey of multivariate stochastic volatility models was given by Ghysels, Harvey & Renault (1996). It may also be worth noting that, in some sense, random coefficient autoregressive models may be regarded as extensions of multivariate ARCH models (see Wong & Li (1997)).

16.8 Exercises

Problem 16.1
Write down BEKK GARCH models explicitly for the following combinations of N and q in (16.3.7):

$$(N, q) = (1,1), (2,1), (1,2), (2,2).$$

Problem 16.2
Write down the factor MGARCH model (16.3.9) explicitly for $N = 2$.

Problem 16.3
Write down in detail all elements of $\Sigma_{t|t-1}$ of a factor MGARCH model as proposed by Vrontos et al. (2003) (see Section 16.3.3) for the case of a bivariate series ($K = 2$).

Problem 16.4
Consider the DEM/USD and GBP/USD exchange rate series from
  www.jmulti.de $\rightarrow$ datasets
(File exrate.dat) and perform the following analysis steps:

(a) Eliminate all rows with missing values from the exchange rate data set.
(b) Determine the VAR order by model selection criteria.
(c) Plot the autocorrelations series and the mean-adjusted squared series. Interpret the plots.
(d) Use ARCH-LM and ARCH-portmanteau tests for the mean-adjusted series and interpret the results. Apply the tests to the bivariate and the two univariate series separately and compare the results.
(e) Fit a bivariate BEKK GARCH(1, 1) model to the bivariate series.
(f) Perform model specification tests based on the residuals of the estimated MGARCH model and interpret the results.

(Hint: A similar data set was analyzed by Herwartz (2004).)

17
Periodic VAR Processes and Intervention Models

17.1 Introduction

In Part II of the book, we have considered cointegrated VAR models and we have seen that they give rise to nonstationary processes with potentially time varying first and second moments. Yet the models have time invariant coefficients. Nonstationarity, that is, time varying first and/or second moments of a process, can also be modelled in the framework of time varying coefficient processes. Suppose, for instance, that the time series under consideration show a seasonal pattern. In that case, a VAR(p) process with different intercept terms for each season may be a reasonable model:

$$y_t = \nu_i + A_1 y_{t-1} + \cdots + A_p y_{t-p} + u_t. \tag{17.1.1}$$

Here ν_i is a ($K \times 1$) intercept vector associated with the i-th season, that is, in (17.1.1), the time index t is assumed to be associated with the i-th season of the year. It is easy to see that such a process has a potentially different mean for each season of the year.

Assuming s seasons, the model (17.1.1) could be written alternatively as

$$y_t = n_{1t} \nu_1 + \cdots + n_{st} \nu_s + A_1 y_{t-1} + \cdots + A_p y_{t-p} + u_t,$$

where

$$n_{it} = 0 \text{ or } 1 \quad \text{and} \quad \sum_{i=1}^{s} n_{it} = 1. \tag{17.1.2}$$

In other words, n_{it} assumes the value of 1 if t belongs to the i-th season and is zero otherwise, that is, n_{it} is a *seasonal dummy variable*.

Of course, the model (17.1.1) is covered, e.g., by the set-up of Chapter 10. In a seasonal context, it is possible, however, that the other coefficients also vary for different seasons. In that case, a more general model may be adequate:

$$y_t = \nu_t + A_{1t}y_{t-1} + \cdots + A_{pt}y_{t-p} + u_t \qquad (17.1.3)$$

with

$$\begin{aligned} B_t &:= [\nu_t, A_{1t}, \ldots, A_{pt}] \\ &= n_{1t}[\nu_1, A_{11}, \ldots, A_{p1}] + \cdots + n_{st}[\nu_s, A_{1s}, \ldots, A_{ps}] \\ &= n_{1t}B_1 + \cdots + n_{st}B_s \end{aligned} \qquad (17.1.4)$$

and

$$\Sigma_t := E(u_t u_t') = n_{1t}\Sigma_1 + \cdots + n_{st}\Sigma_s. \qquad (17.1.5)$$

Here the n_{it} are seasonal dummy variables as in (17.1.2), the $B_i := [\nu_i, A_{1i}, \ldots, A_{pi}]$ are $(K \times (Kp+1))$ coefficient matrices, and the Σ_i are $(K \times K)$ covariance matrices. The model (17.1.3) with periodically varying coefficients as specified in (17.1.4)/(17.1.5) is a general *periodic VAR(p) model*, sometimes abbreviated as PAR(p), with period s. Varying coefficient models of this type will be discussed in Section 17.3.

The model (17.1.3) can also be used in a situation where a stationary, stable data generation process is in operation until period T_1, say, and then some outside intervention occurs after which another VAR(p) process generates the data. This case can be handled within the model class (17.1.3) by defining $s = 2$,

$$n_{1t} = \begin{cases} 1 & \text{for } t \leq T_1, \\ 0 & \text{for } t > T_1, \end{cases}$$

and

$$n_{2t} = \begin{cases} 1 & \text{for } t > T_1, \\ 0 & \text{for } t \leq T_1. \end{cases}$$

Intervention models of this type will be considered in Section 17.4. Interventions in economic systems may, for instance, be due to legislative activities or catastrophic weather conditions. Of course, there could be more than one intervention in the stretch of a time series. The general model (17.1.3) encompasses that situation when the dummy variables are chosen appropriately. In Section 17.2, some properties of the general model (17.1.3) will be given that can be derived without special assumptions regarding the movement of the parameters. These properties are valid for both the periodic and intervention models discussed in Sections 17.3 and 17.4, respectively.

An important characteristic of periodic and intervention models is that only a finite number of regimes exist that are associated with specific, known time periods. In other words, the coefficient variations are systematic. Such a model structure is not realistic in all situations of practical interest. We will therefore discuss models with randomly varying coefficients in Chapter 18.

17.2 The VAR(p) Model with Time Varying Coefficients

In this section, we consider the following general form of a K-dimensional VAR(p) model with time varying coefficients:

$$y_t = \nu_t + A_{1t}y_{t-1} + \cdots + A_{pt}y_{t-p} + u_t, \quad t \in \mathbb{Z}, \qquad (17.2.1)$$

where u_t is a zero mean noise process with covariance matrices $E(u_t u_t') = \Sigma_t$. That is, the u_t may have time varying covariance matrices and, thus, may not be identically distributed. We retain the independence assumption for u_t and u_s, $s \neq t$. Of course, the constant coefficient VAR(p) model considered in previous chapters is a special case of (17.2.1). Further special cases are treated in the next sections. We will now discuss some properties of the general model.

17.2.1 General Properties

To derive general properties, it is convenient to write the model (17.2.1) in VAR(1) form:

$$Y_t = \boldsymbol{\nu}_t + \mathbf{A}_t Y_{t-1} + U_t, \qquad (17.2.2)$$

where

$$Y_t := \begin{bmatrix} y_t \\ \vdots \\ y_{t-p+1} \end{bmatrix}_{(Kp \times 1)}, \quad \boldsymbol{\nu}_t := \begin{bmatrix} \nu_t \\ 0 \\ \vdots \\ 0 \end{bmatrix}_{(Kp \times 1)},$$

$$\mathbf{A}_t := \begin{bmatrix} A_{1,t} & \cdots & A_{p-1,t} & A_{p,t} \\ I_K & & 0 & 0 \\ & \ddots & \vdots & \vdots \\ 0 & \cdots & I_K & 0 \end{bmatrix}_{(Kp \times Kp)}, \quad U_t := \begin{bmatrix} u_t \\ 0 \\ \vdots \\ 0 \end{bmatrix}_{(Kp \times 1)}.$$

By successive substitution we get

$$Y_t = \left(\prod_{j=0}^{h-1} \mathbf{A}_{t-j}\right) Y_{t-h} + \sum_{i=0}^{h-1} \left(\prod_{j=0}^{i-1} \mathbf{A}_{t-j}\right) \boldsymbol{\nu}_{t-i} + \sum_{i=0}^{h-1} \left(\prod_{j=0}^{i-1} \mathbf{A}_{t-j}\right) U_{t-i}. \qquad (17.2.3)$$

Defining the $(K \times Kp)$ matrix $J := [I_K : 0]$ such that $y_t = J Y_t$ and premultiplying (17.2.3) by this matrix gives

$$y_t = J\left(\prod_{j=0}^{h-1} \mathbf{A}_{t-j}\right) Y_{t-h} + \sum_{i=0}^{h-1} \Phi_{it}\nu_{t-i} + \sum_{i=0}^{h-1} \Phi_{it} u_{t-i}, \qquad (17.2.4)$$

where

$$\Phi_{it} := J \left(\prod_{j=0}^{i-1} \mathbf{A}_{t-j} \right) J' \qquad (17.2.5)$$

and it has been used that $J'JU_t = U_t$, $JU_t = u_t$, and similar results hold for ν_t. If

$$\sum_{i=0}^{h-1} \Phi_{it} \nu_{t-i}$$

converges to a constant, say μ_t, for $h \to \infty$, and if the first term on the right-hand side of (17.2.4) converges to zero in mean square and the last term converges in mean square as $h \to \infty$, we get the representation

$$y_t = \mu_t + \sum_{i=0}^{\infty} \Phi_{it} u_{t-i}, \qquad (17.2.6)$$

where $\mu_t = E(y_t)$. In the following, it is assumed without further notice that this representation exists.

It can be used to derive the autocovariance structure of the process. For instance,

$$\begin{aligned} E[(y_t - \mu_t)(y_t - \mu_t)'] &= E\left[\left(\sum_{j=0}^{\infty} \Phi_{jt} u_{t-j} \right) \left(\sum_{i=0}^{\infty} \Phi_{it} u_{t-i} \right)' \right] \\ &= E\left[\sum_{j=0}^{\infty} \sum_{i=0}^{\infty} \Phi_{jt} u_{t-j} u'_{t-i} \Phi'_{it} \right] = \sum_{i=0}^{\infty} \Phi_{it} \Sigma_{t-i} \Phi'_{it} \end{aligned}$$

and

$$\begin{aligned} E[(y_t - \mu_t)(y_{t-1} - \mu_{t-1})'] &= E\left[\left(\sum_{j=0}^{\infty} \Phi_{jt} u_{t-j} \right) \left(\sum_{i=0}^{\infty} \Phi_{i,t-1} u_{t-1-i} \right)' \right] \\ &= E\left[\sum_{j=-1}^{\infty} \sum_{i=0}^{\infty} \Phi_{j+1,t} u_{t-j-1} u'_{t-1-i} \Phi'_{i,t-1} \right] \\ &= \sum_{i=0}^{\infty} \Phi_{i+1,t} \Sigma_{t-1-i} \Phi'_{i,t-1}. \end{aligned}$$

More generally, for some integer h,

$$E[(y_t - \mu_t)(y_{t-h} - \mu_{t-h})'] = \sum_{i=0}^{\infty} \Phi_{i+h,t} \Sigma_{t-h-i} \Phi'_{i,t-h}.$$

Usually these formulas are not very useful for actually computing the autocovariances. They show, however, that the autocovariances generally depend on t and h so that the process y_t is not stationary. In addition, of course, the mean vectors μ_t may be time varying.

Optimal forecasts can be obtained either from (17.2.1) or from (17.2.6). In the former case, the forecasts can be computed recursively as

$$y_t(h) = \nu_{t+h} + A_{1,t+h} y_t(h-1) + \cdots + A_{p,t+h} y_t(h-p), \tag{17.2.7}$$

where $y_t(j) := y_{t+j}$ for $j \leq 0$. Using (17.2.6) gives

$$y_t(h) = \mu_{t+h} + \sum_{i=h}^{\infty} \Phi_{i,t+h} u_{t+h-i} \tag{17.2.8}$$

and the forecast error is

$$y_{t+h} - y_t(h) = \sum_{i=0}^{h-1} \Phi_{i,t+h} u_{t+h-i}. \tag{17.2.9}$$

Hence, the forecast MSE matrices turn out to be

$$\Sigma_t(h) := \mathrm{MSE}[y_t(h)] = \sum_{i=0}^{h-1} \Phi_{i,t+h} \Sigma_{t+h-i} \Phi'_{i,t+h}. \tag{17.2.10}$$

We will discuss some basics of ML estimation for the general model (17.2.1) next.

17.2.2 ML Estimation

Although specific results require specific assumptions, it is useful to establish some general results related to ML estimation for Gaussian processes first. We write the model (17.2.1) as

$$y_t = B_t Z_{t-1} + u_t, \tag{17.2.11}$$

where $B_t := [\nu_t, A_{1t}, \ldots, A_{pt}]$, $Z_{t-1} := (1, Y'_{t-1})'$, and we assume that the $(K \times (Kp+1))$ matrices B_t depend on an $(N \times 1)$ vector $\boldsymbol{\gamma}$ of fixed, time invariant parameters, that is, $B_t = B_t(\boldsymbol{\gamma})$. Furthermore, the Σ_t are assumed to depend on an $(M \times 1)$ vector $\boldsymbol{\sigma}$ of fixed parameters. The vector $\boldsymbol{\sigma}$ is disjoint of and unrelated with $\boldsymbol{\gamma}$. Examples where this situation arises will be seen in the next sections. One example, of course, is a constant coefficient model, where $B_t = B = [\nu, A_1, \ldots, A_p]$ and $\Sigma_t = \Sigma_u$ for all t. Here we may choose $\boldsymbol{\gamma} = \mathrm{vec}(B)$ and $\boldsymbol{\sigma} = \mathrm{vech}(\Sigma_u)$ if no further restrictions are imposed.

Assuming that u_t is a Gaussian noise process, that is, $u_t \sim \mathcal{N}(0, \Sigma_t)$, the log-likelihood function of our general model is

$$\ln l(\boldsymbol{\gamma}, \boldsymbol{\sigma}) = -\frac{KT}{2} \ln 2\pi - \frac{1}{2} \sum_{t=1}^{T} \ln |\Sigma_t| - \frac{1}{2} \sum_{t=1}^{T} u'_t \Sigma_t^{-1} u_t, \tag{17.2.12}$$

where any initial condition terms are ignored. The corresponding normal equations are

$$\begin{aligned}
0 = \frac{\partial \ln l}{\partial \gamma} &= -\sum_{t=1}^{T} \frac{\partial u_t'}{\partial \gamma} \Sigma_t^{-1} u_t \\
&= -\sum_{t=1}^{T} \frac{\partial (y_t - B_t Z_{t-1})'}{\partial \gamma} \Sigma_t^{-1} u_t \\
&= \sum_{t=1}^{T} \frac{\partial \operatorname{vec}(B_t)'}{\partial \gamma} (Z_{t-1} \otimes I_K) \Sigma_t^{-1} u_t \\
&= \sum_{t=1}^{T} \frac{\partial \operatorname{vec}(B_t)'}{\partial \gamma} \Sigma_t^{-1} u_t Z_{t-1}'
\end{aligned} \quad (17.2.13)$$

and

$$\begin{aligned}
0 = \frac{\partial \ln l}{\partial \sigma} &= -\frac{1}{2} \sum_t \left[\frac{\partial \operatorname{vec}(\Sigma_t)'}{\partial \sigma} \frac{\partial \ln |\Sigma_t|}{\partial \operatorname{vec}(\Sigma_t)} \right] \\
&\quad - \frac{1}{2} \sum_t \left[\frac{\partial \operatorname{vec}(\Sigma_t)'}{\partial \sigma} \frac{\partial u_t' \Sigma_t^{-1} u_t}{\partial \operatorname{vec}(\Sigma_t)} \right] \\
&= -\frac{1}{2} \sum_t \left[\frac{\partial \operatorname{vec}(\Sigma_t)'}{\partial \sigma} \operatorname{vec}(\Sigma_t^{-1} - \Sigma_t^{-1} u_t u_t' \Sigma_t^{-1}) \right]. \quad (17.2.14)
\end{aligned}$$

Even if $\partial \operatorname{vec}(B_t)'/\partial \gamma$ is a matrix that does not depend on γ, (17.2.13) is in general a system of equations which is nonlinear in γ and σ because $u_t = y_t - B_t Z_{t-1}$ involves γ. However, we will see in the next sections that in many cases of interest, (17.2.13) reduces to a linear system which is easy to solve. Also, a solution of (17.2.14) is easy to obtain under the conditions of the next sections.

It is furthermore possible to derive an expression for the information matrix associated with the general log-likelihood function (17.2.12). The second partial derivatives with respect to γ are

$$\begin{aligned}
\frac{\partial^2 \ln l}{\partial \gamma \partial \gamma'} &= -\sum_t \left[\frac{\partial \operatorname{vec}(B_t)'}{\partial \gamma} (Z_{t-1} \otimes I_K) \Sigma_t^{-1} (Z_{t-1}' \otimes I_K) \frac{\partial \operatorname{vec}(B_t)}{\partial \gamma'} \right] \\
&\quad + \text{terms with mean zero} \\
&= -\sum_t \left[\frac{\partial \operatorname{vec}(B_t)'}{\partial \gamma} (Z_{t-1} Z_{t-1}' \otimes \Sigma_t^{-1}) \frac{\partial \operatorname{vec}(B_t)}{\partial \gamma'} \right] \\
&\quad + \text{terms with mean zero.} \quad (17.2.15)
\end{aligned}$$

Assuming that $\partial \operatorname{vec}(\Sigma_t)'/\partial \sigma$ does not depend on σ and, thus, the second order partial derivatives of Σ_t with respect to the elements of σ are zero, we get

$$\frac{\partial^2 \ln l}{\partial \sigma \partial \sigma'} = -\frac{1}{2} \sum_t \left(\frac{\partial \operatorname{vec}(\Sigma_t)'}{\partial \sigma} \left[\frac{\partial \operatorname{vec}(\Sigma_t^{-1})}{\partial \sigma'} \right. \right.$$
$$-(I_K \otimes \Sigma_t^{-1} u_t u_t') \frac{\partial \operatorname{vec}(\Sigma_t^{-1})}{\partial \sigma'}$$
$$\left. \left. -(\Sigma_t^{-1} u_t u_t' \otimes I_K) \frac{\partial \operatorname{vec}(\Sigma_t^{-1})}{\partial \sigma'} \right] \right)$$
$$= \frac{1}{2} \sum_t \left[\frac{\partial \operatorname{vec}(\Sigma_t)'}{\partial \sigma} (\Sigma_t^{-1} \otimes \Sigma_t^{-1} - \Sigma_t^{-1} \otimes \Sigma_t^{-1} u_t u_t' \Sigma_t^{-1} \right.$$
$$\left. -\Sigma_t^{-1} u_t u_t' \Sigma_t^{-1} \otimes \Sigma_t^{-1}) \frac{\partial \operatorname{vec}(\Sigma_t)}{\partial \sigma'} \right]. \quad (17.2.16)$$

The assumption of zero second partial derivatives of Σ_t with respect to $\boldsymbol{\sigma}$ will be satisfied in all cases of interest in the following sections. Furthermore, it is easy to see that under the present assumptions

$$E[\partial^2 \ln l / \partial \gamma \partial \sigma'] = 0.$$

Consequently, the information matrix becomes

$$\mathcal{I}(\boldsymbol{\gamma}, \boldsymbol{\sigma}) = E \left[\frac{\partial^2(-\ln l)}{\partial \begin{bmatrix} \gamma \\ \sigma \end{bmatrix} \partial (\gamma', \sigma')} \right] = -E \left[\frac{\partial^2 \ln l}{\partial \begin{bmatrix} \gamma \\ \sigma \end{bmatrix} \partial (\gamma', \sigma')} \right]$$

$$= \begin{bmatrix} \sum_t \left[\frac{\partial \operatorname{vec}(B_t)'}{\partial \gamma} [E(Z_{t-1} Z_{t-1}') \otimes \Sigma_t^{-1}] \frac{\partial \operatorname{vec}(B_t)}{\partial \gamma'} \right] & 0 \\ 0 & \frac{1}{2} \sum_t \left[\frac{\partial \operatorname{vec}(\Sigma_t)'}{\partial \sigma} (\Sigma_t^{-1} \otimes \Sigma_t^{-1}) \frac{\partial \operatorname{vec}(\Sigma_t)}{\partial \sigma'} \right] \end{bmatrix}.$$
(17.2.17)

Although these expressions look a bit unwieldy in their present general form, they are quite handy if special assumptions regarding the time variations of the coefficients are made. We will now turn to such special types of time varying coefficient VAR models.

17.3 Periodic Processes

As we have seen in Section 17.1, in periodic VAR or PAR processes the coefficients vary periodically with period s, say. In other words,

$$y_t = \nu_t + A_t Y_{t-1} + u_t, \quad (17.3.1)$$

where

$$\nu_t = n_{1t}\nu_1 + \cdots + n_{st}\nu_s, \qquad (K \times 1)$$

$$A_t = [A_{1t}, \ldots, A_{pt}] = n_{1t}A_1 + \cdots + n_{st}A_s, \quad (K \times Kp) \qquad (17.3.2)$$

$$\Sigma_t = E(u_t u_t') = n_{1t}\Sigma_1 + \cdots + n_{st}\Sigma_s, \qquad (K \times K)$$

and the n_{it} are seasonal dummy variables which have a value of one if t is associated with the i-th season and zero otherwise. Obviously, the general framework of the previous section encompasses this model. Hence, some properties can be obtained by substituting the expressions from (17.3.2) in the general formulas of the previous section. For periodic processes, however, many properties are more easily derived via another approach which will be introduced and exploited in the next subsection.

Special models arise if only a subset of the parameters vary periodically. For instance, if $\Sigma_i = \Sigma_1$ and $A_i = A_1$ for $i = 1, \ldots, s$, we have a model with seasonal means and otherwise time invariant structure. Simplifications of this kind are useful in practice because they imply a reduction in the number of free parameters to be estimated and thereby result in more efficient estimates and forecasts, at least in large samples. A special case of foremost interest is, of course, a non-periodic, constant coefficient VAR model. If the data generation process turns out to be of that type, the interpretation and analysis is greatly simplified. We will consider estimation and tests of various sets of relevant hypotheses in Subsection 17.3.2.

17.3.1 A VAR Representation with Time Invariant Coefficients

Suppose we have a quarterly process with period $s = 4$ and y_1 belongs to the first quarter. Then we may define an annual process with vectors

$$\mathfrak{y}_1 := \begin{bmatrix} y_4 \\ y_3 \\ y_2 \\ y_1 \end{bmatrix}, \; \mathfrak{y}_2 := \begin{bmatrix} y_8 \\ y_7 \\ y_6 \\ y_5 \end{bmatrix}, \ldots, \mathfrak{y}_\tau := \begin{bmatrix} y_{4\tau} \\ y_{4\tau-1} \\ y_{4\tau-2} \\ y_{4\tau-3} \end{bmatrix}, \ldots$$

This process has a representation with time invariant coefficient matrices. For instance, if the process for each quarter is a VAR(1),

$$\begin{aligned} y_t &= \nu_t + A_{1,t} y_{t-1} + u_t \\ &= \nu_i + A_{1,i} y_{t-1} + u_t, \quad \text{if } t \text{ belongs to the } i\text{-th quarter,} \end{aligned}$$

then the process $\mathfrak{y}_t$ has the representation

$$\begin{bmatrix} I_K & -A_{1,4} & 0 & 0 \\ 0 & I_K & -A_{1,3} & 0 \\ 0 & 0 & I_K & -A_{1,2} \\ 0 & 0 & 0 & I_K \end{bmatrix} \begin{bmatrix} y_{4\tau} \\ y_{4\tau-1} \\ y_{4\tau-2} \\ y_{4\tau-3} \end{bmatrix}$$

$$\begin{bmatrix} \nu_4 \\ \nu_3 \\ \nu_2 \\ \nu_1 \end{bmatrix} + \begin{bmatrix} 0 & 0 & 0 & 0 \\ 0 & 0 & 0 & 0 \\ 0 & 0 & 0 & 0 \\ A_{1,1} & 0 & 0 & 0 \end{bmatrix} \begin{bmatrix} y_{4\tau-4} \\ y_{4\tau-5} \\ y_{4\tau-6} \\ y_{4\tau-7} \end{bmatrix} + \begin{bmatrix} u_{4\tau} \\ u_{4\tau-1} \\ u_{4\tau-2} \\ u_{4\tau-3} \end{bmatrix}. \qquad (17.3.3)$$

More generally, if we have s different regimes (seasons per year) with constant parameters within each regime and if we assume that y_1 belongs to the first season, we may define the sK-dimensional process

$$\mathfrak{y}_\tau := \begin{bmatrix} y_{s\tau} \\ y_{s\tau-1} \\ \vdots \\ y_{s\tau-s+1} \end{bmatrix}_{(sK \times 1)}, \quad \tau = 0, \pm 1, \pm 2, \ldots.$$

This process has the following VAR(P) representation, where P is the smallest integer greater than or equal to p/s:

$$\mathfrak{A}_0 \mathfrak{y}_\tau = \boldsymbol{\nu} + \mathfrak{A}_1 \mathfrak{y}_{\tau-1} + \cdots + \mathfrak{A}_P \mathfrak{y}_{\tau-P} + \mathfrak{u}_\tau, \qquad (17.3.4)$$

where

$$\mathfrak{A}_0 := \begin{bmatrix} I_K & -A_{1,s} & -A_{2,s} & \cdots & -A_{s-1,s} \\ 0 & I_K & -A_{1,s-1} & \cdots & -A_{2,s-1} \\ \vdots & & \ddots & & \vdots \\ \vdots & & & \ddots & \vdots \\ 0 & 0 & 0 & \cdots & I_K \end{bmatrix}_{(sK \times sK)}, \quad \boldsymbol{\nu} := \begin{bmatrix} \nu_s \\ \nu_{s-1} \\ \vdots \\ \nu_1 \end{bmatrix}_{(sK \times 1)},$$

$$\mathfrak{A}_i := \begin{bmatrix} A_{is,s} & A_{is+1,s} & \cdots & A_{(i+1)s-1,s} \\ A_{is-1,s-1} & A_{is,s-1} & \cdots & A_{(i+1)s-2,s-1} \\ \vdots & \vdots & & \vdots \\ A_{is-s+1,1} & A_{is-s+2,1} & \cdots & A_{is,1} \end{bmatrix}_{(sK \times sK)}, \quad i = 1, \ldots, P,$$

$$\mathfrak{u}_\tau := \begin{bmatrix} u_{s\tau} \\ u_{s\tau-1} \\ \vdots \\ u_{s\tau-s+1} \end{bmatrix}_{(sK \times 1)}.$$

All $A_{i,j}$'s with $i > p$ are zero.

The process $\mathfrak{y}_\tau$ is stationary if the y_t's have bounded first and second moments and the VAR operator is stable, that is,

$$\det(\mathfrak{A}_0 - \mathfrak{A}_1 z - \cdots - \mathfrak{A}_P z^P)$$
$$= \det(I_{sK} - \mathfrak{A}_0^{-1} \mathfrak{A}_1 z - \cdots - \mathfrak{A}_0^{-1} \mathfrak{A}_P z^P) \neq 0 \quad \text{for } |z| \leq 1. \qquad (17.3.5)$$

Note that $\det(\mathfrak{A}_0) = 1$.

For the example process (17.3.3), we have

$$\mathfrak{A}_0^{-1} = \begin{bmatrix} I_K & A_{1,4} & A_{1,4}A_{1,3} & A_{1,4}A_{1,3}A_{1,2} \\ 0 & I_K & A_{1,3} & A_{1,3}A_{1,2} \\ 0 & 0 & I_K & A_{1,2} \\ 0 & 0 & 0 & I_K \end{bmatrix}$$

and, thus,

$$\mathfrak{A}_0^{-1}\mathfrak{A}_1 = \begin{bmatrix} A_{1,4}A_{1,3}A_{1,2}A_{1,1} & 0 & 0 & 0 \\ A_{1,3}A_{1,2}A_{1,1} & 0 & 0 & 0 \\ A_{1,2}A_{1,1} & 0 & 0 & 0 \\ A_{1,1} & 0 & 0 & 0 \end{bmatrix}.$$

Hence,

$$\det(\mathfrak{A}_0 - \mathfrak{A}_1 z) = \det(I_K - A_{1,4}A_{1,3}A_{1,2}A_{1,1}z) \neq 0 \quad \text{for } |z| \leq 1$$

is the stability condition for the example process. If this condition is satisfied, we can, for instance, compute the autocovariances of the process $\mathfrak{y}_\tau$ in the usual way. Note, however, that stationarity of $\mathfrak{y}_\tau$ does not imply stationarity of the original process y_t. Even if $\mathfrak{y}_\tau$ has a time invariant mean vector

$$\boldsymbol{\mu} = \begin{bmatrix} \mu_4 \\ \mu_3 \\ \mu_2 \\ \mu_1 \end{bmatrix},$$

for example, the mean vectors μ_4 and μ_3 associated with the fourth and third quarters, respectively, may be different. Similar thoughts apply for other quarters and for the autocovariances associated with different quarters.

The process $\mathfrak{y}_\tau$ corresponding to a periodic process y_t can also be used to determine an upper bound for the order p of the latter. If $\mathfrak{y}_\tau$ is stationary and its order P is selected in the usual way, we know that $p \leq sP$.

Optimal forecasts of a periodic process are easily obtained from the recursions (17.2.7). Assuming that the forecast origin t is associated with the last period of the year, we get

$$\begin{aligned} y_t(1) &= \nu_1 + A_{1,1}y_t + \cdots + A_{p,1}y_{t-p+1} \\ y_t(2) &= \nu_2 + A_{1,2}y_t(1) + \cdots + A_{p,2}y_{t-p+2} \\ &\vdots \\ y_t(s) &= \nu_s + A_{1,s}y_t(s-1) + \cdots + A_{p,s}y_t(s-p) \\ y_t(s+1) &= \nu_1 + A_{1,1}y_t(s) + \cdots + A_{p,1}y_t(s+1-p) \\ &\vdots \end{aligned}$$

17.3.2 ML Estimation and Testing for Time Varying Coefficients

The general framework for ML estimation of the periodic VAR(p) model given in (17.3.1)/(17.3.2), under Gaussian assumptions, is laid out in Section 17.2.2. For the present case, however, a number of simplifications are obtained and closed form expressions can be given for the estimators. In the following, we discuss estimation under various types of restrictions and we consider tests of time invariance of different groups of coefficients. Most of the tests are likelihood ratio (LR) tests, the general form of which is discussed in Appendix C.7. Recall that the LR statistic is

$$\lambda_{LR} = 2[\ln l(\widetilde{\boldsymbol{\delta}}) - \ln l(\widetilde{\boldsymbol{\delta}}_r)], \qquad (17.3.6)$$

where $\widetilde{\boldsymbol{\delta}}$ is the unconstrained ML estimator and $\widetilde{\boldsymbol{\delta}}_r$ is the restricted ML estimator obtained by maximizing the likelihood function under the null hypothesis H_0. If H_0 is true, under general conditions, the LR statistic has an asymptotic χ^2-distribution with degrees of freedom equal to the number of linearly independent restrictions. In the following, we will give the maximum of the likelihood function under various sets of restrictions in addition to the ML estimators. These results will enable us to set up LR tests for different sets of restrictions.

For the present case of a periodic VAR model, the normal equations given in (17.2.13) reduce to

$$\begin{aligned}
0 &= \frac{\partial \ln l}{\partial \gamma} = \sum_{t=1}^{T} \frac{\partial \operatorname{vec}(n_{1t}B_1 + \cdots + n_{st}B_s)'}{\partial \gamma} \Sigma_t^{-1} u_t Z'_{t-1} \\
&= \sum_{i=1}^{s} \sum_{t=1}^{T} n_{it} \frac{\partial \operatorname{vec}(B_i)'}{\partial \gamma} \Sigma_i^{-1} u_t Z'_{t-1},
\end{aligned} \qquad (17.3.7)$$

where $B_i := [\nu_i, A_{1i}, \ldots, A_{pi}] := [\nu_i, A_i]$, $i = 1, \ldots, s$, and

$$\Sigma_t^{-1} = \left(\sum_{i=1}^{s} n_{it} \Sigma_i\right)^{-1} = \sum_i n_{it} \Sigma_i^{-1}$$

has been used. Moreover, (17.2.14) reduces to

$$0 = \frac{\partial \ln l}{\partial \sigma} = -\frac{1}{2} \sum_{i=1}^{s} \sum_{t=1}^{T} n_{it} \left[\frac{\partial \operatorname{vec}(\Sigma_i)'}{\partial \sigma} \operatorname{vec}(\Sigma_i^{-1} - \Sigma_i^{-1} u_t u_t' \Sigma_i^{-1})\right]. \qquad (17.3.8)$$

We will see in the following that the solution of these sets of normal equations is relatively easy in many situations. The discussion follows Lütkepohl (1992).

All Coefficients Time Varying

We begin with a periodic VAR(p) model for which all coefficients are time varying, that is,

$$H_1: B_t = [\nu_t, A_t] = \sum_{i=1}^{s} n_{it} B_i, \quad \Sigma_t = \sum_{i=1}^{s} n_{it} \Sigma_i. \qquad (17.3.9)$$

For this case

$$\boldsymbol{\gamma} = \text{vec}[B_1, \ldots, B_s]$$

and

$$\boldsymbol{\sigma} = [\text{vech}(\Sigma_1)', \ldots, \text{vech}(\Sigma_s)']'.$$

Using a little algebra, the ML estimators can be obtained from (17.3.7) and (17.3.8):

$$\widetilde{B}_i^{(1)} = \left(\sum_{t=1}^{T} n_{it} y_t Z_{t-1}' \right) \left(\sum_{t=1}^{T} n_{it} Z_{t-1} Z_{t-1}' \right)^{-1} \qquad (17.3.10)$$

and

$$\widetilde{\Sigma}_i^{(1)} = \sum_{t} n_{it} (y_t - \widetilde{B}_i^{(1)} Z_{t-1})(y_t - \widetilde{B}_i^{(1)} Z_{t-1})'/T\bar{n}_i, \qquad (17.3.11)$$

for $i = 1, \ldots, s$. Here $\bar{n}_i = \sum_{t=1}^{T} n_{it}/T$. Except for an additive constant, the corresponding maximum of the log-likelihood function is

$$\lambda_1 := -\frac{1}{2} \sum_{t} \ln |\widetilde{\Sigma}_t^{(1)}| = -\frac{1}{2} T(\bar{n}_1 \ln |\widetilde{\Sigma}_1^{(1)}| + \cdots + \bar{n}_s \ln |\widetilde{\Sigma}_s^{(1)}|). \qquad (17.3.12)$$

All Coefficients Time Invariant

The next case we consider is our well-known basic stationary VAR(p) model, where all the coefficients are time invariant:

$$H_2: B_i = B_1, \quad \Sigma_i = \Sigma_1, \quad i = 2, \ldots, s. \qquad (17.3.13)$$

For this case, we know that the ML estimators are

$$\widetilde{B}_1^{(2)} = \left(\sum_{t} y_t Z_{t-1}' \right) \left(\sum_{t} Z_{t-1} Z_{t-1}' \right)^{-1} \qquad (17.3.14)$$

and

$$\widetilde{\Sigma}_1^{(2)} = \sum_t (y_t - \widetilde{B}_1^{(2)} Z_{t-1})(y_t - \widetilde{B}_1^{(2)} Z_{t-1})'/T. \qquad (17.3.15)$$

The maximum log-likelihood is, except for an additive constant,

$$\lambda_2 := -\tfrac{1}{2} T \ln |\widetilde{\Sigma}_1^{(2)}|. \qquad (17.3.16)$$

This case is considered here because H_2 is a null hypothesis of foremost interest in the present context. Of course, if it turns out that H_2 is true, we can proceed with a standard VAR analysis. The slight change of notation relative to previous chapters is useful here to avoid confusion.

Time Invariant White Noise

If just the white noise covariance matrix is time invariant while the other coefficients vary, we have

$$H_3: B_t = [\nu_t, A_t] = \sum_{i=1}^{s} n_{it} B_i \quad \text{and} \quad \Sigma_i = \Sigma_1, \ i = 2, \ldots, s. \qquad (17.3.17)$$

For this case, it follows from (17.3.7) that the ML estimators of the B_i are

$$\widetilde{B}_i^{(3)} = \widetilde{B}_i^{(1)}, \quad i = 1, \ldots, s, \qquad (17.3.18)$$

and (17.3.8) implies

$$\widetilde{\Sigma}_1^{(3)} = \sum_{i=1}^{s} \sum_{t=1}^{T} n_{it}(y_t - \widetilde{B}_i^{(1)} Z_{t-1})(y_t - \widetilde{B}_i^{(1)} Z_{t-1})'/T. \qquad (17.3.19)$$

The resulting maximum log-likelihood turns out to be

$$\lambda_3 := -\tfrac{1}{2} T \ln |\widetilde{\Sigma}_1^{(3)}|, \qquad (17.3.20)$$

where again an additive constant is suppressed.

Time Invariant Covariance Structure

If just the intercept terms and, hence, the means are time varying, we have the conventional case of a model with seasonal dummies and otherwise constant coefficients. In the present framework, this situation may be represented as

$$H_4: \nu_t = \sum_{i=1}^{s} n_{it} \nu_i \quad \text{and} \quad A_i = A_1, \ \Sigma_i = \Sigma_1, \ i = 2, \ldots, s. \qquad (17.3.21)$$

Under this hypothesis, the ML estimators are easily obtained by defining

$$W_{t-1} = \begin{bmatrix} n_{1,t} \\ \vdots \\ n_{s,t} \\ Y_{t-1} \end{bmatrix} \quad \text{and} \quad C = [\nu_1, \ldots, \nu_s, A_1].$$

The ML estimator of C is

$$\widetilde{C} = \left(\sum_t y_t W'_{t-1} \right) \left(\sum_t W_{t-1} W'_{t-1} \right)^{-1} \tag{17.3.22}$$

and that of Σ_1 is

$$\widetilde{\Sigma}_1^{(4)} = \sum_t (y_t - \widetilde{C} W_{t-1})(y_t - \widetilde{C} W_{t-1})'/T. \tag{17.3.23}$$

Dropping again an additive constant, the corresponding maximum of the log-likelihood function is

$$\lambda_4 := -\tfrac{1}{2} T \ln |\widetilde{\Sigma}_1^{(4)}|. \tag{17.3.24}$$

LR Tests

In Table 17.1, the LR tests of some hypotheses of interest are listed. The LR statistics, under general conditions, all have asymptotic χ^2-distributions with the given degrees of freedom. For this result to hold, it is important that the $\bar{n}_i$ are approximately equal for $i = 1, \ldots, s$, as assumed in periodic models. Moreover, the corresponding $\mathfrak{y}_\tau$ process is assumed to be stable. In Chapter 8, Section 8.4.3, we have argued that if integrated variables are involved and VECMs are considered, the degrees of freedom of Chow tests have to be adjusted relative to the stable case. Because Chow tests are formally similar to some of the tests considered here, it is perhaps not surprising that adjustments will also be necessary in the present case if $I(1)$ variables are involved. The reader is invited to check the degrees of freedom listed in Table 17.1 for the case of a stable underlying $\mathfrak{y}_\tau$ process by counting the number of restrictions imposed under the null hypothesis.

In Chapter 4, Section 4.6.1, we have argued that the asymptotic distributions of similar LR tests are poor guides for the actual small sample distributions. Therefore, the same problem must be expected to prevail in the present case. Using bootstrap versions of the present tests may improve the situation (see Appendix D.3).

Testing a Model with Time Varying Error Covariance Matrix Only Against One Where All Coefficients Are Time Varying

ML estimation of models for which all coefficients are time invariant except for the error covariance matrix is complicated by the implied nonlinearity of the normal equations (17.3.7). Thus, LR tests involving the hypothesis

Table 17.1. LR tests for time varying parameters

null hypothesis	alternative hypothesis	LR statistic λ_{LR}	degrees of freedom
H_2	H_1	$2(\lambda_1 - \lambda_2)$	$(s-1)K[K(p+\frac{1}{2}) + \frac{3}{2}]$
H_3	H_1	$2(\lambda_1 - \lambda_3)$	$(s-1)K(K+1)/2$
H_4	H_1	$2(\lambda_1 - \lambda_4)$	$(s-1)K[Kp + (K+1)/2]$
H_2	H_3	$2(\lambda_3 - \lambda_2)$	$(s-1)K(Kp+1)$
H_2	H_4	$2(\lambda_4 - \lambda_2)$	$(s-1)K$

$$H_5: B_i = B_1, \quad i = 2,\ldots,s \quad \text{and} \quad \Sigma_t = \sum_{i=1}^{s} n_{it}\Sigma_i \qquad (17.3.25)$$

are computationally unattractive. If we wish to test H_5 against a model for which all parameters are time varying (H_5 against H_1), estimation under the alternative is straightforward and, therefore, a Wald test may be considered.

Just as a reminder, if the unrestricted estimator $\widetilde{\gamma}$ of a parameter vector γ has an asymptotic normal distribution,

$$\sqrt{T}(\widetilde{\gamma} - \gamma) \xrightarrow{d} \mathcal{N}(0, \Sigma_{\widetilde{\gamma}}),$$

and the restrictions under the null hypothesis are given in the form $R\gamma = 0$, then the Wald statistic is of the form

$$\lambda_W = T\widetilde{\gamma}'R'(R\widetilde{\Sigma}_{\widetilde{\gamma}}R')^{-1}R\widetilde{\gamma}, \qquad (17.3.26)$$

where $\widetilde{\Sigma}_{\widetilde{\gamma}}$ is a consistent estimator of $\Sigma_{\widetilde{\gamma}}$. If $\text{rk}(R) = N$, $R\Sigma_{\widetilde{\gamma}}R'$ is invertible, and the null hypothesis is true, the Wald statistic has an asymptotic $\chi^2(N)$-distribution (see Appendix C.7).

In the case of interest here, the restrictions relate to the VAR coefficients and intercept terms only. Therefore, we consider the $s(K^2p + K)$-dimensional vector $\gamma = \text{vec}[B_1, \ldots, B_s]$. The restrictions under the null hypothesis H_5 can be written as $R\gamma = 0$ with

$$R = \begin{bmatrix} 1 & -1 & & 0 \\ \vdots & & \ddots & \\ 1 & 0 & & -1 \end{bmatrix}_{((s-1)\times s)} \otimes I_{K^2p+K}. \qquad (17.3.27)$$

Denoting the unrestricted ML estimator of γ by $\widetilde{\gamma}$, standard asymptotic theory implies that it has an asymptotic normal distribution with

$$\Sigma_{\widetilde{\gamma}} = \lim_{T \to \infty} T\mathcal{I}(\gamma)^{-1}$$

and

$$\mathcal{I}(\boldsymbol{\gamma}) = -E\left[\frac{\partial^2 \ln l}{\partial \boldsymbol{\gamma} \partial \boldsymbol{\gamma}'}\right]$$

is the upper left-hand block of the information matrix (17.2.17). For the present case, $\mathcal{I}(\boldsymbol{\gamma})$ is seen to be block diagonal with the i-th $((K^2p+K) \times (K^2p+K))$ block on the diagonal being

$$-E\left[\frac{\partial^2 \ln l}{\partial \boldsymbol{\gamma}_i \partial \boldsymbol{\gamma}_i'}\right] = E\left[\sum_t n_{it} Z_{t-1} Z_{t-1}'\right] \otimes \Sigma_i^{-1},$$

where $\boldsymbol{\gamma}_i = \text{vec}(B_i)$. Thus $\Sigma_{\tilde{\boldsymbol{\gamma}}}$ is also block-diagonal and, under standard assumptions, the i-th block is consistently estimated by

$$\left(\frac{1}{T}\sum_t n_{it} Z_{t-1} Z_{t-1}'\right)^{-1} \otimes \widetilde{\Sigma}_i^{(1)}.$$

The resulting estimator $\widetilde{\Sigma}_{\tilde{\boldsymbol{\gamma}}}$ of $\Sigma_{\tilde{\boldsymbol{\gamma}}}$ may be used in (17.3.26). If H_5 is true, λ_W has an asymptotic χ^2-distribution with $(s-1)K(Kp+1)$ degrees of freedom.

Testing a Time Invariant Model Against One with Time Varying Error Covariance

In order to test a stationary constant parameter model (H_2) against one, where the error covariances vary (H_5), an LM (Lagrange multiplier) test is convenient because it requires ML estimation under the null hypothesis only. In Appendix C.7, the general form of the LM statistic is given as

$$\lambda_{LM} = s(\widetilde{\boldsymbol{\gamma}}_r, \widetilde{\boldsymbol{\sigma}}_r)' \mathcal{I}(\widetilde{\boldsymbol{\gamma}}_r, \widetilde{\boldsymbol{\sigma}}_r)^{-1} s(\widetilde{\boldsymbol{\gamma}}_r, \widetilde{\boldsymbol{\sigma}}_r), \qquad (17.3.28)$$

where $\mathcal{I}(\widetilde{\boldsymbol{\gamma}}_r, \widetilde{\boldsymbol{\sigma}}_r)$ is the information matrix of the unrestricted model evaluated at the restricted ML estimators obtained under the null hypothesis and

$$s(\boldsymbol{\gamma}, \boldsymbol{\sigma}) = \begin{bmatrix} \frac{\partial \ln l}{\partial \boldsymbol{\gamma}} \\ \frac{\partial \ln l}{\partial \boldsymbol{\sigma}} \end{bmatrix}$$

is the score vector of first order partial derivatives of the log-likelihood function. In the present case, $\boldsymbol{\gamma} = \text{vec}(B_1)$ is left unrestricted. Thus, $\widetilde{\boldsymbol{\gamma}}_r = \widetilde{\boldsymbol{\gamma}}$ and

$$\left[\frac{\partial \ln l}{\partial \boldsymbol{\gamma}}\bigg|_{\widetilde{\boldsymbol{\gamma}}_r}\right] = 0.$$

Consequently, defining $\boldsymbol{\sigma} = (\text{vech}(\Sigma_1)', \ldots, \text{vech}(\Sigma_s)')'$, the LM statistic reduces to

$$\lambda_{LM} = -\frac{\partial \ln l}{\partial \boldsymbol{\sigma}'} E\left[\frac{\partial^2 \ln l}{\partial \boldsymbol{\sigma} \partial \boldsymbol{\sigma}'}\right]^{-1} \frac{\partial \ln l}{\partial \boldsymbol{\sigma}} \tag{17.3.29}$$

with all derivatives evaluated at the restricted estimator,

$$\widetilde{\boldsymbol{\sigma}}_r = \widetilde{\boldsymbol{\sigma}}^{(2)} := \begin{bmatrix} \text{vech}(\widetilde{\Sigma}_1^{(2)}) \\ \vdots \\ \text{vech}(\widetilde{\Sigma}_1^{(2)}) \end{bmatrix} \quad (\tfrac{1}{2}sK(K+1) \times 1).$$

From (17.3.8), we see that

$$\frac{\partial \ln l}{\partial \text{vech}(\Sigma_i)} = -\frac{1}{2} \sum_{t=1}^{T} n_{it} \mathbf{D}'_K \text{vec}(\Sigma_i^{-1} - \Sigma_i^{-1} u_t u_t' \Sigma_i^{-1}), \tag{17.3.30}$$

where $\mathbf{D}_K = \partial \text{vec}(\Sigma_i)/\partial \text{vech}(\Sigma_i)'$ is the $(K^2 \times \tfrac{1}{2}K(K+1))$ duplication matrix, as usual. Furthermore, for the present case,

$$\frac{\partial \text{vec}(\Sigma_t)'}{\partial \boldsymbol{\sigma}} = \begin{bmatrix} n_{1t}\mathbf{D}'_K \\ \vdots \\ n_{st}\mathbf{D}'_K \end{bmatrix} \quad (\tfrac{1}{2}sK(K+1) \times K^2).$$

Thus, we get from (17.2.17),

$$-E\left[\frac{\partial^2 \ln l}{\partial \boldsymbol{\sigma} \partial \boldsymbol{\sigma}'}\right]$$

$$= \begin{bmatrix} \tfrac{1}{2}T\bar{n}_1 \mathbf{D}'_K(\Sigma_1^{-1} \otimes \Sigma_1^{-1})\mathbf{D}_K & & 0 \\ & \ddots & \\ 0 & & \tfrac{1}{2}T\bar{n}_s \mathbf{D}'_K(\Sigma_s^{-1} \otimes \Sigma_s^{-1})\mathbf{D}_K \end{bmatrix}$$

which implies

$$-E\left[\frac{\partial^2 \ln l}{\partial \boldsymbol{\sigma} \partial \boldsymbol{\sigma}'}\right]^{-1}$$

$$= \begin{bmatrix} 2\mathbf{D}_K^+(\Sigma_1 \otimes \Sigma_1)\mathbf{D}_K^{+\prime}/T\bar{n}_1 & & 0 \\ & \ddots & \\ 0 & & 2\mathbf{D}_K^+(\Sigma_s \otimes \Sigma_s)\mathbf{D}_K^{+\prime}/T\bar{n}_s \end{bmatrix}, \tag{17.3.31}$$

where $\mathbf{D}_K^+$ is the Moore-Penrose inverse of $\mathbf{D}_K$. Using (17.3.30) and (17.3.31) with u_t replaced by $\widetilde{u}_t = y_t - \widetilde{B}_1^{(2)} Z_{t-1}$ and Σ_i replaced by $\widetilde{\Sigma}_1^{(2)}$ the LM statistic in (17.3.29) is easy to evaluate. Under the null hypothesis H_2 and general conditions, it has an asymptotic χ^2-distribution with $(s-1)K(K+1)/2$ degrees of freedom.

17.3.3 An Example

The previously considered theoretical concepts shall now be illustrated by an example from Lütkepohl (1992). We use the first differences of logarithms of quarterly, seasonally unadjusted West German income and consumption data for the years 1960–1987 given in File E4. The two series are plotted in Figure 17.1. Obviously, they exhibit a quite strong seasonal pattern.

There are various problems that may be brought up with respect to the data. For instance, it is possible that the logarithms of the original series are cointegrated (see Part II). In that case, fitting a VAR process to the first differences may be inappropriate. Also, there may be structural shifts during the sample period. We ignore such problems here because we just want to provide an illustrative example for the theoretical results of the previous subsections.

Because we have quarterly data, the period $s = 4$ is given naturally. Stacking the variables for each year in one long 8-dimensional vector $\mathfrak{y}_\tau$, as in Section 17.3.1, we just have 27 observations for each component of $\mathfrak{y}_\tau$. (Note that the first value of the series is lost by differencing.) Thus, the largest full VAR process that can be fitted to the 8-dimensional system is a VAR(3). In such a situation, application of model selection criteria is a doubtful strategy for choosing the order of $\mathfrak{y}_\tau$. Because we want to test the null hypothesis of constant coefficients, it may be reasonable to choose the VAR order under the null hypothesis, that is, to assume a constant coefficient model at the VAR order selection stage. Therefore we have fitted constant coefficient VAR models to the bivariate y_t series consisting of the quarterly income and consumption variables. FPE, AIC, HQ, and SC all have chosen the order $p = 5$ when a maximum of 8 was allowed. Of course, this may not mean too much if the coefficients are actually time varying. The order 5 seems to be a reasonable choice, however, because it means that, for each observation, lags from a whole year and the corresponding quarter of the previous year are included. Therefore, we will work with $p = 5$ in the following.

The first test we carry out is one of H_2 against H_1, that is, a constant coefficient model is tested against one where all the coefficients are time varying. Note that we use the order $p = 5$ also for the model with time varying coefficients. The test value $\lambda_{LR} = 2(\lambda_1 - \lambda_2) = 223.79$ is clearly significant at the 1% level because in this case the number of degrees of freedom of the asymptotic χ^2-distribution is 75. Thus, we conclude that at least some coefficients are not time invariant. To see whether the noise series may be regarded as stationary, we also test H_3 against H_1. The resulting test value is $\lambda_{LR} = 2(\lambda_1 - \lambda_3) = 35.95$ which is also significant at the 1% level because we now have 9 degrees of freedom. Next we use the Wald test described in Section 17.3.2 to see whether the VAR coefficients and intercept terms may be assumed to be constant through time. In other words, we test H_5 against H_1. The test value becomes $\lambda_W = 347$. Comparing this with critical values from the $\chi^2(66)$-distribution, we again reject the null hypothesis H_5 at the

17.3 Periodic Processes   603

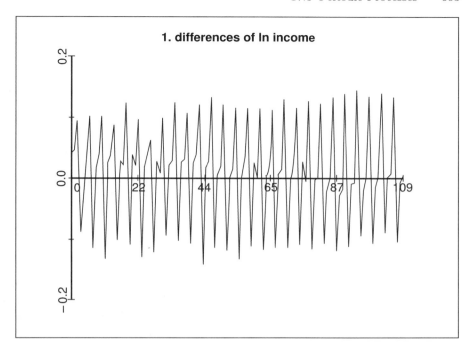

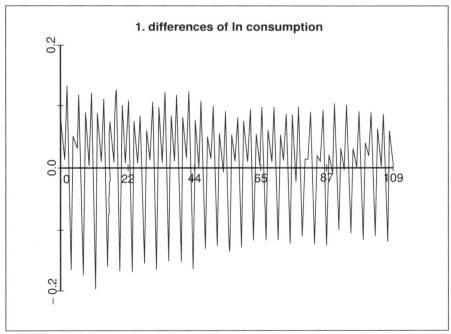

Fig. 17.1. Seasonally unadjusted West German income and consumption series.

1% level. The reader is invited to perform further tests on these data. The tests performed so far support a full periodic model. Notice, however, that our tests are based on asymptotic theory. Their actual distributions in samples as small as the present one are unclear and, in any case, they are not likely to be well approximated by the asymptotic χ^2-distributions. Thus, it is not clear how much evidence in favor of a full periodic model the tests actually provide in this specific case.

Of course, it is possible that a periodic model does not adequately capture the characteristics of the data generating process. In that case, the tests may not have much relevance. To check the adequacy of a periodic model, similar tools may be used as in the stationary nonperiodic case. For instance, a residual analysis could be performed in a similar fashion as for nonperiodic VAR models. The properties of tests for model adequacy may be derived from the stationary representation of the annual process $\mathfrak{y}_\tau$.

17.3.4 Bibliographical Notes and Extensions

Early discussions of periodic time series models include those by Gladyshev (1961) and Jones & Brelsford (1967). Pagano (1978) studied properties of periodic autoregressions while Cleveland & Tiao (1979) considered periodic univariate ARMA models and Tiao & Grupe (1980) explored the consequences of fitting nonperiodic models to data generated by a periodic model. Cipra (1985) discussed inference for periodic moving average processes and Li & Hui (1988) developed an algorithm for ML estimation of periodic ARMA models. A Bayesian analysis of periodic autoregressions was given by Anděl (1983, 1987) and an application of periodic modelling can be found, for instance, in Osborn & Smith (1989). More recently, periodic models for integrated and cointegrated variables were also considered (e.g., Herwartz (1995), Boswijk & Franses (1995, 1996), Boswijk, Franses & Haldrup (1997), Ghysels & Osborn (2001, Chapter 6)). The last publication also includes many more references related to periodic time series models.

17.4 Intervention Models

In Section 17.1, an intervention model was described as one where a particular stationary data generation mechanism is in operation until period T_1, say, and another process generates the data after period T_1. For instance,

$$y_t = \nu_1 + A_1 Y_{t-1} + u_t, \quad E(u_t u_t') = \Sigma_1, \ t \leq T_1 \tag{17.4.1}$$

and

$$y_t = \nu_2 + A_2 Y_{t-1} + u_t, \quad E(u_t u_t') = \Sigma_2, \ t > T_1. \tag{17.4.2}$$

In the present case, it makes a difference whether the intervention is modelled within the intercept form of the process like in (17.4.1)/(17.4.2) or within a mean-adjusted representation. We will consider both cases in turn, following Lütkepohl (1992).

17.4.1 Interventions in the Intercept Model

Before we consider more general situations, it may be useful to study the case described by (17.4.1) and (17.4.2) in a little more detail. For simplicity, suppose that $A_2 = A_1$ and $\Sigma_2 = \Sigma_1$ so that there is just a shift in the intercept terms. Moreover, we assume that the process is stable. In this case, the mean of y_t is

$$E(y_t) = \begin{cases} \sum_{i=0}^{\infty} \Phi_i \nu_1, & t \leq T_1, \\ \sum_{i=0}^{t-T_1} \Phi_i \nu_2 + \sum_{i=t-T_1+1}^{\infty} \Phi_i \nu_1, & t > T_1, \end{cases}$$

where the Φ_i's are the coefficient matrices of the moving average representation of the mean-adjusted process, i.e.,

$$\sum_{i=0}^{\infty} \Phi_i z^i = (I_K - A_{11}z - \cdots - A_{p1}z^p)^{-1}.$$

Hence, after the intervention, the process mean does not reach a fixed new level immediately but only gradually,

$$E(y_t) \xrightarrow[t \to \infty]{} \sum_{i=0}^{\infty} \Phi_i \nu_2.$$

In the more general situation, where all coefficients change due to the intervention, similar results also hold for the autocovariance structure. Of course, such a behavior may be quite plausible in practice because a system may react slowly to an intervention. On the other hand, it is also conceivable that an abrupt change occurs. For the case of a change in the mean, we will discuss this situation in Section 17.4.2.

Before discussing that case, we note that the model setup considered in Section 17.3 may be used for intervention models as well with properly specified n_{it}, as mentioned in Section 17.1. The hypotheses considered in Section 17.3.2 are also of interest in the present context. The test statistics may be computed with the same formulas as in Section 17.3.2 and the tests are often referred to as Chow tests (see also Chapter 4, Section 4.6.1). However, the test statistics do not necessarily have the indicated asymptotic distributions in the present case. The problem is that the ML estimators given in the previous section may not be consistent anymore. To see this, consider, for instance, the hypothesis H_1 (all coefficients time varying) and the model in (17.4.1)/(17.4.2). If T_1 is some fixed finite point and $T > T_1$,

$$\widetilde{B}_1 = [\widetilde{\nu}_1, \widetilde{A}_1] = \left(\sum_{t=1}^{T_1} y_t Z'_{t-1}\right) \left(\sum_{t=1}^{T_1} Z_{t-1} Z'_{t-1}\right)^{-1}$$

will not be consistent because the sample information regarding $B_1 := [\nu_1, A_1]$ does not increase when T goes to infinity. As a way out of this problem it may be assumed that T_1 increases with T. For instance, T_1 may be a fixed proportion of T. Then, under common assumptions,

$$\text{plim } \widetilde{B}_1 = \text{plim} \left(\frac{1}{T_1} \sum_{t=1}^{T_1} y_t Z'_{t-1}\right) \text{plim} \left(\frac{1}{T_1} \sum_{t=1}^{T_1} Z_{t-1} Z'_{t-1}\right)^{-1} = B_1.$$

Also asymptotic normality is easy to obtain in this case and the test statistics have the limiting χ^2-distributions obtained from the results in Section 17.3.2.

A logical problem may arise if more than one intervention is present. In that case, it may not be easy to justify the assumption that all subperiods approach infinity with the sample period T. Whether or not this is a problem of practical relevance must be decided on the basis of the as yet unknown small sample properties of the tests. In any event, the large sample χ^2-distributions are just meant to be a guide for the small sample performance of the tests and as such they may be used if the periods between the interventions are reasonably large. Unfortunately, as mentioned in Chapter 4, Section 4.6.1, the asymptotic χ^2-distributions of the test statistics are not likely to be good approximations to the actual small sample distributions if systems of variables are considered.

17.4.2 A Discrete Change in the Mean

We have seen that in an intercept model like (17.4.1)/(17.4.2) the mean gradually approaches a new level after the intervention. Occasionally, it may be more plausible to assume that there is a one-time jump in the process mean after time T_1. In such a situation, a model in mean-adjusted form,

$$y_t - \mu_t = A_1(y_{t-1} - \mu_{t-1}) + \cdots + A_p(y_{t-p} - \mu_{t-p}) + u_t, \qquad (17.4.3)$$

is easier to work with. Here $\mu_t := E(y_t)$ and, for simplicity, it is assumed that all other coefficients are time invariant and that the process is stable. Therefore, the second subscript is dropped from the VAR coefficient matrices. We also assume Gaussian white noise u_t with time invariant covariance, $u_t \sim \mathcal{N}(0, \Sigma_u)$. Suppose

$$\mu_t = n_{1t}\mu_1 + \cdots + n_{st}\mu_s, \quad n_{it} = 0 \text{ or } 1, \quad \sum_{i=1}^{s} n_{it} = 1. \qquad (17.4.4)$$

In other words, there are s interventions so that for each i, the n_{it}'s, $t = 1, \ldots, T$, are a sequence of zeros and ones, the latter appearing in consecutive positions.

In general, exact ML estimation of the model (17.4.3) results in nonlinear normal equations. To avoid the use of nonlinear optimization algorithms, the μ_i's may be estimated by

$$\widetilde{\mu}_i = \frac{1}{T\bar{n}_i}\sum_{t=1}^{T} n_{it} y_t, \quad i = 1, \ldots, s. \tag{17.4.5}$$

Provided $T\bar{n}_i = \sum_t n_{it}$ approaches infinity with T, it can be shown that under general assumptions, $\widetilde{\mu}_i$ is consistent and

$$\sqrt{T\bar{n}_i}(\widetilde{\mu}_i - \mu_i) \xrightarrow{d} \mathcal{N}(0, \Sigma_{\widetilde{\mu}}), \tag{17.4.6}$$

where

$$\Sigma_{\widetilde{\mu}} = (I_K - A_1 - \cdots - A_p)^{-1} \Sigma_u (I_K - A_1 - \cdots - A_p)'^{-1}$$

(see Chapter 3, Section 3.3, and Problem 17.6). Note that the asymptotic covariance matrix does not depend on i. Furthermore, the $\widetilde{\mu}_i$ are asymptotically independent. Hence, it is quite easy to perform a Wald test of the hypothesis

$$H_6 : \mu_i = \mu_1, \quad i = 2, \ldots, s \quad \text{or} \quad R \begin{bmatrix} \mu_1 \\ \vdots \\ \mu_s \end{bmatrix} = 0, \tag{17.4.7}$$

where R has a similar structure as in (17.3.27). The corresponding Wald statistic is

$$\lambda_W = T[\sqrt{\bar{n}_1}\widetilde{\mu}_1', \ldots, \sqrt{\bar{n}_s}\widetilde{\mu}_s']R'[R(I_s \otimes \Sigma_{\widetilde{\mu}})R']^{-1}R \begin{bmatrix} \sqrt{\bar{n}_1}\widetilde{\mu}_1 \\ \vdots \\ \sqrt{\bar{n}_s}\widetilde{\mu}_s \end{bmatrix}, \tag{17.4.8}$$

where $[R(I_s \otimes \Sigma_{\widetilde{\mu}})R']^{-1}$ reduces to

$$\begin{bmatrix} 2 & 1 & \cdots & 1 \\ 1 & 2 & & 1 \\ \vdots & & \ddots & \vdots \\ 1 & 1 & \cdots & 2 \end{bmatrix}^{-1} \otimes \Sigma_{\widetilde{\mu}}^{-1}$$

and $\Sigma_{\widetilde{\mu}}$ is estimated in the usual way. In other words,

$$\widetilde{A} = \left(\sum_t \widetilde{y}_t \widetilde{Y}_{t-1}'\right)\left(\sum_t \widetilde{Y}_{t-1}\widetilde{Y}_{t-1}'\right)^{-1}$$

and

$$\widetilde{\Sigma}_u = \sum_t (\widetilde{y}_t - \widetilde{A}\widetilde{Y}_{t-1})(\widetilde{y}_t - \widetilde{A}\widetilde{Y}_{t-1})'/T,$$

where $\widetilde{y}_t := y_t - \widetilde{\mu}_t$ and

$$\widetilde{Y}_{t-1} := \begin{bmatrix} y_{t-1} - \widetilde{\mu}_{t-1} \\ \vdots \\ y_{t-p} - \widetilde{\mu}_{t-p} \end{bmatrix}.$$

Under H_6, λ_W has an asymptotic χ^2-distribution with $(s-1)K$ degrees of freedom, if the VAR process is stable.

17.4.3 An Illustrative Example

As an example of testing for structural change in the present framework, we consider again the seasonally adjusted quarterly West German investment, income, and consumption data given in File E1. The data were first used in Chapter 3. As in that and some other chapters, we perform the analysis for the first differences of logarithms (rates of change) of the data. In Chapter 4, tests for a structural break after the year 1978 when the second oil price crisis occurred were already performed. We will now consider different pairs of hypotheses to illustrate the results of this section.

For an event like a drastic oil price increase, a smooth adjustment of the general economic conditions seems more plausible than a discrete change. Therefore the intercept version of an intervention model is chosen with $n_{1t} = 1, n_{2t} = 0$, for $t \leq 1978.4$ and $n_{1t} = 0, n_{2t} = 1$ for $t \geq 1979.1$. Because a VAR(2) model performed reasonably well in Chapter 4 for the period 1960–1978 we use VAR(2) processes for both subperiods. This choice is plausible under the null hypothesis of no structural change after 1978.

We first test a stationary model (H_2) against one where all parameters are allowed to vary (H_1). The resulting value of the LR statistic is $\lambda_{LR} = 64.11$. From Table 17.1 we have

$$(s-1)K\left[K(p+\tfrac{1}{2}) + \tfrac{3}{2}\right] = 27$$

degrees of freedom because $s = 2$, $K = 3$, and $p = 2$. Hence, we can reject the null hypothesis of time invariance at the 1% level of significance ($\chi^2(27)_{.99} = 46.96$). This result, of course, does not necessarily mean that all coefficients are really time varying. For instance, the error covariance matrix may be time invariant while the other coefficients vary. To check this possibility, we test H_3 against H_1 (all coefficients time varying). The value of the LR statistic becomes $\lambda_{LR} = 33.46$ and the number of degrees of freedom for this test is 6. Thus, the test value exceeds the critical value of the χ^2-distribution for a 1% significance level ($\chi^2(6)_{.99} = 16.81$) and we reject the null hypothesis. Further tests on the data are possible and the reader is invited to carry them out.

It is perhaps worth pointing out that we have only four years of data or 16 observations for each variable after the potential structural change. The quality of the χ^2-approximations to the distributions of the LR statistics is therefore doubtful in the present case, as discussed in Chapter 4, Section 4.6.1. In that section, we found that a bootstrap version of a Chow test of H_2 against H_3 may result in a different conclusion than the use of asymptotic critical values. Clearly, similar results are conceivable for the example considered in this section.

17.4.4 Extensions and References

Although we have used the label "intervention" for the type of change that occurs in the models considered in the previous subsections, they could also be regarded as outliers if, for instance, a change in the process mean occurs for a small number of periods only. Tsay (1988) discussed univariate time series models with outliers and structural changes and listed a number of further references. By appropriate choice of the dummy variables n_{it}, it is possible to combine periodic and intervention or outlier models. Extensions of the present framework to VARMA or restricted VAR models are possible in principle. Moreover, cointegrated VAR models with structural shifts were already mentioned in Chapter 8 in the context of testing for the cointegration rank.

More general forms of interventions in the process mean were discussed by Box & Tiao (1975) and Abraham (1980). They assumed that interventions have occurred at $t = T_1, \ldots, T_k$ and they define a vector $I_t = (I_t(T_1), \ldots, I_t(T_k))'$ of dummy variables that may be of the type

$$I_t(T_i) = \begin{cases} 0 & \text{for } t < T_i, \\ 1 & \text{for } t \geq T_i, \end{cases}$$

or of the type

$$I_t(T_i) = \begin{cases} 0 & \text{for } t \neq T_i, \\ 1 & \text{for } t = T_i. \end{cases}$$

They model the interventions as $R(L)I_t$, where $R(L)$ is a matrix of rational functions in the lag operator.

Further complications arise if the time of the break is unknown and has to be estimated in addition to the VAR coefficients. This case was discussed by Bai (1994), Bai, Lumsdaine & Stock (1998), and Lütkepohl et al. (2004), among others.

17.5 Exercises

Problem 17.1
Suppose y_t is a periodic K-dimensional VAR(1) (PAR(1)) process,

$$y_t = \nu_1 + A_{11}y_{t-1} + u_t, \quad E(u_t u_t') = \Sigma_1, \quad \text{if } t \text{ is even},$$

and

$$y_t = \nu_2 + A_{12}y_{t-1} + u_t, \quad E(u_t u_t') = \Sigma_2, \quad \text{if } t \text{ is odd}.$$

(a) Derive explicit expressions for the means μ_t and the matrices Φ_{it}.
(b) Derive the autocovariances $E[(y_t - \mu_t)(y_{t-h} - \mu_{t-h})']$ for $h = 1, 2, 3$, for both cases, t even and t odd. Write down explicitly the assumptions used in deriving the autocovariance matrices.

Problem 17.2
Assume that the process y_t given in Problem 17.1 is bivariate with

$$A_{11} = \begin{bmatrix} .5 & .3 \\ .8 & 1.2 \end{bmatrix}$$

and

$$A_{12} = \begin{bmatrix} .6 & .4 \\ .8 & .5 \end{bmatrix}.$$

Is the corresponding process $\mathfrak{y}_\tau = (y_{2\tau}', y_{2\tau-1}')'$ stable?

Problem 17.3
Give the forecasts $y_t(h)$, $h = 1, 2, 3$, t odd, for the process from Problem 17.1 and derive explicit expressions for the forecast MSE matrices.

Problem 17.4
For the process given in Problem 17.1, construct an LM test of the hypotheses

$$H_0: \nu_1 = \nu_2, \quad A_{11} = A_{12}, \quad \Sigma_1 = \Sigma_2$$

against

$$H_1: \nu_1 = \nu_2, \quad A_{11} = A_{12}, \quad \Sigma_1 \neq \Sigma_2.$$

Provide an explicit expression for the LM statistic.

Problem 17.5
Suppose the process from Problem 17.1 is in operation until period T_1 and after that another periodic VAR(1) process of the same type but with different coefficients generates a set of variables. Define dummy variables in such a way that the complete process can be written in the form (17.1.3)–(17.1.5).

Problem 17.6
Show that (17.4.6) holds. (Hint: See Chapter 3, Section 3.3.)

Problem 17.7
In 1974 the Deutsche Bundesbank officially changed its monetary policy and started targeting the money stock. Use the two interest rate series given in File E5 and test for an intervention after 1974 within the framework discussed in Section 17.4. Use tests for different types of interventions related to shifts in the intercept terms, the VAR coefficients, and the white noise covariances.

18

State Space Models

18.1 Background

State space models may be regarded as generalizations of the models considered so far. They have been used extensively in system theory, the physical sciences, and engineering. The terminology is therefore largely from these fields. The general idea behind these models is that an observed (multiple) time series $y_1, \ldots, y_T$ depends upon a possibly unobserved state z_t which is driven by a stochastic process. The relation between y_t and z_t is described by the *observation* or *measurement equation*

$$y_t = \mathbf{H}_t z_t + v_t, \qquad (18.1.1)$$

where $\mathbf{H}_t$ is a matrix that may also depend on the period of time, t, and v_t is the *observation error* which is typically assumed to be a noise process. The *state vector* or state of nature is generated as

$$z_t = \mathbf{B}_{t-1} z_{t-1} + w_{t-1} \qquad (18.1.2)$$

which is often called the *transition equation* because it describes the transition of the state of nature from period $t-1$ to period t. The matrix $\mathbf{B}_t$ is a coefficient matrix that may depend on t and w_t is an error process. The system (18.1.1)/(18.1.2) is one form of a *state space model*.

The following example from Meinhold & Singpurwalla (1983) may illustrate the related concepts. Suppose we wish to trace a satellite's orbit. The state vector z_t may then consist of the position and the speed of the satellite in period t with respect to the center of the earth. The state cannot be measured directly but, for example, the distance from a certain observatory may be measured. These measurements constitute the observed vectors y_t. As another example, consider the income of an individual which may depend on unobserved factors such as intelligence, special abilities, special interests and so on. In this case, the state vector consists of the variables that describe the abilities of the person and y_t is his or her observed income.

The reader may recall that all the models considered so far have been written in a form similar to (18.1.1)/(18.1.2) at some stage. For instance, our standard (zero mean) VAR(p) model can be written in VAR(1) form as

$$Y_t = \mathbf{A} Y_{t-1} + U_t, \tag{18.1.3}$$

where

$$Y_t := \begin{bmatrix} y_t \\ \vdots \\ y_{t-p+1} \end{bmatrix}, \quad U_t := \begin{bmatrix} u_t \\ 0 \\ \vdots \\ 0 \end{bmatrix}, \quad \text{etc..}$$

Defining $z_t := Y_t$, $\mathbf{B}_t := \mathbf{A}$, and $w_{t-1} := U_t$, Equation (18.1.3) may be viewed as the transition equation of a state space model. The corresponding measurement equation is

$$y_t = [I_K : 0 : \cdots : 0] Y_t$$

with $\mathbf{H}_t := [I_K : 0 : \cdots : 0]$ and $v_t := 0$.

In the next section, we will introduce a slightly more general version of a state space model, we will review many of the previous models, and we will cast them into state space form. As we have seen, the representations of the models used in the previous chapters are useful for many purposes. There are occasions, however, where a state space representation makes life easier. We have actually used state space representations of some models without explicitly mentioning this fact. We will also consider some further models that have been discussed in the literature and which may be set up as special cases of state space models. Thereby we will give an overview of a number of important models that have been considered in the multiple time series literature.

In Section 18.3, we will discuss the Kalman filter which is an extremely useful tool in the analysis of state space models. Given the observable vectors y_t, it provides estimates of the state vectors and measures of the precision of these estimates. In a situation where the state vector consists of unobservable variables, such estimates may be of interest. In a system such as (18.1.1)/(18.1.2), the matrices $\mathbf{B}_t$ and $\mathbf{H}_t$ and the covariance matrices of v_t and w_t will often depend on unknown parameters. The Kalman filter is also helpful in estimating these parameters. This issue will be discussed in Section 18.4.

In this chapter, we will just give a brief introduction to some basic concepts related to state space models and the Kalman filter. Various textbooks exist that provide broader introductions to the topic and a more in-depth discussion. Examples are Jazwinski (1970), Anderson & Moore (1979), Hannan & Deistler (1988), Aoki (1987), and Harvey (1989).

18.2 State Space Models

18.2.1 The Model Setup

As mentioned in the previous section, a state space model consists of a *transition* or *system equation*

$$z_{t+1} = \mathbf{B}_t z_t + \mathbf{F}_t x_t + w_t, \quad t = 0, 1, 2, \ldots,$$

or, equivalently,

$$z_t = \mathbf{B}_{t-1} z_{t-1} + \mathbf{F}_{t-1} x_{t-1} + w_{t-1}, \quad t = 1, 2, \ldots, \quad (18.2.1)$$

and a *measurement* or *observation equation*

$$y_t = \mathbf{H}_t z_t + \mathbf{G}_t x_t + v_t, \quad t = 1, 2, \ldots. \quad (18.2.2)$$

Here

y_t is a $(K \times 1)$ vector of observable *output* or *endogenous variables*,
z_t is an $(N \times 1)$ *state vector* or the *state of nature*,
x_t is an $(M \times 1)$ vector of observable *inputs* or *instruments* or *policy variables*,
v_t is a $(K \times 1)$ vector of *observation* or *measurement errors* or *noise*,
w_t is an $(N \times 1)$ vector of *system* or *transition equation errors* or *noise*,
$\mathbf{H}_t$ is a $(K \times N)$ *measurement matrix*,
$\mathbf{G}_t$ is a $(K \times M)$ *input matrix* of the observation equation,
$\mathbf{B}_t$ is an $(N \times N)$ *transition* or *system matrix*,

and

$\mathbf{F}_t$ is a $(N \times M)$ *input matrix* of the transition equation.

The matrices $\mathbf{H}_t$, $\mathbf{G}_t$, $\mathbf{B}_t$, and $\mathbf{F}_t$ are assumed to be known at time t. Although they are in general allowed to vary, at least some of them will often be time invariant. In practice, at least some of the elements of these matrices are usually unknown and have to be estimated. This issue is deferred to Section 18.4. It is perhaps noteworthy that the process generating the z_t's and, hence, also the y_t's is assumed to be started from an *initial state* z_0 and a given *initial input* x_0.

To complete the description of the model, we make the following stochastic assumptions for the noise processes and the initial state:

The joint process

$$\begin{bmatrix} w_t \\ v_t \end{bmatrix}$$

is a zero mean, serially uncorrelated noise process with possibly time varying covariance matrices

$$\begin{bmatrix} \Sigma_{w_t} & \Sigma_{w_t v_t} \\ \Sigma_{v_t w_t} & \Sigma_{v_t} \end{bmatrix}.$$

The initial state z_0 is uncorrelated with w_t, v_t for all t and has a distribution with mean μ_{z_0} and covariance matrix Σ_{z_0}. The input sequence $x_0, x_1, \ldots$ is assumed to be nonstochastic for simplicity. If the observed inputs are actually stochastic, the analysis is assumed to be conditional on a given sequence of inputs.

With these assumptions we can derive stochastic properties of the states and the system outputs. Successive substitution in (18.2.1) implies

$$z_t = \mathbf{\Phi}_{t,t} z_0 + \sum_{i=1}^{t} \mathbf{\Phi}_{i-1,t}(\mathbf{F}_{t-i} x_{t-i} + w_{t-i}), \tag{18.2.3}$$

where

$$\mathbf{\Phi}_{0,t} := I_N \quad \text{and} \quad \mathbf{\Phi}_{i,t} := \prod_{j=1}^{i} \mathbf{B}_{t-j}, \quad i = 1, 2, \ldots$$

(see also Section 17.2.1). Hence,

$$\mu_{z_t} := E(z_t) = \mathbf{\Phi}_{t,t} \mu_{z_0} + \sum_{i=1}^{t} \mathbf{\Phi}_{i-1,t} \mathbf{F}_{t-i} x_{t-i} \tag{18.2.4}$$

and

$$\begin{aligned} \text{Cov}(z_t, z_{t+h}) &= E[(z_t - \mu_{z_t})(z_{t+h} - \mu_{z_{t+h}})'] \\ &= \mathbf{\Phi}_{t,t} \Sigma_{z_0} \mathbf{\Phi}'_{t+h,t+h} + \sum_{i=1}^{t} \mathbf{\Phi}_{i-1,t} \Sigma_{w_{t-i}} \mathbf{\Phi}'_{h+i-1,t+h}. \end{aligned} \tag{18.2.5}$$

Under the aforementioned stochastic assumptions, it is also easy to derive the means and covariance matrices of the output process:

$$\mu_{y_t} := E(y_t) = \mathbf{H}_t E(z_t) + \mathbf{G}_t x_t$$

and

$$\text{Cov}(y_t, y_{t+h}) = \mathbf{H}_t \text{Cov}(z_t, z_{t+h}) \mathbf{H}'_t \quad \text{for } h \neq 0.$$

Generally, the means and autocovariances of the y_t's are obviously not time invariant. Thus, in general, y_t is a nonstationary process.

We will now consider various special cases of state space models which are obtained by specific definitions of the state vector, the inputs, the noise processes, and the matrices $\mathbf{H}_t, \mathbf{G}_t, \mathbf{B}_t,$ and $\mathbf{F}_t$. These matrices and the noise covariance matrices will often not depend on t, in which case we will suppress the subscript for notational simplicity.

A Finite Order VAR Process

Although we have mentioned earlier how to cast a VAR(p) process,

$$y_t = \nu + A_1 y_{t-1} + \cdots + A_p y_{t-p} + u_t, \tag{18.2.6}$$

in state space form, it may be useful to consider this model again because it illustrates that often different state space models can represent a particular process. One possible state space representation is obtained by defining

$$Y_t := \begin{bmatrix} y_t \\ \vdots \\ y_{t-p+1} \end{bmatrix}, \quad \boldsymbol{\nu} := \begin{bmatrix} \nu \\ 0 \\ \vdots \\ 0 \end{bmatrix},$$

$$\mathbf{A} := \begin{bmatrix} A_1 & \cdots & A_{p-1} & A_p \\ I_K & & 0 & 0 \\ & \ddots & \vdots & \vdots \\ 0 & \cdots & I_K & 0 \end{bmatrix}, \quad U_t := \begin{bmatrix} u_t \\ 0 \\ \vdots \\ 0 \end{bmatrix}. \tag{18.2.7}$$

Hence,

$$Y_t = \mathbf{A} Y_{t-1} + \boldsymbol{\nu} + U_t, \tag{18.2.8}$$

$$y_t = [I_K : 0 : \cdots : 0] Y_t \tag{18.2.9}$$

is a state space model with state vector $z_t := Y_t$, $\mathbf{B} := \mathbf{A}$, $\mathbf{F} := \boldsymbol{\nu}$, $x_t := 1$, $w_t := U_{t+1}$, $\mathbf{H} := [I_K : 0 : \cdots : 0]$, $\mathbf{G} := 0$, $v_t := 0$.

An alternative possibility is to define the state vector as

$$z_t := \begin{bmatrix} 1 \\ y_t \\ \vdots \\ y_{t-p+1} \end{bmatrix}$$

and choose

$$\mathbf{B} := \begin{bmatrix} 1 & 0 & \cdots & 0 & 0 \\ \nu & A_1 & \cdots & A_{p-1} & A_p \\ 0 & I_K & & 0 & 0 \\ \vdots & & \ddots & \vdots & \vdots \\ 0 & 0 & \cdots & I_K & 0 \end{bmatrix} \quad \text{and} \quad w_t := \begin{bmatrix} 0 \\ u_{t+1} \\ 0 \\ \vdots \\ 0 \end{bmatrix},$$

so that

$$z_{t+1} = \mathbf{B} z_t + w_t$$

and

$$y_t = [0 : I_K : 0 : \cdots : 0]z_t,$$

which describes the same process as (18.2.8)/(18.2.9). It may be worth pointing out that in the present framework, the process is assumed to be started at time $t = 1$ with initial values $z_0 = [1, y_0', \ldots, y_{-p+1}']'$, while we have assumed an infinite past of the process in some previous chapters.

A VARMA(p, q) Process

One state space representation of the VARMA(p, q) process

$$y_t = \nu + A_1 y_{t-1} + \cdots + A_p y_{t-p} + u_t + M_1 u_{t-1} + \cdots + M_q u_{t-q} \quad (18.2.10)$$

is known from Chapter 11, Section 11.3.2. It is obtained by choosing a state vector

$$z_t := \begin{bmatrix} y_t \\ \vdots \\ y_{t-p+1} \\ u_t \\ \vdots \\ u_{t-q+1} \end{bmatrix}, \quad \text{transition noise } w_t := \begin{bmatrix} u_{t+1} \\ 0 \\ \vdots \\ 0 \\ u_{t+1} \\ 0 \\ \vdots \\ 0 \end{bmatrix},$$

an input sequence $x_t := 1$ as in (18.2.8), $\mathbf{B} := \mathbf{A}$ from Chapter 11, Equation (11.3.8), $\mathbf{F} := \boldsymbol{\nu}$ defined similarly as in (18.2.8), $\mathbf{H} := [I_K : 0 : \cdots : 0]$, $\mathbf{G} := 0$, and $v_t := 0$. For many purposes, this form is not the most useful state space representation of a VARMA model. Other state space representations are given by Aoki (1987), Hannan & Deistler (1988), and Wei (1990).

The VARX Model

The VARX model

$$y_t = A_1 y_{t-1} + \cdots + A_p y_{t-p} + B_0 x_t + \cdots + B_s x_{t-s} + u_t \quad (18.2.11)$$

considered in Chapter 10 is easily cast in state space form by choosing the state vector

$$z_t := \begin{bmatrix} y_t \\ \vdots \\ y_{t-p+1} \\ x_t \\ \vdots \\ x_{t-s+1} \end{bmatrix},$$

and the transition equation

$$
z_t = \left[\begin{array}{cccc|cccc} A_1 & \cdots & A_{p-1} & A_p & B_1 & \cdots & B_{s-1} & B_s \\ I & & 0 & 0 & 0 & \cdots & 0 & 0 \\ & \ddots & \vdots & \vdots & \vdots & & \vdots & \vdots \\ 0 & \cdots & I & 0 & 0 & \cdots & 0 & 0 \\ \hline & & & & 0 & \cdots & 0 & 0 \\ & & 0 & & I & & 0 & 0 \\ & & & & & \ddots & \vdots & \vdots \\ & & & & 0 & \cdots & I & 0 \end{array}\right] z_{t-1}
$$

$$
+ \left[\begin{array}{c} B_0 \\ 0 \\ \vdots \\ 0 \\ I \\ 0 \\ \vdots \\ 0 \end{array}\right] x_t + \left[\begin{array}{c} u_t \\ 0 \\ \vdots \\ 0 \end{array}\right]. \tag{18.2.12}
$$

The corresponding observation equation is

$$y_t = [I_K : 0 : \cdots : 0] z_t.$$

It is also possible to extend the model so as to allow for a finite order MA(q) error process in (18.2.11) (see Problem 18.1).

Systematic Sampling and Aggregation

Suppose that annual data is available whereas a decision maker is interested in, say, quarterly figures. Let η_{it} be an ($M \times 1$) vector of variables associated with the i-th quarter of year t and suppose the vector of all quarterly variables associated with year t,

$$\eta_t := \left[\begin{array}{c} \eta_{1t} \\ \eta_{2t} \\ \eta_{3t} \\ \eta_{4t} \end{array}\right],$$

is generated by the VAR(p) process

$$\eta_t = A_1 \eta_{t-1} + \cdots + A_p \eta_{t-p} + u_t.$$

Then we may define a state vector

$$z_t := \begin{bmatrix} \eta_t \\ \vdots \\ \eta_{t-p+1} \end{bmatrix}$$

and a transition equation

$$z_t = \mathbf{A} z_{t-1} + U_t,$$

where $\mathbf{A}$ and U_t are the same quantities as in (18.2.7). If the yearly values are obtained by adding (aggregating) the quarterly figures, the observation equation is

$$y_t = [I_M : I_M : I_M : I_M : 0 : \cdots : 0] z_t,$$

where M is the dimension of η_{it}. Alternatively, if the annual figures are obtained by systematic sampling, that is, by taking, say, the fourth quarter values as the annual figures, the observation equation is

$$y_t = [0 : 0 : 0 : I_M : 0 : \cdots : 0] z_t.$$

Extensions of this framework to the case where η_t is generated by a VARMA or VARX process are straightforward. For applications of state space models in aggregation and systematic sampling problems see Nijman (1985), Harvey (1984), Harvey & Pierse (1984), Jones (1980), Ansley & Kohn (1983).

The examples considered so far have in common that the system matrices $\mathbf{H}$, $\mathbf{G}$, $\mathbf{B}$, and $\mathbf{F}$ are all time invariant and the state vector consists of at least some observed or observable variables. In contrast, the state vector is unobservable in the next two examples while the system matrices remain time invariant.

Structural Time Series Models

In a structural time series model, the observed time series is viewed as a sum of unobserved components such as a trend, a seasonal component, and an irregular component (see, e.g., Kitagawa (1981), Harvey & Todd (1983), Harvey (1989)). For instance, for a univariate time series $y_1, \ldots, y_T$, the structural model may have the form

$$y_t = \mu_t + \gamma_t + u_t, \tag{18.2.13}$$

where μ_t is a trend component and γ_t is a seasonal component. Harvey & Todd (1983) assume a local approximation to a linear trend function for which both the level and the slope are shifting. They postulate a process

$$\mu_t = \mu_{t-1} + \beta_{t-1} + \eta_t \quad \text{with } \beta_t = \beta_{t-1} + \xi_t \tag{18.2.14}$$

as the trend generation mechanism. Here η_t and ξ_t are assumed to be white noise processes. This trend model is a mixture of two random walks which are discussed in Chapter 6, Section 6.1. For the seasonal component, it is assumed that the sum over the seasonal factors of a full year is approximately zero,

$$\gamma_t = -\sum_{j=1}^{s-1} \gamma_{t-j} + \omega_t, \qquad (18.2.15)$$

where s is the number of seasons and ω_t is white noise. The three white noise processes η_t, ξ_t, and ω_t are assumed to be independent.

This model can be set up in state space form by defining the state vector to be

$$z_t := \begin{bmatrix} \mu_t \\ \beta_t \\ \gamma_t \\ \vdots \\ \gamma_{t-s+2} \end{bmatrix}$$

and, hence, the transition equation is

$$z_t = \begin{bmatrix} 1 & 1 & | & & & 0 & \\ 0 & 1 & | & & & & \\ \hline & & | & -1 & \cdots & -1 & -1 \\ & & | & 1 & & 0 & 0 \\ & 0 & | & & \ddots & \vdots & \vdots \\ & & | & 0 & \cdots & 1 & 0 \end{bmatrix} z_{t-1} + \begin{bmatrix} \eta_t \\ \xi_t \\ \omega_t \\ 0 \\ \vdots \\ 0 \end{bmatrix}. \qquad (18.2.16)$$

The corresponding measurement equation is

$$y_t = [1, 0, 1, 0, \ldots, 0] z_t + u_t. \qquad (18.2.17)$$

It may be worth noting that these models can be seen to describe special integrated ARMA processes. Hence, it is, of course, not surprising that they can be cast in state space form. Multivariate generalizations of this model are possible (see Harvey (1987) and Proietti (2002)).

Factor Analytic Models

In a classical factor analytic setting, it is assumed that a set of K observed variables y_t depends linearly on $N < K$ unobserved common factors f_t and on individual or idiosyncratic components u_t. In other words,

$$y_t = \mathbf{L} f_t + u_t, \qquad (18.2.18)$$

where $\mathbf{L}$ is a $(K \times N)$ matrix of *factor loadings* and the components of u_t are typically assumed to be uncorrelated, that is, Σ_u is a diagonal matrix (Anderson (1984), Morrison (1976)). One objective of a factor analysis is the construction or estimation of the unobserved factors f_t. We may view (18.2.18) as the measurement equation of a state space model and, if the factors f_t and f_s are independent for $t \neq s$, we may specify a trivial transition equation $f_t = w_{t-1}$.

However, if y_t consists of time series variables, it may be more reasonable to assume that the factors are autocorrelated. For example, they may be generated by a VAR or VARMA process. Also the idiosyncratic components u_t may be autocorrelated. *Dynamic factor analytic models* of this type were considered, for instance, by Sargent & Sims (1977), Geweke (1977), and Engle & Watson (1981). Assuming that

$$f_t = A_1 f_{t-1} + \cdots + A_p f_{t-p} + \eta_t$$

and

$$u_t = C_1 u_{t-1} + \cdots + C_q u_{t-q} + \varepsilon_t,$$

where η_t and ε_t are white noise processes, a state space model can be set up by specifying a state vector,

$$z_t := \begin{bmatrix} f_t \\ \vdots \\ f_{t-p+1} \\ u_t \\ \vdots \\ u_{t-q+1} \end{bmatrix},$$

and a transition equation

$$z_t = \begin{bmatrix} A_1 & \cdots & A_{p-1} & A_p & | & & & & \\ I & & 0 & 0 & | & & & 0 & \\ & \ddots & \vdots & \vdots & | & & & & \\ 0 & \cdots & I & 0 & | & & & & \\ \hline & & & & | & C_1 & \cdots & C_{q-1} & C_q \\ & & & & | & I & & 0 & 0 \\ & & 0 & & | & & \ddots & \vdots & \vdots \\ & & & & | & 0 & \cdots & I & 0 \end{bmatrix} z_{t-1} + \begin{bmatrix} \eta_t \\ 0 \\ \vdots \\ 0 \\ \varepsilon_t \\ 0 \\ \vdots \\ 0 \end{bmatrix}.$$

(18.2.19)

The corresponding measurement equation is

$$y_t = \mathbf{L} f_t + u_t = [\mathbf{L} : 0 : \cdots : 0 : I_K : 0 : \cdots : 0] z_t. \qquad (18.2.20)$$

An extension to the case where f_t and u_t are generated by VARMA processes is left to the reader (see Problem 18.2). If exogenous variables are added to the original model (18.2.18) and, in addition, the factors are dynamic processes, we obtain the dynamic MIMIC models of Engle & Watson (1981). More recent references on dynamic factor models include Stock & Watson (2002a, b) and Forni, Hallin, Lippi & Reichlin (2000).

In all the previous examples the system matrices $\mathbf{H}_t$, $\mathbf{G}_t$, $\mathbf{B}_t$, and $\mathbf{F}_t$ are time invariant. We will now consider models where at least some elements of these matrices vary through time.

VARX Models with Systematically Varying Coefficients

We extend the varying coefficients VAR models of Chapter 17 slightly by adding further "exogenous" variables and assuming that a given multiple time series is generated according to

$$y_t = A_{1,t} y_{t-1} + \cdots + A_{p,t} y_{t-p} + F_t x_t + u_t. \tag{18.2.21}$$

The vector x_t may simply include an intercept term or seasonal dummies. It may also include other deterministic terms and even lags of exogenous variables. Because we are assuming that the input variables of the state space model are nonstochastic, we restrict x_t to be a deterministic sequence, however. Using

$$Y_t := \begin{bmatrix} y_t \\ \vdots \\ y_{t-p+1} \end{bmatrix}, \quad \mathbf{A}_t := \begin{bmatrix} A_{1,t} & \cdots & A_{p-1,t} & A_{p,t} \\ I & & 0 & 0 \\ & \ddots & \vdots & \vdots \\ 0 & \cdots & I & 0 \end{bmatrix},$$

$$\mathbf{F}_t := \begin{bmatrix} F_t \\ 0 \\ \vdots \\ 0 \end{bmatrix}, \quad \text{and} \quad w_{t-1} := \begin{bmatrix} u_t \\ 0 \\ \vdots \\ 0 \end{bmatrix},$$

gives a transition equation

$$Y_t = \mathbf{A}_t Y_{t-1} + \mathbf{F}_t x_t + w_{t-1} \tag{18.2.22}$$

and a measurement equation

$$y_t = [I_K : 0 : \cdots : 0] Y_t. \tag{18.2.23}$$

Obviously, the transition matrix $\mathbf{B}_{t-1} := \mathbf{A}_t$ and the input matrix $\mathbf{F}_t$ of the transition equation may be time varying in this state space model.

Random Coefficients VARX Models

So far all the original models either have time invariant, constant coefficients or, as in the previous example, systematically varying coefficients. We will now consider models with random coefficients and demonstrate how they can be cast in state space form. Let us begin with a simple multivariate regression model of the form

$$y_t = C_t x_t + v_t = (x_t' \otimes I)\text{vec}(C_t) + v_t. \tag{18.2.24}$$

Assuming that the parameter vector $\boldsymbol{\gamma}_t := \text{vec}(C_t)$ is generated by a VAR(q) process,

$$\boldsymbol{\gamma}_t = \nu + B_1 \boldsymbol{\gamma}_{t-1} + \cdots + B_q \boldsymbol{\gamma}_{t-q} + u_t, \tag{18.2.25}$$

we may define the state vector as

$$z_t := \begin{bmatrix} \boldsymbol{\gamma}_t \\ \vdots \\ \boldsymbol{\gamma}_{t-q+1} \end{bmatrix}$$

and get a state space model with the following transition and measurement equations, respectively:

$$z_t = \begin{bmatrix} B_1 & \cdots & B_{q-1} & B_q \\ I & & 0 & 0 \\ & \ddots & \vdots & \vdots \\ 0 & \cdots & I & 0 \end{bmatrix} z_{t-1} + \begin{bmatrix} \nu \\ 0 \\ \vdots \\ 0 \end{bmatrix} + \begin{bmatrix} u_t \\ 0 \\ \vdots \\ 0 \end{bmatrix},$$

$$y_t = [x_t' \otimes I : 0 : \cdots : 0] z_t + v_t.$$

Obviously, the measurement matrix,

$$\mathbf{H}_t := [x_t' \otimes I : 0 : \cdots : 0],$$

may be time varying. It may, in fact, be random if the x_t are stochastic variables. Such an assumption is mandatory if x_t contains lagged y_t variables. To see this point more clearly, let us explicitly introduce lagged y_t's in (18.2.24):

$$\begin{aligned} y_t &= \mathbf{A}_t Y_{t-1} + C_t x_t + v_t \\ &= (Y_{t-1}' \otimes I)\text{vec}(\mathbf{A}_t) + (x_t' \otimes I)\text{vec}(C_t) + v_t, \end{aligned} \tag{18.2.26}$$

where $\mathbf{A}_t := [A_{1t}, \ldots, A_{pt}]$ and $Y_{t-1}' := [y_{t-1}', \ldots, y_{t-p}']$. Now suppose that $\boldsymbol{\gamma}_t := \text{vec}(C_t)$ is generated by the VAR(q) process (18.2.25) and $\boldsymbol{\alpha}_t = \text{vec}(\mathbf{A}_t)$ is driven by a VAR(r) process

$$\boldsymbol{\alpha}_t = D_1 \boldsymbol{\alpha}_{t-1} + \cdots + D_r \boldsymbol{\alpha}_{t-r} + \eta_t, \tag{18.2.27}$$

which is assumed to be independent of γ_t. Defining the state vector as

$$z_t := \begin{bmatrix} \gamma_t \\ \vdots \\ \gamma_{t-q+1} \\ \alpha_t \\ \vdots \\ \alpha_{t-r+1} \end{bmatrix},$$

the following state space model is obtained:

$$z_t = \begin{bmatrix} \begin{array}{cccc|cccc} B_1 & \cdots & B_{q-1} & B_q & & & & \\ I & & 0 & 0 & & & 0 & \\ & \ddots & \vdots & \vdots & & & & \\ 0 & \cdots & I & 0 & & & & \\ \hline & & & & D_1 & \cdots & D_{r-1} & D_r \\ & & 0 & & I & & 0 & 0 \\ & & & & & \ddots & \vdots & \vdots \\ & & & & 0 & \cdots & I & 0 \end{array} \end{bmatrix} z_{t-1}$$

$$+ \begin{bmatrix} \nu \\ 0 \\ \vdots \\ 0 \\ \vdots \\ 0 \end{bmatrix} + \begin{bmatrix} u_t \\ 0 \\ \vdots \\ 0 \\ \eta_t \\ 0 \\ \vdots \\ 0 \end{bmatrix}, \qquad (18.2.28)$$

$$y_t = [x'_t \otimes I : 0 : \cdots : 0 : Y'_{t-1} \otimes I : 0 : \cdots : 0] z_t + v_t. \qquad (18.2.29)$$

Further extensions of this model are possible. For instance, α_t and γ_t may be individually or jointly generated by a VARMA rather than a finite order VAR process. Moreover, input variables with constant coefficients could appear in (18.2.26). These extensions are left to the reader (see Problem 18.3).

The number of publications on random coefficients models is vast in both the econometrics and the time series literature. Famous examples from the earlier econometrics literature on the topic are Hildreth & Houck (1968), Swamy (1971), and Cooley & Prescott (1973, 1976). Surveys of the earlier literature were given by Chow (1984) and Nicholls & Pagan (1985). Both of these articles include extensive reference lists. For a more recent overview see also Swamy & Tavlas (2001). On the time series side, a number of references can be found in the monograph by Nicholls & Quinn (1982). Other important work on the topic includes the article by Doan et al. (1984) who investigated the potential of random coefficients VAR models with Bayesian restrictions for econometric time series analysis.

18.2.2 More General State Space Models

There are also time series models that do not fall into the state space framework considered so far. Therefore it may be worth pointing out that more general nonlinear state space models have been studied in recent publications. A very general setup has the form

$$z_{t+1} = \mathbf{b}_t(z_t, x_t, w_t, \boldsymbol{\delta}_1) \qquad (18.2.30)$$

for the transition equation and

$$y_t = \mathbf{h}_t(z_t, x_t, v_t, \boldsymbol{\delta}_2) \qquad (18.2.31)$$

for the measurement equation. In other words, the functional dependence between the inputs, the states, and the output variables may be of a general nonlinear form and also the transition from one state to the next is described by a more general function than previously. Here $\boldsymbol{\delta}_1$ and $\boldsymbol{\delta}_2$ are vectors of parameters.

Bilinear time series models are examples for which the linear state space framework is too narrow. A very simple univariate bilinear time series model has the form

$$y_t = \alpha y_{t-1} + u_t + \beta y_{t-1} u_{t-1},$$

where u_t is univariate white noise. The product term $\beta y_{t-1} u_{t-1}$ distinguishes this model from a linear specification. Bilinear models have been found useful in modelling nonnormal phenomena (see, e.g., Granger & Andersen (1978)).

A more general multivariate bilinear time series model may be specified as follows:

$$\begin{aligned}y_t &= A_1 y_{t-1} + \cdots + A_p y_{t-p} + u_t + M_1 u_{t-1} + \cdots + M_q u_{t-q} \\ &\quad + \sum_{i=1}^{r} \sum_{j=1}^{s} C_{ij} \operatorname{vec}(y_{t-i} u'_{t-j}).\end{aligned} \qquad (18.2.32)$$

Assuming, without loss of generality, that $p \geq r$ and $q \geq s$ and defining

$$z_t := \begin{bmatrix} y_t \\ \vdots \\ y_{t-p+1} \\ u_t \\ \vdots \\ u_{t-q+1} \end{bmatrix},$$

$$\mathbf{B} := \begin{bmatrix} A_1 & \cdots & A_{p-1} & A_p & | & M_1 & \cdots & M_{q-1} & M_q \\ I & & 0 & 0 & | & 0 & \cdots & 0 & 0 \\ & \ddots & \vdots & \vdots & | & \vdots & & \vdots & \vdots \\ 0 & \cdots & I & 0 & | & 0 & \cdots & 0 & 0 \\ \hline & & & & | & 0 & \cdots & 0 & 0 \\ & & & & | & I & & 0 & 0 \\ & & 0 & & | & & \ddots & \vdots & \vdots \\ & & & & | & 0 & \cdots & I & 0 \end{bmatrix},$$

$$w_t := \begin{bmatrix} u_{t+1} \\ 0 \\ \vdots \\ 0 \\ u_{t+1} \\ 0 \\ \vdots \\ 0 \end{bmatrix},$$

and a matrix $\mathbf{C}$ which contains the elements of the C_{ij} matrices in a suitable arrangement, we get a bilinear state space model of the form

$$z_{t+1} = \mathbf{B} z_t + w_t + \mathbf{C} \text{vec}(z_t z_t'), \tag{18.2.33}$$

$$y_t = [I_K : 0 : \cdots : 0] z_t. \tag{18.2.34}$$

Obviously, the transition equation involves a nonlinear term, namely $\text{vec}(z_t z_t')$. Hence, (18.2.33)/(18.2.34) is an example of a nonlinear state space system.

The work of Granger & Andersen (1978) and others on univariate bilinear models has stimulated investigations in this area. Much of the earlier work is documented in a monograph by Subba Rao & Gabr (1984). More recent work on multivariate bilinear models includes Stensholt & Tjøstheim (1987) and Liu (1989).

With all these examples we have not nearly exhausted the range of models that have been used and studied in the recent time series literature. Important omissions are threshold autoregressive models analyzed by Tong (1983) and exponential autoregressive models introduced by Ozaki (1980) and Haggan & Ozaki (1980). A general nonlinear model class was considered by Priestley (1980) and reviews of many nonlinear models and extensive lists of references were given by Priestley (1988), Anděl (1989), and Granger & Teräsvirta (1993).

18.3 The Kalman Filter

The Kalman filter was originally developed by Kalman (1960) and Kalman & Bucy (1961). It is a tool to recursively estimate the states z_t, given observations $y_1, \ldots, y_T$ of the output variables. Under normality assumptions, the estimator of the state produced by the Kalman filter is the conditional expectation $E(z_t|y_1, \ldots, y_t)$. The Kalman filter also provides the conditional covariance matrix $\text{Cov}(z_t|y_1, \ldots, y_t)$ which may serve as a measure for estimation or prediction uncertainty. Of course, for $t > T$, the estimator $E(z_t|y_1, \ldots, y_T)$ is a forecast or prediction at origin T, in the terminology of the previous chapters. The computation of the estimators $E(z_t|y_1, \ldots, y_t)$, $t = 1, \ldots, T$, is called *filtering* to distinguish it from the forecasting problem.

In some of the examples of Section 18.2, estimation of the state vectors is of obvious interest, for instance, if the state vector consists of time varying coefficients or if the state vector contains the unobserved factors of a dynamic factor analytic model. In other cases, where the state vector is not of foremost interest or where it consists of observable variables, the conditional means and covariance matrices can still be useful in evaluating the likelihood function, for example. We will return to this point in Section 18.4. Now the Kalman filter recursions will be presented.

18.3.1 The Kalman Filter Recursions

Assumptions for the State Space Model

We assume a state space model with transition equation

$$z_t = \mathbf{B} z_{t-1} + \mathbf{F} x_{t-1} + w_{t-1} \tag{18.3.1}$$

and with measurement equation

$$y_t = \mathbf{H}_t z_t + \mathbf{G} x_t + v_t \tag{18.3.2}$$

for $t = 1, 2, \ldots$. Note that both input matrices and the transition matrix are assumed to be time invariant and known. This condition is satisfied in most of the example models of Section 18.2. The measurement matrices $\mathbf{H}_t$ are assumed to be known and nonstochastic at time t. This assumption does not exclude lagged output variables from $\mathbf{H}_t$ because the past output variables are given at time t. The input sequence x_t, $t = 0, 1, \ldots$, is again assumed to be nonstochastic for simplicity. The noise processes w_t and v_t are independent. They are both Gaussian with time invariant covariances,

$$w_t \sim \mathcal{N}(0, \Sigma_w), \quad t = 0, 1, \ldots,$$
$$v_t \sim \mathcal{N}(0, \Sigma_v), \quad t = 1, 2, \ldots.$$

Also the initial state is Gaussian, $z_0 \sim \mathcal{N}(\mu_0, \Sigma_0)$, and it is assumed to be independent of v_t, w_{t-1}, $t = 1, \ldots$. The initial state may be a constant, nonstochastic vector in which case $\Sigma_0 = 0$.

With the exception of the normality assumption, the foregoing conditions are satisfied for most of the example models of Section 18.2 under the usual assumptions entertained for these models. It is possible to derive recursions similar to those given below under more general conditions. If the normality assumption is dropped, the recursions given below can still be justified. We will return to this issue after having presented them.

The Recursions

We will use the following additional notation in stating the Kalman filter recursions:

$$\begin{aligned} z_{t|s} &:= E(z_t|y_1,\ldots,y_s), \\ \Sigma_z(t|s) &:= \mathrm{Cov}(z_t|y_1,\ldots,y_s), \\ y_{t|s} &:= E(y_t|y_1,\ldots,y_s), \\ \Sigma_y(t|s) &:= \mathrm{Cov}(y_t|y_1,\ldots,y_s), \end{aligned} \tag{18.3.3}$$

$(z|y) \sim \mathcal{N}(\mu, \Sigma)$ means that the conditional distribution of z given y is multivariate normal with mean μ and covariance matrix Σ.

Under the previously stated conditions, the normality assumption implies

$$(z_t|y_1,\ldots,y_{t-1}) \sim \mathcal{N}(z_{t|t-1}, \Sigma_z(t|t-1)) \quad \text{for } t = 2,\ldots,T, \tag{18.3.4}$$

$$(z_t|y_1,\ldots,y_t) \sim \mathcal{N}(z_{t|t}, \Sigma_z(t|t)) \quad \text{for } t = 1,\ldots,T, \tag{18.3.5}$$

$$(y_t|y_1,\ldots,y_{t-1}) \sim \mathcal{N}(y_{t|t-1}, \Sigma_y(t|t-1)) \quad \text{for } t = 2,\ldots,T, \tag{18.3.6}$$

and

$$(z_t|y_1,\ldots,y_T) \sim \mathcal{N}(z_{t|T}, \Sigma_z(t|T)), \tag{18.3.7}$$

$$(y_t|y_1,\ldots,y_T) \sim \mathcal{N}(y_{t|T}, \Sigma_y(t|T)) \quad \text{for } t > T. \tag{18.3.8}$$

The conditional means and covariance matrices can be obtained by the following *Kalman filter recursions* which are graphically depicted in Figure 18.1:

Initialization: $z_{0|0} := \mu_0$, $\Sigma_z(0|0) := \Sigma_0$.

Prediction step $(1 \leq t \leq T)$:

$$\begin{aligned} z_{t|t-1} &= \mathbf{B} z_{t-1|t-1} + \mathbf{F} x_{t-1}, \\ \Sigma_z(t|t-1) &= \mathbf{B} \Sigma_z(t-1|t-1) \mathbf{B}' + \Sigma_w, \\ y_{t|t-1} &= \mathbf{H}_t z_{t|t-1} + \mathbf{G} x_t, \\ \Sigma_y(t|t-1) &= \mathbf{H}_t \Sigma_z(t|t-1) \mathbf{H}_t' + \Sigma_v. \end{aligned}$$

Correction step $(1 \leq t \leq T)$:

$$\begin{aligned} z_{t|t} &= z_{t|t-1} + \mathbf{P}_t (y_t - y_{t|t-1}), \\ \Sigma_z(t|t) &= \Sigma_z(t|t-1) - \mathbf{P}_t \Sigma_y(t|t-1) \mathbf{P}_t', \end{aligned}$$

where

$$\mathbf{P}_t := \Sigma_z(t|t-1)\mathbf{H}'_t \Sigma_y(t|t-1)^{-1} \quad (Kalman\ filter\ gain).$$

Although the output variables we have in mind have nonsingular distributions, it may be worth noting that if the inverse of $\Sigma_y(t|t-1)$ does not exist, it may be replaced by a suitable generalized inverse. The recursions proceed by performing the prediction step for $t = 1$. Then the correction step is carried out for $t = 1$. Then the prediction and correction steps are repeated for $t = 2$, and so on.

Forecasting step $(t > T)$:

$$z_{t|T} = \mathbf{B} z_{t-1|T} + \mathbf{F} x_{t-1},$$

$$\Sigma_z(t|T) = \mathbf{B} \Sigma_z(t-1|T)\mathbf{B}' + \Sigma_w,$$

$$y_{t|T} = \mathbf{H}_t z_{t|T} + \mathbf{G} x_t,$$

$$\Sigma_y(t|T) = \mathbf{H}_t \Sigma_z(t|T)\mathbf{H}'_t + \Sigma_v.$$

The forecasting step may be carried out recursively for $t = T+1, T+2, \ldots$.

Computational Aspects and Extensions

In practice, in running through the Kalman filter recursions, computational inaccuracies may accumulate in such a way that the actually computed covariance matrices are not positive semidefinite. These and other computational issues were discussed in Anderson & Moore (1979, Chapter 6) and numerical modifications of the recursions were suggested that may help to overcome the possible difficulties (see also Schneider (1992)).

As mentioned previously, it is possible to justify the Kalman filter recursions even if the initial state and the white noise processes are not Gaussian. In that case, the quantities obtained by the recursions are no longer moments of conditional normal distributions, however. For other interpretations of the quantities see, for example, Schneider (1988).

Sometimes reconstruction of the state vectors, given all the information $y_1, \ldots, y_T$, is of interest. For instance, in the random coefficients models of Section 18.2.1, where the state vector z_t contains the coefficients associated with period t, one may want to estimate the states and, hence, the coefficients, given all the sample information $y_1, \ldots, y_T$. We will see a detailed example in Section 18.5. Recursions are also available to compute $z_{t|T}$ and $\Sigma_z(t|T)$ for $t < T$. The evaluation of $z_{t|T}$ for $t < T$ is known as *smoothing*. Under the previous assumptions (including normality),

$$(z_t|y_1, \ldots, y_T) \sim \mathcal{N}(z_{t|T}, \Sigma_z(t|T))$$

for $t = 0, 1, \ldots, T$. The conditional moments may be obtained recursively, starting at the end of the sample and moving backwards, that is, the recursions proceed for $t = T-1, T-2, \ldots, 0$ as follows.

18.3 The Kalman Filter   629

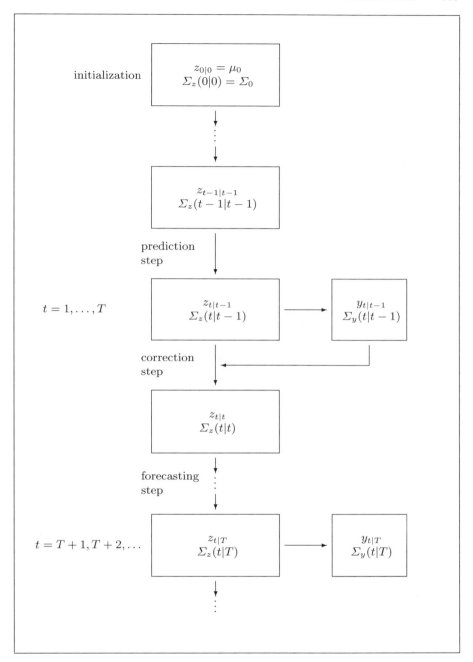

Fig. 18.1. Kalman filter recursions.

Smoothing step ($t < T$):

$$z_{t|T} = z_{t|t} + \mathbf{S}_t(z_{t+1|T} - z_{t+1|t}),$$

$$\Sigma_z(t|T) = \Sigma_z(t|t) - \mathbf{S}_t[\Sigma_z(t+1|t) - \Sigma_z(t+1|T)]\mathbf{S}_t',$$

where

$$\mathbf{S}_t := \Sigma_z(t|t)\mathbf{B}'\Sigma_z(t+1|t)^{-1} \quad (\textit{Kalman smoothing matrix}),$$

(see Anderson & Moore (1979)).

18.3.2 Proof of the Kalman Filter Recursions

The proof follows Anderson & Moore (1979, pp. 39-41) and Meinhold & Singpurwalla (1983). It may be skipped without loss of continuity. We proceed inductively and we use the following properties of multivariate normal distributions (see Propositions B.1 and B.2 of Appendix B):

$y \sim \mathcal{N}(\mu_y, \Sigma_y)$, $z \sim \mathcal{N}(\mu_z, \Sigma_z)$ are independent ($K \times 1$) random vectors

$$\Rightarrow y + z \sim \mathcal{N}(\mu_y + \mu_z, \Sigma_y + \Sigma_z). \tag{18.3.9}$$

If A is a fixed, nonrandom matrix and c a fixed vector,

$$y \sim \mathcal{N}(\mu_y, \Sigma_y) \Rightarrow Ay + c \sim \mathcal{N}(A\mu_y + c, A\Sigma_y A'). \tag{18.3.10}$$

Moreover,

$$\begin{bmatrix} z \\ y \end{bmatrix} \sim \mathcal{N}\left(\begin{bmatrix} \mu_z \\ \mu_y \end{bmatrix}, \begin{bmatrix} \Sigma_z & \Sigma_{zy} \\ \Sigma_{yz} & \Sigma_y \end{bmatrix}\right)$$

$$\Rightarrow (z|y) \sim \mathcal{N}(\mu_z + \Sigma_{zy}\Sigma_y^{-1}(y - \mu_y), \Sigma_z - \Sigma_{zy}\Sigma_y^{-1}\Sigma_{yz}). \tag{18.3.11}$$

Here Σ_y^{-1} may be replaced by a generalized inverse, if Σ_y is singular.

We will now demonstrate the prediction and correction steps for $t = 1$. With that goal in mind, we note that by (18.3.9) and (18.3.10) and the joint normality of w_0 and v_1, the two vectors z_1 and y_1 are jointly normally distributed,

$$\begin{bmatrix} z_1 \\ y_1 \end{bmatrix} = \begin{bmatrix} I \\ \mathbf{H}_1 \end{bmatrix} z_1 + \begin{bmatrix} 0 \\ \mathbf{G}x_1 \end{bmatrix} + \begin{bmatrix} 0 \\ I \end{bmatrix} v_1$$

$$= \begin{bmatrix} I \\ \mathbf{H}_1 \end{bmatrix} \mathbf{B}z_0 + \begin{bmatrix} \mathbf{F}x_0 \\ \mathbf{G}x_1 + \mathbf{H}_1\mathbf{F}x_0 \end{bmatrix} + \begin{bmatrix} I & 0 \\ \mathbf{H}_1 & I \end{bmatrix} \begin{bmatrix} w_0 \\ v_1 \end{bmatrix}$$

$$\sim \mathcal{N}\left(\begin{bmatrix} \mathbf{B}\mu_0 + \mathbf{F}x_0 \\ \mathbf{H}_1(\mathbf{B}\mu_0 + \mathbf{F}x_0) + \mathbf{G}x_1 \end{bmatrix},\right.$$

$$\left. \begin{bmatrix} I & 0 \\ \mathbf{H}_1 & I \end{bmatrix} \begin{bmatrix} \Sigma_w & 0 \\ 0 & \Sigma_v \end{bmatrix} \begin{bmatrix} I & \mathbf{H}_1' \\ 0 & I \end{bmatrix} + \begin{bmatrix} \mathbf{B} \\ \mathbf{H}_1\mathbf{B} \end{bmatrix} \Sigma_0[\mathbf{B}', \mathbf{B}'\mathbf{H}_1']\right).$$

Hence,

$$z_{1|0} := E(z_1) = \mathbf{B}\mu_0 + \mathbf{F}x_0 = \mathbf{B}z_{0|0} + \mathbf{F}x_0,$$

$$\Sigma_z(1|0) := \text{Cov}(z_1) = \Sigma_w + \mathbf{B}\Sigma_0\mathbf{B}' = \mathbf{B}\Sigma_z(0|0)\mathbf{B}' + \Sigma_w,$$

$$y_{1|0} := E(y_1) = \mathbf{H}_1 z_{1|0} + \mathbf{G}x_1,$$

$$\begin{aligned}\Sigma_y(1|0) := \text{Cov}(y_1) &= \mathbf{H}_1\Sigma_w\mathbf{H}_1' + \Sigma_v + \mathbf{H}_1\mathbf{B}\Sigma_0\mathbf{B}'\mathbf{H}_1' \\ &= \mathbf{H}_1\Sigma_z(1|0)\mathbf{H}_1' + \Sigma_v,\end{aligned}$$

which proves the prediction step for $t = 1$. Using these results and (18.3.11), the conditional distribution of z_1 given y_1 is seen to be

$$(z_1|y_1) \sim \mathcal{N}[z_{1|0} + \Sigma_z(1|0)\mathbf{H}_1'\Sigma_y(1|0)^{-1}(y_1 - y_{1|0}),$$
$$\Sigma_z(1|0) - \Sigma_z(1|0)\mathbf{H}_1'\Sigma_y(1|0)^{-1}\mathbf{H}_1\Sigma_z(1|0)],$$

which proves the correction step for $t = 1$.

Now the prediction and correction steps can be shown by induction. Suppose the normal distributions in (18.3.4)–(18.3.6) and the prediction and correction steps are correct for $t-1$. Then, using the transition and measurement equations, z_t and y_t have a joint normal distribution

$$\begin{aligned}\begin{bmatrix} z_t \\ y_t \end{bmatrix} &= \begin{bmatrix} I \\ \mathbf{H}_t \end{bmatrix} z_t + \begin{bmatrix} 0 \\ \mathbf{G} \end{bmatrix} x_t + \begin{bmatrix} 0 \\ I \end{bmatrix} v_t \\ &= \begin{bmatrix} I \\ \mathbf{H}_t \end{bmatrix}(\mathbf{B}z_{t-1} + \mathbf{F}x_{t-1} + w_{t-1}) + \begin{bmatrix} 0 \\ \mathbf{G} \end{bmatrix} x_t + \begin{bmatrix} 0 \\ I \end{bmatrix} v_t.\end{aligned}$$

By the induction assumption and (18.3.9)/(18.3.10), this term has the following conditional normal distribution, given $y_1, \ldots, y_{t-1}$:

$$\mathcal{N}\left(\begin{bmatrix} \mathbf{B}z_{t-1|t-1} + \mathbf{F}x_{t-1} \\ \mathbf{H}_t(\mathbf{B}z_{t-1|t-1} + \mathbf{F}x_{t-1}) + \mathbf{G}x_t \end{bmatrix}, \begin{bmatrix} \Sigma_z(t|t-1) & \bullet \\ \mathbf{H}_t\Sigma_z(t|t-1) & \mathbf{H}_t\Sigma_z(t|t-1)\mathbf{H}_t' + \Sigma_v \end{bmatrix}\right), \quad (18.3.12)$$

where $\Sigma_z(t|t-1) = \mathbf{B}\Sigma_z(t-1|t-1)\mathbf{B}' + \Sigma_w$. This proves the prediction step. Application of (18.3.11) to (18.3.12) gives the conditional distribution of z_t given $y_1, \ldots, y_t$ and proves the correction step.

It remains to prove the forecasting step. Again by induction $(z_t|y_1, \ldots, y_T)$ and $(y_t|y_1, \ldots, y_T)$ both have normal distributions with the first and second moments as stated in the forecasting step.

18.4 Maximum Likelihood Estimation of State Space Models

In this section, we consider ML estimation of the state space system given in Section 18.3.1. We assume that the matrices $\mathbf{B}$, $\mathbf{F}$, $\mathbf{H}_t$, $\mathbf{G}$, Σ_w, Σ_v, Σ_0, and the vector μ_0 depend on a vector of time invariant parameters $\boldsymbol{\delta}$. In other words, $\boldsymbol{\delta}$ is time invariant, even if $\mathbf{H}_t$ is not. For a given $\boldsymbol{\delta}$, the matrices are assumed to be uniquely determined and at least twice continuously differentiable with respect to the elements of $\boldsymbol{\delta}$. For instance, in the state space model (18.2.8)/(18.2.9) which represents the finite order VAR process (18.2.6),

$$\boldsymbol{\delta} = \begin{bmatrix} \text{vec}[\nu, A_1, \ldots, A_p] \\ \text{vech}(\Sigma_u) \end{bmatrix},$$

if no constraints are placed on the VAR coefficients or Σ_u and if the initial conditions $y_{-p+1}, \ldots, y_0$ are assumed to be known and fixed. The objective in this section is to estimate $\boldsymbol{\delta}$. We will set up the log-likelihood function first. Then we discuss its maximization and, finally, the asymptotic properties of the ML estimators are considered.

18.4.1 The Log-Likelihood Function

By Bayes' theorem, the sample density function can be written as

$$\begin{aligned} f(y_1, \ldots, y_T; \boldsymbol{\delta}) &= f(y_1; \boldsymbol{\delta}) f(y_2, \ldots, y_T | y_1; \boldsymbol{\delta}) \\ &\vdots \\ &= f(y_1; \boldsymbol{\delta}) f(y_2 | y_1; \boldsymbol{\delta}) \cdots f(y_T | y_1, \ldots, y_{T-1}; \boldsymbol{\delta}). \end{aligned}$$

Thus, using the notation of the previous section and assuming that y_t has dimension K, the Gaussian log-likelihood for the present case is

$$\begin{aligned} \ln l(\boldsymbol{\delta} | y_1, \ldots, y_T) &= \ln f(y_1, \ldots, y_T; \boldsymbol{\delta}) \\ &= \ln f(y_1; \boldsymbol{\delta}) + \sum_{t=2}^{T} \ln f(y_t | y_1, \ldots, y_{t-1}; \boldsymbol{\delta}) \\ &= -\frac{KT}{2} \ln(2\pi) - \frac{1}{2} \sum_{t=1}^{T} \ln |\Sigma_y(t|t-1)| \\ &\quad - \frac{1}{2} \sum_{t=1}^{T} (y_t - y_{t|t-1})' \Sigma_y(t|t-1)^{-1} (y_t - y_{t|t-1}), \end{aligned}$$

(18.4.1)

where we have used that $y_{1|0} := E(y_1)$, $\Sigma_y(1|0) := \text{Cov}(y_1)$, and

$$(y_t | y_1, \ldots, y_{t-1}) \sim \mathcal{N}(y_{t|t-1}, \Sigma_y(t|t-1)), \quad t = 1, \ldots, T,$$

from Section 18.3.1. Here both $y_{t|t-1}$ and $\Sigma_y(t|t-1)$ depend in general on the parameter vector $\boldsymbol{\delta}$. If a specific vector $\boldsymbol{\delta}$ is given, all the quantities in the log-likelihood function can be computed with the Kalman filter recursions. Thus, the Kalman filter is seen to be a useful tool for evaluating the log-likelihood function of a wide range of models. Note also that we have considered likelihood approximations for VARMA processes in Chapter 12. In the present framework, the exact likelihood may be obtained (see also Solo (1984)).

To simplify the expression for the log-likelihood given in (18.4.1), we use the following notation:

$$e_t(\boldsymbol{\delta}) := y_t - y_{t|t-1} \quad \text{and} \quad \Sigma_t(\boldsymbol{\delta}) := \Sigma_y(t|t-1). \tag{18.4.2}$$

This notation makes the dependence on $\boldsymbol{\delta}$ explicit. Occasionally, we will, however, drop $\boldsymbol{\delta}$. With this notation, the log-likelihood function can be written as

$$\ln l(\boldsymbol{\delta}) = -\frac{KT}{2}\ln(2\pi) - \frac{1}{2}\sum_{t=1}^{T}[\ln|\Sigma_t(\boldsymbol{\delta})| + e_t(\boldsymbol{\delta})'\Sigma_t(\boldsymbol{\delta})^{-1}e_t(\boldsymbol{\delta})]. \tag{18.4.3}$$

18.4.2 The Identification Problem

Recall from the discussion in Chapter 12 that unique maximization of the likelihood function and asymptotic inference require an identified or unique parameterization. Identification is not automatic in the present context because, for instance, VARMA models are not identified without specific restrictions and VARMA processes are just special cases of the presently considered models. Hence, the identification or uniqueness problem is inherent in the general linear state space model, too. We will state the problem here again in sufficient generality to cover the present case.

Let $\mathbf{y} := \text{vec}(y_1, \ldots, y_T)$ be the vector of observed random variables and denote its distribution by $F(\mathbf{y}; \boldsymbol{\delta}_0)$, where $\boldsymbol{\delta}_0$ is the *true parameter vector*. We assume that the true distribution of $\mathbf{y}$ is a member of the parametric family

$$\{F(\mathbf{y}; \boldsymbol{\delta}) | \boldsymbol{\delta} \in \mathbb{D}\},$$

where $\mathbb{D} \subset \mathbb{R}^n$ is the parameter space. The vector $\boldsymbol{\delta}_0$ is said to be *identified* or *identifiable* if it is the only vector in $\mathbb{D}$ which gives rise to the distribution of $\mathbf{y}$. In other words, for any $\boldsymbol{\delta}_1 \in \mathbb{D}$,

$$\boldsymbol{\delta}_1 \neq \boldsymbol{\delta}_0 \Rightarrow F(\mathbf{y}; \boldsymbol{\delta}_1) \neq F(\mathbf{y}; \boldsymbol{\delta}_0) \quad (\text{for at least one } \mathbf{y}). \tag{18.4.4}$$

To compute ML estimators and to derive asymptotic properties it is actually sufficient that $\boldsymbol{\delta}_0$ has a neighborhood in which it is uniquely determined by the true distribution of $\mathbf{y}$. To distinguish this case from one where uniqueness follows for the whole parameter space, the vector $\boldsymbol{\delta}_0$ or the model is often called *locally identified* or *locally identifiable* if there exists a neighborhood $\mathbb{U}(\boldsymbol{\delta}_0)$ of $\boldsymbol{\delta}_0$ such that (18.4.4) holds for any $\boldsymbol{\delta}_1 \in \mathbb{U}(\boldsymbol{\delta}_0)$. In contrast, the model or parameter vector is *globally identifiable* or *globally identified* if (18.4.4) holds for all $\boldsymbol{\delta}_1 \in \mathbb{D}$.

Because the negative log-likelihood function has a locally unique minimum if its Hessian matrix is positive definite, identification conditions for state space models may be formulated via the information matrix. If we are interested in *asymptotic* properties of estimators, it is sufficient to obtain identification in large samples. Hence, under some regularity conditions, the identification assumption may be disguised in the requirement of a positive definite asymptotic information matrix. In a later proposition giving asymptotic properties of the ML estimators, to ensure identification, we will include the condition that the sequence of normalized information matrices, $\mathcal{I}(\boldsymbol{\delta}_0)/T$, is bounded from below by a positive definite matrix, as T goes to infinity. In special case models, other identification conditions are often easier to deal with and are therefore preferred. For example, for VARMA processes the identification conditions given in Section 12.1.2 may be used.

18.4.3 Maximization of the Log-Likelihood Function

From some previous chapters we know that maximization of the log-likelihood function is in general a nonlinear optimization problem. Therefore, numerical methods are required for its solution. One possibility is a gradient algorithm as described, for example, in Section 12.3.2 for iteratively minimizing $-\ln l$. Recall that the general form of the i-th iteration step is

$$\boldsymbol{\delta}_{i+1} = \boldsymbol{\delta}_i - s_i D_i \left[\frac{\partial(-\ln l)}{\partial \boldsymbol{\delta}} \bigg|_{\boldsymbol{\delta}_i} \right], \qquad (18.4.5)$$

where s_i is the step length and D_i is a positive definite direction matrix. The inverse information matrix is one possible choice for this matrix. In that case, the method is called *scoring algorithm*. We will provide the ingredients for this algorithm in the following, that is, we will give expressions for the gradient of $\ln l$ and an estimator of the information matrix. There are various ways to choose the step length s_i. For instance, it could be chosen so as to optimize the progress towards the minimum. Another alternative would be to simply set $s_i = 1$. We will not discuss the step length selection in further detail here because it is of limited importance for the statistical analysis of the model.

The Gradient of the Log-Likelihood

From (18.4.3), we get

$$\frac{\partial \ln l}{\partial \boldsymbol{\delta}'} = -\frac{1}{2} \sum_{t=1}^T \left[\text{vec}\left(\frac{\partial \ln |\Sigma_t|}{\partial \Sigma_t}\right)' \frac{\partial \text{vec}(\Sigma_t)}{\partial \boldsymbol{\delta}'} + \frac{\partial \text{tr}(e_t' \Sigma_t^{-1} e_t)}{\partial \boldsymbol{\delta}'} \right]$$

$$= -\frac{1}{2} \sum_t \left[\text{vec}(\Sigma_t^{-1})' \frac{\partial \text{vec}(\Sigma_t)}{\partial \delta'} + 2e_t' \Sigma_t^{-1} \frac{\partial e_t}{\partial \delta'} \right.$$

$$\left. - \text{vec}(\Sigma_t^{-1} e_t e_t' \Sigma_t^{-1})' \frac{\partial \text{vec}(\Sigma_t)}{\partial \delta'} \right]$$

$$= -\frac{1}{2} \sum_t \left[\text{vec}[\Sigma_t^{-1}(I_K - e_t e_t' \Sigma_t^{-1})]' \frac{\partial \text{vec}(\Sigma_t)}{\partial \delta'} + 2e_t' \Sigma_t^{-1} \frac{\partial e_t}{\partial \delta'} \right], \tag{18.4.6}$$

where $\partial e_t / \partial \delta'$ may be replaced by $-\partial y_{t|t-1} / \partial \delta'$.

The Information Matrix

Using $E(e_t e_t') = \Sigma_t$ $(= \Sigma_y(t|t-1))$ and $E[e_t(\delta_0)] = 0$, straightforward application of the rules for matrix and vector differentiation yields the information matrix,

$$\begin{aligned}
\mathcal{I}(\delta_0) &= -E\left[\frac{\partial^2 \ln l}{\partial \delta \partial \delta'} \bigg|_{\delta_0} \right] \\
&= \frac{1}{2} \sum_{t=1}^{T} \left[\frac{\partial \text{vec}(\Sigma_t)'}{\partial \delta} (\Sigma_t^{-1} \otimes \Sigma_t^{-1}) \frac{\partial \text{vec}(\Sigma_t)}{\partial \delta'} \right. \\
&\quad \left. + 2E\left(\frac{\partial e_t'}{\partial \delta} \Sigma_t^{-1} \frac{\partial e_t}{\partial \delta'} \right) \right].
\end{aligned} \tag{18.4.7}$$

Because the true parameter values involved in this expression are unknown, they are replaced by estimators and the expectation is simply dropped. For instance, in the i-th iteration of the scoring algorithm, δ_i is used as an estimator for δ_0.

Discussion of the Scoring Algorithm

The scoring algorithm may have poor convergence properties far away from the maximum of the log-likelihood function. On the other hand, it has very good convergence properties close to the maximum. Unfortunately, it may be expensive in terms of computation time because it requires (possibly numerical) evaluation of derivatives in each iteration. Therefore, other maximization methods were proposed in the literature. Notably the EM (expectation step-maximization step) algorithm of Dempster, Laird & Rubin (1977) was found to be useful in practice (see Watson & Engle (1983), Schneider (1992)). The EM algorithm is an iterative algorithm which has the advantage of involving much cheaper computations in each iteration step than the scoring algorithm. On the other hand, convergence of the former is slower than that of the latter algorithm. Nicholls & Pagan (1985) and Schneider (1991, 1992) suggested combining the EM and the scoring algorithms. This proposal may be useful if no good initial estimator δ_1 is available from where to start the scoring algorithm. Another alternative is to use the so-called subspace algorithm for getting initial values (e.g., Bauer & Wagner (2002)).

18.4.4 Asymptotic Properties of the ML Estimator

We consider the state space model from Section 18.3.1 with transition equation

$$z_t = \mathbf{B} z_{t-1} + \mathbf{F} x_{t-1} + w_{t-1} \tag{18.4.8}$$

and measurement equation

$$y_t = \mathbf{H}_t z_t + \mathbf{G} x_t + v_t. \tag{18.4.9}$$

All assumptions of Section 18.3.1 are taken to be satisfied. In addition we assume that

(i) the true parameter vector δ_0 is in the interior of the parameter space which is supposed to be compact;
(ii) $\mathbf{H}_t = (x_t \otimes I)J$, where J is a known selection matrix such as $J = [I_K : 0 : \cdots : 0]$ or $\mathbf{H}_t = \mathbf{H}$ is a time invariant nonstochastic matrix;
(iii) the inputs x_t are nonstochastic and uniformly bounded, that is, there exist real numbers c_1 and c_2 such that $c_1 \leq x_t' x_t \leq c_2$ for all $t = 0, 1, 2, \ldots$;
(iv) the sequence of normalized information matrices is bounded from below by a positive definite matrix, that is, there exists a constant c such that $T^{-1} \mathcal{I}(\delta_0) > c I_n$ or, in other words, $T^{-1} \mathcal{I}(\delta_0) - c I_n$ is positive definite, as $T \to \infty$;
(v) all eigenvalues of $\mathbf{B}$ have modulus less than 1.

As we have discussed in Section 18.4.2, (iv) is an identification condition. The last assumption is a stability condition, and (iii) guarantees that the input variables have no trends. We have seen in Chapter 7 that the standard asymptotic theory may not apply for trending variables. Therefore, they are excluded here. With these assumptions, the following proposition can be established.

Proposition 18.1 (*Asymptotic Properties of the ML Estimator*)
With all the assumptions stated in the foregoing, the ML estimator $\widetilde{\delta}$ of δ_0 is consistent and asymptotically normally distributed,

$$\sqrt{T}(\widetilde{\delta} - \delta_0) \xrightarrow{d} \mathcal{N}(0, \Sigma_{\widetilde{\delta}}), \tag{18.4.10}$$

where

$$\Sigma_{\widetilde{\delta}} = \lim T \mathcal{I}(\delta_0)^{-1}$$

is the inverse asymptotic information matrix. It is consistently estimated by substituting the ML estimators for unknown parameters in (18.4.7), dropping the expectation operator and dividing by T. ∎

Pagan (1980) gives a proof of this proposition based on Crowder (1976) (see also Schneider (1988)). Other sets of conditions are possible to accommodate the situation where the inputs x_t are stochastic. They may, in fact, contain lagged y_t's. Moreover, **B** may have eigenvalues on the unit circle if it does not contain unknown parameters. The reader is referred to the articles by Pagan (1980), Nicholls & Pagan (1985), Schneider (1988), and to a book by Caines (1988) for details. When particular models are considered, different sets of assumptions are often preferable for two reasons. First, other sets of conditions may be easier to verify or to understand for special models. Second, the conditions of Proposition 18.1 or the modifications mentioned in the foregoing may not be satisfied. We will see an example of the latter case shortly.

A number of alternatives to ML estimation were suggested, see, e.g., Anderson & Moore (1979), Nicholls & Pagan (1985), Schneider (1988), and Bauer & Wagner (2002) for more details and references.

It may be worth noting that application of the Kalman filter to systems with estimated parameters produces state estimates and precision matrices that do not take into account the estimation variability. Watanabe (1985) and Hamilton (1986) considered the properties of state estimators obtained with estimated parameter Kalman filter recursions. Furthermore, a state space framework for unit root processes was presented by Bauer & Wagner (2003).

18.5 A Real Data Example

As an illustrative example, we consider a dynamic consumption function with time varying coefficients,

$$\begin{aligned} y_t &= \gamma_{0t} + \gamma_{1t}x_t + \gamma_{2t}x_{t-1} + \gamma_{3t}y_{t-1} + \gamma_{4t}x_{t-2} + \gamma_{5t}y_{t-2} + v_t \\ &= X'_t\gamma_t + v_t, \end{aligned} \quad (18.5.1)$$

where

$$X_t := \begin{bmatrix} 1 \\ x_t \\ x_{t-1} \\ y_{t-1} \\ x_{t-2} \\ y_{t-2} \end{bmatrix} \quad \text{and} \quad \gamma_t := \begin{bmatrix} \gamma_{0t} \\ \gamma_{1t} \\ \gamma_{2t} \\ \gamma_{3t} \\ \gamma_{4t} \\ \gamma_{5t} \end{bmatrix}.$$

Here y_t and x_t represent rates of change (first differences of logarithms) of consumption and income, respectively. Suppose that the coefficient vector γ_t differs from γ_{t-1} by an additive random disturbance, that is,

$$\gamma_t = \gamma_{t-1} + w_{t-1}. \quad (18.5.2)$$

In other words, γ_t is driven by a (multivariate) random walk. Clearly, (18.5.1) and (18.5.2) represent the measurement and transition equations of a state space model with $\mathbf{H}_t = X'_t$ and $\mathbf{B} = I_6$. We complete the model by assuming that v_t and w_t are independent Gaussian white noise processes, $v_t \sim \mathcal{N}(0, \sigma_v^2)$ and $w_t \sim \mathcal{N}(0, \Sigma_w)$, where

$$\Sigma_w = \begin{bmatrix} \sigma_{w_0}^2 & & 0 \\ & \ddots & \\ 0 & & \sigma_{w_5}^2 \end{bmatrix} \quad (18.5.3)$$

is a diagonal matrix. Furthermore, the initial state γ_0 is also assumed to be normally distributed, $\gamma_0 \sim \mathcal{N}(\overline{\gamma}_0, \Sigma_0)$, and independent of v_t and w_t. Admittedly, our assumed model is quite simple. Still, it is useful to illustrate some concepts considered in the previous sections.

Assuming that a sample $\mathbf{y} = (y_1, \ldots, y_T)'$ is available, the log-likelihood function of our model is

$$\ln l(\sigma_v^2, \Sigma_w, \overline{\gamma}_0, \Sigma_0 | \mathbf{y})$$
$$= -\frac{T}{2}\ln(2\pi) - \frac{1}{2}\sum_{t=1}^{T}\ln|\Sigma_y(t|t-1)| - \frac{1}{2}\sum_{t=1}^{T}(y_t - y_{t|t-1})^2/\Sigma_y(t|t-1),$$
$$(18.5.4)$$

where $\Sigma_y(t|t-1)$ is a scalar ((1×1) matrix) because y_t is a univariate variable. The log-likelihood function may be evaluated with the Kalman filter recursions for given parameters $\sigma_v^2, \Sigma_w, \overline{\gamma}_0$, and Σ_0. The maximization problem may be solved with an iterative algorithm. Once estimates of the parameters σ_v^2 and Σ_w are available, estimates $\gamma_{t|T}$ of the coefficients of the consumption function (18.5.1) may be obtained with the smoothing recursions given in Section 18.3.1.

Using first differences of logarithms of the quarterly consumption and income data given in File E1 for the years 1960 to 1982, we have estimated the parameters of the state space model (18.5.1)/(18.5.2). The ML estimates of the parameters of interest, namely the variances σ_v^2 and $\sigma_{w_i}^2$, $i = 0, 1, \ldots, 5$, together with estimated standard errors (square roots of the diagonal elements of the estimated inverse information matrix) and corresponding t-ratios are given in Table 18.1.

The interpretation of the standard errors and t-ratios needs caution for various reasons. In Proposition 18.1, where the asymptotic distribution of the ML estimators is given, we have assumed that all eigenvalues of the transition matrix $\mathbf{B}$ have modulus less than 1. This condition is clearly not satisfied in the present example, where $\mathbf{B} = I_6$ and, thus, all six eigenvalues are equal to 1. However, as mentioned in Section 18.4.4, the condition on the eigenvalues of $\mathbf{B}$ is not crucial if $\mathbf{B}$ is a known matrix which does not contain unknown parameters. Of course, setting $\mathbf{B} = I_6$ is just an assumption which may or may not be adequate.

A further deviation from the assumptions of Proposition 18.1 is that the inputs X_t contain lagged endogenous variables and hence are stochastic. Again,

Table 18.1. ML estimates for the example model

parameter	estimate	standard error	t-ratio
σ_v^2	3.91×10^{-5}	1.99×10^{-5}	1.97
$\sigma_{w_0}^2$	2.04×10^{-5}	2.33×10^{-5}	.88
$\sigma_{w_1}^2$	$.14 \times 10^{-2}$	1.08×10^{-2}	.13
$\sigma_{w_2}^2$	$.46 \times 10^{-2}$	$.92 \times 10^{-2}$	.50
$\sigma_{w_3}^2$	$.45 \times 10^{-2}$	1.11×10^{-2}	.41
$\sigma_{w_4}^2$	$.51 \times 10^{-2}$	$.94 \times 10^{-2}$	.54
$\sigma_{w_5}^2$	$.62 \times 10^{-2}$	1.16×10^{-2}	.54

we have mentioned in Section 18.4.4 that this assumption is not necessarily critical. The conditions of Proposition 18.1 could be modified so as to allow for lagged dependent variables.

Another assumption that may be problematic is the normality of the white noise sequences and the initial state. The normality assumption may be checked by computing the skewness and kurtosis of the standardized quantities $(y_t - y_{t|t-1})/\Sigma_y(t|t-1)^{1/2}$. A test for nonnormality may then be based on the χ^2-statistic involving both skewness and kurtosis as described in Chapter 4, Section 4.5. For the present example, the statistic assumes the value 3.00 and has a $\chi^2(2)$-distribution under the null hypothesis of normality. Thus, it is not significant at any conventional level.

Finally, we have assumed in Proposition 18.1 that the true parameter values lie in the interior of the parameter space. Given that the variance estimates are quite small compared to their estimated standard errors, it is possible that at least the $\sigma_{w_i}^2$ are in fact zero and, thus, lie on the boundary of the feasible parameter space. If the $\sigma_{w_i}^2$ are actually zero, the γ_t are time invariant in our model which would be a hypothesis of considerable interest. It would permit us to work with a constant coefficient specification. Unfortunately, if $\sigma_{w_i}^2 = 0$, the corresponding t-ratio does not have an asymptotic standard normal distribution in general. Thus, we cannot use the t-ratios given in Table 18.1 for testing the null hypotheses $\sigma_{w_i}^2 = 0$, $i = 0, 1, \ldots, 5$.

In the present context, we may ignore the problems related to the asymptotic theory for the moment and simply regard the model as a descriptive tool. Using the estimated values of the parameters of the model, we may consider the smoothing estimates $\gamma_{t|T}$ of the states (the coefficients of the consumption function). They are plotted in Figure 18.2. The two-standard error bounds which are also shown in the figure are computed from the $\Sigma_\gamma(t|T)$. These quantities are obtained with the smoothing recursions given in Section 18.3.1. From the plots in Figure 18.2, it can be seen that the intercept term γ_{0t} is the only coefficient that exhibits substantial variation through time. For instance, a considerable downturn is observed in 1966/1967, where the West German economy was in a recession. All the other coefficients show relatively little variation through time, although γ_{3t}, γ_{4t}, and γ_{5t} (the coefficients of

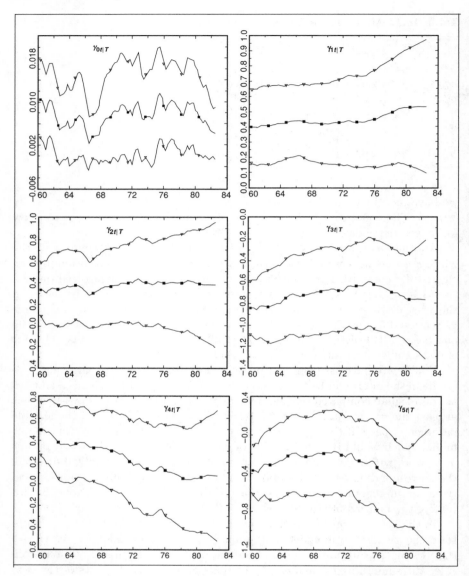

Fig. 18.2. Smoothing estimates of the consumption function coefficients (—■— coefficient estimate, —▽— estimated two-standard error bound).

y_{t-1}, x_{t-2}, and y_{t-2}, respectively) have a tendency to decline in the second half of the 1970s. However, given the estimated two-standard error bounds, overall the results support a specification with constant coefficients of current and lagged income and lagged consumption.

As mentioned previously, this example is quite simplistic. It is just meant to illustrate some of the concepts discussed in this chapter. In a more general model, other lags of income and/or consumption could appear on the right-hand side of the consumption function, the coefficients could be generated by a more general VAR model and the covariance matrix of w_t could be nondiagonal. Moreover, the consumption function may just be a part of a system of equations.

Given that we have discussed different models for the same data in previous chapters, the example also illustrates that there is not just one possible model or model class for the generation process of a multiple time series. The reader may wonder which of the models we have considered in this and the previous chapters is "best". That, however, depends on the questions of interest. In other words, the time series analyst has to decide on the model with the objective of his or her analysis in mind. In this book, we have just tried to introduce some of the possible tools in this venture. With these tools in hand, the analyst is hoped to be able to approach his or her problems of interest in a superior way, with an improved sense of the available possibilities and the potential pitfalls.

18.6 Exercises

Problem 18.1
Write the VARMAX model

$$\begin{aligned} y_t &= A_1 y_{t-1} + \cdots + A_p y_{t-p} + B_0 x_t + \cdots + B_s x_{t-s} \\ &\quad + u_t + M_1 u_{t-1} + \cdots + M_q u_{t-q} \end{aligned}$$

in state space form.

Problem 18.2
Suppose that in the dynamic factor analytic model, $y_t = \mathbf{L} f_t + u_t$, the common factors f_t are generated by the VARMA(p, q) process,

$$f_t = A_1 f_{t-1} + \cdots + A_p f_{t-p} + \eta_t + M_1 \eta_{t-1} + \cdots + M_q \eta_{t-q},$$

and the individual factors u_t are generated by the VARMA(r, s) process

$$u_t = C_1 u_{t-1} + \cdots + C_r u_{t-r} + \varepsilon_t + D_1 \varepsilon_{t-1} + \cdots + D_s \varepsilon_{t-s}.$$

Write the model in state space form.

Problem 18.3
Assume that y_t is generated according to

$$y_t = A_t Y_{t-1} + C_t x_t + v_t,$$

where $A_t := [A_{1t}, \ldots, A_{pt}]$ and $Y_{t-1} := [y'_{t-1}, \ldots, y'_{t-p}]'$. Suppose that $\alpha_t := \text{vec}[A_t : C_t]$ is driven by the VARMA(r, s) process

$$\alpha_t = D_1 \alpha_{t-1} + \cdots + D_r \alpha_{t-r} + \eta_t + M_1 \eta_{t-1} + \cdots + M_s \eta_{t-s}.$$

Write the model in state space form.

Problem 18.4
Write down explicitly the first two steps of the Kalman filter recursions.

Problem 18.5
Suppose the scalar observable variable y_t is generated by the random coefficient regression model

$$y_t = \nu + x_t \beta_t + v_t, \quad t = 1, \ldots, T,$$

where $\beta_t = \alpha \beta_{t-1} + w_t$ is driven by an AR(1) process. Suppose further that v_t and w_t are independent zero mean Gaussian white noise processes with variance 1 and let β_0 be a standard normal random variable.

(a) Determine the conditional distribution of β_t given $y_1, \ldots, y_{t-1}$.
(b) Write down the log-likelihood function of the model and derive its gradient. Find an expression for the information matrix.

Problem 18.6
Consider the K-dimensional Gaussian stable VAR(1) process $y_t = A y_{t-1} + u_t$ with $y_0 \sim \mathcal{N}(0, 0)$ and $u_t \sim \mathcal{N}(0, \Sigma_u)$ for $t = 1, 2, \ldots$. Use the Kalman filter recursions to determine $y_{t|t-1}$.

(a) Show that $y_{t|t-1} = A y_{t-1}$.
(b) Show that the conditions of Proposition 18.1 are satisfied if

$$\delta = \begin{bmatrix} \text{vec}(A) \\ \text{vech}(\Sigma_u) \end{bmatrix}.$$

Problem 18.7
Repeat the analysis of Section 18.5 with the same data and the state space model consisting of the measurement equation

$$y_t = \gamma_{0t} + \gamma_{1t} x_t + \gamma_{2t} y_{t-1} + v_t$$

and the transition equation

$$\begin{bmatrix} \gamma_{0,t} \\ \gamma_{1,t} \\ \gamma_{2,t} \end{bmatrix} = \begin{bmatrix} \gamma_{0,t-1} \\ \gamma_{1,t-1} \\ \gamma_{2,t-1} \end{bmatrix} + w_{t-1}.$$

Appendices

A
Vectors and Matrices

The following summary of matrix and vector algebra is not meant to be an introduction to the subject but is just a brief review of terms and rules used in the text. Most of them can be found in books such as Graybill (1969), Searle (1982), Anderson (1984, Appendix), Magnus & Neudecker (1988), Magnus (1988) or Lütkepohl (1996a). Therefore proofs or further references are only provided in exceptional cases.

A.1 Basic Definitions

A *matrix* is a rectangular array of numbers. For instance,

$$\begin{bmatrix} 3 & -5 \\ .3 & 0 \end{bmatrix}, \quad (0,1,0), \quad \begin{bmatrix} 3 & 5 & .3 & .3 \\ 2 & 2 & 2 & 2 \end{bmatrix}$$

are matrices. More generally,

$$A = (a_{ij}) = \begin{bmatrix} a_{11} & \cdots & a_{1n} \\ \vdots & & \vdots \\ a_{m1} & \cdots & a_{mn} \end{bmatrix} \tag{A.1.1}$$

is a matrix with m rows and n columns. Such a matrix is briefly called ($m \times n$) matrix, m being the *row dimension* and n being the *column dimension*. The numbers a_{ij} are the *elements* or *components* of A. In the following, it is assumed that the elements of all matrices considered are real numbers unless otherwise stated. In other words, we will be concerned with real rather than complex matrices. If the dimensions m and n are clear from the context or if they are of no importance, the notation $A = (a_{ij})$ means that a_{ij} is a *typical element* of A, that is, A consists of elements a_{ij}, $i = 1, \ldots, m, j = 1, \ldots, n$.

A ($1 \times n$) matrix is a *row vector* and an ($m \times 1$) matrix is a *column vector* which is often denoted by a lower case letter in the text. If not otherwise noted, all vectors will be column vectors in the following. Instead of $(m \times 1)$ matrix we sometimes say $(m \times 1)$ vector or simply *m-vector* or *m-dimensional vector*.

An $(m \times m)$ matrix with the number of rows equal to the number of columns is a *square matrix*. An $(m \times m)$ square matrix

$$\begin{bmatrix} a_{11} & 0 & \cdots & 0 \\ 0 & a_{22} & & 0 \\ \vdots & & \ddots & \vdots \\ 0 & 0 & \cdots & a_{mn} \end{bmatrix}$$

with zeros off the main diagonal is a *diagonal matrix*. If all the diagonal elements of a diagonal matrix are one, it is an *identity* or *unit matrix*. An $(m \times m)$ identity matrix is denoted by I_m or simply by I if the dimension is unimportant or obvious from the context. A square matrix with all elements below (above) the main diagonal being zero is called *upper* (*lower*) *triangular* or simply *triangular matrix*. A matrix consisting of zeros only is a *null matrix* or *zero matrix*. Usually, in this text, such a matrix is simply denoted by 0 and its dimensions have to be figured out from the context.

The *transpose* of the $(m \times n)$ matrix A given in (A.1.1) is the $(n \times m)$ matrix

$$A' = \begin{bmatrix} a_{11} & \cdots & a_{m1} \\ \vdots & & \vdots \\ a_{1n} & \cdots & a_{mn} \end{bmatrix},$$

the n rows of A' being the n columns of A. The matrix A is *symmetric* if $A' = A$. For instance,

$$\begin{bmatrix} 3 & 3 & 1 \\ 0 & 1 & 0 \end{bmatrix} \text{ is the transpose of } \begin{bmatrix} 3 & 0 \\ 3 & 1 \\ 1 & 0 \end{bmatrix}$$

and

$$\begin{bmatrix} 2 & -1 \\ -1 & 0 \end{bmatrix}$$

is a symmetric matrix.

A.2 Basic Matrix Operations

Let $A = (a_{ij})$ and $B = (b_{ij})$ be $(m \times n)$ matrices. The two matrices are *equal*, $A = B$, if $a_{ij} = b_{ij}$ for all i, j. The following matrix operations are basic:

$$A + B := (a_{ij} + b_{ij}). \quad \text{(addition)}$$
$$A - B := (a_{ij} - b_{ij}). \quad \text{(subtraction)}$$

For a real constant c,

$$cA = Ac := (ca_{ij}). \quad \text{(multiplication by a scalar)}$$

Let $C = (c_{ij})$ be an $(n \times r)$ matrix, then the product

$$AC := \left(\sum_{j=1}^{n} a_{ij} c_{jk} \right) \quad \text{(multiplication)}$$

is an $(m \times r)$ matrix. For instance,

$$\begin{bmatrix} 3 & 3 \\ 2 & 1 \end{bmatrix} \begin{bmatrix} 2 & 3 & 0 \\ 2 & 4 & -1 \end{bmatrix} = \begin{bmatrix} 3\cdot 2 + 3\cdot 2 & 3\cdot 3 + 3\cdot 4 & 3\cdot 0 - 3\cdot 1 \\ 2\cdot 2 + 1\cdot 2 & 2\cdot 3 + 1\cdot 4 & 2\cdot 0 - 1\cdot 1 \end{bmatrix}$$
$$= \begin{bmatrix} 12 & 21 & -3 \\ 6 & 10 & -1 \end{bmatrix}.$$

If the column dimension of A is the same as the row dimension of C so that A and C can be multiplied, the two matrices are *conformable*. In the product AC the matrix C is *premultiplied* by A and A is *postmultiplied* by C.

Rules: Suppose A, B, and C are matrices with suitable dimensions so that the following operations are defined and c is a scalar.

(1) $A + B = B + A$.
(2) $(A + B) + C = A + (B + C)$.
(3) $A(B + C) = AB + AC$.
(4) $c(A + B) = cA + cB$.
(5) $AB \neq BA$ in general.
(6) $(AB)C = A(BC)$.
(7) $(AB)' = B'A'$.
(8) $AI = IA = A$.
(9) AA' and $A'A$ are symmetric matrices.

A.3 The Determinant

The *determinant* of an $(m \times m)$ square matrix $A = (a_{ij})$ is the sum of all products

$$(-1)^P a_{1i_1} a_{2i_2} \cdots a_{mi_m}$$

consisting of precisely one element from each row and each column multiplied by -1 or 1, depending on the permutation $i_1, \ldots, i_m$ of the subscripts. The -1

is used if the number of inversions of $i_1, \ldots, i_m$ to obtain the order $1, 2, \ldots, m$ is odd and 1 is used otherwise. The sum is taken over all $m!$ permutations of the column subscripts.

For a (1×1) matrix the determinant equals the value of the single element and for $m > 1$ the determinant may be defined recursively as follows. Suppose

$$A = \begin{bmatrix} a_{11} & a_{12} \\ a_{21} & a_{22} \end{bmatrix}$$

is a (2×2) matrix. Then the determinant is

$$\det(A) = |A| = a_{11}a_{22} - a_{12}a_{21}. \qquad (A.3.1)$$

For instance,

$$\det \begin{bmatrix} 3 & 1 \\ 2 & 2 \end{bmatrix} = 4.$$

To specify the determinant of a general $(m \times m)$ matrix $A = (a_{ij})$ we define the *minor* of the ij-th element a_{ij} as the determinant of the $((m-1) \times (m-1))$ matrix that is obtained by deleting the i-th row and j-th column from A. The *cofactor* of a_{ij}, denoted by A_{ij}, is the minor multiplied by $(-1)^{i+j}$. Now

$$\det(A) = |A| = a_{i1}A_{i1} + \cdots + a_{im}A_{im} = a_{1j}A_{1j} + \cdots + a_{mj}A_{mj} \qquad (A.3.2)$$

for any i or $j \in \{1, \ldots, m\}$. It does not matter which row or column is chosen in (A.3.2) because the determinant of a matrix is a unique number.

For example, for the (3×3) matrix

$$A = \begin{bmatrix} 2 & 1 & 3 \\ 0 & 2 & 1 \\ 1 & -1 & 4 \end{bmatrix} \qquad (A.3.3)$$

the minor of the upper right-hand corner element is

$$\det \begin{bmatrix} 0 & 2 \\ 1 & -1 \end{bmatrix} = -2.$$

The cofactor is also -2 because $(-1)^{1+3} = 1$. Developing by the first row gives

$$|A| = 2 \cdot \det \begin{bmatrix} 2 & 1 \\ -1 & 4 \end{bmatrix} - 1 \cdot \det \begin{bmatrix} 0 & 1 \\ 1 & 4 \end{bmatrix} + 3 \cdot \det \begin{bmatrix} 0 & 2 \\ 1 & -1 \end{bmatrix} = 13.$$

The same result is obtained by developing by any other row or column, e.g., using the first column gives

$$|A| = 2 \cdot \det \begin{bmatrix} 2 & 1 \\ -1 & 4 \end{bmatrix} - 0 \cdot \det \begin{bmatrix} 1 & 3 \\ -1 & 4 \end{bmatrix} + 1 \cdot \det \begin{bmatrix} 1 & 3 \\ 2 & 1 \end{bmatrix} = 13.$$

Rules: In the following rules, $A = (a_{ij})$ and $B = (b_{ij})$ are $(m \times m)$ matrices and c is a scalar.

(1) $\det(I_m) = 1$.
(2) If A is a diagonal matrix, $\det(A) = a_{11} \cdot a_{22} \cdots a_{mm}$.
(3) If A is a lower or upper triangular matrix, $|A| = a_{11} \cdots a_{mm}$.
(4) If A contains a row or column of zeros, $|A| = 0$.
(5) If B is obtained from A by adding to one row (column) a scalar multiple of another row (column), then $|A| = |B|$.
(6) If A has two identical rows or columns, then $|A| = 0$.
(7) $\det(cA) = c^m \det(A)$.
(8) $|AB| = |A||B|$.
(9) If C is an $(m \times n)$ matrix, $\det(I_m + CC') = \det(I_n + C'C)$.

A.4 The Inverse, the Adjoint, and Generalized Inverses

A.4.1 Inverse and Adjoint of a Square Matrix

An $(m \times m)$ square matrix A is *nonsingular* or *regular* or *invertible* if there exists a unique $(m \times m)$ matrix B such that $AB = I_m$. The matrix B is denoted by A^{-1}. It is the *inverse* of A,

$$AA^{-1} = A^{-1}A = I_m.$$

For $m > 1$, the $(m \times m)$ matrix of cofactors,

$$A^{adj} = \begin{bmatrix} A_{11} & \cdots & A_{1m} \\ \vdots & \ddots & \vdots \\ A_{m1} & \cdots & A_{mm} \end{bmatrix}'$$

is the *adjoint* of A. For a (1×1) matrix A, we define the adjoint to be 1, that is, $A^{adj} = 1$. To compute the inverse of the $(m \times m)$ matrix A, the relation

$$A^{-1} = |A|^{-1} A^{adj} \tag{A.4.1}$$

is sometimes useful. For this expression to be meaningful, $|A|$ has to be nonzero. Indeed, A is nonsingular if and only if $\det(A) \neq 0$.

As an example consider the matrix given in (A.3.3). Its adjoint is

$$A^{adj} = \begin{bmatrix} \begin{vmatrix} 2 & 1 \\ -1 & 4 \end{vmatrix} & -\begin{vmatrix} 0 & 1 \\ 1 & 4 \end{vmatrix} & \begin{vmatrix} 0 & 2 \\ 1 & -1 \end{vmatrix} \\ -\begin{vmatrix} 1 & 3 \\ -1 & 4 \end{vmatrix} & \begin{vmatrix} 2 & 3 \\ 1 & 4 \end{vmatrix} & -\begin{vmatrix} 2 & 1 \\ 1 & -1 \end{vmatrix} \\ \begin{vmatrix} 1 & 3 \\ 2 & 1 \end{vmatrix} & -\begin{vmatrix} 2 & 3 \\ 0 & 1 \end{vmatrix} & \begin{vmatrix} 2 & 1 \\ 0 & 2 \end{vmatrix} \end{bmatrix}' = \begin{bmatrix} 9 & -7 & -5 \\ 1 & 5 & -2 \\ -2 & 3 & 4 \end{bmatrix}.$$

Consequently,

$$A^{-1} = \frac{1}{13} \begin{bmatrix} 9 & -7 & -5 \\ 1 & 5 & -2 \\ -2 & 3 & 4 \end{bmatrix}.$$

Multiplying this matrix by A is easily seen to result in the (3×3) identity matrix.

Rules:

(1) For an $(m \times m)$ square matrix A, $AA^{adj} = A^{adj}A = |A|I_m$.
(2) An $(m \times m)$ matrix A is nonsingular if and only if $\det(A) \neq 0$.

In the following, $A = (a_{ij})$ and B are nonsingular $(m \times m)$ matrices and $c \neq 0$ is a scalar constant.

(3) $A^{-1} = A^{adj}/|A|$.
(4) $(A')^{-1} = (A^{-1})'$.
(5) $(AB)^{-1} = B^{-1}A^{-1}$.
(6) $(cA)^{-1} = \frac{1}{c}A^{-1}$.
(7) $I_m^{-1} = I_m$.
(8) If A is a diagonal matrix, then A^{-1} is also diagonal with diagonal elements $1/a_{ii}$.
(9) For an $(m \times n)$ matrix C, $(I_m + CC')^{-1} = I_m - C(I_n + C'C)^{-1}C'$.

A.4.2 Generalized Inverses

Let A be an $(m \times n)$ matrix. Any matrix B satisfying $ABA = A$ is a *generalized inverse* of A. For example, if

$$A = \begin{bmatrix} 1 & 0 \\ 0 & 0 \end{bmatrix},$$

the following matrices are generalized inverses of A:

$$\begin{bmatrix} 1 & 0 \\ 0 & 1 \end{bmatrix}, \quad \begin{bmatrix} 1 & 0 \\ 0 & 0 \end{bmatrix}, \quad \begin{bmatrix} 1 & 0 \\ 0 & \frac{1}{2} \end{bmatrix}.$$

Obviously, a generalized inverse is not unique in general. An $(n \times m)$ matrix B is called *Moore-Penrose (generalized) inverse* of A if it satisfies the following four conditions:

$$\begin{aligned} ABA &= A, \\ BAB &= B, \\ (AB)' &= AB, \\ (BA)' &= BA. \end{aligned} \tag{A.4.2}$$

The Moore-Penrose inverse of A is denoted by A^+, it exists for any $(m \times n)$ matrix and is unique.

Rules: (See Magnus & Neudecker (1988, p. 33, Theorem 5).)

(1) $A^+ = A^{-1}$ if A is nonsingular.
(2) $(A^+)^+ = A$.
(3) $(A')^+ = (A^+)'$.
(4) $A'AA^+ = A^+AA' = A'$.
(5) $A'A^{+\prime}A^+ = A^+A^{+\prime}A' = A^+$.
(6) $(A'A)^+ = A^+A^{+\prime}, (AA')^+ = A^{+\prime}A^+$.
(7) $A^+ = (A'A)^+A' = A'(AA')^+$.

A.5 The Rank

Let $x_1, \ldots, x_n$ be $(m \times 1)$ vectors. They are *linearly independent* if, for the constants $c_1, \ldots, c_n$,

$$c_1 x_1 + \cdots + c_n x_n = 0$$

implies $c_1 = \cdots = c_n = 0$. Equivalently, defining the $(n \times 1)$ vector $c = (c_1, \ldots, c_n)'$ and the $(m \times n)$ matrix $X = (x_1, \ldots, x_n)$, the columns of X are linearly independent if $Xc = 0$ implies $c = 0$. The columns of X are *linearly dependent* if $c_1 x_1 + \cdots + c_n x_n = 0$ holds with at least one $c_i \neq 0$. In that case,

$$x_i = d_1 x_1 + \cdots + d_{i-1} x_{i-1} + d_{i+1} x_{i+1} + \cdots + d_n x_n,$$

where $d_j = -c_j/c_i$. In other words, $x_1, \ldots, x_n$ are linearly dependent if at least one of the vectors is a linear combination of the other vectors.

If $n > m$, the columns of X are linearly dependent. Consequently, if $x_1, \ldots, x_n$ are linearly independent, then $n \leq m$.

Let $a_1, \ldots, a_n$ be the columns of the $(m \times n)$ matrix $A = (a_1, \ldots, a_n)$. That is, the a_i are $(m \times 1)$ vectors. The *rank* of A, briefly rk(A), is the maximum number of linearly independent columns of A. Thus, if $n \leq m$ and the $a_1, \ldots, a_n$ are linearly independent, rk$(A) = n$. The maximum number of linearly independent columns of A equals the maximum number of linearly independent rows. Hence, the rank may be defined equivalently as the maximum number of linearly independent rows. If $m \geq n$ ($m \leq n$) then we say that A has *full rank* if rk$(A) = n$ (rk$(A) = m$).

Rules: Let A be an $(m \times n)$ matrix.

(1) rk$(A) \leq \min(m, n)$.
(2) rk$(A) = $ rk(A').
(3) rk$(AA') = $ rk$(A'A) = $ rk(A).
(4) If B is a nonsingular $(n \times n)$ matrix, then rk$(AB) = $ rk(A).

(5) If $\text{rk}(A) = m$, then $A^+ = A'(AA')^{-1}$.
(6) If $\text{rk}(A) = n$, then $A^+ = (A'A)^{-1}A'$.
(7) If B is an $(n \times r)$ matrix, $\text{rk}(AB) \leq \min\{\text{rk}(A), \text{rk}(B)\}$.
(8) If A is $(m \times m)$, then $\text{rk}(A) = m$ if and only if $|A| \neq 0$.

A.6 Eigenvalues and -vectors – Characteristic Values and Vectors

The *eigenvalues* or *characteristic values* or *characteristic roots* of an $(m \times m)$ square matrix A are the roots of the polynomial in λ given by $\det(A - \lambda I_m)$ or $\det(\lambda I_m - A)$. The determinant is sometimes called the *characteristic determinant* and the polynomial is called the *characteristic polynomial* of A. Because the roots of a polynomial are complex numbers, the eigenvalues are also complex in general. A number λ_i is an eigenvalue of A, if the columns of $(A - \lambda_i I_m)$ are linearly dependent. Consequently, there exists an $(m \times 1)$ vector $v_i \neq 0$ such that

$$(A - \lambda_i I_m)v_i = 0 \quad \text{or} \quad Av_i = \lambda_i v_i.$$

A vector with this property is an *eigenvector* or *characteristic vector* of A associated with the eigenvalue λ_i. Of course, any nonzero scalar multiple of v_i is also an eigenvector of A associated with λ_i.

As an example consider the matrix

$$A = \begin{bmatrix} 1 & 0 \\ 1 & 3 \end{bmatrix}.$$

Its eigenvalues are the roots of

$$|A - \lambda I_2| = \det \begin{bmatrix} 1-\lambda & 0 \\ 1 & 3-\lambda \end{bmatrix} = (1-\lambda)(3-\lambda).$$

Hence, $\lambda_1 = 1$ and $\lambda_2 = 3$ are the eigenvalues of A. Associated eigenvectors are obtained by solving

$$\begin{bmatrix} 1 & 0 \\ 1 & 3 \end{bmatrix} \begin{bmatrix} v_{11} \\ v_{21} \end{bmatrix} = \begin{bmatrix} v_{11} \\ v_{21} \end{bmatrix} \quad \text{and} \quad \begin{bmatrix} 1 & 0 \\ 1 & 3 \end{bmatrix} \begin{bmatrix} v_{12} \\ v_{22} \end{bmatrix} = 3 \begin{bmatrix} v_{12} \\ v_{22} \end{bmatrix}.$$

Thus,

$$\begin{bmatrix} v_{11} \\ v_{21} \end{bmatrix} = \begin{bmatrix} 1 \\ -\frac{1}{2} \end{bmatrix} \quad \text{and} \quad \begin{bmatrix} v_{12} \\ v_{22} \end{bmatrix} = \begin{bmatrix} 0 \\ 1 \end{bmatrix}$$

are eigenvectors of A associated with λ_1 and λ_2, respectively.

In the following rules, the *modulus* of a complex number $z = z_1 + iz_2$ is used. Here z_1 and z_2 are the real and imaginary parts of z, respectively, and $i = \sqrt{-1}$. The modulus $|z|$ of z is defined as

$$|z| := \sqrt{z_1^2 + z_2^2}.$$

If $z_2 = 0$ so that z is a real number, the modulus is just the absolute value of z, which justifies the notation.

Rules:

(1) If A is symmetric, then all its eigenvalues are real numbers.
(2) The eigenvalues of a diagonal matrix are its diagonal elements.
(3) The eigenvalues of a triangular matrix are its diagonal elements.
(4) An $(m \times m)$ matrix has at most m eigenvalues.
(5) Let $\lambda_1, \ldots, \lambda_m$ be the eigenvalues of the $(m \times m)$ matrix A, then $|A| = \lambda_1 \cdots \lambda_m$, that is, the determinant is the product of the eigenvalues.
(6) Let λ_i and λ_j be *distinct* eigenvalues of A with associated eigenvectors v_i and v_j. Then v_i and v_j are linearly independent.
(7) All eigenvalues of the $(m \times m)$ matrix A have modulus less than 1 if and only if $\det(I_m - Az) \neq 0$ for $|z| \leq 1$, that is, the polynomial $\det(I_m - Az)$ has no roots in and on the complex unit circle.

A.7 The Trace

The *trace* of an $(m \times m)$ square matrix $A = (a_{ij})$ is the sum of its diagonal elements,

$$\operatorname{tr} A = \operatorname{tr}(A) := a_{11} + \cdots + a_{mm}.$$

For example,

$$\operatorname{tr} \begin{bmatrix} 1 & 0 \\ 1 & 3 \end{bmatrix} = 4.$$

Rules: A and B are $(m \times m)$ matrices and $\lambda_1, \ldots, \lambda_m$ are the eigenvalues of A.

(1) $\operatorname{tr}(A + B) = \operatorname{tr}(A) + \operatorname{tr}(B)$.
(2) $\operatorname{tr} A = \operatorname{tr} A'$.
(3) If C is $(m \times n)$ and D is $(n \times m)$, $\operatorname{tr}(CD) = \operatorname{tr}(DC)$.
(4) $\operatorname{tr} A = \lambda_1 + \cdots + \lambda_m$.

A.8 Some Special Matrices and Vectors

A.8.1 Idempotent and Nilpotent Matrices

An $(m \times m)$ matrix A is *idempotent* if $AA = A^2 = A$. Examples of idempotent matrices are $A = I_m, A = 0$, and

$$A = \begin{bmatrix} \frac{1}{3} & \frac{1}{3} & \frac{1}{3} \\ \frac{1}{3} & \frac{1}{3} & \frac{1}{3} \\ \frac{1}{3} & \frac{1}{3} & \frac{1}{3} \end{bmatrix}.$$

An $(m \times m)$ matrix A is *nilpotent* if there exists a positive integer i such that $A^i = 0$. For instance, the (2×2) matrices

$$A = \begin{bmatrix} 0 & 3 \\ 0 & 0 \end{bmatrix} \quad \text{and} \quad B = \begin{bmatrix} 1 & -1 \\ 1 & -1 \end{bmatrix}$$

are nilpotent because $A^2 = B^2 = 0$.

Rules: In the following rules, A is an $(m \times m)$ matrix.

(1) If A is a diagonal matrix, it is idempotent if and only if all the diagonal elements are either zero or one.
(2) If A is symmetric and idempotent, $\text{rk}(A) = \text{tr}(A)$.
(3) If A is idempotent and $\text{rk}(A) = m$, then $A = I_m$.
(4) If A is idempotent, then $I_m - A$ is idempotent.
(5) If A is symmetric and idempotent, then $A^+ = A$.
(6) If B is an $(m \times n)$ matrix, then BB^+ and B^+B are idempotent.
(7) If A is idempotent, then all its eigenvalues are zero or one.
(8) If A is nilpotent, then all its eigenvalues are zero.

A.8.2 Orthogonal Matrices and Vectors and Orthogonal Complements

Two $(m \times 1)$ *vectors* x and y are *orthogonal* if $x'y = 0$. They are *orthonormal* if they are orthogonal and have unit length, where the length of a vector x is $||x|| := \sqrt{x'x}$.

An $(m \times k)$ matrix B is *orthogonal* to the $(m \times n)$ matrix A if $A'B = 0$. If A is an $(m \times n)$ matrix of full column rank, an *orthogonal complement* of A, denoted by $A_\perp$, is an $(m \times (m-n))$ matrix of full column rank such that $A'A_\perp = 0$. The orthogonal complement of a nonsingular square matrix is zero and the orthogonal complement of a zero matrix is an identity matrix of suitable dimension.

An $(m \times m)$ square *matrix* A is *orthogonal* if its transpose is its inverse, $A'A = AA' = I_m$. In other words, A is orthogonal if its rows and columns are orthonormal vectors.

Examples of orthogonal vectors are

$$x = \begin{bmatrix} 5 \\ 0 \\ 0 \end{bmatrix} \quad \text{and} \quad y = \begin{bmatrix} 0 \\ 2 \\ 0 \end{bmatrix}.$$

The following four matrices are orthogonal matrices:

$$\begin{bmatrix} 0 & 1 \\ 1 & 0 \end{bmatrix}, \quad \begin{bmatrix} \cos\varphi & \sin\varphi \\ -\sin\varphi & \cos\varphi \end{bmatrix},$$

$$\begin{bmatrix} 0 & 1 & 0 \\ 1 & 0 & 0 \\ 0 & 0 & 1 \end{bmatrix}, \quad \begin{bmatrix} 1/\sqrt{3} & 1/\sqrt{3} & 1/\sqrt{3} \\ 1/\sqrt{2} & -1/\sqrt{2} & 0 \\ 1/\sqrt{6} & 1/\sqrt{6} & -2/\sqrt{6} \end{bmatrix}.$$

Suppose

$$A = \begin{bmatrix} 1 & 0 \\ 1 & 1 \\ 0 & 2 \end{bmatrix},$$

then

$$\begin{bmatrix} 1 \\ -1 \\ \frac{1}{2} \end{bmatrix} \quad \text{and} \quad \begin{bmatrix} 2 \\ -2 \\ 1 \end{bmatrix}$$

are orthogonal complements of A.

Rules:

(1) I_m is an orthogonal matrix.
(2) If A is an orthogonal matrix, then $\det(A) = 1$ or -1.
(3) If A and B are orthogonal and conformable matrices, then AB is orthogonal.
(4) If λ_i, λ_j are distinct eigenvalues of a *symmetric* matrix A, then the corresponding eigenvectors v_i and v_j are orthogonal.
(5) For an $(m \times n)$ matrix A of full column rank and $n < m$, the matrix $[A : A_\perp]$ is invertible.

A.8.3 Definite Matrices and Quadratic Forms

Let A be a *symmetric* $(m \times m)$ matrix and x an $(m \times 1)$ vector. The function $x'Ax$ is called a *quadratic form* in x. The symmetric matrix A or the corresponding quadratic form is

(i) *positive definite* if $x'Ax > 0$ for all m-vectors $x \neq 0$;
(ii) *positive semidefinite* if $x'Ax \geq 0$ for all m-vectors x;
(iii) *negative definite* if $x'Ax < 0$ for all m-vectors $x \neq 0$;
(iv) *negative semidefinite* if $x'Ax \leq 0$ for all m-vectors x;
(v) *indefinite* if $x'Ax > 0$ for some x and $x'Ax < 0$ for another x.

Rules: In the following rules, A is a symmetric $(m \times m)$ matrix.

(1) $A = (a_{ij})$ is positive definite if and only if all its principle minors are positive, where

$$\det \begin{bmatrix} a_{11} & \cdots & a_{1i} \\ \vdots & \ddots & \vdots \\ a_{i1} & \cdots & a_{ii} \end{bmatrix}$$

is the i-th principle minor of A.

(2) A is negative definite (semidefinite) if and only if $-A$ is positive definite (semidefinite).
(3) If A is positive or negative definite, it is nonsingular.
(4) All eigenvalues of a positive (negative) definite matrix are greater (smaller) than zero.
(5) A diagonal matrix is positive (negative) definite if and only if all its diagonal elements are positive (negative).
(6) If A is positive definite and B an $(m \times n)$ matrix, then $B'AB$ is positive semidefinite.
(7) If A is positive definite and B an $(m \times n)$ matrix with $\mathrm{rk}(B) = n$, then $B'AB$ is positive definite.
(8) If A is positive definite, then A^{-1} is positive definite.
(9) If A is idempotent, then it is positive semidefinite.

With these rules it is easy to check that

$$\begin{bmatrix} 2 & 1 \\ 1 & 1 \end{bmatrix} \quad \text{and} \quad \begin{bmatrix} 3 & 1 & 0 \\ 1 & 1 & 0 \\ 0 & 0 & 4 \end{bmatrix}$$

are positive definite matrices and

$$\begin{bmatrix} 1 & 1 \\ 1 & 1 \end{bmatrix} \quad \text{and} \quad \begin{bmatrix} 1 & 0 \\ 0 & 0 \end{bmatrix}$$

are positive semidefinite matrices.

A.9 Decomposition and Diagonalization of Matrices

A.9.1 The Jordan Canonical Form

Let A be an $(m \times m)$ matrix with eigenvalues $\lambda_1, \ldots, \lambda_n$. Then there exists a nonsingular matrix P such that

$$P^{-1}AP = \begin{bmatrix} \Lambda_1 & & 0 \\ & \ddots & \\ 0 & & \Lambda_n \end{bmatrix} =: \Lambda \quad \text{or} \quad A = P\Lambda P^{-1}, \tag{A.9.1}$$

where
$$\Lambda_i = \begin{bmatrix} \lambda_i & 1 & 0 & \cdots & 0 \\ 0 & \lambda_i & 1 & & 0 \\ \vdots & & \ddots & \ddots & \vdots \\ 0 & 0 & & \ddots & 1 \\ 0 & 0 & \cdots & \cdots & \lambda_i \end{bmatrix}.$$

This decomposition of A is the *Jordan canonical form*. Because the eigenvalues of A may be complex numbers, Λ and P may be complex matrices. If multiple roots of the characteristic polynomial exist, they may have to appear more than once in the list $\lambda_1, \ldots, \lambda_n$.

The Jordan canonical form has some important implications. For instance, it implies that
$$A^j = (P\Lambda P^{-1})^j = P\Lambda^j P^{-1}$$
and it can be shown that
$$\Lambda_i^j = \begin{bmatrix} \lambda_i^j & \binom{j}{1}\lambda_i^{j-1} & \cdots & \binom{j}{r_i-1}\lambda_i^{j-r_i+1} \\ 0 & \lambda_i^j & \cdots & \binom{j}{r_i-2}\lambda_i^{j-r_i+2} \\ \vdots & & \ddots & \vdots \\ 0 & 0 & \cdots & \lambda_i^j \end{bmatrix},$$
where
$$\binom{p}{q} = \frac{p!}{(p-q)!q!}$$
denotes a binomial coefficient. We have the following rules.

Rules: Suppose A is a real $(m \times m)$ matrix with eigenvalues $\lambda_1, \ldots, \lambda_n$ which have all modulus less than 1, that is, $|\lambda_i| < 1$ for $i = 1, \ldots, n$. Furthermore, let Λ and P be the matrices given in (A.9.1).

(1) $A^j = P\Lambda^j P^{-1} \xrightarrow[j \to \infty]{} 0$.
(2) $\sum_{j=0}^{\infty} A^j = (I_m - A)^{-1}$ exists.
(3) The sequence A^j, $j = 0, 1, 2, \ldots$, is absolutely summable, that is,
$$\sum_{j=0}^{\infty} |\alpha_{kl,j}|$$
is finite for all $k, l = 1, \ldots, m$, where $\alpha_{kl,j}$ is a typical element of A^j. (See Section C.3 regarding the concept of absolute summability.)

A.9.2 Decomposition of Symmetric Matrices

If A is a symmetric $(m \times m)$ matrix, then there exists an orthogonal matrix P such that

$$P'AP = \Lambda = \begin{bmatrix} \lambda_1 & & 0 \\ & \ddots & \\ 0 & & \lambda_m \end{bmatrix} \quad \text{or} \quad A = P\Lambda P', \tag{A.9.2}$$

where the λ_i's are the eigenvalues of A and the columns of P are the corresponding eigenvectors. Here all matrices are real again because the eigenvalues of a symmetric matrix are real numbers. Denoting the i-th column of P by p_i and using that $p_i' p_j = 0$ for $i \neq j$, we get

$$A = P\Lambda P' = \sum_{i=1}^{m} \lambda_i p_i p_i'. \tag{A.9.3}$$

Moreover,

$$A^2 = P\Lambda P' P\Lambda P' = P\Lambda^2 P'$$

and, more generally,

$$A^k = P\Lambda^k P'.$$

If A is a positive definite symmetric $(m \times m)$ matrix, then all eigenvalues are positive so that the notation

$$\Lambda^{1/2} := \begin{bmatrix} \sqrt{\lambda_1} & & 0 \\ & \ddots & \\ 0 & & \sqrt{\lambda_m} \end{bmatrix}$$

makes sense. Defining $Q = P\Lambda^{1/2} P'$, we get $QQ = A$. In generalization of the terminology for positive real numbers, Q may be called a *square root* of A and may be denoted by $A^{1/2}$.

A.9.3 The Choleski Decomposition of a Positive Definite Matrix

If A is a positive definite $(m \times m)$ matrix, then there exists a lower (upper) triangular matrix P with positive main diagonal such that

$$P^{-1} A P'^{-1} = I_m \quad \text{or} \quad A = PP'. \tag{A.9.4}$$

Similarly, if A is positive semidefinite with $\text{rk}(A) = n < m$, then there exists a nonsingular matrix P such that

$$P^{-1}AP'^{-1} = \begin{bmatrix} I_n & 0 \\ 0 & 0 \end{bmatrix}. \tag{A.9.5}$$

Alternatively, $A = QQ'$, where

$$Q = P \begin{bmatrix} I_n & 0 \\ 0 & 0 \end{bmatrix}.$$

For instance,

$$\begin{bmatrix} 26 & 3 & 0 \\ 3 & 9 & 0 \\ 0 & 0 & 81 \end{bmatrix} = \begin{bmatrix} 5 & 1 & 0 \\ 0 & 3 & 0 \\ 0 & 0 & 9 \end{bmatrix} \begin{bmatrix} 5 & 0 & 0 \\ 1 & 3 & 0 \\ 0 & 0 & 9 \end{bmatrix}$$

$$= \begin{bmatrix} \sqrt{26} & 0 & 0 \\ 3/\sqrt{26} & 15/\sqrt{26} & 0 \\ 0 & 0 & 9 \end{bmatrix} \begin{bmatrix} \sqrt{26} & 3/\sqrt{26} & 0 \\ 0 & 15/\sqrt{26} & 0 \\ 0 & 0 & 9 \end{bmatrix}.$$

The decomposition $A = PP'$, where P is lower triangular with positive main diagonal, is sometimes called *Choleski decomposition*. Computer programs are available to determine the matrix P for a given positive definite matrix A. If a lower triangular matrix P is supplied by the program, an upper triangular matrix Q can be obtained as follows: Define an $(m \times m)$ matrix

$$G = \begin{bmatrix} 0 & \cdots & 0 & 1 \\ 0 & \cdots & 1 & 0 \\ \vdots & \ddots & & \vdots \\ 1 & & 0 & 0 \end{bmatrix}$$

with ones on the diagonal from the upper right-hand corner to the lower left-hand corner and zeros elsewhere. Note that $G' = G$ and $G^{-1} = G$. Suppose a decomposition of the $(m \times m)$ matrix A is desired. Then decompose $B = GAG$ as $B = PP'$, where P is lower triangular. Hence,

$$A = GBG = GPGGP'G = QQ',$$

where $Q = GPG$ is upper triangular.

A.10 Partitioned Matrices

Let the $(m \times n)$ matrix A be *partitioned* into *submatrices* $A_{11}, A_{12}, A_{21}, A_{22}$ with dimensions $(p \times q), (p \times (n-q)), ((m-p) \times q)$, and $((m-p) \times (n-q))$, respectively, so that

$$A = \begin{bmatrix} A_{11} & A_{12} \\ A_{21} & A_{22} \end{bmatrix}. \tag{A.10.1}$$

For such a partitioned matrix, a number of useful results hold.

Rules:

(1) $A' = \begin{bmatrix} A'_{11} & A'_{21} \\ A'_{12} & A'_{22} \end{bmatrix}$.

(2) If $n = m$ and $q = p$ and $A, A_{11},$ and A_{22} are nonsingular, then

$$A^{-1} = \begin{bmatrix} D & -DA_{12}A_{22}^{-1} \\ -A_{22}^{-1}A_{21}D & A_{22}^{-1} + A_{22}^{-1}A_{21}DA_{12}A_{22}^{-1} \end{bmatrix}$$
$$= \begin{bmatrix} A_{11}^{-1} + A_{11}^{-1}A_{12}GA_{21}A_{11}^{-1} & -A_{11}^{-1}A_{12}G \\ -GA_{21}A_{11}^{-1} & G \end{bmatrix},$$

where $D := (A_{11} - A_{12}A_{22}^{-1}A_{21})^{-1}$ and $G := (A_{22} - A_{21}A_{11}^{-1}A_{12})^{-1}$.

(3) Under the conditions of (2),

$$(A_{11} - A_{12}A_{22}^{-1}A_{21})^{-1} = A_{11}^{-1} + A_{11}^{-1}A_{12}(A_{22} - A_{21}A_{11}^{-1}A_{12})^{-1}A_{21}A_{11}^{-1}.$$

(4) Under the conditions of (2), if A_{12} and A_{21} are null matrices,

$$A^{-1} = \begin{bmatrix} A_{11}^{-1} & 0 \\ 0 & A_{22}^{-1} \end{bmatrix}.$$

(5) If A is a square matrix ($n = m$) and A_{11} is square and nonsingular, then
$|A| = |A_{11}| \cdot |A_{22} - A_{21}A_{11}^{-1}A_{12}|$.

(6) If A is a square matrix and A_{22} is square and nonsingular, then $|A| = |A_{22}| \cdot |A_{11} - A_{12}A_{22}^{-1}A_{21}|$.

A.11 The Kronecker Product

Let $A = (a_{ij})$ and $B = (b_{ij})$ be $(m \times n)$ and $(p \times q)$ matrices, respectively. The $(mp \times nq)$ matrix

$$A \otimes B := \begin{bmatrix} a_{11}B & \cdots & a_{1n}B \\ \vdots & & \vdots \\ a_{m1}B & \cdots & a_{mn}B \end{bmatrix} \qquad (A.11.1)$$

is the *Kronecker product* or *direct product* of A and B. For example, the Kronecker product of

$$A = \begin{bmatrix} 3 & 4 & -1 \\ 2 & 0 & 0 \end{bmatrix} \quad \text{and} \quad B = \begin{bmatrix} 5 & -1 \\ 3 & 3 \end{bmatrix} \qquad (A.11.2)$$

is

$$A \otimes B = \begin{bmatrix} 15 & -3 & 20 & -4 & -5 & 1 \\ 9 & 9 & 12 & 12 & -3 & -3 \\ 10 & -2 & 0 & 0 & 0 & 0 \\ 6 & 6 & 0 & 0 & 0 & 0 \end{bmatrix}$$

and

$$B \otimes A = \begin{bmatrix} 15 & 20 & -5 & -3 & -4 & 1 \\ 10 & 0 & 0 & -2 & 0 & 0 \\ 9 & 12 & -3 & 9 & 12 & -3 \\ 6 & 0 & 0 & 6 & 0 & 0 \end{bmatrix}.$$

Rules: In the following rules, suitable dimensions are assumed.

(1) $A \otimes B \neq B \otimes A$ in general.
(2) $(A \otimes B)' = A' \otimes B'$.
(3) $A \otimes (B + C) = A \otimes B + A \otimes C$.
(4) $(A \otimes B)(C \otimes D) = AC \otimes BD$.
(5) If A and B are invertible, then $(A \otimes B)^{-1} = A^{-1} \otimes B^{-1}$.
(6) If A and B are square matrices with eigenvalues λ_A, λ_B, respectively, and associated eigenvectors v_A, v_B, then $\lambda_A \lambda_B$ is an eigenvalue of $A \otimes B$ with eigenvector $v_A \otimes v_B$.
(7) If A and B are $(m \times m)$ and $(n \times n)$ square matrices, respectively, then $|A \otimes B| = |A|^n |B|^m$.
(8) If A and B are square matrices,

$$\operatorname{tr}(A \otimes B) = \operatorname{tr}(A)\operatorname{tr}(B).$$

(9) $(A \otimes B)^+ = A^+ \otimes B^+$.

A.12 The vec and vech Operators and Related Matrices

A.12.1 The Operators

Let $A = (a_1, \ldots, a_n)$ be an $(m \times n)$ matrix with $(m \times 1)$ columns a_i. The *vec operator* transforms A into an $(mn \times 1)$ vector by stacking the columns, that is,

$$\operatorname{vec}(A) = \begin{bmatrix} a_1 \\ \vdots \\ a_n \end{bmatrix}.$$

For instance, if A and B are as in (A.11.2), then

$$\operatorname{vec}(A) = \begin{bmatrix} 3 \\ 2 \\ 4 \\ 0 \\ -1 \\ 0 \end{bmatrix} \quad \text{and} \quad \operatorname{vec}(B) = \begin{bmatrix} 5 \\ 3 \\ -1 \\ 3 \end{bmatrix}.$$

Rules: Let A, B, C be matrices with appropriate dimensions.

(1) $\text{vec}(A+B) = \text{vec}(A) + \text{vec}(B)$.
(2) $\text{vec}(ABC) = (C' \otimes A)\text{vec}(B)$.
(3) $\text{vec}(AB) = (I \otimes A)\text{vec}(B) = (B' \otimes I)\text{vec}(A)$.
(4) $\text{vec}(ABC) = (I \otimes AB)\text{vec}(C) = (C'B' \otimes I)\text{vec}(A)$.
(5) $\text{vec}(B')'\text{vec}(A) = \text{tr}(BA) = \text{tr}(AB) = \text{vec}(A')'\text{vec}(B)$.
(6) $\text{tr}(ABC) = \text{vec}(A')'(C' \otimes I)\text{vec}(B)$
$\phantom{\text{tr}(ABC)} = \text{vec}(A')'(I \otimes B)\text{vec}(C)$
$\phantom{\text{tr}(ABC)} = \text{vec}(B')'(A' \otimes I)\text{vec}(C)$
$\phantom{\text{tr}(ABC)} = \text{vec}(B')'(I \otimes C)\text{vec}(A)$
$\phantom{\text{tr}(ABC)} = \text{vec}(C')'(B' \otimes I)\text{vec}(A)$
$\phantom{\text{tr}(ABC)} = \text{vec}(C')'(I \otimes A)\text{vec}(B)$.

The vech operator is closely related to vec. It only stacks the elements on and below the main diagonal of a square matrix. For instance,

$$\text{vech} \begin{bmatrix} \alpha_{11} & \alpha_{12} & \alpha_{13} \\ \alpha_{21} & \alpha_{22} & \alpha_{23} \\ \alpha_{31} & \alpha_{32} & \alpha_{33} \end{bmatrix} = \begin{bmatrix} \alpha_{11} \\ \alpha_{21} \\ \alpha_{31} \\ \alpha_{22} \\ \alpha_{32} \\ \alpha_{33} \end{bmatrix}.$$

In general, if A is an $(m \times m)$ matrix, vech(A) is an $m(m+1)/2$-dimensional vector. The vech operator is usually applied to symmetric matrices to collect the separate elements only.

A.12.2 Elimination, Duplication, and Commutation Matrices

The vec and vech operators are related by the *elimination matrix*, $\mathbf{L}_m$, and the *duplication matrix*, $\mathbf{D}_m$. The former is an $(\frac{1}{2}m(m+1) \times m^2)$ matrix such that, for an $(m \times m)$ square matrix A,

$$\text{vech}(A) = \mathbf{L}_m \text{vec}(A). \tag{A.12.1}$$

Thus, e.g., for $m = 3$,

$$\mathbf{L}_3 = \begin{bmatrix} 1 & 0 & 0 & 0 & 0 & 0 & 0 & 0 & 0 \\ 0 & 1 & 0 & 0 & 0 & 0 & 0 & 0 & 0 \\ 0 & 0 & 1 & 0 & 0 & 0 & 0 & 0 & 0 \\ 0 & 0 & 0 & 0 & 1 & 0 & 0 & 0 & 0 \\ 0 & 0 & 0 & 0 & 0 & 1 & 0 & 0 & 0 \\ 0 & 0 & 0 & 0 & 0 & 0 & 0 & 0 & 1 \end{bmatrix}.$$

The duplication matrix $\mathbf{D}_m$ is $(m^2 \times \frac{1}{2}m(m+1))$ and is defined so that, for any symmetric $(m \times m)$ matrix A,

$$\text{vec}(A) = \mathbf{D}_m \text{vech}(A). \tag{A.12.2}$$

For instance, for $m = 3$,

$$\mathbf{D}_3 = \begin{bmatrix} 1 & 0 & 0 & 0 & 0 & 0 \\ 0 & 1 & 0 & 0 & 0 & 0 \\ 0 & 0 & 1 & 0 & 0 & 0 \\ 0 & 1 & 0 & 0 & 0 & 0 \\ 0 & 0 & 0 & 1 & 0 & 0 \\ 0 & 0 & 0 & 0 & 1 & 0 \\ 0 & 0 & 1 & 0 & 0 & 0 \\ 0 & 0 & 0 & 0 & 1 & 0 \\ 0 & 0 & 0 & 0 & 0 & 1 \end{bmatrix}.$$

Because the rank of $\mathbf{D}_m$ is easily seen to be $m(m+1)/2$, the matrix $\mathbf{D}'_m \mathbf{D}_m$ is invertible. Thus, left-multiplication of (A.12.2) by $(\mathbf{D}'_m \mathbf{D}_m)^{-1} \mathbf{D}'_m$ gives

$$(\mathbf{D}'_m \mathbf{D}_m)^{-1} \mathbf{D}'_m \text{vec}(A) = \text{vech}(A). \tag{A.12.3}$$

Note, however, that $(\mathbf{D}'_m \mathbf{D}_m)^{-1} \mathbf{D}'_m \neq \mathbf{L}_m$ in general because (A.12.3) holds for symmetric matrices A only while (A.12.1) holds for arbitrary square matrices A.

The *commutation matrix*, $\mathbf{K}_{mn}$, is another matrix that is occasionally useful in dealing with the vec operator. $\mathbf{K}_{mn}$ is an $(mn \times mn)$ matrix defined such that, for any $(m \times n)$ matrix A,

$$\text{vec}(A') = \mathbf{K}_{mn} \text{vec}(A)$$

or, equivalently,

$$\text{vec}(A) = \mathbf{K}_{nm} \text{vec}(A').$$

For example,

$$\mathbf{K}_{32} = \begin{bmatrix} 1 & 0 & 0 & 0 & 0 & 0 \\ 0 & 0 & 0 & 1 & 0 & 0 \\ 0 & 1 & 0 & 0 & 0 & 0 \\ 0 & 0 & 0 & 0 & 1 & 0 \\ 0 & 0 & 1 & 0 & 0 & 0 \\ 0 & 0 & 0 & 0 & 0 & 1 \end{bmatrix}$$

because for

$$A = \begin{bmatrix} \alpha_{11} & \alpha_{12} \\ \alpha_{21} & \alpha_{22} \\ \alpha_{31} & \alpha_{32} \end{bmatrix},$$

$$\text{vec}(A') = \begin{bmatrix} \alpha_{11} \\ \alpha_{12} \\ \alpha_{21} \\ \alpha_{22} \\ \alpha_{31} \\ \alpha_{32} \end{bmatrix} = \mathbf{K}_{32} \begin{bmatrix} \alpha_{11} \\ \alpha_{21} \\ \alpha_{31} \\ \alpha_{12} \\ \alpha_{22} \\ \alpha_{32} \end{bmatrix} = \mathbf{K}_{32} \text{vec}(A).$$

Rules:

(7) $\mathbf{L}_m \mathbf{D}_m = I_{m(m+1)/2}$.
(8) $\mathbf{K}_{mm} \mathbf{D}_m = \mathbf{D}_m$.
(9) $\mathbf{K}_{m1} = \mathbf{K}_{1m} = I_m$.
(10) $\mathbf{K}'_{mn} = \mathbf{K}_{mn}^{-1} = \mathbf{K}_{nm}$.
(11) $\operatorname{tr} \mathbf{K}_{mm} = m$.
(12) $\det(\mathbf{K}_{mn}) = (-1)^{mn(m-1)(n-1)/4}$.
(13) $\operatorname{tr}(\mathbf{D}'_m \mathbf{D}_m) = m^2$, $\operatorname{tr}(\mathbf{D}'_m \mathbf{D}_m)^{-1} = m(m+3)/4$.
(14) $\det(\mathbf{D}'_m \mathbf{D}_m) = 2^{m(m-1)/2}$.
(15) $\operatorname{tr}(\mathbf{D}_m \mathbf{D}'_m) = m^2$.
(16) $|\mathbf{D}'_m (A \otimes A) \mathbf{D}_m| = 2^{m(m-1)/2} |A|^{m+1}$, where A is an $(m \times m)$ matrix.
(17) $(\mathbf{D}'_m (A \otimes A) \mathbf{D}_m)^{-1} = (\mathbf{D}'_m \mathbf{D}_m)^{-1} \mathbf{D}'_m (A^{-1} \otimes A^{-1}) \mathbf{D}_m (\mathbf{D}'_m \mathbf{D}_m)^{-1}$, if A is a nonsingular $(m \times m)$ matrix.
(18) $\mathbf{L}_m \mathbf{L}'_m = I_{m(m+1)/2}$.
(19) $\mathbf{L}_m \mathbf{L}'_m$ and $\mathbf{L}_m \mathbf{K}_{mm} \mathbf{L}'_m$ are idempotent.

Let A and B be lower triangular $(m \times m)$ matrices. Then we have the following rules:

(20) $\mathbf{L}_m (A \otimes B) \mathbf{L}'_m$ is lower triangular.
(21) $\mathbf{L}'_m \mathbf{L}_m (A' \otimes B) \mathbf{L}'_m = (A' \otimes B) \mathbf{L}'_m$.
(22) $[\mathbf{L}_m (A' \otimes B) \mathbf{L}'_m]^s = \mathbf{L}_m ((A')^s \otimes B^s) \mathbf{L}'_m$ for $s = 0, 1, \ldots$ and for $s = \ldots, -2, -1$, if A^{-1} and B^{-1} exist.

Let G be $(m \times n)$, F $(p \times q)$, and $\mathbf{b}$ $(p \times 1)$. Then the following results hold:

(23) $\mathbf{K}_{pm}(G \otimes F) = (F \otimes G) \mathbf{K}_{qn}$.
(24) $\mathbf{K}_{pm}(G \otimes F) \mathbf{K}_{nq} = F \otimes G$.
(25) $\mathbf{K}_{pm}(G \otimes \mathbf{b}) = \mathbf{b} \otimes G$.
(26) $\mathbf{K}_{pm}(\mathbf{b} \otimes G) = G \otimes \mathbf{b}$.
(27) $\operatorname{vec}(G \otimes F) = (I_n \otimes \mathbf{K}_{qm} \otimes I_p)(\operatorname{vec}(G) \otimes \operatorname{vec}(F))$.
(28) $(\mathbf{D}'_m \mathbf{D}_m)^{-1} \mathbf{D}'_m \mathbf{K}_{mm} = (\mathbf{D}'_m \mathbf{D}_m)^{-1} \mathbf{D}'_m$.

A.13 Vector and Matrix Differentiation

In the following, it will be assumed that all derivatives exist and are continuous. Let $f(\beta)$ be a scalar function that depends on the $(n \times 1)$ vector $\beta = (\beta_1, \ldots, \beta_n)'$.

$$\frac{\partial f}{\partial \beta} := \begin{bmatrix} \frac{\partial f}{\partial \beta_1} \\ \vdots \\ \frac{\partial f}{\partial \beta_n} \end{bmatrix}, \quad \frac{\partial f}{\partial \beta'} := \begin{bmatrix} \frac{\partial f}{\partial \beta_1}, \ldots, \frac{\partial f}{\partial \beta_n} \end{bmatrix}$$

are $(n \times 1)$ and $(1 \times n)$ vectors of first order partial derivatives, respectively, and

$$\frac{\partial^2 f}{\partial \beta \partial \beta'} := \left[\frac{\partial^2 f}{\partial \beta_i \partial \beta_j}\right] = \begin{bmatrix} \frac{\partial^2 f}{\partial \beta_1 \partial \beta_1} & \cdots & \frac{\partial^2 f}{\partial \beta_1 \partial \beta_n} \\ \vdots & & \vdots \\ \frac{\partial^2 f}{\partial \beta_n \partial \beta_1} & \cdots & \frac{\partial^2 f}{\partial \beta_n \partial \beta_n} \end{bmatrix}$$

is the $(n \times n)$ *Hessian matrix* of second order partial derivatives. If $f(A)$ is a scalar function of an $(m \times n)$ matrix $A = (a_{ij})$, then

$$\frac{\partial f}{\partial A} := \left[\frac{\partial f}{\partial a_{ij}}\right]$$

is an $(m \times n)$ matrix of partial derivatives. If the $(m \times n)$ matrix $A = (a_{ij})$ depends on the scalar β, then

$$\frac{\partial A}{\partial \beta} := \left[\frac{\partial a_{ij}}{\partial \beta}\right]$$

is an $(m \times n)$ matrix. If $y(\beta) = (y_1(\beta), \ldots, y_m(\beta))'$ is an $(m \times 1)$ vector that depends on the $(n \times 1)$ vector β, then

$$\frac{\partial y}{\partial \beta'} := \begin{bmatrix} \frac{\partial y_1}{\partial \beta_1} & \cdots & \frac{\partial y_1}{\partial \beta_n} \\ \vdots & & \vdots \\ \frac{\partial y_m}{\partial \beta_1} & \cdots & \frac{\partial y_m}{\partial \beta_n} \end{bmatrix}$$

is an $(m \times n)$ matrix and

$$\frac{\partial y'}{\partial \beta} := \left(\frac{\partial y}{\partial \beta'}\right)'.$$

For example, if $\beta = (\beta_1, \beta_2)'$ and $f(\beta) = \beta_1^2 - 2\beta_1\beta_2$, then

$$\frac{\partial f}{\partial \beta} = \begin{bmatrix} \frac{\partial f}{\partial \beta_1} \\ \frac{\partial f}{\partial \beta_2} \end{bmatrix} = \begin{bmatrix} 2\beta_1 - 2\beta_2 \\ -2\beta_1 \end{bmatrix}.$$

If

$$y(\beta) = \begin{bmatrix} \beta_1^3 + \beta_2 \\ e^{\beta_1} \end{bmatrix}, \quad \text{then} \quad \frac{\partial y}{\partial \beta'} = \begin{bmatrix} 3\beta_1^2 & 1 \\ e^{\beta_1} & 0 \end{bmatrix}.$$

The following two propositions are useful for deriving rules for vector and matrix differentiation.

Proposition A.1 (*Chain Rule for Vector Differentiation*)
Let α and β be $(m \times 1)$ and $(n \times 1)$ vectors, respectively, and suppose $h(\alpha)$ is $(p \times 1)$ and $g(\beta)$ is $(m \times 1)$. Then, with $\alpha = g(\beta)$,

$$\frac{\partial h(g(\beta))}{\partial \beta'} = \frac{\partial h(\alpha)}{\partial \alpha'} \frac{\partial g(\beta)}{\partial \beta'} \qquad (p \times n).$$

∎

Proposition A.2 (*Product Rules for Vector Differentiation*)
(a) Suppose β is $(m \times 1)$, $a(\beta) = (a_1(\beta), \ldots, a_n(\beta))'$ is $(n \times 1)$, $c(\beta) = (c_1(\beta), \ldots, c_p(\beta))'$ is $(p \times 1)$ and $A = (a_{ij})$ is $(n \times p)$ and does not depend on β. Then

$$\frac{\partial [a(\beta)'Ac(\beta)]}{\partial \beta'} = c(\beta)'A'\frac{\partial a(\beta)}{\partial \beta'} + a(\beta)'A\frac{\partial c(\beta)}{\partial \beta'}.$$

(b) If β is a (1×1) scalar, $A(\beta)$ is $(m \times n)$ and $B(\beta)$ is $(n \times p)$, then

$$\frac{\partial AB}{\partial \beta} = \frac{\partial A}{\partial \beta}B + A\frac{\partial B}{\partial \beta}.$$

(c) If β is an $(m \times 1)$ vector, $A(\beta)$ is $(n \times p)$ and $B(\beta)$ is $(p \times q)$, then

$$\frac{\partial \operatorname{vec}(AB)}{\partial \beta'} = (I_q \otimes A)\frac{\partial \operatorname{vec}(B)}{\partial \beta'} + (B' \otimes I_n)\frac{\partial \operatorname{vec}(A)}{\partial \beta'}.$$

∎

Proof:

(a)

$$\begin{aligned}\frac{\partial (a'Ac)}{\partial \beta'} &= \frac{\partial \left(\sum_{i,j} a_i a_{ij} c_j\right)}{\partial \beta'} \\ &= \sum_{i,j}\left[\frac{\partial a_i}{\partial \beta'}a_{ij}c_j + a_i a_{ij}\frac{\partial c_j}{\partial \beta'}\right] \\ &= c'A'\frac{\partial a}{\partial \beta'} + a'A\frac{\partial c}{\partial \beta'}.\end{aligned}$$

(b)

$$AB = \left[\sum_j a_{ij}b_{jk}\right] \text{ and }$$

$$\frac{\partial \left(\sum_j a_{ij} b_{jk}\right)}{\partial \beta} = \sum_j \left[\frac{\partial a_{ij}}{\partial \beta} b_{jk} + a_{ij} \frac{\partial b_{jk}}{\partial \beta}\right].$$

(c) Follows from (b) by stacking the columns of AB and writing the resulting columns $\partial \text{vec}(AB)/\partial \beta_i$ for $i = 1, \ldots, m$ in one matrix. ∎

The following rules are now easy to verify.

Rules:

(1) Let A be an $(m \times n)$ matrix and β be an $(n \times 1)$ vector. Then

$$\frac{\partial A\beta}{\partial \beta'} = A \quad \text{and} \quad \frac{\partial \beta' A'}{\partial \beta} = A'.$$

Proof: This result is a special case of Proposition A.2(a). ∎

(2) Let A be $(m \times m)$ and β be $(m \times 1)$. Then

$$\frac{\partial \beta' A \beta}{\partial \beta} = (A + A')\beta \quad \text{and} \quad \frac{\partial \beta' A \beta}{\partial \beta'} = \beta'(A' + A).$$

Proof: See Proposition A.2(a). ∎

(3) If A is $(m \times m)$ and β is $(m \times 1)$, then

$$\frac{\partial^2 \beta' A \beta}{\partial \beta \partial \beta'} = A + A'.$$

Proof: Follows from (1) and (2). ∎

(4) If A is a symmetric $(m \times m)$ matrix and β an $(m \times 1)$ vector then

$$\frac{\partial^2 \beta' A \beta}{\partial \beta \partial \beta'} = 2A.$$

Proof: See (3). ∎

(5) Let Ω be a symmetric $(n \times n)$ matrix and $c(\beta)$ an $(n \times 1)$ vector that depends on the $(m \times 1)$ vector β. Then

$$\frac{\partial c(\beta)' \Omega c(\beta)}{\partial \beta'} = 2c(\beta)' \Omega \frac{\partial c(\beta)}{\partial \beta'}$$

and

$$\frac{\partial^2 c(\beta)' \Omega c(\beta)}{\partial \beta \partial \beta'} = 2\left[\frac{\partial c(\beta)'}{\partial \beta} \Omega \frac{\partial c(\beta)}{\partial \beta'} + [c(\beta)' \Omega \otimes I_m] \frac{\partial \, \text{vec}(\partial c(\beta)'/\partial \beta)}{\partial \beta'}\right].$$

In particular, if y is an $(n \times 1)$ vector and X an $(n \times m)$ matrix,

$$\frac{\partial (y - X\beta)'\Omega(y - X\beta)}{\partial \beta'} = -2(y - X\beta)'\Omega X$$

and

$$\frac{\partial^2 (y - X\beta)'\Omega(y - X\beta)}{\partial \beta \partial \beta'} = 2X'\Omega X.$$

Proof: Follows from Proposition A.2(a). ∎

(6) Suppose β is $(m \times 1)$, $B(\beta)$ is $(n \times p)$, A is $(k \times n)$, and C is $(p \times q)$ and the latter two matrices do not depend on β. Then

$$\frac{\partial \operatorname{vec}(ABC)}{\partial \beta'} = (C' \otimes A)\frac{\partial \operatorname{vec}(B)}{\partial \beta'}.$$

Proof: Follows from Rule (2), Section A.12, and Proposition A.1. ∎

(7) Suppose β is $(m \times 1)$, $A(\beta)$ is $(n \times p)$, $D(\beta)$ is $(q \times r)$, and C is $(p \times q)$ and does not depend on β. Then

$$\frac{\partial \operatorname{vec}(ACD)}{\partial \beta'} = (I_r \otimes AC)\frac{\partial \operatorname{vec}(D)}{\partial \beta'} + (D'C' \otimes I_n)\frac{\partial \operatorname{vec}(A)}{\partial \beta'}.$$

Proof: Follows from Proposition A.2(c) by setting $B = CD$ and noting that $\partial \operatorname{vec}(CD)/\partial \beta' = (I_r \otimes C)\partial \operatorname{vec}(D)/\partial \beta'$. ∎

(8) If β is $(m \times 1)$ and $A(\beta)$ is $(n \times n)$, then, for any positive integer h,

$$\frac{\partial \operatorname{vec}(A^h)}{\partial \beta'} = \left[\sum_{i=0}^{h-1}(A')^{h-1-i} \otimes A^i\right]\frac{\partial \operatorname{vec}(A)}{\partial \beta'}.$$

Proof: Follows inductively from Proposition A.2(c). The result is evident for $h = 1$. Assuming it holds for $h - 1$ gives

$$\frac{\partial \operatorname{vec}(AA^{h-1})}{\partial \beta'} = (I_n \otimes A)\left[\sum_{i=0}^{h-2}(A')^{h-2-i} \otimes A^i\right]\frac{\partial \operatorname{vec}(A)}{\partial \beta'}$$
$$+ ((A')^{h-1} \otimes I_n)\frac{\partial \operatorname{vec}(A)}{\partial \beta'}.$$

∎

(9) If A is a nonsingular $(m \times m)$ matrix, then

$$\frac{\partial \operatorname{vec}(A^{-1})}{\partial \operatorname{vec}(A)'} = -(A^{-1})' \otimes A^{-1}.$$

Proof: Using Proposition A.2(c),

$$0 = \frac{\partial \operatorname{vec}(I_m)}{\partial \operatorname{vec}(A)'} = \frac{\partial \operatorname{vec}(A^{-1}A)}{\partial \operatorname{vec}(A)'} = (I_m \otimes A^{-1})\frac{\partial \operatorname{vec}(A)}{\partial \operatorname{vec}(A)'}$$
$$+ (A' \otimes I_m)\frac{\partial \operatorname{vec}(A^{-1})}{\partial \operatorname{vec}(A)'}.$$

■

(10) Let A be a symmetric positive definite $(m \times m)$ matrix and let P be a lower triangular $(m \times m)$ matrix with positive elements on the main diagonal such that $A = PP'$. Moreover, let $\mathbf{L}_m$ be an $(\frac{1}{2}m(m+1) \times m^2)$ elimination matrix such that $\mathbf{L}_m \operatorname{vec}(A) = \operatorname{vech}(A)$ consists of the elements on and below the main diagonal of A only. Then

$$\frac{\partial \operatorname{vech}(P)}{\partial \operatorname{vech}(A)'} = \{\mathbf{L}_m[(I_m \otimes P)\mathbf{K}_{mm} + (P \otimes I_m)]\mathbf{L}_m'\}^{-1}$$
$$= \{\mathbf{L}_m(I_{m^2} + \mathbf{K}_{mm})(P \otimes I_m)\mathbf{L}_m'\}^{-1},$$

where $\mathbf{K}_{mm}$ is an $(m^2 \times m^2)$ commutation matrix such that $\mathbf{K}_{mm}\operatorname{vec}(P) = \operatorname{vec}(P')$.

Proof: See Lütkepohl (1989a). ■

(11) If $A = (a_{ij})$ is an $(m \times m)$ matrix, then

$$\frac{\partial \operatorname{tr}(A)}{\partial A} = I_m.$$

Proof: $\operatorname{tr}(A) = a_{11} + \cdots + a_{mm}$. Hence,

$$\frac{\partial \operatorname{tr}(A)}{\partial a_{ij}} = \begin{cases} 0 & \text{if } i \neq j, \\ 1 & \text{if } i = j. \end{cases}$$

■

(12) If $A = (a_{ij})$ is $(m \times n)$ and $B = (b_{ij})$ is $(n \times m)$, then

$$\frac{\partial \operatorname{tr}(AB)}{\partial A} = B'.$$

Proof: Follows because $\operatorname{tr}(AB) = \sum_{j=1}^n a_{1j}b_{j1} + \cdots + \sum_{j=1}^n a_{mj}b_{jm}$. ■

(13) Suppose A is an $(m \times n)$ matrix and B, C are $(m \times m)$ and $(n \times m)$, respectively. Then

$$\frac{\partial \operatorname{tr}(BAC)}{\partial A} = B'C'.$$

Proof: Follows from Rule (12) because $\operatorname{tr}(BAC) = \operatorname{tr}(ACB)$. ■

(14) Let A, B, C, D be $(m \times n)$, $(n \times n)$, $(m \times n)$, and $(n \times m)$ matrices, respectively. Then

$$\frac{\partial \operatorname{tr}(DABA'C)}{\partial A} = CDAB + D'C'AB'.$$

670   A Vectors and Matrices

Proof: See Murata (1982, Appendix, Theorem 6a). ∎

(15) Let A, B, and C be $(m \times m)$ matrices and suppose A is nonsingular. Then
$$\frac{\partial \operatorname{tr}(BA^{-1}C)}{\partial A} = -(A^{-1}CBA^{-1})'.$$

Proof: By Rule (6) of Section A.12,
$$\operatorname{tr}(BA^{-1}C) = \operatorname{vec}(B')'(C' \otimes I_m) \operatorname{vec}(A^{-1}).$$

Hence, using (9),
$$\begin{aligned}\frac{\partial \operatorname{tr}(BA^{-1}C)}{\partial \operatorname{vec}(A)'} &= -\operatorname{vec}(B')'(C' \otimes I_m)((A^{-1})' \otimes A^{-1}) \\ &= -\left[(A^{-1}C \otimes A^{-1'}) \operatorname{vec}(B')\right]' \\ &= -\left[\operatorname{vec}(A^{-1'}B'C'A^{-1'})\right]'\end{aligned}$$

by Rule (2) of Section A.12. ∎

(16) Let $A = (a_{ij})$ be an $(m \times m)$ matrix. Then
$$\frac{\partial |A|}{\partial A} = (A^{adj})',$$
where A^{adj} is the adjoint of A.

Proof: Developing by the i-th row of A gives
$$|A| = a_{i1}A_{i1} + \cdots + a_{im}A_{im},$$
where A_{ij} is the cofactor of a_{ij}. Hence,
$$\frac{\partial |A|}{\partial a_{ij}} = A_{ij}$$
because A_{ij} does not contain a_{ij}. ∎

(17) If A is a nonsingular $(m \times m)$ matrix with $|A| > 0$, then
$$\frac{\partial \ln |A|}{\partial A} = (A')^{-1}.$$

Proof: Using Proposition A.1 (chain rule),
$$\frac{\partial \ln |A|}{\partial A} = \frac{\partial \ln |A|}{\partial |A|} \cdot \frac{\partial |A|}{\partial A} = \frac{1}{|A|}(A^{adj})' = (A')^{-1}.$$
∎

Proposition A.3 (*Taylor's Theorem*)
Let $f(\beta)$ be a scalar valued function of the $(m \times 1)$ vector β. Suppose $f(\beta)$ is at least twice continuously differentiable on an open set $\mathbb{S}$ that contains β_0, β, and the entire line segment between β_0 and β. Then there exists a point $\overline{\beta}$ on the line segment such that

$$f(\beta) = f(\beta_0) + \frac{\partial f(\beta_0)}{\partial \beta'}(\beta - \beta_0) + \frac{1}{2}(\beta - \beta_0)' \frac{\partial^2 f(\overline{\beta})}{\partial \beta \partial \beta'}(\beta - \beta_0), \quad (A.13.1)$$

where $\partial f(\beta_0)/\partial \beta' := (\partial f/\partial \beta'|_{\beta_0})$. ∎

The expansion of f given in (A.13.1) is a *second order Taylor expansion* at or around β_0.

A.14 Optimization of Vector Functions

Suppose $f(\beta)$ is a real valued (scalar) differentiable function of the $(m \times 1)$ vector β. A necessary condition for a local optimum (minimum or maximum) at $\widetilde{\beta}$ is that

$$\frac{\partial f}{\partial \beta} = 0 \quad \text{for } \beta = \widetilde{\beta}, \quad \text{that is,} \quad \frac{\partial f(\widetilde{\beta})}{\partial \beta} := \left[\frac{\partial f}{\partial \beta}\bigg|_{\widetilde{\beta}}\right] = 0.$$

In other words, $f(\cdot)$ has a *stationary point* at $\widetilde{\beta}$. If this condition is satisfied and the *Hessian matrix* of second order partial derivatives

$$\frac{\partial^2 f}{\partial \beta \partial \beta'}$$

is negative (positive) definite for $\beta = \widetilde{\beta}$, then $\widetilde{\beta}$ is a local maximum (minimum).

If a set of constraints is given in the form

$$\varphi(\beta) = (\varphi_1(\beta), \ldots, \varphi_n(\beta))' = 0,$$

that is, $\varphi(\beta)$ is an $(n \times 1)$ vector, then a local optimum, subject to these constraints, is obtained at a stationary point of the *Lagrange function*

$$\mathcal{L}(\beta, \lambda) = f(\beta) - \lambda' \varphi(\beta),$$

where λ is an $(n \times 1)$ vector of *Lagrange multipliers*. In other words, a necessary condition for a constrained local optimum is that

$$\frac{\partial \mathcal{L}}{\partial \beta} = 0 \quad \text{and} \quad \frac{\partial \mathcal{L}}{\partial \lambda} = 0$$

hold simultaneously.

The following results are useful in some optimization problems.

Proposition A.4 (*Maximum of* $\operatorname{tr}(B'\Omega B)$)
Let Ω be a positive semidefinite symmetric $(K \times K)$ matrix with eigenvalues $\lambda_1 \geq \lambda_2 \geq \cdots \geq \lambda_K$ and corresponding orthonormal $(K \times 1)$ eigenvectors $v_1, v_2, \ldots, v_K$. Moreover, let B be a $(K \times r)$ matrix with $B'B = I_r$. Then the maximum of $\operatorname{tr}(B'\Omega B)$ with respect to B is obtained for

$$B = \widehat{B} = [v_1, \ldots, v_r]$$

and

$$\max_B \operatorname{tr}(B'\Omega B) = \lambda_1 + \cdots + \lambda_r.$$

∎

Proof: The proposition follows from Theorem 6, p. 205, of Magnus & Neudecker (1988) by induction. For $r = 1$, our result is just a special case of that theorem. For $r > 1$, assuming that the proposition holds for $r-1$ and denoting the columns of B by $b_1, \ldots, b_r$,

$$\begin{aligned}
\operatorname{tr}(B'\Omega B) &= \operatorname{tr} \begin{bmatrix} b_1' \\ \vdots \\ b_r' \end{bmatrix} \Omega [b_1, \ldots, b_r] = \operatorname{tr} \begin{bmatrix} b_1'\Omega b_1 & & * \\ & \ddots & \\ * & & b_r'\Omega b_r \end{bmatrix} \\
&= b_1'\Omega b_1 + \cdots + b_{r-1}'\Omega b_{r-1} + b_r'\Omega b_r \\
&= \lambda_1 + \cdots + \lambda_{r-1} + b_r'\Omega b_r
\end{aligned}$$

and $\max b_r'\Omega b_r = v'\Omega v_r = \lambda_r$, under the conditions of the proposition, by the aforementioned theorem from Magnus & Neudecker (1988). ∎

The next proposition may be regarded as a corollary of Proposition A.4.

Proposition A.5 (*Minimum of* $\operatorname{tr}(Y - BCX)'\Sigma_u^{-1}(Y - BCX)$)
Let Y, X, Σ_u, B, and C be matrices of dimensions $(K \times T), (Kp \times T), (K \times K), (K \times r)$, and $(r \times Kp)$, respectively, with Σ_u positive definite, $\operatorname{rk}(B) = \operatorname{rk}(C) = r$, $\operatorname{rk}(X) = Kp$, and $\operatorname{rk}(Y) = K$. Then a minimum of

$$\operatorname{tr}[(Y - BCX)'\Sigma_u^{-1}(Y - BCX)] \tag{A.14.1}$$

with respect to B and C is obtained for

$$B = \widehat{B} = \Sigma_u^{1/2}\widehat{V} \quad \text{and} \quad C = \widehat{C} = \widehat{V}'\Sigma_u^{-1/2}YX'(XX')^{-1}, \tag{A.14.2}$$

where $\widehat{V} = [\widehat{v}_1, \ldots, \widehat{v}_r]$ is the $(K \times r)$ matrix of the orthonormal eigenvectors corresponding to the r largest eigenvalues of

$$\frac{1}{T}\Sigma_u^{-1/2}YX'(XX')^{-1}XY'\Sigma_u^{-1/2}$$

in nonincreasing order. ∎

Proof: We first assume $\Sigma_u = I_K$.

$$\begin{aligned}&\operatorname{tr}[(Y - BCX)'(Y - BCX)] \\ &= \operatorname{tr}[(Y - BCX)(Y - BCX)'] \\ &= [\operatorname{vec}(Y) - \operatorname{vec}(BCX)]'[\operatorname{vec}(Y) - \operatorname{vec}(BCX)] \\ &= [\operatorname{vec}(Y) - (X' \otimes B)\operatorname{vec}(C)]'[\operatorname{vec}(Y) - (X' \otimes B)\operatorname{vec}(C)].\end{aligned} \quad (A.14.3)$$

A derivation similar to that in Section 3.2.1 shows that this sum of squares is minimized with respect to $\operatorname{vec}(C)$ when this vector is chosen to be

$$\begin{aligned}\operatorname{vec}(\widehat{C}) &= [(X \otimes B')(X' \otimes B)]^{-1}(X \otimes B')\operatorname{vec}(Y) \\ &= (XX' \otimes B'B)^{-1}\operatorname{vec}(B'YX') \\ &= \operatorname{vec}[(B'B)^{-1}B'YX'(XX')^{-1}].\end{aligned}$$

Because we may normalize the columns of B, we choose $B'B = I_r$ without loss of generality. Hence,

$$\widehat{C} = B'YX'(XX')^{-1}. \quad (A.14.4)$$

Substituting for C in (A.14.3) gives

$$\begin{aligned}&\operatorname{tr}[(Y - BB'YX'(XX')^{-1}X)(Y - BB'YX'(XX')^{-1}X)'] \\ &= \operatorname{tr}(YY') - \operatorname{tr}(BB'YX'(XX')^{-1}XY') - \operatorname{tr}(YX'(XX')^{-1}XY'BB') \\ &\quad + \operatorname{tr}(BB'YX'(XX')^{-1}XX'(XX')^{-1}XY'BB') \\ &= \operatorname{tr}(YY') - \operatorname{tr}(B'YX'(XX')^{-1}XY'B),\end{aligned}$$

where again $B'B = I_r$ has been used. This expression is minimized with respect to B, where

$$\frac{1}{T}\operatorname{tr} B'YX'(XX')^{-1}XY'B$$

assumes its maximum. By Proposition A.4, the maximum is attained if B consists of the eigenvectors corresponding to the r largest eigenvalues of

$$\frac{1}{T}YX'(XX')^{-1}XY'$$

which proves the proposition for $\Sigma_u = I_K$.

If $\Sigma_u \neq I_K$,

$$\operatorname{tr}[(Y - BCX)'\Sigma_u^{-1}(Y - BCX)] = \operatorname{tr}[(Y^{\#} - B^{\#}CX)'(Y^{\#} - B^{\#}CX)]$$

has to be minimized with respect to $B^{\#}$ and C. Here $Y^{\#} = \Sigma_u^{-1/2}Y$ and $B^{\#} = \Sigma_u^{-1/2}B$. From the above derivation the solution is $\widehat{B}^{\#} = \widehat{V}$ and

$$\widehat{C} = \widehat{B}^{\#\prime}Y^{\#}X'(XX')^{-1} = \widehat{V}'\Sigma_u^{-1/2}YX'(XX')^{-1},$$

where the columns of $\widehat{V}$ are the eigenvectors corresponding to the r largest eigenvalues of

$$\frac{1}{T} Y^{\#} X'(XX')^{-1} XY^{\#\prime} = \frac{1}{T} \Sigma_u^{-1/2} Y X'(XX')^{-1} XY' \Sigma_u^{-1/2}.$$

Hence, $\widehat{B} = \Sigma_u^{1/2} \widehat{B}^{\#} = \Sigma_u^{1/2} \widehat{V}$. ∎

A result similar to that in Proposition A.4 also holds for the maximum and minimum of a determinant. The following proposition is a slight modification of Theorem 15 of Magnus & Neudecker (1988, Chapter 11).

Proposition A.6 (*Maximum and Minimum of* $|C\Omega C'|$)
Let Ω be a positive definite symmetric $(K \times K)$ matrix with eigenvalues $\lambda_1 \geq \lambda_2 \geq \cdots \geq \lambda_K$ and corresponding orthonormal $(K \times 1)$ eigenvectors $v_1, \ldots, v_K$. Furthermore, let C be an $(r \times K)$ matrix with $CC' = I_r$. Then

$$\max_C |C\Omega C'| = \lambda_1 \cdots \lambda_r$$

and the maximum is attained for

$$C = \widehat{C} = [v_1, \ldots, v_r]'.$$

Moreover,

$$\min_C |C\Omega C'| = \lambda_K \lambda_{K-1} \cdots \lambda_{K-r+1}$$

and the minimum is attained for

$$C = \widehat{C} = [v_K, \ldots, v_{K-r+1}]'.$$

∎

An important implication of this proposition is used in Chapter 7 and is stated next.

Proposition A.7 (*Minimum of* $|T^{-1}(Y - BCX)(Y - BCX)'|$)
Let Y and X be $(K \times T)$ matrices of rank K and let B and C be of rank r and dimensions $(K \times r)$ and $(r \times K)$, respectively. Furthermore, let $\lambda_1 \geq \cdots \geq \lambda_K$ be the eigenvalues of

$$(XX')^{-1/2} XY'(YY')YX'(XX')^{-1/2\prime}$$

and the corresponding orthonormal eigenvectors are $v_1, \ldots, v_K$. Here

$$(XX')^{-1/2}$$

is some matrix satisfying

$$(XX')^{-1/2}(XX')(XX')^{-1/2\prime} = I_K.$$

Then

$$\min_{B,C} |T^{-1}(Y - BCX)(Y - BCX)'| = |T^{-1}YY'|(1 - \lambda_1) \cdots (1 - \lambda_r)$$

and the minimum is attained for

$$C = \widehat{C} = [v_1, \ldots, v_r]'(XX')^{-1/2}$$

and

$$B = \widehat{B} = YX'\widehat{C}'(\widehat{C}XX'\widehat{C}')^{-1}.$$

∎

A proof of this proposition can be found in Tso (1981). It should be noted that the minimizing matrices $\widehat{B}$ and $\widehat{C}$ are not unique. Any nonsingular $(r \times r)$ matrix F leads to another set of minimizing matrices $F\widehat{C}, \widehat{B}F^{-1}$.

A.15 Problems

The following problems refer to the matrices

$$A = \begin{bmatrix} 5 & 2 \\ -1 & 1 \end{bmatrix}, \quad B = \begin{bmatrix} 6 & 0 & 0 \\ -6 & 1 & 0 \end{bmatrix}, \quad C = \begin{bmatrix} 1 & 4 & 0 \\ 2 & 2 & 2 \\ 1 & 2 & 0 \end{bmatrix},$$

$$D = \begin{bmatrix} 5 & 2 \\ 2 & 1 \end{bmatrix}, \quad H(\beta) = \begin{bmatrix} 4\beta_1 & 2\beta_1 + \beta_2 \\ 1 + \beta_2 & 3 \end{bmatrix}.$$

Problem A.1
Determine $A + D$, $A - 2D$, A', AB, BC, $B'A$, $B'A'$, $A \otimes D$, $B \otimes D$, $D \otimes B$, $B' \otimes D'$, $B + BC$, tr A, tr D, det A, $|D|$, $|C|$, vec(B), vec(B'), vech(C), $\mathbf{K}_{33}$, A^{-1}, D^{-1}, $(A \otimes D)^{-1}$, rk(C), rk(B), det($A \otimes D$), tr($A \otimes D$), C^{-1} (use the rules for the partitioned inverse).

Problem A.2
Determine the eigenvalues of A, D, and $A \otimes D$.

Problem A.3
Find an upper triangular matrix Q such that $D = QQ'$ and find an orthogonal matrix P such that $D = P\Lambda P'$, where Λ is a diagonal matrix with the eigenvalues of D on the main diagonal. Compute D^5.

Problem A.4
Is $F = I_2 - BB'$ idempotent? Is BB' positive definite?

Problem A.5
Determine the following derivatives.

$$\frac{\partial \det(H)}{\partial \beta}, \quad \frac{\partial^2 \det(H)}{\partial \beta \partial \beta'}, \quad \frac{\partial \operatorname{tr}(H)}{\partial \beta},$$

$$\frac{\partial \operatorname{vec}(H)}{\partial \beta'}, \quad \frac{\partial \operatorname{vec}(H^2)}{\partial \beta'}, \quad \frac{\partial H(\beta)\beta}{\partial \beta'},$$

where $\beta = (\beta_1, \beta_2)'$.

Problem A.6
Determine the stationary points of $|H|$ with respect to β. Are they local extrema?

Problem A.7
Give a second order Taylor expansion of $\det(H)$ around $\beta = (0,0)'$.

B
Multivariate Normal and Related Distributions

B.1 Multivariate Normal Distributions

A K-dimensional vector of continuous random variables $y = (y_1, \ldots, y_K)'$ has a *multivariate normal distribution* with mean vector $\mu = (\mu_1, \ldots, \mu_K)'$ and covariance matrix Σ, briefly

$$y \sim \mathcal{N}(\mu, \Sigma),$$

if its distribution has the probability density function (p.d.f.)

$$f(y) = \frac{1}{(2\pi)^{K/2}} |\Sigma|^{-1/2} \exp\left[-\frac{1}{2}(y-\mu)' \Sigma^{-1} (y-\mu)\right]. \tag{B.1.1}$$

Alternatively, $y \sim \mathcal{N}(\mu, \Sigma)$, if for any K-vector c for which $c'\Sigma c \neq 0$ the linear combination $c'y$ has a univariate normal distribution, that is, $c'y \sim \mathcal{N}(c'\mu, c'\Sigma c)$ (see Rao (1973, Chapter 8)). This definition of a multivariate normal distribution is useful because it carries over to the case where Σ is positive semidefinite and singular, while the multivariate density in (B.1.1) is only meaningful, if Σ is positive definite and, hence, nonsingular. It must be emphasized, however, that the two definitions are equivalent, if Σ is positive definite rather than just positive semidefinite. Another possibility to define a multivariate normal distribution with singular covariance matrix may be found in Anderson (1984).

The following results regarding the multivariate normal and related distributions are useful. Many of them are stated in Judge et al. (1985, Appendix A). Proofs can be found in Rao (1973, Chapter 8) and Hogg & Craig (1978, Chapter 12).

Proposition B.1 (*Marginal and Conditional Distributions of a Multivariate Normal*)
Let y_1 and y_2 be two random vectors such that

$$\begin{bmatrix} y_1 \\ y_2 \end{bmatrix} \sim \mathcal{N}\left(\begin{bmatrix} \mu_1 \\ \mu_2 \end{bmatrix}, \begin{bmatrix} \Sigma_{11} & \Sigma_{12} \\ \Sigma_{21} & \Sigma_{22} \end{bmatrix}\right),$$

where the partitioning of the mean vector and covariance matrix corresponds to that of the vector $(y_1', y_2')'$. Then,

$$y_1 \sim \mathcal{N}(\mu_1, \Sigma_{11})$$

and the conditional distribution of y_1 given $y_2 = c$ is also multivariate normal,

$$(y_1|y_2 = c) \sim \mathcal{N}(\mu_1 + \Sigma_{12}\Sigma_{22}^{-1}(c - \mu_2), \Sigma_{11} - \Sigma_{12}\Sigma_{22}^{-1}\Sigma_{21}).$$

If Σ_{22} is singular, the inverse can be replaced by a generalized inverse. Moreover, y_1 and y_2 are independent if and only if $\Sigma_{12} = \Sigma_{21}' = 0$. ∎

Proposition B.2 (*Linear Transformation of a Multivariate Normal Random Vector*)
Suppose $y \sim \mathcal{N}(\mu, \Sigma)$ is $(K \times 1)$, A is an $(M \times K)$ matrix and c an $(M \times 1)$ vector. Then

$$x = Ay + c \sim \mathcal{N}(A\mu + c, A\Sigma A').$$

∎

B.2 Related Distributions

Suppose $y \sim \mathcal{N}(0, I_K)$. The distribution of $z = y'y$ is a (central) *chi-square distribution* with K degrees of freedom,

$$z \sim \chi^2(K).$$

Proposition B.3 (*Distributions of Quadratic Forms*)
(1) Suppose $y \sim \mathcal{N}(0, I_K)$ and A is a symmetric idempotent $(K \times K)$ matrix with $\text{rk}(A) = n$. Then $y'Ay \sim \chi^2(n)$.
(2) If $y \sim \mathcal{N}(0, \Sigma)$, where Σ is a positive definite $(K \times K)$ matrix, then $y'\Sigma^{-1}y \sim \chi^2(K)$.
(3) Let $y \sim \mathcal{N}(0, QA)$, where Q is a symmetric idempotent $(K \times K)$ matrix with $\text{rk}(Q) = n$ and A is a positive definite $(K \times K)$ matrix. Then $y'A^{-1}y \sim \chi^2(n)$.
(4) Suppose $y \sim \mathcal{N}(0, \Sigma)$, where Σ is a nonsingular $(K \times K)$ covariance matrix. Furthermore, let A be a $(K \times K)$ matrix with $\text{rk}(A) = n$. Then

$$y'Ay \sim \chi^2(n) \quad \Rightarrow \quad A\Sigma A = A$$

and

$$A\Sigma A = A \quad \Rightarrow \quad y'Ay \sim \chi^2(n).$$

∎

B.2 Related Distributions

Proposition B.4 (*Independence of a Normal Vector and a Quadratic Form*)
Suppose $y \sim \mathcal{N}(\mu, \sigma^2 I_K)$, A is a symmetric, idempotent $(K \times K)$ matrix, B is an $(M \times K)$ matrix and $BA = 0$. Then By is stochastically independent of the random variable $y'Ay$. ■

Proposition B.5 (*Independence of Quadratic Forms*)
Suppose $y \sim \mathcal{N}(\mu, \sigma^2 I_K)$ and A and B are symmetric, idempotent $(K \times K)$ matrices with $AB = 0$, then $y'Ay$ and $y'By$ are stochastically independent. ■

If $z \sim \mathcal{N}(0,1)$ and $u \sim \chi^2(m)$ are stochastically independent, then

$$T = \frac{z}{\sqrt{u/m}}$$

has a *t-distribution* with m degrees of freedom, $T \sim t(m)$. If $u \sim \chi^2(m)$ and $v \sim \chi^2(n)$ are independent, then

$$\frac{u/m}{v/n} \sim F(m,n),$$

that is, the ratio of two independent χ^2 random variables, each divided by its degrees of freedom, has an *F-distribution* with m and n degrees of freedom. The numbers m and n indicate the numerator and denominator degrees of freedom, respectively.

Proposition B.6 (*Distributions of Ratios of Quadratic Forms*)

(1) Suppose $x \sim \mathcal{N}(0, I_m)$ and $y \sim \mathcal{N}(0, I_n)$ are independent. Then

$$\frac{x'x/m}{y'y/n} \sim F(m,n).$$

(2) If $y \sim \mathcal{N}(0, I_K)$ and A and B are symmetric, idempotent $(K \times K)$ matrices with $\operatorname{rk}(A) = m$, $\operatorname{rk}(B) = n$ and $AB = 0$, then

$$\frac{y'Ay/m}{y'By/n} \sim F(m,n).$$

(3) $z \sim F(m,n) \Rightarrow \dfrac{1}{z} \sim F(n,m)$.

■

If $y \sim \mathcal{N}(\mu, I_K)$, then $y'y$ has a *noncentral χ^2-distribution* with K degrees of freedom and *noncentrality parameter* (or simply *noncentrality*) $\tau = \mu'\mu$. Briefly,

$$y'y \sim \chi^2(K; \tau).$$

The noncentrality parameter is sometimes defined differently in the literature. For instance, $\lambda = \frac{1}{2}\mu'\mu$ is sometimes called noncentrality parameter. Let $w \sim \chi^2(m; \tau)$ and $v \sim \chi^2(n)$ be independent random variables, then

$$\frac{w/m}{v/n} \sim F(m, n; \tau),$$

that is, the ratio has a *noncentral F-distribution* with m and n degrees of freedom and noncentrality parameter τ.

Proposition B.7 (*Quadratic Form with Noncentral χ^2-Distribution*)
If $y \sim \mathcal{N}(\mu, \Sigma)$ with positive definite $(K \times K)$ covariance matrix Σ, then $y'\Sigma^{-1}y \sim \chi^2(K; \mu'\Sigma^{-1}\mu)$. ∎

C
Stochastic Convergence and Asymptotic Distributions

It is often difficult to derive the exact distributions of estimators and test statistics. In that case, their asymptotic or limiting properties, when the sample size gets large, are of interest. The limiting properties are then regarded as approximations to the properties for the sample size available. In order to study the limiting properties, some concepts of convergence of sequences of random variables and vectors are useful. They are discussed in Sections C.1 and C.2. Infinite sums of random variables are treated in Section C.3. Laws of large numbers and central limit theorems are given in Section C.4. Asymptotic properties of estimators are considered in Section C.5. Maximum likelihood estimators and their asymptotic properties are discussed in Section C.6 and some common testing principles are treated in Section C.7. Finally, asymptotic properties of nonstationary processes with unit roots are dealt with in Section C.8.

This appendix contains a brief summary of results used in the text. Many of these results can be found in Judge et al. (1985, Section 5.8). A more complete discussion and proofs are provided in Fuller (1976), Roussas (1973), Serfling (1980), Davidson (1994, 2000) and other more advanced books on statistics. Further references will be given in the following.

C.1 Concepts of Stochastic Convergence

Let $x_1, x_2, \ldots$ or $\{x_T\}$, $T = 1, 2, \ldots$, be a sequence of scalar random variables which are all defined on a common probability space $(\Omega, \mathcal{F}, \Pr)$. The sequence $\{x_T\}$ *converges in probability* to the random variable x (which is also defined on $(\Omega, \mathcal{F}, \Pr)$) if for every $\epsilon > 0$,

$$\lim_{T \to \infty} \Pr(|x_T - x| > \epsilon) = 0$$

or, equivalently,

$$\lim_{T \to \infty} \Pr(|x_T - x| < \epsilon) = 1.$$

This type of stochastic convergence is abbreviated as

$$\text{plim } x_T = x \quad \text{or} \quad x_T \xrightarrow{p} x.$$

The limit x may be a fixed, nonstochastic real number which is then regarded as a degenerate random variable that takes on one particular value with probability one.

The sequence $\{x_T\}$ *converges almost surely* (a.s.) or *with probability one* to the random variable x if for every $\epsilon > 0$,

$$\Pr\left(\lim_{T \to \infty} |x_T - x| < \epsilon\right) = 1.$$

This type of stochastic convergence is often written as $x_T \xrightarrow{a.s.} x$ and is sometimes called *strong convergence*.

The sequence $\{x_T\}$ *converges in quadratic mean* or *mean square error* to x, briefly $x_T \xrightarrow{q.m.} x$, if

$$\lim_{T \to \infty} E(x_T - x)^2 = 0.$$

This type of convergence requires that the mean and variance of the x_T's and x exist.

Finally, denoting the distribution functions of x_T and x by F_T and F, respectively, the sequence $\{x_T\}$ is said to *converge in distribution* or *weakly* or *in law* to x, if for all real numbers c for which F is continuous,

$$\lim_{T \to \infty} F_T(c) = F(c).$$

This type of convergence is abbreviated as $x_T \xrightarrow{d} x$. It must be emphasized that we do not require the convergence of the sequence of p.d.f.s of the x_T's to the p.d.f. of x. In fact, we do not even require that the distributions of the x_T's have p.d.f.s. Even if they do have p.d.f.s, convergence in distribution does not imply their convergence to the p.d.f. of x.

All these concepts of stochastic convergence can be extended to sequences of random vectors (multivariate random variables). Suppose $\{x_T = (x_{1T}, \ldots, x_{KT})'\}$, $T = 1, 2, \ldots$, is a sequence of K-dimensional random vectors and $x = (x_1, \ldots, x_K)'$ is a K-dimensional random vector. Then the following definitions are used:

$\text{plim } x_T = x \quad \text{or} \quad x_T \xrightarrow{p} x \quad \text{if} \quad \text{plim } x_{kT} = x_k \text{ for } k = 1, \ldots, K.$

$x_T \xrightarrow{a.s.} x \quad \text{if} \quad x_{kT} \xrightarrow{a.s.} x_k \text{ for } k = 1, \ldots, K.$

$x_T \xrightarrow{q.m.} x \quad \text{if} \quad \lim E[(x_T - x)'(x_T - x)] = 0.$

$x_T \xrightarrow{d} x \quad \text{if} \quad \lim F_T(c) = F(c) \text{ for all continuity points of } F.$

Here F_T and F are the joint distribution functions of x_T and x, respectively. Almost sure convergence and convergence in probability can be defined for matrices in the same way in terms of convergence of the individual elements. Convergence in quadratic mean and in distribution is easily extended to sequences of random matrices by vectorizing them. In the following proposition, the relationships between the different modes of convergence are given.

Proposition C.1 (*Convergence Properties of Sequences of Random Variables*)
Suppose $\{x_T\}$ is a sequence of K-dimensional random variables. Then the following relations hold:

(1) $x_T \xrightarrow{a.s.} x \Rightarrow x_T \xrightarrow{p} x \Rightarrow x_T \xrightarrow{d} x$.
(2) $x_T \xrightarrow{q.m.} x \Rightarrow x_T \xrightarrow{p} x \Rightarrow x_T \xrightarrow{d} x$.
(3) If x is a fixed, nonstochastic vector, then

$$x_T \xrightarrow{q.m.} x \quad \Leftrightarrow \quad [\lim E(x_T) = x \text{ and } \lim E\{(x_T - Ex_T)'(x_T - Ex_T)\} = 0].$$

(4) If x is a fixed, nonstochastic random vector, then

$$x_T \xrightarrow{p} x \quad \Leftrightarrow \quad x_T \xrightarrow{d} x.$$

(5) (Slutsky's Theorem) If $g : \mathbb{R}^K \to \mathbb{R}^m$ is a continuous function, then

$$x_T \xrightarrow{p} x \quad \Rightarrow \quad g(x_T) \xrightarrow{p} g(x) \quad [\text{plim } g(x_T) = g(\text{plim } x_T)],$$

$$x_T \xrightarrow{d} x \quad \Rightarrow \quad g(x_T) \xrightarrow{d} g(x),$$

and

$$x_T \xrightarrow{a.s.} x \quad \Rightarrow \quad g(x_T) \xrightarrow{a.s.} g(x).$$

∎

Proposition C.2 (*Properties of Convergence in Probability and in Distribution*)
Suppose $\{x_T\}$ and $\{y_T\}$ are sequences of $(K \times 1)$ random vectors, $\{A_T\}$ is a sequence of $(K \times K)$ random matrices, x is a $(K \times 1)$ random vector, c is a fixed $(K \times 1)$ vector, and A is a fixed $(K \times K)$ matrix.

(1) If plim x_T, plim y_T, and plim A_T exist, then
   (a) plim $(x_T \pm y_T)$ = plim x_T ± plim y_T;
   (b) plim $(c'x_T) = c'(\text{plim } x_T)$;
   (c) plim $x_T'y_T = (\text{plim } x_T)'(\text{plim } y_T)$;
   (d) plim $A_T x_T$ = plim (A_T)plim (x_T).
(2) If $x_T \xrightarrow{d} x$ and plim $(x_T - y_T) = 0$, then $y_T \xrightarrow{d} x$.
(3) If $x_T \xrightarrow{d} x$ and plim $y_T = c$, then

(a) $x_T \pm y_T \xrightarrow{d} x \pm c$;

(b) $y_T' x_T \xrightarrow{d} c'x$.

(4) If $x_T \xrightarrow{d} x$ and plim $A_T = A$, then $A_T x_T \xrightarrow{d} Ax$.

(5) If $x_T \xrightarrow{d} x$ and plim $A_T = 0$, then plim $A_T x_T = 0$.

∎

Proposition C.3 (*Limits of Sequences of t and F Random Variables*)

(1) $t(T) \xrightarrow{d}_{T \to \infty} \mathcal{N}(0,1)$

(that is, a sequence of random variables with t-distributions with T degrees of freedom converges to a standard normal distribution as the degrees of freedom go to infinity).

(2) $JF(J,T) \xrightarrow{d}_{T \to \infty} \chi^2(J)$.

∎

C.2 Order in Probability

Let $\{a_T\}$ be a sequence of real numbers and $\{b_T\}$ a sequence of positive real numbers. Then a_T is said to be *of smaller order than* b_T ($a_T = o(b_T)$) if $\lim_{T \to \infty} a_T/b_T = 0$ and a_T is said to be *at most of order* b_T ($a_T = O(b_T)$) if there exists a number c such that for all T, $|a_T|/b_T \leq c$.

Proposition C.4 (*Order of Convergence Results*)
For sequences of real numbers $\{a_T\}$, $\{b_T\}$ and sequences of positive real numbers $\{c_T\}$, $\{d_T\}$, the following results hold:

(1) $a_T = o(c_T), b_T = o(d_T) \Rightarrow a_T b_T = o(c_T d_T)$, $a_T + b_T = o(\max[c_T, d_T])$ and $|a_T|^s = o(c_T^s)$ for $s > 0$.

(2) $a_T = O(c_T), b_T = O(d_T) \Rightarrow a_T b_T = O(c_T d_T)$, $a_T + b_T = O(\max[c_T, d_T])$ and $|a_T|^s = O(c_T^s)$ for $s > 0$.

(3) $a_T = o(c_T), b_T = O(d_T) \Rightarrow a_T b_T = o(c_T d_T)$.

∎

Let $\{A_T = (a_{ij,T})\}$ be a sequence of random ($m \times n$) matrices and $\{b_T\}$ a sequence of positive real numbers. Then A_T is said to be *of smaller order in probability than* b_T ($A_T = o_p(b_T)$) if plim $_{T \to \infty} A_T/b_T = 0$ and A_T is said to be *at most of order in probability* b_T or *bounded in probability by* b_T ($A_T = O_p(b_T)$) if, for every $\epsilon > 0$, there exists a number c_ϵ such that for all T, $\Pr\{|a_{ij,T}| \geq c_\epsilon b_T\} \leq \epsilon$ for $i = 1, \ldots, m$, $j = 1, \ldots, n$. The following results hold for sequences of random matrices.

Proposition C.5 (*Order in Probability Results*)
For sequences of random matrices of suitable fixed dimensions $\{A_T\}$, $\{B_T\}$ and sequences of positive real numbers $\{c_T\}$, $\{d_T\}$ the following results hold:

(1) $A_T = o_p(c_T), B_T = o_p(d_T) \Rightarrow A_T B_T = o_p(c_T d_T)$ and $A_T + B_T = o_p(\max[c_T, d_T])$.
(2) $A_T = O_p(c_T), B_T = O_p(d_T) \Rightarrow A_T B_T = O_p(c_T d_T)$ and $A_T + B_T = O_p(\max[c_T, d_T])$.
(3) $A_T = o_p(c_T), B_T = O_p(d_T) \Rightarrow A_T B_T = o_p(c_T d_T)$.

∎

For the next result see, e.g., Fuller (1976, p. 192).

Proposition C.6 (*Taylor's Theorem for Functions of Random Vectors*)
Let $y_T = (y_{1T}, \ldots, y_{KT})' = a + O_p(r_T)$ be a K-dimensional random vector sequence, where $r_T = o(1)$, and let $g : \mathbb{R}^K \to \mathbb{R}$ be a function with continuous partial derivatives of order two at $a = (a_1, \ldots, a_K)'$. Then

$$g(y_T) = g(a) + \frac{\partial g(a)}{\partial y'}(y_T - a) + O_p(r_T^2).$$

If g has continuous partial derivatives of order three,

$$g(y_T) = g(a) + \frac{\partial g(a)}{\partial y'}(y_T - a) + \frac{1}{2}(y_T - a)'\frac{\partial^2 g(a)}{\partial y \partial y'}(y_T - a) + O_p(r_T^3).$$

∎

C.3 Infinite Sums of Random Variables

The MA representation of a VAR process is often an infinite sum of random vectors. As in the study of infinite sums of real numbers, we must specify what we mean by such an infinite sum. The concept of absolute convergence is basic in the following. A doubly infinite sequence of real numbers $\{a_i\}$, $i = 0, \pm 1, \pm 2, \ldots$, is *absolutely summable* if

$$\lim_{n \to \infty} \sum_{i=-n}^{n} |a_i|$$

exists and is finite. The limit is usually denoted by

$$\sum_{i=-\infty}^{\infty} |a_i|.$$

The following theorem provides a justification for working with infinite sums of random variables. A proof may be found in Fuller (1976, pp. 29-31).

Proposition C.7 (*Existence of Infinite Sums of Random Variables*)
Suppose $\{a_i\}$ is an absolutely summable sequence of real numbers and $\{z_t\}$, $t = 0, \pm 1, \pm 2, \ldots$, is a sequence of random variables satisfying

$$E(z_t^2) \leq c, \quad t = 0, \pm 1, \pm 2, \ldots,$$

for some finite constant c. Then there exists a sequence of random variables $\{y_t\}$, $t = 0, \pm 1, \pm 2, \ldots$, such that

$$\sum_{i=-n}^{n} a_i z_{t-i} \xrightarrow[n \to \infty]{q.m.} y_t$$

and, thus,

$$\plim_{n \to \infty} \sum_{i=-n}^{n} a_i z_{t-i} = y_t.$$

The random variables y_t are uniquely determined except on a set of probability zero. If, in addition, the z_t are independent random variables, then

$$\sum_{i=-n}^{n} a_i z_{t-i} \xrightarrow{a.s.} y_t.$$

∎

This theorem makes precise what we mean by a (univariate) infinite MA

$$y_t = \sum_{i=0}^{\infty} \Phi_i u_{t-i},$$

where u_t is univariate zero mean white noise with variance $\sigma_u^2 < \infty$. Defining $a_i = 0$ for $i < 0$ and $a_i = \Phi_i$ for $i \geq 0$ and assuming that $\{a_i\}$ is absolutely summable, the proposition guarantees that the process y_t is uniquely defined as a limit in mean square, except on a set of probability zero. The latter qualification may be ignored for practical purposes because we may always change a random variable on a set of probability zero without changing its probability characteristics. The requirement for the MA coefficients to be absolutely summable is satisfied if y_t is a stable AR process. For instance, if $y_t = \alpha y_{t-1} + u_t$ is an AR(1) process, $\Phi_i = \alpha^i$ which is an absolutely summable sequence for $|\alpha| < 1$. With respect to the moments of an infinite sum of random variables the following result holds:

Proposition C.8 (*Moments of Infinite Sums of Random Variables*)
Suppose z_t satisfies the conditions of Proposition C.7, $\{a_i\}$ and $\{b_i\}$ are absolutely summable sequences of real numbers,

$$y_t = \sum_{i=-\infty}^{\infty} a_i z_{t-i}, \quad \text{and} \quad x_t = \sum_{i=-\infty}^{\infty} b_i z_{t-i}.$$

Then

$$E(y_t) = \lim_{n \to \infty} \sum_{i=-n}^{n} a_i E(z_{t-i})$$

and

$$E(y_t x_t) = \lim_{n \to \infty} \sum_{i=-n}^{n} \sum_{j=-n}^{n} a_i b_j E(z_{t-i} z_{t-j})$$

and, in particular,

$$E(y_t^2) = \lim_{n \to \infty} \sum_{i=-n}^{n} \sum_{j=-n}^{n} a_i a_j E(z_{t-i} z_{t-j}).$$

■

Proof: Fuller (1976, pp. 32-33). ■

All these concepts and results may be extended to vector processes. A sequence of $(K \times K)$ matrices $\{A_i = (a_{mn,i})\}$, $i = 0, \pm 1, \pm 2, \ldots$, is *absolutely summable* if each sequence $\{a_{mn,i}\}$, $m, n = 1, \ldots, K$; $i = 0, \pm 1, \pm 2, \ldots$, is absolutely summable. Equivalently, $\{A_i\}$ may be defined to be absolutely summable if the sequence $\{\|A_i\|\}$ is summable, where

$$\|A_i\| = [\text{tr}(A_i A_i')]^{1/2} = \left(\sum_m \sum_n a_{mn,i}^2 \right)^{1/2}$$

is the Euclidean norm of A_i. To see the equivalence of the two definitions, note that

$$|a_{mn,i}| \leq \|A_i\| \leq \sum_m \sum_n |a_{mn,i}|.$$

Hence,

$$\sum_{i=-\infty}^{\infty} |a_{mn,i}| \qquad (C.3.1)$$

exists and is finite if

$$\sum_{i=-\infty}^{\infty} \|A_i\| \qquad (C.3.2)$$

is finite. In turn, if (C.3.1) is finite for all m, n, then, for all h,

$$\sum_{i=-h}^{h} \|A_i\| \leq \sum_{i=-h}^{h} \sum_m \sum_n |a_{mn,i}|$$

so that (C.3.2) is finite. Thus, the two definitions are indeed equivalent.

Proposition C.9 (*Existence of Infinite Sums of Random Vectors*)
Suppose $\{A_i\}$ is an absolutely summable sequence of real $(K \times K)$ matrices and $\{z_t\}$ is a sequence of K-dimensional random variables satisfying

$$E(z_t' z_t) \leq c, \quad t = 0, \pm 1, \pm 2, \ldots,$$

for some finite constant c. Then there exists a sequence of K-dimensional random variables $\{y_t\}$ such that

$$\sum_{i=-n}^{n} A_i z_{t-i} \xrightarrow[n \to \infty]{q.m.} y_t.$$

The sequence is uniquely determined except on a set of probability zero. ∎

Proof: Analogous to Fuller (1976, pp. 29-31); replace the absolute value by $\|\cdot\|$. ∎

This proposition ensures that the infinite MA representations of the VAR processes considered in this text are well-defined because it can be shown that the MA coefficient matrices of a stable VAR process form an absolutely summable sequence. With respect to moments of infinite sums, we have the following result.

Proposition C.10 (*Moments of Infinite Sums of Random Vectors*)
Suppose z_t satisfies the conditions of Proposition C.9, $\{A_i\}$ and $\{B_i\}$ are absolutely summable sequences of $(K \times K)$ matrices,

$$y_t = \sum_{i=-\infty}^{\infty} A_i z_{t-i} \quad \text{and} \quad x_t = \sum_{i=-\infty}^{\infty} B_i z_{t-i}.$$

Then

$$E(y_t) = \lim_{n \to \infty} \sum_{i=-n}^{n} A_i E(z_{t-i})$$

and

$$E(y_t x_t') = \lim_{n \to \infty} \sum_{i=-n}^{n} \sum_{j=-n}^{n} A_i E(z_{t-i} z_{t-j}') B_j',$$

where the limit of the sequence of matrices is the matrix of limits of the sequences of individual elements. ∎

Proof: Along similar lines as the proof of Fuller (1976, Theorem 2.2.2, pp. 32-33). ∎

While we have restricted the discussion to absolutely summable sequences of coefficients, it may be worth mentioning that infinite sums of random variables and vectors can be defined in more general terms.

C.4 Laws of Large Numbers and Central Limit Theorems

The derivation of asymptotic properties of estimators and test statistics is largely based on *laws of large numbers* (LLNs) and *central limit theorems* (CLTs) some examples of which are listed in the following. So-called *weak LLNs* specify conditions under which a sample mean converges in probability to the population mean and *strong LLNs* state the corresponding results for almost sure convergence.

In stating some of the results, martingale difference processes are useful tools. Suppose $\{x_t\}$ $(t = 1, 2, \ldots)$ is a sequence of zero mean random variables and let Ω_t be an information set available at time t which includes at least $\{x_1, \ldots, x_t\}$ and possibly other random variables. The sequence $\{x_t\}$ is said to be a *martingale difference sequence with respect to the sequence* Ω_t if $E(x_t|\Omega_{t-1}) = 0$ for all $t = 2, 3, \ldots$. It is simply referred to as *martingale difference sequence* if $E(x_t) = 0$ for $t = 1, 2, \ldots$, and $E(x_t|x_{t-1}, \ldots, x_1) = 0$ for $t = 2, 3, \ldots$. More generally, a sequence $\{x_t\}$ of K-dimensional vector random variables satisfying $E(x_t) = 0$ for all t and $E(x_t|x_{t-1}, \ldots, x_1) = 0$ for $t = 2, 3, \ldots$, is a *vector martingale difference sequence*.

It is sometimes useful to allow the x_t's to depend on the sample size. This way a different sequence for each sample size T is obtained. Denoting by $x_{T,t}$ the t-th element of the T-th sequence, not just a sequence but an *array* of random variables $\{x_{T,t}\}$ $(t = 1, 2, \ldots, T; T = 1, 2, \ldots)$ is obtained. Such an array is called a *martingale difference array* if $E(x_{T,t}) = 0$ for all t and T and $E(x_{T,t}|x_{T,t-1}, \ldots, x_{T,1}) = 0$ for all t and $T > 1$. This definition also applies for vector arrays.

The following inequality is a useful device for deriving asymptotic results. It is therefore presented here (see, e.g., Fuller (1976, Theorem 5.1.1)).

Proposition C.11 (*Chebyshev's Inequality*)
Given $r \in \mathbb{N}$, $r > 0$, let x be a random variable such that $E(|x|^r)$ exists. Then, for any $c \in \mathbb{R}$ and $\epsilon > 0$,

$$\Pr\{|x - c| \geq \epsilon\} \leq \frac{E(|x - c|^r)}{\epsilon^r}.$$

∎

The next proposition collects some weak LLNs (see, e.g., Davidson (1994, Part IV)).

Proposition C.12 (*Weak Laws of Large Numbers*)

(1) (Khinchine's Theorem) (Rao (1973, p. 112))
Let $\{x_t\}$ be a sequence of i.i.d. random variables with $E(x_t) = \mu < \infty$. Then
$$\overline{x}_T := \frac{1}{T} \sum_{t=1}^{T} x_t \xrightarrow{p} \mu.$$

(2) Let $\{x_t\}$ be a sequence of independent random variables with $E(x_t) = \mu < \infty$ and $E|x_t|^{1+\epsilon} \leq c < \infty$ ($t = 1, 2, \ldots$) for some $\epsilon > 0$ and a finite constant c. Then $\overline{x}_T \xrightarrow{p} \mu$.

(3) (Chebyshev's Theorem) (Rao (1973, p. 112))
Let $\{x_t\}$ be a sequence of uncorrelated random variables with $E(x_t) = \mu < \infty$ and $\lim_{T \to \infty} E(\overline{x}_T - \mu)^2 = 0$. Then $\overline{x}_T \xrightarrow{p} \mu$.

(4) (Corollary to Chebyshev's Theorem)
Let $\{x_t\}$ be a sequence of independent random variables with $E(x_t) = \mu < \infty$ and $\text{Var}(x_t) \leq c < \infty$ ($t = 1, 2, \ldots$) for some finite constant c. Then $\overline{x}_T \xrightarrow{p} \mu$.

(5) (LLN for Martingale Differences)
Let $\{x_t\}$ be a strictly stationary martingale difference sequence with $E|x_t| < \infty$ ($t = 1, 2, \ldots$). Then $\overline{x}_T \xrightarrow{p} 0$.

(6) (LLN for Martingale Difference Arrays)
Let $\{x_{T,t}\}$ be a martingale difference array with $E|x_{T,t}|^{1+\epsilon} \leq c < \infty$ for all t and T for some $\epsilon > 0$ and a finite constant c. Then $\overline{x}_T := T^{-1} \sum_{t=1}^{T} x_{T,t} \xrightarrow{p} 0$.

(7) (Stationary Processes) (Hamilton (1994, Proposition 7.5))
Let $\{x_t\}$ be a stationary stochastic process with $E(x_t) = \mu < \infty$ and $E[(x_t - \mu)(x_{t-j} - \mu)] = \gamma_j$ ($t = 1, 2, \ldots$) such that $\sum_{j=0}^{\infty} |\gamma_j| < \infty$. Then $\overline{x}_T \xrightarrow{q.m.} \mu$ and, hence, $\overline{x}_T \xrightarrow{p} \mu$, and $\lim_{T \to \infty} T E(\overline{x}_T - \mu)^2 = \sum_{j=-\infty}^{\infty} \gamma_j$. ∎

Notice that the i.i.d. assumption in Khinchine's theorem may be replaced by the requirement that moments exist of order larger than one. In fact, Chebyshev's theorem even requires the existence of second order moments. It is actually sufficient that the variances of the x_t are bounded. It may be worth noting that heterogenous variances are allowed for the weak LLN to hold, if the variances are bounded. The last result in the proposition shows that uncorrelated elements of the sequence under consideration are not required. Actually, a martingale difference sequence does not necessarily have independent elements so that for most of the above results independence of the sequence elements is not assumed.

Notice, that the proposition generalizes straightforwardly to sequences of random vectors because convergence in probability for a sequence of random vectors is defined in terms of convergence of the sequences of the individual

C.4 Laws of Large Numbers and Central Limit Theorems

elements. The following CLTs are stated for vector sequences and, of course, hold for univariate sequences as special cases.

Proposition C.13 (*Central Limit Theorems*)

(1) (Lindeberg-Levy CLT)
Let $\{x_t\}$ be a sequence of K-dimensional i.i.d. random vectors with mean μ and covariance matrix Σ_x. Then
$$\sqrt{T}(\overline{x}_T - \mu) \xrightarrow{d} \mathcal{N}(0, \Sigma_x).$$

(2) (CLT for Martingale Difference Arrays) (see Hamilton (1994, Proposition 7.9))
Let $\{x_{T,t} = (x_{1T,t}, \ldots, x_{KT,t})'\}$ be a K-dimensional martingale difference array with covariance matrices $E(x_{T,t}x'_{T,t}) = \Sigma_{Tt}$ such that $T^{-1}\sum_{t=1}^{T} \Sigma_{Tt} \to \Sigma$, where Σ is positive definite. Moreover, suppose that $T^{-1}\sum_{t=1}^{T} x_{T,t}x'_{T,t} \xrightarrow{p} \Sigma$ and $E(x_{iT,t}x_{jT,t}x_{kT,t}x_{lT,t}) < \infty$ for all t and T and all $1 \le i, j, k, l \le K$. Then
$$\sqrt{T}\,\overline{x}_T \xrightarrow{d} \mathcal{N}(0, \Sigma).$$

(3) (CLT for Stationary Processes)
Let $x_t = \mu + \sum_{j=0}^{\infty} \Phi_j u_{t-j}$ be a K-dimensional stationary stochastic process with $E(x_t) = \mu < \infty$, $\sum_{j=0}^{\infty} \|\Phi_j\| < \infty$ and $u_t \sim (0, \Sigma_u)$ i.i.d. white noise. Then
$$\sqrt{T}(\overline{x}_T - \mu) \xrightarrow{d} \mathcal{N}\left(0, \sum_{j=-\infty}^{\infty} \Gamma_x(j)\right),$$
where $\Gamma_x(j) := E[(x_t - \mu)(x_{t-j} - \mu)']$. ∎

The results in Proposition C.13 are just examples of useful CLTs. A variety of similar results exists for different sets of conditions. More discussion of CLTs and proofs can be found in Davidson (1994, Part V). For the CLT for stationary processes see Anderson (1971, Chapters 7 and 8).

To derive the asymptotic distribution of a vector sequence it is actually sufficient to consider univariate series. This is a consequence of the following result.

Proposition C.14 (*Cramér-Wold Device*) (Rao (1973, p. 123))
Let x_T be a K-dimensional sequence of random vectors and x a K-dimensional random vector. If $c'x_T \xrightarrow{d} c'x$ for all $c \in \mathbb{R}^K$, then $x_T \xrightarrow{d} x$. ∎

Therefore, to show asymptotic normality of a sequence, $\sqrt{T}(\widehat{\beta}_T - \beta) \overset{d}{\to} \mathcal{N}(0, \Sigma)$, it suffices to show for all K-vectors c with $c'\Sigma c \neq 0$,

$$\frac{\sqrt{T} c'(\widehat{\beta}_T - \beta)}{(c'\Sigma c)^{1/2}} \overset{d}{\to} \mathcal{N}(0, 1).$$

Hence, CLTs for univariate series can in fact be used to show multivariate results.

C.5 Standard Asymptotic Properties of Estimators and Test Statistics

Suppose we have a sequence of $(m \times n)$ estimators $\{\widehat{B}_T\}$ for an $(m \times n)$ parameter matrix B, where T denotes the sample sizes (time series lengths) on which the estimators are based. For simplicity we will delete the subscript T in the following and we will mean the sequence of estimators when we use the term "estimator".

The estimator $\widehat{B}$ is *consistent* if plim $\widehat{B} = B$. In the related literature, this type of consistency is sometimes called *weak consistency*. However, in this text, we simply use the term consistency instead. The estimator is *strongly consistent* if $\widehat{B} \overset{a.s.}{\to} B$, and the estimator is *mean square consistent* if $\widehat{B} \overset{q.m.}{\to} B$. By Proposition C.1, both strong consistency and mean square consistency imply consistency.

Let $\widehat{\beta}$ be an estimator (a sequence of estimators) of a $(K \times 1)$ vector β. The estimator is said to have an *asymptotic normal distribution* if $\sqrt{T}(\widehat{\beta} - \beta)$ converges in distribution to a random vector with multivariate normal distribution $\mathcal{N}(0, \Sigma)$, that is,

$$\sqrt{T}(\widehat{\beta} - \beta) \overset{d}{\to} \mathcal{N}(0, \Sigma). \tag{C.5.1}$$

In that case, for large T, $\mathcal{N}(\beta, \Sigma/T)$ is usually used as an approximation to the distribution of $\widehat{\beta}$. Equivalently, by the Cramér-Wold device (Proposition C.14), (C.5.1) may be defined by requiring that

$$\frac{\sqrt{T} c'(\widehat{\beta} - \beta)}{(c'\Sigma c)^{1/2}} \overset{d}{\to} \mathcal{N}(0, 1),$$

for any $(K \times 1)$ vector c for which $c'\Sigma c \neq 0$. The following proposition provides some useful rules for determining the asymptotic distributions of estimators and test statistics.

Proposition C.15 (*Asymptotic Properties of Estimators*)
Suppose $\widehat{\beta}$ is an estimator of the $(K \times 1)$ vector β with $\sqrt{T}(\widehat{\beta} - \beta) \overset{d}{\to} \mathcal{N}(0, \Sigma)$. Then the following rules hold:

(1) If plim $\widehat{A} = A$, then $\sqrt{T}\widehat{A}(\widehat{\beta} - \beta) \xrightarrow{d} \mathcal{N}(0, A\Sigma A')$ (see Schmidt (1976, p. 251)).
(2) If $R \neq 0$ is an $(M \times K)$ matrix, then $\sqrt{T}(R\widehat{\beta} - R\beta) \xrightarrow{d} \mathcal{N}(0, R\Sigma R')$.
(3) (Delta method)
If $g(\beta) = (g_1(\beta), \ldots, g_m(\beta))'$ is a vector-valued continuously differentiable function with $\partial g/\partial \beta' \neq 0$ at β, then

$$\sqrt{T}[g(\widehat{\beta}) - g(\beta)] \xrightarrow{d} \mathcal{N}\left(0, \frac{\partial g(\beta)}{\partial \beta'} \Sigma \frac{\partial g(\beta)'}{\partial \beta}\right).$$

If $\partial g/\partial \beta' = 0$ at β, $\sqrt{T}[g(\widehat{\beta}) - g(\beta)] \xrightarrow{p} 0$. (See Serfling (1980, pp. 122-124)).
(4) If Σ is nonsingular, $T(\widehat{\beta} - \beta)'\Sigma^{-1}(\widehat{\beta} - \beta) \xrightarrow{d} \chi^2(K)$.
(5) If Σ is nonsingular and plim $\widehat{\Sigma} = \Sigma$, then $T(\widehat{\beta} - \beta)'\widehat{\Sigma}^{-1}(\widehat{\beta} - \beta) \xrightarrow{d} \chi^2(K)$.
(6) If $\Sigma = QA$, where Q is symmetric, idempotent of rank n and A is positive definite, then $T(\widehat{\beta} - \beta)'A^{-1}(\widehat{\beta} - \beta) \xrightarrow{d} \chi^2(n)$.

∎

C.6 Maximum Likelihood Estimation

Suppose $y_1, y_2, \ldots$ is a sequence of K-dimensional random vectors, the first T of which have a joint probability density function $f_T(y_1, \ldots, y_T; \delta_0)$, where δ_0 is an unknown $(M \times 1)$ vector of parameters that does not depend on T. It is assumed to be from a subset $\mathbb{D}$ of the M-dimensional Euclidean space $\mathbb{R}^M$. Suppose further that $f_T(\cdot; \delta)$ has a known functional form and one wishes to estimate δ_0.

For a fixed realization $y_1, \ldots, y_T$, the function

$$l(\delta) = l(\delta|y_1, \ldots, y_T) = f_T(y_1, \ldots, y_T; \delta),$$

viewed as a function of δ, is the *likelihood function*. Its natural logarithm $\ln l(\delta|\cdot)$ is the *log-likelihood function*. A vector $\widetilde{\delta}$, maximizing the likelihood function or log-likelihood function, is called a *maximum likelihood* (ML) *estimate*, that is, if

$$l(\widetilde{\delta}) = \sup_{\delta \in \mathbb{D}} l(\delta),$$

then $\widetilde{\delta}$ is an ML estimate. Here sup denotes the supremum, that is, the least upper bound, which may exist even if the maximum does not. In general, $\widetilde{\delta}$ depends on $y_1, \ldots, y_T$, that is, $\widetilde{\delta} = \widetilde{\delta}(y_1, \ldots, y_T)$. Replacing the fixed values $y_1, \ldots, y_T$ by their corresponding random vectors, $\widetilde{\delta}$ is an *ML estimator* of δ_0 if the functional dependence on $y_1, \ldots, y_T$ is such that $\widetilde{\delta}$ is a random vector.

If $l(\delta)$ is a differentiable function of δ, the vector of first order partial derivatives of $\ln l(\delta)$, that is,

$$s(\delta) = \partial \ln l(\delta)/\partial \delta,$$

regarded as a random vector (a function of the random vectors $y_1, \ldots, y_T$), is the *score vector*. It vanishes at $\delta = \widetilde{\delta}$ if the maximum of $\ln l(\delta)$ is attained at an interior point of the parameter space $\mathbb{D}$. The *information matrix* for δ_0 is minus the expectation of the matrix of second order partial derivatives of $\ln l$, evaluated at the true parameter vector δ_0,

$$\mathcal{I}(\delta_0) = -E\left[\left.\frac{\partial^2 \ln l}{\partial \delta \partial \delta'}\right|_{\delta_0}\right].$$

The matrix

$$\mathcal{I}_a(\delta_0) = \lim_{T \to \infty} \mathcal{I}(\delta_0)/T,$$

if it exists, is the *asymptotic information matrix* for δ_0. If it is nonsingular, its inverse is a lower bound for the covariance matrix of the asymptotic distribution of any consistent estimator with asymptotic normal distribution. In other words, if $\widehat{\delta}$ is a consistent estimator of δ_0 with

$$\sqrt{T}(\widehat{\delta} - \delta_0) \xrightarrow{d} \mathcal{N}(0, \Sigma_{\widehat{\delta}}),$$

then $\mathcal{I}_a(\delta_0)^{-1} \leq \Sigma_{\widehat{\delta}}$, that is, $\Sigma_{\widehat{\delta}} - \mathcal{I}_a(\delta_0)^{-1}$ is positive semidefinite. Under quite general regularity conditions, an ML estimator $\widetilde{\delta}$ for δ_0 is consistent and

$$\sqrt{T}(\widetilde{\delta} - \delta_0) \xrightarrow{d} \mathcal{N}(0, \mathcal{I}_a(\delta_0)^{-1}).$$

Thus, in large samples, $\widetilde{\delta}$ is approximately distributed as $\mathcal{N}(\delta_0, \mathcal{I}_a(\delta_0)^{-1}/T)$.

C.7 Likelihood Ratio, Lagrange Multiplier, and Wald Tests

Three principles for constructing tests of statistical hypotheses are employed frequently in the text. We consider testing of

$$H_0 : \varphi(\delta_0) = 0 \quad \text{against} \quad H_1 : \varphi(\delta_0) \neq 0, \tag{C.7.1}$$

where δ_0 is the true $(M \times 1)$ parameter vector, as in the previous section, and $\varphi : \mathbb{R}^M \to \mathbb{R}^N$ is a continuously differentiable function so that $\varphi(\delta)$ is of dimension $(N \times 1)$. We assume that $[\partial \varphi/\partial \delta'|_{\delta_0}]$ has rank N. This condition implies that $N \leq M$ and the N restrictions for the parameter vector are

C.7 Likelihood Ratio, Lagrange Multiplier, and Wald Tests

distinguishable in a neighborhood of δ_0. Often the hypotheses can be written alternatively as

$$H_0 : \delta_0 = g(\gamma_0) \quad \text{against} \quad H_1 : \delta_0 \neq g(\gamma_0), \tag{C.7.2}$$

where γ_0 is an $(M - N)$-dimensional vector and $g : \mathbb{R}^{M-N} \to \mathbb{R}^M$ is a continuously differentiable function in a neighborhood of γ_0 (see Gallant (1987, pp. 57-58)).

The *likelihood ratio* (LR) *test* of (C.7.1) or (C.7.2) is based on the statistic

$$\lambda_{LR} = 2[\ln l(\widetilde{\delta}) - \ln l(\widetilde{\delta}_r)],$$

where $\widetilde{\delta}$ denotes the unconstrained ML estimator and $\widetilde{\delta}_r$ is the restricted ML estimator of δ_0, subject to the restrictions specified under H_0, that is, $\widetilde{\delta}_r$ is obtained by maximizing $\ln l$ over the parameter space restricted by the conditions stated in H_0. Under suitable regularity conditions, we have

$$\lambda_{LR} \xrightarrow{d} \chi^2(N). \tag{C.7.3}$$

The *Lagrange multiplier* (LM) *statistic* for testing (C.7.1) or (C.7.2) is of the form

$$\lambda_{LM} = s(\widetilde{\delta}_r)' \mathcal{I}(\widetilde{\delta}_r)^{-1} s(\widetilde{\delta}_r), \tag{C.7.4}$$

where $s(\delta)$ denotes the score vector and $\mathcal{I}(\delta)$ the information matrix, as before. In the LM statistic, both functions are evaluated at the restricted estimator of δ_0. Under H_0, λ_{LM} has an asymptotic $\chi^2(N)$-distribution, if weak regularity conditions are satisfied. The name derives from the fact that it can be written as

$$\lambda_{LM} = \widetilde{\lambda}' \left[\frac{\partial \varphi}{\partial \delta'} \bigg|_{\widetilde{\delta}_r} \right] \mathcal{I}(\widetilde{\delta}_r)^{-1} \left[\frac{\partial \varphi'}{\partial \delta} \bigg|_{\widetilde{\delta}_r} \right] \widetilde{\lambda}, \tag{C.7.5}$$

where $\widetilde{\lambda}$ is the vector of Lagrange multipliers for which the Lagrange function has a stationary point corresponding to the constrained estimator (see Appendix A.14).

The equivalence of (C.7.4) and (C.7.5) can be seen by recalling that the constrained minimum of $-\ln l$ is attained at a stationary point of the Lagrange function

$$\mathcal{L}(\delta, \lambda) = -\ln l(\delta) + \lambda' \varphi(\delta).$$

In other words, $\widetilde{\delta}_r$ satisfies

$$0 = \left[\frac{\partial \mathcal{L}}{\partial \delta'} \bigg|_{\widetilde{\delta}_r, \widetilde{\lambda}} \right] = -\left[\frac{\partial \ln l}{\partial \delta'} \bigg|_{\widetilde{\delta}_r} \right] + \widetilde{\lambda}' \left[\frac{\partial \varphi}{\partial \delta'} \bigg|_{\widetilde{\delta}_r} \right] = -s(\widetilde{\delta}_r)' + \widetilde{\lambda}' \left[\frac{\partial \varphi}{\partial \delta'} \bigg|_{\widetilde{\delta}_r} \right].$$

The LM statistic is often computed via an auxiliary regression. To see how this can be done, consider a normal regression model of the form

$$\mathbf{y} = X\boldsymbol{\beta} + Z\boldsymbol{\gamma} + \mathbf{u},$$

where $\mathbf{y}$ and $\mathbf{u}$ are $(T \times 1)$ vectors, X and Z are $(T \times M)$ and $(T \times N)$ regressor matrices, respectively, $\boldsymbol{\beta}$ and $\boldsymbol{\gamma}$ are $(M \times 1)$ and $(N \times 1)$ parameter vectors and $\mathbf{u} \sim \mathcal{N}(0, \sigma_u^2 I_T)$. Suppose we wish to test the pair of hypotheses

$$H_0 : \boldsymbol{\gamma} = 0 \quad \text{versus} \quad H_1 : \boldsymbol{\gamma} \neq 0.$$

In this case, the score vector is

$$s\begin{pmatrix} \boldsymbol{\beta} \\ \boldsymbol{\gamma} \end{pmatrix} = \frac{1}{\sigma_u^2} \begin{bmatrix} X' \\ Z' \end{bmatrix} (\mathbf{y} - X\boldsymbol{\beta} - Z\boldsymbol{\gamma}),$$

the inverse information matrix is

$$\sigma_u^2 \begin{bmatrix} X'X & X'Z \\ Z'X & Z'Z \end{bmatrix}^{-1},$$

and the restricted estimator is

$$\begin{bmatrix} \widehat{\boldsymbol{\beta}} \\ 0 \end{bmatrix} = \begin{bmatrix} (X'X)^{-1} X'\mathbf{y} \\ 0 \end{bmatrix}.$$

Notice that the first order conditions for computing this estimator imply

$$X'(\mathbf{y} - X\widehat{\boldsymbol{\beta}}) = X'\widehat{\mathbf{u}} = 0.$$

Here $\widehat{\mathbf{u}} := \mathbf{y} - X\widehat{\boldsymbol{\beta}}$ is the residual vector of the restricted estimation. Hence, the score vector evaluated at the restricted estimator is

$$s\begin{pmatrix} \widehat{\boldsymbol{\beta}} \\ 0 \end{pmatrix} = \frac{1}{\sigma_u^2} \begin{bmatrix} X'(\mathbf{y} - X\widehat{\boldsymbol{\beta}}) \\ Z'(\mathbf{y} - X\widehat{\boldsymbol{\beta}}) \end{bmatrix} = \frac{1}{\sigma_u^2} \begin{bmatrix} 0 \\ Z'\widehat{\mathbf{u}} \end{bmatrix}$$

and the LM statistic becomes

$$\begin{aligned}
\lambda_{LM} &= [0 : \widehat{\mathbf{u}}'Z] \begin{bmatrix} X'X & X'Z \\ Z'X & Z'Z \end{bmatrix}^{-1} \begin{bmatrix} 0 \\ Z'\widehat{\mathbf{u}} \end{bmatrix} / \sigma_u^2 \\
&= \widehat{\mathbf{u}}'Z(Z'Z - Z'X(X'X)^{-1}X'Z)^{-1} Z'\widehat{\mathbf{u}} / \sigma_u^2,
\end{aligned}$$

where the rules for the partitioned inverse have been used (see Appendix A.10).

The same statistic is obtained by using the usual χ^2-statistic for testing $\boldsymbol{\gamma} = 0$ in the auxiliary regression model

$$\widehat{\mathbf{u}} = X\boldsymbol{\beta} + Z\boldsymbol{\gamma} + \mathbf{e},$$

where **e** is an error vector. The LS estimator from this model is

$$\begin{bmatrix} \widetilde{\beta} \\ \widetilde{\gamma} \end{bmatrix} = \begin{bmatrix} X'X & X'Z \\ Z'X & Z'Z \end{bmatrix}^{-1} \begin{bmatrix} X'\widehat{\mathbf{u}} \\ Z'\widehat{\mathbf{u}} \end{bmatrix}.$$

Using $X'\widehat{\mathbf{u}} = 0$ and the rules for the partitioned inverse gives

$$\begin{aligned} \widetilde{\gamma} &= (Z'Z - Z'X(X'X)^{-1}X'Z)^{-1}Z'\widehat{\mathbf{u}} \\ &\sim \mathcal{N}(\gamma, \sigma_u^2(Z'Z - Z'X(X'X)^{-1}X'Z)^{-1}). \end{aligned}$$

Hence, the χ^2-statistic

$$\widetilde{\gamma}'(Z'Z - Z'X(X'X)^{-1}X'Z)\widetilde{\gamma}/\sigma_u^2$$

is easily seen to be identical to the previously obtained expression for λ_{LM}. Of course, algebraically the same result is obtained if σ_u^2 is replaced by an estimator. Using the usual modifications, the statistic has an F-distribution in this case. More precisely,

$$\frac{\widetilde{\gamma}'(Z'Z - Z'X(X'X)^{-1}X'Z)\widetilde{\gamma}}{N\widehat{\sigma}_u^2} \sim F(N, T - M - N).$$

Although we have used a normal regression model with nonstochastic regressors in this illustration, a similar reasoning often applies for more general situations and it implies an auxiliary regression model from which the LM statistic can be obtained. The reason is that much of the derivation rests on the algebraic properties of the quantities involved. Therefore, similar arguments can be used, for example, if the regressors are stochastic or a GLS estimation is used. In Chapters 4 and 5, the LM statistics for residual autocorrelation in VAR models are, for instance, derived in this way.

The *Wald statistic* is based on an unconstrained estimator which is asymptotically normal,

$$\sqrt{T}(\widetilde{\delta} - \delta_0) \xrightarrow{d} \mathcal{N}(0, \Sigma_{\widetilde{\delta}}).$$

By Proposition C.15(3), it follows that

$$\sqrt{T}[\varphi(\widetilde{\delta}) - \varphi(\delta_0)] \xrightarrow{d} \mathcal{N}\left(0, \left[\frac{\partial \varphi}{\partial \delta'}\bigg|_{\delta_0}\right] \Sigma_{\widetilde{\delta}} \left[\frac{\partial \varphi'}{\partial \delta}\bigg|_{\delta_0}\right]\right).$$

Thus, by Proposition C.15(5), if $H_0 : \varphi(\delta_0) = 0$ is true and the covariance matrix is invertible,

$$\lambda_W = T\varphi(\widetilde{\delta})' \left(\left[\frac{\partial \varphi}{\partial \delta'}\bigg|_{\widetilde{\delta}}\right] \widetilde{\Sigma}_{\widetilde{\delta}} \left[\frac{\partial \varphi'}{\partial \delta}\bigg|_{\widetilde{\delta}}\right]\right)^{-1} \varphi(\widetilde{\delta}) \xrightarrow{d} \chi^2(N), \qquad \text{(C.7.6)}$$

where $\widetilde{\Sigma}_{\widetilde{\delta}}$ is a consistent estimator of $\Sigma_{\widetilde{\delta}}$. The statistic λ_W is the Wald statistic. For further discussion of the three test statistics and proofs of their asymptotic distributions see also Hayashi (2000, Chapter 7).

In summary, we have three test statistics with equivalent asymptotic distributions under the null hypothesis. The LR statistic involves both the restricted and the unrestricted ML estimators, the LM statistic is based on the restricted estimator only, and the Wald statistic requires just the unrestricted estimator. The choice among the three statistics is often based on computational convenience. Wald tests have the disadvantage that they are not invariant under transformations of the restrictions. In other words, if the restrictions can be written in two equivalent ways (e.g., $\delta_i = 0$ and $\delta_i^2 = 0$) the corresponding Wald tests may have different small sample properties. Their small sample power may be low (see Gregory & Veall (1985), Breusch & Schmidt (1988)).

C.8 Unit Root Asymptotics

C.8.1 Univariate Processes

In deriving asymptotic results for processes with unit roots, it is helpful to consider also continuous stochastic processes. An important example is a *standard Brownian motion* or a *standard Wiener process* $\mathbf{W}(\cdot)$ which is a function defined on the unit interval $[0, 1]$ and it assigns a random variable $\mathbf{W}(t)$ to each $t \in [0, 1]$ such that the following conditions hold:

(1) $\mathbf{W}(0) = 0$ with probability one.
(2) $\mathbf{W}(t)$ is continuous in t with probability one.
(3) For any partitioning of the unit interval, $0 \leq t_1 < t_2 < \cdots < t_k \leq 1$, the vector

$$\begin{bmatrix} \mathbf{W}(t_2) - \mathbf{W}(t_1) \\ \vdots \\ \mathbf{W}(t_k) - \mathbf{W}(t_{k-1}) \end{bmatrix} \sim \mathcal{N}\left(\begin{bmatrix} 0 \\ \vdots \\ 0 \end{bmatrix}, \begin{bmatrix} t_2 - t_1 & & 0 \\ & \ddots & \\ 0 & \cdots & t_k - t_{k-1} \end{bmatrix}\right),$$

that is, the differences have a multivariate normal distribution with independent components, means of zero, and variances $t_i - t_{i-1}$.

Wiener processes play an important role in the asymptotic theory for unit root processes. Nonstandard versions of the type $Z(t) = \sigma \mathbf{W}(t)$ are often encountered. Their increments are still independent but $Z(t) - Z(s) \sim \mathcal{N}(0, \sigma^2(t-s))$ for $s < t$. Notice also that $Z(t) \sim \mathcal{N}(0, \sigma^2 t)$.

In developing unit root asymptotics, we are often interested in quantities of the form

$$X_T(r) = \frac{1}{T} \sum_{t=1}^{[Tr]} w_t,$$

where w_t is a stationary stochastic process, $r \in [0, 1]$ denotes a fraction and $[Tr]$ signifies the largest integer less than or equal to Tr. If the $w_t = u_t$ are i.i.d. $(0, \sigma_u^2)$, we know from a central limit theorem (see Proposition C.13) that

C.8 Unit Root Asymptotics

$$\sqrt{T}X_T(r) = \frac{\sqrt{[Tr]}}{\sqrt{T}} \frac{1}{\sqrt{[Tr]}} \sum_{t=1}^{[Tr]} u_t \xrightarrow{d} \mathcal{N}(0, r\sigma_u^2),$$

for every $r \in [0,1]$, because $\sqrt{[Tr]}/\sqrt{T} \to \sqrt{r}$. Moreover,

$$\sqrt{T}[X_T(r_2) - X_T(r_1)]/\sigma_u \xrightarrow{d} \mathcal{N}(0, r_2 - r_1),$$

for $r_1 < r_2$. For nonoverlapping partitions of the unit interval, the partial sums will be made up of independent terms and they are therefore independent. Hence, it is plausible to write

$$\sqrt{T}X_T(\cdot)/\sigma_u \xrightarrow{d} \mathbf{W}(\cdot). \tag{C.8.1}$$

This notation and result generalizes the previously defined concept of convergence in distribution because now convergence is stated for a sequence of continuous time stochastic processes. The result is often referred to as a *functional central limit theorem* (FCLT) or *invariance principle* or *Donsker's theorem*.

Giving a precise definition of the related concept of convergence in distribution is simplified by considering convergence of probability measures. A sequence of probability measures $\Pr_T$ is said to converge to the probability measure $\Pr$ or $\Pr_T$ *converges weakly* to $\Pr$, if $\Pr_T(\mathbb{A}) \to \Pr(\mathbb{A})$ for all measurable sets $\mathbb{A}$, with the exception of sets for which the boundary points have nonzero probability mass. Instead of considering the distribution functions, we may define convergence in distribution via weak convergence of the corresponding sequence of probability measures. Thus, constructing a probability space on a suitable space of functions defined on the unit interval, the convergence in (C.8.1) can be defined rigorously. Although this type of convergence is more properly called weak convergence, we will still use the symbol $\xrightarrow{d}$ for signifying it. For more precise discussions see, for example, Davidson (1994, 2000) or Johansen (1995).

We may also generalize the concept of convergence in probability to the case of sequences of random functions. For a sequence $G_T(\cdot)$ and a random function $G(\cdot)$ we write $G_T \xrightarrow{p} G$ if

$$\sup_{t \in [0,1]} |G_T(t) - G(t)| \xrightarrow{p} 0.$$

Another useful tool in dealing with unit root process is the *continuous mapping theorem* which states that, given a sequence of stochastic functions $\{G_T(\cdot)\}$, a stochastic function $G(\cdot)$ and a continuous functional $g(\cdot)$ (a function defined on a space of functions), we have

$$G_T \xrightarrow{d} G \quad \Rightarrow \quad g(G_T) \xrightarrow{d} g(G).$$

Using the FCLT, this theorem implies, for instance, that

$$\int_0^1 \sqrt{T}X_T(r)dr \xrightarrow{d} \sigma_u \int_0^1 \mathbf{W}(r)dr$$

because the integral is a continuous functional.

These tools are useful in proving the following proposition from Hamilton (1994, Proposition 17.1) which summarizes a number of helpful results from the literature, many of which were derived, e.g., by Phillips (1987).

Proposition C.16 (*Properties of Random Walks and Related Quantities*)
Suppose $x_t = x_{t-1} + u_t$ is a random walk with i.i.d. white noise, $u_t \sim (0, \sigma_u^2)$, and $x_0 = 0$. Then the following results hold:

(1) $T^{-1/2} \sum_{t=1}^T u_t \xrightarrow{d} \sigma_u \mathbf{W}(1) = \mathcal{N}(0, \sigma_u^2)$.
(2) $T^{-1} \sum_{t=1}^T x_{t-1} u_t \xrightarrow{d} \frac{1}{2}\sigma_u^2 [\mathbf{W}(1)^2 - 1] = \frac{1}{2}\sigma_u^2 [\chi^2(1) - 1]$.
(3) $T^{-3/2} \sum_{t=1}^T t u_t \xrightarrow{d} \sigma_u \mathbf{W}(1) - \sigma_u \int_0^1 \mathbf{W}(r)dr = \mathcal{N}(0, \sigma_u^2/3)$.
(4) $T^{-3/2} \sum_{t=1}^T x_{t-1} \xrightarrow{d} \sigma_u \int_0^1 \mathbf{W}(r)dr = \mathcal{N}(0, \sigma_u^2/3)$.
(5) $T^{-2} \sum_{t=1}^T x_{t-1}^2 \xrightarrow{d} \sigma_u^2 \int_0^1 \mathbf{W}(r)^2 dr$.
(6) $T^{-5/2} \sum_{t=1}^T t x_{t-1} \xrightarrow{d} \sigma_u \int_0^1 r\mathbf{W}(r)dr$.
(7) $T^{-3} \sum_{t=1}^T t x_{t-1}^2 \xrightarrow{d} \sigma_u^2 \int_0^1 r\mathbf{W}(r)^2 dr$.
(8) $T^{-(n+1)} \sum_{t=1}^T t^n \to 1/(n+1)$ for $n = 0, 1, \ldots$.

■

From these results, the following asymptotic distributions of Dickey-Fuller (DF) statistics for unit roots can be derived. For details see, e.g., Hamilton (1994, Section 17.4). It is assumed that estimation is based on a sample $y_1, \ldots, y_T$ and a presample value y_0 is also available.

Proposition C.17 (*Asymptotic Distributions of Dickey-Fuller Test Statistics*)

(1) Suppose $\widehat{\rho} = \sum_{t=1}^T y_{t-1} y_t / \sum_{t=1}^T y_{t-1}^2$ is the LS estimator of the coefficient ρ of the AR(1) process $y_t = \rho y_{t-1} + u_t$, $t = 1, 2, \ldots$, where $u_t \sim (0, \sigma_u^2)$ is i.i.d. white noise. Here y_0 is a fixed starting value or a stochastic variable with a given fixed distribution (which does not depend on the sample size). Then, if $\rho = 1$,

$$T(\widehat{\rho} - 1) \xrightarrow{d} \frac{\frac{1}{2}[\mathbf{W}(1)^2 - 1]}{\int_0^1 \mathbf{W}(r)^2 dr}$$

and the t-statistic

$$t_{\widehat{\rho}-1} = \frac{\widehat{\rho} - 1}{\widehat{\sigma}_{\widehat{\rho}}} \xrightarrow{d} \frac{\frac{1}{2}[\mathbf{W}(1)^2 - 1]}{\left[\int_0^1 \mathbf{W}(r)^2 dr\right]^{1/2}},$$

where $\widehat{\sigma}_{\widehat{\rho}}^2 = T^{-1} \sum_{t=1}^T (y_t - \widehat{\rho} y_{t-1})^2 / \sum_{t=1}^T y_{t-1}^2$ is the usual LS estimator of the variance of $\widehat{\rho}$.

(2) Suppose $y_t = \mu + x_t$, $t = 1, 2, \ldots$, with $x_t = \rho x_{t-1} + u_t$, where $u_t \sim (0, \sigma_u^2)$ is i.i.d. white noise and μ is a fixed mean term. Moreover, let $x_0 = 0$ and y_0 be a fixed starting value or a stochastic variable with a given fixed distribution. Furthermore, $\widehat{\rho}$ is the LS estimator of ρ from a regression $y_t = \nu + \rho y_{t-1} + u_t$. Then, if $\rho = 1$,

$$T(\widehat{\rho} - 1) \xrightarrow{d} \frac{\frac{1}{2}[\mathbf{W}(1)^2 - 1] - \mathbf{W}(1) \int_0^1 \mathbf{W}(r) dr}{\int_0^1 \mathbf{W}(r)^2 dr - \left[\int_0^1 \mathbf{W}(r) dr\right]^2}$$

and the t-statistic

$$t_{\widehat{\rho}-1} = \frac{\widehat{\rho} - 1}{\widehat{\sigma}_{\widehat{\rho}}} \xrightarrow{d} \frac{\frac{1}{2}[\mathbf{W}(1)^2 - 1] - \mathbf{W}(1) \int_0^1 \mathbf{W}(r) dr}{\left\{\int_0^1 \mathbf{W}(r)^2 dr - \left[\int_0^1 \mathbf{W}(r) dr\right]^2\right\}^{1/2}},$$

where $\widehat{\sigma}_{\widehat{\rho}}^2$ is the usual LS estimator of the variance of $\widehat{\rho}$.

(3) Suppose $y_t = \nu + y_{t-1} + u_t$, $t = 1, 2, \ldots$, where $u_t \sim (0, \sigma_u^2)$ is i.i.d. white noise and $\nu \neq 0$ is a constant term. Moreover, let y_0 be a fixed starting value or a stochastic variable with a given fixed distribution. Furthermore, $\widehat{\rho}$ is the LS estimator of ρ from a regression $y_t = \nu + \rho y_{t-1} + u_t$. Then, if $\rho = 1$,

$$T^{3/2}(\widehat{\rho} - 1) \xrightarrow{d} \mathcal{N}(0, 12\sigma_u^2/\nu^2)$$

and the t-statistic

$$t_{\widehat{\rho}-1} = \frac{\widehat{\rho} - 1}{\widehat{\sigma}_{\widehat{\rho}}} \xrightarrow{d} \mathcal{N}(0, 1),$$

where $\widehat{\sigma}_{\widehat{\rho}}^2$ is the usual LS estimator of the variance of $\widehat{\rho}$.

(4) Suppose $y_t = \mu_0 + \mu_1 t + x_t$, $t = 1, 2, \ldots$, with $x_t = \rho x_{t-1} + u_t$, where $u_t \sim (0, \sigma_u^2)$ is i.i.d. white noise and μ_0 and μ_1 are fixed intercept and trend slope terms. Moreover, let $x_0 = 0$ and y_0 be a fixed starting value or a stochastic variable with a given fixed distribution. Furthermore, $\widehat{\rho}$ is the LS estimator of ρ from a regression $y_t = \nu_0 + \nu_1 t + \rho y_{t-1} + u_t$. Then, if $\rho = 1$,

$$T(\widehat{\rho} - 1) \xrightarrow{d} a/b$$

and

$$t_{\widehat{\rho}-1} = \frac{\widehat{\rho} - 1}{\widehat{\sigma}_{\widehat{\rho}}} \xrightarrow{d} a/\sqrt{b},$$

where

$$a = \int_0^1 \mathbf{W}(r)d\mathbf{W}(r)$$
$$+12\left[\int_0^1 r\mathbf{W}(r)dr - \frac{1}{2}\int_0^1 \mathbf{W}(r)dr\right]\left[\int_0^1 \mathbf{W}(r)dr - \frac{1}{2}\mathbf{W}(1)\right]$$
$$-\mathbf{W}(1)\int_0^1 \mathbf{W}(r)dr$$

and

$$b = \int_0^1 \mathbf{W}(r)^2 dr - 12\left(\int_0^1 r\mathbf{W}(r)dr\right)^2$$
$$+12\int_0^1 \mathbf{W}(r)dr \int_0^1 r\mathbf{W}(r)dr - 4\left(\int_0^1 \mathbf{W}(r)dr\right)^2.$$

Furthermore, $\widehat{\sigma}_{\widehat{\rho}}^2$ is the usual LS estimator of the variance of $\widehat{\rho}$.

∎

Obviously, most of the asymptotic distributions obtained for $\widehat{\rho}$ are nonstandard if $\rho = 1$. In fact, even the convergence rate of the estimator is nonstandard. It converges at a much faster rate to its true value of 1 than usual estimators based on stationary processes. More precisely, $\widehat{\rho} - \rho = O_p(T^{-1})$ if $\rho = 1$ in Cases 1, 2, and 4 in the proposition, whereas in the stationary case of an AR(1) process $y_t = \rho y_{t-1} + u_t$, say, we have for the LS estimator of ρ, $\widehat{\rho} - \rho = O_p(T^{-1/2})$. The latter rate also holds if y_t is stationary and has a nonzero mean term. In Case 3 of Proposition C.17, the convergence rate of $\widehat{\rho}$ is even larger because in that case the estimator is dominated by the linear trend which is generated by the drift term.

It is important to note that the limiting distributions in Cases 1, 2, and 4 are free of unknown nuisance parameters. Therefore, it is easy to compute percentage points of the limiting distributions by simulation methods. To do that, it is strictly speaking not even necessary to know the exact form of the asymptotic distributions of the estimators. It is sufficient to know that well-defined asymptotic distributions are obtained which do not depend on unknown nuisance parameters. Of course, there are also situations when a more detailed knowledge of the asymptotic distributions and closed form expressions are helpful.

The results of Proposition C.16 can be generalized in different ways. First of all, the process x_t may have a more complicated dependence structure. In particular, the error process may be a stationary process. Consider, for instance, a process $x_t = x_{t-1} + w_t$, where $w_t = \sum_{j=0}^{\infty} \theta_j u_{t-j} = \theta(L)u_t$ is a stationary process with $\sum_{j=0}^{\infty} j|\theta_j| < \infty$ and $u_t \sim (0, \sigma_u^2)$ is white noise, then x_t can be rewritten as

$$x_t = x_0 + w_1 + \cdots + w_t = x_0 + \theta(1)(u_1 + \cdots + u_t) + \sum_{j=0}^{\infty} \theta_j^* u_{t-j} - w_0^*,$$

where $\theta(1) = \sum_{j=0}^{\infty} \theta_j$, $\theta_j^* = -\sum_{i=j+1}^{\infty} \theta_i$, $j = 0, 1, \ldots$, and $w_0^* = \sum_{j=0}^{\infty} \theta_j^* u_{-j}$ contains initial values. Thus, x_t is the sum of a random walk, a stationary process and initial values. Note that the condition $\sum_{j=0}^{\infty} j|\theta_j| < \infty$ ensures that $\sum_{j=0}^{\infty} |\theta_j^*| < \infty$, so that $\sum_{j=0}^{\infty} \theta_j^* u_{t-j}$ is indeed well-defined according to Proposition C.7. Although the condition for the θ_j is stronger than absolute summability, it is satisfied for many processes of practical interest. For example, the MA representation of a stable AR or ARMA process satisfies the condition. The decomposition of x_t in a random walk, a stationary component, and initial values is known as the *Beveridge-Nelson decomposition*. It is a convenient tool in generalizing the results in Propositions C.16 and C.17.

In fact, if y_t is a finite order AR process, $y_t = \alpha_1 y_{t-1} + \cdots + \alpha_p y_{t-p} + u_t$, where u_t is again white noise, y_t can be rewritten as

$$y_t = \rho y_{t-1} + \gamma_1 \Delta y_{t-1} + \cdots + \gamma_{p-1} \Delta y_{t-p+1} + u_t$$

or, subtracting y_{t-1} on both sides,

$$\Delta y_t = (\rho - 1)y_{t-1} + \gamma_1 \Delta y_{t-1} + \cdots + \gamma_{p-1} \Delta y_{t-p+1} + u_t.$$

Estimating ρ or $\rho-1$ from these equations by LS, it turns out that the resulting estimators have the same asymptotic properties as in Proposition C.17 (see, e.g., Hamilton (1994)).

Another possible generalization of these results may be obtained by considering multivariate processes. We will tackle both generalizations at once in the following.

C.8.2 Multivariate Processes

For the present purposes, multivariate Brownian motions or Wiener processes are of central importance. The univariate definition can be generalized as follows. A K-dimensional *standard Brownian motion* or *standard Wiener process* $\mathbf{W}(\cdot)$ is a function defined on the unit interval $[0, 1]$ which assigns a K-dimensional random vector $\mathbf{W}(t)$ to each $t \in [0, 1]$ such that:

(1) $\mathbf{W}(0) = 0$ with probability one.
(2) A realization $\mathbf{W}(t)$ is a continuous function in t on the unit interval with probability one.
(3) For any partitioning of the unit interval, $0 \leq t_1 < t_2 < \cdots < t_k \leq 1$, the vector

$$\begin{bmatrix} \mathbf{W}(t_2) - \mathbf{W}(t_1) \\ \vdots \\ \mathbf{W}(t_k) - \mathbf{W}(t_{k-1}) \end{bmatrix} \sim \mathcal{N}\left(\begin{bmatrix} 0 \\ \vdots \\ 0 \end{bmatrix}, \begin{bmatrix} (t_2 - t_1)I_K & & 0 \\ & \ddots & \vdots \\ 0 & \cdots & (t_k - t_{k-1})I_K \end{bmatrix} \right),$$

that is, the differences have multivariate normal distributions with independent components, means of zero, and variances of the form $t_i - t_{i-1}$, depending on their difference in time.

Again, for any nonsingular $(K \times K)$ matrix P, a nonstandard version of a Wiener process $Z(t) := P\mathbf{W}(t)$ is obtained for which the increments are still independent but $Z(t) - Z(s) \sim \mathcal{N}(0, (t-s)PP')$ for $s < t$. Moreover, $Z(t) \sim \mathcal{N}(0, tPP')$.

For a sequence $G_T(\cdot)$ of multivariate random functions, we define convergence in probability to a random function G, $G_T \xrightarrow{p} G$, to hold if

$$\sup_{t \in [0,1]} \|G_T(t) - G(t)\| \xrightarrow[T \to \infty]{p} 0.$$

Also, the continuous mapping theorem remains valid in the multivariate case.

As in the univariate case, it is of interest to consider quantities of the form

$$X_T(r) = \frac{1}{T} \sum_{t=1}^{[Tr]} w_t,$$

where w_t is a stationary stochastic process, $r \in [0, 1]$ denotes a fraction and $[Tr]$ signifies the largest integer less than or equal to Tr. If $w_t = u_t \sim (0, \Sigma_u)$ is i.i.d. white noise, it follows from a multivariate version of a suitable CLT (see Proposition C.13) that

$$\sqrt{T}[X_T(r_2) - X_T(r_1)] \xrightarrow{d} \mathcal{N}(0, (r_2 - r_1)\Sigma_u)$$

for $r_1 < r_2$. Hence, using the same ideas as in the univariate case,

$$\sqrt{T}\Sigma_u^{-1/2} X_T(\cdot) \xrightarrow{d} \mathbf{W}(\cdot),$$

which is a multivariate version of the previously stated FCLT also referred to as *invariance principle* or *Donsker's theorem*.

If $x_t = x_{t-1} + w_t$, where

$$w_t = \Xi(L)u_t = \sum_{j=0}^{\infty} \Xi_j u_{t-j}, \quad \text{with} \quad \sum_{j=0}^{\infty} j\|\Xi_j\| < \infty,$$

and $u_t \sim (0, \Sigma_u = (\sigma_{ij}))$ is white noise, then a *multivariate Beveridge-Nelson decomposition* is available,

$$x_t = x_0 + w_1 + \cdots + w_t = x_0 + \Xi(1) \sum_{s=1}^{t} u_s + \sum_{j=0}^{\infty} \Xi_j^* u_{t-j} - w_0^*,$$

where $\Xi(1) = \sum_{j=0}^{\infty} \Xi_j$, $\Xi_j^* = -\sum_{i=j+1}^{\infty} \Xi_i$, $j = 0, 1, \ldots$, and $w_0^* = \sum_{j=0}^{\infty} \Xi_j^* u_{-j}$ contains initial values. Now x_t is a sum of a multivariate random walk, a stationary process, and initial values (see also Proposition 6.1). Using these concepts, the following generalized version of Proposition C.16 can be established. It also goes back to Phillips and others (see Phillips & Durlauf (1986), Park & Phillips (1988, 1989), Phillips & Solo (1992), Sims et al. (1990), Johansen (1995)) and may be found, e.g., in Hamilton (1994, Proposition 18.1).

Proposition C.18 (*Properties of Multivariate Unit Root Processes*)
Suppose $x_t = x_{t-1} + w_t$, $t = 1, 2, \ldots$, is a K-dimensional generalized random walk with initial vector $x_0 = 0$ and stationary error term

$$w_t = \Xi(L)u_t = \sum_{j=0}^{\infty} \Xi_j u_{t-j}, \qquad t \in \mathbb{Z},$$

where

$$\sum_{j=0}^{\infty} j\|\Xi_j\| < \infty,$$

and $u_t \sim (0, \Sigma_u = (\sigma_{ij}))$, $t \in \mathbb{Z}$, is i.i.d. white noise with finite fourth moments. Let P be a lower triangular matrix such that $\Sigma_u = PP'$,

$$\Gamma_w(h) := E(w_t w'_{t-h}) = \sum_{j=0}^{\infty} \Xi_{j+h} \Sigma_u \Xi'_j, \qquad h = 0, 1, 2, \ldots,$$

for an arbitrary positive integer n, $W_t := (w'_{t-1}, \ldots, w'_{t-n})'$ is a Kn-dimensional vector with $\Sigma_W := E(W_t W'_t)$, and the $(K \times K)$ matrix $\Lambda := \Xi(1)P$. Then the following results hold:

(1) $T^{-1/2} \sum_{t=1}^{T} w_t \xrightarrow{d} \Lambda \mathbf{W}(1)$.
(2) $T^{-1/2} \sum_{t=1}^{T} W_t u_{it} \xrightarrow{d} \mathcal{N}(0, \sigma_{ii} \Sigma_W)$ for $i = 1, \ldots, K$.
(3) $T^{-1} \sum_{t=1}^{T} w_t w'_{t-h} \xrightarrow{p} \Gamma_w(h)$ for $h = 0, 1, 2, \ldots$.
(4) $T^{-1} \sum_{t=1}^{T} (x_{t-1} w'_{t-h} + w_{t-h} x'_{t-1})$

$$\xrightarrow{d} \begin{cases} \Lambda \mathbf{W}(1)\mathbf{W}(1)'\Lambda' - \Gamma_w(0) & \text{for } h = 0, \\ \Lambda \mathbf{W}(1)\mathbf{W}(1)'\Lambda' - \Gamma_w(0) + \sum_{j=-h+1}^{h-1} \Gamma_w(j) & \text{for } h = 1, 2, \ldots. \end{cases}$$

(5) $T^{-1} \sum_{t=1}^{T} x_{t-1} w'_t \xrightarrow{d} \Lambda \left\{ \int_0^1 \mathbf{W}(r) d\mathbf{W}(r)' \right\} \Lambda' + \sum_{j=1}^{\infty} \Gamma_w(j)$.
(6) $T^{-1} \sum_{t=1}^{T} x_{t-1} u'_t \xrightarrow{d} \Lambda \left\{ \int_0^1 \mathbf{W}(r) d\mathbf{W}(r)' \right\} P'$.
(7) $T^{-3/2} \sum_{t=1}^{T} x_{t-1} \xrightarrow{d} \Lambda \int_0^1 \mathbf{W}(r) dr$.
(8) $T^{-3/2} \sum_{t=1}^{T} t w_{t-h} \xrightarrow{d} \Lambda \left\{ \mathbf{W}(1) - \int_0^1 \mathbf{W}(r) dr \right\}$ for $h = 0, 1, 2, \ldots$.
(9) $T^{-2} \sum_{t=1}^{T} x_{t-1} x'_{t-1} \xrightarrow{d} \Lambda \left\{ \int_0^1 \mathbf{W}(r)\mathbf{W}(r)' dr \right\} \Lambda'$.
(10) $T^{-5/2} \sum_{t=1}^{T} t x_{t-1} \xrightarrow{d} \Lambda \int_0^1 r\mathbf{W}(r) dr$.
(11) $T^{-3} \sum_{t=1}^{T} t x_{t-1} x'_{t-1} \xrightarrow{d} \Lambda \left\{ \int_0^1 r\mathbf{W}(r)\mathbf{W}(r)' dr \right\} \Lambda'$.

■

These results are the basis for much of the asymptotic theory related to multivariate VAR processes with unit roots. Extensions exist for more general processes w_t and u_t.

D

Evaluating Properties of Estimators and Test Statistics by Simulation and Resampling Techniques

If asymptotic theory is difficult or only small samples are available, properties of estimators and test statistics are sometimes investigated by heavy use of the computer. The idea is to simulate the distribution (or some of its properties) of the random variables of interest by artificially sampling from some known distribution. Generally, if the random variable or vector of interest, say $q = q(z)$, is a function of a random vector z with a known distribution F_z, then samples $z_1, \ldots, z_n$ are drawn from F_z and the empirical distribution of q given by $q_n = q(z_n)$, $n = 1, \ldots, N$, is determined. The characteristics of the actual distribution of q are then inferred from the empirical distribution.

Often the statistics of interest in this book are functions of multiple time series generated by VAR(p) processes. Therefore, we will briefly describe in the next section how to simulate such time series. Afterwards, some more details are given on simulation and resampling techniques for evaluating estimators and test statistics.

D.1 Simulating a Multiple Time Series with VAR Generation Process

To simulate a multiple time series of dimension K and length T, we first generate a series of (often independent) disturbance vectors $u_{-s}, \ldots, u_0, u_1, \ldots, u_T$. If a series of Gaussian disturbances is desired, i.e., $u_t \sim \mathcal{N}(0, \Sigma_u)$, we may choose K independent univariate standard normal variates $v_1, \ldots, v_K$ and multiply by a $(K \times K)$ matrix P for which $PP' = \Sigma_u$, that is,

$$u_t = P \begin{bmatrix} v_1 \\ \vdots \\ v_K \end{bmatrix}.$$

This process is repeated $T + s + 1$ times until we have the desired series of disturbances. Programs for generating (pseudo) standard normal variates are available on most computers. Also facilities for generating random numbers from other distributions are usually available and may be used in a similar manner to obtain disturbances from other distributions of interest.

For a given set of parameters $\nu, A_1, \ldots, A_p$, where ν is $(K \times 1)$ and the A_i are $(K \times K)$, and a given set of starting values $y_{-p+1}, \ldots, y_0$, the u_t may be used to simulate a time series $y_1, \ldots, y_T$ with VAR(p) generation process recursively as

$$y_t = \nu + A_1 y_{t-1} + \cdots + A_p y_{t-p} + u_t$$

starting with $t = 1, t = 2$, etc. until $t = T$. There are different ways to obtain the initial values. Assuming that the desired process is stable, they may be set to zero or to the process mean $\mu = (I_K - A_1 - \cdots - A_p)^{-1}\nu$. Because the choice of initial values has some impact on the generated time series, a number of presample values $y_t, t = -s, \ldots, 0$, is often generated and then discarded in the subsequent analysis.

A possible way to ensure the same correlation structure for the initial values and the rest of the time series is to determine the covariance matrix of p consecutive y_t vectors, say Σ_Y. Using the results of Chapter 2, Section 2.1, that matrix may be obtained from

$$\text{vec}(\Sigma_Y) = (I_{(Kp)^2} - \mathbf{A} \otimes \mathbf{A})^{-1} \text{vec}(\Sigma_U),$$

where

$$\mathbf{A} = \begin{bmatrix} A_1 & A_2 & \cdots & A_{p-1} & A_p \\ I_K & 0 & \cdots & 0 & 0 \\ 0 & I_K & & 0 & 0 \\ \vdots & & \ddots & \vdots & \vdots \\ 0 & 0 & \cdots & I_K & 0 \end{bmatrix}_{(Kp \times Kp)} \quad \text{and} \quad \Sigma_U = \begin{bmatrix} \Sigma_u & 0 & \cdots & 0 \\ 0 & 0 & \cdots & 0 \\ \vdots & \vdots & \ddots & \vdots \\ 0 & 0 & \cdots & 0 \end{bmatrix}_{(Kp \times Kp)}.$$

Then a $(Kp \times Kp)$ matrix Q is chosen such that $QQ' = \Sigma_Y$ and p initial starting vectors are obtained as

$$\begin{bmatrix} y_0 \\ \vdots \\ y_{-p+1} \end{bmatrix} = Q \begin{bmatrix} v_1 \\ \vdots \\ v_{Kp} \end{bmatrix} + \begin{bmatrix} \mu \\ \vdots \\ \mu \end{bmatrix},$$

where the v_i are independent variates with mean zero and unit variance.

D.2 Evaluating Distributions of Functions of Multiple Time Series by Simulation

Suppose we are interested in the function $q_T = q(y_1, \ldots, y_T)$ of some VAR(p) process y_t, where q_T is of dimension $(M \times 1)$. The quantity q_T may be some estimator or test statistic. To investigate the distribution F_T of q_T, we generate a large number, say N, of independent multiple time series of length T and compute the corresponding values of q_T, say $q_T(n)$, $n = 1, \ldots, N$. The properties of F_T are then estimated from the empirical distribution of the $q_T(n)$. For instance, the mean vector of q_T is estimated as

$$\frac{1}{N} \sum_{n=1}^{N} q_T(n).$$

Analogously, we may estimate the variances, standard deviations, quantiles or other characteristics of F_T.

D.3 Resampling Methods

If the distribution of the disturbances of a VAR model under consideration is unknown, so-called *bootstrap* or *resampling* methods may be applied to investigate the distributions of functions of stochastic processes or multiple time series. Suppose a time series $y_1, \ldots, y_T$ and the presample values required for estimation are available. Fitting a VAR(p) model to this time series, we get coefficient estimates $\widehat{\nu}, \widehat{A}_1, \ldots, \widehat{A}_p$, and a series of residuals $\widehat{u}_1, \ldots, \widehat{u}_T$. An estimator of a quantity of interest, say $q = q(A_1, \ldots, A_p)$, is then obtained as

$$\widehat{q} = q(\widehat{A}_1, \ldots, \widehat{A}_p). \tag{D.3.1}$$

The properties of $\widehat{q}$ follow from those of $\widehat{A}_1, \ldots, \widehat{A}_p$. To assess the sampling uncertainty of $\widehat{q}$, confidence intervals are often established, based on the asymptotic distribution of $\widehat{q}$. Alternatively, if q is a test statistic, its p-value may be of interest which can be approximated on the basis of the asymptotic distribution. Unfortunately, this distribution is often a rather poor approximation of the actual distribution for a given finite sample. In some of these cases, bootstrap methods provide a better small sample approximation. The theoretical justification for the bootstrap also rests on asymptotic theory, however. In particular, it can usually be justified if the quantity of interest has a normal limiting distribution (Horowitz (2001)).

A *residual based bootstrap* is often used in this context. Assuming that a sample $y_1, \ldots, y_T$ plus presample values as required are available, it proceeds as follows:

(1) The parameters of the model under consideration are estimated. Let $\widehat{u}_t$, $t = 1, \ldots, T$, be the estimation residuals.
(2) Centered residuals $\widehat{u}_1 - \overline{u}_., \ldots, \widehat{u}_T - \overline{u}_.$ are computed. Here $\overline{u}_. = T^{-1} \sum \widehat{u}_t$ denotes the usual average. Bootstrap residuals $u_1^*, \ldots, u_T^*$ are then obtained by randomly drawing with replacement from the centered residuals.

(3) Bootstrap time series are computed recursively as

$$y_t^* = \widehat{\nu} + \widehat{A}_1 y_{t-1}^* + \cdots + \widehat{A}_p y_{t-p}^* + u_t^*, \quad t = 1, \ldots, T,$$

where the same initial values may be used for each generated series, $(y_{-p+1}^*, \ldots, y_0^*) = (y_{-p+1}, \ldots, y_0)$.
(4) Based on the bootstrap time series, the parameters $A_1, \ldots, A_p$ are reestimated.
(5) Using the parameter estimates obtained in the previous stage, a bootstrap version of the statistic of interest, say $\widehat{q}^*$, is calculated.
(6) These steps are repeated N times, where N is a large number.

There is now a range of other bootstrap methods which may have advantages in certain situations. For example, rather than using a residual-based bootstrap, a *block bootstrap* may be applied which is based on the original observations rather than the model residuals (see, e.g., Li & Maddala (1996) for details). It may be preferable if there is uncertainty regarding specific aspects of the model like, for instance, the VAR order. These methods are not discussed here because residual based bootstraps are still the most popular methods in the present context.

In the following, the symbol q denotes the quantity of interest for which a confidence interval is desired. Its estimator implied by the estimators of the model coefficients and the corresponding bootstrap estimator are denoted by $\widehat{q}$ and $\widehat{q}^*$, respectively. The following bootstrap confidence intervals are examples that have been considered in the literature in the context of impulse response analysis (see, e.g., Benkwitz, Lütkepohl & Wolters (2001)):

- *Standard percentile interval*
  Denoting by $s_{\gamma/2}^*$ and $s_{(1-\gamma/2)}^*$ the $\gamma/2$- and $(1-\gamma/2)$-quantiles, respectively, of the N bootstrap versions of $\widehat{q}^*$, the interval

  $$CI_S = \left[s_{\gamma/2}^*, s_{(1-\gamma/2)}^* \right],$$

  may be set up. It is the percentile confidence interval discussed, e.g., by Efron & Tibshirani (1993).
- *Hall's percentile interval*
  Hall (1992) uses the result that asymptotically the distribution of $\sqrt{T}(\widehat{q}-q)$ corresponds to that of $\sqrt{T}(\widehat{q}^* - \widehat{q})$, to derive the interval

  $$CI_H = \left[\widehat{q} - t_{(1-\gamma/2)}^*, \widehat{q} - t_{\gamma/2}^* \right].$$

  Here $t_{\gamma/2}^*$ and $t_{(1-\gamma/2)}^*$ are the $\gamma/2$- and $(1-\gamma/2)$-quantiles, respectively, of $(\widehat{q}^* - \widehat{q})$ and the interval is obtained by pretending that these are the quantiles of $(\widehat{q} - q)$.
- *Hall's studentized interval*
  A studentized statistic $(\widehat{q} - q)/(\widehat{\text{Var}(\widehat{q})})^{1/2}$ often results in more precise

confidence intervals at least in theory. Using bootstrap quantiles $t^{**}_{\gamma/2}$ and $t^{**}_{(1-\gamma/2)}$ from the distribution of $(\widehat{q}^* - \widehat{q})/(\widehat{\text{Var}}(\widehat{q}^*))^{1/2}$, an interval

$$CI_{SH} = \left[\widehat{q} - t^{**}_{(1-\gamma/2)}\sqrt{\widehat{\text{Var}}(\widehat{q})},\ \widehat{q} - t^{**}_{\gamma/2}\sqrt{\widehat{\text{Var}}(\widehat{q})}\right]$$

can be constructed by using these quantities in conjunction with $(\widehat{q} - q)/(\widehat{\text{Var}}(\widehat{q}))^{1/2}$. Here the variance $\text{Var}(\widehat{q})$ may be estimated from the bootstrap estimates of q,

$$\widehat{\text{Var}}(\widehat{q}) = \frac{1}{N-1} \sum_{i=1}^{N} \left(\widehat{q}^{*,i} - \overline{\widehat{q}^*}\right)^2,$$

where N is the number of bootstrap replications and $\widehat{q}^{*,i}$ denotes the value of the statistic of interest obtained in the i-th bootstrap replication. Moreover, the variances $\widehat{\text{Var}}(\widehat{q}^*)$ may be estimated by a bootstrap within each bootstrap replication. In other words,

$$\widehat{\text{Var}}(\widehat{q}^*) = \frac{1}{N^*-1} \sum_{i=1}^{N^*} \left(\widehat{q}^{**,i} - \overline{\widehat{q}^{**}}\right)^2,$$

where $\widehat{q}^{**,i}$ is obtained by a double bootstrap, that is, pseudo-data are generated according to a process obtained on the basis of the bootstrap systems parameters and N^* is the number of bootstrap replications within each bootstrap replication.

A number of refinements and modifications of these intervals exist (see Hall (1992)).

The bootstrap confidence intervals have the property that they attain the nominal confidence content at least asymptotically under general conditions. Roughly speaking, if $\sqrt{T}(\widehat{q} - q)$ converges as $T \to \infty$, $\sqrt{T}(\widehat{q}^* - \widehat{q})$ converges to the same limit distribution under suitable conditions (e.g., Hall (1992)). Therefore CI_H has the correct size asymptotically, that is, $\Pr(q \in CI_H) \to 1 - \gamma$ as $T \to \infty$, under general conditions, and, hence, Hall's percentile method is asymptotically precise. The same holds for the CI_{SH} interval. On the other hand, to obtain such a result for the standard percentile interval CI_S, the limiting distribution of $\sqrt{T}(\widehat{q} - q)$ has to be symmetric about zero. For example, this result holds if it is zero mean normal. Roughly speaking, CI_S works with an implicit asymptotic unbiasedness assumption for $\widehat{q}$. If the distribution of $\widehat{q}$ is not centered at q, CI_S will generally not have the desired confidence content even asymptotically (see also Efron & Tibshirani (1993) and Benkwitz et al. (2000) for a more detailed discussion of this point).

If $\widehat{q}$ is a statistic for which a p-value is desired, the following method may be used. Recall that the p-value of a test is the probability of obtaining a value of the test statistic greater than the observed one, if the null hypothesis holds.

Hence, the p-value may be estimated by the proportion of bootstrap values $\widehat{q}^*$ exceeding the value of the test statistic $\widehat{q}$. Again, under general assumptions, this estimator is consistent.

References

Abraham, B. (1980). Intervention analysis and multiple time series, *Biometrika* **67**: 73–78.

Ahn, S. K. (1988). Distribution for residual autocovariances in multivariate autoregressive models with structured parameterization, *Biometrika* **75**: 590–593.

Ahn, S. K. & Reinsel, G. C. (1988). Nested reduced-rank autoregressive models for multiple time series, *Journal of the American Statistical Association* **83**: 849–856.

Ahn, S. K. & Reinsel, G. C. (1990). Estimation of partially nonstationary multivariate autoregressive models, *Journal of the American Statistical Association* **85**: 813–823.

Akaike, H. (1969). Fitting autoregressive models for prediction, *Annals of the Institute of Statistical Mathematics* **21**: 243–247.

Akaike, H. (1971). Autoregressive model fitting for control, *Annals of the Institute of Statistical Mathematics* **23**: 163–180.

Akaike, H. (1973). Information theory and an extension of the maximum likelihood principle, *in* B. N. Petrov & F. Csáki (eds), *2nd International Symposium on Information Theory*, Académiai Kiadó, Budapest, pp. 267–281.

Akaike, H. (1974). A new look at the statistical model identification, *IEEE Transactions on Automatic Control* **AC-19**: 716–723.

Akaike, H. (1976). Canonical correlation analysis of time series and the use of an information criterion, *in* R. K. Mehra & D. G. Lainiotis (eds), *Systems Identification: Advances and Case Studies*, Academic Press, New York, pp. 27–96.

Amisano, G. & Giannini, C. (1997). *Topics in Structural VAR Econometrics*, 2nd edn, Springer, Berlin.

Anděl, J. (1983). Statistical analysis of periodic autoregression, *Aplikace Mathematiky* **28**: 364–385.

Anděl, J. (1987). On multiple periodic autoregression, *Aplikace Mathematiky* **32**: 63–80.

Anděl, J. (1989). On nonlinear models for time series, *Statistics* **20**: 615–632.

Anderson, B. D. O. & Moore, J. B. (1979). *Optimal Filtering*, Prentice-Hall, Englewood Cliffs, NJ.

Anderson, T. W. (1971). *The Statistical Analysis of Time Series*, John Wiley, New York.

Anderson, T. W. (1984). *An Introduction to Multivariate Statistical Analysis*, 2nd edn, John Wiley, New York.

Anderson, T. W. (1999). Asymptotic distribution of the reduced rank regression estimator under general conditions, *Annals of Statistics* **27**: 1141–1154.

Anderson, T. W. (2002). Canonical correlation analysis and reduced rank regression in autoregressive models, *Annals of Statistics* **30**: 1134–1154.

Andrews, D. W. K. (1987). Asymptotic results for generalized Wald tests, *Econometric Theory* **3**: 348–358.

Ansley, C. F. & Kohn, R. (1983). Exact likelihood of vector autoregressive-moving average process with missing or aggregated data, *Biometrika* **70**: 275–278.

Aoki, M. (1987). *State Space Modeling of Time Series*, Springer, Berlin.

Baba, Y., Engle, R. F., Kraft, D. F. & Kroner, K. F. (1990). Multivariate simultaneous generalized ARCH, *mimeo*, Department of Economics, University of California, San Diego.

Bai, J. (1994). Least squares estimation of a shift in linear processes, *Journal of Time Series Analysis* **15**: 453–472.

Bai, J., Lumsdaine, R. L. & Stock, J. H. (1998). Testing for and dating common breaks in multivariate time series, *Review of Economic Studies* **65**: 395–432.

Baillie, R. T. (1981). Prediction from the dynamic simultaneous equation model with vector autoregressive errors, *Econometrica* **49**: 1331–1337.

Baillie, R. T. & Bollerslev, T. (1992). Prediction in dynamic models with time-dependent conditional variances, *Journal of Econometrics* **52**: 91–113.

Banerjee, A., Dolado, J. J., Galbraith, J. W. & Hendry, D. F. (1993). *Co-integration, Error-Correction, and the Econometric Analysis of Non-stationary Data*, Oxford University Press, Oxford.

Baringhaus, L. & Henze, N. (1988). A consistent test for multivariate normality based on the empirical characteristic function, *Metrika* **35**: 339–348.

Barone, P. (1987). A method for generating independent realizations of a multivariate normal stationary and invertible ARMA(p,q) process, *Journal of Time Series Analysis* **8**: 125–130.

Bartel, H. & Lütkepohl, H. (1998). Estimating the Kronecker indices of cointegrated echelon form VARMA models, *Econometrics Journal* **1**: C76–C99.

Basu, A. K. & Sen Roy, S. (1986). On some asymptotic results for multivariate autoregressive models with estimated parameters, *Calcutta Statistical Association Bulletin* **35**: 123–132.

Basu, A. K. & Sen Roy, S. (1987). On asymptotic prediction problems for multivariate autoregressive models in the unstable nonexplosive case, *Calcutta Statistical Association Bulletin* **36**: 29–37.

Bauer, D. & Wagner, M. (2002). Estimating cointegrated systems using subspace algorithms, *Journal of Econometrics* **111**: 47–84.

Bauer, D. & Wagner, M. (2003). A canonical form for unit root processes in the state space framework, Diskussionsschriften 03-12, Universität Bern.

Bauwens, L., Laurent, S. & Rombouts, J. V. K. (2004). Multivariate GARCH models: A survey, CORE Discussion Paper 2003/31.

Benkwitz, A., Lütkepohl, H. & Neumann, M. (2000). Problems related to bootstrapping impulse responses of autoregressive processes, *Econometric Reviews* **19**: 69–103.

Benkwitz, A., Lütkepohl, H. & Wolters, J. (2001). Comparison of bootstrap confidence intervals for impulse responses of German monetary systems, *Macroeconomic Dynamics* **5**: 81–100.

Bera, A. K. & Higgins, M. L. (1993). ARCH models: Properties, estimation and testing, *Journal of Economic Surveys* **7**: 305–366.

Berk, K. N. (1974). Consistent autoregressive spectral estimates, *Annals of Statistics* **2**: 489–502.

Bernanke, B. (1986). Alternative explanations of the money–income correlation, *Carnegie-Rochester Conference Series on Public Policy*, North-Holland, Amsterdam.

Bewley, R. & Yang, M. (1995). Tests for cointegration based on canonical correlation analysis, *Journal of the American Statistical Association* **90**: 990–996.

Bhansali, R. J. (1978). Linear prediction by autoregressive model fitting in the time domain, *Annals of Statistics* **6**: 224–231.

Bierens, H. J. (1981). *Robust Methods and Asymptotic Theory in Nonlinear Econometrics*, Springer, Berlin.

Bierens, H. J. (1997). Nonparametric cointegration analysis, *Journal of Econometrics* **77**: 379–404.

Black, F. (1976). Studies in stock price volatility changes, *Proceedings of the 1976 Meeting of the Business and Economic Statistics Section*, American Statistical Association, pp. 177–181.

Blanchard, O. & Quah, D. (1989). The dynamic effects of aggregate demand and supply disturbances, *American Economic Review* **79**: 655–673.

Bollerslev, T. (1986). Generalized autoregressive conditional heteroskedasticity, *Journal of Econometrics* **31**: 307–327.

Bollerslev, T. (1990). Modeling the coherence in short-run nominal exchange rates: A multivariate generalized ARCH model, *Review of Economics and Statistics* **72**: 498–505.

Bollerslev, T., Chou, R. Y. & Kroner, K. F. (1992). ARCH modelling in finance: A review of the theory and empirical evidence, *Journal of Econometrics* **52**: 5–59.

Bollerslev, T., Engle, R. F. & Nelson, D. B. (1994). ARCH models, in R. Engle & D. McFadden (eds), *Handbook of Econometrics*, Vol. 4, Elsevier, Amsterdam, pp. 2959–3038.

Bollerslev, T., Engle, R. F. & Wooldridge, J. M. (1988). A capital asset pricing model with time-varying covariances, *Journal of Political Economy* **96**: 116–131.

Bossaerts, P. (1988). Common nonstationary components of asset prices, *Journal of Economic Dynamics and Control* **12**: 347–364.

Boswijk, H. P. & Doornik, J. A. (2002). Identifying, estimating and testing restricted cointegrated systems: An overview, *Paper presented at the Henri Theil Memorial Conference*, Universiteit van Amsterdam.

Boswijk, H. P. & Franses, P. H. (1995). Periodic cointegration: Representation and inference, *Review of Economics and Statistics* **77**: 436–454.

Boswijk, H. P. & Franses, P. H. (1996). Unit roots in periodic autoregressions, *Journal of Time Series Analysis* **17**: 221–245.

Boswijk, H. P., Franses, P. H. & Haldrup, N. (1997). Multiple unit roots in periodic autoregression, *Journal of Econometrics* **80**: 167–193.

Boswijk, H. P. & Lucas, A. (2002). Semi-nonparametric cointegration testing, *Journal of Econometrics* **108**: 253–280.

Box, G. E. P. & Jenkins, G. M. (1976). *Time Series Analysis: Forecasting and Control*, Holden-Day, San Francisco.

Box, G. E. P. & Tiao, G. C. (1975). Intervention analysis with applications to economic and environmental problems, *Journal of the American Statistical Association* **70**: 70–79.

Braun, P. A., Nelson, D. B. & Sunier, A. M. (1995). Good news, bad news, volatility, and betas, *Journal of Finance* **50**: 1575–1603.

Breitung, J., Brüggemann, R. & Lütkepohl, H. (2004). Structural vector autoregressive modeling and impulse responses, *in* H. Lütkepohl & M. Krätzig (eds), *Applied Time Series Econometrics*, Cambridge University Press, Cambridge, pp. 159–196.

Breitung, J. & Hassler, U. (2002). Inference on the cointegration rank in fractionally integrated processes, *Journal of Econometrics* **110**: 167–185.

Breusch, T. S. (1978). Testing for autocorrelation in dynamic linear models, *Australian Economic Papers* **17**: 334–355.

Breusch, T. S. & Schmidt, P. (1988). Alternative forms of the Wald test: How long is a piece of string, *Communications in Statistics, Theory and Methods* **17**: 2789–2795.

Brockwell, P. J. & Davis, R. A. (1987). *Time Series: Theory and Methods*, Springer, New York.

Brooks, C. & Burke, S. (2003). Information criteria for GARCH model selection, *European Journal of Finance* **9**: 557–580.

Brooks, C., Burke, S. & Persand, G. (2003). Multivariate GARCH models: Software choice and estimation issues, *Journal of Applied Econometrics* **18**: 725–734.

Brüggemann, R. (2004). *Model Reduction Methods for Vector Autoregressive Processes*, Springer, Berlin.

Brüggemann, R. & Lütkepohl, H. (2001). Lag selection in subset VAR models with an application to a U.S. monetary system, *in* R. Friedmann, L. Knüppel & H. Lütkepohl (eds), *Econometric Studies: A Festschrift in Honour of Joachim Frohn*, LIT Verlag, Münster, pp. 107–128.

Brüggemann, R. & Lütkepohl, H. (2004). Practical problems with reduced rank ML estimators for cointegration parameters and a simple alternative, *Discussion paper*, European University Institute, Florence.

Brüggemann, R., Lütkepohl, H. & Saikkonen, P. (2004). Residual autocorrelation testing for vector error correction models, *Discussion paper*, European University Institute, Florence.

Caines, P. E. (1988). *Linear Stochastic Systems*, John Wiley, New York.

Candelon, B. & Lütkepohl, H. (2001). On the reliability of Chow-type tests for parameter constancy in multivariate dynamic models, *Economics Letters* **73**: 155–160.

Caner, M. (1998). Tests for cointegration with infinite variance errors, *Journal of Econometrics* **86**: 155–175.

Caporale, G. M. & Pittis, N. (2004). Estimator choice and Fisher's paradox: A Monte Carlo study, *Econometric Reviews* **23**: 25–52.

Cheung, Y.-W. & Ng, L. K. (1996). A causality-in-variance test and its application to financial market prices, *Journal of Econometrics* **72**: 33–48.

Chitturi, R. V. (1974). Distribution of residual autocorrelations in multiple autoregressive schemes, *Journal of the American Statistical Association* **69**: 928–934.

Choi, I. (1994). Residual-based tests for the null of stationarity with applications to U.S. macroeconomic time series, *Econometric Theory* **10**: 720–749.

Chow, G. C. (1975). *Analysis and Control of Dynamic Economic Systems*, John Wiley, New York.

Chow, G. C. (1981). *Econometric Analysis by Control Methods*, John Wiley, New York.

Chow, G. C. (1984). Random and changing coefficient models, *in* Z. Griliches & M. D. Intriligator (eds), *Handbook of Econometrics*, Vol. 2, North-Holland, Amsterdam, pp. 1213–1245.

Christiano, L. J., Eichenbaum, M. & Evans, C. (1996). The effects of monetary policy shocks: Evidence from the flow of funds, *Review of Economics and Statistics* **78**: 16–34.

Cipra, T. (1985). Periodic moving average process, *Aplikace Matematiky* **30**: 218–229.

Cleveland, W. P. & Tiao, G. C. (1979). Modeling seasonal time series, *Economie Appliquée* **32**: 107–129.

Comte, F. & Lieberman, O. (2000). Second-order noncausality in multivariate GARCH processes, *Journal of Time Series Analysis* **21**: 535–557.

Comte, F. & Lieberman, O. (2003). Asymptotic theory for multivariate GARCH processes, *Journal of Multivariate Analysis* **84**: 61–84.

Cooley, T. F. & Prescott, E. (1973). Varying parameter regression: A theory and some applications, *Annals of Economic and Social Measurement* **2**: 463–474.

Cooley, T. F. & Prescott, E. (1976). Estimation in the presence of stochastic parameter variation, *Econometrica* **44**: 167–184.

Cooper, D. M. & Wood, E. F. (1982). Identifying multivariate time series models, *Journal of Time Series Analysis* **3**: 153–164.

Crowder, M. J. (1976). Maximum-likelihood estimation for dependent observations, *Journal of the Royal Statistical Society* **B38**: 45–53.

Davidson, J. (1994). *Stochastic Limit Theory*, Oxford University Press, Oxford.

Davidson, J. (2000). *Econometric Theory*, Blackwell, Oxford.

Davidson, J. E. H., Hendry, D. F., Srba, F. & Yeo, S. (1978). Econometric modelling of the aggregate time-series relationship between consumer's expenditure and income in the United Kingdom, *Economic Journal* **88**: 661–692.

Davies, N., Triggs, C. M. & Newbold, P. (1977). Significance levels of the Box-Pierce portmanteau statistic in finite samples, *Biometrika* **64**: 517–522.

Deistler, M. & Hannan, E. J. (1981). Some properties of the parameterization of ARMA systems with unknown order, *Journal of Multivariate Analysis* **11**: 474–484.

Deistler, M. & Pötscher, B. M. (1984). The behaviour of the likelihood function for ARMA models, *Advances in Applied Probability* **16**: 843–865.

Dempster, A. P., Laird, N. M. & Rubin, D. B. (1977). Maximum-likelihood from incomplete data via the EM-algorithm, *Journal of the Royal Statistical Society* **B39**: 1–38.

Diebold, F. X. & Nerlove, M. (1989). Dynamic exchange rate volatility: A multivariate latent factor ARCH model, *Journal of Applied Econometrics* **4**: 1–21.

Doan, T., Litterman, R. B. & Sims, C. A. (1984). Forecasting and conditional projection using realistic prior distributions, *Econometric Reviews* **3**: 1–144.

Dolado, J. J. & Lütkepohl, H. (1996). Making Wald tests work for cointegrated VAR systems, *Econometric Reviews* **15**: 369–386.

Doornik, J. A. (1996). Testing vector error autocorrelation and heteroscedasticity, unpublished paper, Nuffield College.

Doornik, J. A. & Hansen, H. (1994). A practical test of multivariate normality, unpublished paper, Nuffield College.

Doornik, J. A. & Hendry, D. F. (1997). *Modelling Dynamic Systems Using PcFiml 9.0 for Windows*, International Thomson Business Press, London.

Doornik, J. A., Hendry, D. F. & Nielsen, B. (1998). Inference in cointegrating models: UK M1 revisited, *Journal of Economic Surveys* **12**: 533–572.

Dufour, J.-M. (1985). Unbiasedness of predictions from estimated vector autoregressions, *Econometric Theory* **1**: 387–402.

Dufour, J.-M. & Renault, E. (1998). Short run and long run causality in time series: Theory, *Econometrica* **66**: 1099–1125.

Dufour, J.-M. & Roy, R. (1985). Some robust exact results on sample autocorrelations and tests of randomness, *Journal of Econometrics* **29**: 257–273.

Dunsmuir, W. T. M. & Hannan, E. J. (1976). Vector linear time series models, *Advances in Applied Probability* **8**: 339–364.

Edgerton, D. & Shukur, G. (1999). Testing autocorrelation in a system perspective, *Econometric Reviews* **18**: 343–386.

Efron, B. & Tibshirani, R. J. (1993). *An Introduction to the Bootstrap*, Chapman & Hall, New York.

Eichenbaum, M. & Evans, C. (1995). Some empirical evidence on the effects of shocks to monetary policy on exchange rates, *Quarterly Journal of Economics* **110**: 975–1009.

Engle, R. F. (1982). Autoregressive conditional heteroscedasticity with estimates of the variance of United Kingdom inflation, *Econometrica* **50**: 987–1007.

Engle, R. F. (2002). Dynamic conditional correlation: A simple class of multivariate generalized autoregressive conditional heteroscedasticity models, *Journal of Business & Economic Statistics* **20**: 339–350.

Engle, R. F. & Granger, C. W. J. (1987). Co-integration and error correction: Representation, estimation and testing, *Econometrica* **55**: 251–276.

Engle, R. F., Granger, C. W. J. & Kraft, D. (1986). Combining competing forecasts of inflation using a bivariate ARCH model, *Journal of Economic Dynamics and Control* **8**: 151–165.

Engle, R. F., Hendry, D. F. & Richard, J. F. (1983). Exogeneity, *Econometrica* **51**: 277–304.

Engle, R. F. & Kroner, K. F. (1995). Multivariate simultaneous generalized GARCH, *Econometric Theory* **11**: 122–150.

Engle, R. F., Lilien, D. M. & Robins, R. P. (1987). Estimating time varying risk premia in the term structure: The ARCH-M model, *Econometrica* **55**: 391–407.

Engle, R. F. & Ng, V. G. (1993). Measuring and testing the impact of news on volatility, *Journal of Finance* **48**: 1749–1778.

Engle, R. F. & Watson, M. (1981). A one-factor multivariate time series model of metropolitan wage rates, *Journal of the American Statistical Association* **76**: 774–781.

Engle, R. F. & Yoo, B. S. (1987). Forecasting and testing in co-integrated systems, *Journal of Econometrics* **35**: 143–159.

Ericsson, N. R. (1994). Testing exogeneity: An introduction, *in* N. R. Ericsson (ed.), *Testing Exogeneity*, Oxford University Press, Oxford, pp. 3–38.

Fiorentini, G., Sentana, E. & Calzolari, G. (2004). On the validity of the Jarque-Bera normality test in conditionally heteroskedastic dynamic regression models, *Economics Letters* **83**: 307–312.

Fisher, L. A. & Huh, H. (1999). Weak exogeneity and long-run and contemporaneous identifying restrictions in VEC models, *Economics Letters* **63**: 159–165.

Forni, M., Hallin, M., Lippi, M. & Reichlin, L. (2000). The generalized dynamic factor model: Identification and estimation, *Review of Economics and Statistics* **82**: 540–552.

Friedmann, R. (1981). The reliability of policy recommendations and forecasts from linear econometric models, *International Economic Review* **22**: 415–428.
Fuller, W. A. (1976). *Introduction to Statistical Time Series*, John Wiley, New York.
Galí, J. (1992). How well does the IS-LM model fit postwar U.S. data?, *Quarterly Journal of Economics* **107**: 709–738.
Galí, J. (1999). Technology, employment, and the business cycle: Do technology shocks explain aggregate fluctuations?, *American Economic Review* **89**: 249–271.
Gallant, A. R. (1987). *Nonlinear Statistical Models*, John Wiley, New York.
Gallant, A. R., Rossi, P. E. & Tauchen, G. (1993). Nonlinear dynamic structures, *Econometrica* **61**: 871–907.
Gallant, A. R. & White, H. (1988). *A Unified Theory of Estimation and Inference for Nonlinear Dynamic Models*, Blackwell, Oxford.
Gasser, T. (1975). Goodness-of-fit tests for correlated data, *Biometrika* **62**: 563–570.
Geweke, J. (1977). The dynamic factor analysis of economic time-series models, *in* D. J. Aigner & A. S. Goldberger (eds), *Latent Variables in Socio-Economic Models*, North-Holland, New York, pp. 365–383.
Geweke, J. (1982). Causality, exogeneity, and inference, *in* W. Hildenbrand (ed.), *Advances in Econometrics*, Cambridge University Press, Cambridge, pp. 209–235.
Geweke, J. (1984). Inference and causality in economic time series models, *in* Z. Griliches & M. D. Intriligator (eds), *Handbook of Econometrics*, Vol. 2, North-Holland, Amsterdam, pp. 1101–1144.
Geweke, J., Meese, R. & Dent, W. (1983). Comparing alternative tests of causality in temporal systems: Analytic results and experimental evidence, *Journal of Econometrics* **21**: 161–194.
Ghysels, E., Harvey, A. & Renault, E. (1996). Stochastic volatility, *in* G. S. Maddala & C. R. Rao (eds), *Handbook of Statistics*, Elsevier, Amsterdam, pp. 119–191.
Ghysels, E. & Osborn, D. R. (2001). *The Econometric Analysis of Seasonal Time Series*, Cambridge University Press, Cambridge.
Giannini, C. (1992). *Topics in Structural VAR Econometrics*, Springer, Heidelberg.
Gladyshev, E. G. (1961). Periodically correlated random sequences, *Soviet Mathematics* **2**: 385–388.
Glosten, L., Jagannathan, R. & Runkle, D. (1993). Relationship between the expected value and the volatility of the nominal excess return on stocks, *Journal of Finance* **48**: 1779–1801.
Godfrey, L. G. (1978). Testing for higher order serial correlation in regression equations when the regressors include lagged dependent variables, *Econometrica* **46**: 1303–1313.
Godfrey, L. G. (1988). *Misspecification Tests in Econometrics*, Cambridge University Press, Cambridge.
Gonzalo, J. (1994). Five alternative methods of estimating long-run equilibrium relationships, *Journal of Econometrics* **60**: 203–233.
Gonzalo, J. & Lee, T.-H. (1998). Pitfalls in testing for long run relationships, *Journal of Econometrics* **86**: 129–154.
Gonzalo, J. & Ng, S. (2001). A systematic framework for analyzing the dynamic effects of permanent and transitory shocks, *Journal of Economic Dynamics and Control* **25**: 1527–1546.
Granger, C. W. J. (1969a). Investigating causal relations by econometric models and cross-spectral methods, *Econometrica* **37**: 424–438.

Granger, C. W. J. (1969b). Prediction with a generalized cost of error function, *Operations Research Quarterly* **20**: 199–207.
Granger, C. W. J. (1981). Some properties of time series data and their use in econometric model specification, *Journal of Econometrics* **16**: 121–130.
Granger, C. W. J. (1982). Generating mechanisms, models, and causality, *in* W. Hildenbrand (ed.), *Advances in Econometrics*, Cambridge University Press, Cambridge, pp. 237–253.
Granger, C. W. J. & Andersen, A. P. (1978). *An Introduction to Bilinear Time Series Models*, Vandenhoeck & Ruprecht, Göttingen.
Granger, C. W. J. & Newbold, P. (1986). *Forecasting Economic Time Series*, 2nd edn, Academic Press, New York.
Granger, C. W. J., Robins, R. P. & Engle, R. F. (1986). Wholesale and retail prices: Bivariate time series modeling with forecastable error variances, *in* D. A. Belsley & E. Kuh (eds), *Model Reliability*, MIT Press, Cambridge, pp. 1–17.
Granger, C. W. J. & Teräsvirta, T. (1993). *Modelling Nonlinear Economic Relationships*, Oxford University Press, Oxford.
Graybill, F. A. (1969). *Introduction to Matrices with Applications in Statistics*, Wadsworth, Belmont, CA.
Gregory, A. W. & Veall, M. R. (1985). Formulating Wald tests of nonlinear restrictions, *Econometrica* **53**: 1465–1468.
Guo, L., Huang, D. W. & Hannan, E. J. (1990). On ARX(∞) approximation, *Journal of Multivariate Analysis* **32**: 17–47.
Hafner, C. M. & Herwartz, H. (1998a). Structural analysis of portfolio risk using beta impulse response functions, *Statistica Neerlandica* **52**: 336–355.
Hafner, C. M. & Herwartz, H. (1998b). Time-varying market price of risk in the CAPM. Approaches, empirical evidence and implications, *Finance* **19**: 93–112.
Haggan, V. & Ozaki, T. (1980). Amplitude-dependent exponential AR model fitting for non-linear random vibrations, *in* O. D. Anderson (ed.), *Time Series*, North-Holland, Amsterdam, pp. 57–71.
Hall, P. (1992). *The Bootstrap and Edgeworth Expansion*, Springer, New York.
Hamilton, J. D. (1986). A standard error for the estimated state vector of a state space model, *Journal of Econometrics* **33**: 387–397.
Hamilton, J. D. (1994). *Time Series Analysis*, Princeton University Press, Princeton, New Jersey.
Hannan, E. J. (1969). The identification of vector mixed autoregressive moving average systems, *Biometrika* **56**: 223–225.
Hannan, E. J. (1970). *Multiple Time Series*, John Wiley, New York.
Hannan, E. J. (1976). The identification and parameterization of ARMAX and state space forms, *Econometrica* **44**: 713–723.
Hannan, E. J. (1979). The statistical theory of linear systems, *in* P. R. Krishnaiah (ed.), *Developments in Statistics*, Academic Press, New York, pp. 83–121.
Hannan, E. J. (1981). Estimating the dimension of a linear system, *Journal of Multivariate Analysis* **11**: 459–473.
Hannan, E. J. & Deistler, M. (1988). *The Statistical Theory of Linear Systems*, John Wiley, New York.
Hannan, E. J. & Kavalieris, L. (1984). Multivariate linear time series models, *Advances in Applied Probability* **16**: 492–561.
Hannan, E. J. & Quinn, B. G. (1979). The determination of the order of an autoregression, *Journal of the Royal Statistical Society* **B41**: 190–195.

Hannan, E. J. & Rissanen, J. (1982). Recursive estimation of mixed autoregressive moving average order, *Biometrika* **69**: 81–94.
Hansen, G., Kim, J. R. & Mittnik, S. (1998). Testing cointegrating coefficients in vector autoregressive error correction models, *Economics Letters* **58**: 1–5.
Hansen, H. & Johansen, S. (1999). Some tests for parameter constancy in cointegrated VAR-models, *Econometrics Journal* **2**: 306–333.
Hansen, P. R. (2003). Structural changes in the cointegrated vector autoregressive model, *Journal of Econometrics* **114**: 261–295.
Harbo, I., Johansen, S., Nielsen, B. & Rahbek, A. (1998). Asymptotic inference on cointegrating rank in partial systems, *Journal of Business & Economic Statistics* **16**: 388–399.
Harvey, A. C. (1981). *The Econometric Analysis of Time Series*, Allan, Oxford.
Harvey, A. C. (1984). A unified view of statistical forecasting procedures, *Journal of Forecasting* **3**: 245–275.
Harvey, A. C. (1987). Applications of the Kalman filter in econometrics, *in* T. F. Bewley (ed.), *Advances in Econometrics – Fifth World Congress*, Vol. I, Cambridge University Press, Cambridge, pp. 285–313.
Harvey, A. C. (1989). *Forecasting, Structural Time Series Models, and the Kalman Filter*, Cambridge University Press, Cambridge.
Harvey, A. C. & Pierse, R. G. (1984). Estimating missing observations in economic time series, *Journal of the American Statistical Association* **79**: 125–131.
Harvey, A. C. & Todd, P. H. J. (1983). Forecasting economic time series with structural and Box-Jenkins models: A case study, *Journal of Business & Economic Statistics* **1**: 299–315.
Haug, A. A. (1996). Tests for cointegration: A Monte Carlo comparison, *Journal of Econometrics* **71**: 89–115.
Haugh, L. D. & Box, G. E. P. (1977). Identification of dynamic regression (distributed lag) models connecting two time series, *Journal of the American Statistical Association* **72**: 121–130.
Hausman, J. A. (1983). Specification and estimation of simultaneous equation models, *in* Z. Griliches & M. D. Intriligator (eds), *Handbook of Econometrics*, Vol. 1, North-Holland, Amsterdam, pp. 391–448.
Hayashi, F. (2000). *Econometrics*, Princeton University Press, Princeton.
Hendry, D. F. (1995). *Dynamic Econometrics*, Oxford University Press, Oxford.
Hendry, D. F. & Krolzig, H.-M. (2001). *Automatic Econometric Model Selection with PcGets*, Timberlake Consultants Press, London.
Hendry, D. F. & von Ungern-Sternberg, T. (1981). Liquidity and inflation effects on consumer's expenditure, *in* A. S. Deaton (ed.), *Essays in the Theory and Measurement of Consumer's Behavior*, Cambridge University Press, Cambridge.
Herwartz, H. (1995). *Analyse saisonaler Zeitreihen mit Hilfe periodischer Zeitreihenmodelle*, Josef Eul, Bergisch Gladbach.
Herwartz, H. (2004). Conditional heteroskedasticity, *in* H. Lütkepohl & M. Krätzig (eds), *Applied Time Series Econometrics*, Cambridge University Press, Cambridge, pp. 197–221.
Herwartz, H. & Lütkepohl, H. (2000). Multivariate volatility analysis of VW stock prices, *International Journal of Intelligent Systems in Accounting, Finance & Management* **9**: 35–54.
Hildreth, C. & Houck, J. P. (1968). Some estimators for a linear model with random coefficients, *Journal of the American Statistical Association* **63**: 584–595.

Hillmer, S. C. & Tiao, G. C. (1979). Likelihood function of stationary multiple autoregressive moving average models, *Journal of the American Statistical Association* **74**: 652–660.

Ho, M. S. & Sørensen, B. E. (1996). Finding cointegration rank in high dimensional systems using the Johansen test: An illustration using data based Monte Carlo simulations, *Review of Economics and Statistics* **78**: 726–732.

Hogg, R. V. & Craig, A. T. (1978). *Introduction to Mathematical Statistics*, Macmillan, New York.

Hong, Y. (2001). A test for volatility spillover with application to exchange rates, *Journal of Econometrics* **103**: 183–224.

Horowitz, J. L. (2001). The bootstrap, *in* J. J. Heckman & E. Leamer (eds), *Handbook of Econometrics*, Vol. 5, North-Holland, Amsterdam, pp. 3159–3228.

Hosking, J. R. M. (1980). The multivariate portmanteau statistic, *Journal of the American Statistical Association* **75**: 602–608.

Hosking, J. R. M. (1981a). Equivalent forms of the multivariate portmanteau statistic, *Journal of the Royal Statistical Society* **B43**: 261–262.

Hosking, J. R. M. (1981b). Lagrange-multiplier tests of multivariate time series models, *Journal of the Royal Statistical Society* **B43**: 219–230.

Hsiao, C. (1979). Autoregressive modeling of Canadian money and income data, *Journal of the American Statistical Association* **74**: 553–560.

Hsiao, C. (1982). Time series modelling and causal ordering of Canadian money, income and interest rates, *in* O. D. Anderson (ed.), *Time Series Analysis: Theory and Practice 1*, North-Holland, Amsterdam, pp. 671–699.

Huang, D. W. & Guo, L. (1990). Estimation of nonstationary ARMAX models based on the Hannan-Rissanen method, *Annals of Statistics* **18**: 1729–1756.

Hubrich, K., Lütkepohl, H. & Saikkonen, P. (2001). A review of systems cointegration tests, *Econometric Reviews* **20**: 247–318.

Hylleberg, S. (1986). *Seasonality in Regression*, Academic Press, Orlando.

Inder, B. (1993). Estimating long run relationships in economics: A comparison of different approaches, *Journal of Econometrics* **57**: 53–68.

Jacobson, T., Vredin, A. & Warne, A. (1997). Common trends and hysteresis in Scandinavian unemployment, *European Economic Review* **41**: 1781–1816.

Jarque, C. M. & Bera, A. K. (1987). A test for normality of observations and regression residuals, *International Statistical Review* **55**: 163–172.

Jazwinski, A. H. (1970). *Stochastic Processes and Filtering Theory*, Academic Press, New York.

Jeantheau, T. (1998). Strong consistency of estimators for multivariate ARCH models, *Econometric Theory* **14**: 70–86.

Jenkins, G. M. & Alavi, A. S. (1981). Some aspects of modelling and forecasting multivariate time series, *Journal of Time Series Analysis* **2**: 1–47.

Johansen, S. (1988). Statistical analysis of cointegration vectors, *Journal of Economic Dynamics and Control* **12**: 231–254.

Johansen, S. (1991). Estimation and hypothesis testing of cointegration vectors in Gaussian vector autoregressive models, *Econometrica* **59**: 1551–1580.

Johansen, S. (1992). Cointegration in partial systems and the efficiency of single-equation analysis, *Journal of Econometrics* **52**: 389–402.

Johansen, S. (1994). The role of the constant and linear terms in cointegration analysis of nonstationary time series, *Econometric Reviews* **13**: 205–231.

Johansen, S. (1995). *Likelihood-based Inference in Cointegrated Vector Autoregressive Models*, Oxford University Press, Oxford.

Johansen, S. & Juselius, K. (1990). Maximum likelihood estimation and inference on cointegration–with applications to the demand for money, *Oxford Bulletin of Economics and Statistics* **52**: 169–210.

Johansen, S., Mosconi, R. & Nielsen, B. (2000). Cointegration analysis in the presence of structural breaks in the deterministic trend, *Econometrics Journal* **3**: 216–249.

Jones, R. H. (1980). Maximum likelihood fitting of ARMA models to time series with missing observations, *Technometrics* **22**: 389–395.

Jones, R. H. & Brelsford, W. M. (1967). Time series with periodic structure, *Biometrika* **54**: 403–408.

Judge, G. G., Griffiths, W. E., Hill, R. C., Lütkepohl, H. & Lee, T.-C. (1985). *The Theory and Practice of Econometrics*, 2nd edn, John Wiley, New York.

Kalman, R. E. (1960). A new approach to linear filtering and prediction problems, *Journal of Basic Engineering* **82**: 35–45.

Kalman, R. E. & Bucy, R. S. (1961). New results in linear filtering and prediction theory, *Journal of Basic Engineering* **83**: 95–108.

Kapetanios, G. (2003). A note on an iterative least-squares estimation method for ARMA and VARMA models, *Economics Letters* **79**: 305–312.

Kauppi, H. (2004). On the robustness of hypothesis testing based on fully modified vector autoregression when some roots are almost one, *Econometric Theory* **20**: 341–359.

Kilian, L. (1998). Small-sample confidence intervals for impulse response functions, *Review of Economics and Statistics* **80**: 218–230.

Kilian, L. (1999). Finite-sample properties of percentile and percentile-t bootstrap confidence intervals for impulse responses, *Review of Economics and Statistics* **81**: 652–660.

Kilian, L. & Demiroglu, U. (2000). Residual-based tests for normality in autoregressions: Asymptotic theory and simulation evidence, *Journal of Business & Economic Statistics* **18**: 40–50.

King, R. G., Plosser, C. I., Stock, J. H. & Watson, M. W. (1991). Stochastic trends and economic fluctuations, *American Economic Review* **81**: 819–840.

Kitagawa, G. (1981). A nonstationary time series model and its fitting by a recursive filter, *Journal of Time Series Analysis* **2**: 103–116.

Kleibergen, F. & Paap, R. (2002). Priors, posteriors and bayes factors for a Bayesian analysis of cointegration, *Journal of Econometrics* **111**: 223–249.

Kleibergen, F. & van Dijk, H. K. (1994). On the shape of the likelihood/posterior in cointegration models, *Econometric Theory* **10**: 514–551.

Kohn, R. (1979). Asymptotic estimation and hypothesis testing results for vector linear time series models, *Econometrica* **47**: 1005–1030.

Kohn, R. (1981). A note on an alternative derivation of the likelihood of an autoregressive moving average process, *Economics Letters* **7**: 233–236.

Koop, G. (1992). Aggregate shocks and macroeconomic fluctuations: A Bayesian approach, *Journal of Applied Econometrics* **7**: 395–411.

Koop, G., Pesaran, M. H. & Potter, S. M. (1996). Impulse response analysis in nonlinear multivariate models, *Journal of Econometrics* **74**: 119–147.

Koop, G., Strachan, R. W., van Dijk, H. K. & Villani, M. (2005). Bayesian approaches to cointegration, *in* K. Patterson & T. C. Mills (eds), *Palgrave Handbook of Econometrics, Volume 1: Econometric Theory*, Palgrave Macmillan, Houndmills.

Koreisha, S. G. & Pukkila, T. M. (1987). Identification of nonzero elements in the polynomial matrices of mixed VARMA processes, *Journal of the Royal Statistical Society* **B49**: 112–126.

Lee, K. C., Pesaran, M. H. & Pierse, R. G. (1992). Persistence of shocks and its sources in a multisectoral model of UK output growth, *Economic Journal* **102**: 342–356.

Lewis, R. & Reinsel, G. C. (1985). Prediction of multivariate time series by autoregressive model fitting, *Journal of Multivariate Analysis* **16**: 393–411.

Li, H. & Maddala, G. S. (1996). Bootstrapping time series models, *Econometric Reviews* **15**: 115–158.

Li, W. K. & Hui, Y. V. (1988). An algorithm for the exact likelihood of periodic autoregressive moving average models, *Communications in Statistics–Simulation and Computation* **17**: 1483–1494.

Li, W. K. & McLeod, A. I. (1981). Distribution of the residual autocorrelations in multivariate ARMA time series models, *Journal of the Royal Statistical Society* **B43**: 231–239.

Lin, W.-L. (1992). Alternative estimators of factor GARCH models – A Monte Carlo comparison, *Journal of Applied Econometrics* **7**: 259–279.

Lin, W.-L. (1997). Impulse response functions for conditional volatility in GARCH models, *Journal of Business & Economic Statistics* **15**: 15–25.

Ling, S. & McAleer, M. (2003). Asymptotic theory for a new vector ARMA GARCH model, *Econometric Theory* **19**: 280–310.

Litterman, R. B. (1986). Forecasting with Bayesian vector autoregressions - five years of experience, *Journal of Business & Economic Statistics* **4**: 25–38.

Liu, J. (1989). On the existence of a general multiple bilinear time series, *Journal of Time Series Analysis* **10**: 341–355.

Liu, L.-M. & Hanssens, D. M. (1982). Identification of multiple-input transfer function models, *Communications in Statistics – Theory and Methods* **11**: 297–314.

Ljung, G. M. & Box, G. E. P. (1978). On a measure of lack of fit in time series models, *Biometrika* **65**: 297–303.

Lomnicki, Z. A. (1961). Tests for departure from normality in the case of linear stochastic processes, *Metrika* **4**: 37–62.

Lucas, A. (1997). Cointegration testing using pseudo likelihood ratio tests, *Econometric Theory* **13**: 149–169.

Lucas, A. (1998). Inference on cointegrating ranks using LR and LM tests based on pseudo-likelihoods, *Econometric Reviews* **17**: 185–214.

Lucas, Jr., R. E. & Sargent, T. J. (1981). *Rational Expectations and Econometric Practice*, Vol. 1 and 2, University of Minnesota Press, Minneapolis.

Lundbergh, S. & Teräsvirta, T. (2002). Evaluating GARCH models, *Journal of Econometrics* **110**: 417–435.

Lütkepohl, H. (1984). Linear transformations of vector ARMA processes, *Journal of Econometrics* **26**: 283–293.

Lütkepohl, H. (1985). Comparison of criteria for estimating the order of a vector autoregressive process, *Journal of Time Series Analysis* **6**: 35–52, Correction, **8** (1987), 373.

Lütkepohl, H. (1986). *Prognose aggregierter Zeitreihen*, Vandenhoeck & Ruprecht, Göttingen.

Lütkepohl, H. (1987). *Forecasting Aggregated Vector ARMA Processes*, Springer, Berlin.

Lütkepohl, H. (1988a). Asymptotic distribution of the moving average coefficients of an estimated vector autoregressive process, *Econometric Theory* **4**: 77–85.
Lütkepohl, H. (1988b). Prediction tests for structural stability, *Journal of Econometrics* **39**: 267–296.
Lütkepohl, H. (1989a). A note on the asymptotic distribution of impulse response functions of estimated VAR models with orthogonal residuals, *Journal of Econometrics* **42**: 371–376.
Lütkepohl, H. (1989b). Prediction tests for structural stability of multiple time series, *Journal of Business & Economic Statistics* **7**: 129–135.
Lütkepohl, H. (1990). Asymptotic distributions of impulse response functions and forecast error variance decompositions of vector autoregressive models, *Review of Economics and Statistics* **72**: 116–125.
Lütkepohl, H. (1991). *Introduction to Multiple Time Series Analysis*, Springer, Berlin.
Lütkepohl, H. (1992). Testing for time varying parameters in vector autoregressive models, *in* W. E. Griffiths, H. Lütkepohl & M. E. Bock (eds), *Readings in Econometric Theory and Practice*, North-Holland, Amsterdam, pp. 243–264.
Lütkepohl, H. (1993). Testing for causation between two variables in higher dimensional VAR models, *in* H. Schneeweiß & K. F. Zimmermann (eds), *Studies in Applied Econometrics*, Physica, Heidelberg, pp. 75–91.
Lütkepohl, H. (1996a). *Handbook of Matrices*, John Wiley, Chichester.
Lütkepohl, H. (1996b). Testing for nonzero impulse responses in vector autoregressive processes, *Journal of Statistical Planning and Inference* **50**: 1–20.
Lütkepohl, H. (1997). Statistische Modellierung von Volatilitäten, *Allgemeines Statistisches Archiv* **81**: 62–84.
Lütkepohl, H. (2004). Vector autoregressive and vector error correction models, *in* H. Lütkepohl & M. Krätzig (eds), *Applied Time Series Econometrics*, Cambridge University Press, Cambridge, pp. 86–158.
Lütkepohl, H. & Burda, M. M. (1997). Modified Wald tests under nonregular conditions, *Journal of Econometrics* **78**: 315–332.
Lütkepohl, H. & Claessen, H. (1997). Analysis of cointegrated VARMA processes, *Journal of Econometrics* **80**: 223–239.
Lütkepohl, H. & Krätzig, M. (eds) (2004). *Applied Time Series Econometrics*, Cambridge University Press, Cambridge.
Lütkepohl, H. & Poskitt, D. S. (1991). Estimating orthogonal impulse responses via vector autoregressive models, *Econometric Theory* **7**: 487–496.
Lütkepohl, H. & Poskitt, D. S. (1996). Testing for causation using infinite order vector autoregressive processes, *Econometric Theory* **12**: 61–87.
Lütkepohl, H. & Poskitt, D. S. (1998). Consistent estimation of the number of cointegration relations in a vector autoregressive model, *in* R. Galata & H. Küchenhoff (eds), *Econometrics in Theory and Practice. Festschrift for Hans Schneeweiß*, Physica, Heidelberg, pp. 87–100.
Lütkepohl, H. & Reimers, H.-E. (1992a). Granger-causality in cointegrated VAR processes: The case of the term structure, *Economics Letters* **40**: 263–268.
Lütkepohl, H. & Reimers, H.-E. (1992b). Impulse response analysis of cointegrated systems, *Journal of Economic Dynamics and Control* **16**: 53–78.
Lütkepohl, H. & Saikkonen, P. (1999a). A lag augmentation test for the cointegrating rank of a VAR process, *Economics Letters* **63**: 23–27.

Lütkepohl, H. & Saikkonen, P. (1999b). Order selection in testing for the cointegrating rank of a VAR process, *in* R. F. Engle & H. White (eds), *Cointegration, Causality, and Forecasting. A Festschrift in Honour of Clive W.J. Granger*, Oxford University Press, Oxford, pp. 168–199.

Lütkepohl, H. & Saikkonen, P. (2000). Testing for the cointegrating rank of a VAR process with a time trend, *Journal of Econometrics* **95**: 177–198.

Lütkepohl, H., Saikkonen, P. & Trenkler, C. (2001). Maximum eigenvalue versus trace tests for the cointegrating rank of a VAR process, *Econometrics Journal* **4**: 287–310.

Lütkepohl, H., Saikkonen, P. & Trenkler, C. (2004). Testing for the cointegrating rank of a VAR process with level shift at unknown time, *Econometrica* **72**: 647–662.

Lütkepohl, H. & Schneider, W. (1989). Testing for nonnormality of autoregressive time series, *Computational Statistics Quarterly* **5**: 151–168.

MacKinnon, J. G., Haug, A. A. & Michelis, L. (1999). Numerical distribution functions of likelihood ratio tests for cointegration, *Journal of Applied Econometrics* **14**: 563–577.

Magnus, J. R. (1988). *Linear Structures*, Charles Griffin, London.

Magnus, J. R. & Neudecker, H. (1988). *Matrix Differential Calculus with Applications in Statistics and Econometrics*, John Wiley, Chichester.

Mann, H. B. & Wald, A. (1943). On the statistical treatment of linear stochastic difference equations, *Econometrica* **11**: 173–220.

Mardia, K. V. (1980). Tests for univariate and multivariate normality, *in* P. R. Krishnaiah (ed.), *Handbook of Statistics*, Vol. 1, North-Holland, Amsterdam, pp. 279–320.

Meinhold, R. J. & Singpurwalla, N. D. (1983). Understanding the Kalman-filter, *American Statistician* **37**: 123–127.

Mittnik, S. (1990). Computation of theoretical autocovariance matrices of multivariate autoregressive moving average time series, *Journal of the Royal Statistical Society* **B52**: 151–155.

Mood, A. M., Graybill, F. A. & Boes, D. C. (1974). *Introduction to the Theory of Statistics*, 3rd edn, McGraw-Hill, Auckland.

Morrison, D. F. (1976). *Multivariate Statistical Methods*, 2nd edn, McGraw-Hill, New York.

Murata, Y. (1982). *Optimal Control Methods for Linear Discrete-Time Economic Systems*, Springer, New York.

Muth, J. F. (1961). Rational expectations and the theory of price movements, *Econometrica* **29**: 315–335.

Nankervis, J. C. & Savin, N. E. (1988). The Student's t approximation in a stationary first order autoregressive model, *Econometrica* **56**: 119–145.

Nelson, D. (1991). Conditional heteroskedasticity in asset returns: A new approach, *Econometrica* **59**: 347–370.

Nicholls, D. F. & Pagan, A. R. (1985). Varying coefficient regression, *in* E. J. Hannan, P. R. Krishnaiah & M. M. Rao (eds), *Handbook of Statistics*, Vol. 5, North-Holland, Amsterdam, pp. 413–449.

Nicholls, D. F. & Pope, A. L. (1988). Bias in the estimation of multivariate autoregressions, *Australian Journal of Statistics* **30A**: 296–309.

Nicholls, D. F. & Quinn, B. C. (1982). *Random Coefficient Autoregressive Models: An Introduction*, Springer, New York.

Nijman, T. E. (1985). *Missing Observations in Dynamic Macroeconomic Modeling*, Free University Press, Amsterdam.
Osborn, D. R. & Smith, J. P. (1989). The performance of periodic autoregressive models in forecasting seasonal U.K. consumption, *Journal of Business & Economic Statistics* **7**: 117–127.
Ozaki, T. (1980). Non-linear time series models for nonlinear random vibrations, *Journal of Applied Probability* **17**: 84–93.
Pagan, A. (1980). Some identification and estimation results for regression models with stochastically varying coefficients, *Journal of Econometrics* **13**: 341–363.
Pagan, A. (1995). Three econometric methodologies: An update, *in* L. Oxley, D. A. R. George, C. J. Roberts & S. Sayer (eds), *Surveys in Econometrics*, Blackwell, Oxford.
Pagan, A. (1996). The econometrics of financial markets, *Journal of Empirical Finance* **3**: 15–102.
Pagano, M. (1978). On periodic and multiple autoregressions, *Annals of Statistics* **6**: 1310–1317.
Pantelidis, T. & Pittis, N. (2004). Testing for Granger causality in variance in the presence of causality in mean, *Economics Letters* **85**: 201–207.
Park, J. Y. & Phillips, P. C. B. (1988). Statistical inference in regressions with integrated processes: Part 1, *Econometric Theory* **4**: 468–497.
Park, J. Y. & Phillips, P. C. B. (1989). Statistical inference in regressions with integrated processes: Part 2, *Econometric Theory* **5**: 95–131.
Paulsen, J. (1984). Order determination of multivariate autoregressive time series with unit roots, *Journal of Time Series Analysis* **5**: 115–127.
Paulsen, J. & Tjøstheim, D. (1985). On the estimation of residual variance and order in autoregressive time series, *Journal of the Royal Statistical Society* **B47**: 216–228.
Penm, J. H. W., Brailsford, T. J. & Terrell, R. D. (2000). A robust algorithm in sequentially selecting subset time series systems using neural networks, *Journal of Time Series Analysis* **21**: 389–412.
Penm, J. H. W. & Terrell, R. D. (1982). On the recursive fitting of subset autoregressions, *Journal of Time Series Analysis* **3**: 43–59.
Penm, J. H. W. & Terrell, R. D. (1984). Multivariate subset autoregressive modelling with zero constraints for detecting 'overall causality', *Journal of Econometrics* **24**: 311–330.
Penm, J. H. W. & Terrell, R. D. (1986). The 'derived' moving average model and its role in causality, *in* J. Gani & M. B. Priestley (eds), *Essays in Time Series and Allied Processes*, Applied Probability Trust, Sheffield, pp. 99–111.
Pesaran, M. H. (1987). *The Limits to Rational Expectations*, Blackwell, Oxford.
Pesaran, M. H. & Shin, Y. (1996). Cointegration and speed of convergence to equilibrium, *Journal of Econometrics* **71**: 117–143.
Phillips, P. C. B. (1987). Time series regression with a unit root, *Econometrica* **55**: 277–301.
Phillips, P. C. B. (1988). Multiple regression with integrated time series, *Contemporary Mathematics* **80**: 79–105.
Phillips, P. C. B. (1991). Optimal inference in cointegrated systems, *Econometrica* **59**: 283–306.
Phillips, P. C. B. (1994). Some exact distribution theory for maximum likelihood estimators of cointegrating coefficients in error correction models, *Econometrica* **62**: 73–93.

Phillips, P. C. B. (1995). Fully modified least squares and vector autoregression, *Econometrica* **63**: 1023–1078.

Phillips, P. C. B. & Durlauf, S. N. (1986). Multiple time series regression with integrated processes, *Review of Economic Studies* **53**: 473–495.

Phillips, P. C. B. & Hansen, B. E. (1990). Statistical inference in instrumental variables regression with I(1) processes, *Review of Economic Studies* **57**: 99–125.

Phillips, P. C. B. & Ouliaris, S. (1990). Asymptotic properties of residual based tests for cointegration, *Econometrica* **58**: 165–193.

Phillips, P. C. B. & Park, J. Y. (1988). Asymptotic equivalence of ordinary least squares and generalized least squares in regressions with integrated regressors, *Journal of the American Statistical Association* **83**: 111–115.

Phillips, P. C. B. & Solo, V. (1992). Asymptotics for linear processes, *Annals of Statistics* **20**: 971–1001.

Poskitt, D. S. (1987). A modified Hannan-Rissanen strategy for mixed autoregressive-moving average order determination, *Biometrika* **74**: 781–790.

Poskitt, D. S. (1989). A method for the estimation of transfer function models, *Journal of the Royal Statistical Society* **B51**: 29–46.

Poskitt, D. S. (1992). Identification of echelon canonical forms for vector linear processes using least squares, *Annals of Statistics* **20**: 196–215.

Poskitt, D. S. (2003). On the specification of cointegrated autoregressive moving-average forecasting systems, *International Journal of Forecasting* **19**: 503–519.

Poskitt, D. S. (2004). On the identification and estimation of partially nonstationary ARMAX systems, *Discussion paper*, Monash University, Australia.

Poskitt, D. S. & Lütkepohl, H. (1995). Consistent specification of cointegrated autoregressive moving average systems, *Discussion Paper 54*, SFB 373, Humboldt-Universität zu Berlin.

Poskitt, D. S. & Tremayne, A. R. (1981). An approach to testing linear time series models, *Annals of Statistics* **9**: 974–986.

Poskitt, D. S. & Tremayne, A. R. (1982). Diagnostic tests for multiple time series models, *Annals of Statistics* **10**: 114–120.

Priestley, M. B. (1980). State-dependent models: A general approach to non-linear time series analysis, *Journal of Time Series Analysis* **1**: 47–71.

Priestley, M. B. (1988). *Non-Linear and Non-Stationary Time Series Analysis*, Academic Press, London.

Proietti, T. (2002). Forecasting with structural time series models, *in* M. P. Clements & D. F. Hendry (eds), *A Companion to Economic Forecasting*, Blackwell, Oxford, pp. 105–132.

Quenouille, M. H. (1957). *The Analysis of Multiple Time-Series*, Griffin, London.

Quinn, B. (1980). Order determination for a multivariate autoregression, *Journal of the Royal Statistical Society* **B42**: 182–185.

Rao, C. R. (1973). *Linear Statistical Inference and Its Applications*, 2nd edn, John Wiley, New York.

Reimers, H.-E. (1991). *Analyse kointegrierter Variablen mittels vektorautoregressiver Modelle*, Physica, Heidelberg.

Reinsel, G. C. (1983). Some results on multivariate autoregressive index models, *Biometrika* **70**: 145–156.

Reinsel, G. C. (1993). *Elements of Multivariate Time Series Analysis*, Springer, New York.

Reinsel, G. C. & Ahn, S. K. (1992). Vector autoregressive models with unit roots and reduced rank structure: Estimation, likelihood ratio test, and forecasting, *Journal of Time Series Analysis* **13**: 353–375.
Reinsel, G. C. & Velu, R. P. (1998). *Multivariate Reduced-Rank Regression, Theory and Applications*, Springer, New York.
Rohatgi, V. K. (1976). *An Introduction to Probability Theory and Mathematical Statistics*, John Wiley, New York.
Rothenberg, T. J. (1971). Identification in parametric models, *Econometrica* **39**: 577–591.
Roussas, G. G. (1973). *A First Course in Mathematical Statistics*, Addison-Wesley, Reading, MA.
Runkle, D. E. (1987). Vector autoregressions and reality, *Journal of Business & Economic Statistics* **5**: 437–442.
Saikkonen, P. (1992). Estimation and testing of cointegrated systems by an autoregressive approximation, *Econometric Theory* **8**: 1–27.
Saikkonen, P. (2005). Stability results for nonlinear error correction models, *Journal of Econometrics*.
Saikkonen, P. & Lütkepohl, H. (1996). Infinite order cointegrated vector autoregressive processes: Estimation and inference, *Econometric Theory* **12**: 814–844.
Saikkonen, P. & Lütkepohl, H. (1999). Local power of likelihood ratio tests for the cointegrating rank of a VAR process, *Econometric Theory* **15**: 50–78.
Saikkonen, P. & Lütkepohl, H. (2000a). Asymptotic inference on nonlinear functions of the coefficients of infinite order cointegrated VAR processes, *in* W. A. Barnett, D. F. Hendry, S. Hylleberg, T. Teräsvirta, D. Tjøstheim & A. Würtz (eds), *Nonlinear Econometric Modeling in Time Series Analysis*, Cambridge University Press, Cambridge, pp. 165–201.
Saikkonen, P. & Lütkepohl, H. (2000b). Testing for the cointegrating rank of a VAR process with an intercept, *Econometric Theory* **16**: 373–406.
Saikkonen, P. & Lütkepohl, H. (2000c). Testing for the cointegrating rank of a VAR process with structural shifts, *Journal of Business & Economic Statistics* **18**: 451–464.
Saikkonen, P. & Lütkepohl, H. (2000d). Trend adjustment prior to testing for the cointegrating rank of a vector autoregressive process, *Journal of Time Series Analysis* **21**: 435–456.
Saikkonen, P. & Luukkonen, R. (1997). Testing cointegration in infinite order vector autoregressive processes, *Journal of Econometrics* **81**: 93–129.
Salmon, M. (1982). Error correction mechanisms, *Economic Journal* **92**: 615–629.
Samaranayake, V. A. & Hasza, D. P. (1988). Properties of predictors for multivariate autoregressive models with estimated parameters, *Journal of Time Series Analysis* **9**: 361–383.
Sargent, T. J. & Sims, C. A. (1977). Business cycle modeling without pretending to have too much a priori economic theory, *in* C. A. Sims (ed.), *New Methods in Business Cycle Research: Proceedings from a Conference*, Federal Reserve Bank of Minneapolis, Minneapolis, pp. 45–109.
Schmidt, P. (1973). The asymptotic distribution of dynamic multipliers, *Econometrica* **41**: 161–164.
Schmidt, P. (1976). *Econometrics*, Marcel Dekker, New York.
Schneider, W. (1988). Analytical uses of Kalman filtering in econometrics – a survey, *Statistical Papers* **29**: 3–33.

Schneider, W. (1991). Stability analysis using Kalman filtering, scoring, EM, and an adaptive EM method, *in* P. Hackl & A. Westlund (eds), *Economic Structural Change, Analysis and Forecasting*, Springer, Berlin, pp. 191–221.

Schneider, W. (1992). Systems of seemingly unrelated regression equations with time-varying coefficients – an interplay of Kalman filtering, scoring, EM- and MINQUE-method, *Computers and Mathematics with Applications* **24**: 1–16.

Schwarz, G. (1978). Estimating the dimension of a model, *Annals of Statistics* **6**: 461–464.

Searle, S. R. (1982). *Matrix Algebra Useful for Statistics*, John Wiley, New York.

Serfling, R. J. (1980). *Approximation Theorems of Mathematical Statistics*, John Wiley, New York.

Shapiro, M. & Watson, M. W. (1988). Sources of business cycle fluctuations, *NBER Macroeconomics Annual* **3**: 111–156.

Shibata, R. (1980). Asymptotically efficient selection of the order of the model for estimating parameters of a linear process, *Annals of Statistics* **8**: 147–164.

Shin, Y. (1994). A residual-based test of the null of cointegration against the alternative of no cointegration, *Econometric Theory* **10**: 91–115.

Sims, C. A. (1980). Macroeconomics and reality, *Econometrica* **48**: 1–48.

Sims, C. A. (1981). An autoregressive index model for the U.S. 1948-1975, *in* J. Kmenta & J. B. Ramsey (eds), *Large-Scale Macro-Econometric Models*, North-Holland, Amsterdam, pp. 283–327.

Sims, C. A. (1986). Are forecasting models usable for policy analysis?, *Quarterly Review, Federal Reserve Bank of Minneapolis* **10**: 2–16.

Sims, C. A., Stock, J. H. & Watson, M. W. (1990). Inference in linear time series models with some unit roots, *Econometrica* **58**: 113–144.

Sims, C. A. & Zha, T. (1999). Error bands for impulse responses, *Econometrica* **67**: 1113–1155.

Snell, A. (1999). Testing for r versus $r-1$ cointegrating vectors, *Journal of Econometrics* **88**: 151–191.

Solo, V. (1984). The exact likelihood for a multivariate ARMA model, *Journal of Multivariate Analysis* **15**: 164–173.

Stensholt, B. K. & Tjøstheim, D. (1987). Multiple bilinear time series models, *Journal of Time Series Analysis* **8**: 221–233.

Stock, J. H. (1987). Asymptotic properties of least squares estimators of cointegrating vectors, *Econometrica* **55**: 1035–1056.

Stock, J. H. & Watson, M. W. (1988). Testing for common trends, *Journal of the American Statistical Association* **83**: 1097–1107.

Stock, J. H. & Watson, M. W. (1993). A simple estimator of cointegrating vectors in higher order integrated systems, *Econometrica* **61**: 783–820.

Stock, J. H. & Watson, M. W. (2002a). Forecasting using principal components from a large number of predictors, *Journal of the American Statistical Association* **97**: 1167–1179.

Stock, J. H. & Watson, M. W. (2002b). Macroeconomic forecasting using diffusion indexes, *Journal of Business & Economic Statistics* **20**: 147–162.

Strachan, R. W. (2003). Valid Bayesian estimation of the cointegrating error correction model, *Journal of Business & Economic Statistics* **21**: 185–195.

Strachan, R. W. & Inder, B. (2004). Bayesian analysis of the error correction model, *Journal of Econometrics* **123**: 307–325.

Subba Rao, T. & Gabr, M. M. (1984). *An Introduction to Bispectral Analysis and Bilinear Time Series Models*, Springer, New York.

Swamy, P. A. V. B. (1971). *Statistical Inference in Random Coefficient Regression Models*, Springer, Berlin.
Swamy, P. A. V. B. & Tavlas, G. S. (2001). Random coefficient models, *in* B. H. Baltagi (ed.), *A Companion to Theoretical Econometrics*, Blackwell, Oxford, pp. 410–428.
Tay, A. S. & Wallis, K. F. (2002). Density forecasting: A survey, *in* M. P. Clements & D. F. Hendry (eds), *Companion to Economic Forecasting*, Blackwell, Oxford, pp. 45–68.
Taylor, S. J. (1986). *Modelling Financial Time Series*, John Wiley, Chichester.
Theil, H. (1971). *Principles of Econometrics*, Wiley, Santa Barbara.
Tiao, G. C. & Box, G. E. P. (1981). Modeling multiple time series with applications, *Journal of the American Statistical Association* **76**: 802–816.
Tiao, G. C. & Grupe, M. R. (1980). Hidden periodic autoregressive-moving average models in time series data, *Biometrika* **67**: 365–373.
Tiao, G. C. & Tsay, R. S. (1989). Model specification in multivariate time series (with discussion), *Journal of the Royal Statistical Society* **B51**: 157–213.
Tjøstheim, D. & Paulsen, J. (1983). Bias of some commonly-used time series estimates, *Biometrika* **70**: 389–399.
Toda, H. Y. & Phillips, P. C. B. (1993). Vector autoregressions and causality, *Econometrica* **61**: 1367–1393.
Toda, H. Y. & Yamamoto, T. (1995). Statistical inference in vector autoregressions with possibly integrated processes, *Journal of Econometrics* **66**: 225–250.
Tong, H. (1983). *Threshold Models in Non-linear Time Series Analysis*, Springer, New York.
Tsay, R. S. (1984). Order selection in nonstationary autoregressive models, *Annals of Statistics* **12**: 1425–1433.
Tsay, R. S. (1985). Model identification in dynamic regression (distributed lag) models, *Journal of Business & Economic Statistics* **3**: 228–237.
Tsay, R. S. (1988). Outliers, level shifts and variance changes in time series, *Journal of Forecasting* **7**: 1–20.
Tsay, R. S. (1989a). Identifying multivariate time series models, *Journal of Time Series Analysis* **10**: 357–372.
Tsay, R. S. (1989b). Parsimonious parameterization of vector autoregressive moving average models, *Journal of Business & Economic Statistics* **7**: 327–341.
Tse, Y. K. & Tsui, A. K. C. (1999). A note on diagnosing multivariate conditional heteroscedasticity models, *Journal of Time Series Analysis* **20**: 679–691.
Tse, Y. K. & Tsui, A. K. C. (2002). A multivariate generalized autoregressive conditional heteroscedasticity model with time-varying correlations, *Journal of Business & Economic Statistics* **20**: 351–362.
Tso, M. K.-S. (1981). Reduced-rank regression and canonical analysis, *Journal of the Royal Statistical Society* **B43**: 183–189.
Uhlig, H. (1994). What are the effects of monetary policy on output? Results from an agnostic identification procedure, mimeo. Department of Economics, Humboldt University Berlin.
van der Weide, R. (2002). GO-GARCH: A multivariate generalized orthogonal GARCH model, *Journal of Applied Econometrics* **17**: 549–564.
van Overschee, P. & DeMoor, B. (1994). N4sid: Subspace algorithms for the identification of combined deterministic-stochastic systems, *Automatica* **30**: 75–93.
Velu, R. P., Reinsel, G. C. & Wichern, D. W. (1986). Reduced rank models for multiple time series, *Biometrika* **73**: 105–118.

Vlaar, P. J. G. (2004). On the asymptotic distribution of impulse response functions with long-run restrictions, *Econometric Theory* **20**: 891–903.

Vrontos, I. D., Dellaportas, P. & Politis, D. N. (2003). A full-factor multivariate GARCH model, *Econometrics Journal* **6**: 312–334.

Wallis, K. F. (1977). Multiple time series analysis and the final form of econometric models, *Econometrica* **45**: 1481–1497.

Watanabe, N. (1985). Note on the Kalman filter with estimated parameters, *Journal of Time Series Analysis* **6**: 269–278.

Watson, M. W. & Engle, R. F. (1983). Alternative algorithms for the estimation of dynamic factor, MIMIC and varying coefficient regression models, *Journal of Econometrics* **23**: 385–400.

Weber, C. E. (1995). Cyclical output, cyclical unemployment, and Okun's coefficient: A new approach, *Journal of Applied Econometrics* **10**: 433–445.

Wei, W. W. S. (1990). *Time Series Analysis: Univariate and Multivariate Methods*, Addison-Wesley, Redwood City, CA.

White, H. & MacDonald, G. M. (1980). Some large-sample tests for nonnormality in the linear regression model, *Journal of the American Statistical Association* **75**: 16–28.

Wickens, M. R. & Breusch, T. S. (1988). Dynamic specification, the long run and the estimation of transformed regression models, *Economic Journal* **98**: 189–205.

Wold, H. (1938). *A Study in the Analysis of Stationary Time-Series*, Almqvist and Wiksells, Uppsala.

Wong, H. & Li, W. K. (1997). On a multivariate conditional heteroscedastic model, *Biometrika* **84**: 111–123.

Yamamoto, T. (1980). On the treatment of autocorrelated errors in the multiperiod prediction of dynamic simultaneous equation models, *International Economic Review* **21**: 735–748.

Yang, M. & Bewley, R. (1996). On cointegration tests for VAR models with drift, *Economics Letters* **51**: 45–50.

Yap, S. F. & Reinsel, G. C. (1995). Estimation and testing for unit roots in a partially nonstationary vector autoregressive moving average model, *Journal of the American Statistical Association* **90**: 253–267.

Young, P. C., Jakeman, A. J. & McMurtrie, R. (1980). An instrumental variable method for model order identification, *Automatica* **16**: 281–294.

Zellner, A. (1962). An efficient method of estimating seemingly unrelated regressions and tests of aggregation bias, *Journal of the American Statistical Association* **57**: 348–368.

Zellner, A. & Palm, F. (1974). Time series analysis and simultaneous equation econometric models, *Journal of Econometrics* **2**: 17–54.

Index of Notation

Most of the notation is clearly defined in the text where it is used. The following list is meant to provide some general guidelines. Occasionally, in the text a symbol has a meaning which differs from the one specified in this list when confusion is unlikely. For instance, A usually stands for a VAR coefficient matrix whereas in the Appendix it is often a general matrix.

General Symbols

$=$	equals
$:=$	equals by definition
$\Rightarrow$	implies
$\Leftrightarrow$	is equivalent to
$\sim$	is distributed as
$\in$	element of
$\subset$	subset of
$\cup$	union
$\cap$	intersection
$\sum$	summation sign
$\prod$	product sign
$\rightarrow$	converges to, approaches
$\xrightarrow{p}$	converges in probability to
$\xrightarrow{a.s.}$	converges almost surely to
$\xrightarrow{q.m.}$	converges in quadratic mean to
$\xrightarrow{d}$	converges in distribution to
i.i.d.	independently, identically distributed
lim	limit
plim	probability limit
max	maximum
min	minimum
sup	supremum, least upper bound
ln	natural logarithm
exp	exponential function
$\|z\|$	absolute value or modulus of z
K	dimension of a stochastic process or time series
T	sample size, time series length
$\mathbb{R}$	real numbers
$\mathbb{R}^m$	m-dimensional Euclidean space
$\mathbb{C}$	complex numbers
$\mathbb{Z}$	integers
$\mathbb{N}$	positive integers
$\mathbb{I}(\cdot)$	indicator function
L	lag operator
Δ	differencing operator
E	expectation
Var	variance
Cov	covariance, covariance matrix
MSE	mean squared error (matrix)
Pr	probability

$l(\cdot)$	likelihood function
$\ln l$	log-likelihood function
$l_0(\cdot)$	approximate likelihood function
$\ln l_0$	approximate log-likelihood function
λ_{LM}	Lagrange multiplier statistic
λ_{LR}	likelihood ratio statistic
λ_W	Wald statistic
Q_h	portmanteau statistic
$\bar{Q}_h$	modified portmanteau statistic
d.f.	degrees of freedom
AIC	Akaike information criterion
FPE	final prediction error (criterion)
HQ	Hannan-Quinn (criterion)
SC	Schwarz criterion

Distributions and Stochastic Processes

$\mathcal{N}(\mu, \Sigma)$	(multivariate) normal distribution with mean (vector) μ and variance (covariance matrix) Σ
$\chi^2(m)$	χ^2-distribution with m degrees of freedom
$F(m, n)$	F-distribution with m numerator and n denominator degrees of freedom
$t(m)$	t-distribution with m degrees of freedom
AR	autoregressive (process)
AR(p)	autoregressive process of order p
ARCH	autoregressive conditional heteroskedasticity
ARMA	autoregressive moving average (process)
ARMA(p,q)	autoregressive moving average process of order (p, q)
ARMA$_E$	echelon form VARMA model
ARMA$_E(p_1, \ldots, p_K)$	echelon form VARMA model with Kronecker indices $(p_1, \ldots, p_K)$
EC-ARMA$_{RE}$	error correction echelon form VARMA model
GARCH	generalized autoregressive conditional heteroskedasticity
MA	moving average (process)
MA(q)	moving average process of order q
MGARCH	multivariate generalized autoregressive conditional heteroskedasticity
PAR	periodic (vector) autoregression
VAR	vector autoregressive (process)
VAR(p)	vector autoregressive process of order p
VARMA	vector autoregressive moving average (process)
VARMA(p,q)	vector autoregressive moving average process of order (p, q)
VECM	vector error correction model

Index of Notation

Vector and Matrix Operations

M'	transpose of M
M^{adj}	adjoint of M
M^{-1}	inverse of M
M^+	Moore-Penrose generalized inverse of M
$M_\perp$	orthogonal complement of M
$M^{1/2}$	square root of M
M^k	k-th power of M
MN	matrix product of M and N
$+$	plus
$-$	minus
$\otimes$	Kronecker product
$\det(M)$, $\det M$	determinant of M
$|M|$	determinant of M
$\|M\|$	Euclidean norm of M
$\text{rk}(M)$, $\text{rk } M$	rank of M
$\text{tr}(M)$, $\text{tr } M$	trace of M
vec	column stacking operator
vech	column stacking operator for symmetric matrices (stacks the elements on and below the main diagonal only)
$\dfrac{\partial \varphi}{\partial \beta'}$	vector or matrix of first order partial derivatives of φ with respect to β
$\dfrac{\partial^2 \varphi}{\partial \beta \partial \beta'}$	Hessian matrix of φ, matrix of second order partial derivatives of φ with respect to β

General Matrices

$\mathbf{D}_m$	$(m^2 \times \tfrac{1}{2}m(m+1))$ duplication matrix
I_m	$(m \times m)$ unit or identity matrix
$\mathcal{I}(\cdot)$	information matrix
$\mathcal{I}_a(\cdot)$	asymptotic information matrix
J	$:= [I_K : 0 : \cdots : 0]$
$\mathbf{K}_{mn}$	$(mn \times mn)$ commutation matrix
$\mathbf{L}_m$	$(\tfrac{1}{2}m(m+1) \times m^2)$ elimination matrix
0	zero or null matrix or vector

Vectors and Matrices Related to Stochastic Processes and Multiple Time Series

u_t K-dimensional white noise process
u_{kt} k-th element of u_t

$$u_{(k)} := \begin{bmatrix} u_{k1} \\ \vdots \\ u_{kT} \end{bmatrix}$$

$U := [u_1, \ldots, u_T]$
$\mathbf{u} := \text{vec}(U)$

$$U_t := \begin{bmatrix} u_t \\ 0 \\ \vdots \\ 0 \end{bmatrix} \quad \text{or} \quad \begin{bmatrix} u_t \\ 0 \\ \vdots \\ 0 \\ u_t \\ 0 \\ \vdots \\ 0 \end{bmatrix}$$

y_t K-dimensional stochastic process
y_{kt} k-th element of y_t

$$y_{(k)} := \begin{bmatrix} y_{k1} \\ \vdots \\ y_{kT} \end{bmatrix}$$

$\bar{y} := \sum_{t=1}^{T} y_t/T$, sample mean (vector)

$y_t(h)$ h-step forecast of y_{t+h} at origin t
$Y := [y_1, \ldots, y_T]$
$\mathbf{y} := \text{vec}(Y)$

$$Y_t := \begin{bmatrix} y_t \\ \vdots \\ y_{t-p+1} \end{bmatrix} \quad \text{or} \quad \begin{bmatrix} y_t \\ \vdots \\ y_{t-p+1} \\ u_t \\ \vdots \\ u_{t-q+1} \end{bmatrix} \quad \text{or} \quad \begin{bmatrix} y_t \\ \vdots \\ y_{t-p+1} \\ x_t \\ \vdots \\ x_{t-s+1} \end{bmatrix}$$

$$Z_t := \begin{bmatrix} 1 \\ y_t \\ \vdots \\ y_{t-p+1} \end{bmatrix}$$

Index of Notation

Matrices and Vectors Related to VAR and VARMA Representations and VECMs (Parts I, II, III, IV)

A_i VAR coefficient matrix

$A \;\; := [A_1, \ldots, A_p]$

$\boldsymbol{\alpha} \;\; := \text{vec}(A)$

$$\mathbf{A} \;\; := \begin{bmatrix} A_1 & \cdots & A_{p-1} & A_p \\ I_K & & 0 & 0 \\ & \ddots & \vdots & \vdots \\ 0 & \cdots & I_K & 0 \end{bmatrix} \quad \text{or} \quad \begin{bmatrix} \mathbf{A}_{11} & \mathbf{A}_{12} \\ \mathbf{A}_{21} & \mathbf{A}_{22} \end{bmatrix}$$

$$\mathbf{A}_{11} := \begin{bmatrix} A_1 & \cdots & A_{p-1} & A_p \\ I_K & & 0 & 0 \\ & \ddots & \vdots & \vdots \\ 0 & \cdots & I_K & 0 \end{bmatrix} \quad (Kp \times Kp)$$

$$\mathbf{A}_{12} := \begin{bmatrix} M_1 & \cdots & M_{q-1} & M_q \\ 0 & \cdots & 0 & 0 \\ \vdots & & \vdots & \vdots \\ 0 & \cdots & 0 & 0 \end{bmatrix} \quad (Kp \times Kq)$$

$\mathbf{A}_{21} := 0 \quad (Kq \times Kp)$

$$\mathbf{A}_{22} := \begin{bmatrix} 0 & 0 \\ I_{K(q-1)} & 0 \end{bmatrix} \quad (Kq \times Kq)$$

M_i MA coefficient matrix

$\mathbf{m} \;\; := \text{vec}[M_1, \ldots, M_q]$

$$\mathbf{M} \;\; := \begin{bmatrix} \mathbf{M}_{11} & \mathbf{M}_{12} \\ \mathbf{M}_{21} & \mathbf{M}_{22} \end{bmatrix} \quad (K(p+q) \times K(p+q))$$

$$\mathbf{M}_{11} := \begin{bmatrix} -M_1 & \cdots & -M_{q-1} & -M_q \\ I_K & & 0 & 0 \\ & \ddots & \vdots & \vdots \\ 0 & \cdots & I_K & 0 \end{bmatrix} \quad (Kq \times Kq)$$

$$\mathbf{M}_{12} := \begin{bmatrix} -A_1 & \cdots & -A_{p-1} & -A_p \\ 0 & \cdots & 0 & 0 \\ \vdots & & \vdots & \vdots \\ 0 & \cdots & 0 & 0 \end{bmatrix} \quad (Kq \times Kp)$$

$\mathbf{M}_{21} := 0 \quad (Kp \times Kq)$

$$\mathbf{M}_{22} := \begin{bmatrix} 0 & 0 \\ I_{K(p-1)} & 0 \end{bmatrix} \quad (Kp \times Kp)$$

Φ_i coefficient matrix of canonical MA representation
Π_i coefficient matrix of pure VAR representation
α loading matrix of VECM
β cointegration matrix
Π $:= \alpha\beta'$
Γ_i short-run coefficient matrix of VECM

Impulse Responses and Related Quantities

Φ_i matrix of forecast error impulse responses
Ψ_m $:= \sum_{i=0}^{m} \Phi_i$, matrix of accumulated forecast error impulse responses
Ψ_∞ $:= \sum_{i=0}^{\infty} \Phi_i$, matrix of total or long-run forecast error impulse responses
Θ_i matrix of orthogonalized impulse responses
Ξ_m $:= \sum_{i=0}^{m} \Theta_i$, matrix of accumulated orthogonalized impulse responses
Ξ_∞ $:= \sum_{i=0}^{\infty} \Theta_i$, matrix of total or long-run orthogonalized impulse responses
$\omega_{jk,h}$ proportion of h-step forecast error variance of variable j, accounted for by innovations in variable k
Ξ matrix of long-run effects
Ξ_j^* matrix of transitory effects

Moment Matrices

Γ := plim ZZ'/T
$\Gamma_y(h)$:= $\text{Cov}(y_t, y_{t-h})$ for a stationary process y_t
$R_y(h)$ correlation matrix corresponding to $\Gamma_y(h)$
Σ_u := $E(u_t u_t') = \text{Cov}(u_t)$, white noise covariance matrix
Σ_y := $E[(y_t - \mu)(y_t - \mu)'] = \text{Cov}(y_t)$, covariance matrix of a stationary process y_t
P lower triangular Choleski decomposition of Σ_u
$\Sigma_{\widehat{\alpha}}$ covariance matrix of the asymptotic distribution of $\sqrt{T}(\widehat{\alpha} - \alpha)$
$\Omega(h)$ correction term for MSE matrix of h-step forecast
$\Sigma_y(h)$ MSE or forecast error covariance matrix of h-step forecast of y_t
$\Sigma_{\widehat{y}}(h)$ approximate MSE matrix of h-step forecast of estimated process y_t

Author Index

Abraham, B., 609
Ahn, S.K., 157, 165, 170, 222, 271, 272, 280, 290, 292–294, 297, 316, 323
Akaike, H., 146, 147, 507
Alavi, A.S., 507
Amisano, G., 358, 374
Anděl, J., 604, 625
Andersen, A.P., 624, 625
Anderson, B.D.O., 612, 628, 630, 637
Anderson, T.W., 104, 186, 222, 397, 619, 645, 677, 691
Andrews, D.W.K., 318
Ansley, C.F., 618
Aoki, M., 507, 612, 616

Baba, Y., 564
Bai, J., 609
Baillie, R.T., 406, 562
Banerjee, A., 301, 343
Baringhaus, L., 177
Barone, P., 30
Bartel, H., 515, 523
Basu, A.K., 95, 316
Bauer, D., 636, 637
Bauwens, L., 559, 568, 571, 577
Benkwitz, A., 112, 113, 129, 710, 711
Bera, A.K., 175, 177, 559
Berk, K.N., 532
Bernanke, B., 357
Bewley, R., 343
Bhansali, R.J., 532
Bierens, H.J., 343, 398
Black, F., 568
Blanchard, O., 357, 367, 376, 385

Boes, D.C., IX
Bollerslev, T., 559, 560, 562, 564, 565
Bossaerts, P., 300
Boswijk, H.P., 307, 342, 604
Box, G.E.P., 171, 400, 401, 493, 495, 507, 609
Brailsford, T.J., 212
Braun, P.A., 568
Breitung, J., 343, 357, 364, 366, 372, 386
Brelsford, W.M., 604
Breusch, T.S., 173, 301, 698
Brockwell, P.J., 86, 434
Brooks, C., 573, 583
Brüggemann, R., 211, 212, 297, 345, 346, 348, 357, 364, 366, 372, 375, 386
Bucy, R.S., 625
Burda, M.M., 106–108
Burke, S., 573, 583

Caines, P.E., 637
Calzolari, G., 578
Candelon, B., 184
Caner, M., 342
Caporale, G.M., 301
Cheung, Y.W., 579
Chitturi, R.V., 157
Choi, I., 343
Chou, R.Y., 559
Chow, G.C., 410, 411, 413, 623
Christiano, L.J., 362
Cipra, T., 604
Claessen, H., 515, 518, 526, 529

Author Index

Cleveland, W.P., 604
Comte, F., 569, 579
Cooley, T.F., 623
Cooper, D.M., 507
Craig, A.T., IX, 677
Crowder, M.J., 637

Davidson, J., 246, 400, 681, 689, 691, 699
Davies, N., 171
Davis, R.A., 86, 434
Deistler, M., 397, 400, 422, 455, 467, 476, 477, 479, 499, 500, 502, 504, 505, 612, 616
Dellaportas, P., 567, 584
Demiroglu, U., 181, 348
DeMoor, B., 479
Dempster, A.P., 635
Dent, W., 104
Diebold, F.X., 562
Doan, T., 225, 311, 623
Dolado, J.J., 301, 318–320, 343
Doornik, J.A., 173, 181, 307, 341, 576
Dufour, J.M., 49, 50, 94, 161
Dunsmuir, W.T.M., 479
Durlauf, S.N., 290, 302, 323, 704

Edgerton, D., 173
Efron, B., 710, 711
Eichenbaum, M., 362
Engle, R.F., 245, 246, 316, 343, 388, 559, 560, 562, 564–566, 568, 577, 579, 581–583, 620, 621, 635
Ericsson, N.R., 388
Evans, C., 362

Fiorentini, G., 578
Fisher, L.A., 372
Forni, M., 621
Franses, P.H., 604
Friedmann, R., 411
Fuller, W.A., VIII, 35, 73, 84, 159, 546, 681, 685, 687–689

Gabr, M.M., 625
Galbraith, J.W., 301, 343
Galí, J., 364, 376
Gallant, A.R., 398, 580, 695
Gasser, T., 177

Geweke, J., 51, 104, 388, 620
Ghysels, E., 583, 604
Giannini, C., 358, 374, 375
Gladyshev, E.G., 604
Glosten, L., 568
Godfrey, L.G., 173
Gonzalo, J., 300, 343, 357, 370, 372
Granger, C.W.J., 33, 41, 51, 245, 246, 343, 507, 562, 579, 624, 625
Graybill, F.A., IX, 645
Gregory, A.W., 698
Griffiths, W.E., 229, 398, 400, 470–472, 677, 681
Grupe, M.R., 604
Guo, L., 523

Hafner, C.M., 568, 582
Haggan, V., 625
Haldrup, N., 604
Hall, P., 710, 711
Hallin, M., 621
Hamilton, J.D., VIII, 637, 690, 691, 700, 703, 704
Hannan, E.J., 25, 141, 149, 150, 159, 397, 400, 422, 455, 476, 477, 479, 495, 498–500, 502, 504, 505, 523, 612
Hansen, B.E., 301, 302
Hansen, G., 297
Hansen, H., 181, 351
Hansen, P.R., 349
Hanssens, D.M., 401
Harbo, I., 401
Harvey, A.C., 397, 583, 612, 618, 619
Hassler, U., 343
Hasza, D.P., 95
Haug, A.A., 343, 401
Haugh, L.D., 401
Hausman, J.A., 398
Hayashi, F., 398, 697
Hendry, D.F., 212, 246, 301, 341, 343, 388, 576
Henze, N., 177
Herwartz, H., 568, 571, 582, 584, 604
Higgins, M.L., 559
Hildreth, C., 623
Hill, R.C., 229, 398, 400, 470–472, 677, 681
Hillmer, S.C., 463, 466

Ho, M.S., 343
Hogg, R.V., IX, 677
Hong, Y., 579
Horowitz, J.L., 709
Hosking, J.R.M., 157, 171, 508, 510
Houck, J.P., 623
Hsiao, C., 212
Huang, D.W., 523
Hubrich, K., 336, 341, 343
Huh, H., 372
Hui, Y.V., 604
Hylleberg, S., 411

Inder, B., 301, 312

Jacobson, T., 357
Jagannathan, R., 568
Jakeman, A.J., 401
Jarque, C.M., 175, 177
Jazwinski, A.H., 612
Jeantheau, T., 570
Jenkins, G.M., 400, 493, 495, 507
Johansen, S., 251, 280, 283, 290, 294, 296, 300, 302, 329–332, 334–336, 340, 341, 351, 400, 401, 699, 704
Jones, R.H., 604, 618
Judge, G.G., 229, 398, 400, 470–472, 677, 681
Juselius, K., 302

Kalman, R.E., 625
Kapetanios, G., 479
Kauppi, H., 318
Kavalieris, L., 400, 476, 477, 498, 504
Kilian, L., 126, 181, 348
Kim, J.R., 297
King, R.G., 357, 370, 371, 377
Kitagawa, G., 618
Kleibergen, F., 312
Kohn, R., 463, 479, 508, 618
Koop, G., 312, 383, 581
Koreisha, S.G., 479
Kraft, D.F., 562, 564
Krätzig, M., IX
Krolzig, H.M., 212
Kroner, K.F., 559, 564–566, 577, 583

Laird, N.M., 635
Laurent, S., 559, 568, 571, 577

Lee, K.C., 383
Lee, T.C., 229, 398, 400, 470–472, 677, 681
Lee, T.H., 343
Lewis, R., 532, 536
Li, H., 710
Li, W.K., 157, 510, 583, 604
Lieberman, O., 569, 579
Lilien, D.M., 583
Lin, W.L., 567, 582
Ling, S., 570
Lippi, M., 621
Litterman, R.B., 225, 311, 315, 623
Liu, J., 625
Liu, L.M., 401
Ljung, G.M., 171
Lomnicki, Z.A., 175
Lucas, A., 342
Lucas, R.E., 394
Lumsdaine, R.L., 609
Lundbergh, S., 577
Lütkepohl, H., 49, 63, 106–108, 110, 112, 113, 116, 129, 138, 156, 178, 184, 186, 188, 189, 211, 212, 229, 263, 297, 302, 303, 309, 318, 320, 327, 334, 336, 337, 339–341, 343, 345, 346, 348, 357, 364, 366, 372, 386, 398, 400, 435, 438, 440, 441, 470–472, 515, 518, 523–526, 529, 534–536, 541, 543, 546, 549–552, 561, 568, 571, 582, 595, 602, 605, 645, 669, 677, 681, 710, 711
Luukkonen, R., 525

MacDonald, G.M., 177
MacKinnon, J.G., 401
Maddala, G.S., 710
Magnus, J.R., 645, 651, 672, 674
Mann, H.B., 73
Mardia, K.V., 177
McAleer, M., 570
McLeod, A.I., 157, 510
McMurtrie, R., 401
Meese, R., 104
Meinhold, R.J., 611, 630
Michelis, L., 401
Mittnik, S., 297, 431
Mood, A.M., IX
Moore, J.B., 612, 628, 630, 637

Morrison, D.F., 619
Mosconi, R., 335
Murata, Y., 410, 411, 670
Muth, J.F., 393

Nankervis, J.C., 82
Nelson, D.B., 559, 568
Nerlove, M., 562
Neudecker, H., 645, 651, 672, 674
Neumann, M., 112, 113, 129, 711
Newbold, P., 33, 171, 507
Ng, L.K., 579
Ng, S., 357, 370, 372
Ng, V.G., 581, 582
Nicholls, D.F., 82, 623, 635, 637
Nielsen, B., 335, 341, 401
Nijman, T.E., 618

Osborn, D.R., 604
Ouliaris, S., 343
Ozaki, T., 625

Paap, R., 312
Pagan, A.R., 364, 366, 559, 623, 635, 637
Pagano, M., 604
Palm, F., 494, 496
Pantelidis, T., 579
Park, J.Y., 251, 280, 290, 302, 704
Paulsen, J., 82, 86, 144, 149, 150, 326
Penm, J.H.W., 207, 212
Persand, G., 583
Pesaran, M.H., 383, 394, 581
Phillips, P.C.B., 251, 280, 290, 297, 300–302, 318, 323, 343, 548, 700, 704
Pierse, R.G., 383, 618
Pittis, N., 301, 579
Plosser, C.I., 357, 370, 371, 377
Politis, D.N., 567, 584
Pope, A.L., 82, 623
Poskitt, D.S., 327, 400, 401, 477, 495, 498, 499, 505–508, 510, 515, 517, 518, 522–525, 534, 535, 541, 543
Pötscher, B.M., 467
Potter, S.M., 581
Prescott, E., 623
Priestley, M.B., 625
Proietti, T., 619

Pukkila, T.M., 479

Quah, D., 357, 367, 376, 385
Quenouille, M.H., 507
Quinn, B.G., 149, 150, 153

Rahbek, A., 401
Rao, C.R., 173, 677, 690, 691
Reichlin, L., 621
Reimers, H.E., 302, 316, 320
Reinsel, G.C., 86, 222, 271, 272, 280, 290, 292–294, 297, 316, 323, 522, 525, 532, 536
Renault, E., 49, 50, 583
Richard, J.F., 388
Rissanen, J., 495
Robins, R.P., 579, 583
Rohatgi, V.K., IX
Rombouts, J.V.K., 559, 568, 571, 577
Rossi, P.E., 580
Rothenberg, T.J., 360, 375
Roussas, G.G., 681
Roy, R., 161
Rubin, D.B., 635
Runkle, D., 126, 568

Saikkonen, P., 254, 292, 294, 334, 336, 337, 339–341, 343, 345, 346, 348, 525, 546, 549–553, 609
Salmon, M., 246
Samaranayake, V.A., 95
Sargent, T.J., 394, 620
Savin, N.E., 82
Schmidt, P., 407, 693, 698
Schneider, W., 178, 628, 635, 637
Schwarz, G., 150
Searle, S.R., 645
Sen Roy, S., 95, 316
Sentana, E., 578
Serfling, R.J., 681, 693
Shapiro, M., 357
Shibata, R., 151
Shin, Y., 343, 383
Shukur, G., 173
Sims, C.A., 66, 116, 225, 289, 311, 320, 357, 383, 620, 623, 704
Singpurwalla, N.D., 611, 630
Smith, J.P., 604
Snell, A., 343
Solo, V., 548, 633, 704

Sørensen, B.E., 343
Srba, F., 246
Stensholt, B.K., 625
Stock, J.H., 289, 300–302, 320, 343, 357, 370, 371, 377, 609, 621, 704
Strachan, R.W., 312
Subba Rao, T., 625
Sunier, A.M., 568
Swamy, P.A.V.B., 623

Tauchen, G., 580
Tavlas, G.S., 623
Tay, A.S., 41
Taylor, S.J., 560
Teräsvirta, T., 577, 625
Terrell, R.D., 207, 212
Theil, H., 59
Tiao, G.C., 463, 466, 507, 604, 609
Tibshirani, R.J., 710, 711
Tjøstheim, D., 82, 86, 144, 150, 625
Toda, H.Y., 318
Todd, P.H.J., 618
Tong, H., 625
Tremayne, A.R., 401, 508, 510
Trenkler, C., 334, 339, 340, 609
Triggs, C.M., 171
Tsay, R.S., 326, 401, 507, 609
Tse, Y.K., 568, 577
Tso, M.K.S., 222, 675
Tsui, A.K.C., 568, 577

Uhlig, H., 383
Ungern-Sternberg, T. von, 246

van der Weide, R., 568
van Dijk, H.K., 312

van Overschee, P., 479
Veall, M.R., 698
Velu, R.P., 222
Villani, M., 312
Vlaar, P.J.G., 376, 377
Vredin, A., 357
Vrontos, I.D., 567, 584

Wagner, M., 636, 637
Wald, A., 73
Wallis, K.F., 41, 494
Warne, A., 357
Watanabe, N., 637
Watson, M.W., 289, 300, 301, 320, 343, 357, 370, 371, 377, 620, 621, 635, 704
Weber, C.E., 385
Wei, W.W.S., 616
White, H., 177, 398
Wichern, D.W., 222
Wickens, M.R., 301
Wold, H., 25
Wolters, J., 710
Wong, H., 583
Wood, E.F., 507
Wooldridge, J.M., 562, 564, 565

Yamamoto, T., 318, 406
Yang, M., 343
Yap, S.F., 522, 525
Yeo, S., 246
Yoo, B.S., 316
Young, P.C., 401

Zellner, A., 71, 494, 496
Zha, T., 116, 383

Subject Index

Accumulated forecast error impulse responses, 55, 108
Accumulated impulse responses, 55, 108
Accumulated orthogonalized impulse responses, 108
Adjoint of a matrix, 649
Aggregation
— of MA processes, 435–436
— of VARMA processes, 440–441
contemporaneous —, 434–435, 440
temporal —, 434–435, 440–441, 616–618
AIC, 147, 208
Akaike information criterion, see AIC
Analysis
Bayesian —, 222–229, 309–315
causality —, 261–262, 316–321, 441–444
impulse response —, 205–206, 262–264, 321–322, 444, 490
multiplier —, 392, 406–408
ARCH model, 557–584
generalized —, 559
interpretation of —, 579–582
multivariate —, 563–564
univariate —, 559–562
ARCH process, 557–584
forecasting of —, 561–562
generalized —, 559
ARCH-in-mean, 583
ARCH-LM test, 576–577
ARCH-M, 583
ARCH-portmanteau test, 576–577

ARMA$_{RE}$ form
— of VARMA process, 518–519
estimation of —, 521–522
Asymmetric GARCH model, 568
Asymptotic distribution
— of EGLS estimator, 197
— of GLS estimator, 197
— of LM statistic, 510
— of LR statistic, 140
— of LS estimator, 73, 532
— of ML estimator, 93, 200, 296, 479–487, 636–637
— of Wald statistic, 102, 321
— of Yule-Walker estimator, 86
— of autocorrelations, 159
— of autocovariances, 159
— of forecast error variance decomposition, 110–112, 541–543
— of impulse responses, 110–112, 205–206, 321–322, 541–543
— of kurtosis, 175, 178
— of lag augmented Wald statistic, 318
— of multivariate LS estimator, 73, 532
— of residual autocorrelations, 166, 213
— of residual autocovariances, 165, 213
— of skewness, 175, 178
Asymptotic properties
— of LM statistic, 138–143
— of LR statistic, 508–510

- of LS estimator, 72–77, 531–536
- of ML estimator, 90–93, 200, 296, 479–487, 636–637
- of Wald statistic, 102, 321
- of Yule-Walker estimator, 86
- of estimated forecasts, 93–102, 405–406, 536–538
- of impulse responses, 110–112, 321, 490
- of lag augmented Wald statistic, 318
- of sample mean, 84
- of white noise covariance matrix estimator, 75–76, 200, 296, 479–481

Asymptotically stationary process, 241

Autocorrelation function
- of VAR process, 30–31
- of VARMA process, 430
- of residuals, 161–169, 212–214
  asymptotic distribution of —, 157–171
  estimation of —, 157–171

Autocorrelation matrix
- of VARMA process, 430
- of residuals, 161–169, 212–214
  estimation of —, 157–171

Autocovariance function
- of MA process, 422
- of VAR process, 21, 26–30
- of VARMA process, 429–432
- of residuals, 161–169, 212–214
  asymptotic distribution of —, 157–171
  estimation of —, 157–169

Autoregressive conditional heteroskedasticity, 559

Autoregressive process, 4

Autoregressive representation of VARMA process
  infinite order —, 425
  pure —, 425

Bayesian estimation
- of Gaussian VAR process, 222–229
- of integrated systems, 309–315
- with normal priors, 222–225, 309–312
  basics of —, 222

BEKK model, 565–567

Beveridge-Nelson decomposition, 242, 703
  multivariate —, 252, 704

Bilinear state space model, 624

Bilinear time series model, 624

Block bootstrap, 710

Bootstrap, 709–712
- Hall's percentile interval, 710
- Hall's studentized interval, 710
- confidence interval, 710
- standard percentile interval, 710
  block —, 710
  residual based —, 709

Bootstrap confidence interval
  Hall's percentile —, 710
  Hall's studentized —, 710
  standard percentile —, 710

Bottom-up specification of subset VAR model, 210–211

Bounded in probability, 684

Box-Jenkins methodology, 493, 495

Breusch-Godfrey test for autocorrelation, 171–174

Brownian bridge, 338

Brownian motion, 698
  multivariate —, 703

Causality, 41–51, 102–108, 261–262, 316–321, 441–444
- in r-th moment, 579
- in variance, 579–580
  Granger —, 41–51
  instantaneous —, 41–51
  multi-step —, 41–51
  Wold —, 359

CCC GARCH model, 568

Central limit theorem, 689–692
- for martingale difference arrays, 691
- for stationary processes, 691
  Donsker's —, 699
  functional —, 699
  invariance principle, 699
  Lindeberg-Levy —, 691

Chain rule for vector differentiation, 665

Characteristic
- determinant, 652

— value of a matrix, 652
— vector of a matrix, 652
polynomial
— of a matrix, 652
Chebyshev's inequality, 689
Chebyshev's theorem, 690
Checking the adequacy
— of VAR models, 157–189
— of VARMA models, 508–510
— of cointegrated systems, 344–351
— of dynamic SEMs, 400–401
— of subset VAR models, 212–217
Checking the whiteness
— of VAR residuals, 157–174
— of VARMA residuals, 510
Chi-square distribution, 678
noncentral —, 679
Choleski decomposition, 362
— of a positive definite matrix, 659
Chow forecast test, 183
Chow test, 182
Closed-loop control strategy, 388
CLT, 689
Cofactor of an element of a square matrix, 648
Cointegrated process, 244–256
— of order (d, b), 245
Cointegrated system, 244–256
Granger representation theorem for —, 251
vector error correction model of —, 244–256
Cointegrated VAR process, 256
checking the adequacy of —, 344–351
estimation of —, 269–309
forecasting of —, 258–261, 315–316
GLS estimation of —, 291–294
Granger-causality in —, 261, 316
impulse response analysis of —, 262–264, 321–322
least squares estimation of —, 286–291
LS estimation of —, 286–291
ML estimation of —, 294–300
structural analysis of —, 261–264, 316–322
two-stage estimation of —, 301–302
Cointegrated VARMA process, 515–521
$ARMA_{RE}$ form of —, 518–519

EC-$ARMA_{RE}$ form of —, 519–521
error correction echelon form of —, 519–521
estimation of —, 521–522
reverse echelon form of —, 518–519
specification of cointegrating rank of —, 525–526
Cointegrating
— matrix, 256
— vector, 245
Cointegrating rank, see cointegration rank
Cointegration matrix, 248
Cointegration rank, 248
— of VAR process, 248
— of VARMA process, 525–526
LR test for —, 327–335, 551–552
maximum eigenvalue test for —, 329
testing for —, 327–343, 551–552
trace test for —, 329
Column vector, 645
Common trend, 245
Commutation matrix, 663
Complex matrix, 657
Complex number
modulus of —, 652
Conditional forecast, 402
Conditional likelihood function, 464
Conditional model, 387
Conditional moment profiles, 580–582
Confidence interval
— for forecast error variance components, 114
— for impulse responses, 112
Hall's percentile —, 710
Hall's studentized —, 710
standard percentile —, 710
Consistency
super —, 288, 301
Consistent estimation
— of Kronecker indices, 501
— of VAR order, 148–150, 326
— of white noise covariance matrix, 76
Constant conditional correlation GARCH model, 568
Constrained VAR models
linear constraints, 194–221
nonlinear constraints, 221–222

Subject Index

Contemporaneous aggregation of VARMA process, 440
Continuous mapping theorem, 699
Convergence
— almost surely, 682
— in distribution, 682
— in law, 682
— in mean square error, 682
— in probability, 681
— in quadratic mean, 682
— with probability one, 682
strong —, 682
weak —, 682
Cramér-Wold device, 691

Data generation process, 4
DCC GARCH model, 568
Decomposition of matrices, 656–659
Choleski —, 659
Jordan —, 656–657
Definite
— matrix, 655–656
— quadratic form, 655–656
Degree
McMillan —, 453
Degrees of freedom, 678
Determinant of a matrix, 647
Deterministic trend, 238
DGP, 4
Diagnostic checking
— of VAR models, 157–189
— of VARMA models, 508–510
— of cointegrated systems, 344–351
— of dynamic SEMs, 400–401
— of restricted VAR models, 212–214
Diagonal matrix, 646
Diagonalization of a matrix, 657–658
Dickey-Fuller test, 700
Difference operator, 242
Differencing, 242
Differentiation of vectors and matrices, 664–671
Direction matrix, 471
Discrete stochastic process, 3
Distributed lag model, 387, 391–392
rational —, 391–392
Distribution
— multivariate normal, 677–678
— normal, 677–678

— of quadratic form, 678
chi-square —, 678
F —, 679
noncentral F —, 680
noncentral chi-square —, 679
posterior —, 222
prior —, 222
Distribution function, 3
joint —, 3
Donsker's theorem, 699, 704
Drift of a random walk, 238
Dummy variable, 585
seasonal —, 585
Duplication matrix, 662
Dynamic
— MIMIC model, 621
— factor analytic model, 620
— multipliers, 392
Dynamic conditional correlation GARCH model, 568
Dynamic SEM
checking the adequacy of —, 400–401
estimation of —, 394–400
final equations of —, 392
final form of —, 391
forecasting of —, 401–406
conditional —, 402
unconditional —, 402
multipliers of —, 406–408
optimal control of —, 408–411
rational expectations in —, 392–394
reduced form of —, 390
specification of —, 400–401
structural form of —, 390

EC-ARMA$_{RE}$ form
— of VARMA process, 519–521
estimation of —, 522
specification of —, 523–526
Echelon form
— VARMA representation, 452–453
— of a VARMA process, 452–453
specification of —, 498–507
Effect of linear transformations
— on MA process, 435
— on VARMA orders, 436
— on forecast efficiency, 439
Efficiency
— of estimators, 198–200

— of forecasts, 439
EGARCH model, 568
EGLS estimation
   — of parameters arranged equation-wise, 200–201
   asymptotic properties of —, 197–200
   implied restricted —, 197
   restricted —, 197–201
Eigenvalue of a matrix, 652
Eigenvector of a matrix, 652
Elimination matrix, 662
EM algorithm, 635
Empirical distribution, 707
   generation of —, 707
Endogenous variable, 387–390, 613
Equilibrium, 244–245
Equilibrium relation, 245
Error correction echelon form, 519–521
   estimation of —, 522
   specification of —, 523–526
Error correction model, 246
Error process
   — of the measurement equation of a state space model, 613
   — of the transition equation of a state space model, 613
Estimated generalized least squares estimation, see EGLS estimation
Estimation
   — of AB-model, 372–375
   — of $ARMA_{RE}$ form, 521–522
   — of Blanchard-Quah model, 376
   — of SVAR, 372–376
   — of SVAR with long-run restrictions, 376
   — of SVECM, 376–377
   — of VAR models, 69–93, 531–536
   — of VARMA models, 447–487
   — of autocorrelations, 157–169
   — of autocovariances, 157–169
   — of cointegrated VARMA process, 521–522
   — of cointegrated systems, 269–309
   — of dynamic SEMs, 394–400
   — of error correction echelon form, 522
   — of integrated VAR processes, 309–315
   — of multivariate GARCH model, 569–571
   — of periodic models, 594–598
   — of process mean, 83–85
   — of reverse echelon form, 521–522
   — of state space models, 631–637
   — of time varying coefficient models, 589–591
   — of white noise covariance matrix, 75–77, 197–198
   — with linear restrictions, 195–204
   — with nonlinear restrictions, 222
   — with unknown process mean, 85
   Bayesian —, 222–229
   EGLS —, 197
   generalized least squares —, 291–294
   GLS —, 195
   least squares —, 286–291
   LS —, 69–82, 197
   maximum likelihood —, 87–93, 200, 294–300, 589–591
   multivariate least squares —, 69–82, 531–536
   preliminary —, 474–477
   restricted
      EGLS —, 195–200
      GLS —, 195–200
   restricted —, 195–204, 222
   two-stage —, 301–302
   Yule-Walker —, 85–86
Exact likelihood function, 458–461
Exogenous variable, 387–390
   strictly —, 389
   strongly —, 388
   super —, 388
   systems with —, 388–390
   weakly —, 388
Expectation
   rational —, 392

F-distribution, 679
   noncentral, 680
Factor analytic model
   dynamic —, 620
Factor GARCH model, 567
Factor loadings, 620
FCLT, 699
Filter
   Kalman —, 625–631

752   Subject Index

Filtering, 625
Final equations form
— VARMA representation, 452
— of a dynamic simultaneous equations model, 392
specification of —, 494–498
Final form of a dynamic simultaneous equations system, 391
Final prediction error criterion, see FPE criterion
Finite order MA process, 420–423
Forecast
— interval, 39–41
— of VAR process, 93–102, 536–540
— of VARMA process, 432–434, 487–490
— region, 39–41
conditional —, 402
estimated —, 93–102, 487–490, 536–540
loss function for —, 32–33
minimum MSE —, 35–39
point —, 33–39
unconditional —, 402
Forecast error, 93–95
Forecast error impulse responses, 51–56
Forecast error variance component, 108
Forecast error variance decomposition, 63–66, 540–545
— of VAR process, 63–66
— of cointegrated system, 264
asymptotic distribution of —, 108–118, 205–206, 541–543
structural —, 381–382
Forecast interval, 98
Forecast MSE matrix, 434
approximate —, 96–98, 489–490, 536
Forecast region, 98
Forecasting
— of ARCH process, 561–562
— of GARCH model, 561–562
— of VAR process, 31–41
— of VARMA process, 432–434
— of cointegrated system, 258–261, 315–316
— of dynamic SEM, 401–406
— of estimated VAR process, 93–102, 204–205

— of estimated VARMA process, 487–490
— of infinite order VAR process, 536–540
— of integrated system, 258–261, 315–316
— of restricted VAR process, 204–205
FPE criterion, 146
Fully modified VAR estimation, 318
Functional central limit theorem, 699
multivariate —, 704

GARCH model, 557–584
asymmetric —, 568
CCC —, 568
constant conditional correlation —, 568
dynamic conditional correlation —, 568
exponential —, 568
factor —, 567
generalized orthogonal —, 568
interpretation of —, 579–582
multivariate —, 562–584
univariate —, 559–562
GARCH process, 557–584
forecasting of —, 561–562
Gaussian likelihood function
— of MA process, 458–463
— of VAR process, 87–89
— of VARMA process, 463–467
— of cointegrated process, 294
— of state space model, 631–633
Gaussian process
VAR, 16
VARMA, 423
white noise, 75
Generalized autoregressive conditional heteroskedasticity, 559
Generalized impulse responses, 580–582
Generalized inverse of a matrix, 650
Generalized orthogonal GARCH model, 568
Generating process of a time series, 4
Generation process of a time series, 4
Global identification, 633–634
Globally identified model, 633–634
GLS estimation, 195–200
— of cointegrated system, 291–294

asymptotic properties of —, 197
Gradient
— algorithm, 469
— of log-likelihood function, 635
— of vector function, 469
Granger representation theorem, 251
Granger-causality
— in VAR models, 41–51, 102–104
— in VARMA models, 441–444
— in cointegrated system, 261–262
— in cointegrated systems, 316–321
characterization of —, 316
lag augmentation test for —, 318
lag augmented Wald test for —, 318
test for —, 102–104, 316–321
Wald test for —, 102–104, 316–321

Hall's percentile confidence interval, 710
Hall's studentized confidence interval, 710
Hannan-Kavalieris procedure, 503–505
Hannan-Quinn criterion, *see* HQ criterion
Hessian matrix, 665
HQ criterion, 150, 208

Idempotent matrix, 653
Identification
— of VARMA model, 447–458
— of VARX model, 400
— of dynamic simultaneous equations system, 400
— of state space model, 633–634
global —, 634
local —, 634
Identification problem, 447–458
Identified model
globally —, 634
locally —, 634
state space, 633–634
VARMA, 447–458
Identity matrix, 646
Impact multiplier, 61
Impulse response analysis, 377–382
— of VAR model, 51–63
— of VARMA model, 444, 490
— of cointegrated system, 262–264, 321–322
Impulse responses, 51–63, 377–382

— of VARMA model, 444, 490
— of cointegrated system, 262–264, 321–322
accumulated —, 55
asymptotic distribution of —, 108–118, 205–206, 541–543
estimation of —, 108, 205–206
forecast error —, 51–56
generalized —, 580–582
orthogonalized —, 56–62, 359
structural —, 359, 377–382
total —, 56
Indefinite
— matrix, 656
— quadratic form, 656
Index model, 222
Infinite order MA representation
— of a VARMA process, 423
— of a time varying coefficient process, 587
Infinite order VAR representation
— of a VARMA process, 425
— of an MA process, 420
Information matrix
— of VAR process, 90
— of VARMA process, 472–474
— of state space model, 635
— of time varying coefficient VAR model, 591
Initial
— input, 613
— state, 613
Innovations
structural —, 359
Input
— matrix of a state space model, 613
— variables, 388
observable —, 388, 613
unobservable —, 388
Inputs of a state space model, 613
Instantaneous causality
— in VAR models, 41–51
tests for —, 104–108
Instrument
— variable, 388, 613
observable —, 388, 613
Integrated
— of order d, 242
— process, 237–244

— time series, 237–244
— variable, 237–244
Integration
 order of —, 242
Interim multipliers, 56, 392
Interpretation
 — of ARCH model, 579–582
 — of GARCH model, 579–582
 — of VARMA model, 441–444
 classical versus Bayesian —, 228–229
Interval forecast, 39–41, 98
Intervention
 — in intercept model, 604–606
 testing for —, 605–606
Intervention model, 586
 estimation of —, 604–608
 specification of —, 604–608
Invariance principle, 699
 multivariate —, 704
Inverse of a matrix, 649
Invertible
 — MA operator, 422
 — MA process, 420–422
 — VARMA process, 425
 — matrix, 649
IS-LM model, 366
Iterative optimization algorithm
 EM algorithm, 635
 Newton algorithm, 471
 scoring algorithm, 472, 634–636

Jarque-Bera test, 175
Jordan canonical form, 657

Kalman filter, 625–631
 — correction step, 627
 — forecasting step, 627
 — gain, 627
 — initialization, 627
 — prediction step, 627
 — recursions, 626–630
 — smoothing step, 630
Kalman gain, 627
Kalman smoothing matrix, 630
Khinchine's theorem, 690
Kronecker indices, 453
 — of VARMA process, 453
 — of cointegrated VARMA process, 518

— of echelon form, 453
— of reverse echelon form, 518
 determination of —, 498–507
 estimation of —, 498–507
 specification of —, 498–507
Kronecker product, 660
Kurtosis
 asymptotic distribution of —, 175, 178
 measure of multivariate —, 174–180

Lagrange function, 671, 695
Lagrange multiplier statistic, 508–510, 600–601
 asymptotic distribution of —, 510, 601
Lagrange multiplier test, 508–510, 600–601, 694–698
Lagrange multipliers, 671
Law of large numbers, 689–692
 — for martingale difference arrays, 690
 — for martingale difference sequence, 690
 — for stationary processes, 690
 strong —, 689
 weak —, 689
Least squares estimation
 — of VAR process, 69–82, 531–536
 — of cointegrated VAR process, 286–291
 — with mean-adjusted data, 82–85
 asymptotic properties of —, 72–77, 197–200, 532–533
 multivariate —, 69–82, 531–536
 restricted —, 197–200
 small sample properties of —, 80–82
Least squares estimator of white noise covariance matrix, 75–77, 535–536
 asymptotic properties of —, 75, 535–536
Left-coprime operator, 452
Leptokurtosis, 560
Leverage effect, 568
Likelihood function, 693
 — of MA process, 458–463
 — of VAR process, 87–89
 — of VARMA process, 463–467
 — of cointegrated process, 294

— of state space model, 631–633
— of time varying coefficient VAR model, 589
conditional —, 464
Likelihood ratio statistic
 asymptotic distribution of —, 140
 definition of —, 138
Likelihood ratio test, 694–698
— for cointegration rank, 327–343, 551–552
— of linear restrictions, 138–143
— of periodicity, 598
— of varying coefficients, 595–598
— of zero restrictions, 138–143
Lindeberg-Levy central limit theorem, 691
Linear constraints
— for VAR coefficients, 194–195
Linear system, 387
Linear transformation
— of MA process, 435–436
— of VARMA process, 436–440
— of multivariate normal distribution, 678
Linearly dependent vectors, 652
Linearly independent vectors, 652
Litterman prior
— for nonstationary process, 310–315
— for stationary process, 225–227
LLN, 689
LM test, 695
— for autocorrelation, 171–174
Loading matrix, 248
Locally identified model, 634
Log-likelihood function, 693
Lomnicki-Jarque-Bera test, 175
Long-run
— effect, 392
— multiplier, 392
Loss function, 32–33
 quadratic —, 409
LR test, 695
LS estimation, see least squares estimation

MA operator, 422
MA process
 autocovariances of —, 422
 finite order —, 420–423
 invertible —, 420–422
 likelihood function of —, 458–463
MA representation
— of a VARMA process, 423
canonical —, 426
forecast error —, 426
prediction error —, 426
MA representation of VAR process, 18–24
Martingale difference array, 689
 law of large numbers for —, 690
Martingale difference sequence, 689
 law of large numbers for —, 690
 vector —, 689
Matrix, 645
— addition, 646
— differentiation, 664–671
— multiplication, 646
— multiplication by a scalar, 646
— operations, 646–647
— rules, 645–675
— subtraction, 646
operator
 left-coprime —, 452
 unimodular —, 452
adjoint of —, 649
characteristic determinant of —, 652
characteristic polynomial of —, 652
characteristic root of —, 652
characteristic value of —, 652
characteristic vector of —, 652
Choleski decomposition of —, 659
cofactor of an element of —, 648
column dimension of —, 645
commutation —, 663
conformable —, 647
decomposition of —, 656–659
determinant of —, 647
diagonal —, 646
diagonalization of —, 657–658
duplication —, 662
eigenvalue of —, 652
eigenvector of —, 652
element of —, 645
elimination —, 662
full rank —, 652
generalized inverse of —, 650
Hessian —, 665
idempotent —, 653

identity —, 646
indefinite —, 656
information —, 694
inverse of —, 649
invertible —, 649
Jordan canonical form of —, 657
lower triangular —, 646
minor of an element of —, 648
Moore-Penrose inverse of —, 650
negative definite —, 656
negative semidefinite —, 656
nilpotent —, 653
nonsingular —, 649
null —, 646
orthogonal —, 654
orthogonal complement of —, 654
partitioned —, 659
positive definite —, 655
positive semidefinite —, 655
rank of —, 652
regular —, 649
row dimension of —, 645
square —, 645
square root of —, 658
symmetric —, 646
trace of —, 653
transpose of —, 646
triangular —, 646
typical element of —, 645
unit —, 646
upper triangular —, 646
zero —, 646
Maximum eigenvalue test for cointegration rank, 329
Maximum likelihood estimation, *see* ML estimation
McMillan degree
— of VARMA process, 453
— of echelon form, 453
Mean squared error matrix, *see* MSE matrix
Mean vector of a VAR process, 82
Mean-adjusted
— VAR process, 82
— process, 82
Measurement
— equation of state space model, 611, 613
— errors, 613

— matrix, 613
MGARCH, 562–584
MIMIC models, 621
Minimization
— algorithms, 469–472
iterative —, 469–472
numerical —, 469–472
Minimum MSE forecast, 35–39
Minor of an element of a square matrix, 648
ML estimates
computation of —, 89–90, 467–477, 631–637
ML estimation, 693
— of AB-model, 372–375
— of Blanchard-Quah model, 376
— of SVAR, 372–376
— of SVECM, 376–377
— of VAR process, 87–93
— of VAR process with time varying coefficients, 589–591
— of VARMA process, 458–487
— of cointegrated system, 294–300
— of periodic VAR process, 594–598
— of restricted VAR process, 200
— of state space model, 631–637
quasi —, 140
Model checking
— of VAR models, 157–189
— of VARMA models, 508–510
— of cointegrated systems, 344–351
— of dynamic SEMs, 400–401
— of restricted VAR models, 212–217
— of state space models, 639
— of subset VAR models, 212–217
Model selection
— of VAR models, 135–157
— of VARMA models, 493–508
— of cointegrated processes, 325–344
— of subset VAR models, 206–212
Model specification
— of VAR models, 135–157
— of VARMA models, 493–508
— of cointegrated processes, 325–344
— of dynamic SEMs, 400–401
— of periodic VAR models, 594–604
— of subset VAR models, 206–212
Model specification criteria
AIC, 147, 208

FPE, 146
HQ, 150, 208
SC, 150, 208
Modified portmanteau statistic, 174, 214
  approximate distribution of —, 174, 214
Modified portmanteau test, 174, 214, 510
Modulus of a complex number, 652
Moore-Penrose (generalized) inverse, 650
Moving average process, see MA process
Moving average representation of VAR process, 18–24
MSE matrix, 434
  approximate —, 96–98, 489–490, 536
MSE of forecast, 96–98, 434, 489–490, 536
Multi-step causality
  — in VAR models, 41–51
  tests for —, 105–108
Multiplicative operator, 221–222
Multiplier
  — analysis, 392, 406–408
  dynamic —, 392
  impact —, 61
  interim —, 392
  long-run —, 392
  total —, 392
Multivariate ARCH model, 563–564
  interpretation of —, 579–582
Multivariate Beveridge-Nelson decomposition, 252
Multivariate GARCH model, 562–584
  BEKK, 565–567
  estimation of —, 569–571
  interpretation of —, 579–582
Multivariate least squares estimation, 69–86
  — of VAR process, 69–82
  — of infinite order VAR process, 531–536
Multivariate normal distribution, 677–678
  linear transformation of —, 678
Multivariate stochastic process
  discrete —, 3

Negative definite
  — matrix, 656
  — quadratic form, 656
Negative semidefinite matrix, 656
Newton algorithm, 471
Newton-Raphson algorithm, 471
Nilpotent matrix, 653
Noncentral F-distribution, 680
Noncentral chi-square distribution, 679
Noncentrality parameter, 679
Nonlinear
  — parameter restrictions, 221–222
  — state space model, 623–625
Nonnormality
  tests for —, 174–180
Nonsingular matrix, 649
Nonstationary
  — VAR process, 242, 256, 585–586
  — process, 237, 585–586, 614, 621–623
  — time series, 237
Normal distribution
  — multivariate, 677–678
Normal equations
  — for VAR coefficient estimates, 71
  — for VAR process with time varying coefficients, 589
  — for VARMA estimation, 467–469
Normal prior, 222–225
  — p.d.f., 222
Normal prior for Gaussian VAR process, 222–225, 309

Observable
  — input, 388
  — output, 388
  — variables, 388
Observation
  — equation of state space model, 611, 613
  — error, 611, 613
  — noise, 613
Open-loop strategy, 411
Operator
  left-coprime —, 452
  MA —, 422
  unimodular —, 452
Optimal control, 408–411
  closed-loop —, 411

Subject Index

open-loop —, 411
problem of —, 410
Optimization
— algorithms, 469–472
— of vector functions, 671–675
Order determination
— for VAR process, 135–157
— for cointegrated process, 325–327
criteria for —, 146–157
tests for —, 136–145
Order estimation
— for cointegrated processes, 325–327
— of VAR process, 146–157
consistent —, 148–150
criteria for —, 146–157
Order in probability, 684–685
Order of
— MA process, 420
— VAR process, 136
— VARMA process, 423
Orthogonal
— matrix, 654
— vectors, 654
Orthogonal complement of a matrix, 654
Orthogonalized impulse responses, 56–62
accumulated —, 108
Orthonormal vectors, 654
Outlier, 609
Output
— of a state space system, 613
observable —, 388

Partial model, 387
Partitioned matrix, 659
rules for —, 659–660
Period of a stochastic process, 591
Periodic VAR process
— definition of, 591–594
— estimation of, 594–598
— specification of, 594–604
Permanent shock, 369
Point forecast, 33–39
Policy
— simulation, 406
— variable, 613
Portmanteau statistic, 169–171, 214, 510

approximate distribution of —, 169, 214, 510
modified —, 171, 214, 510
Portmanteau test, 169–171, 214, 510
modified —, 171, 214, 510
Positive definite
— matrix, 655
— quadratic form, 655
Positive semidefinite matrix, 655
Poskitt's procedure, 505–507
Posterior
— density, 222
— mean, 222
— p.d.f., 222
Postmultiplication, 647
Predetermined variable, 388
Prediction tests for structural change
— based on one forecast, 184–186
— based on several forecasts, 186–188
— for cointegrated systems, 349–351
— of VAR processes, 184–189
— of VARMA processes, 510
Preliminary estimation of VARMA process, 474–477
Preliminary estimator of VARMA process, 475–477
Premultiplication, 647
Probability space, 2
Process
cointegrated —, 244–256
invertible MA —, 420–422
invertible VARMA —, 425
periodic —, 586
stable VARMA —, 423
VAR —, 5
VARMA —, 423–426
Product rule for vector differentiation, 665
Pure MA representation of a VARMA process, 423
Pure VAR representation of a VARMA process, 425

Quadratic form, 655
distribution of —, 678
indefinite —, 656
independence of —, 679
negative definite —, 656
negative semidefinite —, 656

positive definite —, 655
positive semidefinite —, 655
Quasi ML estimator, 140

Random coefficient VARX model, 621–623
Random variable, 2
Random vector, 3
Random walk, 237
Random walk with drift, 238
Rank of a matrix, 652
Rank of cointegration
  LR test for —, 327–343, 551–552
  testing for —, 327–343, 551–552
Rational
  — distributed lag model, 391–392
  — expectations, 392
  — transfer function, 392
  — transfer function model, 392
Real matrices, 645
Recursions
  Kalman filter —, 626–630
Recursive computation
  — of derivatives, 467–468
  — of residuals, 478
Reduced form of a dynamic SEM, 390
Regular matrix, 649
Resampling, 709–712
Resampling technique, 709–712
Residual autocorrelation
  — of VAR process, 161–169, 212–213
  — of VARMA process, 510
  asymptotic properties of —, 166, 212–213
  estimation of —, 161–169, 212–213
Residual autocovariance
  — of VAR process, 161–169, 212–213
  — of VARMA process, 510
  asymptotic properties of —, 165, 212–213
  estimation of —, 161–169, 212–213
Residual based bootstrap, 709
Residuals of VAR process
  checking the whiteness of —, 157–174, 214
Residuals of VARMA process
  checking the whiteness of —, 510
  estimation of —, 475

Restricted estimation of VAR models, 195–204
  asymptotic properties of —, 197–201
  EGLS, 197–200
  GLS, 195–197
  LS, 197
  ML, 200
Restrictions for VAR coefficients
  — for individual equations, 200–201
  linear —, 194–195
  nonlinear —, 221–222
  tests of —, 104–108, 138–143
  Wald test of —, 104–108
  zero —, 206–212
Restrictions for VARMA coefficients
  Granger-causality —, 441–444
  identifying —, 452–454
  linear —, 464
  LM test of —, 508–510
  tests of —, 508–510
Restrictions on white noise covariance, 202–204
Reverse echelon form, 518–519
  estimation of —, 521–522
Row vector, 645

Sample
  — autocorrelations, 159
  — autocovariances, 157
  — mean, 83–85
SC, 150, 208
Schwarz criterion, see SC
Score vector, 694
Scoring algorithm, 374, 472, 634–636
Seasonal
  — dummies, 585
  — model, 585
  — operator, 221
  — process, 585
  — time series, 585
Second order Taylor expansion, 671
SEM, 387
Sequential elimination of regressors
  specification of subset VAR model, 211
Shock
  permanent —, 369
  transitory —, 369
Simulation techniques

evaluating properties of estimators by
—, 707–709
evaluating properties of test statistics
by —, 707–709
Simultaneous equations model, *see*
SEM
Skewness
asymptotic distribution of —, 175,
178
measure of multivariate —, 174–180
Slutsky's theorem, 683
Small sample properties
— of LS estimator, 80–82
— of VAR order selection criteria,
151–157
— of estimated forecasts, 100–102
— of estimators, 707–709
— of test statistics, 707–709
investigation of —, 80, 707–709
Smoothing, 630
Smoothing matrix
Kalman —, 630
Smoothing step, 630
Specification of
— EC-ARMA$_{RE}$ form, 523–526
— VAR models, 135–157
— VARMA models, 493–508
— cointegrated systems, 325–344
— dynamic SEMs, 400–401
— echelon form, 498–507
— error correction echelon form,
523–526
— final equations form, 494–498
— subset VAR models, 206–212
Specification of cointegrating rank
— of EC-ARMA$_{RE}$ form, 525–526
— of VAR process, 327–343
— of error correction echelon form,
525–526
Square root of a matrix, 658
Stability condition, 15, 16
Stability of a VARMA process, 423
Stable
— VAR process, 13–18
— VARMA process, 423
Standard percentile confidence interval,
710
Standard VARMA representation, 448
Standard white noise, 73

State space model
estimation of —, 631–637
global identification of —, 634
identification of —, 633–634
local identification of —, 634
log-likelihood function of —, 631–633
ML estimation of —, 631–637
nonlinear —, 623–625
State space representation
— of VAR process, 614–616
— of VARMA process, 616
— of VARX process, 616
— of VARX process with systematically varying coefficients,
621
— of factor analytic model, 619–621
— of random coefficient VARX
model, 621–623
State vector, 611, 613
Stationarity
asymptotic —, 241
strict —, 24
Stationarity condition for VAR process,
25
Stationary point of a function, 671
Stationary stochastic process, 24–26
strictly —, 24
Stationary VAR process, 24–26
Step direction, 469
Stochastic convergence, 681–684
— almost surely, 682
— in distribution, 682
— in law, 682
— in mean square error, 682
— in probability, 681
— in quadratic mean, 682
— with probability one, 682
strong —, 682
weak —, 682
Stochastic process
cointegrated —, 244–256
discrete —, 3
MA, 420–423
multivariate —, 3
nonstationary —, 237, 585–586, 614,
621–623
periodic —, 591–594
VAR, 13–18
VARMA, 423–426

VARX, 387, 616, 621–623
Stochastic trend, 238
Stochastic volatility model, 583
Strictly exogenous variable, 389
Strictly stationary stochastic process, 24
Strong law of large numbers, 689
Strongly exogenous variable, 388
Structural analysis
— of VARMA models, 441–444
— of cointegrated system, 261–264
— of cointegrated systems, 316–322
— of dynamic SEMs, 406–408
— of subset VAR models, 205–206, 221
Structural change, 182
  Chow test for —, 182–184, 348–349
  prediction test for —, 349–351
  testing for —, 182–189, 348–351, 510, 598–601, 608
Structural form
— of a VAR process, 358
— of a dynamic SEM, 390
Structural impulse responses, 359, 377–382
Structural innovation, 359
  permanent —, 369
  transitory —, 369
Structural models
  VAR, 357–386
  VECM, 357–386
Structural time series model, 618–619
Structural VAR, 358–368
— with Blanchard-Quah restrictions, 367–368
— with long-run restrictions, 367–368
AB-model, 364–367
A-model, 358–362
B-model, 362–364
Structural vector autoregression, *see* structural VAR
Structural vector error correction model, 368–372
Submatrix, 659
Subset model
  bottom-up procedure for —, 344
  full search procedure for —, 344
  sequential elimination of regressors, 344
  top-down procedure for —, 344
Subset VAR model, 206–221
  checking of —, 212–217
  specification of —, 206–212
    bottom-up strategy, 211
    sequential elimination of regressors, 211
    top-down strategy, 208–210
  structural analysis of —, 221
Super-exogenous variable, 388
Superconsistent estimator, 288, 301
SVAR, 357–368
— with Blanchard-Quah restrictions, 367–368
— with long-run restrictions, 367–368
AB-model, 364–367
A-model, 358–362
B-model, 362–364
Blanchard-Quah —, 367–368
concentrated likelihood function, 373
estimation of —, 372–376
ML estimation of —, 372–376
SVECM, 368–372
  estimation of —, 376–377
  ML estimation of —, 376–377
Symmetric matrix, 646
System equation, 611
System matrix, 613
System with exogenous variables, 388–412
Systematic sampling, 616–618
Systematically varying coefficients
— of VAR models, 585–589
— of VARX models, 621

Taylor expansion, 671
  second order —, 671
Taylor's theorem, 670, 685
Temporal aggregation, 434–435, 440–441, 616–618
Testing for
— Granger-causality, 102–104, 316–321
— causal relations, 102–108, 316–321
— instantaneous causality, 104–108
— multi-step causality, 105–108
— nonnormality, 174–180
— periodicity, 598–604

762   Subject Index

— rank of cointegration, 327–343, 551–552
— structural change, 181–189, 348–351, 510, 598–601, 608
— whiteness of residuals, 169–174, 214, 510
nonnormality
— of VAR process, 177–180
— of white noise process, 174–177
residual autocorrelation
— of VAR process, 169–174
— of VARMA process, 510
— of subset VAR model, 214
— of white noise process, 157–161
structural change
— based on one forecast period, 182–186
— based on several forecast periods, 186–188
Tests of parameter restrictions
linear restrictions, 102, 138–143
nonlinear restrictions, 508–510
Threshold models, 625
Time invariant
— autocovariances, 597
— coefficients, 596
Time series
nonstationary—, 237
seasonal —, 585
Time varying
— coefficients, 585–591
randomly —, 621–623
systematically —, 585–591
Top-down strategy for subset VAR specification, 208–210
Total forecast error impulse responses, 56
Total impact matrix, 367
Total impulse responses, 56
Total multiplier, 392
Trace of a matrix, 653
Trace test for cointegration rank, 329
Transfer function, 392
Transfer function model, 387, 392
rational —, 392
Transformation
— of MA process, 435–436
— of VARMA process, 436–440
linear —, 435–440

Transition equation
— errors, 613
— noise, 613
— of a state space model, 611
Transition matrix, 613
Transitory shock, 369
Transpose of a matrix, 646
Trend
deterministic —, 238
stochastic —, 238
Triangular matrix
lower —, 646
upper —, 646
Triangular representation of cointegrated system, 251
Two-stage estimation
— of cointegrated system, 301–302
asymptotic properties of —, 301

Unconditional forecast, 402
Unimodular operator, 452
Univariate ARCH model, 559–562
Univariate GARCH model, 559–562
Unmodelled variable, 387–390

VAR estimation
fully modified —, 318
VAR order estimator
consistent —, 148
small sample properties of —, 151–157
strongly consistent —, 148
VAR order selection
AIC criterion for —, 147
comparison of criteria for —, 150–157
consistent —, 148–150
criteria for —, 146–157
FPE criterion for —, 146
HQ criterion for —, 150
SC criterion for —, 150
sequence of tests for —, 136–145
testing scheme for —, 143–144
VAR process, 5
— with linear parameter restrictions, 194–221
— with nonlinear parameter restrictions, 221–222
— with parameter constraints, 193–231

Subject Index   763

— with time varying coefficients, 586–591
autocorrelations of —, 30–31
autocovariances of —, 21, 26–30
checking the adequacy of —, 157–189
estimation of —, 69–93, 531–536
forecast error variance decomposition of —, 63–66
forecasting of —, 31–41, 93–102, 536–540
impulse response analysis of —, 108–129, 540–545
infinite order —, 531–552
LS estimation of —, 69–86, 531–536
MA representation of —, 18–24
mean-adjusted —, 82
nonstationary —, 256, 586–594
order determination of —, 135–157
order estimation of —, 146–157
specification of —, 135–157
stable —, 13–18
state space representation of —, 614–616
stationarity condition for —, 25
structural —, 358–368
subset —, 206–221
unstable —, 256
Yule-Walker estimation of —, 85–86
VAR representation of a VARMA process
infinite order —, 425
pure —, 425
VARMA process
— for integrated variables, 515–521
— in standard form, 448
— representation in standard form, 448
aggregation of —, 440–441
$ARMA_{RE}$ form of —, 518–519
autocorrelations of —, 430
autocovariances of —, 429–432
checking the adequacy of —, 508–510
definition of —, 423–426
EC-$ARMA_{RE}$ form of —, 519–521
echelon form of —, 452–453
error correction echelon form of —, 519–521
estimation of —, 447–487, 521–522
final equations form of —, 452

forecasting of —, 432–434, 487–490
Granger-causality in —, 441–444
identifiability of —, 447–458
identification of —, 447–458
impulse response analysis of —, 444, 490
interpretation of —, 441–444
invertible —, 425
linear transformation of —, 435–440
MA representation of —, 423
ML estimation of —, 458–487
nonuniqueness of —, 447–452
preliminary estimation of —, 474–477
reverse echelon form of —, 518–519
specification of —, 493–508
stable —, 423
standard form —, 448
state space representation of —, 616
transformation of —, 434–441
VAR representation of —, 425
VAR(1) representation of —, 426–429
VARX model, 387, 616, 621
random coefficient —, 621–623
Vec operator, 661
Vech operator, 662
VECM, 244–256
Vector autoregressive moving average process, see VARMA process
Vector autoregressive process, see VAR process
Vector differentiation, 664–671
Vector error correction model, 244–256

Wald statistic, 102, 321, 598
asymptotic distribution of —, 102, 321, 598
Wald test, 694–698
— for Granger-causality, 102, 316–321
— for instantaneous causality, 104–108
— for multi-step causality, 105–108
— of linear constraints, 102, 316–321
— of zero constraints, 104–108, 598–600
Weak law of large numbers, 689
Weakly exogenous variable, 388
White noise
Gaussian —, 75
standard —, 73

testing for —, 157–161, 169–174, 214, 510
White noise assumption
  checking of —, 157–161
  testing of —, 169–174, 214, 510
White noise covariance matrix estimator
  asymptotic properties of —, 75–76, 200, 296, 479–481
Whiteness of residuals
  checking the —, 157–174, 214, 510
  testing for —, 169–174, 214, 510

Wiener process, 698
  multivariate —, 703
  standard —, 698
Wold causal ordering, 61, 359
Wold's decomposition theorem, 25

Yule-Walker estimation of VAR process, 85–86

Zero mean VARMA process, 429

Printing and Binding: Strauss GmbH, Mörlenbach